sustainabili

Sustainability in Construction Engineering

Edited by
Edmundas Kazimieras Zavadskas, Jonas Šaparauskas and
Jurgita Antuchevičienė

Printed Edition of the Special Issue Published in *Sustainability*

www.mdpi.com/journal/sustainability

MDPI

Sustainability in Construction Engineering

Sustainability in Construction Engineering

Special Issue Editors

Edmundas Kazimieras Zavadskas
Jonas Šaparauskas
Jurgita Antuchevičienė

MDPI • Basel • Beijing • Wuhan • Barcelona • Belgrade

MDPI

Special Issue Editors
Edmundas Kazimieras Zavadskas, Jonas Šaparauskas and Jurgita Antuchevičienė
Vilnius Gediminas Technical University
Lithuania

Editorial Office
MDPI
St. Alban-Anlage 66
Basel, Switzerland

This is a reprint of articles from the Special Issue published online in the open access journal *Sustainability* (ISSN 2071-1050) from 2017 to 2018 (available at: http://www.mdpi.com/journal/sustainability/special issues/Sustainability Construction Engineering)

For citation purposes, cite each article independently as indicated on the article page online and as indicated below:

LastName, A.A.; LastName, B.B.; LastName, C.C. Article Title. *Journal Name* **Year**, *Article Number*, Page Range.

ISBN 978-3-03897-166-5 (Pbk)
ISBN 978-3-03897-167-2 (PDF)

Contents

About the Special Issue Editors

Edmundas Kazimieras Zavadskas, PhD, DSc, Professor at the Department of Construction Management and Real Estate, Chief Research Fellow at the Research Institute of Sustainable Construction, Vilnius Gediminas Technical University, Lithuania. Professor Zavadskas received his PhD in Building Structures (1973), Dr Sc. (1987) in Building Technology and Management. He is a member of Lithuanian and several foreign Academies of Sciences, received the Doctor Honoris Causa from Poznan, Saint Petersburg and Kiev Universities, and is an Honorary International Chair Professor in the National Taipei University of Technology. He was acknowledged by the International Association of Grey System and Uncertain Analysis (GSUA) for his huge input in Grey System field, and has been elected as an Honorary Fellowship of International Association of Grey System and Uncertain Analysis, which is a part of IEEE (2016); awarded by the Neutrosophic Science—International Association for distinguished achievements in neutrosophics and has been conferred an honorary membership (2016); awarded a Thomson Reuters certificate for access to the list of the most highly cited scientists (2014). His main research interests are: multi-criteria decision-making, operations research, decision support systems, multiple-criteria optimization in construction technology and management. He has published over 400 publications in Clarivate Analytics Web of Science, h = 48, and a number of monographs in Lithuanian, English, German and Russian. He is the Editor in Chief of *Technological and Economic Development of Economy and the Journal of Civil Engineering and Management*, and Guest Editor of over ten Special Issues related to decision-making in civil engineering.

Jonas Šaparauskas is an Associated Professor at the Department of Construction Management and Real Estate at Vilnius Gediminas Technical University, Lithuania. He received his PhD in Civil Engineering in 2004. His research interests include construction technology and organization, project management, quality management, multiple criteria decision-making, sustainable development. He has published 29 papers in Clarivate Analytics Web of Science journals. He is a member of the EURO Working Group OR in Sustainable Development and Civil Engineering (EWG-ORSDCE). He is the Deputy Editor-in-Chief of *Technological and Economic Development of Economy*, an Editorial Board Member of *E&M Ekonomie a Management, Baltic Journal of Real Estate Economics and Construction Management* and *Journal of Engineering, Project, and Production Management* (EPPM-Journal). He is the Guest Editor of the following Special Issues: "Sustainability in Construction Engineering" (2018) in *Sustainability* and "Economic Growth as a Consequence of the Industry 4.0 Concept" (2018) in *Economies*.

Jurgita Antuchevičienė is a Professor at the Department of Construction Management and Real Estate at Vilnius Gediminas Technical University, Lithuania. She received her PhD in Civil Engineering in 2005. Her research interests include multiple-criteria decision-making theory and applications, sustainable development, construction technology and management, construction investment. Professor Antucheviciene has published over 70 papers in Clarivate Analytics Web of Science journals. She is a member of IEEE SMC, Systems Science and Engineering Technical Committee: Grey Systems and of two EURO Working Groups: Multicriteria Decision Aiding (EWG MCDA) and Operations Research in Sustainable Development and Civil Engineering (EWG ORSDCE). She is a Deputy Editor in Chief of the *Journal of Civil Engineering and Management,* an Editorial Board Member of *Applied Soft Computing* and *Sustainability*, and a Guest Editor of several Special Issues: "Decision Making Methods and Applications in Civil Engineering" (2015) and "Mathematical Models for Dealing with Risk in Engineering" (2016) in *Mathematical Problems in Engineering*, "Managing Information Uncertainty and Complexity in Decision-Making" (2017) in *Complexity*, "Civil Engineering and Symmetry" (2018) in *Symmetry*, as well as "Sustainability in Construction Engineering" (2018) in *Sustainability*.

sustainability

MDPI

Editorial

Sustainability in Construction Engineering

Edmundas Kazimieras Zavadskas [1,2], Jonas Šaparauskas [1] and Jurgita Antucheviciene [1,*]

[1]   Department of Construction Management and Real Estate, Vilnius Gediminas Technical University,
      Sauletekio al. 11, LT-10223 Vilnius, Lithuania; edmundas.zavadskas@vgtu.lt (E.K.Z.);
      jonas.saparauskas@vgtu.lt (J.Š.)
[2]   Institute of Sustainable Construction, Vilnius Gediminas Technical University, Sauletekio al. 11,
      LT-10223 Vilnius, Lithuania
[*]   Correspondence: jurgita.antucheviciene@vgtu.lt; Tel.: +370-5-274-5233

Received: 27 June 2018; Accepted: 27 June 2018; Published: 29 June 2018

Abstract: The concept of sustainability has been expanding to all areas of economic activity, including construction engineering. Construction engineering is a complex discipline that involves designing, planning, constructing and managing infrastructures. In this Special Issue, 27 selected and peer-reviewed papers contribute to sustainable construction by offering technological, economic, social and environmental benefits through a variety of methodologies and tools, including fundamental decision-making models and methods as well as advanced multi-criteria decision-making (MCDM) methods and techniques. The papers are mainly concentrated in five areas: Sustainable architecture; construction/reconstruction technology and sustainable construction materials; construction economics, including investments, supply, contracting and costs calculation; infrastructure planning and assessment; project risk perception, analysis and assessment, with an emphasis on sustainability.

Keywords: construction engineering; sustainable construction; construction building technology; construction economics; project risk assessment; multi-criteria decision-making (MCDM)

---

1. Introduction

The concept of sustainability has been increasingly applied in construction engineering. Construction engineering involves all stages of the life cycle of building, including the design of a building or structure, construction planning and management, construction works, maintenance and the rehabilitation of buildings or infrastructure objects.

A large number of alternative solutions must be analyzed to obtain the most effective and sustainable decisions in the life cycle of building. Decision-making methods can facilitate making these decisions. Sustainable decision-making in construction engineering can be supported by fundamental models or modern multiple-criteria decision-making (MCDM) techniques.

In recent years, a large number of research papers have been published dealing with the achievements of fundamental sciences applied in construction. Many of these research papers were summarized in several review papers [1–6]. Meanwhile, besides the fundamental methods, MCDM developments and novel applications dealing with construction problems have been constantly growing. In particular, this group of methods can effectively support sustainable decisions when we are faced with the necessity to evaluate the performance of a large number and, in most cases, contradictory criteria. A great variety of mixed information can be successfully managed by applying multi-criteria decision-making methods [7]. The findings of the review paper [8] show that MCDM applications in civil engineering and construction have been constantly growing, and the increase is correlated with an increase of interest in sustainable development. A number of papers on sustainability problems in construction increased 7.6 times in the last decade [8]. The number of papers related to MCDM developments and applications was 7.3 times higher in the last decade [8]. Correspondingly, the number of MCDM applications in construction was 5.5 times higher in the same period [8]. These findings confirm a great potential for the research of sustainable decision-making in construction problems. A great interest for readers should attract review papers devoted to MCDM applications in construction engineering. A comprehensive review was prepared in 2014 [9], presenting an overview of popular MCDM techniques and their applications for construction problems. In 2015, the next two papers summarized applications of the methods in particular areas of civil engineering, including construction building technology and management [10,11]. Subsequently, more advanced hybrid multi-criteria decision-making (HMCDM) methods gained more popularity. The application of hybrid methods for engineering was analyzed in 2016 [12] and, in the next review paper, applications of HMCDM methods for sustainability problems, were overviewed [13]. In addition, several review papers devoted to MCDM or HMCDM applications for sustainable development problems are worth mentioning: Making sustainable decisions in architectural and engineering design [14,15]; sustainable supply chains [16]; green technologies; green building; sustainable design; and energy related problems [17–19]. A great amount of attention is devoted to risk assessment, or dealing with other uncertainties in construction engineering, by applying mathematical models and methods [20].

The above-mentioned items highlight the topicality of the issue and the need to provide a possibility for researchers to disseminate their new ideas and findings related to sustainable decisions in construction. Therefore, the current Special Issue received a great number of submissions from different institutions, countries and continents. The number of papers that were positively evaluated by qualified reviewers and editors was 27. The next chapter discusses the main research areas of submissions and a contribution of each paper to the aim of the Special Issue in terms of analyzed problems and applied methods.

2. Contributions

The Special Issue collects 23 research papers, 3 review papers and 1 case report paper. The papers contribute to sustainable construction by offering technological, economic, social and environmental benefits through a variety of methodologies and tools, including fundamental decision-making models and methods, as well as advanced multi-criteria decision-making (MCDM) methods and techniques.

The topics of the Special Issue gained attention all over the World. The paper from all four Continents have been submitted (Figure 1).

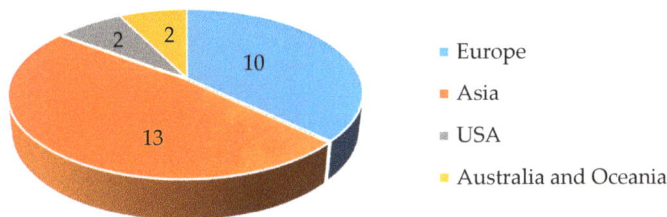

Figure 1. Number of publications from different Continents.

Regarding the origin of papers, the papers from 14 countries have been published in the Special Issue. The distribution of papers according to the authors' affiliation is presented in Table 1. Authors and co-authors from Lithuania contributed to 14 papers, those from Poland, 6 papers, those from Iran, 4 papers, and those from Korea, 3 papers. The authors from other countries, listed in Table 1, contributed to 1 or 2 papers.

Table 1. Publications by countries.

Countries	Number of Papers
Lithuania	4
Poland	4
Korea	3
Iran and Lithuania	4
Poland and Lithuania	2
Australia and Lithuania	1
USA and Lithuania	1
Turkey and Lithuania	1
Cambodia, New Zeland and Lithuania	1
USA	1
China	1
Taiwan	1
China and Taiwan	1
Qatar	1
Italy	1

The papers are mainly concentrated in five areas: Sustainable architecture; construction/reconstruction technology and sustainable construction materials; construction economics, including investments, supply, contracting and costs calculation; infrastructure planning and assessment; project risk perception, analysis and assessment, with an emphasis on sustainability.

As can be seen in Figure 2, the most numerous research areas are construction building technology and materials, including reconstruction and rehabilitation, as well as construction economics, covering a wide spectrum of problems related to investments and costs calculation, contracting and supply chains.

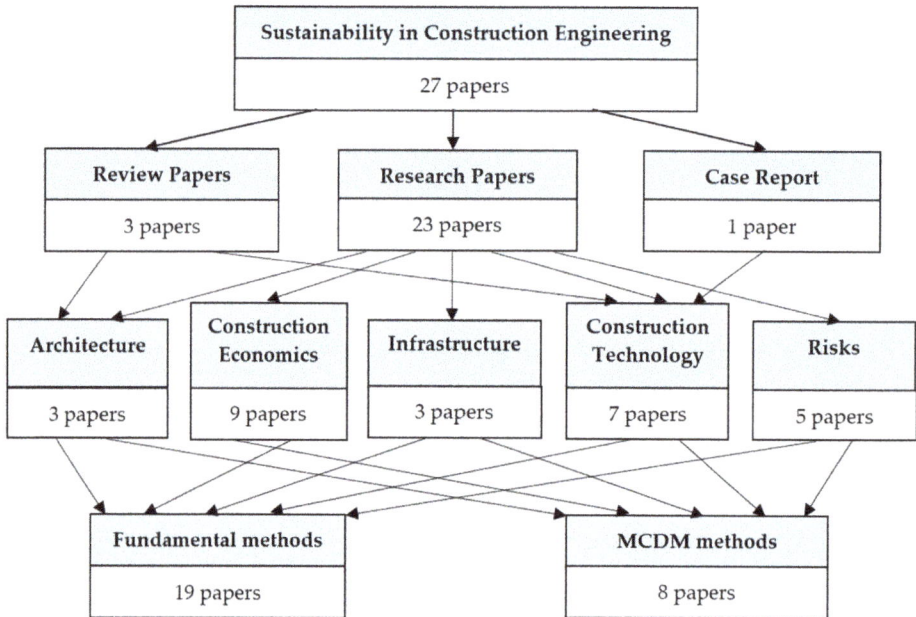

Figure 2. Types and research areas of publications.

A number of papers are very closely related to topical issues of sustainability in construction. The Issue received a paper related to Green Building, suggesting a building evaluation system by applying a combination of a Decision Making Trial and Evaluation Laboratory (DEMATEL) and an Analytic Network Process (ANP), called DEMATEL-based ANP (DANP) [21]. The other paper presents a construction project assessment according to sustainable development criteria and also suggests the application of a combination of MCDM methods—a Fuzzy Analytic Hierarchy Process and an improved Grey Relational Analysis (GRA) model [22]. Summarizing the applications of MCDM methods for various civil engineering and construction building technology problems is summarized in the review paper [8].

Tall residential building construction is analyzed [23], while sustainable renovation of buildings, with an emphasis on key performance indicators, is provided [24].

A couple of papers deal with construction materials and techniques for the construction or reconstruction/rehabilitation of buildings or infrastructure objects. A tool for CO_2 emission evaluation in the life cycle of concrete is presented [25]. The emission of Volatile Organic Compounds (VOCs) from dispersion and cementitious waterproofing products is analyzed [26].

In papers related to infrastructure objects, sustainable techniques for gravel road rehabilitation [27] and problems related to the displacement measures of bridges [28] are discussed.

As a part of a sustainable built environment, infrastructure for electric vehicles in cities and resorts is evaluated [29].

The next group of papers (9 papers) is related to construction economics and involves different topics. The construction industry is analyzed in terms of sustainability in Poland [30] and in Cambodia [31]. Promoting sustainability in construction investments through Building Information Modeling (BIM) implementation [32] or proper judicial conflict-resolution is suggested [33]. Two papers analyze the construction project cost calculation, considering the requirements of sustainable development [34,35], and the next paper is focused on the prices of implemented housing projects [36]. A new group decision-making model for contractor assessment based on a combination of MCDM methods under uncertainty is presented [37]. A topical issue of sustainable supply chains is reflected in a single paper in the current Issue [38].

It is worthwhile talking about risk-related papers individually. Two papers apply advanced MCDM methods for project risk assessment [39] and occupational risks on a construction site [40]. The paper [41] suggests the use of Bayesian Networks for project portfolio risk identification. Two more papers are devoted to sustainable construction risk perception [42,43].

Finally, the Special Issue touched a very unique topic that has been little studied heretofore in connection with sustainability—architecture. One research paper, focussing on the analysis of the key factor of sustainable architecture by the fuzzy HMCDM method, has been published [44]. Additionally, two very comprehensive review papers are elaborated [45,46], which are expected to attract a large interest from the architecture community.

3. Conclusions

The scope of the Special Issue raised the interest of researchers all over the World. Papers from 14 countries, located in four Continents, were published.

The main topics of the papers published in the Issue mainly cover five research areas: Construction/reconstruction technology and materials, construction economics, risk analysis, sustainable infrastructure and sustainable architecture.

The papers contribute to sustainable construction by offering technological, economic, social and environmental benefits through a variety of methodologies and tools. Multi-criteria decision-making techniques proved to be very suitable for sustainability assessment. Almost one third of papers (8 papers from 27) apply MCDM methods.

Author Contributions: E.K.Z., J.Š. and J.A. contributed equally to this work.

Funding: This research received no external funding.

Acknowledgments: Authors express their gratitude to the journal, Sustainability, for offering an academic platform for researchers to contribute and exchange their recent findings concerning sustainable construction.

Conflicts of Interest: The authors declare no conflict of interest.

References

1. Siddique, N.; Adeli, H. Nature-Inspired Chemical Reaction Optimisation Algorithms. *Cogn. Comput.* **2017**, *9*, 411–422. [CrossRef] [PubMed]
2. Siddique, N.; Adeli, H. Physics-based search and optimization: Inspirations from nature. *Expert Syst.* **2016**, *33*, 607–623. [CrossRef]
3. Siddique, N.; Adeli, H. Brief History of Natural Sciences for Natural-Inspired Computing in Engineering. *J. Civ. Eng. Manag.* **2016**, *22*, 287–301. [CrossRef]
4. Siddique, N.; Adeli, H. Applications of Gravitational Search Algorithm in Engineering. *J. Civ. Eng. Manag.* **2016**, *22*, 981–990. [CrossRef]
5. Yeganeh-Fallah, A.; Taghikhany, T. A Modified Sliding Mode Fault Tolerant Control for Large Scale Civil Infrastructures. *Comput.-Aided Civ. Infrastruct. Eng.* **2016**, *31*, 550–561. [CrossRef]
6. Ghaedi, K.; Ibrahim, Z.; Adeli, H.; Javanmardi, A. Invited Review: Recent developments in vibration control of building and bridge structures. *J. Vibroeng.* **2017**, *19*, 3564–3580.
7. Cinelli, M.; Coles, S.R.; Kirwan, K. Analysis of the potentials of multi criteria decision analysis methods to conduct sustainability assessment. *Ecol. Indic.* **2014**, *46*, 138–148. [CrossRef]
8. Zavadskas, E.K.; Antucheviciene, J.; Vilutiene, T.; Adeli, H. Sustainable decision-making in civil engineering, construction and building technology. *Sustainability* **2018**, *10*, 14. [CrossRef]
9. Jato-Espino, D.; Castillo-Lopez, E.; Rodriguez-Hernandez, J.; Canteras-Jordana, J.C. A review of application of multi-criteria decision making methods in construction. *Autom. Constr.* **2014**, *45*, 151–162. [CrossRef]
10. Zavadskas, E.K.; Antuchevičienė, J.; Kapliński, O. Multi-criteria decision making in civil engineering: Part I—A state-of-the-art survey. *Eng. Struct. Technol.* **2015**, *7*, 103–113. [CrossRef]
11. Zavadskas, E.K.; Antuchevičienė, J.; Kapliński, O. Multi-criteria decision making in civil engineering. Part II—Applications. *Eng. Struct. Technol.* **2015**, *7*, 151–167. [CrossRef]
12. Zavadskas, E.K.; Antucheviciene, J.; Turskis, Z.; Adeli, H. Hybrid multiple-criteria decision-making methods: A review of applications in engineering. *Sci. Iran.* **2016**, *23*, 1–20.
13. Zavadskas, E.K.; Govindan, K.; Antucheviciene, J.; Turskis, Z. Hybrid multiple criteria decision-making methods: A review of applications for sustainability issues. *Econ. Res.-Ekon. Istraz.* **2016**, *29*, 857–887. [CrossRef]
14. Pons, O.; de la Fuente, A.; Aguado, A. The Use of MIVES as a Sustainability Assessment MCDM Method for Architecture and Civil Engineering Applications. *Sustainability* **2016**, *8*, 460. [CrossRef]
15. Penades-Pla, V.; Garcia-Segura, T.; Marti, J.V.; Yepes, V. A Review of Multi-Criteria Decision-Making Methods Applied to the Sustainable Bridge Design. *Sustainability* **2016**, *8*, 1295. [CrossRef]
16. Keshavarz Ghorabaee, M.; Amiri, M.; Zavadskas, E.K.; Antucheviciene, J. Supplier evaluation and selection in fuzzy environments: A review of MADM approaches. *Econ. Res.-Ekon. Istraz.* **2016**, *30*, 1073–1118. [CrossRef]
17. Si, J.; Marjanovic-Halburd, L.; Nasiri, F.; Bell, S. Assessment of building-integrated green technologies: A review and case study on applications of Multi-Criteria Decision Making (MCDM) method. *Sustain. Cities Soc.* **2016**, *27*, 106–115. [CrossRef]
18. Streimikiene, D.; Balezentis, T. Multi-criteria assessment of small scale CHP technologies in buildings. *Renew. Sustain. Energy Rev.* **2013**, *26*, 183–189. [CrossRef]
19. Mardani, A.; Jusoh, A.; Zavadskas, E.K.; Cavallaro, F.; Khalifah, Z. Sustainable and Renewable Energy: An Overview of the Application of Multiple Criteria Decision Making Techniques and Approaches. *Sustainability* **2015**, *7*, 13947–13984. [CrossRef]

20. Antuchevičienė, J.; Kala, Z.; Marzouk, M.; Vaidogas, E.R. Solving civil engineering problems by means of fuzzy and stochastic MCDM methods: Current state and future research. *Math. Probl. Eng.* **2015**, *2015*, 362579. [CrossRef]

21. Shao, Q.-G.; Liou, J.J.H.; Weng, S.-S.; Chuang, Y.-C. Improving the Green Building Evaluation System in China Based on the DANP Method. *Sustainability* **2018**, *10*, 1173. [CrossRef]

22. Hatefi, S.M.; Tamošaitienė, J. Construction Projects Assessment Based on the Sustainable Development Criteria by an Integrated Fuzzy AHP and Improved GRA Model. *Sustainability* **2018**, *10*, 991. [CrossRef]

23. Kim, T.; Lim, H.; Kim, C.-W.; Lee, D.; Cho, H.; Kang, K.-I. The Accelerated Window Work Method Using Vertical Formwork for Tall Residential Building Construction. *Sustainability* **2018**, *10*, 456. [CrossRef]

24. Vilutiene, T.; Ignatavičius, Č. Towards sustainable renovation: Key performance indicators for quality monitoring. *Sustainability* **2018**, *10*, 1840. [CrossRef]

25. Kim, T.; Lee, S.; Chae, C.U.; Jang, H.; Lee, K. Development of the CO_2 Emission Evaluation Tool for the Life Cycle Assessment of Concrete. *Sustainability* **2017**, *9*, 2116. [CrossRef]

26. Kozicki, M.; Piasecki, M.; Goljan, A.; Deptuła, H.; Niesłochowski, A. Emission of Volatile Organic Compounds (VOCs) from Dispersion and Cementitious Waterproofing Products. *Sustainability* **2018**, *10*, 2178. [CrossRef]

27. Vaitkus, A.; Vorobjovas, V.; Tuminienė, F.; Gražulytė, J.; Čygas, D. Soft Asphalt and Double Otta Seal—Self-Healing Sustainable Techniques for Low-Volume Gravel Road Rehabilitation. *Sustainability* **2018**, *10*, 198. [CrossRef]

28. Jo, B.-W.; Lee, Y.-S.; Jo, J.H.; Khan, R.M.A. Computer Vision-based Bridge Displacement Measurements using Rotation-Invariant Image Processing Technique. *Sustainability* **2018**, *10*, 1785. [CrossRef]

29. Palevičius, V.; Podviezko, A.; Sivilevičius, H.; Prentkovskis, O. Decision-Aiding Evaluation of Public Infrastructure for Electric Vehicles in Cities and Resorts of Lithuania. *Sustainability* **2018**, *10*, 904. [CrossRef]

30. Hoła, B.; Nowobilski, T. Classification of Economic Regions with Regards to Selected Factors Characterizing the Construction Industry. *Sustainability* **2018**, *10*, 1637. [CrossRef]

31. Durdyev, S.; Zavadskas, E.K.; Thurnell, D.; Banaitis, A.; Ihtiyar, A. Sustainable Construction Industry in Cambodia: Awareness, Drivers and Barriers. *Sustainability* **2018**, *10*, 392. [CrossRef]

32. Reizgevičius, M.; Ustinovičius, L.; Cibulskienė, D.; Kutut, V.; Nazarko, L. Promoting Sustainability through Investment in Building Information Modeling (BIM) Technologies: A Design Company Perspective. *Sustainability* **2018**, *10*, 600. [CrossRef]

33. Bugajev, A.; Šostak, O.R. An Algorithm for Modelling the Impact of the Judicial Conflict-Resolution Process on Construction Investment. *Sustainability* **2018**, *10*, 182. [CrossRef]

34. Leśniak, A.; Zima, K. Cost Calculation of Construction Projects Including Sustainability Factors Using the Case Based Reasoning (CBR) Method. *Sustainability* **2018**, *10*, 1608. [CrossRef]

35. Gunduz, M.; Fahmi Naser, A. Cost Based Value Stream Mapping as a Sustainable Construction Tool for Underground Pipeline Construction Projects. *Sustainability* **2017**, *9*, 2184. [CrossRef]

36. Trojanek, R.; Tanas, J.; Raslanas, S.; Banaitis, A. The Impact of Aircraft Noise on Housing Prices in Poznan. *Sustainability* **2017**, *9*, 2088. [CrossRef]

37. Hashemi, H.; Mousavi, S.M.; Zavadskas, E.K.; Chalekaee, A.; Turskis, Z. A New Group Decision Model Based on Grey-Intuitionistic Fuzzy-ELECTRE and VIKOR for Contractor Assessment Problem. *Sustainability* **2018**, *10*, 1635. [CrossRef]

38. Dallasega, P.; Rauch, E. Sustainable Construction Supply Chains through Synchronized Production Planning and Control in Engineer-to-Order Enterprises. *Sustainability* **2017**, *9*, 1888. [CrossRef]

39. Wu, S.; Wang, J.; Wei, G.; Wei, Y. Research on Construction Engineering Project Risk Assessment with Some 2-Tuple Linguistic Neutrosophic Hamy Mean Operators. *Sustainability* **2018**, *10*, 1536. [CrossRef]

40. Seker, S.; Zavadskas, E.K. Application of Fuzzy DEMATEL Method for Analyzing Occupational Risks on Construction Sites. *Sustainability* **2017**, *9*, 2083. [CrossRef]

41. Ghasemi, F.; Sari, M.H.M.; Yousefi, V.; Falsafi, R.; Tamošaitienė, J. Project Portfolio Risk Identification and Analysis, Considering Project Risk Interactions and Using Bayesian Networks. *Sustainability* **2018**, *10*, 1609. [CrossRef]

42. Ploywarin, S.; Song, J.; Sun, D. Research on Factors Affecting Public Risk Perception of Thai High-speed Railway Projects Based on "Belt and Road Initiative". *Sustainability* **2018**, *10*, 1978. [CrossRef]

43. Ismael, D.; Shealy, T. Sustainable Construction Risk Perceptions in the Kuwaiti Construction Industry. *Sustainability* **2018**, *10*, 1854. [CrossRef]

44. Mahdiraji, H.A.; Arzaghi, S.; Stauskis, G.; Zavadskas, E.K. A Hybrid Fuzzy BWM-COPRAS Method for Analyzing Key Factors of Sustainable Architecture. *Sustainability* **2018**, *10*, 1626. [CrossRef]
45. Martek, I.; Hosseini, M.R.; Shrestha, A.; Zavadskas, E.K.; Seaton, S. The Sustainability Narrative in Contemporary Architecture: Falling Short of Building a Sustainable Future. *Sustainability* **2018**, *10*, 981. [CrossRef]
46. Bonenberg, W.; Kapliński, O. The Architect and the Paradigms of Sustainable Development: A Review of Dilemmas. *Sustainability* **2018**, *10*, 100. [CrossRef]

sustainability

MDPI

Review

Sustainable Decision-Making in Civil Engineering, Construction and Building Technology

Edmundas Kazimieras Zavadskas [1,2,*] **, Jurgita Antucheviciene** [1] **, Tatjana Vilutiene** [1] **and Hojjat Adeli** [3]

[1] Department of Construction Management and Real Estate, Vilnius Gediminas Technical University, Sauletekio al. 11, LT-10223 Vilnius, Lithuania; jurgita.antucheviciene@vgtu.lt (J.A.); tatjana.vilutiene@vgtu.lt (T.V.)
[2] Institute of Sustainable Construction, Vilnius Gediminas Technical University, Sauletekio al. 11, LT-10223 Vilnius, Lithuania
[3] College of Engineering, The Ohio State University, 470 Hitchcock Hall, 2070 Neil Avenue, Columbus, OH 43210, USA; adeli.1@osu.edu
* Correspondence: edmundas.zavadskas@vgtu.lt; Tel.: +370-5-274-5233

Received: 13 November 2017; Accepted: 20 December 2017; Published: 22 December 2017

Abstract: Sustainable decision-making in civil engineering, construction and building technology can be supported by fundamental scientific achievements and multiple-criteria decision-making (MCDM) theories. The current paper aims at overviewing the state of the art in terms of published papers related to theoretical methods that are applied to support sustainable evaluation and selection processes in civil engineering. The review is limited solely to papers referred to in the Clarivate Analytic Web of Science core collection database. As the focus is on multiple-criteria decision-making, it aims at reviewing how the papers on MCDM developments and applications have been distributed by period of publishing, by author countries and institutions, and by journals. Detailed analysis of 2015–2017 journal articles from two Web of Science categories (engineering civil and construction building technology) is presented. The articles are grouped by research domains, problems analyzed and the decision-making approaches used. The findings of the current review paper show that MCDM applications have been constantly growing and particularly increased in the last three years, confirming the great potential and prospects of applying MCDM methods for sustainable decision-making in civil engineering, construction and building technology.

Keywords: civil engineering; construction building technology; sustainability; decision-making; MCDM; literature review

---

1. Introduction

Civil engineering is based on fundamental scientific achievements. In the design and construction of engineering structures and buildings, theoretical methods are applied that are based on the fundamental sciences, such as mathematics, physics and chemistry. Over the last five years (2013–2017) a number of review articles have been prepared dealing with the achievements in these areas of fundamental sciences and their application in civil engineering as well as in building and construction. Optimizations "inspired by nature" based on chemistry [1], physics [2] and other natural sciences [3] were described. Applications of gravitational search algorithms [4], simulated annealing [5] and central force metaheuristic optimization [6] as nature-inspired conceptual frameworks in engineering are presented. Much attention is being paid to vibration control and the health monitoring of buildings and engineering structures [7–11], including bridges [12,13] and high-rise buildings [14–16]. A comprehensive review of tuned mass dampers for the vibration control of structures was provided [17].

Continuing our overview of review articles, a number of review articles have been published to address specific civil engineering issues and information technology applications to assist in solving engineering problems. The usage of support vector machines in structural engineering was presented [18,19]. Neurocomputing, in terms of the application of artificial neural networks for civil infrastructure optimization, monitoring and control is reviewed [20]. A review of how automation in construction operations was applied and automated equipment was incorporated in building construction phases [21] is presented. Transportation systems and transport technologies are systematically assessed in [22].

As sustainable development is becoming more relevant, more and more publications are being published related to sustainability in construction (Figure 1). Sustainable, innovative and efficient structural design [23,24], sustainable building design [25], including sustainability in high-rise building design [26], and integrated planning for sustainable building [27] is acknowledged, as well as a model for the structural health monitoring of high-rise buildings [28], and the vibration control of smart structures [29] were discussed, including sustainability aspects. Sustainable urban design [30] is no less important for assuring overall sustainability. Ceravolo et al. [31] describe a methodology for assessing the time-dependent structural performance of electric road infrastructures. Katsigarakis et al. [32] present a sense–think–act methodology for intelligent building energy management. Wang and Szeto [33] present a multiobjective environmentally sustainable road network design using Pareto optimization. Wang et al. [34] present a multi-objective path optimization for critical infrastructure links with consideration of seismic resilience. Bozza et al. [35] advocate alternative resilience indices for city ecosystems subject to natural hazards. Cahill et al. [36] study the effect of road surface, vehicle and device characteristics on energy harvesting from bridge–vehicle interactions.

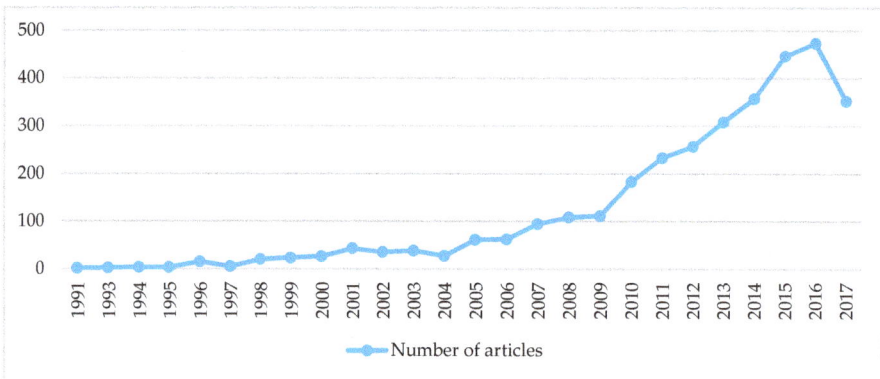

Figure 1. Number of publications on the topic "sustainability" in civil engineering and construction building technology Web of Science categories (Web of Science core collection database, 15 October 2017).

In applying the principles of sustainability, besides technological and economic aspects, environmental and social aspects also need to be considered. Accordingly, when choosing the most effective project decisions, everyone is faced with the need to evaluate the performance of a number of criteria. Mixed information and a wide variety of information types can be managed by applying multi-criteria decision analysis methods [37]. Multiple-criteria decision-making (MCDM) methods cover a wide range of somewhat distinct approaches. The methods can be broadly classified into two categories: discrete MADM (multi-attribute decision-making) methods and continuous MODM (multi-objective decision-making) methods. This classification has risen from two schools of thought regarding what human choice is based on: a French school and an American school. The French school mainly promotes the outranking concept for evaluating discrete alternatives. The American school is based on multi-attribute value functions and multi-attribute utility theory. Lately multiple-criteria decision-making (MCDM) methods have been increasingly applied (Figure 2), combining MODM and MADM techniques.

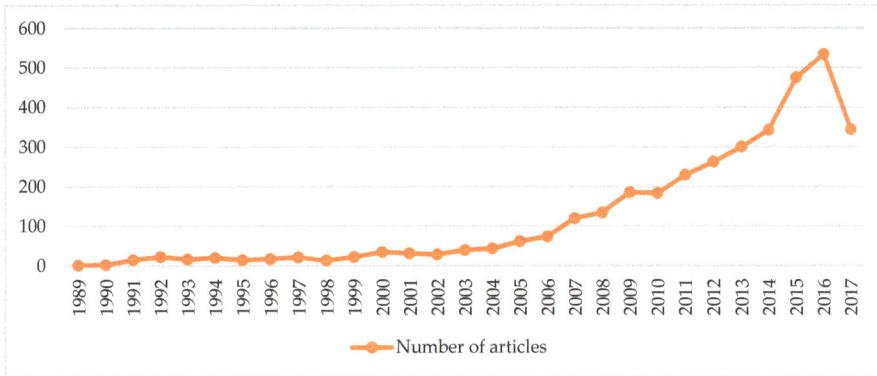

Figure 2. Number of publications on the topic "MCDM" in Web of Science core collection database (15 October 2017).

There are not many review articles aimed at analyzing MCDM (including MODM and MADM) for civil engineering applications. A very comprehensive paper was published by Kabir et al. [38]. Jato-Espino et al. [39] published a review article where an overview of the most widely-applied multi-criteria techniques and the main applications of the techniques to construction was provided.

Zavadskas et al. [40,41] reviewed the development of MCDM methods from 1772 to 2015. The first publication on multiple-criteria methods is considered a letter written by Franklin [42]. Pareto [43] publications played a particularly important role. Several Nobel prizes were awarded to Debrese (1959), Frisch (1969), Samuelson (1970), Arrow (1972), Nash (1994), Kantorovich and Koopmans (1975), Dantzig (1976), Sen (1998). The work of Simon (1978) [44] played a special role in the most up-to-date MCDM theory. Other important contributions were made by Saaty [45], Zeleny [46], Zadeh [47]. Zadeh [47] introduced the fuzzy sets theory. In 2015, Herrera-Viedma, a well-known scholar in the field of MCDM, prepared a special issue [48] devoted to the fifty-year theory of Zadeh. Kou and Ergu [49] prepared a special issue devoted to Satty's 90th anniversary and an overview article for pairwise comparison matrixes in multi-criteria decision-making [50]. Later Zavadskas et al. reviewed applications of MCDM methods in civil engineering until 2015 [40,41]. Applications in particular civil engineering areas were summarized in a number of papers. In 2016, Zavadskas et al. [51] reviewed the application of hybrid MCDM (HMCDM) methods in engineering. This article also gave an overview of the historical development of MCDM and the main publications on this topic. The focus of the article was on a broad overview, i.e., engineering applications on the whole, not focusing on building and construction. In another review article, Zavadskas et al. [52] presented a comprehensive analysis of the application of HMCDM methods for sustainability problems, including technology or product development/selection, personnel selection and company management, site selection, supply chains, etc. Yi and Wang [53] presented a multi-objective mathematical programming approach for construction laborer assignment with equity consideration. Pons et al. [54] published an article devoted to the application of MCDM methods for the assessment of sustainability in architectural and engineering design; Penades-Pla et al. [55] overviewed the sustainable design of bridges. Keshavarz Ghorabaee et al. [56] provided a broad overview of the application of MCDM methods in supply chains. Si et al. [57] reviewed the application of MCDM methods for the assessment of green technologies. Decision-making for green building, sustainable design, and energy related problems were overviewed [58,59]. Cerveira et al. [60] discussed wind farm distribution network optimization. These published review articles well illustrate the current state of the art in solving sustainability issues in civil engineering by applying MCDM, including MADM and MODM, methods. The whole and continuously increasing number of publications applying MCDM in civil engineering, construction and building technology is presented in Figure 3.

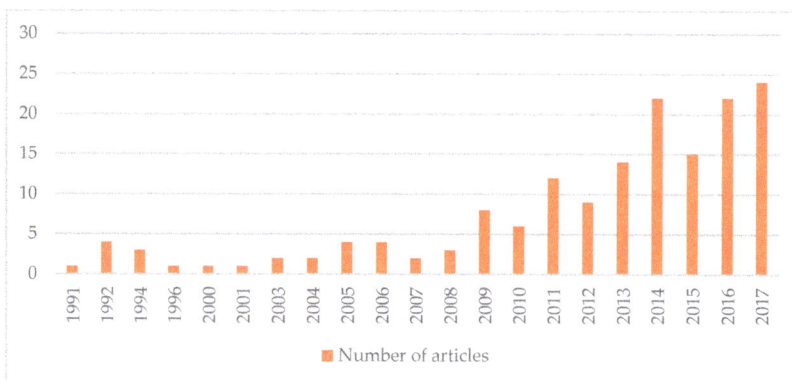

Figure 3. Number of publications on the topic "MCDM" in civil engineering and construction building technology Web of Science categories (Web of Science core collection database, 15 October 2017).

In addition, review articles focusing on the development of MCDM techniques also play an important role. Broad MADM and fuzzy MADM overviews were presented by Mardani et al. [61,62], Kahraman et al. [63], Antucheviciene et al. [64]. Zavadskas et al. [65] and Behzadian et al. [66] reviewed TOPSIS (technique for order of preference by similarity to ideal solution) method. Development of the VIKOR (VIseKriterijumska Optimizacija I Kompromisno Resenje (in Serbian), which means multicriteria optimization and compromise solution) method was presented by Mardani et al. [67]. Balezentis and Balezentis [68] reviewed the MULTIMOORA (multiobjective optimization by ratio analysis plus full multiplicative form) method. Behzadian et al. [69] discussed numerous developments and applications in various areas of PROMETHEE (preference ranking organization method for enrichment evaluations) method. Yang et al. [70] discussed multiobjective inventory routing with uncertain demand using population-based metaheuristics. Pan et al. [71] presented a region division-based diversity maintaining approach for many-objective optimization. Marttunen et al. [72] reviewed combinations of methods and structuring challenges for multiple-criteria decision analysis in practice. The books, summarizing the MCDM methodology, and published by Tzeng [73–75], Kou et al. [76], Bisdorff et al. [77], and Liu [78] also play an important role. These publications facilitate the work of researchers and the ability to orient themselves in choosing the methods.

In the next section, MCDM applications in civil engineering and construction building technology are reviewed in terms of the distribution of papers by years, countries, institutions, and journals. Detailed analysis of the latest articles (2015–2017) in terms of application domains, analyzed problems and MCDM methods applied is provided in Section 3.

2. Research Methodology and Preliminary Results

The search in the online Web of Science core collection database was made on 15 October 2017. The research procedure is presented in Figure 4. The search was made on the topic "MCDM". From all the documents identified (3571), a total of 2605 articles were identified, including research articles and reviews, and excluding proceedings papers and book chapters.

A search for MCDM applications in civil engineering and construction building technology Web of Science categories identified 195 papers, including 160 research articles and review articles.

This search revealed that MCDM methods were applied by researchers from over 100 institutions in 91 countries all over the world and in more than 100 research areas, while MCDM in the Web of Science categories of civil engineering and construction building technology was applied by researchers from 34 countries (Figure 5). It is worth mentioning that the leading three countries were Iran, the USA and Lithuania. The leading university was Vilnius Gediminas Technical University, Lithuania (30 articles), followed by the University of Tehran, Iran (17 articles), as presented in Table 1.

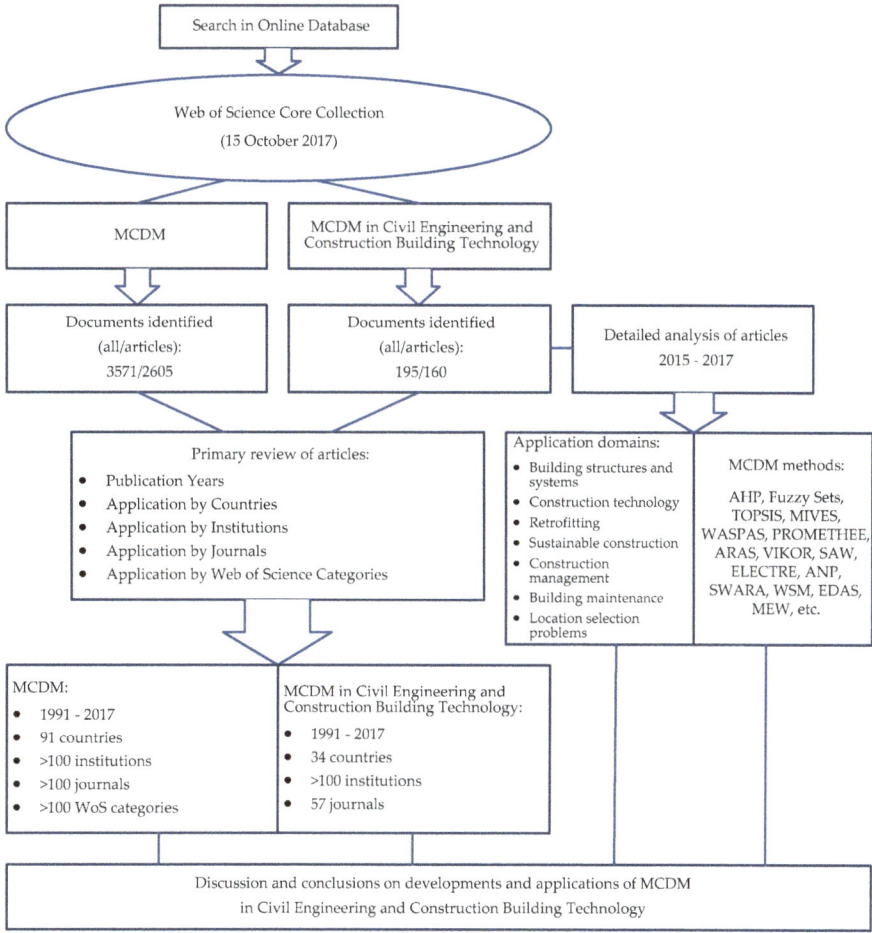

Figure 4. The research procedure and preliminary results.

The numerous papers on MCDM developments and applications were published in more than 100 different journals, often in operations research and computer science journals, while applications of MCDM for civil engineering, construction and building technology were published in 57 journals (Table 2), and mostly related to engineering. The leading journal was Journal of Civil Engineering and Management (24 articles).

As MCDM applications in civil engineering up to 2015 have been reviewed earlier [40,41], the current review focuses on detailed analysis of the papers published in 2015–2017. The changes during the period from 2014 until now are summarized in Table 3. It was found that the number of publications on MCDM methods increased by 56 percent in the three-year period. The geography of the authors expanded from 72 to 91 countries. A number of publications on the application of MCDM methods for civil engineering problems increased by 41 percent and the geography of the authors expanded from 28 to 34 countries in three year period. Researchers from the leading institution, Vilnius Gediminas Technical University, published 84 papers on the topic of MCDM in 2015–2017, including nine in the category of civil engineering and construction building technology.

Table 1. Publications by institutions on the topic "MCDM" civil engineering and construction building technology Web of Science categories (Web of Science core collection, 15 October 2017).

Institutions	Number of Articles
Vilnius Gediminas Technical University	30
University of Tehran	17
Amirkabir University of Technology	7
University of Naples Federico II	5
University of Arizona	5
Polytechnic University of Catalonia	5
Iran University Science Technology	5
Hong Kong Polytechnic University	5
University of British Columbia	4
Seoul National University of Science Technology	4
Istanbul Teknik University	4
Indian Institute of Technology IIT	4
Texas A&M University System	3
Royal Institute of Technology	3
Kaunas University of Technology	3
Islamic Azad University	3
Hohai University	3

Note: The names of institutions are presented as given in the Web of Science core collection database. The table does not represent the institutions that published less than three articles on the topic. Those are: two articles—Yonsei University; Yildiz Technical University; University of North Carolina; University of Nebraska System; University of Nebraska Lincoln; University of Illinois System; University of California System; Universiti Teknologi Malaysia; Universiti Malaya; Univ Mohaghegh Ardabili; Tsinghua University; Tennessee Technological University; Telecom Italia; Poznan University of Technology; Pontificia Universidad Catolica De Chile; Parthenope University Naples; Pacific Century Premium Dev Ltd.; National Central University; Nan Kai University Technology; Munzur Univ; Indian Institute of Technology IIT Roorkee; Engref; Engn Resinst Nat Disaster Shakhes Pajouh; Birla Institute of Technology Science. One article—Lublin University of Technology; Laval University;Lasbela Univ Agr; Kunsan Natl Univ; Korea University; Korea Environment Institute Kei; Korea Advanced Institute of Science Technology Kaist; Kocaeli University; Klaipeda University; Karlsruhe Inst Technol; Izmir Katip Celebi University; Istanbul Bilgi University; Isfahan University of Technology; Higher Institute of Applied Biological Sciences of Tunis; Inst Land Reclam Grass Farm; Indian Institute of Technology IIT Kharagpur; Indian Institute of Technology Iit Kanpur; Indian Institute of Science Iisc Bangalore; Indian Inst Remote Sensing; Imperial College London; Imam Khomeini Int Univ; Iett; Hyundai Inst Construct Technol; Hyundai Engn Construct Co., Ltd.; Huafan University; Hong Kong University of Science Technology; Heriot Watt University; Hellen Inst Transport; Harp Akademileri Komutanligi; George Mason University; Gaziosmanpasa University; Firat University; Feng Chia University; Fed Univ Petr Resources; Fateh Res Grp; Eskisehir Osmangazi University; El Paso Metropolitan Planning Or; Ecole Natl Genie Rural; East Carolina University; Dogus University; Dalian University of Technology; Council of Scientific Industrial Research Csir India; Concordia University Canada; Chongqing Jiaotong University; Chia Nan Univ Pharm Sci; Centre National De La Recherche Scientifique Cnrs; Canik Basari University; Calif Dept Transp; Cairo University; Bursa Technical University; Bur Rech Geol Minieres; Brandon University; Bialystok Tech Univ; Beijing University of Technology; Beijing Normal University; Asian Institute of Technology; Aristotle University of Thessaloniki; Akdeniz University; Acad Sci Innovat Res Acsir.

Table 2. Publications by journal on the topic "MCDM" in civil engineering and construction building technology Web of Science categories (Web of Science core collection, 15 October 2017).

Title of Journal	Number of Articles
Journal of Civil Engineering and Management	24
Water Resources Management	23
Archives of Civil and Mechanical Engineering	7
Stochastic Environmental Research and Risk Assessment	6
Journal of Hydroinformatics	6
Energy and Buildings	6
Water Resources Bulletin	5
Journal of Construction Engineering and Management	5
Tunnelling and Underground Space Technology	4
Ocean Engineering	4
Journal of Advanced Transportation	4
Automation in Construction	4
Transportation	3
Sustainable Cities and Society	3
Structure and Infrastructure Engineering	3
Building and Environment	3

Note: The table does not represent the journals, which published less than 3 articles on the topic. Those are: two articles—Transportation Research Record; Transportation Research Part E Logistics and Transportation Review; Journal of Water Resources Planning and Management; Journal of Performance of Constructed Facilities; Journal of Irrigation and Drainage Engineering; Journal of Hydrology; Journal of Computing in Civil Engineering; Computer Aided Civil and Infrastructure Engineering; Civil Engineering and Environmental Systems; Baltic Journal of Road and Bridge Engineering; one article—Water International; Thin Walled Structures; Stochastic Hydrology and Hydraulics; Smart Structures and Systems; Proceedings of The Institution of Mechanical Engineers Part F Journal of Rail and Rapid Transit; Preservation of Roadway Structures and Pavements; Latin American Journal of Solids and Structures; KSCE Journal of Civil Engineering; Journal of Water Supply Research and Technology Aqua; Journal of Water Resources Planning and Management ASCE; Journal of Urban Planning and Development ASCE; Journal of Urban Planning and Development; Journal of Transportation Engineering ASCE; Journal of Structural Engineering ASCE; Journal of Management in Engineering; Journal of Information Technology in Construction; Journal of Hydrologic Engineering; Journal of Earthquake Engineering; Journal of Construction Engineering and Management ASCE; Journal of Building Engineering; Journal of Aerospace Engineering; Iranian Journal of Science and Technology, Transactions of Civil Engineering; International Journal of Geomate; International Journal of Concrete Structures and Materials; International Journal of Civil Engineering; Gradevinar; European Journal of Environmental and Civil Engineering; Earthquakes and Structures; Construction and Building Materials; Computers Structures; Advances in Structural Engineering.

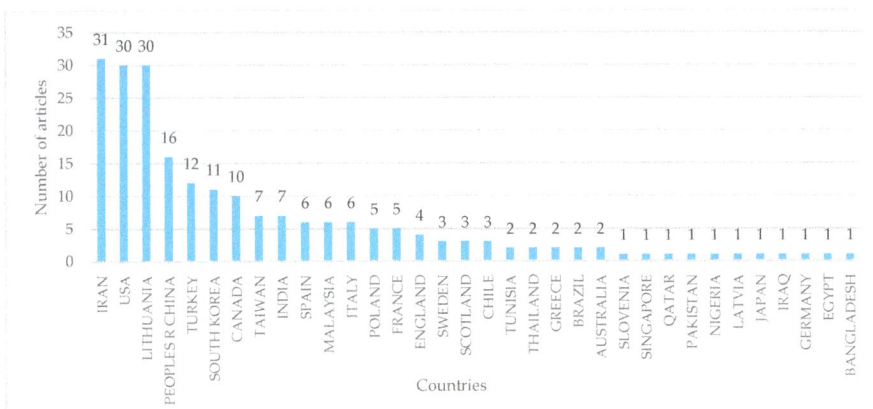

Figure 5. Publications by country on the topic of "MCDM" in civil engineering and construction building technology Web of Science categories (Web of Science core collection database, 15 October 2017).

Table 3. Changes in number of publications on the topic of "MCDM" in Web of Science core collection database.

Publications	Number of Publications	
	1991–2014	1991–2017 (15 October)
Publications on MCDM methods		
All	2290	3571
Articles:	1589	2605
• Countries	72	91
• Institutions	>100	>100
• Journals	>100	>100
Publications on MCDM in Civil Engineering and Construction Building Technology		
All	138	195
Articles:	113	160
• Countries	28	34
• Institutions	>100	>100
• Journals	38	57

Articles on the topic of "MCDM" in the Web of Science categories of civil engineering and construction building technology, published 1 January 2015–15 October 2017, are further analyzed in detail in Section 3.

3. Detailed Analysis of Articles Published in the Period of 2015–2017

The detailed review of the articles published in the research areas of civil engineering and construction building technology was made grouping journal articles by seven application domains (Figure 6) that cover separate aspects of the sustainable building life cycle. It was decided to analyze the most resent papers, specifically those published in the period 2015–2017. According to the research areas set out in the search, 61 papers formed the sample. After detailed review, papers published in the Web of Science categories "Water resources", "Environmental sciences", "Engineering mechanical", "Engineering marine", "Engineering aerospace" and "Engineering industrial" were removed from the sample, because their scope was not related to the object of this research. Research papers and review articles were only assigned to the analyzed research areas, in a total of 36 documents. This set was analyzed by scrutinizing the application areas (domains), problems solved and the MCDM methods applied.

The results of the detailed review presented in Table 4 revealed that the most important application domains in last three years were sustainable construction and construction technology. The number of articles in these domains was, respectively, 28% and 22%, of all set of sample articles. The next three important domains were building structures and systems, construction management and retrofitting, each comprising about 11% of the total number. Building maintenance and location selection problems as research domains were each about 8% in total.

Different problems related to sustainability in the construction sector were analyzed by the application of different MCDM approaches and methods. Naubi et al. [79] developed the watershed sustainability index (WSI) for the identification of the problematic areas within the watersheds. De la Fuente et al. [80] developed a method for analyzing the sustainability of different concrete and reinforcement configurations for segmental linings of tunnels. Arroyo et al. [81,82] compared different decision-making approaches applicable for design decisions involving sustainability factors in architecture, engineering and the construction industry. Ignatius et al. [83] proposed a novel integrated method for assessing green buildings realistically, based on stakeholders' fuzzy preferences. Hosseini et al. [84] presented a method for assessing the sustainability of post-disaster temporary housing. Chen & Pan [85] integrated Building Information Modeling (BIM) with MCDM and developed a BIM-aided variable fuzzy MCDM model for selecting low-carbon building (LCB) measures.

Jalaei et al. [86] proposed a methodology that integrates BIM with decision-making problem-solving approaches in order to optimize efficiently the selection of sustainable building components at the conceptual design stage of building projects. Medineckiene et al. [87] presented a new multi-criteria decision-making technique to select criteria for building sustainability assessment. Nakhaei et al. [88] presented the model to assess the vulnerability of buildings against explosion.

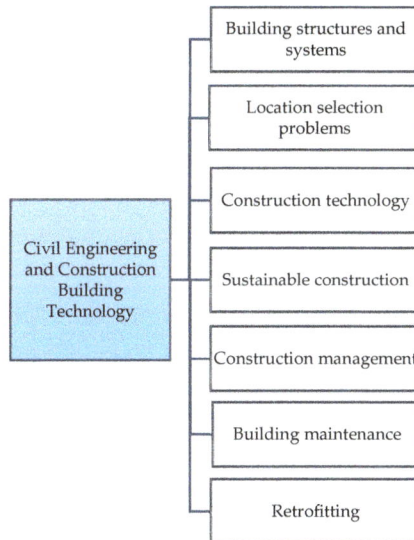

Figure 6. Application domains in research areas of civil engineering and construction building technology.

The MCDM approach was used very often for the solution of problems related to the construction technology area of research. Yousefi et al. [89] used the MCDM approach for the selection of combined building cooling, heating and power (CCHP) technologies. Kalibatas and Kovaitis [90] compared waterproofing alternatives for multifunctional inverted flat roofs by the use of multi-criteria decision-making techniques. Turskis et al. [91] presented a MCDM model for the evaluation and selection of building foundation alternatives. Leonavičiūtė et al. [92] analyzed personal protection devices for the prevention of falls from elevations using the new MCDM method. Turskis and Juodagalvienė [93] proposed a decision-making model to assess stairs for dwellings. Ebrahimian et al. [94] studied the selection of the most suitable construction method in urban storm water collection systems using a systematic and structured hybrid multi-criteria decision making approach. Nezarat et al. [95] performed a risk assessment of a mechanized tunneling project by the application of MCDM techniques. Shariati et al. [96] proposed a new multi-criteria decision-making model to evaluate the critical factors for the application of nanotechnology in the construction industry.

A multi-criteria decision-making approach was used for the solution of building structures and systems-related problems, like the evaluation and selection of waste materials for recovery and reuse in concrete [97], the selection of a best practicable decommissioning method [98], the optimization of the fire protection of structures [99] and the selection of an appropriate fan for an underground coal mine [100]. Fewer applications were found in the construction management [101–104], retrofitting [57,105–107], building maintenance [108–110] and location selection [111–113] application domains.

Table 4. Multiple-criteria decision-making (MCDM) applications in the domains of civil engineering and construction building technology research areas.

Application Domain and Problem Solved	MCDM Method(s) Applied *	Publication
Sustainable construction (27.78%)		
Watershed sustainability	PROMETHEE	Naubi et al. (2017) [79]
Sustainability based-approach to determine the concrete type and reinforcement configuration	AHP, MIVES	de la Fuente et al. (2017) [80]
Sustainable building design	CBA, WRC approaches	Arroyo et al. (2016) [81]
Choosing problem in building detailed design	AHP, CBA	Arroyo et al. (2015) [82]
Approach for green building assessment	fuzzy ANP	Ignatius et al. (2016) [83]
Method for assessing the sustainability of post-disaster temporary housing	MIVES	Hosseini et al. (2016a) [84]
BIM-aided variable fuzzy MCDM model for selecting Low-carbon building (LCB) measures	fuzzy PROMETHEE	Chen & Pan (2016a) [85]
Selection of Sustainable Building Components	TOPSIS	Jalaei et al. (2015) [86]
Sustainable building assessment/certification	AHP, ARAS	Medineckiene et al. (2015) [87]
Assessment of vulnerability of office buildings to blast	SMART, SWARA	Nakhaei et al. (2016) [88]
Construction technology (22.22%)		
Integration of a hybrid CCHP system into a commercial building	AHP	Yousefi et al. (2017) [89]
Selecting the most effective alternative of waterproofing membranes for multifunctional inverted flat roofs	SAW, Hurwicz, Laplace and Bayes rules	Kalibatas & Kovaitis (2017) [90]
Multicriteria evaluation of building foundation alternatives	AHP, WASPAS-G	Turskis et al. (2016) [91]
Analysis and prevention of construction site accidents	WASPAS-G	Leonavičiūtė et al. (2016) [92]
Decision-making model to assess a stairs shape for dwelling houses	AHP, SAW, MEW, TOPSIS, EDAS, ARAS, Laplace Rule, Bayes Rule, FM	Turskis & Juodagalvienė (2016) [93]
Selection of urban storm water construction method	Fuzzy AHP and CP	Ebrahimian et al. (2015) [94]
Ranking of geological risks in mechanized tunneling	Fuzzy AHP	Nezarat et al. (2015) [95]
The critical factors of the application of nanotechnology in construction	IFS, ANP	Shariati et al. (2017) [96]
Building structures and systems (11.11%)		
Evaluation and selection of waste materials for recovery and reuse in concrete	Choquet integral based fuzzy approach	Onat & Celik (2017) [97]
Selection of a best practicable decommissioning method	AHP	Na et al. (2017) [98]
Optimization of fire protection of cultural heritage structures	AHP	Naziris et al. (2016) [99]
Selection of an appropriate fan for an underground coal mine	AHP	Kursunoglu & Onder (2015) [100]
Construction management (11.11%)		
Selecting the best bidder during a tendering procedure	WRC, BVS, CBA approaches	Schöttle & Arroyo (2017) [101]
Dispute resolution method for disputes in construction projects	Laplace, Hurwicz, Hodges-Lehmann rules for games with grey numbers	Khanzadi et al. (2017) [102]
Introduction of private sectors into major projects. A hybrid model for evaluation and selection of the private sector for partnership projects	SWOT, Fuzzy VIKOR, PROMEEHTE	Dadpour & Shakeri (2017) [103]
Supply vendor selection model. Development of liquefied natural gas (LNG) plants as megaprojects	Fuzzy TOPSIS	Jang et al. (2016) [104]

Table 4. *Cont.*

Application Domain and Problem Solved	MCDM Method(s) Applied *	Publication
Retrofitting (11.11%)		
Optimal seismic upgrading of a reinforced concrete school building with metal-based devices	TOPSIS	Formisano et al. (2017) [105]
Assessment of building-integrated green technologies	AHP	Si et al. (2016) [57]
The optimal system for seismic retrofitting and vertical addition of existing buildings	TOPSIS, ELECTRE, VIKOR	Formisano & Mazzolani (2015) [106]
The analysis of vertical addition systems for energetic retrofitting of existing masonry buildings	TOPSIS	Terracciano et al. (2015) [107]
Building maintenance (8.33%)		
Quantitative tool to support the definition of a maintenance-inspection policy	Multicriteria model based on the delay-time concept	Cavalcante et al. (2016) [108]
Assessment of existing telecommunication towers	Fuzzy TOPSIS	Verma et al. (2015) [109]
Decision-making framework for procurement strategy selection in building maintenance work	AHP	Lin et al. (2015) [110]
Location selection problems (8.33%)		
Location selection of distribution centers	ELECTRE I	Agrebi (2017) [111]
Garage location selection for residential house	AHP, WASPAS-SVNS	Baušys & Juodagalvienė (2017) [112]
Model to support decision makers in choosing site locations for temporary housing	AHP, MIVES	Hosseini et al. (2016b) [113]

* PROMETHEE—Preference Ranking Organization METHod for Enrichment of Evaluations; AHP—Analytic Hierarchy Process; MIVES—The Integrated Value Model for Sustainable Assessment; CBA—Choosing By Advantages; WRC—Weighting Rating and Calculating; ANP—Analytic Network Process; TOPSIS—Technique for Order of Preference by Similarity to Ideal Solution; ARAS—Additive Ratio ASsessment; SMART—Simple Multi Attribute Ranking Technique; SAW—Simple Additive Weighting; WASPAS-G—Weighted Aggregated Sum Product ASsessment with grey numbers; MEW—Multiplicative Exponential Weighting; EDAS—The Method of Evaluation Based on Distance from Average Solution; CP—Compromise Programming; IFS—Intuitionistic Fuzzy Set; BVS—Best Value Selection; VIKOR—VIseKriterijumska Optimizacija I Kompromisno Resenje (that means Multicriteria Optimization and Compromise Solution); ELECTRE—ELimination Et Choix Traduisant la REalité (that means ELimination and Choice Expressing Reality); WASPAS-SVNS—Weighted Aggregated Sum Product ASsessment with Single-Valued Neutrosophic Set; FM—Full Multiplicative Utility function.

The question of how one method is better suited than another to address a specific problem is constantly scrutinized by researches and practitioners. This is also typical for the reports from the last three years. Schöttle and Arroyo [101] applied a value-based decision-making method conditionally named weighting rating and calculating (WRC), best value selection (BVS), and choosing by advantages (CBA) approaches to illustrate and compare the impact of these methods on the results of a tendering procedure. Khanzadi et al. [102] developed a hybrid MCDM model of discrete, zero-sum, two-person matrix games with grey numbers as a framework to solve dispute problems in the construction industry. The proposed approach was based on the application of several game theory methods—Laplace, Hurwicz, Bayes and Hodges-Lehmann rules. The integration of different methods in one model allows the minimization of the influence of shortcomings and the use of the advantages of separate methods, and at the same time to compare the obtained results. Dadpour and Shakeri [103] introduced a hybrid model that combines SWOT (Strengths, Weaknesses, Opportunities, and Threats) and fuzzy VIKOR methods for the evaluation and selection of the private sector for partnership projects. The case study compares the results obtained with the proposed approach and the PROMETHEE method.

Cases where the single method was applied are also common in the analyzed sample. Si et al. [57] employed AHP to derive the relative performance scores for technologies to be retrofitted in existing buildings to reduce carbon emissions and energy consumption. Lin et al. [110] presented the procedure based on AHP for the selection of the procurement method. Formisano et al. [105], with the TOPSIS method, assessed the five alternative upgrading techniques against cost, structural and environmental criteria. A final outcome of this study was the consideration that the applied TOPSIS method is a reliable technique to solve problems of different retrofitting solutions for existing buildings. The conclusion was complementary to the findings of earlier research, where Formisano and Mazzolani [106] used a combined application of TOPSIS, ELECTRE (ELimination Et Choix Traduisant la REalité (that means Elimination and Choice Expressing Reality)) and VIKOR methods for the analysis of retrofitting scenarios. Terracciano et al. [107] analyzed solutions for the vertical addition of existing masonry buildings and applied the TOPSIS method for the validation of the numerical results of the calculations. Through a sensitivity analysis, it was checked that the final results were not influenced by the decision-maker judgments. Agrebi [111] proposed a new approach for location selection based on the ELECTRE I method.

The combination of different approaches with the inclusion of MCDM methods is not a rare phenomenon. Cavalcante et al. [108] proposed a multi-criteria model based on the delay-time concept to provide the builder with a quantitative tool to support the decision-making process in building maintenance. The model proposed by Verma et al. [109] is a modified version of fuzzy TOPSIS applied in order to minimize the vagueness of visual inspection. This model ranks the alternative solutions based on similarity with fuzzy positive ideal solutions rather than the distance from fuzzy positive and negative ideal solutions. Baušys and Juodagalvienė [112], for the selection of garage locations, applied the AHP and an extension of the Weighted Aggregated Sum Product ASsessment (WASPAS) approach, namely Weighted Aggregated Sum Product ASsessment with Single-Valued Neutrosophic Set (WASPAS-SVNS), constructed based on the single-valued neutrosophic set. Hosseini et al. [113] assessed the sustainability of technologies using a newly designed sustainability model based on AHP and MIVES (Modelo Integrado de Valor para una Evaluación Sostenible (that means the Integrated Value Model for Sustainable Assessment), including a simplified life-cycle assessment (LCA).

Table 5 presents a summary of multi-criteria decision-making methods applied in articles published in the civil engineering and construction building technology application domains. The table show that the AHP [114], fuzzy approach [47] and TOPSIS method [115] were mostly used in the analyzed period. The most popular and commonly applied method in the analyzed sample was AHP, which was developed in 1980 [114]. AHP was used most commonly for the weighting of criteria as a single method or in integration with others for selection and decision-making [58,80,82,87,89,91,93–95,98–100,110,113]. Fuzzy sets were the second most popular approach in the analyzed sample. It was applied for different assessments as a single method and combined with others, i.e., ANP [83,96], PROMETHEE [85],

AHP [94,95], VIKOR [103], and TOPSIS [104,109]. The latter have been also intensively applied as a single method [86,105,107] or in integration with others [93,104,106,109].

Table 5. Methods applied in articles on civil engineering and construction building technology.

Methods	Articles
AHP, Saaty 1980 [114]	15
Fuzzy Sets, Zadeh 1965 [47]	12
TOPSIS, Hwang, Yoon 1981 [115]	7
MIVES, San-José, Cuadrado 2010 [116]	3
WASPAS-G, Zavadskas et al., 2015b [117]	2
PROMETHEE, Mareschal, Brans 1992 [118]	2
ARAS, Zavadskas, Turskis 2010 [119]	2
VIKOR, Opricovic 1998 [120]	2
SAW, MacCrimon 1968 [121]	2
Laplace Rule, Laplace 1814 [122]	2
Hurwicz Rule, Hurwicz 1951 [123]	2
Bayes Rule, Bayes 1763 [124]	2
WASPAS, Zavadskas et al., 2012 [125]	2
ELECTRE, Roy 1968 [126]	1
ANP, Saaty 1996 [127]	1
SWARA, Kersuliene et al., 2010 [128]	1
WSM, MacCrimon 1968 [121]	1
EDAS, Keshavarz Ghorabaee et al., 2015 [129]	1
MEW, Yoon, Hwang 1995 [130]	1
Hodges-Lehmann Rule, Hodges, Lehmann 1952 [131]	1
FM, Bridgman 1922 [132]	1

4. Discussion

A pressing task facing the world today is the sustainable development of cities and urban infrastructure addressed through the constructive interaction of environmental, economic and social factors. Sustainability priorities encompass integrated problems that address environmental protection, energy efficiency, optimized mobility, e-city technology and other fostering issues, including those appearing throughout all building life cycles, and deal with various levels of management and interest groups with different goals. From the mathematical point of view, these are multi-criteria group decision-making problems. In other words, the multi-criteria problems came from the multidimensionality paradigm conditioned by the ideology of sustainable development.

The most important advantage of the multi-criteria decision-making methods is their capability to address the problems that are characterized by conflicting goals. Therefore, the article was focused on the MCDM techniques and approaches that are being employed for decision-making in sustainability issues, particularly those related to the construction sector.

Usually, the selection of the most effective solution in construction-related problems is not such a simple task. The methods used in structural engineering do not allow for the assessment of the sustainability of alternative solutions. It has been noticed, that often alternative solutions and the results of numerical calculations have been validated by applying a MCDM method [91,105–107]. In particular, a sensitivity analysis was usually applied as a complementary approach to check that the results were not influenced by the judgments of decision-makers [83,86,99,101,105–107,109,111].

The results of the in-depth analysis revealed that AHP, fuzzy sets and TOPSIS methods are among the most well-known, not only during the last three decades, but also during last three years, and thus prevail in scientific articles. A rapid growth of AHP and TOPSIS applications was also recorded in Zyoud and Fuchs-Hanusch [133].

The TOPSIS approach is popular and employed for several main reasons most often referred to in analyzed articles [66,86,104–107,109]. First, the TOPSIS technique has a rational and understandable logic. Second, the computation process is straightforward; and the concept is depicted in a simple mathematical form. In several sources, TOPSIS is mentioned as one of the most popular MCDM techniques thanks to its easy application [105,106], as well as consistency and reliability [107]. Working with vast numbers of alternatives and attributes, TOPSIS is more efficient and faster compared with other MCDM methods, e.g., ELECTRE [86]. The technique creates two additional positive and negative ideal alternatives as a basis that guides the decision-maker to choose the optimal alternative among those considered. Subsequently, the solution of a problem is represented by the alternative with the minimum distance from positive ideal and the maximum distance from the negative ideal in a geometrical sense.

Generally, MCDM methods help the decision-maker to select objective solutions not influenced by the evaluation process. Real world problems are normally not defined exactly due to the uncertainty of human judgment; therefore, the extension of the classic methods enabling decision-making in uncertain environments has appeared, e.g., fuzzy TOPSIS. The popularity of fuzzy TOPSIS could be explained by one of the key advantages mentioned by Zavadskas et al. [66], i.e., the ability to deal with different types of values: crisp, interval, fuzzy or linguistic. Starting from the ideas presented in Zadeh's "Fuzzy Sets", published in 1965 [47], the fuzzy logic theory has proved to have numerous applications and developments until now [48,134]. Thus, the integration of fuzzy logic into classic methods provides a solution to handle subjective uncertain data and strengthens the comprehensiveness of the decision-making process.

The weights of the criteria in many papers in the analyzed sample were determined through the AHP method developed by Saaty (1980) [114]. Si et al. [57] emphasized that in the AHP method, the hierarchy between criteria influences weight value allocations. When more criteria are taken, the interrelations between the criteria can be changed, and alternative hierarchies may influence the difference in the allocation of weights. Consequently, different allocation of weight values will change the final ranking results. This leads to the necessity to set up an agreement regarding the development of a criteria hierarchy. Thus, sensitivity analysis is strongly recommended to identify the desirable weight values for decision makers. Despite the clear drawbacks, this method remains one of the most popular in technological and economic development, including multiple-criteria decision-making [49]. The superiority of AHP was proved by its predominance in the research and evidenced through the huge number of publications [133].

The analysis also revealed that research uses a combined approach instead of a single method. Particularly, [106] stated that the combined application of MCDM methods would allow decision-makers to find with the highest probability an objective optimal solution under different points of view. The diversity of the applied MCDM methods has also increased because of a growing trend to combine different MCDM methods and a need to integrate MCDM with other methods [62,72].

The use of MCDM methods reached various subareas of civil engineering and construction/building technology, proving that researchers in many fields are becoming increasingly aware of the importance of considering multiple aspects of reality when it comes to sustainable decision-making in civil engineering and construction/building technology. This review is also of political relevance because it shows that sustainability is a complex, manifold task, which urges political decision-makers to consider aspects that go beyond financials and implement solutions that make a balance between cost and benefits to all stakeholders.

This manuscript summarized carefully the papers that were available in the Web of Science core collection database, although a number of relevant works may have remained outside the scope of this study. However, the authors believe that this sample is representative, as the Web of Science core collection database is presented as the most accurate, objective, and complete resource available, and the articles included in it have passed a rigorous selection process inherent to high quality articles. Moreover, the authors limited the research on purpose; otherwise, the volume of the article would have increased significantly. On the other hand, the limitation specified above allows others in the future to get deeper into the subject, expand the sample and review those papers that are not mentioned in this article.

5. Conclusions

Sustainable decision-making in civil engineering, construction and building technology is based on fundamental scientific achievements and can be supported by multiple-criteria decision-making approaches. The current research justifies the need and usefulness of the application of MCDM methods for sustainable decision-making. It was identified that the number of publications on the topic of "sustainability" significantly increased in 2010. The number of publications on the topic "MCDM" began to grow starting from 2010. An analogous growth trend in publications applying MCDM methods has been observed in civil engineering and construction building technology Web of Science categories.

The main contribution of the current research is that it analyses in detail the newest publications for the last three year period. It was found that the number of publications on the application of MCDM methods for civil engineering and construction building technology problems increased by over 40 percent, the geography of the authors expanded from 28 to 34 countries in three years, 19 new journals were additionally added to the existing 38, and papers by researchers from over 100 institutions were published in 57 different journals by the current date.

The detailed analysis revealed that sustainable construction and construction technology were the most important application domains in last three years in the research areas of civil engineering and construction building technology. The number of articles in these two domains makes up to 50% of the sample. The next three important domains are building structures and systems, construction management and retrofitting, each comprising about 11% of total number. The analysis shows that the analytical hierarchy process (AHP), fuzzy approach and TOPSIS method were mostly used in the analyzed period. The results of the current research show that the MCDM approach is very useful for decision support in the construction industry in assessing design and technological alternatives with regard to sustainability, vulnerability and other important aspects.

The limitations of the current research are that only two Web of Science categories (engineering civil and construction building technology) were analyzed, while MCDM methods are applied in over 100 Web of Science categories, including some other categories also related to civil engineering. Accordingly, the research methodology is versatile and could be applied to analyzing publications including more Web of Science categories in a future.

The aim of the article was to introduce the thematic issue, to summarize the latest research in the field under study. As a result, the paper provides a better understanding of recent research directions in topics of sustainable development and construction engineering and can assist in conducting further research and seeking information. The study shows that decision-making methods have been developing in the last three years and their application has had a positive effect. The inclusion of multi-criteria decision-making methods as a robust and flexible tool for assessing possible alternatives provides the possibility to select a rational solution more precisely, taking into account the trade-offs that inevitably exist between the various candidate solutions. The obvious efforts to combine several methods show that the scientific community is still searching for the proper combination of decision-making methods for the solution of concrete problems. Thus, this analysis helps to anticipate future directions for the development of multi-criteria decision-making methods. Thus, the authors intend to make a comparative analysis and a more rigorous investigation of the existing methods, such as a comparison of previous approaches in terms of pros and cons, in the near future. In the light of the above, expectedly, this study can be employed by scholars as a basis for further research.

Acknowledgments: The authors express their gratitude to the reviewers for their comments and valuable suggestions.

Author Contributions: All authors contributed equally to this work.

Conflicts of Interest: The authors declare no conflict of interest.

References

1. Siddique, N.; Adeli, H. Nature-Inspired Chemical Reaction Optimisation Algorithms. *Cogn. Comput.* **2017**, *9*, 411–422.
2. Siddique, N.; Adeli, H. Physics-based search and optimization: Inspirations from nature. *Expert Syst.* **2016**, *33*, 607–623. [CrossRef]
3. Siddique, N.; Adeli, H. Brief History of Natural Sciences for Natural-Inspired Computing in Engineering. *J. Civ. Eng. Manag.* **2016**, *22*, 287–301. [CrossRef]
4. Siddique, N.; Adeli, H. Applications of Gravitational Search Algorithm in Engineering. *J. Civ. Eng. Manag.* **2016a**, *22*, 981–990. [CrossRef]
5. Siddique, N.; Adeli, H. Simulated annealing, its variants and engineering applications. *Int. J. Artif. Intell. Tools* **2016**, *25*, 1630001. [CrossRef]
6. Siddique, N.; Adeli, H. Central force metaheuristic optimisation. *Sci. Iran.* **2015**, *22*, 1941–1953.
7. Amezquita-Sanchez, J.P.; Adeli, H. Feature extraction and classification techniques for health monitoring of structures. *Sci. Iran.* **2015**, *22*, 1931–1940.
8. Qarib, H.; Adeli, H. Recent advances in health monitoring of civil structures. *Sci. Iran.* **2014**, *21*, 1733–1742.
9. Soto, M.G.; Adeli, H. Placement of control devices for passive, semi-active, and active vibration control of structures. *Sci. Iran.* **2013**, *20*, 1567–1578.
10. El-Khoury, O.; Adeli, H. Recent Advances on Vibration Control of Structures under Dynamic Loading. *Arch. Comput. Methods Eng.* **2013**, *20*, 353–360. [CrossRef]
11. Yeganeh-Fallah, A.; Taghikhany, T. A Modified Sliding Mode Fault Tolerant Control for Large Scale Civil Infrastructures. *Comput. Aided Civ. Infrastruct. Eng.* **2016**, *31*, 550–561. [CrossRef]
12. Ghaedi, K.; Ibrahim, Z.; Adeli, H.; Javanmardi, A. Invited Review: Recent developments in vibration control of building and bridge structures. *J. Vibroeng.* **2017**, *19*, 3564–3580.
13. Amezquita-Sanchez, J.P.; Adeli, H. Signal processing techniques for vibration-based health monitoring of smart structures. *Arch. Comput. Methods Eng.* **2016**, *23*, 1–15. [CrossRef]
14. Aldwaik, M.; Adeli, H. Advances in optimization of highrise building structures. *Struct. Multidiscip. Optim.* **2014**, *50*, 899–919. [CrossRef]
15. Soto, M.G.; Adeli, H. Tuned Mass Dampers. *Arch. Comput. Methods Eng.* **2013**, *20*, 419–431. [CrossRef]
16. Bakule, L.; Rehák, B.; Papík, M. Decentralized Networked Control of Building Structures. *Comput. Aided Civ. Infrastruct. Eng.* **2016**, *31*, 871–886. [CrossRef]
17. Karami, K.; Akbarabadi, S. Developing a smart structure using integrated subspace-based damage detection and semi-active control. *Comput. Aided Civ. Infrastruct. Eng.* **2016**, *31*, 887–902. [CrossRef]
18. Chou, J.S.; Pham, A.D. Smart Artificial Firefly Colony-based Support Vector Regression for Enhanced Forecasting in Civil Engineering. *Comput. Aided Civ. Infrastruct. Eng.* **2015**, *30*, 715–732. [CrossRef]
19. Amezquita-Sanchez, J.P.; Valtierra-Rodriguez, M.; Aldwaik, M.; Adeli, H. Neurocomputing in civil infrastructure. *Sci. Iran.* **2016**, *23*, 2417–2428. [CrossRef]
20. Vaha, P.; Heikkila, T.; Kilpelainen, P.; Jarviluoma, M.; Gambao, E. Extending Automation of Building Construction—Survey on Potential Sensor Technologies and Robotic Applications. *Autom. Constr.* **2013**, *36*, 168–178. [CrossRef]
21. Streimikiene, D.; Balezentis, T.; Balezentiene, L. Comparative assessment of road transport technologies. *Renew. Sustain. Energy Rev.* **2013**, *20*, 611–618. [CrossRef]
22. Pongiglione, M.; Calderini, C. Sustainable Structural Design: Comprehensive Literature Review. *J. Struct. Eng.* **2016**, *142*, 04016139. [CrossRef]
23. Dai, H. A wavelet support vector machine-based neural network meta model for structural reliability assessment. *Comput. Aided Civ. Infrastruct. Eng.* **2017**, *32*, 344–357. [CrossRef]
24. Asadi, E.; Adeli, H. Diagrid: An innovative, sustainable, and efficient structural system. *Struct. Des. Tall Spec. Build.* **2017**, *26*, e1358. [CrossRef]
25. Wang, N.M.; Adeli, H. Sustainable Building Design. *J. Civ. Eng. Manag.* **2014**, *20*, 1–10. [CrossRef]
26. Rafiei, M.H.; Adeli, H. Sustainability in highrise building design and construction. *Struct. Des. Tall Spec. Build.* **2016**, *25*, 643–658. [CrossRef]
27. Mikaelsson, L.A.; Larsson, J. Integrated Planning for Sustainable Building—Production an Evolution Over Three Decades. *J. Civ. Eng. Manag.* **2017**, *23*, 319–326. [CrossRef]

28. Oh, B.K.; Kim, K.J.; Kim, Y.; Park, H.S.; Adeli, H. Evolutionary learning based sustainable strain sensing model for structural health monitoring of high-rise buildings. *Appl. Soft Comput.* **2017**, *58*, 576–585. [CrossRef]
29. Soto, M.G.; Adeli, H. Multi-agent replicator controller for sustainable vibration control of smart structures. *J. Vibroeng.* **2017**, *19*, 4300–4322. [CrossRef]
30. Akbari, H.; Cartalis, C.; Kolokotsa, D.; Muscio, A.; Pisello, A.L.; Rossi, F.; Santamouris, M.; Synnefa, A.; Wong, N.H.; Zinzi, M. Local Climate Change and Urban Heat Island Mitigation Techniques—The State of the Art. *J. Civ. Eng. Manag.* **2016**, *22*, 1–16. [CrossRef]
31. Ceravolo, R.; Miraglia, G.; Surace, C.; Zanotti-Fragonara, L. A computational methodology for assessing the time-dependent structural performance of electric road infrastructures. *Comput. Aided Civ. Infrastruct. Eng.* **2016**, *31*, 701–716. [CrossRef]
32. Katsigarakis, K.; Kontes, G.D.; Giannakis, G.I.; Rovas, D.V. Sense-think-act Methodology for Intelligent Building Energy Management. *Comput. Aided Civ. Infrastruct. Eng.* **2016**, *31*, 50–64. [CrossRef]
33. Wang, Y.; Szeto, W.Y. Multiobjective environmentally sustainable road network design using Pareto optimization. *Comput. Aided Civ. Infrastruct. Eng.* **2017**, *32*, 964–987. [CrossRef]
34. Wang, Z.; Wang, Q.; Zukerman, M.; Guo, J.; Wang, Y.; Wang, G.; Yang, J.; Moran, B. Multiobjective Path Optimization for Critical Infrastructure Links with Consideration to Seismic Resilience. *Comput. Aided Civ. Infrastruct. Eng.* **2017**, *32*, 836–855. [CrossRef]
35. Bozza, A.; Napolitano, R.; Asprone, D.; Parisi, F.; Manfredi, G. Alternative resilience indices for city ecosystems subjected to natural hazards. *Comput. Aided Civ. Infrastruct. Eng.* **2017**, *32*, 527–545. [CrossRef]
36. Cahill, P.; Jaksic, V.; John Keane, J.; O'Sullivan, A.; Mathewson, A.; Ali, S.F.; Pakrashi, V. Effect of Road Surface, Vehicle and Device Characteristics on Energy Harvesting from Bridge-Vehicle Interactions. *Comput. Aided Civ. Infrastruct. Eng.* **2016**, *31*, 921–935. [CrossRef]
37. Cinelli, M.; Coles, S.R.; Kirwan, K. Analysis of the potentials of multi criteria decision analysis methods to conduct sustainability assessment. *Ecol. Indic.* **2014**, *46*, 138–148. [CrossRef]
38. Kabir, G.; Sadiq, R.; Tesfamariam, S. A review of multi-criteria decision-making methods for infrastructure management. *Struct. Infrastruct. Eng.* **2014**, *10*, 1176–1210. [CrossRef]
39. Jato-Espino, D.; Castillo-Lopez, E.; Rodriguez-Hernandez, J.; Canteras-Jordana, J.C. A review of application of multi-criteria decision making methods in construction. *Autom. Constr.* **2014**, *45*, 151–162. [CrossRef]
40. Zavadskas, E.K.; Antuchevičienė, J.; Kapliński, O. Multi-criteria decision making in civil engineering: Part I—A state-of-the-art survey. *Eng. Struct. Technol.* **2015**, *7*, 103–113. [CrossRef]
41. Zavadskas, E.K.; Antuchevičienė, J.; Kapliński, O. Multi-criteria decision making in civil engineering. Part II—Applications. *Eng. Struct. Technol.* **2015**, *7*, 151–167. [CrossRef]
42. Franklin, B. *Letter to Joseph Priesley, 1772*; Reprinted in the Benjamin Franklin Sampler; Fawcett: New York, NY, USA, 1956.
43. Pareto, V. *Cours E-Economic*; Universite de Lausanne: Lausanne, Switzerland, 1896/1897.
44. Simon, H.A. A behaviour model of rational choice. *Q. J. Econom.* **1955**, *69*, 99–118. [CrossRef]
45. Saaty, T.L. *Decision Making for Leaders: the Analytical Hierarchy Process for Decisions in a Complex World*; Lifetime Learning Publications: Belmont, CA, USA, 1982.
46. Zeleny, M. *Multiple Criteria Decision Making*; McGraw-Hill: New York, NY, USA, 1982.
47. Zadeh, L.A. Fuzzy sets. *Inf. Control* **1965**, *8*, 338–353. [CrossRef]
48. Herrera-Viedma, E. Fuzzy Sets and Fuzzy Logic in Multi-Criteria Decision Making. The 50th Anniversary of Prof. Lotfi Zadeh's Theory: Introduction. *Technol. Econ. Dev. Econ.* **2015**, *21*, 677–683. [CrossRef]
49. Kou, G.; Ergu, D. AHP/ANP Theory and Its Application in Technological and Economic Development: The 90th Anniversary of Thomas L. Saaty. *Technol. Econ. Dev. Econ.* **2016**, *22*, 649–650. [CrossRef]
50. Kou, G.; Ergu, D.; Lin, C.S.; Chen, Y. Pairwise Comparison Matrix in Multiple Criteria Decision Making. *Technol. Econ. Dev. Econ.* **2016**, *22*, 738–765. [CrossRef]
51. Zavadskas, E.K.; Antucheviciene, J.; Turskis, Z.; Adeli, H. Hybrid multiple-criteria decision-making methods: A review of applications in engineering. *Sci. Iran.* **2016**, *23*, 1–20.
52. Zavadskas, E.K.; Govindan, K.; Antucheviciene, J.; Turskis, Z. Hybrid multiple criteria decision-making methods: A review of applications for sustainability issues. *Econ. Res. Ekon. Istraz.* **2016**, *29*, 857–887. [CrossRef]
53. Yi, W.; Wang, S. Multi-objective mathematical programming approach to construction laborer assignment with equity consideration. *Comput. Aided Civ. Infrastruct. Eng.* **2016**, *31*, 954–965. [CrossRef]

54. Pons, O.; de la Fuente, A.; Aguado, A. The Use of MIVES as a Sustainability Assessment MCDM Method for Architecture and Civil Engineering Applications. *Sustainability* **2016**, *8*, 460. [CrossRef]

55. Penades-Pla, V.; Garcia-Segura, T.; Marti, J.V.; Yepes, V. A Review of Multi-Criteria Decision-Making Methods Applied to the Sustainable Bridge Design. *Sustainability* **2016**, *8*, 1295. [CrossRef]

56. Keshavarz Ghorabaee, M.; Amiri, M.; Zavadskas, E.K.; Antucheviciene, J. Supplier evaluation and selection in fuzzy environments: A review of MADM approaches. *Econ. Res. Ekon. Istraz.* **2016**, *30*, 1073–1118. [CrossRef]

57. Si, J.; Marjanovic-Halburd, L.; Nasiri, F.; Bell, S. Assessment of building-integrated green technologies: A review and case study on applications of Multi-Criteria Decision Making (MCDM) method. *Sustain. Cities Soc.* **2016**, *27*, 106–115. [CrossRef]

58. Streimikiene, D.; Balezentis, T. Multi-criteria assessment of small scale CHP technologies in buildings. *Renew. Sustain. Energy Rev.* **2013**, *26*, 183–189. [CrossRef]

59. Mardani, A.; Jusoh, A.; Zavadskas, E.K.; Cavallaro, F.; Khalifah, Z. Sustainable and Renewable Energy: An Overview of the Application of Multiple Criteria Decision Making Techniques and Approaches. *Sustainability* **2015**, *7*, 13947–13984. [CrossRef]

60. Cerveira, A.; Baptista, J.; Solteiro Pires, E.J. Wind Farm Distribution Network Optimization. *Integr. Comput. Aided Eng.* **2016**, *23*, 69–79. [CrossRef]

61. Mardani, A.; Jusoh, A.; Nor, K.M.D.; Khalifah, Z.; Zakwan, N.; Valipour, A. Multiple criteria decision-making techniques and their applications—A review of the literature from 2000 to 2014. *Econ. Res. Ekon. Istraz.* **2015**, *28*, 516–571. [CrossRef]

62. Mardani, A.; Jusoh, A.; Zavadskas, E.K. Fuzzy multiple criteria decision-making techniques and applications—Two decades review from 1994 to 2014. *Expert Syst. Appl.* **2015**, *42*, 4126–4148. [CrossRef]

63. Kahraman, C.; Onar, S.C.; Oztaysi, B. Fuzzy Multicriteria Decision-Making: A Literature Review. *Int. J. Comput. Intell. Syst.* **2015**, *8*, 637–666. [CrossRef]

64. Antucheviciene, J.; Kala, Z.; Marzouk, M.; Vaidogas, E.R. Solving Civil Engineering Problems by Means of Fuzzy and Stochastic MCDM Methods: Current State and Future Research. *Math. Probl. Eng.* **2015**, *2015*, 362579. [CrossRef]

65. Zavadskas, E.K.; Mardani, A.; Turskis, Z.; Jusoh, A.; Nor, K.M.D. Development of TOPSIS Method to Solve Complicated Decision-Making Problems: An Overview on Developments from 2000 to 2015. *Int. J. Inf. Technol. Decis. Mak.* **2016**, *15*, 645–682. [CrossRef]

66. Behzadian, M.; Otaghsara, S.K.; Yazdani, M.; Ignatius, J. A state-of the-art survey of TOPSIS applications. *Expert Syst. Appl.* **2012**, *39*, 13051–13069. [CrossRef]

67. Mardani, A.; Zavadskas, E.K.; Govindan, K.; Senin, A.A.; Jusoh, A. VIKOR Technique: A Systematic Review of the State of the Art Literature on Methodologies and Applications. *Sustainability* **2016**, *8*, 37. [CrossRef]

68. Balezentis, T.; Balezentis, A. A Survey on Development and Applications of the Multi-criteria Decision Making Method MULTIMOORA. *J. Multi-Criteria Decis. Anal.* **2014**, *21*, 209–222. [CrossRef]

69. Behzadian, M.; Kazemadeh, R.B.; Albadvi, A.; Aghdasi, M. PROMETHEE: A comprehensive literature review on methodologies and applications. *Eur. J. Oper. Res.* **2010**, *200*, 198–215. [CrossRef]

70. Yang, Z.; Emmerich, M.; Baeck, T.; Kok, J. Multiobjective Inventory Routing with Uncertain Demand Using Population-based Metaheuristics. *Integr. Comput. Aided Eng.* **2016**, *23*, 205–220. [CrossRef]

71. Pan, L.; He, C.; Tian, Y.; Su, Y.; Zhang, X. A Region Division Based Diversity Maintaining Approach for Many-Objective Optimization. *Integr. Comput. Aided Eng.* **2017**, *24*, 279–296. [CrossRef]

72. Marttunen, M.; Lienert, J.; Belton, V. Structuring problems for Multi-Criteria Decision Analysis in practice: A literature review of method combinations. *Eur. J. Oper. Res.* **2017**, *263*, 1–17. [CrossRef]

73. Tzeng, G.-H.; Huang, J.J. *Multiple Attribute Decision Making: Methods and Applications*; CRC Press, Taylor and Francis Group: Boca Raton, FL, USA, 2011; 349p.

74. Tzeng, G.-H.; Huang, J.J. *Fuzzy Multiple Objective Decision Making*; CRC Press, Taylor and Francis Group: Boca Raton, FL, USA, 2014; 313p.

75. Tzeng, G.-H.; Shen, K.-Y. *New Concepts and Trends of Hybrid Multiple Criteria Decision Making*; CRC Press, Taylor and Francis Group: Boca Raton, FL, USA, 2017.

76. Kou, G.; Ergu, D.; Peng, Y.; Shi, Y. Data Processing for the AHP/ANP. In *Quantitative Management*; Springer: Berlin/Heidelberg, Germany, 2013; Volume 1.

77.   Bisdorff, R.; Dias, L.C.; Meyer, P.; Mousseau, V.; Pirlot, M. (Eds.) Evaluation and Decision Models with Multiple Criteria: Case Studies. In *International Handbooks on Information Systems*; Springer: Berlin/Heidelberg, Germany, 2015.

78.   Liu, H.-C. *FMEA Using Uncertainty Theories and MCDM Methods*; Springer: Singapore, 2016.

79.   Naubi, I.; Zardari, N.H.; Shirazi, S.M.; Roslan, N.A.; Yusop, Z.; Haniffah, M.R.B.M. Ranking of Skudai river sub-watersheds from sustainability indices application of PROMETHEE method. *Int. J.* **2017**, *12*, 124–131. [CrossRef]

80.   De la Fuente, A.; Blanco, A.; Armengou, J.; Aguado, A. Sustainability based-approach to determine the concrete type and reinforcement configuration of TBM tunnels linings. Case study: Extension line to Barcelona Airport T1. *Tunn. Undergr. Space Technol.* **2017**, *61*, 179–188. [CrossRef]

81.   Arroyo, P.; Fuenzalida, C.; Albert, A.; Hallowell, M.R. Collaborating in decision making of sustainable building design: An experimental study comparing CBA and WRC methods. *Energy Build.* **2016**, *128*, 132–142. [CrossRef]

82.   Arroyo, P.; Tommelein, I.D.; Ballard, G. Comparing AHP and CBA as decision methods to resolve the choosing problem in detailed design. *J. Constr. Eng. Manag.* **2015**, *141*, 04014063. [CrossRef]

83.   Ignatius, J.; Rahman, A.; Yazdani, M.; Šaparauskas, J.; Haron, S.H. An integrated fuzzy ANP–QFD approach for green building assessment. *J. Civ. Eng. Manag.* **2016**, *22*, 551–563. [CrossRef]

84.   Hosseini, S.A.; de la Fuente, A.; Pons, O. Multicriteria decision-making method for sustainable site location of post-disaster temporary housing in urban areas. *J. Constr. Eng. Manag.* **2016**, *142*, 04016036. [CrossRef]

85.   Chen, L.; Pan, W. BIM-aided variable fuzzy multi-criteria decision making of low-carbon building measures selection. *Sustain. Cities Soc.* **2016**, *27*, 222–232. [CrossRef]

86.   Jalaei, F.; Jrade, A.; Nassiri, M. Integrating Decision Support System (DSS) and Building Information Modeling (BIM) to Optimize the Selection of Sustainable Building Components. *J. Inf. Technol. Constr.* **2015**, *20*, 399–420.

87.   Medineckiene, M.; Zavadskas, E.K.; Björk, F.; Turskis, Z. Multi-criteria decision-making system for sustainable building assessment/certification. *Arch. Civ. Mech. Eng.* **2015**, *15*, 11–18. [CrossRef]

88.   Nakhaei, J.; Bitarafan, M.; Lale Arefi, S.; Kapliński, O. Model for rapid assessment of vulnerability of office buildings to blast using SWARA and SMART methods (a case study of swiss re tower). *J. Civ. Eng. Manag.* **2016**, *22*, 831–843. [CrossRef]

89.   Yousefi, H.; Ghodusinejad, M.H.; Noorollahi, Y. GA/AHP-based optimal design of a hybrid CCHP system considering economy, energy and emission. *Energy Build.* **2017**, *138*, 309–317. [CrossRef]

90.   Kalibatas, D.; Kovaitis, V. Selecting the most effective alternative of waterproofing membranes for multifunctional inverted flat roofs. *J. Civ. Eng. Manag.* **2017**, *23*, 650–660. [CrossRef]

91.   Turskis, Z.; Daniūnas, A.; Zavadskas, E.K.; Medzvieckas, J. Multicriteria evaluation of building foundation alternatives. *Comput. Aided Civ. Infrastruct. Eng.* **2016**, *31*, 717–729. [CrossRef]

92.   Leonavičiūtė, G.; Dėjus, T.; Antuchevičienė, J. Analysis and prevention of construction site accidents. *Građevinar* **2016**, *68*, 399–410.

93.   Turskis, Z.; Juodagalvienė, B. A novel hybrid multi-criteria decision-making model to assess a stairs shape for dwelling houses. *J. Civ. Eng. Manag.* **2016**, *22*, 1078–1087. [CrossRef]

94.   Ebrahimian, A.; Ardeshir, A.; Rad, I.Z.; Ghodsypour, S.H. Urban stormwater construction method selection using a hybrid multi-criteria approach. *Autom. Constr.* **2015**, *58*, 118–128. [CrossRef]

95.   Nezarat, H.; Sereshki, F.; Ataei, M. Ranking of geological risks in mechanized tunneling by using Fuzzy Analytical Hierarchy Process (FAHP). *Tunn. Undergr. Space Technol.* **2015**, *50*, 358–364. [CrossRef]

96.   Shariati, S.; Abedi, M.; Saedi, A.; Yazdani-Chamzini, A.; Tamošaitienė, J.; Šaparauskas, J.; Stupak, S. Critical factors of the application of nanotechnology in construction industry by using ANP technique under fuzzy intuitionistic environment. *J. Civ. Eng. Manag.* **2017**, *23*, 914–925. [CrossRef]

97.   Onat, O.; Celik, E. An integral based fuzzy approach to evaluate waste materials for concrete. *Smart Struct. Syst.* **2017**, *19*, 323–333. [CrossRef]

98.   Na, K.L.; Lee, H.E.; Liew, M.S.; Zawawi, N.W.A. An expert knowledge based decommissioning alternative selection system for fixed oil and gas assets in the South China Sea. *Ocean Eng.* **2017**, *130*, 645–658. [CrossRef]

99.   Naziris, I.A.; Lagaros, N.D.; Papaioannou, K. Optimized fire protection of cultural heritage structures based on the analytic hierarchy process. *J. Build. Eng.* **2016**, *8*, 292–304. [CrossRef]

100. Kursunoglu, N.; Onder, M. Selection of an appropriate fan for an underground coal mine using the Analytic Hierarchy Process. *Tunn. Undergr. Space Technol.* **2015**, *48*, 101–109. [CrossRef]

101. Schöttle, A.; Arroyo, P. Comparison of Weighting-Rating-Calculating, Best Value, and Choosing by Advantages for Bidder Selection. *J. Constr. Eng. Manag.* **2017**, *143*, 05017015. [CrossRef]

102. Khanzadi, M.; Turskis, Z.; Ghodrati Amiri, G.; Chalekaee, A. A model of discrete zero-sum two-person matrix games with grey numbers to solve dispute resolution problems in construction. *J. Civ. Eng. Manag.* **2017**, *23*, 824–835. [CrossRef]

103. Dadpour, M.; Shakeri, E. A Hybrid Model Based on Fuzzy Approach Type II to Select Private Sector in Partnership Projects. *Iran. J. Sci. Technol. Trans. Civ. Eng.* **2017**, *41*, 175–186. [CrossRef]

104. Jang, W.; Hong, H.U.; Han, S.H.; Baek, S.W. Optimal supply vendor selection model for LNG plant projects using fuzzy-TOPSIS theory. *J. Manag. Eng.* **2016**, *33*, 04016035. [CrossRef]

105. Formisano, A.; Castaldo, C.; Chiumiento, G. Optimal seismic upgrading of a reinforced concrete school building with metal-based devices using an efficient multi-criteria decision-making method. *Struct. Infrastruct. Eng.* **2017**, *13*, 1373–1389. [CrossRef]

106. Formisano, A.; Mazzolani, F.M. On the selection by MCDM methods of the optimal system for seismic retrofitting and vertical addition of existing buildings. *Comput. Struct.* **2015**, *159*, 1–13. [CrossRef]

107. Terracciano, G.; Di Lorenzo, G.; Formisano, A.; Landolfo, R. Cold-formed thin-walled steel structures as vertical addition and energetic retrofitting systems of existing masonry buildings. *Eur. J. Environ. Civ. Eng.* **2015**, *19*, 850–866. [CrossRef]

108. Cavalcante, C.A.V.; Alencar, M.H.; Lopes, R.S. Multicriteria Model to Support Maintenance Planning in Residential Complexes under Warranty. *J. Constr. Eng. Manag.* **2016**, *143*, 04016110. [CrossRef]

109. Verma, M.; Rajasankar, J.; Anandavalli, N.; Prakash, A.; Iyer, N.R. Fuzzy similarity approach for ranking and health assessment of towers based on visual inspection. *Adv. Struct. Eng.* **2015**, *18*, 1399–1414. [CrossRef]

110. Lin, S.C.J.; Ali, A.S.; Bin Alias, A. Analytic hierarchy process decision-making framework for procurement strategy selection in building maintenance work. *J. Perform. Constr. Facil.* **2015**, *29*, 04014050. [CrossRef]

111. Agrebi, M.; Abed, M.; Omri, M.N. ELECTRE I based relevance decision-makers feedback to the location selection of distribution centers. *J. Adv. Transp.* **2017**, *2017*, 7131094. [CrossRef]

112. Baušys, R.; Juodagalvienė, B. Garage location selection for residential house by WASPAS-SVNS method. *J. Civ. Eng. Manag.* **2017**, *23*, 421–429. [CrossRef]

113. Hosseini, S.A.; de la Fuente, A.; Pons, O. Multi-criteria decision-making method for assessing the sustainability of post-disaster temporary housing units technologies: A case study in Bam, 2003. *Sustain. Cities Soc.* **2016b**, *20*, 38–51. [CrossRef]

114. Saaty, T.L. *The Analytic Hierarchy Process*; McGraw-Hill: New York, NY, USA, 1980.

115. Hwang, C.L.; Yoon, K. *Multiple Attributes Decision Making Methods and Applications*; Springer: Berlin/Hedelberg, Germany, 1981.

116. San-José, J.T.; Cuadrado, J. Industrial building design stage based on a system approach to their environmental sustainability. *Constr. Build. Mater.* **2010**, *24*, 438–447. [CrossRef]

117. Zavadskas, E.K.; Turskis, Z.; Antucheviciene, J. Selecting a contractor by using a novel method for multiple attribute analysis: Weighted Aggregated Sum Product Assessment with grey values (WASPAS-G). *Stud. Inform. Control* **2015**, *24*, 141–150. [CrossRef]

118. Mareschal, B.; Brans, J.P. *PROMETHEE V: MCDM Problems with Segmentation Constrains*; Universite Libre de Brusells: Brussels, Belgium, 1992.

119. Zavadskas, E.K.; Turskis, Z. A new additive ratio assessment (ARAS) method in multicriteria decision-making. *Technol. Econ. Dev. Econ.* **2010**, *16*, 159–172. [CrossRef]

120. Opricovic, S. *Multicriteria Optimization of Civil Engineering Systems*; University of Belgrade: Belgrade, Serbia, 1998.

121. MacCrimmon, K.R. *Decision Makingamong Multipleattribute Alternatives: A Survey and Consolidated Approach*; RAND Memorandum, RM-4823-ARPA; RAND Corporation: Santa Monica, CA, USA, 1968.

122. Laplace, P.-S. *Essai Philosophique sur les Probabilités*; Courcier: Paris, France, 1814.

123. Hurwicz, L. Optimality Criteria for Decision-Making under Ignorance: Cowles Commission Paper. *Statistics* **1951**, *370*, 45–52.

124. Bayes, T. An Essay towards solving a Problem in the Doctrine of Chances. *Philos. Trans.* **1763**, *53*, 370–418. [CrossRef]

125. Zavadskas, E.K.; Turskis, Z.; Antuchevičienė, J.; Zakarevičius, A. Optimization of weighted aggregated sum product assessment. *Electron. Electr. Eng.* **2012**, *122*, 3–6. [CrossRef]

126. Roy, B. La methode ELECTRE. *Rev. Inform. Rech. Oper. RIRO* **1968**, *8*, 57–75.

127. Saaty, T.L. *Decision Making with Dependence and Feedback. The Analytic Network Process*; RWS Publications: Pitsburg, PA, USA, 1996; 370p.

128. Keršulienė, V.; Zavadskas, E.K.; Turskis, Z. Selection of rational dispute resolution method by applying new stepwise weight assessment ratio analysis (SWARA). *J. Bus. Econ. Manag.* **2010**, *11*, 243–258. [CrossRef]

129. Keshavarz Ghorabaee, M.; Zavadskas, E.K.; Laya, O.; Turskis, Z. Multi-Criteria Inventory Classification Using a New Method of Evaluation Based on Distance from Average Solution (EDAS). *Informatica* **2015**, *26*, 435–451. [CrossRef]

130. Yoon, K.; Hwang, C. *Multiple Attribute Decision Making: An Introduction*; Sage Publications: London, UK, 1995.

131. Hodges, J.L.; Lehmann, E.L. The Use of Previous Experience in Reaching Statistical Decision. *Ann. Math. Stud.* **1952**, *23*, 396–407. [CrossRef]

132. Bridgman, P.W. *Dimensional Analysis*; Yale University Press: New Haven, CT, USA, 1922.

133. Zyoud, S.H.; Fuchs-Hanusch, D. A bibliometric-based survey on AHP and TOPSIS techniques. *Expert Syst. Appl.* **2017**, *78*, 158–181. [CrossRef]

134. Dzitac, I.; Filip, F.G.; Manolescu, M.J. Fuzzy Logic Is Not Fuzzy: World-renowned Computer Scientist Lotfi A. Zadeh. *Int. J. Comput. Commun. Control* **2017**, *12*, 748–789. [CrossRef]

sustainability

MDPI

Review

The Sustainability Narrative in Contemporary Architecture: Falling Short of Building a Sustainable Future

Igor Martek [1], M. Reza Hosseini [1,*], Asheem Shrestha [1], Edmundas Kazimieras Zavadskas [2] and Stewart Seaton [1]

[1]   School of Architecture and Built Environment, Deakin University, Geelong, Victoria 3220, Australia;
     igor.martek@Deakin.edu.au (I.M.); asheem.shrestha@deakin.edu.au (A.S.); s.seaton@deakin.edu.au (S.S.)
[2]   Institute of Sustainable Construction, Faculty of Civil Engineering, Vilnius Gediminas Technical University,
     Sauletekio Ave. 11, Vilnius LT-10223, Lithuania; edmundas.zavadskas@vgtu.lt
*   Correspondence: reza.hosseini@Deakin.edu.au

Received: 28 February 2018; Accepted: 25 March 2018; Published: 27 March 2018

Abstract: Sustainability has emerged, arguably, as the premiere mission of contemporary architecture. Green assessment tools abound, consultancy services flourish, buildings are marketed on the basis of sustainability performance, and government, media, and corporations seem preoccupied with assessing the quality of the built environment through a green lens. Yet for all the effort, and indeed for all the progress made, fundamental issues resistant to the structural change that is essential for genuine sustainability remain. This paper reviews the state of play of sustainability across the urban landscape. It considers the road travelled so far, and points out some of the major challenges that lie ahead.

Keywords: sustainability; green buildings; green development; green environment; rating tools; low carbon living; livable cities; sustainable cities; sustainability assessment tools

---

1. Introduction

Who would dare suggest that sustainability is anything but a good thing? With this simple test, it should be clear that sustainability is an ideology [1]. Nobody can deny its worth. However, in order to complete the test fairly, we need to ask "What does sustainability really mean?" The devil is in the details.

This study provides a broad overview of the sustainability narrative. It begins by answering the question of what sustainability means, noting that it is a trade-off between three competing ambitions: environmental protection, economic growth, and fairness for people [2]. It charts the historic development of sustainability consciousness in society. The rise of human civilisation has long been seen as synonymous with the conquest of nature and of overcoming natural barriers. Our confidence in controlling nature underlies both our belief that we are generating incalculable damage, and that we can and should do something about it [3].

Against this backdrop, the study goes on to chart the mission of architecture over the ages. Mostly, it was the preserve of tradition, encoding ritual, ceremony and value systems in built form. However, with the advent of industrialisation and the rise of the Modern Movement, architecture relinquished the past in favour of the future. The aims of 20th century architecture—in their various manifestations—was to generate utopias that embraced technology. The greatest legacy to arise out this was the "industrial city"—a concept that is still alive today, a century after its conception, and is largely indistinguishable from its manifestation as when first introduced. Here, utopia is a world full of factories, power plants, and railroads. The industrial city, then and now, seeks economic growth above all else [2,4].

Green sustainability rating tools arise out of the collision of these two agendas: continued economic growth, and redressing the environmental harm that such growth causes [5]. However, where is the correct balance to be found? How much economic benefit should we sacrifice in order to slow environmental damage? What economic practices must go, and what environmental goals should we prioritise? The study goes on to plot the rise and development of rating tools, noting that there are some 600 in use [6], each touting a different balance of give-and-take to the question [7].

Indeed, the academic world's contribution to the debate is to offer yet more varieties of rating tools. Rating "rating tools" is the new game in sustainability research. Over time, we see rating tools expanding vertically over industries and horizontally across value chains; expanding holistically to encompass whole regions; and collapsing microscopically to explore parameters in ever more refined nuances. However, we also see some pushback. Research also shows that rating tools may be failing to measure what they claim [8–10], either because they are off-plan and unverified, or because they actually get it wrong. Sometimes, the embodied energy of the components used in sustainable buildings generate a larger "carbon footprint" than the energy they purportedly save.

In the end, sustainability is a forum of ideas. It is a place where contesting interests struggle over what matters [1,11,12]. Rating tools are merely the physical artifacts of values generated by their proponents. What is most instructive is how more and more issues are being considered sustainability concerns. Sustainability is expanding to include so much that it risks being about nothing at all. This paper closes by reviewing the uncertainty of that legacy.

2. Defining Sustainability

The etymology of the word "sustainability" originates from the Latin *sustinēre* in which the words *sub*—from below—and *tenēre*—held up—combine to generate the idea of something that supports, maintains, or endures. According to the *Oxford English Dictionary*, the adjective form entered English usage in 1965 as an economics term—sustainable growth—and transformed into a developmental term in a 1972 report simulating population growth projections, which was entitled *The limits of growth* [13]. Terms such as "sustainable development" and "sustainable lifestyle" are now so commonplace in everyday parlance that it is hard to imagine that as recently as the 1980s the concept remained virtually unknown. Bill McKibben wrote the first ever book on climate change, which was published in 1989 [14]. Seven years later, in 1996, writing in the *New York Times*, he felt compelled to lament that "sustainability" would never catch on and would never become part of our social consciousness [15]. Yet beginning this century, there has been an explosion of attention afforded to the topic of sustainability. Today, a Google search of the term will return over 100 million hits.

Notwithstanding the newness of the term, "sustainability" has evolved significantly in meaning. It initially referred to practices that curbed environmental damage, and in that capacity, was the preserve of government regulators. However, the curtailment of resource exploitation and pollution came with a cost, and a second iteration of sustainability emerged which prioritised economic efficiencies in resource management. Then, in 2009, Mazmanian and Kraft observed that sustainability principles had begun to permeate societal values at large, and were influencing decisions made by individuals on a mass, personal scale, including what people bought, how they managed waste, and generally how they lived their lives [16].

At the institutional level, modern understandings of sustainability generally begin with the Brundtland Commission definition: *Sustainable development is the development that meets the needs of the present without compromising the needs of future generations* [17]. It describes sustainability as comprising an intersection of three elements: the environment, the economy, and equity. The premise is that all three elements must be preserved, and that any one element cannot be promoted at the expense of the others; sustainability is achieved when these are balanced. Sometimes, a fourth "e", education, is added, reflecting the importance of community buy-in. Others argue that the relationship between elements is hierarchical, with the environment paramount, and thus best depicted as concentric circles, as shown in Figure 1. There are also a great many other definitions and conceptualisations.

The important point here is that the fluidity of the term impacts its adoption, interpretation, and execution [18,19]. An extreme scientific modelling approach was proposed by Albert Bartlett, concluding that sustainable growth was an oxymoron, where: *Modern agriculture is the use of land to convert petroleum into food* [20]. At the other end of the spectrum, John Dryzek contends that sustainability can never be more than a platform for debate [21]. Superficially, while sustainability may be conveniently summarised as the parallel pursuit of the three objectives of the "triple bottom line"—economy, environment, and society—it is plainly evident that no specific point of arrival can be articulated. In this regard, sustainability aligns more with abstract ideals such as justice, democracy, and virtue. Ultimately, sustainability is a philosophical problem that is concerned more with the conditions under which we should be living, rather than with the apprehension of a specific set of quantifiable performance measures [22].

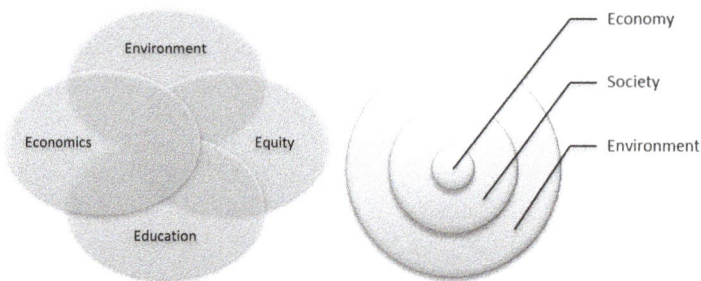

Figure 1. Depictions of the concept of sustainability [23].

3. Attitudes to Nature across History

In the far, distant past of pre-history, the world must have seemed to be a threatening and menacing place to our ancestors. Food sources were uncertain, natural phenomenon such as flood and drought hovered to unleash their fury, and predators, enemy tribes, and disease lurked at every turn. In the words of Thomas Hobbes, the political theorist, life was *"solitary, poor, nasty, brutish, and short"* [24]. Mankind's engagement with nature has shaped our attitude to it. For almost all of history, the challenge was to resist nature's onslaught, overcome the threat, and conquer it.

Early philosophical thought sought to reconcile mankind's place in nature. "Salvation" in primitive religion was communal. Deities were entreated to bless the crops and bring a bountiful harvest so that the whole tribe might survive to the next year. The conviction that developed was an anthropocentric view of the world in which the divine destiny of human beings was to rise over nature and dominate. For the Greeks, as exemplified by Socrates, the whole purpose of animals was to nourish humans. To the Roman Cicero, humans were on a higher plane than animals, which was evident in their ability to impose their will on raw creation, and in so doing improve it. With the emergence of cities and the accompanying transformation of the locus of salvation from group to the individual, mankind's ascendant position over nature consolidated. Perhaps one of the best-known ancient injunctions can be found in the Bible. God, speaking to Adam, instructs him: *Be fruitful and multiply, and replenish the earth, and subdue it: and have dominion over the fish of the sea, and over the fowl of the air, and over every living thing that moveth upon the earth. (Genesis 1:28).*

Through the Renaissance and Enlightenment periods, the capacity, and duty even, of mankind to perfect nature took on a secular flavor. Great influencers, such as Renè Descartes in his *Discourse on Method*, not only elevated the confidence of scientists and thinkers in the ability of human achievement, but the very act of introducing scientific methods to the exploration of the physical world, along with the reductionist approach to physical enquiry, diluted the mystery and enigma of nature into that of mere mechanisms. Viewing nature in terms of assembled components glossed the complexity of the holistic whole. Indeed, Newton's spectacular success in predicting the motion of planets and objects through physical laws transformed popular views of the universe as being just one big machine. The philosopher Herbert Spencer, writing in solidarity with Darwin's *Origin of the Species*, argued that survival of the fittest meant struggling against nature. Immanuel Kant went further, proclaiming that mankind alone has "understanding", giving him the right to rule over nature and relinquishing him of any moral responsibility with regards to his actions.

As distasteful as this amorality may seem to us today, when couched in the economic terminology driving current values, the message becomes familiar. Adam Smith asserted that individuals left to freely trade in consumer goods, and acting only out of their own self-interest, will ultimately generate the greatest enrichment for society as a whole; resources will be allocated efficiently. Through science, nature bends to human will, and human will arcs inevitably towards progress and the greater good. Even Marxism sees resources as nothing more than commodities, without recognising them as finite and depletable. For Marx, capital was the ultimate civilising force, which in the hands of the proletariat would see the renunciation of nature. In Pokrovsky's *History of Russia*, he proclaimed a criterion by which the Soviet struggle with the West would be measured: *In the future, when science and technique have attained to a perfection which we are unable to visualise, nature will become soft wax in his [man's] hands which he will be able to cast into whatever form he chooses.* Quoted in Ponting [25].

The lesson here is that while we may now accept that human activity has in fact damaged the environment to an extent that would have been unimaginable to the ancients, the belief remains that it is right and proper that we continue to intervene in nature in order to shape it to our will.

4. Architectural Theory and Sustainability

Serious architects, unlike construction managers, engineers, or perhaps any other professional, are prone to speak of their work in conceptual terms. Any respectable architecture program at university level will have a history and theory component. Styles from the ancient Classical, Romanesque, Norman, and Gothic; through Mannerism, the Baroque, Georgian, Victorian; to the Modernism, Bauhaus and International movements of the last century, all in their turn capture a "zeitgeist", or spirit of the age in which the aspirations of human civilisation are immortalised in the built form. The commonalities linking architectural forms speak to the perennial boundedness of the human condition: the need for shelter and for spaces to carry out both the mundane activities of everyday life, as well as the ceremonies and rituals that lift the soul. The differences between these forms speak to the evolutionary nature of human progress and the iterations in our efforts to lift ourselves into a better tomorrow. The comparison of architectural images, such as the two shown in Figure 2—Stonehenge, United Kingdom (UK), and the Marina Bay Sands Resort, Singapore—is an activity familiar to architects. What would they see? There is of course the remarkable technological differential that separates the 3800-year divide in construction dates. Yet, that divide is bridged in the realisation that the makers of both structures shared a driving ambition to impose their will on the landscape and bring their architectural vision into reality. The gesture of lintel and columns that are so evident in both structures would not be lost on Jungian psychologists, nor indeed on architects themselves.

However, a further question with regard to the images that is relevant to the discussion at hand would be to ask: which of the two structures represents a more sustainable solution? Answering is no mean feat, since in order to choose, a raft of parameters must be identified and evaluated. Essentially, to choose the ancient is to renounce current achievements, while choosing the modern is to qualify away current problems. This intractability must be met head on if the global sustainability project is to achieve genuine momentum.

The Gothic style lasted 500 years, from the 12th to 16th centuries, not to mention the period of revival it enjoyed in the 19th century. Its churches, too, took decades, if not a century to complete. These were communal projects; their artisans and architects were largely unknown. Soaring, vaulted ceilings reached for the heavens, and high stained-glass windows filtered light back down from the angels. Similar to much ancient architecture, the message in the light-as-air stonework spoke of a divine destiny for mankind on earth, the ultimate transcendence of the material world, and hope for the unspeakable beauty of the next. While we lived in this world, its stewardship was not a priority. We were only passing through; our reward lay in the next life.

Figure 2. Stonehenge, United Kingdom (UK), (**left**) and the Marina Sands Beach Resort, Singapore (**right**) [26].

By contrast, the 20th century was forward-looking and materialistic. It was an epoch of rapid change, uncertainty, and anxiety, bringing in its wake a cascade of architectural movements. Functionalism took on what was being learned in the new sciences, and brought them into architecture. These were the biological, mechanical, gastronomic, and linguistic analogies [27]. Frank Lloyd Wright may be its best-known purveyor through his many organically designed masterpieces, while Louis Sullivan may have coined architecture's best known cliché: *Form follows function* [28]. By contrast, Revivalism was a nostalgic paradigm, a strategy for retreating from an uncertain and tumultuous present while looking for the imagined past idyllic. These were the Roman, Greek, Renaissance, above-mentioned Gothic, and other revivals, as well as experiments in polychromy and eclecticism. Other approaches attempted in their own various ways to reconcile the old and the new. Rationalism sought to synthesise past traditions into a machine age. Expressionism adopted new materials and technical innovations, but aimed at socialist and utopian, if not utterly fantastic ideals.

Historians will assess these movements and their contributions in any manner of ways, and have done so [29]. Ultimately, for all their perceived impact, they were relatively short-lived. The rationalisations that propelled them into existence all too quickly dried up or were superseded. The seminal works of these movements can be summarised in the key projects of their respective champions. However, individual architectural heroes aside, the tradition that has left the heaviest imprint on the built environment in terms of sustainability, and whose impact lingers despite the demise of the name under which it flourished, is undoubtedly the Modern Movement. An appreciation of this contention begins with a sober assessment of the degrading impact of the world population on the environment. Just how extreme population growth has been over the last few generations is made plain when considering Figure 3. For almost all of history, the total number of humans on the entire planet may have hovered at not much more than a few million. With the abandonment of hunting and gathering, and the adoption of settled agriculture, populations began to edge upwards. At 1000 BC, it may have jumped to 50 million, reaching one billion in the 1800s. By 1960, it was three billion, and today it is eight billion. Currently, 100 million people are added to the planet every year, which was the entire world population of ancient times. The sudden, recent, and exponential growth in population would not have been possible without the associated magnitudinal leap in industrialisation. While this link is well established, it is less appreciated that industrialisation itself was energised and spurred forward by the Modern Movement.

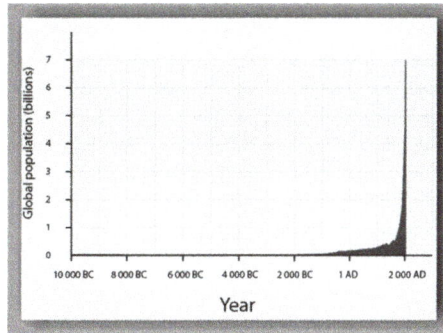

Figure 3. World population—10,000 BC to the present [30].

If the Modern Movement stood for anything, it stood for the belief that science and engineering, through the power of the human intellect, could solve any problem. An early advocate of the Modern Movement was Tony Garnier. He is known, firstly, for using concrete inside and outside buildings, and for doing so without concealing its inherent raw character; without ornamentation or reference to previous styles. He wrote: *Ancient architecture is an error. Truth alone is beautiful. In architecture, truth is the result of calculations made to satisfy known necessities with known materials* [31]. This again is the familiar cry of the primacy of human intellect and its ascendancy over nature. His great work, secondly, was in formulating the concept of the "industrial city". For all the subsequent criticisms that have been levelled against it, here, for the first time, was a manifesto on urban life that may be more familiar today than when it was conceptualised back in 1901. It was planned around the needs of factories and the workers that would fuel an economy dependent on productive output. An infrastructure grid of railways, roads, and communications networks ensured logistical efficiency. Surrounding dams and power stations provided the energy needed to keep the city alive. Here, for the first time, was a vision of the future that arguably has not dated. It is not only recognisable in today's cities, it also espouses values that even now permeate policy-making and social discourse, as shown in Figure 4. Critically, it is also dependent on the premise of an infinite, open system, where nature retains the capacity to supply an endless stream of input resources, while being able to absorb an endless waste output. The most significant environmental threat over the last 100 years is not so much how we build cities, but rather that we continue to build in much the same way, despite now knowing their limitations with regard to resource depletion and waste production.

Figure 4. Tony Garnier's industrial city, 1901—wholly recognisable in the cities of today [32].

5. Green Buildings and the Rise of Rating Tools

Against this backdrop of rising concern regarding the impact of humans on the planet, building construction looms large as generating the greatest environmental damage [33]. Buildings consume one-third of all resources [34]. This breaks down as one-sixth of all freshwater usage, one-quarter of world wood harvests, and four-tenths of all other raw materials [35]. Moreover, close to 50% of all global energy usage and green-house gas emissions arise from building operations, with a further 10% of energy consumed in the manufacturing of building materials [36]. On top of all this, 40% of all waste produced worldwide is generated in the initial construction and later demolition of buildings [37,38].

The industrialisation process of the 19th and 20th century saw buildings move from being evaluated as communal assets in cultural terms, to private assets that primarily serve an economic function. Indeed, the evolution of building performance measures has a long history, shifting only in recent times from concerns of financial return and occupant health and safety, to accommodate environmental performance [39]. Nevertheless, while sustainability, greening, energy efficiency, and ecolabelling are now arguably the emergent priority of building and planning regulators, sustainability performance measures remain embedded in traditional assessment paradigms. Building codes and regulations, at least in the West, emerged as a result of the Great Fire of London in 1666, setting standards in construction that enhanced fire resistance. Later, these codes centered on structural design to ensure minimum standards of load-bearing capacity for anticipated dead and live loads. Health and safety were also incorporated, looking to ensure the well-being of occupants. Then, in the 1970s, the world was hit with the oil crisis. As a result, regulations were expanded to encompass considerations of energy efficiency [8,40]. When public interest was aroused in climate change in the 1990s, these codes were again extended. In Australia, energy efficiency was introduced into the residential building code in 2005, and then into the commercial building code in 2006. At first, this was limited to new buildings, but it was then brought across to include the refurbishment of existing structures [41].

What is important to understand here is that compliance with all of the relevant codes is a prerequisite to obtaining a building permit. That is, compliance is assessed prior to the building actually being built—it is assessed off the plan. Following this precedent, sustainability assessment, too, is done in this way—off the plan. The difficulty is that while the engineering design of a component is reliably predictive of the engineering components' performance, sustainability design is not [9]. For one thing, there is the problem of measurement. It is relatively easy to comprehend and observe structural failure. However, sustainability performance unfolds as a matter of degrees, and is not objectively quantifiable. Moreover, even more fundamentally, there is the problem of what it is that is actually being measured. There is agreement on what constitutes structural failure; there is far more controversy over what sustainable performance entails [6].

6. Green Buildings and the Rise of Rating Tools

Given the ingredients of a recent but rocketing concern over the environmental impact of the built environment, combined with a regulatory legacy steeped in off-plan compliance assessment and a culture of outsourcing assessment to commercial bodies vying in an open market, we have fostered the emergence of the competitive green-rating assessment regimes that we see today. Just as with any fast-growing market for a new high-demand product, green building rating tools have exploded onto the scene, with a plethora of providers offering all sorts of rating tool variations. At last count, there were over 600 such sustainability rating tools competing in the market [42]. Similar to any emerging market, the market for green rating tools is characterised by early entrants that have captured market share, followed by niche entrants that have tweaked initial offerings to serve specialist sectors or adapt to different regional conditions. The Building Research Establishment Assessment Method (BREEAM) dominates Europe, the Leadership in Energy and Environmental Design (LEED) is preeminent in America, the Comprehensive Assessment System for Built Environment Efficiency

(CASBEE) is centered on Asia, and Green Star is the main provider in Australasia. A schematic of their evolutionary development is shown in Figure 5.

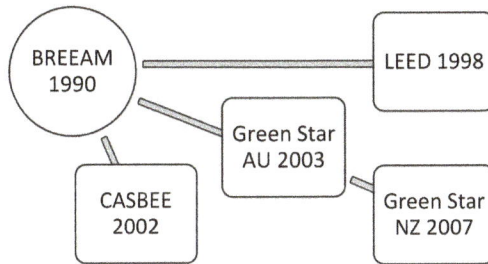

Figure 5. The product development pathway of the major global green rating tools [43]. BREEAM: Building Research Establishment Assessment Method; CASBEE: Comprehensive Assessment System for Built Environment Efficiency; LEED: Leadership in Energy and Environmental Design.

The Building Research Establishment Assessment Method, or BREEAM, is regarded as the first green building rating system. It was developed by the UK Building Research Establishment in 1990, and was revised in 1993 to assess new offices. Later, it was expanded again to include existing offices, supermarkets, and light industrial buildings. Though adapted to local codes, it has been applied internationally, and encompasses the whole building life cycle, including: design, operation, and refurbishment. BREEAM evaluates the environmental performance of buildings across eight factors: land use, transport, energy, water, materials, waste, pollution, and management. With the first-mover advantage into the marketplace, BREEAM accounts for 80% of the sustainable building certifications in Europe, which totalled some 560,000 buildings in 2017. By 2017, BREEAM had expanded to 75 countries [44]. Its market penetration success is attributed to the support it provides building professionals by way of certification manuals, and the attractive rating system it offers: fair, good, very good, and excellent—with three, if not all four, of the awards reading as positive [45]. At its core, BREEAM measures energy usage, with certification available to buildings achieving as little as a 6% energy cost reduction over non-certified buildings [46].

LEED stands for Leadership in Energy and Environmental Design, and was developed in the United States (US). Though introduced well after BREEAM, in 1998, it is now perhaps the mostly widely adopted rating system. It operates in over 160 countries, with over 15 billion square feet of building space certified. LEED's assessment categories are: site, water, energy, materials, resources, and indoor environmental quality [47]. CASBEE, on the other hand, is largely limited to its country of origin, Japan—though a worldwide version was piloted in 2015, and is expected to be offered globally in the near future. CASBEE stands for Comprehensive Assessment System for Built Environment Efficiency; it was developed by a consortium of government, industry and academia. While presently limited in market distribution, it does evaluate a much broader combination of sustainability factors, extending its scope of assessment from buildings to embrace city and urban contexts in which the deeper issues of sustainability are grounded. In light of this, CASBEE takes on a different approach to assessment. Rather than summing credits across a range of categories, as do most systems, it evaluates the trade-off between the positive improvements achieved with regard to a building's internal "Built Environment Quality" against the negative impacts that result on the building's surrounding context in the form of the "Built Environment Load". The scheme recognises that positive benefits achieved within a building may come at the expense of its surroundings, and that these externalities are overlooked in other rating schemes [48].

Green Star was developed in Australia in 2003, drawing on the BREEAM model, while adapting it to Australian conditions. Green Star NZ (New Zealand) is a further adaptation, extrapolating the Australian Green Star to New Zealand requirements; it was launched in 2007. The Building

Sustainability Index (BASIX), Evaluation Manual for Green Buildings (EMGB), and National Australian Building Environmental Rating System (NABERS) are further rating permutations that have been developed by government, adding or subtracting various factors, while adjusting the weightings given to each. Most other rating systems are privately developed, and again tailored to different markets and different client needs. A summary of some of the better-known rating tools are provided in Table 1.

In fact, in just about every locality, you will find a plethora of additional software programs, excel spreadsheets, and other instruments offered by maverick operators that can be downloaded and utilised to generate a tailored sustainability rating. For example, the Cooperative Research Centres Program in Australia hosts a 'Low-Carbon Living' website on which 13 tools are listed, along with the developers' contact information. These tools promise all sorts of measures: energy cost predictions from weather data, carbon value engineering, degree of comfort loss in raising house energy efficiency, embodied carbon flows in supply chains, predicted energy load changes as a product of design variation, sustainability gains from co-opting community groups in design processes, behavioural change programme effectiveness, and, psychological readiness to engage in sustainability [49].

Table 1. Summary of major built environment sustainability performance assessment instruments (compiled by authors—Refer Ding [45], Jackson [50]). UN: United Nations; USA: United States of America.

Instrument	Source	Measurement Domain
ABGR	Australian Building Greenhouse Rating; Depart. of Commerce, Australia, 2005	❖ National approach to benchmarking performance ❖ Based on 12 months' energy consumption
AccuRate	CSIRO, 2006	❖ Revised from NatHERS ❖ Uses extensive materials database
BASIX	Department of Infrastructure Planning and Natural Resources, 2004	❖ Web-based planning too ❖ Assesses water and energy in new residences
BEPAC	Canada, 1993	❖ More comprehensive version of BREEAM ❖ Voluntary tool
BREEAM	UK, 1990, 2012	❖ First system developed—Predictive ❖ Expanded into master-planning
CASBEE	Japan, 2004	❖ Developed by government and industry ❖ Based on the concept of closed ecosystems
CEPAS	Hong Kong, 2001	❖ Developed as a benchmark tool with eight criteria ❖ Flexible across both new and existing buildings
CoS	UN Global Cities Program, 2014	❖ Extends across metropolitan regions Measurement based
CPA	Comprehensive project evaluation, UK, 2001	❖ Assesses projects during development ❖ Uses financial and economic approach
DGNB	German Sustainable Building Council, 2013	❖ Precinct assessment ❖ Predicted for five years
DQI	Design Quality Indicator, UK	❖ Aims to help clients define project success ❖ Assesses functionality, build quality, and impact
EcoDistricts	USA, 2015	❖ Neighborhood assessment ❖ Predicted and measured
Eco-Quantum	The Netherlands	❖ Based on life cycle assessment ❖ Only applies to single residences
ED	Urban Development Institute of Australia, 2006	❖ Master-planning assessment ❖ Predicted for 12 months
EMGB	Evaluation Manual for Green Buildings, Taiwan, 1998	❖ Only assesses quantifiable criteria ❖ Single tool for all types of buildings
EPGB	Environmental Performance Guide for Buildings—Public Works, Australia	❖ Measures resource loading ❖ Buildings rated on one indicator

Table 1. *Cont.*

Instrument	Source	Measurement Domain
GBTool	Green Building Challenge, Worldwide—over 20 countries, 1995	❖ Comprehensive framework ❖ Over 90 performance criteria
GHEM	Green home evaluation, China, 2001	❖ Measures environmental performance standards ❖ No weightings of assessment criteria
Green Mark	Building and Construction Authority of Singapore	❖ Master-planning assessment ❖ Predicted
Green Star	Green Building Council, Australia, 2012	❖ Measures commercial buildings only ❖ Ratings from 0 to 6 stars
HKBEAM	Hong Kong building environmental assessment method, HK, 1996	❖ Based on BREEAM ❖ Separate systems for new and existing buildings
LCC	Living Community Challenge—International Future Institute, USA, 2014	❖ District assessment ❖ Independent third-party certification
LEED	Leadership in Energy and Environmental Design, USA, 2000	❖ Developed to create an industry standard ❖ Voluntary self-assessment tool
NABERS	Department of Environment and Heritage, Australia, 2001	❖ For existing commercial and residential buildings ❖ Voluntary self-assessment tool
NatHERS	CSIRO, Australia	❖ Computer-based house energy rating system ❖ Considers design, insulation, and climate
OPC	One Planet Communities—Bioregional Development Group, UK, 2004	❖ Neighborhood design and as built ❖ Predicted and measured verification
SBAT	Sustainable building assessment tool, South Africa	❖ Based on building life cycle ❖ Considers social and economic issues
SPeAR	Sustainable project appraisal routine—Ove Arup	❖ Used for rapid review of project sustainability ❖ Graphical format to present sustainable design

7. Academic Contribution to Sustainability Tool Development

Mirroring this expansion of rating tools has been a burst of academic research on the subject. A recent review study by Li et al. of papers published between 2004 and 2016 identified a total of 57 peer-reviewed articles that made direct comparisons between building assessment methods. Coincidently, these 57 papers examined and compared exactly 57 assessment methods [51]; 74% of the papers made general comparisons between assessment methods, while 70% made category comparisons, 51% made criterion comparisons, and 33% made indicator comparisons between assessment methods [51].

The agenda behind much of the academic research is to foster further variation and embellishment to the list of sustainability factors. The hope is to advance what we already have with ever more sophisticated, all-encompassing, and nuanced schemes. That is, not only is there strong competition among the many sustainability tools on offer, there is also strong pressure to develop even more. Building on the core notion of sustainability as the equilibrium of economic, environmental, and social objectives, Silvius et al., writing in the specific context of project management, parses these into 36 sub-categories, citing disparate concepts such as freedom of association, organisational learning, and diversity as recommended project sustainability checklist items [52]. Swarup et al. identify 40 metrics for ascertaining building performance, including project team selection process, onerous contract clauses, and deviation in actual project cost over projected project cost [53]. At times, it would seem that these studies simply attempt to capture everything that can be thought of; measures and weighting are necessarily subjective. Moreover, the very practice of ballooning the range sustainability factors blurs the boundaries between sustainability performance with that of the building project itself: they become one and the same.

Complementing this interest in the fine-grained performance measurement of building projects is a parallel effort to expand the scope of sustainability beyond the singular point at which the building comes into operation. In assessing the sustainability of projects, Marjaba and Chidiac take the core criteria of economic, environmental, and social objectives, and then lay them out over the three phases of a project's life cycle: manufacturing and construction, occupancy and use, and demolition and reuse. Sustainability performance thus becomes the cumulative performance across many decades of the building's existence. In this regard, site quality, process quality, and technical quality are added dimensions to traditional sustainability metrics [54]. Sustainability measurement has spread horizontally across the value chain, from raw material extraction, through material fabrication, component manufacturing, transportation, construction, use, maintenance, refurbishment, decommissioning, demolition, and recycling. Characteristically, academics have filled these stages with a plethora of sustainability measures; Shen et al. recommended some 112 checklist items covering a project's life cycle [55]. It has also spread vertically across construction sectors, from residential, commercial, industrial, manufacturing, petroleum, energy, transport, and into various other infrastructure domains; each with their own finely tailored sustainability criteria [56].

If that were not enough, even within these many, increasingly narrow criteria, further refinements and qualifications are recommended. In considering just the matter of housing affordability within the broader context of sustainability, Pullen et al. proposed an assessment framework with some 29 considerations [57]. While adding more and more layers of assessment, with greater and greater numbers of performance criteria, may appear cumbersome, if not simply disingenuous, the approach does recognise that a "systems thinking method", in which components must be considered in terms of their relationships with all others, is fundamental to making meaningful headway. The problem of course remains as to what to do with this ever-expanding list of criteria, once identified. What is the strength of the relationship between factors, and how can these be addressed as a system, rather than as merely a list?

8. The Limits of Sustainability Rating Tools

Building environmental assessment methods are considered one of the most potent and effective means to improve the performance of buildings [51]. This quotation sums up the prevailing architectural narrative on sustainability. Liu et al. went on to add: *Defining new assessment items is important . . . due to the fact that existing assessment tools may not always be able to objectively or accurately reflect a building's sustainability . . .* In other words, the recommended solution to the problem with rating tools, is more rating tools.

Indeed, it has become the fashion to rate rating tools. In a comparative study conducted by Doan et al., BREEAM was deemed the strongest rating system, and Green Star NZ the weakest [34], even though Green Star NZ was derived from BREEAM, but adapted to particular local requirements, as shown in Figure 5. Whole sustainability regimes have come under the spotlight. Bond et al. compare the sustainability assessment practices of England, Western Australia, Canada, and South Africa, pointing out again that sustainability is a game of trade-offs, and that ultimately everybody has different priorities as to what they are willing to give up in order to gain something else. Underlying this narrative is not so much that sustainability is subjective, and that each should be left to his own in determining where the right balance might be between economic, environmental, and social objectives, but rather that sustainability has a deeper, political agenda to shape public opinion on where that subjective balance should lie. Competing interests lobby to introduce their aspirations into the ratings game; once included, everybody else becomes compelled to serve this agenda, conforming to ideals that may not be their own, but doing so in order to gain the sustainability star ratings they need.

Resistance to the sustainability movement has not gone unnoticed, but those that are slow to take it up are typically cast as complacent, if not characterised in condescending terms. The study by Swarup et al. comes to the rather obvious conclusion that better sustainability outcomes are achieved when all of the stakeholders are committed: owner, construction contractor, and project team—the earlier their involvement, the better [53]. However, engagement is the product of incentive. Investing in green buildings is costly, both in terms of additional consultants and the construction itself [58]. These green costs can inflate project costs by as much as 25% [59]. Compounding this disincentive are the additional challenges of increased technical difficulty, lengthy approval processes, more complex contractual arrangements, higher project delivery risk, and longer duration to completion.

However, the problem runs deeper. If a sustainability rating system is to work, it must be evidence-based [60]. Yet, despite broad assumptions that the metrics used stand up, there is no consistent environmental data anywhere on the market [61]. In the words of Marjaba and Chidiac, *Certification systems, such as LEED, BREEAM and DGNB are found to be useful and successful at meeting their purpose, however they fail to address all of sustainability's requirements. Moreover, these certification systems have yet to produce metrics that are repeatable, reproducible, and a true reflection of the building performance* [54]. An emerging body of research is finding that participants in environmental rating performance schemes sometimes show no better performance than non-participants [9].

What'a more, it could be worse than that. Despite the intuitive appeal of solar panels as a vehicle for reducing dependency on carbon-based energy and green-house gas emissions, it turns out that the embodied energy in manufacturing solar panels, along with the contaminants released into the environment at the time of their disposal, may outweigh the actual environmental credits gained from their use [62]. This counterproductive side effect of going green extends to the building project itself. Current Australian building energy efficiency regulations do not consider embodied energy. Yet, a study by Crawford et al. showed that while increasing the thermal energy rating of a building envelope can improve the energy performance of a building, that gain may in fact be offset by the increased embodied energy required to make insulating materials and windows more energy efficient. Specifically, Crawford determined that with an embodied energy of +572 gigajoules (GJs) for rated Melbourne houses, the life cycle energy benefit was negligible, while for rated Brisbane houses, with an embodied energy of +422 GJs, the life cycle energy demands actually increased [63]. A recent study conducted by the University of South Australia used the AccuRate energy simulation software program to compare the energy performance of six and eight star-rated homes in Sydney and Adelaide with so-called traditional energy inefficient buildings. While NatHERS endorsed the star-rated homes, their over-reliance on air-conditioning in fact made them far less heat stress resistant than unrated traditionally designed houses [8].

In sum, rating tools differ in what they measure; the range and their weightings are largely subjective, reflecting political aspirations more than scientific accuracy. While systems that require energy monitoring—such as Singapore's Green Mark—have emerged, for the most part, rating tools depend heavily of project documentation in making those measurements, leaving imputed ratings effectively unverified in practice. They tend to be market-oriented, relying on voluntary compliance, with users doing so more to win public relations advances than for anything else. Finally, since most rating systems have not been independently scrutinised, they may not be delivering the environmental benefits they claim [64].

9. Discussion

Pre-industrial cities were, for the most part, relatively small islands of human habitation in a sea of natural landscape. Almost all had walls. These provided a defensive barrier, but also demarcated the limits of the city. Within this circumscribed domain, all of its services could be accessed on foot [25]. This has changed entirely. Megacities are becoming the new normal, with dozens of cities carrying populations in excess of 10 million springing up all around the world [65]. Australia, the least populated continent (excluding Antarctica), is 97% urban, with close to half of all Australians living in or around its two key cities of Melbourne and Sydney, as shown in Figure 6.

The point has been made that this remarkable development arose out of the industrial revolution, with the accompanying urban vision springing from Modernist concepts of progress that themselves were spawned from architectural manifestos such as the "industrial city". However, there is more to be said. In rejecting the past, and past value systems, Modernism as a philosophy sowed the seeds of its own destruction. Once the past was overturned, Modernism had no further foundation on which to establish an enduring value system. Consider the Modernist composers of the early twentieth century—Arnold Schönberg, Alban Berg, Anton Webern, and others. They rejected the past traditions of setting music to a key, or composing with a tonal center. Music was instead, atonal, without a recognisable key—indeed composed to be unidentifiable as sitting in any traditional genre. Their compositions were an intellectual rejection of the past. In 1952, John Cage performed a musical piece entitled 4′33″. In the performance, Cage sat in front of a grand piano, and played not a single note. He is quoted as saying: *I have nothing to say, and I am saying it* [66]. The same can be said for Modernist painters, such as Mark Rothko (blocks of monochromatic paint on canvas) or Jackson Pollock (erratic brush-drippings on canvas). Their works are much lauded, and sell for millions, but the net effect is that they have left art at a dead end. The elephant in the room with regard to all of the traditional arts taken over by Modernism is that, except for the few lingering elite, such art has gravitated to nihilism, and has ceased to be of relevance to mainstream society.

Figure 6. *Herald Sun* newspaper headline: "Melbourne will need another 720,000 homes by 2031" [67].

The same is true of architecture. Where it once spoke to whole communities, now it is largely limited to the patronage of elite clientele looking to build brand awareness through awards, photo displays, news headlines, or simple notoriety. Good architecture produces a dramatic statement, but often the statement made is empty and narcissistic. Where once students attended architecture schools to be mentored in particular ideals and traditions, now, schools attempt to provide the accreditation that students are looking for in as value-neutral an environment as possible.

Nature abhors a vacuum, and into the vacuum has fallen the cause of sustainability. Sustainability is the new cause célèbre in architecture. It is in this light and against this historic background that the sustainability movement is best understood. As such, sustainability in the built environment is less about how the many variables that comprise a green rating tool are measured, than about what those variables should be. Sustainability is less the science of measurement, and more the philosophy of what matters. Stepping into this space are new evolving iterations of the ratings paradigm. The 'Well Building Standard' is a not-for-profit initiative attempting to explore the effects of buildings on human well-being. Through sustainability, architecture has found a road out of its commercial marketing role and back into the fray of debate about the kind of utopia we want to build for ourselves. Sustainability is a catch all term. It is all things to all people. It is the ground into which new seeds of architectural debate are being sown. The debate is far from settled, but through sustainability, architecture is again venturing towards something meaningful to say and do.

10. Conclusions

Sustainability is a narrative; it is about telling stories. Government, architects, academics, consultants, builders, and the community, each in their many ways, are telling us where they see the right balance between the environment, the economy, and social equity. For mainstream sustainability pundits, the focus has been on environmental issues: improved energy efficiency, better water management, responsible waste treatment, and the like. "Green" proponents hold that resources are finite, and that they are being consumed at an ever-increasing rate. This is where the green rating schemes have their origin. An examination of the best-known rating tools—BREEAM, LEED, Green Star, etc.—reveal their purpose; our "carbon footprint" must be reduced.

A criticism arising is that the most these tools can hope to achieve is to slow the inevitable. As long as populations grow, as long as we take up more land to build, and as long as we consume beyond the capacity for the planet to regenerate, environmental degradation can only be slowed, but never reversed. In this doomsday view, the call is made to increase the range of controls, raise the sustainability bar, and be ever more stringent in demanding compliance. This, again, is the view among those that endorse the mainstream sustainability narrative.

However, there are dissenting views. For one, sustainability is meant to be a balance between the environment and economic growth—not merely a struggle to supplant economic development with environmental idealism. This position has credibility, and is not limited only to the so-called fringe "global warming deniers". In 1798, the mathematician Robert Malthus famously predicted that the increasing population, combined with increasing consumption, would inevitably lead to a collapse of civilisation. Populations grow geometrically, he argued, while food production could only grow arithmetically. There was a time, not far off he said, when there simply would not be enough food to feed everybody [68]. Two things should be noted here. Firstly, this did not happen (as an aside—all famines in history were man-made), yet secondly, the premise to his argument continues to drive Green logic.

Malthus's mistake was to presume that resources are finite; they are not. Technological advances outstrip even population growth, putting us in reach of more and more resources that previously were inaccessible. In ancient times, almost everybody was a farmer—the year-round labor of an individual fed little more than one person. Now, in the US, only 2% of the population are farmers, and the US is a net exporter of food. In 1940, an acre of land yielded 30 bushels of corn; in 2016, the same acre yielded 170 bushels. Yields continue to increase at the rate of 1.9 bushels, per acre, per year [69]. The variable at play is technology.

Currently, we see the beginnings of a trend away from fossil-fuel powered cars to electric cars. Possibly, within a generation, these will be driverless. However, it should be noted that this is not because oil is running out. In a now oft-quoted statement to the *New York Times* Magazine, Saudi Arabia's oil minister Sheik Ahmed Zaki Yamani said, *The Stone Age didn't end for lack of stone, and the oil age will end long before the world runs out of oil* [70]. We are in fact finding oil faster than we are using it. In the last 30 years, proven extractable reserves have grown from 680 billion barrels by some 1000 billion barrels [71], as shown in Figure 7. The shift away from oil—such as it is—is not an expedient, it is value-driven.

Then, there is the third side of the triangle: equity. Sustainability is not just about the environment or economy; it is about fairness. Housing should be affordable, amenities such as proximity to schools, parks, and transport must be considered, and in the workplace, health, well-being and the quality of the work environment must be maintained. These dimensions, too, are finding their way into the rating tools framework. Extreme positions on equity are held. The feminist Hayden, writing on the ideal of a non-sexist city, posits . . . *the private suburban house was the stage set for the effective sexual division of labor. It was the commodity par excellence, a spur for male paid labor and a container for female unpaid labor* [72]. For her, the very idea of the suburban house is unsustainable. Other positions are more normative; calling for fair working conditions, fair-trade practices, and a fair distribution of outcomes.

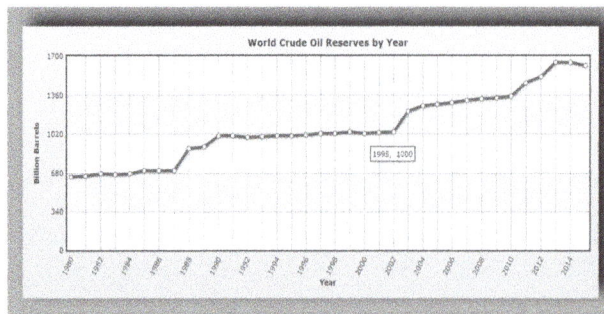

Figure 7. World retrievable crude oil reserves increasing year-on-year, despite consumption rates [71].

In summary, the sustainability project brings together debates about what is economically viable, what is environmentally sound, and what is fair to all. Where that balance ought to be is what is really behind the rise of rating tools. These are physical artifacts of that debate. To some, rating tools cannot be relied on to measure all that is important to sustainability. Returning to Bill McKibben, the world's first recognised environmentalist, we would do well to consider his injunction: *Winning slowly is the same as losing* [73]. Others go further, suggesting that the sustainability project has merely become a meal ticket to its new army of consultants and rating tool purveyors, or a convenient new forum by which lobby groups promote narrow self-interests. However, it is also fair to say that sustainability has provided a platform by which architects can explore their passion for bequeathing us all a better built tomorrow.

Acknowledgments: This paper provides a background to the successful Integral Design Futures (IDF) funding scheme's 2017 program: Charting Pre-Design Sustainability Indicators (School of Architecture and Built Environment, Deakin University).

Conflicts of Interest: The authors declare no conflict of interest.

References

1. Zavadskas, E.K.; Antucheviciene, J.; Vilutiene, T.; Adeli, H. Sustainable decision-making in civil engineering, construction and building technology. *Sustainability* **2017**, *10*, 14. [CrossRef]
2. Bonenberg, W.; Kapliński, O. The architect and the paradigms of sustainable development: A review of dilemmas. *Sustainability* **2018**, *10*, 100. [CrossRef]
3. Čereška, A.; Zavadskas, E.K.; Cavallaro, F.; Podvezko, V.; Tetsman, I.; Grinbergienė, I. Sustainable assessment of aerosol pollution decrease applying multiple attribute decision-making methods. *Sustainability* **2016**, *8*, 586. [CrossRef]
4. Hosseini, M.R.; Banihashemi, S.; Rameezdeen, R.; Golizadeh, H.; Arashpour, M.; Ma, L. Sustainability by information and communication technology: A paradigm shift for construction projects in iran. *J. Clean. Prod.* **2017**, *168*, 1–13. [CrossRef]
5. Foong, D.; Mitchell, P.; Wagstaff, N.; Duncan, E.; McManus, P. Transitioning to a more sustainable residential built environment in sydney? *Geogr. Environ.* **2017**, *4*. [CrossRef]
6. Reed, R.; Wilkinson, S.; Bilos, A.; Schulte, K.-W. A Comparison of International Sustainable Building Tools—An Update. In Proceedings of the 17th Annual Pacific Rim Real Estate Society Conference, Gold Coast, Australia, 16–19 January 2011; pp. 16–19.
7. Mattoni, B.; Guattari, C.; Evangelisti, L.; Bisegna, F.; Gori, P.; Asdrubali, F. Critical review and methodological approach to evaluate the differences among international green building rating tools. *Renew. Sustain. Energy Rev.* **2018**, *82*, 950–960. [CrossRef]
8. Hatvani-Kovacs, G.; Belusko, M.; Pockett, J.; Boland, J. Heat stress-resistant building design in the australian context. *Energy Build.* **2018**, *158*, 290–299. [CrossRef]

9. van der Heijden, J. On the potential of voluntary environmental programmes for the built environment: A critical analysis of leed. *J. Hous. Built Environ.* **2015**, *30*, 553–567. [CrossRef]

10. Van der Heijden, J. The new governance for low-carbon buildings: Mapping, exploring, interrogating. *Build. Res. Inf.* **2016**, *44*, 575–584. [CrossRef]

11. Zavadskas, E.K.; Cavallaro, F.; Podvezko, V.; Ubarte, I.; Kaklauskas, A. Mcdm assessment of a healthy and safe built environment according to sustainable development principles: A practical neighborhood approach in vilnius. *Sustainability* **2017**, *9*, 702. [CrossRef]

12. Zavadskas, E.K.; Bausys, R.; Kaklauskas, A.; Ubarte, I.; Kuzminske, A.; Gudiene, N. Sustainable market valuation of buildings by the single-valued neutrosophic mamva method. *Appl. Soft Comput.* **2017**, *57*, 74–87. [CrossRef]

13. Meadows, D. *Limits of Growth*; Club of Rome, Chelsea Green Publishing: New York, NY, USA, 1972.

14. McKibben, W. *The End of Nature*; Penguin, Random House: New York, NY, USA, 1989.

15. McKibben, W. Buzzless buzzwords. *New York Times*, 10 April 1996.

16. Mazmanian, D.; Kraft, M. *Towards Sustainable Communities: Transition and Transformation in Environmental Policy*; MIT Press: Cambridge, MA, USA, 2009.

17. WCED World Commission on Environment and Development. *Our Common Future*; Oxford University Press: Oxford, UK, 1987.

18. Aarseth, W.; Ahola, T.; Aaltonen, K.; Økland, A.; Andersen, B. Project sustainability strategies: A systematic literature review. *Int. J. Proj. Manag.* **2017**, *35*, 1071–1083. [CrossRef]

19. Zuofa, T.; Ochieng, E. Sustainability in construction project delivery: A study of experienced project managers in nigeria. *Proj. Manag. J.* **2016**, *47*, 44–55. [CrossRef]

20. Bartlett, A. Reflections on sustainability, population growth and the environment—Revisited. *Renew. Resour. J.* **1998**, *15*, 6–23.

21. Dryzek, J. *The Politics of the Earth: Environmental Discourses*, 2nd ed.; Oxford University Press: Oxford, UK, 2005.

22. Caradonna, J. *Sustainability: A History*; Oxford University Press: New York, NY, USA, 2016.

23. Portney, K. *Sustainability*; MIT Press: Cambridge, MA, USA, 2015.

24. Hobbes, T. *Leviathan*; Penguin Books: Baltimore, MD, USA, 1968.

25. Ponting, C. *A Green History of the World*; Penguin: London, UK, 2011.

26. Craven, J. Architecture Timeline—Historic Periods and Styles of the West. Available online: https://www.thoughtco.com/architecture-timeline-historic-periods-styles-175996 (accessed on 28 January 2018).

27. Collins, P. *Changing Ideals in Modern Architecture*; Faber and Faber: London, UK, 2006.

28. Stinson, L. Remembering the Legend behind form Follows Function. Available online: https://www.wired.com/2015/09/man-coined-form-follows-function-born-today/ (accessed on 27 March 2018).

29. Pevsner, N. *The Sources of Modern Architecture and Design*; Cambridge University Press: Cambridge, UK, 2010.

30. World Population Balance. The Global Population Situation—An Overview. Available online: http://www.worldpopulationbalance.org/global_population (accessed on 27 March 2018).

31. Garnier, T.; Pawlowski, C. *Une Cite Industrielle*; Il Balcone: Paris, France, 1967.

32. Wiebenson, D. Utopian aspects of tony garnier's cité industrielle. *J. Soc. Arch. Hist.* **1960**, *19*, 16–24. [CrossRef]

33. Banihashemi, S.; Tabadkani, A.; Hosseini, M.R. Integration of parametric design into modular coordination: A construction waste reduction workflow. *Autom. Constr.* **2018**, *88*, 1–12. [CrossRef]

34. Doan, D.T.; Ghaffarianhosseini, A.; Naismith, N.; Zhang, T.; Ghaffarianhosseini, A. A critical compariosn of green building rating systems. *Build. Environ.* **2017**, *123*, 243–260. [CrossRef]

35. Dixit, M.K.; Culp, C.H.; Fernandez-Solís, J.L. System boundary for embodied energy in buildings: A conceptual model for definition. *Sustain. Energy Rev.* **2013**, *21*, 153–164. [CrossRef]

36. Wong, J.K.W.; Zhou, J. Enhancing environmental sustainability over building life cycles through green bim. *Autom. Constr.* **2015**, *57*, 156–165. [CrossRef]

37. Udawatta, N.; Zuo, J.; Chiveralls, K.; Zillante, G. Attitudinal and behavioural approaches to improving waste management on construction projects in australia. *Int. J. Constr. Manag.* **2015**, *15*, 137–147.

38. Nikmehr, B.; Hosseini, M.R.; Rameezdeen, R.; Chileshe, N.; Ghoddousi, P.; Arashpour, M. An integrated model for factors affecting construction and demolition waste management in Iran. *Eng. Constr. Archit. Manag.* **2017**, *24*, 1246–1268. [CrossRef]

39. Cooper, I. Transgressing discipline boundaries. *Build. Res. Inf.* **2002**, *27*, 321–331. [CrossRef]

40. Mardani, A.; Jusoh, A.; Zavadskas, E.K.; Cavallaro, F.; Khalifah, Z. Sustainable and renewable energy: An overview of the application of multiple criteria decision making techniques and approaches. *Sustainability* **2015**, *7*, 13947–13984. [CrossRef]

41. Wilkinson, S. *Report for Royal Institution of Chartered Surveyors*; Royal Institution: London, UK, 2014.

42. Vierra, S. *Green Building Standards and Certification Systems*; National Institute of Building Sciences: Washington, DC, USA, 2011.

43. Mao, X.; Lu, H.; Li, Q. A comparison study of mainstream sustainable/green building rating tools in the world. In Proceedings of the International Conference on Management and Service Science, Wuhan, China, 20–22 September 2009; pp. 1–5.

44. BREEAM. Breeam Homepage. Available online: http://www.breeam.com/ (accessed on 12 February 2018).

45. Ding, G. Sustainable construction—The role of environmental assessment tools. *J. Environ. Manag.* **2008**, *86*, 451–464. [CrossRef] [PubMed]

46. Vimpari, J.; Junnila, S. Value influencing mechanism of green certificates in the discounted cash flow valuation. *Int. J. Strateg. Prop. Manag.* **2014**, *18*, 238–252. [CrossRef]

47. LEED. Leed Homepage. Available online: https://new.usgbc.org/leed (accessed on 12 February 2018).

48. CASBEE. Casbee Homepage. Available online: http://www.ibec.or.jp/CASBEE/english/ (accessed on 12 February 2018).

49. Low Carbon Living. Crclcl Tools Catalogue. Available online: http://www.lowcarbonlivingcrc.com.au/resources/crc-publications/fact-sheet/crclcl-tools-catalogue (accessed on 12 February 2018).

50. Jackson, S. A summary of urban assessment tools for application in australia. In *Environment Design Guide*; Australian Institute of Architects: Melbourne, Australia, 2016.

51. Li, Y.; Chen, X.; Wang, X.; Xu, Y.; Chen, P. A review of studies on green building assessment methods by comparative analysis. *Energy Build.* **2017**, *146*, 152–159. [CrossRef]

52. Silvius, G.; Schipper, R.; Nedeski, S. Consideration of sustainability in projects and project management: An empirical study. In *Sustainability Integration for Effective Project Management*; IGI Global: Hershey, PA, USA, 2013.

53. Swarup, L.; Korkmaz, S.; Riley, D. Project delivery metrics for sustainable, high-performance buildings. *J. Constr. Eng. Manag.* **2011**, *137*, 1043–1051. [CrossRef]

54. Marjaba, G.E.; Chidiac, S.E. Sustainability and resiliency metrics for buildings—Critical review. *Build. Environ.* **2016**, *101*, 116–125. [CrossRef]

55. Shen, L.Y.; Hao, J.L.; Tam, V.; Yao, H. A checklist for assessing sustainability performance of construction projects. *J. Civ. Eng. Manag.* **2010**, *13*, 273–281.

56. Ortiz, O.; Castells, F.; Sonnemann, G. Sustainability in the construction industry: A review of recent developments based on lca. *Constr. Build. Mater.* **2009**, *23*, 28–39. [CrossRef]

57. Pullen, S.; Arman, M.; Zillante, G.; Zo, J.; Chileshe, N.; Wilson, L. Developing an assessment framework for affordable and sustainable housing. *Aust. J. Constr. Econ. Build.* **2010**, *10*, 48–64.

58. Metibogun, L.; Baird, G. Integrating green building index consultancy with residential design. In *Fifty Years Later: Revisiting the Role of Architectural Science in Design and Practice: 50th International Conference of the Architectural Science Association*; Zuo, J., Daniel, L., Soebarto, V., Eds.; The Architectural Science Association: Adelaide, Australia, 2016.

59. Hwang, B.-G.; Ng, W.J. Project management knowledge and skills for green construction: Overcoming challenges. *Int. J. Proj. Manag.* **2013**, *31*, 272–284. [CrossRef]

60. Low Carbon Living. *Viable Integrated Systems for Zero Carbon Housing*; University of South Australia: Adelaide, Australia, 2014.

61. Tam, V.; Zeng, S.X. Sustainable performance indicators for australian residential buildings. *J. Leg. Aff. Disput. Resolut. Eng. Constr.* **2013**, *5*, 168–179. [CrossRef]

62. Mulvaney, D. Solar's Green Dilemma: Must Cheaper Photovoltaics Come with a Higher Environmental Price Tag? Available online:. Available online: http://spectrum.ieee.org/solar0914 (accessed on 12 February 2018).

63. Crawford, R.; Bartak, E.; Stephan, A.; Jensen, C. Does current policy on building energy efficiency reduce a building's life-cycle energy demand? In Proceedings of the 49th International Conference of the Architectural Science Association Living and Learning: Research for a Better Built Environment, Melbourne, Australia, 2–4 December 2015; pp. 332–341.

64. Wong, S.Y.; Susilawanti, C.; Miller, W.; Mardiasmo, D. A comparison of international and australian rating tools for sustainability elements of residential property. In *COBRA*; RICS: London, UK, 2015.

65. Khanna, P. *Connectography—Mapping the Future of Global Civilization*; Random House: London, UK, 2016.

66. Jones, J. The Curious Score for John Cage's "Silent" Zen Composition 4'33''. Available online: http://www.openculture.com/2013/10/see-the-curious-score-for-john-cages-silent-zen-composition-433.html (accessed on 23 March 2018).

67. Masanauskas, J. *Melbourne's Continuing Population Boom Means Another 720,000 Homes Will Be Needed by 2031*; Herals Sun: Melbourne, Australia, 2014.

68. Sachs, J. Are Malthus's Predicted 1798 Food Shortages Coming True? *Scientific American*, September 2008.

69. Nielsen, B. Historical Corn Grain Yields for the U.S. Available online: https://www.agry.purdue.edu/ext/corn/news/timeless/yieldtrends.html (accessed on 23 March 2018).

70. Maass, P. The breaking point. *New York Times Magazine*, 21 August 2005.

71. United States Energy Information Administration. World Crude Oil Reserves by Year. Available online: https://www.indexmundi.com/energy/?product=oil&graph=reserves (accessed on 23 March 2018).

72. Hayden, D. *What would a non-sexist city be like? In Urban Communities in the 21st Century—From Industrialization to Sustainability*; Hutson, M., Ed.; University of California: Berkeley, CA, USA, 2010; pp. 219–235.

73. McKibben, W. Winning Slowly Is the Same as Losing. Available online: https://www.rollingstone.com/politics/news/bill-mckibben-winning-slowly-is-the-same-as-losing-w512967 (accessed on 23 March 2018).

Review

The Architect and the Paradigms of Sustainable Development: A Review of Dilemmas

Wojciech Bonenberg and Oleg Kapliński *

Faculty of Architecture, Poznań University of Technology, Nieszawska Str. 13C, 60-965 Poznań, Poland; wojciech.bonenberg@put.poznan.pl
* Correspondence: oleg.kaplinski@put.poznan.pl; Tel.: +48-61-665-3260

Received: 20 December 2017; Accepted: 30 December 2017; Published: 3 January 2018

Abstract: This article presents the architect's attitude towards the paradigms of sustainable development. The place and role of the architect in the implementation of the multidimensional processes of sustainable design are presented. Basic dilemmas and antinomies are presented. The analysis of architects' attitudes towards these problems is performed in various contexts, examining the architect's awareness and his/her environment in view of changes under way. The article draws attention to the status of knowledge, changes in design paradigms, legislative and organizational requirements. The importance of architectural culture level, the need for training and ways to support the implementation of new design paradigms through integrated activities are indicated. The research results, regarding public awareness of architecture and sustainable development, are illustrated, with examples from Poland.

Keywords: architect; sustainable architecture; paradigms of design; knowledge; society; Poland

---

1. Introduction

The results of developing sustainable architecture are founded on the symbiosis of ecologists and architects. It began with these two professional groups proposing a change in the function of the building, i.e., a transition from a linear approach to a closed circulation plan. Therefore, from an ecological point of view, the plan of the building function has become a paradigm.

In a linear pattern, the building is treated as a "place of processing natural resources into waste". For example, energy is "converted" into heat losses, clean water into sewage, fresh air is converted into used air, materials and consumer goods into classical waste.

In a closed circulation plan, a building may change from a voracious consumer of energy and all other resources, into a more self-sufficient unit. It will be possible to use much less energy for heating in winter, and cooling and ventilation in the summer (part of the energy will be recovered). Part of the water can not only be saved, but also re-used. Generally, a large amount of waste can be avoided altogether, or used again.

The transition from one plan to another is evolutionarily. The first step in this design trend was passive, low energy buildings. The next step was friendly buildings—friendly not only for people, but also for the environment. Today, we are talking about almost zero energy buildings, autonomous buildings, and IQ architecture. There are numerous examples of such buildings. The shape of *The Edge*, a new office building in Amsterdam, described as the most modern and the most "green commercial building" in the world, is quite unusual. It was included in the category of intelligent buildings and, as part of the BREEAM (Building Research Establishment Environmental Assessment Method) certificate; in 2016, the building then scored a record value of 98.36% (on a scale from 0 to 100%). Only two years after its construction, two more office buildings scored even higher: Bloomberg's new European headquarters in London (which scored 98.5%) and the Geelen Counterflow Office in Haelen (The Netherlands) which scored 99.94% [1–3]—of course, all in the Offices—New category.

However, there are also examples of achieving spectacular success in the area of designing sustainable settlements or cities. Such an example is the city of Masdar in the Abu Dhabi emirate, together with the Masdar Institute of Science and Technology, which is autonomous in terms of energy. Moreover, it meets all other criteria of sustainable development.

Arriving at sustainable design is a continuous process. What consequently changes is the architect's attitude towards the design paradigms, which are particularly noticeable in the context of the intellectual and ecological revolution.

Society was, or has been a witness to, three revolutions, which have significantly influenced architecture. The industrial revolution in the late nineteenth century (replacing physical labour with machines) was a foundation of two other revolutions: the information and administration revolution (from the mid-twentieth century: information processing, strengthening mental abilities) and sustainable development type revolution, embracing aspects of ecology, economics and social/cultural values. This has been accompanied by enormous progress in the field of digitisation. All these revolutions are developmental in character, but one can also say that their derivative is a new term: architectural IQ. The dynamics of new examples/practical implementation of designs is amazing. One perhaps cannot understand the essence of the revolution—especially the sustainable development type—but, in design, it is obligatory to adapt to applicable law [4]. Unfortunately, the understanding of the paradigms of sustainable development and design paradigms in some social groups, including some architects, still produces some difficulties—dilemmas arise; there are contradictions in the interpretation of rules.

According to the authors, the key to solving these dilemmas is knowledge (its acquisition, broadening the scope), which will strengthen the understanding and application of the above-mentioned paradigms. Sometimes resistance to sustainable design is evident. The source of reluctance is limited awareness by part of the public as well as a certain group of architects.

The article deals with the architect's attitude towards these problems, examining the architect's awareness in different contexts, and his/her environment in the light of the changes under way. The results of research on the public awareness have been illustrated with examples from Poland. In addition, a review of the literature [5] indicates that there exists abundant studies in the sphere of sustainable development, but few in the area of the analysis of architects' attitudes in the face of changes in design, and ways of design (see also [6–8]).

2. Explicit and Tacit Knowledge

Many renown architects began their education and professional careers under the influence of modernism, which, in the 1980s, significantly impacted the approach to architectural and urban design. The difference in the level of knowledge prevailing in that period, and the knowledge necessary to understand and implement the principles of sustainable development, is enormous. The condensation of problems and new elements of knowledge faced by the architect today is quoted, in part, in Figure 1.

The quoted set of knowledge elements indicates that design is a team game, that, in order to cope with, you need a competent team at hand and new organizational methods should be used (e.g., integrated design), new tools (e.g., Building Information Modeling).

In designing, and from the architect's point of view (and his/her office), two types of knowledge are distinguished: formal (*explicit knowledge*) and hidden (*tacit knowledge*). *Explicit knowledge* is acquired; it stems from standards, technical design conditions, and can even be obtained from technical specifications. It also includes design paradigms. Tacit knowledge refers to individual skills, creativity, and is used on a regular basis in designing. Both types of knowledge constitute intellectual and creative capital [9].

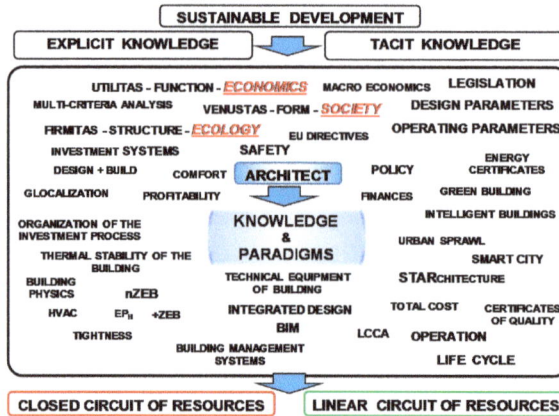

Figure 1. Typical dilemmas in the interaction of knowledge—design paradigms—sustainable development.

An architect who is unable to use the elements of knowledge supporting sustainable design (Figure 1) loses not only his/her prestige, but can also make technical mistakes and become vulnerable to conflicts with the participants of the investment process [10–12]. Figure 1 has been supplemented with two contrasting patterns of the function of buildings (closed and linear circulation), which are at the basis of ecological design.

3. Basic Legislative Requirements

Design paradigms require permanent extension of knowledge. Legal acts accompany this process. As early as 1987, under the UNESCO patronage, a specific message for the architect was formulated, i.e., suggesting that the architect—taking into account the three criteria below—turn his/her attention to the beneficiary; that is, that he contributes to a fully balanced model of life. These criteria consisted of emphasizing the integration of activities in three areas: economic growth and an even distribution of profits, natural resources and protection of the environment and social development.

Further initiatives in this area were taken over by the European Parliament and the Council of Europe, in particular through the Commité Européen de Normalisation (CEN/TC 350), (cf. [13]). Subsequent acts tried to structure a number of issues related to sustainable development in construction; however, it has quickly turned out that designing buildings in accordance with these principles is not easy and, above all, requires a holistic, integrated approach. This results from the multitude of parameters defining the impact of the building on the three pillars of sustainable development: environment, society and economy. Therefore, a number of documents, developed by CEN/TC 350, have seeked to organize matters. The architect has been advised to design buildings that would not be burdensome for the natural environment, would meet conditions of comfort for users, and are at the lowest possible costs during the entire life cycle of the structure.

According to CEN/TC 350 and national standardization committees (e.g., in Poland, KT PKN 307, cf. [14]), the architect should take into account three basic areas, supported by the following standards:

- the area of environmental assessment,
- the area of social evaluation,
- the area of economic evaluation.

This section is divided by subheadings. It should provide a concise and precise description of the experimental results, their interpretation as well as the experimental conclusions that can be drawn.

The standards for assessing the impact of a building on the natural environment are determined by four groups of impacts, i.e.,

- parameters defining environmental interactions (including six quantifiable parameters),
- parameters defining the consumption of resources (including 10 quantifiable parameters),
- parameters regarding the amount of waste generated during the life cycle of a product or building (including three quantifiable parameters),
- parameters included in the LCA (Life Cycle Assessment) assessment, so-called output streams (including three quantifiable parameters).

This multitude of regulations favours the idea of sustainable architecture; however, it raises anxiety among designers. There are, however, indications which have the traits of imperatives. Indeed, in the last few years, the architect (and other entities of the investment process) collided with two very important requirements. They have a clear social and economic dimension.

The first of these results in requirements from the introduction of decisive instruments regarding sustainable policy, such as buildings with a demand for energy close to zero. This is a result of the 2002/91/CE EU Directive, amended in May 2010. This directive entered into force, for example in Poland, on 8 July 2010. The reality is that after 31 December 2018, all new buildings, used and owned by state administration, are to be designed/erected as buildings with almost zero energy consumption; however, after 31 December 2020, all new buildings are to be almost zero energy buildings ("nZEB").

The second very important requirement results from the 2010/31/EU Directive and national standards (e.g., PN-EN 15459). The requirement applies to the calculation period of the building's value (not as of today, but in a time perspective). A calculation period of 30 years has been introduced for residential and public buildings, and 20 years for non-residential buildings used for commercial purposes. This forces designers to become acquainted with such terms as Life Cycle Cost (LCC), Life-Cycle Cost Analysis (LCCA) and, above all, costs generated in the operation phase.

Both the 2010/31/UE Directive and the (EU) Commission Delegated Regulation No 244/2012 clearly indicate that the procedure for determining the "nZEB" standard must be based on economic calculations including the total cost. Total costs account for the sum of the net present value of the initial investment costs (Net Present Value, NPV), the sum reflecting maintenance and operating costs, as well as the costs of removal (liquidation of the investment). In order for the building assessment results to be comparable, a discount rate of 3% was adopted. This means the end of using the term "low investment costs", which, unfortunately, still functions in some countries as a criterion for evaluation, especially in public tenders. Incidentally, a discount rate of 3% is appropriate for EU countries with stable economies. It is at least twice as high in Polish conditions.

The introduction of the abovementioned legislative requirements has highlighted two characteristic phenomena.

Firstly, there is a permanent paradox to be observed, especially in Central and Eastern Europe: in the coming years: the "nZEB" standard of designing will apply, while the energy standard of about 10 million apartments in Poland is lower than 240 kWh/m^2. This information is sensitive, indicating a low level of environmental culture. However, at the same time, one can take an optimistic view of this problem: there will be jobs for our graduates owing to the expected modernization of many buildings.

Secondly, the introduction of the sustainable development paradigm resulted in a significant increase in demand for knowledge of physics of buildings. Only in this way can one explain myths and common opinions about, for example, the so-called glass houses (glass architecture), thermos buildings, breathing buildings, CO_2 emission, tightness, etc. Unfortunately, what is needed on top of that, is detailed knowledge about heat transfer (another aspect is the accumulation mass of the building), knowledge about air exchange, moisture sorption, water vapour diffusion, the role of ventilation, etc., which are more and more often noticed during various conferences (cf. [15–19]).

4. Integration of Creative Processes on the Path to Attaining Integrated Design

It is standard to start teaching architecture from a presentation of the Vitruvian triad, which presents the perfection of architectural work as the balance between durability, usability and beauty (*Firmitas, Utilitas, Venustas*). Meanwhile, it is L.H. Sullivan's motto which is closer to our times: *Form follows Function*, nowadays transformed into *Form follows Energy*—the perspective of perceiving architecture changes, especially in the context of sustainable development. Another maxim results from three premises: ecology, economy and society. This triad is shown on the right hand side of Figure 2, as the green triad. The drawing also presents the pathological (though contemporary) black triad, namely: space, politics and money.

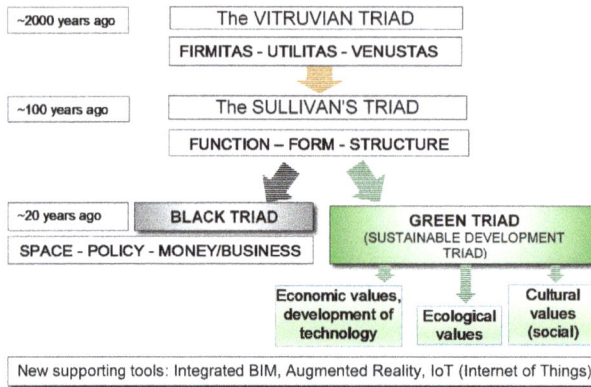

Figure 2. Architectural design paradigm triads.

One could ask the ironic question: which path to choose? Of course, none of the architects will officially dare to support any other triad but the one on the right (the green triad), on the other hand, why are there so many cases in Poland resulting from the black triad? Examples can be found in almost every country, where architectural services do not honour their obligations. One of the drastic examples is the gigantic "Univermag" in Zakopane in the historic Krupówki Street, blocking the view of the Tatra Mountains. How could an architect possibly match the notion of *genius loci* or *context of the place* with such a situation?

Several paradigms can be mentioned in the framework of sustainable design. One of them is a paradigm based on characteristic (target) parameters, which is a function of design (permanent) parameters and operational (variable) parameters. This approach is emphasized in the reverse aspects of design [20], and within *architechnology*. This intellectual trend gave rise to the paradigms of designing energy-efficient buildings, and then, "smart buildings". Today, the term *architectural IQ* already exists, and architectural IQ may be measured by an additional criterion, i.e., the building's capability of adapting changes occurring not only in the internal, but also in the external environment.

Within the framework of environment-friendly approach to architecture, the building analysis should be applied throughout its entire life cycle. Architects call this "from cradle to grave".

Under the new regulations, the application of the rule of low investment costs has been eliminated in favour of total costs. No documentation which does not specify the value of the building in the time span of 20 or 30 years can leave the architect's design office. This dimension was also adopted in Poland; the authors of this article believe that this period is too short, because at this time there may yet be no repairs, replacement of doors and windows, etc.

The next paradigm of sustainable policy mandates that, within either a year or three years, buildings must be designed with a demand for energy close to zero. The dates depend on the form of ownership.

The architect entered the twenty-first century with the changing paradigm of designing from linear (traditional) to integrated. The linear process is characterized by the division into industry disciplines, joining in, one after another, to the implementation of project documentation. In retrospect, it can be determined that such mode of design certainly gives a visual effect (aesthetic, [21]), but it often leads to ineffective energy performance of the created works, not to mention high operating costs and debatable impact on the environment.

The American Institute of Architects (AIA) has created a new type of contract, used for the implementation (design) of construction projects, called the *Integrated Project Delivery* (IPD), [22]. Integrated design is an iterative and interactive process, a way of implementing the entire investment process, which, in a rational (almost optimal, [5,23–27]) manner, in terms of cooperation of the project team—which includes the architect, the industry, the investor, the contractor and the user—allows creation of a balanced object from the viewpoint of construction and operating costs.

When comparing both design processes, it can be determined that the linear process is characterized by the separation of creative processes, while the integrated design is characterized by the integration of creative processes. The next stages of integration of creative processes—i.e., departing from the separation of these processes—are presented in a synthetic way, in Figure 3. *Integrated Project Delivery* (IPD) clearly promotes sustainable architecture.

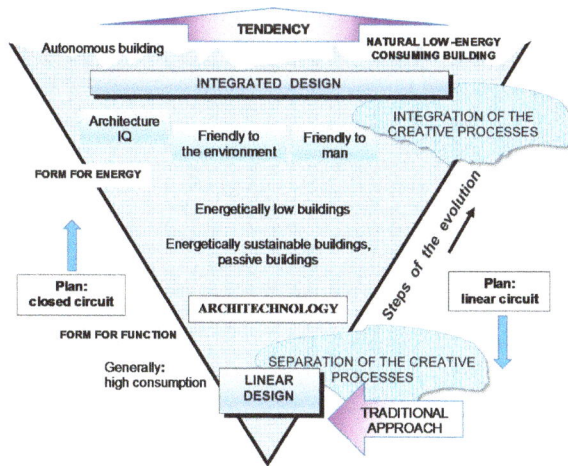

Figure 3. A reverse pyramid of the integration of creative processes in the process of achieving integrated design.

In some countries, this way of investing is underestimated. The most important limitation is inadequate legislation governing the organization of the investment process. Independent research [28,29] points directly to two important limitations: the shortage of specialists and experts from many disciplines who would understand architecture and the design process and the shortage of architects who would be prepared to run an integrated project team and to act as moderators/leaders of such a team. In countries such as Poland, there is a structural constraint, which is still important: the 2015 BCMM's research [30] shows that currently small-scale offices, employing up to three architects, dominate the structure of architectural office management in Poland, taking the share of up to three-quarters of the market. The following questions have arisen: How can we implement digital design technologies in such small teams? How can these small teams adapt to the canons of integrated design? How can we prevail over broadly-understood competition?

5. The Architect's Attitude towards Antinomic Situations

In his/her professional practice, the architect encounters various conflicts, not only administrative and substantive, but also squabbles with contracting parties and users. In the aspect of sustainable development, first and foremost, the architectural versus energy conflict is visible. It is a functional–spatial, usage related, financial and aesthetic conflict.

Architectural design is known for the need of compromise, especially in the context of energy criteria—functional and architectural needs—and increasing the insulation of external partitions. A typical, though a more specific conflict, is the relationship between the surface to volume of the building (A/V ratio) and a flat's or office's access to natural light, including the issue of terms of self-shading. It is known that the smaller the A/V ratio, the more compact the building is and the lower the heat loss. Unfortunately, in our Eastern European climate, it is not possible to achieve a passive standard for an A/V ratio exceeding 0.7. Studies have shown that an increase in the value of the A/V ratio by 0.1 means an increase in the external surface of the building in relation to its volume, and higher heat losses. To compensate for this, it is necessary to increase the thickness of the thermal insulation of an opaque barrier by around 3 cm [31]. This can explain why this coefficient is close to one in this part of Europe.

All this meant that originally passive buildings erected in some European countries, including Poland, were designated as "*zeroarchitectural*". Practice has shown something different, and splendid constructions were built. Moreover, the additional costs of the passive standard in Austria and Germany no longer exceed 4–6%, so they have dropped from around 20%, according to the Passive House Institute (PHI), in Darmstadt [32]. Most architects seek mitigation and even elimination of architectural and energy conflicts in *Integrated Project Delivery* (see e.g., [33–36]).

In this context, one of the architect's characteristic attitudes is quite noticeable. If there are no explicitly defined endpoints in the contract (in the requirements laid out by the contracting authority, or in the technical specifications), the architect will only prefer functional and architectural solutions, and is not guided by the energy prerogatives resulting from sustainable development. In this case, such an attitude does not result from the architect's ignorance of, but from, conformism. In the age of sustainable development, should an architect not promote innovative solutions as part of his/her mission? (See [4,6–9]).

Almost simultaneously with the concept of sustainable architecture, the concept of STARchitecture or STAR-architecture appeared. This describes emotional, star and media oriented architecture, in need of publicity (and the architect's success) and, unfortunately, often lacking the features of pro-ecological architecture. From the point of view of sustainable development, in most cases, this is a poorly-understood success. These types of buildings can be assessed (and compared on the backdrop of already-developed criteria used in certificates, e.g., BREEAM, LEED.) Unfortunately, data from this field are reluctantly disclosed. A positive example is the *Edge* building, mentioned above.

In the last few years, structures have emerged expansively, which can be attributed to the definitions of biomorphic, bionic or biomimetic architecture. The examples are excellent: the Zayed National Museum—called the desert sculpture in the United Arab Emirates (Foster & Partners, Figure 4a); Callebaut's stone mounds in China (Figure 4b) and the Ordos Museum—called the high tech bionic dune in Inner Mongolia (Chinese MAN design office, Figure 4c). The question arises whether bionic or biomorphic architecture is, by definition, sustainable architecture? We have to be very careful providing the answer, because bio-objects must meet the classic conditions assigned to sustainable architecture. Some see this type of architectural solution as a vehicle securing the status of sustainable architecture. Bio is not, by definition, a fully sustainable architecture, but it is certainly interesting. In STAR architecture, anti-smog architecture has emerged in a good context, for example "Anti Smog: An Innovation Centre in Sustainable Development" in Paris. V. Callebaut's building, is now an example of *sustainable design* (Figure 4d). The building, equipped with vertical axis wind turbines, due to financial reasons, has, unfortunately, not been built.

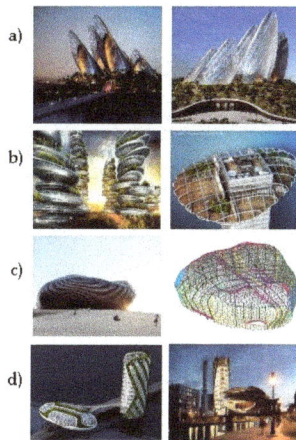

Figure 4. Examples of bio-type architecture (sources [37–43]). (**a**) The Zayed National Museum; (**b**) Callebaut's stone mounds; (**c**) The Ordos Museum; (**d**) Anti Smog: An Innovation Centre in Sustainable Development.

There is one more area of misunderstanding. The truth is that many designers limit sustainability to just designing a building. Spatial planning and urban planning are less often discussed in the context of sustainable development. From scanning through literature using web browsers, one can conclude that the number of articles on sustainable buildings is significantly higher than articles on sustainable cities. This ratio is 5:1. However, Smart City is more than just a sustainable building. The building is only part of the space. All the three pillars of sustainable development should be fully interpreted and applied, and, above all, the socio-cultural pillar [44,45].

There is also a counter-phenomenon, dangerous for sustainable development. It is the *urban sprawl* or sub- or ex-urbanization. This phenomenon is now recognized as the opposite of sustainable development. *Urban sprawl* is treated as a threat to a city's identity [46]. The reaction of the architectural community should be unambiguous.

6. Professional Prestige and Identity of the Architect

The symbolic scope of knowledge, as outlined in Figure 1, is indispensable for the architect, to allow conscious implementation of the paradigms of designing sustainable structures. However, it is extensive interdisciplinary knowledge, organizational skills and creative skills that predispose the architect to be a moderator in Integrated Project Delivery (IPD). These factors will determine the architect's competence. In turn, professional competences strongly affect the prestige of an architect. The role of universities and professional self-government is of great importance here. In all universities, academic curricula have been, or are being, modified (see good examples [47,48]). The reactions of professional self-government organizations of local architects' engineers, supported by editorial boards of trade magazines are satisfactory. Examples include the Architects' Council of Europe (ACE) and the National Chamber of Polish Architects (IARP), see [49].

Apart from the need for intensive acquisition of knowledge, the architect is required to lobby for changes in legislation (adapted to changing paradigms and organization of design), even to become involved in marketing addressed to the general public in the field of architectural culture, and influencing the image of the profession. This also affects the professional prestige and identity of the architect, as illustrated in Figure 5. Explicit knowledge—increase of competences—of social interaction skills, means an increase in professional identity and a stronger impact on society and even on legislation. Hence, there are guarantees of high-quality of architectural space, and the practical use of sustainable development paradigms.

A graduate of the architecture department at the beginning of the course of studies should be aware that technical knowledge is rapidly aging. It requires continuous self-education: one cannot follow the routine. Broadly understood design paradigms change.

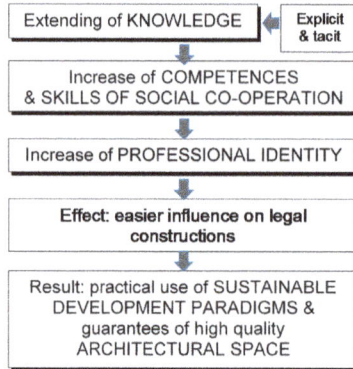

Figure 5. Prestige of the profession and identity of the architect in the context of sustainable development.

Unexpectedly, the architect's professional prestige is raised by BIM. BIM is an instrument which is expected to link sustainable construction with integrated design, and with a building's life cycle. M in this abbreviation means model, modelling and management. It is said that it is almost a revolution in managing large project projects. BIM technologies allow all participants of the process to access the same model of the building being designed at the same time. Therefore, designing in BIM is not only a change in software, but a change in the whole approach to the organization of investment and, above all, the design process. Conferences and even summits (e.g., [50]) take place in almost every country, concluding that the use of BIM definitely increases the prestige of design offices and contractors.

Tardiness is nothing out of the ordinary, though. Here is an example: as results from the research carried out in 2015, and the report [51], the awareness and use of BIM in Poland, compared to other countries, is not favourable. Nevertheless, awareness of its existence is noticeably higher in larger offices employing more than 10 people (56.5%) and among younger people more focused on new design techniques, and this amounts to about 60 percent. At the same time, the respondents say that BIM awareness is higher among architects (65.4%) than in other industries. Such a high level of awareness, though, is not the same as active use of this tool.

According to these studies, among the factors that slow down the development of BIM (in Poland), respondents mention the existence of a small number of specialists working with BIM (71.4%), low awareness of benefits for investors (68.9%), lack of shared operating standards (68.9%), project prices that are too low (83.9%) and, above all, reluctance to make changes in the methodology of design (61.5%).

The element accelerating the implementation of BIM is the 2014/24/EU Directive of the European Parliament and of the Council of 26 February 2014. The Member States were supposed to implement this directive into their national legal systems by 18 April 2016. Some countries, including Poland, followed the British experience [52–54]. In Great Britain the preparations, lasting from 2008, were divided into four stages, and from 2016, BIM became obligatory for public investments [55]. Interesting remarks are also presented in [56,57].

One thing is almost univocal: hope in the implementation of all these paradigms and instruments should be placed in architectural youth, better prepared for the profession. Moreover, the obligation to introduce BIM changes the structure of the design/investment market to a large extent.

7. Knowledge and Public Awareness

The architect alone will not solve the above issues. Knowledge and public awareness are important for the success of the sustainable development program. Unfortunately, the international Dodge Data & Analytics report [58] says that there are countries with diverse public awareness about the benefits of sustainable construction. The group of countries with the lowest awareness includes Colombia, Brazil, India and Poland. A similar contestation was provided by the RenoValue report. [59]. Most respondents indicated a lack of knowledge and support from the central administration or local governments as the main obstacle on the path of development of the sustainable market. Awareness alone will not help if there are no proper legal regulations [60].

Since the level of public awareness in individual countries cannot be consolidated, detailed considerations are focused on Poland. In 2015, the Public Opinion Research Centre (TNS OBOP) surveyed the knowledge of Poles about energy saving. It turned out that 86 percent of the subjects associated energy saving only with the reduction of electricity consumption, while as much as 71 percent of the energy is used by heating [61–63].

Architecture is not the subject of interest for the average Pole (though they declare it is), or, with few exceptions, of the leading media.

The low architectural culture of Poland undoubtedly results from complex, long-term historical processes, as well as the lack of basic knowledge about architecture and the importance of space for the community. The SARP Report (Association of Polish Architects) "Space of Polish life" [64] indicates that the reason for this state is also the lack of basic architectural education, which should be taught as part of the knowledge about the environment from kindergarten, and then enriched by incorporating architectural topics into art education in schools.

The opinions about green/sustainable buildings are characteristic. For most market participants, the most important characteristic of green buildings is energy saving: 95%. A low percentage of respondents have noticed that sustainable construction does not only mean energy saving, but also improvement of well-being, health and reduction of absenteeism. The detailed Construction Marketing Group (CMG) report is uncompromising and exposes this situation (see [65], compare also [66–69]).

The research has shown how low the level of knowledge on the market is about the impact of sustainable buildings on the health and productivity of employees (CMG 2013) [67]. The perception of benefits of sustainable construction from the viewpoint of developers, investors and major tenants was examined. Fifty-four percent of respondents (the same number of companies) disagree or are uncertain about the statement that their company might be willing to incur higher costs of purchasing or renting a green building, in order to provide their tenants or employees with a healthy workplace.

In the United States, absences due to health problems cost companies several hundred to even 2500 dollars per employee per year [70]. Green solutions help to significantly reduce the number of absences caused by illness and offset these losses by up to 40%. Also, according to European Union research, sustainable construction significantly improves the health of employees, which is evidenced by a smaller number of sick leaves (35–55%), [71]. However, sustainable development aims to ensure a proper balance between concentration and communication. This is important, because of the efficiency of employees and their well-being.

CMG also conducted studies in the area of cost awareness. The results are rather optimistic. Only 8% of respondents pointed to an increase in the cost of sustainable construction by 15–20%, while the majority (38%) indicated a 5–10% increase in costs. As to the reasons for the higher costs of erecting a sustainable building, developers, first of all, referred to costs related to the use of modern technologies (according to 87% of responses, this is a "strong" or "fairly large" impact). However, as many as three-quarters of the surveyed developers believed that the excess costs associated with the design and construction of a green building will decrease in the future. This shows that developers are convinced that these are prospective investments.

The above review does not yet provide a full picture of the public awareness of sustainable development and architecture. Therefore, it is worth quoting excerpts from two CBOS (Public Opinion

Research Center, Poland) surveys (2013 and 2014) [72] about the awareness of architecture and the architects' work. The results are surprising:

- Twenty-four percent of adult Polish residents have not seen any city abroad, and together with those who saw one or two cities, this makes one-third of the population of respondents;
- One-third acknowledge that they are not interested in it at all (i.e., architecture);
- The answer to the question of how the knowledge of architecture is obtained prodded answers which the architects have never taken into consideration, namely, TV series and other TV programs. After rejection of negative votes, the sum of responses regarding TV series and TV accounted for 23%, but it was the Internet that prevailed (25%). This case is illustrated in Figure 6.

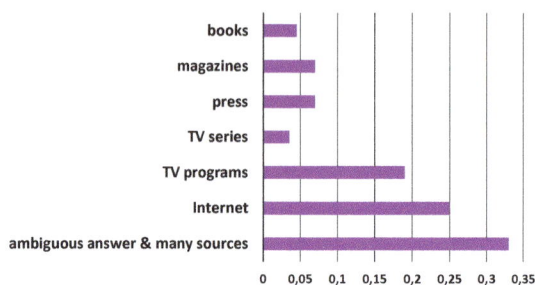

Figure 6. Answers to the question of 'Where do you get the knowledge about architecture (after rejecting negative answers?'

According to earlier CBOS surveys, performed in 2013, as many as 62% of respondents believe that "everyone should be able to build a house that they like" [73]. From these studies, the inability to verbalise feelings and assessments of architecture came to light [74,75].

The quoted surveys and reports rise concern about architectural culture. At the same time, the hermeticity of knowledge is emphasized and there is a lack of discussion about the social effects of the designed structures [36,64]. A clear position is contained in the "The Poles Living Space" SARP Report [64], emphasizing the urgent need for education, also in the area of sustainable development, so that raising public awareness of architecture and sustainable development does not take place under the dictate of TV series.

In this context, the attitude of architecture faculty students becomes extremely important [76]. Designers/practitioners place their hopes in graduates who they hire, expecting that they have basic knowledge of building physics and energy efficiency. Unfortunately, as they have stated in the professional press, these hopes are shattered [77].

The issues of low architectural culture cannot be taken lightly. Shaping public awareness is primarily a mission of architects, but their own knowledge and awareness are becoming a strategic element of the success of the idea of sustainable development.

8. Conclusions

Implementation of the sustainable development process entails changes in design standards, and is dependent on both the public's, and the architect's, awareness. The significance of knowledge is increased intentionally, including the understanding of changes in design paradigms.

Unfortunately, the presented surveys indicate a low level of architectural culture (based on the example of Polish residents), as well as insufficient knowledge about the essence of sustainable development.

This phenomenon must not be underrated. Shaping public awareness is, above all, a mission of architects; their knowledge and awareness become a very important strategic element of the success of the whole idea, because the architect's perspective must always be wider and multidimensional.

The synergy of action for legislative changes, implementation of new technologies, raising the level of knowledge and, a change in public awareness will guarantee the success of architecture in the process of sustainable development.

Acknowledgments: The authors wish to thank the Institute of Architecture and Spatial Planning at PUT for its financial support.

Author Contributions: All authors contributed equally to this work.

Conflicts of Interest: The authors declare no conflict of interest.

References

1. Wang, L. Bloomberg's New London HQ Rated World's Most Sustainable Office. Available online: https://inhabitat.com/bloombergs-new-london-hq-rated-worlds-most-sustainable-office/ (accessed on 15 November 2017).
2. Best of BREEAM 2017 Exceptional Sustainable Places and Project Teams from the BREEAM Awards 2017. Available online: http://www.breeam.com/filelibrary/BREEAM%20Awards/BREEAM-Awards-2017/BREEAM_Awards_Brochure_-1061-.pdf (accessed on 15 November 2017).
3. Best of BREEAM 2016 Outstanding Sustainable Buildings from the BREEAM Awards 2016. Available online: http://www.breeam.com/filelibrary/BREEAM%20Awards/109611_Best_of_BREEAM_Awards_2016_WEB.pdf (accessed on 15 November 2017).
4. Bonenberg, W.; Kapliński, O. Postawa architekta wobec zrównoważonego rozwoju. In *Architektura Wobec Wyzwań Zrównoważonego Rozwoju*; Poznan University of Technology Press: Poznań, Poland, 2016; Volume 2, pp. 11–30. ISBN 978-83-7775-438-2.
5. Ubarte, I.; Kapliński, O. Review of the sustainable built environment in 1998–2015. *Eng. Struct. Technol.* **2016**, *8*, 41–51. [CrossRef]
6. Baranowski, A. *Projektowanie Zrównoważone w Architekturze*; Wyd. Politechniki Gdańskiej: Gdańsk, Poland, 1998.
7. Bać, A. *Zrównoważenie w Architekturze. Od Idei do Realizacji na tle Dokonań Kanadyjskich*; Oficyna Wydawnicza Polit. Wrocł.: Wrocław, Poland, 2016.
8. Kronenberg, J.; Bergier, T. *Wyzwania Zrównoważonego Rozwoju w Polsce*; Fundacja Sendzimira: Kraków, Poland, 2010.
9. Bonenberg, W.; Kapliński, O. Knowledge is the key to innovation in architectural design. *Procedia Eng.* **2017**, *208*, 2–7. [CrossRef]
10. Belniak, S.; Leśniak, A.; Plebankiewicz, E.; Zima, K. The influence of the building shape on the cost of its construction. *J. Financ. Manag. Prop. Constr.* **2013**, *18*, 90–102. [CrossRef]
11. Dziadosz, A. The influence of solutions adopted at the stage of planning the building investment on the accuracy of cost estimation. *Procedia Eng.* **2013**, *54*, 625–635. [CrossRef]
12. Kapliński, O. An important contribution to the discussion on research methods and techniques in designing. *Eng. Struct. Technol.* **2015**, *7*, 50–53. [CrossRef]
13. Directive 2010/31/EU of the European Parliament and of the Council of 19 May 2010 on the Energy Performance of Buildings (Recast). 2010. Available online: http://eur-lex.europa.eu/legal-content/en/TXT/?uri=CELEX:32010L0031 (accessed on 29 December 2017).
14. Polski Komitet Normalizacyjny. *Energy Performance of Buildings—Economic Evaluation Procedure for Energy Systems in Buildings—Part 1: Calculation Procedures, Module M1-14*; PN-EN 15459-1:2017-07E; PKN: Warszawa, Poland, 2017.
15. Bonenberg, W. Requirements engineering as a tool for sustainable architectural design. In *Advances in Human Factors, Sustainable Urban Planning and Infrastructure*; Charytonowicz, J., Ed.; Springer: Cham, Switzerland, 2018; Volume 600, pp. 218–227. ISBN 978-3-319-60449-7.
16. Januchta-Szostak, A.; Banach, M. *Architektura Wobec Wyzwań Zrównoważonego Rozwoju*; Faculty of Architecture, Poznan University of Technology: Poznań, Poland, 2016.

17. Kapliński, O. Sustainable development and intelligent buildings as elements of humanization of technological civilization. In *Technikos Humanizavimas, Humanizacja Techniki, Humanisation of Technology*; Technika: Vilnius, Lithuania, 2001; pp. 225–231.

18. Koczyk, H.; Basińska, M. Optimum energetyczno-ekonomiczne rozwiązań instalacyjnych budynków energooszczędnych. In *Budownictwo Energetyczne w Polsce—Stan i Perspektywy, Proceedings of the KILiW PAN & KN PZITB Conference, Krynica, Poland, 27–28 May 2015*; UTP-Bydgoszcz Press: Bydgoszcz, Poland, 2015; pp. 241–252.

19. Szczechowiak, E. Parametry budynków niemal zero-energetycznych w warunkach polskich. In *Budownictwo Energetyczne w Polsce—Stan i Perspektywy, Proceedings of the KILiW PAN & KN PZITB Conference, Krynica, Poland, 27–28 May 2015*; UTP-Bydgoszcz Press: Bydgoszcz, Poland, 2015; pp. 57–68.

20. Rosolski, S. *Projektowanie Architektoniczne a Zagadnienia Odwrotne*; Exemplum: Poznań, Poland, 2012.

21. Crouch, C.; Kaye, N.; Crouch, J. *An Introduction to Sustainability and Aesthetics: The Arts and Design for the Environment*; Brown Walker Press: Irvine, CA, USA, 2015.

22. The American Institute of Architects; American Institute of Architects California Council. *Integrated Project Delivery: A Guide*; version 1National California Council: San Francisco, CA, USA, 2007.

23. Kaklauskas, A.; Zavadskas, E.K.; Dargis, R.; Bardauskiene, D. *Sustainable Development of Real Estate*; Technika: Vilnius, Lithuania, 2015.

24. Zavadskas, E.K.; Antucheviciene, J. Development of an indicator model and ranking of sustainable revitalization alternatives of derelict property: A Lithuanian case study. *Sustain. Dev.* **2006**, *5*, 287–299. [CrossRef]

25. Zavadskas, E.K.; Antuchevicienė, J.; Kaplinski, O. Multi-criteria decision making in civil engineering. Part 1—A state-of-the-art survey. *Eng. Struct. Technol.* **2015**, *7*, 103–113. [CrossRef]

26. Zavadskas, E.K.; Antuchevicienė, J.; Kaplinski, O. Multi-criteria decision making in civil engineering. Part 2—Applications. *Eng. Struct. Technol.* **2015**, *7*, 151–167. [CrossRef]

27. Gajzler, M. Usefulness of mining methods in knowledge source analysis in the construction industry. *Arch. Civ. Eng.* **2016**, *62*, 127–142. [CrossRef]

28. Majerska-Pałubicka, B. *Zintegrowane Projektowanie Architektoniczne w Kontekście Zrównoważonego Rozwoju. Doskonalenie Procesu*; Silesia Technical University Press: Gliwice, Poland, 2014.

29. Perspektywy Rozwoju Budownictwa Energooszczędnego w Polsce. Zespół Go4Energy. Available online: http://g4e.pl/wp-content/uploads/2014/05/Go4Energy_raport_wiosna_2014.pdf (accessed on 30 April 2016).

30. BCMM. Nastroje Architektów 2015. Wyniki Badania Sondażowego Wśród Architektów, Kwiecień 2015. Available online: www.bcmm.com.pl (accessed on 4 April 2016).

31. Laskowski, L. *Ochrona Cieplna i Charakterystyka Energetyczna Budynku*; Oficyna Wydawnicza Pol. Warsz.: Warszawa, Poland, 2005.

32. The Passive House Institute (PHI). Available online: http://www.passiv.de/en/01_passivehouseinstitute/01_passivehouseinstitute.htm (accessed on 15 October 2017).

33. Bickert, E. Green Building, Sustainable Real Estate Investment, Sustainable Valuation & Efficiency Assessment: Neue Projekte, Berichte und Leitfäden. Forum Nachhaltige Immobilien. Available online: https://forumnachhaltigeimmobilien.com/2016/06/06/green-building-sustainable-real-estate-investment-sustainable-valuation-efficiency-assessment-neue-projekte-berichte-und-leitfaeden/ (accessed on 1 June 2017).

34. Hegger, M.; Fuchs, M.; Stark, T.; Zeumer, M. *Energy Manual. Sustainable Architecture*; Birkhäuser: Basel, Switzerland, 2008.

35. Marchwiński, J. Konflikty architektoniczno-energetyczne w projektowaniu miejskich budynków wielorodzinnych. *Ciepłownictwo, Ogrzewnictwo, Wentylacja* **2014**, *45*, 10–14.

36. Niezabitowska, E.; Masły, D. *Oceny Jakości Środowiska Zbudowanego i Ich Znaczenie dla Rozwoju Koncepcji Budynku Zrównoważonego*; Silesia Technical University Press: Gliwice, Poland, 2007.

37. Zayed National Museum. Available online: https://www.e-architect.co.uk/dubai/zayed-national-museum (accessed on 1 June 2017).

38. Warmann, C. Zayed National Museum by Foster + Partners. Available online: https://www.dezeen.com/2010/11/25/zayed-national-museum-by-foster-partners/ (accessed on 1 June 2017).

39. Sikora, A. Asian Cairns. Sustainable Farmscrapers for Rural Urbanity—Vincent Callebaut Architectes. 2013. Available online: http://progg.eu/kamienne-kopce-callebauta-biomorficzna-architektura-w-bionicznym-miescie/ (accessed on 1 June 2017).

40. Kalogeropoulos, S. Vincent Callebaut Architectures: Asian Cairns, Shenzhen, China. Available online: http://mydesignstories.com/vincent-callebaut-architectures-asian-cairns-shenzhen-china/ (accessed on 1 June 2017).

41. Frearson, A. Ordos Museum by MAD. Available online: https://www.dezeen.com/2011/12/13/ordos-museum-by-mad/ (accessed on 1 June 2017).

42. Ordos Museum. Available online: https://en.wikiarquitectura.com/building/ordos-museum/ (accessed on 1 June 2017).

43. Kriscenski, A. Anti Smog Architecture: A Catalyst for Cleaner Air in Paris. Available online: https://inhabitat.com/anti-smog-architecture-a-catalyst-for-cleaner-air-in-paris/ (accessed on 1 June 2017).

44. Bonenberg, A. *Cityscape in the Era of Information and Communication Technologies*; Springer: Cham, Switzerland, 2018; ISBN 978-3-319-69541-9.

45. Zhou, M.; Bonenberg, W. Application of the green roof system in small and medium urban cities. In *Advances in Human Factors and Sustainable Infrastructure, Proceedings of the AHFE 2016 International Conference on Human Factors and Sustainable Infrastructure, Orlando, FL, USA, 27–31 July 2016*; Walt Disney World: Orlando, FL, USA, 2016; pp. 125–136.

46. Bonenberg, W. Urban sprawl jako zagrożenie tożsamości miasta. *Zeszyty Naukowe Politechniki Poznańskiej. Architektura i Urbanistyka* **2011**, *23*, 7–14.

47. Alvarez, S.P.; Lee, K.; Park, J.; Rieh, S.-Y. A comparative study on sustainability in architectural education in Asia—With a focus on professional degree curricula. *Sustainability* **2016**, *8*, 290. [CrossRef]

48. Hassanpour, B.; Atun, R.A.; Ghaderi, S. From words to action: Incorporation of sustainability in architectural education. *Sustainability* **2017**, *9*, 1790. [CrossRef]

49. Architects' Council of Europe (ACE). The Architectural Profession in Europe 2016—ACE Sector Study. Available online: http://www.ace-cae.eu/837/?L=0 (accessed on 3 November 2017).

50. The European Summit on BIM. Available online: http://europeanbimsummit.com/the-summit/ (accessed on 11 November 2017).

51. Raport BIM—Polska Perspektywa. Available online: www.ebuilder.pl/index.php?act=article&sub=save&id=9902 (accessed on 30 November 2015).

52. The British Standards Institution. *Sustainability of Construction Works. Assessment of Environmental Performance of Buildings. Calculation Method*; BS EN 15978:2011; The British Standards Institution: London, UK, 2011.

53. Bewa, M.; Richards, M. *UK Government BIM Roadmap*; RIBA: London, UK, 2008.

54. The British Standards Institution. *Specification for Information Management for the Capital/Delivery Phase of Construction Projects Using Building Information Modeling*; PAS 1192-2:2013, Incorporating Corrigendum No. 1; The British Standards Institution: London, UK, 2013.

55. Salamonowicz, M. BIM: Wdrożenie w Wielkiej Brytanii. *Zawód:Architekt* **2015**, *11/12*, 132–140.

56. Zima, K.; Leśniak, A. Limitations of cost estimation using building information modeling in Poland. *J. Civ. Eng. Archit.* **2013**, *7*, 545–554.

57. Juszczyk, M.; Vyskala, M.; Zima, K. Prospects for the use of BIM in Poland and the Czech Republic—Preliminary research results. *Procedia Eng.* **2015**, *123*, 250–259. [CrossRef]

58. Dodge Data & Analytics. *World Green Building Trends 2016*; Smart Market Report; DODGE Data & Analytics: Bedford, MA, USA, 2016.

59. *Reno Value Report: Drivers for Change: Strengthening the Role of Valuation Professionals in Market Transition*; Market Insights Report. EU: Brussels, Belgium, 2015. Available online: https://ec.europa.eu/easme/sites/easme-site/files/RENOVALUE%20report%20JUne%202015.pdf (accessed on 1 November 2017).

60. Eko Budownictwo w Polsce Wciąż bez Wystarczającego Wsparcia Rządowego. Available online: http://beta.oswbz.org/wp-content/uploads/2015/12/Eko-budownictwo-w-Polsce-propertyjournal.pdf (accessed on 1 February 2016).

61. Competency Based Occupational Standards. Polacy o Oszczędzaniu Energii i Energetyce Obywatelskiej. Komunikat z Badań nr 36/2016. 2016. Available online: http://www.cbos.pl/SPISKOM.POL/2016/K_036_16.PDF (accessed on 10 November 2017).

62. Raport RWE Polska 2013 Świadomość Energetyczna Polaków. Available online: http://www.postawnaslonce.pl/pliki/wokol_energii_i_fotowoltaiki/Badanie_swiadomosci_Polakow/Raport_Swiadomosc_Energetyczna_Polakow.pdf (accessed on 1 January 2017).
63. Analiza TNS Polska dla Ministerstwa Środowiska. Raport z Analizy Badań Świadomości, Postaw i Zachowań Ekologicznych Polaków Przeprowadzonych w Polsce w Latach 2009–2015. 2015. Available online: http://www.3xsrodowisko.pl/uploads/media/badanie_dr_ekologia_ministerstwo_srodowiska.pdf (accessed on 1 November 2017).
64. SARP Warszawa. *Raport SARP: Przestrzeń życia Polaków*; SARP Warszawa: Warsaw, Poland, 2014; Chapter I; pp. 145–170.
65. Colliers International. *Zdrowie i Produktywność w Zrównoważonych Budynkach*; Construction Marketing Group: Warsaw, Poland, 2015.
66. Colliers International. *Zielone Budynki w Polsce 2015*; Building Consultancy Services, Green Building Certification: Warsaw, Poland, 2015.
67. *Analiza Rynku Zrównoważonego Budownictwa w Polsce. Badanie Percepcji Rynku*; Construction Marketing Group: Warsaw, Poland, 2013.
68. Polacy o Oszczędzaniu Energii w Budownictwie—Architekci, Inwestorzy. 6paliwo-Raport-2. Available online: http://6paliwo.pl/wp-content/uploads/2012/02/6paliwo-raport-2.pdf (accessed on 6 April 2017).
69. Raport RWE Polska 2013. *Świadomość Energetyczna Polaków*; Biuro Prasowe RWE: Warsaw, Poland, 2013.
70. Health, Wellbeing and Productivity in Offices: The Next Chapter for Green Building. Available online: http://www.worldgbc.org/news-media/health-wellbeing-and-productivity-offices-next-chapter-green-building (accessed on 24 September 2014).
71. Sick Pay and Sickness Benefit Schemes in the European Union. Background Report for the Social Protection Committee's. In-Depth Review on Sickness Benefits Brussels. Available online: Ec.europa.eu/social/BlobServlet?docId=16969&langId=en[KE-02-17-045-EN-N(1).pdf] (accessed on 17 October 2016).
72. Competency Based Occupational Standards. Polacy o Architektach. Komunikat z Badań nr 161/2014. Available online: http://www.cbos.pl/SPISKOM.POL/2014/K_161_14.PDF (accessed on 10 November 2017).
73. Prośniewski, B. *Gust Nasz Pospolity*; Fundacja Bęc Zmiana: Warsaw, Poland, 2004.
74. Prośniewski, B. Anioł Najbardziej Przykuwa Moją Uwagę—Czyli Polski Gust Architektoniczny. *Zawód:Architekt* **2015**, *43*, 39–45.
75. Osowski, S. Polacy o architektach w badaniu CBOS. *Zawód:Architekt* **2015**, *43*, 46–48.
76. Bać, A. Idea zrównoważenia i jej wybrane przejawy. *Architectus* **2014**, *2*, 3–14.
77. Mielczyński, T. Forma podąża za energią. *Zawód:Architekt* **2014**, *38*, 92–94.

sustainability

MDPI

Article

A Hybrid Fuzzy BWM-COPRAS Method for Analyzing Key Factors of Sustainable Architecture

Hannan Amoozad Mahdiraji [1], Sepas Arzaghi [2], Gintaras Stauskis [3,*] and Edmundas Kazimieras Zavadskas [4]

[1] Faculty of Management, University of Tehran, Tehran 1417614418, Iran; h.amoozad@ut.ac.ir
[2] Department of Management, University of Tehran, Kish International Campus, Kish 7941655655, Iran; arzaghi.sepas@ut.ac.ir
[3] Faculty of Architecture, Department of Urban Design, Vilnius Gediminas Technical University, Vilnius 10223, Lithuania
[4] Faculty of Civil Engineering, Institute of Sustainable Construction, Vilnius Gediminas Technical University, Vilnius 10223, Lithuania; edmundas.zavadskas@vgtu.lt
*   Correspondence: gintaras.stauskis@vgtu.lt; Tel.: +370-6879-0312

Received: 4 April 2018; Accepted: 16 May 2018; Published: 18 May 2018

Abstract: Sustainable development by emphasizing on satisfying the current needs of the general public without threating their futures, alongside with taking the environment and future generations under consideration, has become one of the prominent issues in different societies. Therefore, identifying and prioritizing the key factors of sustainable architecture according to regional and cultural features could be the first step in sustaining the architecture as a process and an outcome. In this paper, the key indicators of the environmental sustainability in contemporary architecture of Iran has been identified and prioritized. This study has been performed in three phases. First, identifying key factors of environmental sustainability according to the experts' point of view and transforming the collected data to triangular fuzzy numbers. Subsequently, the best-worst multi-criteria decision-making method (henceforth BWM) under grey system circumstances has determined the weights and priority of the identified criteria. Eventually, identified key factors were prioritized by the complex proportional assessment method (hereafter COPRAS) under the condition of fuzzy sets. The results indicate that the key factors of creating engagement between buildings and other urban systems has the highest priority in the built environment sustainability in contemporary architecture and proving building management systems has the lowest.

Keywords: sustainable development; sustainable architecture; best-worst method (BWM); complex proportional assessment method (COPRAS); grey system; fuzzy sets

---

1. Introduction

Sustainability is not a new concern; since the introduction of the concept in 1987, there has been a proliferation of competing notions of sustainability to the extent that it has become an empty box, a fragmented concept. It seems that sustainability is what you make of it [1]. The etymology of the word "sustainability" originates from the Latin sustin_ere in which the words sub—from below—and ten_ere—held up—combine to generate the idea of something that supports, maintains, or endures [2]. However, the term 'sustainability' is increasingly used in the context of ecological, economic, and social studies. In green economics it is often used interchangeably with the term 'sustainable development', defined by the World Commission on Environment and Development as, "development which meets the needs of the present without compromising the ability of future generations to meet their own needs" [3]. The concept of sustainable development refers to the ideas of "our common future" based upon a report published by the United Nations World Commission on Environment and Development [4]. In fact, sustainable development is all about ensuring a better quality of life for everyone, now and for generations to come [5]. The World Commission of Environment and Development identified sustainable development as a development that can fulfill requirements, including two concepts:

- The concept of needs, especially poor populace basic needs, as a top priority; and
- Sustainable development, which includes the ideas that limit every country because of ecological, social, and environmental situations (Figure 1). It means that each country should identify the sustainable development purposes operationally inside itself, according to human soberness of himself and natural resources of the Earth, and it wants a unique and sustainable lifestyle for everyone. It is against the overuse and the dissipation of resources, disregard for future generations, and disconnection with the past [6]. Several models have been constructed for the identification of sustainable development, on the basis of its understanding. The best known is the three-pillar model, generally considered that there are three distinct sectors via which sustainability can be affected and enhanced, encompassing environmental sustainability, economic sustainability, and social sustainability [7]. Accordingly, the achievement of a sustainable built environment leads to holistic design methods capable of balancing the varied demands of environmental, social, and economic issues [8].

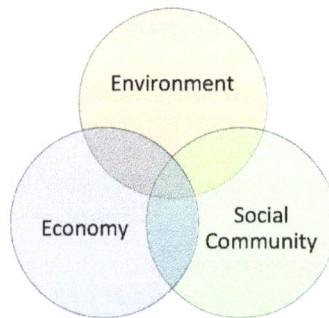

Figure 1. Venn diagram showing that sustainability consists of a balance between environmental quality (Environmental), sociocultural quality (Social), and economic quality (Economic) [9].

Based on a 1987 Brundtland report, sustainable development is rooted in sustainable forest and environmental issues in the 20th century. 'Sustainability' should be considered as humanity's target goal of human-ecosystem equilibrium, while sustainable development refers to the holistic approach and temporal processes [10]. Furthermore, it can be considered as the practice of maintaining processes of productivity indefinitely by replacing resources used with resources of equal or greater value without degrading or endangering natural biotic systems [11].

In the territory of environmental sustainability, sustainable architecture has been introduced as the most prominent component. Environment and its changes, reduction in energy consumption, and green building are the most important factors in shaping different approaches of sustainable architecture [12]. A building, energy and resource efficient to sustain the lifecycle of its operations, meanwhile conducive to the health and comfort of its occupants, could be considered a green building [13]. Sustainable architecture is a way of prolonging the aging process of existing architecture [14]. Furthermore, is the inclusive part of the Green Building paradigm—the comprehensive professional philosophy, design methodology, and assessment toolkit that came into use in the 1990s along with the introduction of the BREEAM (Building Research Establishment Environmental Assessment Method) sustainability assessment system that has identified ten performance areas to measure sustainability of new or in-use buildings [15]. Recently, many similar systems elaborate analogous assessment criteria and are used for creating sustainable architecture internationally, regionally, or nationally.

Architectural sustainability in developing countries such as Iran, is still in the early stages compared with developed countries. Although, in recent decades, construction has been one of the most beneficial industries in Iran, but, unfortunately, most of the buildings suffer from a lack of sustainability. In Iran, the running architect's education programs still poorly address the questions of sustainable architecture, there is no sustainability assessment tradition, and for this reason there is a great need for scientifically-based methods for prioritizing the sustainable architecture indicators. Zarghami et al., in 2018, published a report investigating the opportunities to customize the well-known sustainability assessment systems for the Iranian architecture and construction environment and underline the need for a method that explicitly displays the key factors of sustainable architecture [16]. In 2017, Shareef and Altan published research analyzing the sustainability assessment systems that are in use in the Middle East and globally, to determine qualitative and quantitative ways to weight the selected sustainability indicators and use this for developing the national sustainability codes, including the digital calculation systems [17]. The scope of sustainability theories, methodologies, and assessment systems often unreasonably limit themselves to the scale of a separately-taken building with its internal systems, as Hashemkhani and Zolfani note in 2018 [18]. At the same moment, architects and planners seamlessly neglect the necessity of an integrated approach with the superior urban structures as blocks and districts and urban systems as mobility and technical infrastructure. There, urban planners and architects make the strategic decisions for sustainability and clients harvest the major benefits, and we aim to check this for the Iranian environment.

Statistics indicate that, in the developing countries, approximately 40% of the energy is consumed in the construction sector [19]. It is predicted that this percentage will rise up to 50% by the year 2050 [20]. Consequently, each year a great amount of energy, and financial and non-financial resources are wasted during the process of building, and even during the regular use of the built buildings. The main aim of this particular research is to increase the pace of sustaining Persian architecture by applying innovative analysis methods and relevant conceptual tools. Moreover, the results of the following research could cause a reduction of energy, financial, and non-financial resources that the construction industry annually uses in Iran. For this reason, key factors of environmental sustainability in the contemporary architecture of Iran have been introduced and prioritized. Prioritizing these key factors will help the architecture and the other related companies to put more emphasis on necessary factors that will lead to sustainable buildings in the context of sustainable city spaces.

In the following, related subjects to this paper, such as sustainable architecture, Persian contemporary architecture, best-worst, and the complex proportional assessment (hereafter COPRAS) methods have been demonstrated.

2. Literature Review

2.1. Sustainable Architecture

Sustainability has become one of the most important and progressive trends in architecture over the last two decades [21]. The environmental awareness of professionals has put sustainability in the center of the profession of architecture and has resulted in introducing and implementing ecological designs both in the scale of buildings and cities [22]. Multiple investors with different viewpoints and, often, conflicting interests are involved in the sustainable design of buildings [23]. In fact, sustainability is a broad field relating to many areas in architecture as education, design, construction, and others. Therefore, sustainable architectural education is encouraged to be integrated throughout the program across the disciplines, where more opportunities will enable architects to acquire the knowledge effectively and transform it practically [24]. The concept of sustainable architecture is upon an undeniable reality that ecological conditions influence economic and social activities, including the idea of creating a logical environment [6].

Architects took over the criteria of green building as to achieve more sustainable architecture outcomes whereas the new approach put more focus on building performance in a longer time. By trying to cover all the important environmental, social, and economic aspects, still the most emphasis went to the environmental dimension and partly to the social aspects, while economic sustainability was a hard nut for architects before the German Sustainable and Building Council (DGNB) system brought that into use in 2007. The implemented green building assessment schemes definitely increased attention to the quality of architect's work and brought more interdisciplinary design work in sustainable architecture [25]; nonetheless, their impact on sustainable development and greenhouse gas emissions was lower than expected [26].

Sustainable architecture employs a conscious approach to energy and ecological conservation in the design of the built environment (Dublin Institute of Technology). The results of sustainable architecture are founded on the symbiosis of ecologist's and architects common work. These two professional groups could propose a change in the function of the building, such as a transition from a linear approach to a closed circulation plan. A building in a linear pattern would be treated as a place of processing natural resources into waste. For instance, energy is converted into heat losses. On the other hand, a building in a closed circulation plan approach could change from a voracious consumer of energy into a more self-sufficient unit [27]. Although sustainable design and construction might have a higher capital cost, it would provide savings through lower operating costs over the life cycle of the building in the long term [17].

Furthermore, sustainable architecture defines an understanding of environmentally-friendly architecture and contains [28] many characteristics, as listed below:

- Designing efficient ventilation infrastructures;
- Energy-efficient lighting and appliances;
- Facilities for water-saving plumbing systems;
- Increasing passive solar energy by landscape scheduling;
- Diminishing harm to natural habitat;
- Employing other possible power sources;
- Non-synthetic, non-toxic materials;
- Locally-obtained and responsibly-harvested woods and stone;
- Compatible usage of former constructions;
- Use of recycled architectural salvage; and
- Efficient use of space.

While most green buildings do not have all of these features, the highest goal of green architecture is to be fully sustainable [29]. Green building involves considering four main areas including site development (to reduce the impact of development on the natural environment); material selection and minimization, contemporary sustainable building practice and looking for a reliable way of assessing and certifying materials. Critical review underline the need to upgrade the sustainability assessment requirements in the "Materials" section of different green building assessment schemes [21].

Based on this idea, the use of materials is progressively detected through ecological specifications that refer to various interaction between a material and the environment including embodied energy, pollution, waste generation and recycling possibilities [30]. Architectural sustainability must go down to standards that underscore the use of renewable energy, the maintenance and renewal of energy without contamination, contrary to minimum energy consumption [31]. Intelligent building systems monitor various parts of the building and create the right conditions to provide concurrent services, which can lead to the optimization of energy consumption, improvement in efficiency, productivity of devices, value- added, and the facilities in the building [32]. Take advantage of the natural elements and technologies engenders conservation of resources and increases occupant comfort/productivity, while lowering long-term operational costs and pollutants [33]. Furthermore, design for high indoor air quality to promote occupant health and productivity, minimizes the waste in construction, and demolition processes by recovering materials and reusing or recycling [34].

While seeking for sustainable solutions architects try to reuse the traditional construction and crafts techniques as vernacular climatic strategies (VCS) that focus on the natural ventilation, external walls insulation, and integrated shading. These vernacular low-cost solutions can reduce energy consumption and CO_2 emissions as well as have positive impact on the aesthetics of modern architecture by minimizing the volumes of externally and internally-located equipment [35].

Several paradigms could be defined in the framework of sustainable architecture. The first paradigm focuses on characteristic parameters, which could be mentioned as a function of design and operational variables (reverse aspect). The next paradigm is the framework of environment-friendly approach to architecture, being known as "from cradle to grave" among architects, the building analysis should be applied throughout its entire life cycle. The final paradigm of sustainable policy indicates that, between one to three years, buildings should be designed with a demand for energy close to zero [27].

2.2. Contemporary Architecture of Iran

Modern architecture of Iran is a period of Persian architecture that started during the empire of QAJARS and continues to the present. At first, modern architecture of Iran was deeply affected by European architecture. Hence, more spacing creativity could be seen in the buildings of that era. Concurrently, new spaces are created and were faced with more various spaces. In other words, Persian architecture during Qajars Empire can be named as the evolution phase of Persian traditional architecture. Qajars architecture improved principles and patterns of Persian architecture and provided space innovations. Construction of castles and palaces is the most important component of Persian architecture during this period. The Sepahsalar School is known as the first modern building of Iran designed by the first Iranian architect who had been trained in Europe (Figure 2) [36].

Figure 2. Sepahsalar School [36].

During the Pahlavi Empire, transformation of European architecture, such as using concrete, steel, and glass, instead of traditional materials, is observable in Persian architecture. Since this era, European decoration and elements are presented in the houses and interior design of the buildings. In Pahlavis architecture, some building elements, such as, porches, doors, and windows became more look like European forms. In fact, Persian architecture during this time was a combination of Qajars architecture and some imported elements. Modern architecture of Iran was affected by expressionism architecture of Germany before 1930. In addition to forming public and residential architecture, during this era, industrial architecture, which is one of the components of modern architecture, were detected. Sa'dabad Palace is one of the best-known buildings of this era. (Figure 3) [36].

After the Islamic revolution, architecture in Iran changed to a combination of modern architecture, Persian architecture, and Islamic architecture. Since then, Iranian architecture consisted of a wide range of tendencies embracing postmodern classicism (Figures 4 and 5). Among the mentioned tendencies,

Iranian-Islamic tendency was strongly supported by the government. In fact, the main purpose of Persian architects during this time was to provide a balance between Persian traditional architecture and modern architecture [36].

Figure 3. Sa'dabad Palace [36].

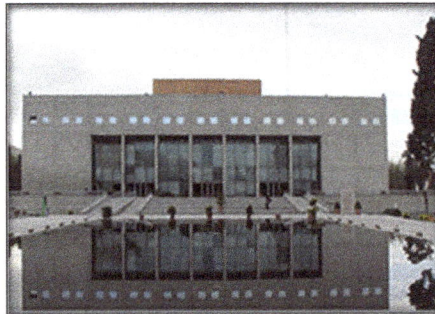

Figure 4. National Library of Shiraz [36].

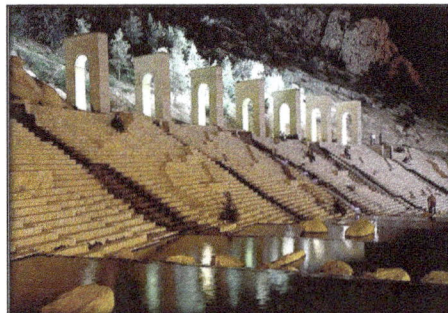

Figure 5. Shiraz Northern Entrance [36].

In Table 1 case studies which have been introduced in Figures 2–4, have been investigated based on typology, material, and technology.

Table 1. Investigation of case studies, source: own.

Figure No.	Typology	Material	Technology
2	introverted	traditional	traditional
3	extroverted	traditional and modern	traditional and modern
4	extroverted	modern	modern
5	extroverted	traditional	modern

In recent years, a few studies have been performed in the field of sustainability related to the contemporary architecture of Iran. For example, in 2016, Alidadi and Zadeh indicated the effect of sustainable architecture in designing a five-star hotel in Iran [37]. Karimi and Zandieh compared modern architectural principles of sustainability in Persian architecture with five modern sustainable cities in the world [38]. Moreover, in 2016, Makari Faraji and Mokhtari Taleghani studied regarding the role of sustainable architecture in valuable historical districts of Tehran [39]. Table 2 demonstrates a classification of recent studies related to sustainable architecture.

Table 2. Previous studies related to sustainability and contemporary architecture of Iran and internationally, source: own.

Researcher/s	Year	Iranian Architecture/ International Architecture	Aim and Focus
[27]	2018	International	Presenting the architect's attitude towards the paradigms of sustainable development;
			Presenting the place and role of the architect in the implementation of the multidimensional processes of sustainable design;
[6]	2017	Iranian	Investigating sustainable architecture and reasons of its usage;
			Introducing the unseen principles of sustainable architecture in vernacular architecture.
[30]	2017	International	Examining the sustainable design strategies of the Balkan vernacular architecture in the example of the traditional house.
[32]	2017	International	Considers Intelligent Buildings with sustainable architecture Approach;
			Defining the concepts of computerizing, intelligent buildings, sustainable development and architecture.
[31]	2017	Iranian	Maxing use of natural light out in mountainous climates based on sustainable architecture.
[9]	2017	International	Examining the integration of sustainability principles into architectural education programs in South Korean Universities;
			Identifying eight sustainability-related SPCs that can be utilized to teach sustainability;
			Identifying the average number of credits per sustainability-related SPC in different course types.
[37]	2016	Iranian	Designing a hotel obtaining practical needs beside its aesthetic aspects.
[38]	2016	Iranian and International	The objectives of human communities resorting to sustainable development have been studied in three separate domains, environmental, social and economic;
			Reasons and features of sustainability of ancient Iranian cities are considered and compared to five cases of sustainable cities of the modern world.
[39]	2016	Iranian	Modern architecture in valuable city textures is one of challengeable issues in modern architecture and renovation in Iran;
			The old textures of Tehran are faced to severe structural erosion and social decline.
[5]	2016	International	Considering the construction strategies in the fields of "Efficacy", "maintenance" and "end of life" issues;
			Concentrating the considerations on the structural members and materials in the buildings;
			Proposing some outlines in design of futuristic structures.
[35]	2017	Iranian	Reuse of vernacular climatic strategies VCS in modern sustainable architecture, application of dynamic modelling systems.

The innovative aspect of the current research in comparison with the aforementioned studies is argued as follow: First, we used the new hybrid approach of MCDM methods and sustainability for the first time to analyze the contemporary architecture of Iran. Furthermore, this research presents a follow-up discussion and builds upon the Bonenberg's and Kaplinski's research [27].

The next, for the novelty of our proposed approach based upon the literature review, the authors found only four articles related to MCDM methods that were applied in the field of sustainable architecture in the Web of Science (Clarivate Analytic) database [16,18,23,40]. Furthermore, based upon review articles [41–44], many studies focusing on sustainable decision-making in civil engineering, construction, and building have been made, but nearly no applications in architecture were identified. Moreover, even fewer hybrid MCDM methods were applied for sustainability and engineering problems, though it was proved that they are extremely suitable [42,44]. Accordingly, they argued that sustainability studies could be considered in the best way by applying MCDM methods when evaluating several criteria groups. Thus, our proposed method considering hybrid MCDM models and their application in sustainable architecture is novel according to the studied literature. Moreover, a new hybrid method, including BMW and COPRAS methods under a fuzzy environment, has been developed and applied for the first time.

3. Research Methodology

In this section, at first, best-worst multi-criteria decision-making method (BWM), complex proportional assessment method (COPRAS), fuzzy sets theory, and a defuzzification technique are explained in detail. Afterwards, the stages of multi-criteria analysis of prioritizing key factors of sustainability in Persian contemporary architecture are described.

3.1. Best-Worst Method (BWM)

BWM is one of the newest methods for solving multi-criteria decision-making problems, and was introduced by Rezaei [45] in 2015. In this method, first, the best (e.g., most desirable, most important) and the worst (e.g., least desirable, least important) criteria are chosen by the decision-maker. Afterwards, upon the BWM questionnaire, pairwise comparisons should be performed between each of these two criteria (best and worst) and the other criteria. For determining the weights of decision-making criteria a maximum problem is formulated. In order to check the reliability of the comparisons a consistency ratio is proposed for the BWM [45]. In our proposed approach, this method was employed in order to weight the decision-making criteria, considering each one's advantages in comparison with other existing MCDM methods, being the requirement of less comparison data in conjunction with more consistent comparisons and more reliable results.

In 2017, Gupta and Barua applied BWM in order to select suppliers among SMEs on the basis of their green innovation ability [46]. Likewise, Ahmadi et al. performed BWM in favor of assessing the social sustainability of supply chains [47]. Van de Kaa, Kamp, and Rezaei practiced the mentioned method for the selection of biomass thermochemical conversion technology in the Netherlands [48]. BWM was also tested to analyze the barriers to humanitarian supply chain management by Sahebi et al. [49]. In addition, a supplier selection life cycle approach integrating traditional and environmental criteria by using BWM, was presented by Rezaei et al. [50]. Eventually, Mokhtarzadeh et al., presented a hybrid model including BWM to analyze the technology portfolio selection problem [51]. In this part, the steps of BWM for deriving the weights of the criteria are described [45].

Step 1. A set of decision criteria $\{c_1, c_2, \cdots, c_n\}$ have to be determined.

Step 2. The best and the worst criteria in general are identified by the decision-maker.

Step 3. Based on the BWM questionnaire, the preference of the best criterion over all the other criteria using a number between 1 and 9 should be determined. The resulting best-to-others vector would be:

$$A_B = (a_{B1}, a_{B2}, \cdots, a_{Bn}) \tag{1}$$

where a_{Bj} indicates the preference of the best criterion B over criterion j. It is clear that $a_{BB} = 1$.

Step 4. According to the BWM questionnaire the preference of all the criteria over the worst criterion using a number between 1 and 9 must be determined. The resulting others-to-worst vector would be:

$$A_w = (a_{1W}, a_{2W}, \cdots, a_{nW}) \tag{2}$$

where a_{jW} indicates the preference of the criterion j over the worst criterion W. It is clear that $a_{WW} = 1$.

Step 5. Finding the optimal weight $\{W_1^*, W_2^*, \cdots, W_n^*\}$ by solving Equation (3). It should be mentioned that the optimal weight for the criteria is the one where for each pair of $\frac{W_B}{W_j}$ and $\frac{W_j}{W_W}$, we have $\frac{W_B}{W_j} = a_{Bj}$ and $\frac{W_j}{W_W} = a_{jW}$. Considering the non-negativity and sum condition for the weights, the following problem results:

$$min \; \varepsilon$$
$$s.t.$$
$$\left| \frac{W_B}{W_j} - a_{Bj} \right| \le \varepsilon, \; for \; all \; j$$
$$\left| \frac{W_j}{W_W} - a_{jW} \right| \le \varepsilon, \; for \; all \; j \tag{3}$$
$$\sum W_j = 1$$
$$W_j \ge 0, \; for \; all \; j$$

By using ε^*, the consistency ratio can be calculated. It is clear that the greater the ε^*, the higher the consistency ratio, and the less reliable the comparisons are [45]. The consistency ratio can be obtained from the following formula (Table 3):

$$Consistency \; Ratio = \frac{\varepsilon^*}{Consistency \; Index} \tag{4}$$

Table 3. Consistency index [45].

a_{BW}	1	2	3	4	5	6	7	8	9
Consistency Index	0	0.44	1	1.63	2.3	3.00	3.73	4.47	5.23

Considering uncertain circumstances, the following two models are used to calculate the lower and upper bounds of the weights of criterion j based on the grey systems [52]:

$$min \; W_j$$
$$s.t.$$
$$\left| \frac{W_B}{W_j} - a_{Bj} \right| \le \varepsilon^*, \; for \; all \; j$$
$$\left| \frac{W_j}{W_W} - a_{jW} \right| \le \varepsilon^*, \; for \; all \; j \tag{5}$$
$$\sum W_j = 1$$
$$W_j \ge 0, \; for \; all \; j$$

$$max \; W_j$$
$$s.t.$$
$$\left| \frac{W_B}{W_j} - a_{Bj} \right| \le \varepsilon^*, \; for \; all \; j$$
$$\left| \frac{W_j}{W_W} - a_{jW} \right| \le \varepsilon^*, \; for \; all \; j \tag{6}$$
$$\sum W_j = 1$$
$$W_j \ge 0, \; for \; all \; j$$

Solving these two models for each criterion, the optimal weights as interval values can be determined. For ranking the criteria the center of the intervals can be used. However, another option is to rank the criteria based on the interval weights by the help of a matrix of preferences [52].

3.2. Complex Proportional Assessment Method (COPRAS)

The COPRAS method was introduced by Zavadskas et al. [53]. The reliability and accuracy of the COPRAS method is acknowledged by several scholars and, nowadays, it is used to solve different engineering and management multi-attribute problems [54–58]. Moreover, the accuracy of performance measures in the COPRAS method assumes direct and proportional dependence of the significance and utility degree of investigated alternatives on a system of criteria [59].

COPRAS with fuzzy sets information is a developed method for solving decision-making problems under uncertain situations, introduced by Zavadskas and Antucheviciene [60]. In this paper, the assessment and prioritizing of the key factors of environmental sustainability in contemporary rural buildings is analyzed by the fuzzy COPRAS method.

In 2016 Beheshti et al. performed the COPRAS method for strategy portfolio optimization [53]. Pitchipoo et al. applied the COPRAS method in order to optimize blind spots in heavy vehicles [61]. In addition, the mentioned method was used to assess the neglected areas in Vilnius by Bielinskas et al. [62]. By applying the COPRAS method, the evaluation of construction projects of hotels based on environmental sustainability was practiced by Hashemkhani Zolfani et al. [18]. Moreover, Polat et al. applied the COPRAS method as a tool for mechanical designer selection [63].

Lithuanian scientists Zavadskas and Kaklauskas presented a method of multi-criteria complex proportional evaluation for formulating construction and engineering multi-objectives and multi-attribute problems since 1996 [64]. The four stages of this method are presented as below.

Stage 1: Calculate the normalized matrix by the following formula [18,65,66]:

$$d_{ij} = \frac{x_{ij}q_i}{\sum_{j=1}^{n} x_{ij}}, \; i = 1, \cdots, m, \; j = 1, \cdots, n \tag{7}$$

Remark that x_{ij} demonstrates the ith criterion in the jth alternative, m presents the number of criteria, where n stands for alternatives. Moreover, q_i illustrates the weight of ith criteria and D_{ij} is the normalized weighted value of each criterion. Note that:

$$\sum_{i=1}^{m} q_i = 1 \tag{8}$$

The values of weight q_i are usually determined based on the experts' point of view [18]. The influence of weight q_i on a_j distributes in proportion to the values of the investigated criterion x_{ij}:

$$q_i = \sum_{j=1}^{n} d_{ij}, \; i = 1, \cdots, m, \; j = 1, \cdots, n \tag{9}$$

Stage 2: Following Equation (7), $d_{ij}, \; i = 1, \cdots, m, \; j = 1, \cdots, n$ indicates the normalized weighted value of the ith criteria, which could be a benefit criteria (+) or cost (−). Thus, the jth alternative would be indicated by maximizing $d_{ij}^+, \; i = 1, \cdots, m$, where i is a benefit and minimizing $d_{ij}^-, \; i = 1, \cdots, m$, where i is a cost [67]. Maximizing the higher value and minimizing the lower value would be more desirable. Minimizing indices (d_{ij}^-) and maximizing indices (d_{ij}^+) are calculated for each jth alternative. The sum of weighted normalized minimizing and maximizing indices S_j^- and S_j^+, are calculated by [18,67]:

$$\tilde{S}_i^+ = \sum d_{ij}^+, \; i = 1, 2, \cdots, m \tag{10}$$

$$\tilde{S}_i^- = \sum d_{ij}^-, \; i = 1, 2, \cdots, m \tag{11}$$

In all the cases, S_j^+ is the sum of maximizing values from j row's alternative and S_j^- is the sum of minimizing values from j row's alternative [68]:

$$S^+ = \sum_{j=1}^{n} s_i^+ = \sum_{i=1}^{m} \sum_{j=1}^{n} d_{ij}^+ \tag{12}$$

$$S^- = \sum_{j=1}^{n} s_i^- = \sum_{i=1}^{m} \sum_{j=1}^{n} d_{ij}^- \tag{13}$$

$$i = 1, \cdots, m, \ j = 1, \cdots, n$$

Stage 3: The relative significance ($\tilde{Q}_i$) of each alternative aj should be determined according to positive $\tilde{S}_j^+$ and negative $\tilde{S}_j^-$. It could be calculated by the following formula [69]:

$$\tilde{Q}_i = \frac{s_{min}^- \sum s_i^-}{s_i^- \sum \frac{s_{min}^-}{s_i^-}} + s_i^+, \ i = 1, 2, \cdots, m \tag{14}$$

Stage 4: After stage 3, the priority of alternatives would be determined. The assessment results of alternatives reflect the initial data submitted by experts [68]. Based on Equation (7), the normalized weighted value of each ith criterion d_{ij}, $i = 1, \cdots, m$, $j = 1, \cdots, n$; would have a direct and proportional relationship with the variables x_{ij} and q_i. According to Equations (10) and (11), it is clear that the sums of $\tilde{S}_j^+$ and $\tilde{S}_j^-$ are linear functions of d_{ij}. In addition, based on Equation (14), the generalizing criterion Q_j has a direct linear relationship with the values and weights of the investigated criteria [60]. As a result, the greater the value of the generalizing criterion Q_j, the more effective the alternative will be. The satisfaction degree of demands and goals pursued by experts would be indicated by Q_j of a_j. The significance Q_{max} will always be the highest [60].

In order to visually assess the efficiency of alternative the utility degree N_i can be calculated. The degree of utility is determined by comparing the alternative analyzed with the most efficient alternative from the set of alternatives [70]. By comparing the variant which is analyzed with Q_{max}, the degree of the variant utility (N_i) can be determined [60]:

$$N_i = \frac{Q_i}{Q_{max}} \times 100 \tag{15}$$

All the utility degree values related to the alternatives analyzed range from 0% to 100%.

3.3. Fuzzy Numbers

Considering the fuzziness of the available data, hereby, the decision matrix can be converted into a fuzzy decision matrix and a weighted normalized fuzzy decision matrix will be constructed [71]. In this paper the triangular fuzzy numbers were applied [72]. A triangular fuzzy number f can be defined by a triplet (f_1, f_2, f_3) and is shown in Figure 6.

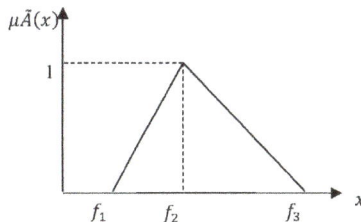

Figure 6. Membership function of TFN [71].

The membership function μf of f is defined as [67]:

$$\mu\tilde{A}(x) = \begin{cases} 0, x < f_1 \\ \dfrac{x - f_1}{f_2 - f_1}, f_1 \le x \le f_2 \\ \dfrac{x - f_3}{f_2 - f_3}, f_2 \le x < f_3 \\ 0, x > f_3 \end{cases}$$

The operations on fuzzy triangular numbers used in this research are defined as follows [72,73].

$$\tilde{f} + \tilde{f} = (f_1 + f_1, \ f_2 + f_2, \ f_3 + f_3) \tag{16}$$

$$\tilde{f} \times \tilde{f} = (f_1 \times f_1, \ f_2 \times f_2, \ f_3 \times f_3) \tag{17}$$

$$\tilde{f} \div \tilde{f} = (f_1 \div f_3, \ f_2 \div f_2, \ f_3 \div f_1) \tag{18}$$

3.3.1. A Linguistic Variable

According to Zadeh, it is complicated for a conventional quantification to reasonably express those situations being complex to describe. Accordingly, a linguistic variable is useful in such circumstances [71,74]. Fuzzy numbers can represent linguistic variables. The relationship between linguistic variables and TFN are presented in Tables 4 and 5.

Table 4. Relationships between linguistic variables and triangular fuzzy numbers [75].

Linguistic Variable	Triangular Fuzzy Numbers
Equal	(1, 1, 1)
a little better	(1/2, 1, 3/2)
mostly better	(1, 3/2, 2)
Better	(3/2, 2, 5/2)
much better	(2, 5/2, 3)
completely better	(5/2, 3, 7/2)

Table 5. Relationships between linguistic variables and triangular fuzzy numbers [60].

Linguistic Variable	Triangular Fuzzy Numbers
very poor/very light	(0, 1, 2)
poor/light	(1, 2, 3)
mostly poor/mostly light	(2, 3.5, 5)
Fair	(4, 5, 6)
mostly good/mostly difficult	(5, 6.5, 8)
good/difficult	(7, 8, 9)
very good/very difficult	(8, 9, 10)

In this paper, the ratings of qualitative criteria and the weights, and evaluating key factors of environmental sustainability in contemporary architecture of Iran, are considered as linguistic variables.

3.3.2. Defuzzification

The results of fuzzy decisions are fuzzy numbers. As a result, a problem of ranking fuzzy numbers may appear in MCDM. In order to solve this problem a defuzzification should be performed. The procedure of defuzzification is to locate the best non-fuzzy performance (BNP) value [72]. Several methods of defuzzification are available, such as mean-of-maximum, center-of-area, and a-cut methods [76,77]. In this research the center-of-area method is used. The defuzzified value of a fuzzy number would be obtained by applying the following equation [74]:

$$Q_i = f_1 + \frac{[(f_3 - f_1) + (f_2 - f_1)]}{3} \tag{19}$$

where BNP is the best non-fuzzy performance value, f_2 is a mode, and f_1 and f_3 are the lower and the upper limits of fuzzy triangular number f, respectively.

3.4. Proposed Approach

The process of multiple-criteria analysis in identifying and prioritizing key factors of environmental sustainability of Iranian contemporary architecture by performing the fuzzy set approach is performed in several stages, being presented in Figure 7.

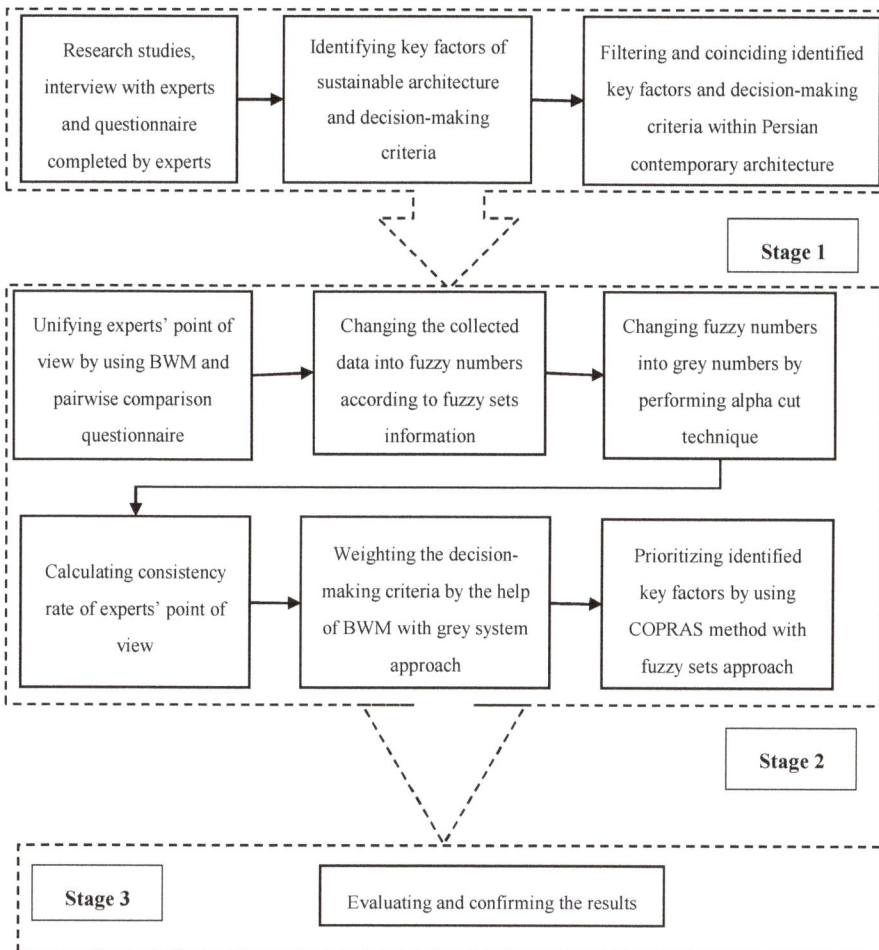

Figure 7. Stages of multi-criteria analysis of prioritizing key factors of sustainability in Persian contemporary architecture, source: own.

Stage 1. According to the method of multiple-criteria decision-making, the key factors of environmental sustainability should be determined. For this reason, the initial information was collected by the help of research studies, interviews, and questionnaires.

Stage 2.

    Step 1

- After identifying key factors and criteria of the decision-making problem according to the experts, their opinions should be unified. For this reason, at first, the best and the worst criteria are resolved. Thereupon, the experts determine the preference of the best criterion over all the other criteria. Moreover, preference of the criteria over the worst criterion is determined accordingly.

    Step 2

- The results of pairwise comparisons are transferred into triangular fuzzy numbers. By means of unification, the average of triangular fuzzy numbers is calculated.

    Step 3

- Since the mathematical model of BWM is a deterministic nonlinear model, in this step, the average of the triangular fuzzy numbers should be transformed into grey numbers through alpha cut technique by using the following equation:

$$(l(\alpha), u(\alpha)) = (f_1 + \alpha(f_2 - f_1), f_3 + \alpha(f_2 - f_3)) \tag{20}$$

    Step 4

- After changing triangular fuzzy numbers into grey numbers by the help of the alpha cut technique, A_B (the preference of the best criterion over all the other criteria) and A_W (the preference of all the criteria over the worst criterion) can be created based on the different amounts of alpha between zero and one.
- Since the center of the grey numbers for different amounts of alpha is the same, based on Equation (3), a mathematical model could be defined. By solving the model, ξ^* is obtained. According to ξ^* and Table 3, the consistency ratio can be calculated.

    Step 5

- In order to determine the lower and upper bounds of the weight of criterion j, two mathematical models should be proposed for different amounts of alpha based on Equations (5) and (6). By solving these two models for all the criteria, the optimal weights of the criteria can be determined as interval values.

    Step 6

- In this part, each expert fills out the evaluation questionnaire of alternatives over decision-making criteria based on fuzzy numbers. Next, by means of unification, the average of triangular fuzzy numbers is calculated.
- In order to prioritizing alternatives with COPRAS method, first of all, the decision- making matrix with triangular numbers must be created. Afterwards, the decision-making matrix should be turned into normalized weighted matrix.
- After creating the normalized weighted decision-making matrix, the values S_j^+ and S_j^- have to be figured out for all the alternatives.
- In the last step, the relative significance of each alternative $a_j(Q_j)$ is determined according to positive S_j^+ and negative S_j^-. Then, the determined $\tilde{Q}_i$ should be defuzzied (Q_j) and, finally, the degree of the variant utility (N_i) is calculated by a comparison of the variant that is analyzed with the most efficient one.

Stage 3.

- Eventually the results are emanated and the comparison between different alpha cuts is presented for more consideration.

4. Results and Case Study

4.1. Identifying Alternatives and Criteria

Based on the explored principles of the sustainable architecture, including energy saving, harmony with climate, reduction in the use of new resources, satisfying the needs of residents, harmony with the site and unification [32], and according to the experts' point of view, six factors were chosen as the key factors of environmental sustainability in the contemporary architecture of Iran. Experts were chosen among academic and executive professionals in the fields of architecture, civil engineering, and project management. Moreover, they were divided into three different groups based on their professions. Table 6 shows three expert groups in detail.

Table 6. Three group of experts, source: own.

Group	Profession	No. of Academic Experts	No. of Executive Experts
1	Architecture	4	3
2	Civil engineering	3	1
3	Project manager	1	2

The mentioned six factors of environmental sustainability in contemporary architecture of Iran, emanated from the aforementioned approach, are illustrated as follows:

1. Proving building management systems (a_1);
2. Applying rules of continental design (a_2);
3. Practicing renewal sources (a_3);
4. Performing human design (a_4);
5. Adopting ecological rules in design (a_5); and
6. Creating engagement between buildings and other urban systems (a_6).

Similarly, the following six items were determined as the criteria of the decision-making problem, to evaluate the key factors of environmental sustainability in contemporary architecture of Iran:

1. Execution cost (c_1);
2. Maintenance cost (c_2);
3. Easiness of execution (c_3);
4. Reducing current costs (c_4);
5. Adaptability with built buildings (c_5); and
6. Accessibility to required knowledge (c_6).

4.2. Weighting Criteria with BWM

As it was mentioned in Section 3.1, BWM was chosen for weighting criteria because of its advantages, which require less comparison data in conjunction with more consistent comparisons and more reliable results.

Step 1. At first, c_6 (accessibility to required knowledge) was determined as the best criteria and c_3 (easiness of execution) was determined as the worst criteria by the brain-storming among the experts. Subsequently, the preference of the best criterion (c_6) over all the other criteria was determined by the experts. Table 7 shows the results of this step.

Table 7. Pairwise comparison vector for the best criterion, source: own.

	c_1	c_2	c_3	c_4	c_5	c_6
			Group expert 1			
c_6	$(5/2, 3, 7/2)$	$(3/2, 2, 5/2)$	$(5/2, 3, 7/2)$	$(1, 1, 1)$	$(1, 3/2, 2)$	$(1, 1, 1)$
			Group expert 2			
c_6	$(3/2, 2, 5/2)$	$(1, 3/2, 2)$	$(5/2, 3, 7/2)$	$(1, 3/2, 2)$	$(3/2, 2, 5/2)$	$(1, 1, 1)$
			Group expert 3			
c_6	$(3/2, 2, 5/2)$	$(3/2, 2, 5/2)$	$(2, 5/2, 3)$	$(1/2, 1, 3/2)$	$(1/2, 1, 3/2)$	$(1, 1, 1)$
Average	$(1.83, 2.33, 2.83)$	$(1.33, 1.83, 2.33)$	$(2.33, 2.83, 3.33)$	$(0.83, 1.16, 1.5)$	$(1, 1.5, 2)$	$(1, 1, 1)$

Preference of the criteria over the worst criterion was determined according to the experts' opinion. Table 8 indicates the results of this step.

Table 8. Pairwise comparison vector for the worst criterion, source: own.

	Group expert 1	Group expert 2	Group expert 3	Average
	c_3	c_3	c_3	
c_1	$(1/2, 1, 3/2)$	$(1, 1, 1)$	$(1, 1, 1)$	$(0.83, 1, 1.16)$
c_2	$(1/2, 1, 3/2)$	$(1, 1, 1)$	$(1, 3/2, 2)$	$(0.83, 1.16, 1.5)$
c_3	$(1, 1, 1)$	$(1, 1, 1)$	$(1, 1, 1)$	$(1, 1, 1)$
c_4	$(2, 5/2, 3)$	$(3/2, 2, 5/2)$	$(2, 5/2, 3)$	$(1.83, 2.33, 2.83)$
c_5	$(3/2, 2, 5/2)$	$(3/2, 2, 5/2)$	$(2, 5/2, 3)$	$(1.66, 2.16, 2.66)$
c_6	$(5/2, 3, 7/2)$	$(5/2, 3, 7/2)$	$(2, 5/2, 3)$	$(2.33, 2.83, 3.33)$

Step 2. The average of the triangular fuzzy numbers has transformed into grey numbers through the alpha cut technique and by Equation (20). Table 9 shows the results of changing triangular fuzzy numbers into grey numbers according to different amounts of alpha.

Table 9. Pairwise comparison vector for the best criterion with grey numbers, source: own.

α	c_6c_1	c_6c_2	c_6c_3	c_6c_4	c_6c_5	c_6c_6
0	$[1.83, 2.83]$	$[1.33, 2.33]$	$[2.33, 3.33]$	$[0.83, 1.5]$	$[1, 2]$	$[1, 1]$
0.1	$[1.88, 2.78]$	$[1.38, 2.28]$	$[2.38, 3.28]$	$[0863, 1.466]$	$[1.05, 1.95]$	$[1, 1]$
0.2	$[1.93, 2.73]$	$[1.43, 2.23]$	$[2.43, 3.23]$	$[0896, 1.432]$	$[1.1, 1.9]$	$[1, 1]$
0.3	$[1.98, 2.68]$	$[1.48, 2.18]$	$[2.48, 3.18]$	$[0.929, 1.398]$	$[1.15, 1.85]$	$[1, 1]$
0.4	$[2.03, 2.63]$	$[1.53, 2.13]$	$[2.53, 3.13]$	$[0.962, 1.364]$	$[1.2, 1.8]$	$[1, 1]$
0.5	$[2.08, 2.58]$	$[1.58, 2.08]$	$[2.58, 3.08]$	$[0.995, 1.33]$	$[1.25, 1.75]$	$[1, 1]$
0.6	$[2.13, 2.53]$	$[1.63, 2.03]$	$[2.63, 3.03]$	$[1.028, 1.296]$	$[1.3, 1.7]$	$[1, 1]$
0.7	$[2.18, 2.48]$	$[1.68, 1.98]$	$[2.68, 2.98]$	$[1.061, 1.262]$	$[1.35, 1.65]$	$[1, 1]$
0.8	$[2.33, 2.43]$	$[1.73, 1.93]$	$[2.73, 2.93]$	$[1.094, 1.228]$	$[1.4, 1.6]$	$[1, 1]$
0.9	$[2.28, 2.38]$	$[1.78, 1.88]$	$[2.78, 2.88]$	$[1.127, 1.194]$	$[1.45, 1.55]$	$[1, 1]$
1	$[2.33, 2.33]$	$[1.83, 1.83]$	$[2.83, 2.83]$	$[1.16, 1.16]$	$[1.5, 1.5]$	$[1, 1]$

Moreover, Table 10 indicates the results of pairwise comparison vector for the worst criterion with grey numbers according to different amounts of alpha.

Table 10. Pairwise comparison vector for the worst criterion with grey numbers, source: own.

α	c_1c_3	c_2c_3	c_3c_3	c_4c_3	c_5c_3	c_6c_3
0	[0.83, 1.16]	[0.83, 1.5]	[1, 1]	[1.83, 2.83]	[1.66, 2.66]	[2.33, 3.33]
0.1	[0847, 1.144]	[0863, 1.466]	[1, 1]	[1.88, 2.78]	[1.71, 2.61]	[2.38, 3.28]
0.2	[0.864, 1.128]	[0.896, 1.432]	[1, 1]	[1.93, 2.73]	[1.76, 2.56]	[2.43, 3.23]
0.3	[0.881, 1.112]	[0.929, 1.398]	[1, 1]	[1.98, 2.68]	[1.81, 2.51]	[2.48, 3.18]
0.4	[0.898, 1.096]	[0.962, 1.364]	[1, 1]	[2.03, 2.63]	[1.86, 2.46]	[2.53, 3.13]
0.5	[0.915, 1.08]	[0.995, 1.33]	[1, 1]	[2.08, 2.58]	[1.91, 2.41]	[2.58, 3.08]
0.6	[0.932, 1.064]	[1.028, 1.296]	[1, 1]	[2.13, 2.53]	[1.96, 2.36]	[2.63, 3.03]
0.7	[0.949, 1.048]	[1.061, 1.262]	[1, 1]	[2.18, 2.48]	[2.01, 2.31]	[2.68, 2.98]
0.8	[0.966, 1.032]	[1.094, 1.228]	[1, 1]	[2.23, 2.43]	[2.06, 2.26]	[2.73, 2.93]
0.9	[0.983, 1.016]	[1.127, 1.194]	[1, 1]	[2.28, 2.38]	[2.11, 2.21]	[2.78, 2.88]
1	[1, 1]	[1.16, 1.16]	[1, 1]	[2.33, 2.33]	[2.16, 2.16]	[2.83,2.83]

In this step, according to Equations (1) and (2), A_B (the preference of the best criterion over all the other criteria) and A_W (the preference of all the criteria over the worst criterion) were created based on the different amounts of alpha.

Step 3. Since the center of the grey numbers for different amounts of alpha is the same, according to Equation (3), the following problem could be defined:

$$Min\,Max\left\{\left|\frac{W_6}{W_1}-2.33\right|,\left|\frac{W_6}{W_2}-1.83\right|,\left|\frac{W_6}{W_3}-2.83\right|,\left|\frac{W_6}{W_4}-1.16\right|,\left|\frac{W_6}{W_5}-1.5\right|,\left|\frac{W_1}{W_3}-1\right|,\left|\frac{W_2}{W_3}\right.\right.$$
$$\left.\left.-1.16\right|,\left|\frac{W_4}{W_3}-2.33\right|,\left|\frac{W_5}{W_3}-2.16\right|\right\}$$

$$s.t:$$
$$W_1+W_2+W_3+W_4+W_5+W_6=1$$
$$W_j\geq 0,\ \forall j\in n$$

This problem can be transferred to the following problem:

$$Min\,\varepsilon$$
$$S.t.$$
$$\left|\frac{W_6}{W_1}-2.33\right|\leq\varepsilon$$
$$\left|\frac{W_6}{W_2}-1.83\right|\leq\varepsilon$$
$$\left|\frac{W_6}{W_3}-2.83\right|\leq\varepsilon$$
$$\left|\frac{W_6}{W_4}-1.16\right|\leq\varepsilon$$
$$\left|\frac{W_6}{W_5}-1.5\right|\leq\varepsilon$$
$$\left|\frac{W_1}{W_3}-1\right|\leq\varepsilon$$
$$\left|\frac{W_2}{W_3}-1.16\right|\leq\varepsilon$$
$$\left|\frac{W_4}{W_3}-2.33\right|\leq\varepsilon$$
$$\left|\frac{W_5}{W_3}-2.16\right|\leq\varepsilon$$
$$W_1+W_2+W_3+W_4+W_5+W_6=1$$
$$W_j\geq 0,\ \forall j\in n$$

By solving the problem with Lingo 17.0 software (Chicago, IL, USA), ε^* is obtained (0.2611469). According to ε^*, Table 2, and Equation (4), the consistency ratio can be calculated as below, indicating the appropriate consistency of results obtained from average opinion:

$$CR = \frac{0.261}{3} = 0.087$$

Step 4. Based on Equations (5) and (6), the following two models can be proposed for different amounts of alpha, in order to determine the upper and lower weights of criterion j:

Lower bound of w_1

Min W_1

S.t.

$\left| \frac{W_6}{W_1} - 2.33 \right| \leq 0.261$

$\left| \frac{W_6}{W_2} - 1.83 \right| \leq 0.261$

$\left| \frac{W_6}{W_3} - 2.83 \right| \leq 0.261$

$\left| \frac{W_6}{W_4} - 1.16 \right| \leq 0.261$

$\left| \frac{W_6}{W_5} - 1.5 \right| \leq 0.261$

$\left| \frac{W_1}{W_3} - 1 \right| \leq 0.261$

$\left| \frac{W_2}{W_3} - 1.16 \right| \leq 0.261$

$\left| \frac{W_4}{W_3} - 2.33 \right| \leq 0.261$

$\left| \frac{W_5}{W_3} - 2.16 \right| \leq 0.261$

$W_1 + W_2 + W_3 + W_4 + W_5 + W_6 = 1$

$W_j \geq 0, \forall j \in n$

Upper bound of w_1

Max W_1

S.t.

$\left| \frac{W_6}{W_1} - 2.33 \right| \leq 0.261$

$\left| \frac{W_6}{W_2} - 1.83 \right| \leq 0.261$

$\left| \frac{W_6}{W_3} - 2.83 \right| \leq 0.261$

$\left| \frac{W_6}{W_4} - 1.16 \right| \leq 0.261$

$\left| \frac{W_6}{W_5} - 1.5 \right| \leq 0.261$

$\left| \frac{W_1}{W_3} - 1 \right| \leq 0.261$

$\left| \frac{W_2}{W_3} - 1.16 \right| \leq 0.261$

$\left| \frac{W_4}{W_3} - 2.33 \right| \leq 0.261$

$\left| \frac{W_5}{W_3} - 2.16 \right| \leq 0.261$

$W_1 + W_2 + W_3 + W_4 + W_5 + W_6 = 1$

$W_j \geq 0, \forall j \in n$

By solving these two models for all the criteria, we can determine the optimal weights of the criteria as interval values. The center of intervals can be used to rank the criteria or alternatives. Table 11 shows the optimal weights of the criteria.

Table 11. Optimal interval weights, source: own.

Optimal Weight	Lower Bound	Upper Bound	Center	Width
w^*_1	0.099473	0.113167	0.10632	0.013694
w^*_2	0.122048	0.128845	0.125447	0.00668
w^*_3	0.08588	0.090663	0.088272	0.004783
w^*_4	0.187646	0.224733	0.206189	0.037087
w^*_5	0.206014	0.217487	0.21175	0.011473
w^*_6	0.255221	0.269434	0.262328	0.014213

Figure 8 illustrates the width, lower, and upper bounds of the optimal weights.

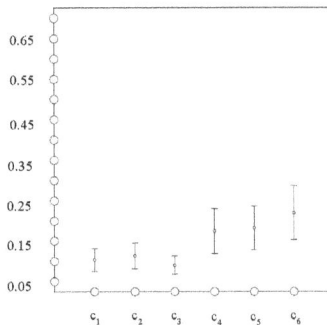

Figure 8. Optimal interval weights, source: own.

However, another option is to rank the criteria or alternatives based on the interval weights. The matrix of the degree of preference and the matrix of preferences were applied for final prioritization. The results are presented in Table 12 [42].

Table 12. Results of preferences matrix, source: own.

		w_1	w_2	w_3	w_4	w_5	w_6			w_1	w_2	w_3	w_4	w_5	w_6	sum
	w_1	0.5	0	1	0	0	0		w_1	0	0	1	0	0	0	1
	w_2	1	0.5	1	0	0	0		w_2	1	0	1	0	0	0	2
$DP_{ij} =$	w_3	0	0	0.5	0	0	0	$P_{ij}=$	w_3	0	0	0	0	0	0	0
	w_4	1	1	1	0.5	0.38	0		w_4	1	1	1	0	0	0	3
	w_5	1	1	1	0.62	0.5	0		w_5	1	1	1	1	0	0	4
	w_6	1	1	1	1	1	0.5		w_6	1	1	1	1	1	0	5

Based on the sum of the rows, the following ranking emerges as below, being compatible with the initial weights emanated from nonlinear model:

$$c_6 > c_5 > c_4 > c_2 > c_1 > c_3 \tag{}$$

4.3. Prioritizing Alternatives with the COPRAS Method

In order to prioritize alternatives, the COPRAS method was selected according to its superiority compared to other existing MCDM methods, including easiness of the procedure, considering every aspect of each criteria, and separation of negative and positive criteria.

Step 1. In the first step, evaluation of alternatives over decision-making criteria was applied by the experts. Table 13 demonstrated the results of this step.

Step 2. In order to prioritize alternatives with the fuzzy COPRAS method, the decision-making matrix with triangular fuzzy numbers was created. It should be mentioned that, in this matrix in order to use the calculated weights of the criteria, the center of the grey numbers has been applied. Table 14 indicates the decision-making matrix with fuzzy numbers; (+) presents positive criteria and (−) illustrates negative criteria.

Step 3. The decision-making matrix has been transferred into a normalized weighted matrix using Equations (7)–(9). Table 15 denotes the normalized weighted decision-making matrix with fuzzy numbers.

Table 13. Evaluation of alternatives over decision-making criteria, source: own.

Criteria / Alternative	c_1	c_2	c_3	c_4	c_5	c_6
a_1	(7, 8, 9)	(7, 8, 9)	(4, 5, 6)	(7, 8, 9)	(1, 2, 3)	(7, 8, 9)
a_2	(8, 9, 10)	(8, 9, 10)	(5, 6.5, 8)	(7, 8, 9)	(4, 5, 6)	(7, 8, 9)
a_3	(5, 6.5, 8)	(4, 5, 6)	(4, 5, 6)	(7, 8, 9)	(1, 2, 3)	(7, 8, 9)
a_4	(2, 3.5, 5)	(7, 8, 9)	(2, 3.5, 5)	(2, 3.5, 5)	(4, 5, 6)	(7, 8, 9)
a_5	(8, 9, 10)	(8, 9, 10)	(5, 6.5, 8)	(5, 6.5, 8)	(4, 5, 6)	(7, 8, 9)
a_6	(4, 5, 6)	(7, 8, 9)	(5, 6.5, 8)	(7, 8, 9)	(4, 5, 6)	(7, 8, 9)
Group Expert 1						
a_1	(8, 9, 10)	(8, 9, 10)	(4, 5, 6)	(8, 9, 10)	(2, 3.5, 5)	(5, 6.5, 8)
a_2	(5, 6.5, 8)	(8, 9, 10)	(4, 5, 6)	(7, 8, 9)	(4, 5, 6)	(7, 8, 9)
a_3	(7, 8, 9)	(7, 8, 9)	(5, 6.5, 8)	(7, 8, 9)	(0, 1, 2)	(5, 6.5, 8)
a_4	(4, 5, 6)	(5, 6.5, 8)	(2, 3.5, 5)	(4, 5, 6)	(5, 6.5, 8)	(7, 8, 9)
a_5	(5, 6.5, 8)	(7, 8, 9)	(4, 5, 6)	(7, 8, 9)	(1, 2, 3)	(7, 8, 9)
a_6	(1, 2, 3)	(5, 6.5, 8)	(7, 8, 9)	(4, 5, 6)	(7, 8, 9)	(8, 9, 10)
Group Expert 2						
a_1	(7, 8, 9)	(8, 9, 10)	(1, 2, 3)	(8, 9, 10)	(4, 5, 6)	(8, 9, 10)
a_2	(7, 8, 9)	(7, 8, 9)	(7, 8, 9)	(8, 9, 10)	(7, 8, 9)	(8, 9, 10)
a_3	(7, 8, 9)	(5, 6.5, 8)	(7, 8, 9)	(8, 9, 10)	(2, 3.5, 5)	(4, 5, 6)
a_4	(1, 2, 3)	(8, 9, 10)	(5, 6.5, 8)	(4, 5, 6)	(5, 6.5, 8)	(8, 9, 10)
a_5	(8, 9, 10)	(7, 8, 9)	(5, 6.5, 8)	(8, 9, 10)	(2, 3.5, 5)	(5, 6.5, 8)
a_6	(5, 6.5, 8)	(5, 6.5, 8)	(7, 8, 9)	(7, 8, 9)	(5, 6.5, 8)	(8, 9, 10)
Group Expert 3						
a_1	(7.33, 8.33, 9.33)	(7.66, 8.66, 9.66)	(3, 4, 5)	(7.66, 8.66, 9.66)	(2.33, 3.5, 4.66)	(6.66, 7.83, 9)
a_2	(6.66, 7.83, 9)	(7.66, 8.66, 9.66)	(5.33, 6.5, 7.66)	(7.33, 8.33, 9.33)	(5, 6, 7)	(7.33, 8.33, 9.33)
a_3	(6.33, 7.5, 8.66)	(5.33, 6.5, 7.66)	(5.33, 6.5, 7.66)	(7.33, 8.33, 9.33)	(1, 2.16, 3.33)	(5.33, 6.5, 7.66)
a_4	(2.33, 3.5, 4.66)	(6.66, 7.83, 9)	(3, 4.5, 6)	(3.33, 4.5, 5.66)	(4.66, 6, 7.33)	(7.33, 7.33, 9.33)
a_5	(7, 8.16, 9.33)	(7.33, 8.33, 9.33)	(4.66, 6, 7.33)	(6.66, 7.83, 9)	(2.33, 3.5, 4.66)	(6.33, 7.5, 8.66)
a_6	(3.33, 4.5, 5.66)	(5.66, 7, 8.33)	(6.33, 7.5, 8.66)	(6, 7, 8)	(5.33, 6.5, 7.66)	(7.66, 8.66, 9.66)

Table 14. Decision-making matrix with fuzzy numbers, source: own.

	c_1	c_2	c_3	c_4	c_5	c_6
	(+)	(−)	(+)	(+)	(+)	(+)
a_1	(7.33, 8.33, 9.33)	(7.66, 8.66, 9 9.66)	(3, 4, 5)	(7.66, 8.66, 9.66)	(2.33, 3.5, 4.66)	(6.66, 7.83, 9)
a_2	(6.66, 7.83, 9)	(7.66, 8.66, 9.66)	(5.33, 6.5, 7.66)	(7.33, 8.33, 9.33)	(5, 6, 7)	(7.33, 8.33, 9.33)
a_3	(6.33, 7.5, 8.66)	(5.33, 6.5, 7.66)	(5.33, 6.5, 7.66)	(7.33, 8.33, 9.33)	(1, 2.16, 3.33)	(5.33, 6.5, 7.66)
a_4	(2.33, 3.5, 4.66)	(6.66, 7.83, 9)	(3, 4.5, 6)	(3.33, 4.5, 5.66)	(4.66, 6, 7.33)	(7.33, 8.33, 9.33)
a_5	(7, 8.16, 9.33)	(7.33, 8.33, 9.33)	(4.66, 6, 7.33)	(6.66, 7.83, 9)	(2.33, 3.5, 4.66)	(6.33, 7.5, 8.66)
a_6	(3.33, 4.5, 5.66)	(5.66, 7, 8.33)	(6.33, 7.5, 8.66)	(6, 7, 8)	(5.33, 6.5, 7.66)	(7.66, 8.66, 9.66)
Total	(32.98, 39.82, 46.64)	(40.3, 46.98, 53.64)	(27.65, 35, 42.31)	(38.31, 44.65, 50.98)	(20.65, 27.66, 34.64)	(40.64, 47.15, 53.64)
w_j	0.11	0.12	0.09	0.2	0.22	0.26

Table 15. Normalized weighted decision-making matrix with fuzzy numbers, source: own.

	c_1	c_2	c_3	c_4	c_5	c_6
	(+)	(−)	(+)	(+)	(+)	(+)
a_1	(0.017, 0.023, 0.031)	(0.017, 0.022, 0.028)	(0.006, 0.01, 0.016)	(0.003, 0.004, 0.005)	(0.014, 0.027, 0.049)	(0.032, 0.043, 0.057)
a_2	(0.015, 0.021, 0.030)	(0.017, 0.022, 0.028)	(0.011, 0.016, 0.024)	(0.002, 0.003, 0.004)	(0.031, 0.047, 0.074)	(0.035, 0.045, 0.059)
a_3	(0.014, 0.020, 0.028)	(0.011, 0.016, 0.022)	(0.011, 0.016, 0.024)	(0.002, 0.003, 0.004)	(0.006, 0.017, 0.035)	(0.025, 0.035, 0.049)
a_4	(0.005, 0.009, 0.015)	(0.014, 0.020, 0.026)	(0.006, 0.011, 0.019)	(0.001, 0.002, 0.003)	(0.029, 0.047, 0.078)	(0.035, 0.045, 0.059)
a_5	(0.016, 0.022, 0.031)	(0.016, 0.021, 0.027)	(0.009, 0.015, 0.023)	(0.002, 0.003, 0.004)	(0.014, 0.027, 0.049)	(0.030, 0.041, 0.055)
a_6	(0.007, 0.012, 0.018)	(0.012, 0.017, 0.024)	(0.013, 0.019, 0.028)	(0.002, 0.003, 0.004)	(0.033, 0.051, 0.081)	(0.037, 0.047, 0.061)

Step 4. Subsequently, the values $\widetilde{S}_i^+$ and $\widetilde{S}_i^-$ have been determined for all the alternatives according to Equations (11) and (12). The results are uncovered in Table 16.

Table 16. Matrix of $\widetilde{S}_i^+$ and $\widetilde{S}_i^-$, source: own.

	$\widetilde{S}_i^+$	$\widetilde{S}_i^-$
a_1	(0.069, 0.107, 0.158)	(0.017, 0.022, 0.028)
a_2	(0.094, 0.132, 0.191)	(0.017, 0.022, 0.028)
a_3	(0.058, 0.091, 0.14)	(0.011, 0.016, 0.022)
a_4	(0.076, 0.114, 0.174)	(0.014, 0.020, 0.026)
a_5	(0.071, 0.108, 0.162)	(0.016, 0.021, 0.027)
a_6	(0.092, 0.132, 0.192)	(0.012, 0.017, 0.024)

Step 5. The relative significance of each alternative $a_j(\widetilde{Q}_i)$ was determined according to Equation (14). Afterwards, the determined $\widetilde{Q}_i$ has been defuzzied (Q_i) by Equation (19). Finally, the degree of the variant utility (N_i) was calculated by Equation (15). The results are illustrated in Table 17.

Table 17. Matrix of $\widetilde{Q}_i$, Q_i, and N_i, source: own.

	$\widetilde{Q}_i$	Q_i	N_i	Priority
a_1	(0.073, 0.125, 0.237)	0.145	78.37%	6
a_2	(0.098, 0.15, 0.27)	0.172	92.97%	2
a_3	(0.063, 0.116, 0.261)	0.146	78.91%	5
a_4	(0.08, 0.134, 0.268)	0.16	86.48%	3
a_5	(0.075, 0.127, 0.245)	0.149	80.54%	4
a_6	(0.096, 0.155, 0.305)	0.185	100%	1

As Table 17 reveals, the priority of key factors of environmental sustainability in Iranian contemporary architecture is presented as below:

1. Creating engagement between buildings and other urban systems;
2. Applying rules of continental design;
3. Performing human design;
4. Adopting ecological rules in design;
5. Practicing renewable resources; and
6. Proving building management systems.

5. Conclusions

In this research, key factors of environmental sustainability were analyzed with an emphasis on Iranian contemporary architecture, and by combining economic benefits of sustainable architecture, environmental potential, and social interest. Six possible alternatives for sustaining Persian architecture were suggested, including using building management systems, applying rules of continental design, using renewable energy and construction resources, applying human design, applying ecological rules in design, and creating engagement between buildings and other urban systems. Ranking of alternatives was performed on the mathematical statistical calculations and was based on the criteria system, as developed by the experts.

Calculations by applying fuzzy COPRAS and BWM were suggested, which took into consideration the uncertainty caused by incomplete and inconsistent information that related to sustainable development. The consistency ratio was calculated and the priority of alternatives was determined (Table 17). It has been concluded that among the alternatives, creating engagement between buildings and other urban systems has the first priority, and proving building management systems has the last one. The alternative of applying rules of continental design has the second, performing human design has the third, adopting ecological rules in design has the fourth, and practicing renewable resources presents the fifth priority.

In order to complete this research, and based on the limitations that the authors had been faced with, some suggestions could be mentioned for future studies. Instead of using a lower and upper bounded grey model, the grey model could be solved with other approaches. Approaches possible to perform in uncertain conditions, such as the interval-valued intuitionistic fuzzy approach, could be applied instead of fuzzy and grey approaches. Moreover, one could observe and evaluate other components of sustainable development, like economic sustainability and social sustainability, next to environmental sustainability.

The findings are important for further driving the development of sustainable architecture in Iran, mainly by giving the grounds for weighting different sustainability criteria for the decision-making in the strategy, design, implementation, and impact assessment phases. It is important to note that the number and the list of sustainability criteria may change depending on the country, the region, or the urban setting analyzed, and the method will still work. The demonstrated method allows the architects and urban planners to compose the individually-shaped set of sustainability indicators and weight them accordingly. In particular, both the authorities and the specialists may adapt the findings of this paper to the national sustainability standards and design guidelines, as well as for the sustainability assessment tools in Iran and beyond. The wider question remains: how the research findings can find their way to the regulatory and professional development environments, and this may be the topic for coming research.

Author Contributions: The individual contributions of the authors are indicated as follows: The initial idea was proposed by H.A.M.; The real world case data gathering was prepared by S.A.; applying the proposed approach and analyzing the data was carried out by H.A.M. and S.A.; E.K.Z. and G.S. revised the manuscript completely regarding the abstract, introduction, research design, research methodology, and conclusions; and all authors have read and approved the final manuscript.

References

1. Zaib Khan, A.; Vandevyvere, H.; Allacker, K. Design for the Ecological Age: Rethinking the Role. *J. Archit. Educ.* **2013**, *67*, 175–185. [CrossRef]
2. Martek, I.; Hosseini, M.R.; Shrestha, A.; Zavadskas, E.K.; Seaton, S. The Sustainability Narrative in Contemporary Architecture: Falling Short of Building a Sustainable Future. *Sustainability* **2018**, *10*, 981. [CrossRef]
3. Grierson, D.; Salama, A. Forging advances in sustainable architecture and urbanism. *Open House Int.* **2016**, *41*, 4–6.
4. Hamiti, S.W.; Wydler, H. Supporting the Integration of Sustainability into Higher Education Curricula—A Case Study from Switzerland. *Sustainability* **2014**, *6*, 3291–3300. [CrossRef]
5. Rahaei, O.; Derakhshan, S.; Shirgir, N. Environmental architecture: The role of sustainable structures in futuristic buildings. *J. Fundam. Appl. Sci.* **2016**, *8*, 559–613. [CrossRef]
6. Amiri, N.; Vatandoost, M. The Study of the Relationship between Sustainable Architecture and Vernacular Architecture in the North of Iran. *J. Hist. Cult. Art Res.* **2017**, *6*, 436–450. [CrossRef]
7. Maywald, C.; Riesser, F. Sustainability-the art of modern architecture. *Procedia Eng.* **2016**, *155*, 238–248. [CrossRef]
8. Farmer, G. From Differentiation to Concretisation: Integrative Experiments in Sustainable Architecture. *Societies* **2017**, *7*, 35. [CrossRef]
9. Rieh, S.-Y.; Lee, B.-Y.; Oh, J.-G.; Schuetze, T.; Álvarez, S.P.; Lee, K.; Park, J. Integration of Sustainability into Architectural Education at Accredited Korean Universities. *Sustainability* **2017**, *9*, 1121. [CrossRef]
10. Shaker, R.R. The spatial distribution of development in Europe and its underlying sustainability correlations. *Appl. Geogr.* **2015**, *63*, 304–314. [CrossRef]

11. Kahle, L.; Gurel-Atay, E. *Communicating Sustainability for the Green Economy*; M.E. Sharpe: New York, NY, USA, 2014.

12. Hoseynof, E.; Fikret, O. Planning of sustainable cities in view of green architecture. *Procedia Eng.* **2011**, *21*, 534–542. [CrossRef]

13. Wang, X.; Altan, H.; Kang, J. Parametric study on the performance of green residential buildings in China. *Front. Archit. Res.* **2015**, *4*, 56–67. [CrossRef]

14. Zebari, H.; Ibrahim, R. Methods and strategies for sustainable architecture in Kurdistan region, Iraq. *Procedia Environ. Sci.* **2016**, *34*, 202–211. [CrossRef]

15. BREEAM. Available online: http://www.breeam.org (accessed on 16 March 2018).

16. Zarghami, E. Azemati, H. Fatourehchi, D. Karamloob, M. Customizing well-known sustainability assessment tools for Iranian residential buildings using Fuzzy Analytic Hierarchy Process. *Build. Environ.* **2018**, *128*, 107–128. [CrossRef]

17. Shareef, S.L.; Altan, H. Building sustainability rating systems in the Middle East. *Proc. Inst. Civ. Eng. Eng. Sustain.* **2017**, *170*, 283–293. [CrossRef]

18. Hashemkhani Zolfani, S.; Pourhossein, M.; Yazdani, M.; Zavadskas, E.K. Evaluating construction projects of hotels based on environmental sustainability with MCDM framework. *Alex. Eng. J.* **2018**, *57*, 357–365. [CrossRef]

19. Mazraeh, H.M.; Pazhouhanfar, M. Effects of vernacular architecture structure on urban sustainability case study: Qeshm Island, Iran. *Front. Archit. Res.* **2018**, *7*, 11–24. [CrossRef]

20. Sahebzadeh, S.; Heidari, A.; Kamelnia, H.; Baghbani, A. Sustainability Features of Iran's Vernacular Architecture: A Comparative Study between the Architecture of Hot–Arid and Hot–Arid–Windy Regions. *Sustainability* **2017**, *9*, 749. [CrossRef]

21. Park, J.; Yoon, J.; Kim, K.-H. Critical Review of the Material Criteria of Building Sustainability Assessment Tools. *Sustainability* **2017**, *9*, 186. [CrossRef]

22. Celiker, A. Sustainable housing: A conceptual approach. *Open House Int.* **2017**, *42*, 49–57.

23. Arroyo, P.; Fuenzalida, C.; Albert, A.; Hallowell, M.R. Collaborating in decision making of sustainable building design: An experimental study comparing CBA and WRC methods. *Energy Build.* **2016**, *128*, 132–142. [CrossRef]

24. Ismail, M.; Keumala, N.; Dabdoob, R. Review on integrating sustainability knowledge into architectural education: Practice in the UK and the USA. *J. Clean. Prod.* **2017**, *140*, 1542–1552. [CrossRef]

25. Lowe, J.; Watts, N. An evaluation of a Breem case study project. *Built Environ. Res. Trans.* **2011**, *3*, 42–53.

26. Yudelson, J. *Reinventing Green Building. Why Certification Systems Aren't Working and What We Can Do About It*; New Society Publishers: Gabriola Island, BC, Canada, 2016.

27. Bonenberg, W.; Kapliński, O. The Architect and the Paradigms of Sustainable Development: A Review of Dilemmas. *Sustainability* **2018**, *10*, 100. [CrossRef]

28. Burcu, G. Sustainability Education by sustainable School Design. *Procedia Soc. Behav. Sci.* **2015**, *186*, 868–873. [CrossRef]

29. Ragheb, A.; El-Shimy, H.; Ragheb, G. Green architecture: A concept of sustainability. *Procedia Soc. Behav. Sci.* **2016**, *216*, 778–787. [CrossRef]

30. Tomovska, R.; Radivojevic, A. tracing sustainable design strategies in the example of the traditional. *J. Clean. Prod.* **2017**, *147*, 10–24. [CrossRef]

31. Safi, S.; Rahro Mehrabani, S. Design based on sustainable environmental architecture emphasizing maximum natural light output in mountainous climates. *Rev. QUID* **2017**, *1*, 593–596.

32. Nastarani, S.; Khaksari, P.; Mohammadi, M. Intelligent buildings with sustainable architecture approach. *Rev. QUID* **2017**, *1*, 912–920.

33. CBFEE. *Skylighting and Retail Sales: An Investigation into the Relationship between Daylighting and Human Performance*; The Heschong Group on behalf of California Board for Energy Efficiency Third Party Program; Pacific Gas and Electric Co.: San Francisco, CA, USA, 1999.

34. CGB, C. Building the GREEN Garden State. *New Jersey Municipalities Magazine*, 2009; 86.

35. Mohammadi, A.; Saghafi, M.; Tahbaz, M.; Nasrollahi, A. Effects of Vernacular Climatic Strategies (VCS) on Energy Consumption in Common Residential Buildings in Southern Iran: The Case Study of Bushehr City. *Sustainability* **2017**, *9*, 1950. [CrossRef]

36. Bani Masoud, A. *Iranian Contemporary Architecture*; Honar-E-Memari Gharn: Tehran, Iran, 2009.

37. Alidadi, S.; Yousefi Zadeh, Z. The effect of sustainable architecture in designing five star hotel. *Turk. Online J. Des. Art Commun. TOJDAC* **2016**, *6*, 2765–2771. [CrossRef]

38. Karimi, S.; Zandieh, M. The development of modern architectural principles of sustainability in order to compare Iran traditional architecture with five modern sustainable city in the world (case study: Vancouver, Copenhagen, Oslo, Curitiba and Masdar). *Turk. Online J. Des. Art Commun. TOJDAC* **2016**, *6*, 3141–3157. [CrossRef]

39. Makari Faraji, M.; Mokhtari Taleghani, E. The role of sustainable architecture in valuable historical districts of tehran (a case study of sustainable residential development in Sanglaj district). *Turk. Online J. Des. Art Commun. TOJDAC* **2016**, *6*, 1860–1869. [CrossRef]

40. Pons, O.; de la Fuente, A.; Aguado, A. The Use of MIVES as a sustainability assessment MCDM method for architecture and civil engineering applications. *Sustainability* **2016**, *8*, 460. [CrossRef]

41. Zavadskas, E.K.; Antucheviciene, J.; Vilutiene, T.; Adeli, H. Sustainable Decision-Making in Civil Engineering, Construction and Building Technology. *Sustainability* **2017**, *10*, 14. [CrossRef]

42. Zavadskas, E.K.; Govindan, K.; Antucheviciene, J.; Turskis, Z. Hybrid multiple criteria decision-making methods: A review of applications for sustainability issues. *Econ. Res. Ekon. Istraž.* **2016**, *29*, 857–887. [CrossRef]

43. Kaplinski, O. Innovative solutions in construction industry: Review of 2016–2018 events and trends. *Sustainability* **2018**, *10*, 27–33. [CrossRef]

44. Zavadskas, E.K.; Antucheviciene, J.; Turskis, Z.; Adeli, H. Hybrid multiple-criteria decision-making methods: A review of applications in engineering. *Sci. Iran.* **2016**, *23*, 1–20.

45. Rezaei, J. Best-worst multi-criteria decision-making method. *Omega* **2015**, *53*, 49–57. [CrossRef]

46. Gupta, H.; Barua, M. Supplier selection among SMEs on the basis of their green innovation ability using BWM and fuzzy TOPSIS. *J. Clean. Prod.* **2017**, *152*, 242–258. [CrossRef]

47. Badri Ahmadi, H.; Kusi-Sarpong, S.; Rezaei, J. Assessing the social sustainability of supply chains using Best Worst Method. *Resour. Conserv. Recycl.* **2017**, *126*, 99–106. [CrossRef]

48. Van de Kaa, G.; Kamp, L.; Rezaei, J. Selection of biomass thermochemical conversion technology in the Netherlands: A best worst method approach. *J. Clean. Prod.* **2017**, *166*, 32–39. [CrossRef]

49. Ghasemian Sahebi, I.; Arab, A.; Sadeghi Moghadam, M. analyzing the barriers to humanitarian supply chain management: A case study of the Tehran Red Crescent Societies. *Int. J. Disaster Risk Reduct.* **2017**, *24*, 232–241. [CrossRef]

50. Rezaei, J.; Nispeling, T.; Sarkis, J.; Tavasszy, L. A supplier selection life cycle approach integrating traditional and environmental criteria using the best worst method. *J. Clean. Prod.* **2016**, *135*, 577–588. [CrossRef]

51. Mokhtarzadeh, N.; Amoozad Mahdiraji, H.; Beheshti, M.; Zavadskas, E. A Novel Hybrid Approach for Technology Selection in the Information Technology Industry. *Technologies* **2018**, *6*, 34. [CrossRef]

52. Rezaei, J. Best-worst multi-criteria decision-making method: Some properties and a linear model. *Omega* **2016**, *64*, 126–130. [CrossRef]

53. Beheshti, M.; Amoozad Mahdiraji, H.; Zavadskas, E.K. Strategy Portfolio Optimization: A COPRAS G-MODM Hybrid approach. *Transform. Bus. Econ.* **2016**, *15*, 500–519. [CrossRef]

54. Akahvan, P.; Barak, S.; Maghsoudlou, H.; Antucheviciene, J. FQSPM-SWOT for strategic alliance planning and partner selection; case study in a holding car manufacturer company. *Technol. Econ. Dev. Econ.* **2015**, *21*, 165–185. [CrossRef]

55. Rasiulis, R.; Ustinovichius, L.; Vilutiene, T.; Popov, V. Decision model for selection of modernization measures: Public building case. *J. Civ. Eng. Manag.* **2016**, *22*, 124–133. [CrossRef]

56. Ecer, F. A Hybrid banking websites quality evaluation model using AHP and COPRAS-G: A Turkey case. *Technol. Econ. Dev. Econ.* **2014**, *20*, 758–782. [CrossRef]

57. Cereska, A.; Podvezko, V.; Zavadskas, E.K. Operating Characteristics Analysis of Rotor Systems Using MCDM Methods. *Stud. Inform. Control* **2016**, *25*, 59–68. [CrossRef]

58. Cereska, A.; Zavadskas, E.; Cavallaro, F.; Podvezko, V.; Tetsman, I.; Grinbergiene, I. Sustainable Assessment of Aerosol Pollution Decrease Applying Multiple Attribute Decision-Making Methods. *Sustainability* **2016**, *8*, 586. [CrossRef]

59. Tamošaitienė, J.; Gaudutis, E. Complex assessment of structural systems. *J. Civ. Eng. Manag.* **2013**, *19*, 305–317. [CrossRef]

60. Zavadskas, E.; Antucheviciene, J. Multiple criteria evaluation of rural buildings. *Build. Environ.* **2007**, *42*, 436–451. [CrossRef]
61. Pitchipoo, P.; Vincent, D.; Rajini, N.; Rajakarunakaran, S. COPRAS Decision Model to Optimize Blind Spot in Heavy Vehicles: A Comparative Perspective. *Procedia Eng.* **2014**, *97*, 1049–1059. [CrossRef]
62. Bielinskas, V.; Burinskienė, M.; Palevičius, V. Assessment of Neglected Areas in Vilnius City Using MCDM and COPRAS Methods. *Procedia Eng.* **2015**, *122*, 29–38. [CrossRef]
63. Polat, G.; Bingol, B.; Var, O. An Integrated Multi-criteria-decision-making Tool for Mechanical Designer Selection. *Procedia Eng.* **2017**, *196*, 278–285. [CrossRef]
64. Zavadskas, E.; Kaklauskas, A. *Multiple Criteria Evaluation of Buildings*; Technika: Vilnius, Lithuania, 1996.
65. Kvederyte, E.; Zavadskas, E.K.; Kaklauskas, A. Multiple Criteria Analysis of a Dwelling Life Cycle. *J. Statyb.* **2000**, *6*, 179–192. [CrossRef]
66. Zavadskas, E.K.; Kaklauskas, A.; Banaitiene, N. *Multiple Criteria Analysis of a Building's Life Cycle*; Technika: Vilnius, Lithuania, 2001.
67. Zavadskas, E.; Kaklauskas, A.; Banaitis, A.; Kvederyte, N. Housing credit access model: The case for Lithuania. *Eur. J. Oper. Res.* **2004**, *155*, 335–352. [CrossRef]
68. Staniunas, M.; Medineckiene, M.; Zavadskas, E.; Kalibatas, D. To modernize or not: Ecological–economical assessment of multi-dwelling houses modernization. *Arch. Civ. Mech. Eng.* **2013**, *13*, 88–98. [CrossRef]
69. Šiožinytė, E.; Antuchevičienė, J. solving the problems of daylighting and tradition continuity in a reconstructed vernacular building. *J. Civ. Eng. Manag.* **2013**, *19*, 873–882. [CrossRef]
70. Tupenaite, L.; Zavadskas, E.; Kaklauskas, A.; Turskis, Z.; Seniut, M. Multiple criteria assessment of alternatives for Built and Human Environment Renovation. *J. Civ. Eng. Manag.* **2010**, *16*, 257–266. [CrossRef]
71. Chen, C. Extensions of the TOPSIS for group decision-making under fuzzy environment. *Fuzzy Sets Syst.* **2000**, *114*, 1–9. [CrossRef]
72. Triantaphyllou, E.; Lin, C. Development and evaluation of five fuzzy multi-attribute decision-making methods. *Int. J. Approx. Reason.* **1996**, *14*, 281–310. [CrossRef]
73. Hwang, C.; Yoon, K. *Multiple Attribute Decision-Making Methods and Applications*; Springer: Berlin/Heidelberg, Germany; New York, NY, USA, 1981.
74. Tsaur, S.; Chang, T.; Yen, C. The evaluation of airline service quality by fuzzy MCDM. *Tour. Manag.* **2002**, *23*, 107–115. [CrossRef]
75. Barbosa de Santis, R.; Golliat, L.; Pestana de Aguiar, E. Multi-criteria supplier selection using fuzzy analytic hierarchy process. *Braz. J. Oper. Prod. Manag.* **2017**, *14*, 428–437. [CrossRef]
76. Van Leekwijck, W.; Kerre, E. Defuzzification: Criteria and classification. *Fuzzy Sets Syst.* **1999**, *108*, 159–178. [CrossRef]
77. Bardossy, A.; Duckstein, L. *Fuzzy Rule-Based Modeling with Applications to Geophysical, Biological and Engineering Systems*; CRC Press: Boca Raton, FL, USA, 1995.

sustainability

MDPI

Article

Improving the Green Building Evaluation System in China Based on the DANP Method

Qi-Gan Shao [1,3], James J. H. Liou [2,*], Sung-Shun Weng [3] and Yen-Ching Chuang [2]

[1] School of Economics & Management, Xiamen University of Technology, Xiamen 361024, China; qgshao@xmut.edu.cn
[2] Department of Industrial Engineering and Management, National Taipei University of Technology, Taipei 10608, Taiwan; yenching.chuang@gmail.com
[3] Department of Information and Finance Management, National Taipei University of Technology, Taipei 10608, Taiwan; wengss@ntut.edu.tw
* Correspondence: jamesjhliou@gmail.com; Tel.: +886-2-2771-2171; Fax: +886-2-2731-7168

Received: 19 March 2018; Accepted: 9 April 2018; Published: 13 April 2018

Abstract: Against the background of sustainable development, green building practices could be part of the strategy for solving environmental and energy problems in developing countries. The aim of this paper is to explore a system for the assessment of green buildings in China that provides the government and stakeholders with ways to improve their strategies for green building development. We apply a hybrid model, developed by integrating the Decision-Making Trial and Evaluation Laboratory and Analytical Network Process (called DANP) method, to build an influential network relationship map (INRM) between assessment systems and to derive the criterion weights. The INRM and derived weights can help us to understand this complex assessment system and to set improvement priorities for green building development. The results demonstrate that indoor environment, materials, and smart facilities are the top three critical factors for green building evaluation. Finally, we discuss some management implications based on an actual case study with solutions provided using this model.

Keywords: green building; sustainability; Decision-Making Trial and Evaluation Laboratory (DEMATEL); Analytical Network Process (ANP); DEMATEL-based ANP (DANP)

---

1. Introduction

In today's world, there is a need to construct a large number of buildings in response to rapid urbanization, to truly improve the economic strength of a developing country. However, this building boom can contribute to the problems of climate change, global warming and environmental pollution to some extent. More than one third of total worldwide greenhouse gas emissions and energy exhaust are related to buildings [1]. At the same time, a sixth of the world's freshwater usage, 40% of the raw materials and one fourth of wood harvested, are consumed by the construction industry [2,3]. Buildings are not only responsible for economic but also sustainable environmental development. In addition to construction, during the life cycle of the buildings, energy and water consumption, technological innovations and garbage disposal, are all related to environmental performance and these factors affect human health and sustainability. In other words, the construction of green buildings has become a multi-dimensional issue that can be viewed from multiple perspectives [4]. This makes the building of a suitable evaluation system for green building construction and operations, and how to find the factors that effect the implementation of green building assessment, a difficult but significant subject of study.

Numerous different levels of certification systems have been founded for green building evaluation, owing to different situations in different countries and regions. The primary green building assessment tools used around the world include the Leadership in Energy and Environmental Design (LEED) guidelines developed in the USA, the Building Research Establishment Environmental Assessment Method (BREEAM) of the UK, and the Green Building Labeling Assessment Standard for Green Buildings (CGBL) in China [1,5,6]. Currently, China is the largest construction market, with up to 2 billion square meters of annual construction and predicted to account for about 50% of new construction globally in the coming decade, it is the largest contributor of CO_2 emissions in the world. Although the popularity of green buildings has grown in China, they only make up about 4% of the buildings in the world's largest construction market. Furthermore, low levels of management and a lack of transparency during the design, construction and operation of the buildings make it difficult to truly implement a green building practice. With the government's "One Belt, One Road" policy and continued urbanization, there will be more opportunities for China to accelerate its infrastructure construction. As a result of these factors and an increasing environmental awareness in China, a rational and effective building evaluation system needs to be constructed and utilized.

Many researchers have improved upon the existing rating tools and evaluation systems for green building performance assessment. For example, Balaban and Puppim de Oliveira [7] argued that a co-benefits approach could address energy and CO_2 reduction. Their results demonstrate that green buildings could obtain a significant level of effectiveness in terms of energy and CO_2 reduction, and improve health conditions for building users. Wong and Abe [8] noted that, although the Comprehensive Assessment System for Built Environment Efficiency (CASBEE) developed in Japan has been in use since 2002, it lacks a reward system to promote the adoption of the CASBEE guidelines. Based on the situation in Oman, Al-Jebouri et al. [4] developed a framework for green building construction and proposed an evaluation system for the construction industry. Dwaikat and Ali [9] proposed an earned value method to monitor a green building's energy life cycle cost. Kang [10] produced a systematic model for green building assessment standards by using the Environmental Impact Assessment (EIA) framework. In his survey of 104 expert opinions he found that energy cost, environmental problems, water withdrawal, health, and company reputation to be the five drivers for operating Green Building (GB) technologies [11]. However, there appears to be a void in the literature on how to develop and improve a valid green building assessment tool within a developing country or area by applying a systemic decision model. For example, how to systematically determine the most important factors, such as operation management, material quality, indoor environment quality, and energy efficiency in green building performance evaluation and how to identify the relationship between these factors is currently lacking in most rating systems. Thus these factors fail to be implemented by construction departments and governments. Therefore, the existing system needs to be checked and improved upon by underlining green building related management and technology innovation [12].

The development of the Chinese national green building system construction is relativity young. For the evaluation and implementation of sustainable green building practices, China built a national 3-star Green Building rating system in 2006 by learning from the international rating systems, such as LEED and BREEAM. However, Ye et al. [6] argued that there have been about 17 national level and about 50 or more provincial level green building assessment standards practiced in China. However, a lack of understanding of green building incentive imbalances is the main reason for the slowdown in green building construction. Builders usually project long-term savings based on short-term costs, such as materials and labor costs, rather than energy efficiency or green building technologies. The increasing urbanization in China and growing awareness of environmental protection issues has encouraged the government to give priority to infrastructure development. Finding the factors of influence for a green building system may provide the answer as to how to promote the development of green buildings in China. However, few studies have focused on this topic.

This paper seeks to fill this gap in the research by proposing a causal relationship framework for improving green building assessment capabilities. This study proposes a hybrid Multiple Criteria Decision Making (MCDM) model that combines Decision-Making Trial and Evaluation

Laboratory (DEEMATEL) and the Analytical Network Process (DANP) methods to construct interdependent connections among the assessment dimension and criteria [13–16]. Based on graph theory, the DEMATEL technology is a powerful tool to use the knowledge of experienced experts to arrange the structural model of a system [17]. The effect on each category and criterion is confirmed, making such complex systems easy to understand [18]. The derived influential network relationship map (INRM) can help decision makers understand the complex relationships in green building systems. The DANP method, based on the results of DEMATEL, can then be used to calculate the weights of the green building assessment criteria. The strength of the suggested model is that it can be used to build an orderly plan of complex systems, showing the cause-and-effect relationships and obtaining the influence weights of factors within a green building system. Consequently, differing from existing studies focusing on green building evaluation, this paper contributes to the literature by trying to construct a cause-and-effect system for green building assessment and environmental management that can not only assist construction companies and governments to identify the key factors for Green Buildings, but also provide direction for improvement.

The rest of this study is organized as follows. Section 2 gives a brief review of the relevant literature. Section 3 describes the DANP method used to build the INRM and find the weights of the criteria. The green building weighting system is introduced in Section 4. The results are discussed in Section 5, and some concluding remarks are presented in Section 6.

2. Literature Review

A review of the literature on green or sustainable building evaluation systems indicate that a structural assessment model for the production and management of effective green building is lacking. Most of the research in the past has been limited to a discussion about green building certification tools and agreements to evaluate energy efficiency, environmental problems, materials and resources and economic topics in the building industry [19–21]. Several of the studies related to green building or sustainable building evaluation and management published over the last decade are discussed below. The Building Research Establishment Environmental Assessment Method (BREEAM) is considered the first green building rating assessment in the world. It was created by the UK [22,23]. The Leadership in Energy and Environmental Design (LEED), China Green Building Labeling (CGBL), Comprehensive Assessment System for Built Environment efficiency (CASBEE) and other main systems were developed based on BREEAM. However, it is difficult to judge which evaluation system has the most complete assessment criteria for the certification of green buildings in a specific area. Some have compared green building rating tools in different areas and countries, including Shad et al. [1] who compared BREEAM, LEED, Green Star of Australia, CASBEE of Japan and proposed a new system for Iranian green building evaluation. Ali and Al Nsairat [5] discussed a green building rating system for residential buildings in Jordan based on a comparison of BREEAM, LEED, and the GB Tool.

Several academics [6,24–26] have proposed some suggestions regarding the development of a Chinese green building evaluation system and standards, but we found no research aimed at finding the relationship between the factors of the green building rating systems. Ye et al. [6] suggested an evaluation system for green building practices by proposing a three layer basis for general and specialized standards. Guo et al. [24] discussed the enforcement of civil building efficiency codes for the monitoring of energy consumption in China. Their results indicated an obvious increase in energy saving. Hong et al. [25] updated the design standards for energy efficiency in public buildings and compared the GB 50189-2014 standards for Chinese public buildings with USA standards. Yu et al. [26] proposed a rating method for green store buildings in China. They considered eight dimensions, landscape, energy efficiency, water efficiency, material and resources, indoor environment, construction management, and operation management, proposing 23 criteria for assessing green store buildings. The expert group decision analysis hierarchy process (AHP) method was used for deriving the relative importance of the dimensions.

Furthermore, Si et al. [27] also applied the AHP method for the selection of green technologies for assessing buildings. Their results indicated that social criteria, including occupant satisfaction, should be considered as part of the list. By studying LEED, BREEAM and other main international green building assessment tools, Ali and Al Nsairat [5] defined a new assessment system by considering the local context of Jordan, then used the AHP to weigh the importance of the criteria. Banani et al. [28] proposed a structure for sustainable non-residential building assessment for Saudi Arabia, and also applied the AHP method to weigh the proposed 36 criteria. Liu et al. [29] proposed a model of an evaluation system for green university assessment based on Fuzzy AHP (FAHP). The FAHP was also used to develop an expert system to assess green building performance considering such factors as environmental management, pollution, and energy management [30]. Sabaghi et al. [31] introduced a hybrid model combining FAHP with entropy to evaluate green products. With the exception of the AHP method, no other MCDM methods have been used for weighing the dimensions or criteria for a green building assessment system.

Although AHP is a widely used MCDM technology, it assumes the dimensions or criteria of the system to be independent with a unidirectional hierarchical relationship. However, the dimensions or criteria of an evaluation system are seldom independent in the real world. In contrast with previous studies that applied AHP to weight the green building rating system, this paper utilizes the DEMATEL-based ANP (DANP) method to explore the network relationship of green building criteria and the influential weights of the criteria in a complex evaluation system. This method can not only be used to build the network relationship of the evaluation system but also understand the cause-effect relationships between criteria for the construction of a better green building assessment system. The procedures are described in detail in the next section.

3. Methodology

In this section, we introduce the DANP model that combines DEMATEL with ANP to establish the interdependent structure and receive the weights of the dimensions and criteria. Government and the construction industry can figure out the complex relationship between green building management and the cause-effect within the criteria through the derived influenced network relationship map (INRM). The obtained weights of the criteria and the INRM can help governments and the construction industry to set improved priorities for bettering the green buildings in China. The detailed procedures are illustrated as follows [32–38].

3.1. DEMATEL Method

The DEMATEL method is used to establish the interrelationship between factors used to construct an INRM. The method is summarized as follows:

*Step 1: Identify the direct relation average matrix **M** on a scale of 0–4 ranging from "no influence (0)" to "very high influential (4)".* Using the aforementioned scale, k respondents are asked to judge the extent of direct influence between two pairwise criteria, denoted by m_{ij}. Then, the direct relation average matrix **M** is acquired through the mean of the same criteria in the k matrices for the respondents. Matrix **M** is shown as Equation (1), where is n the number of criteria. Thus,

$$\mathbf{M} = \begin{bmatrix} m_{11} & \cdots & m_{1j} & \cdots & m_{1n} \\ \vdots & \cdots & \vdots & & \vdots \\ m_{i1} & \cdots & m_{ij} & \cdots & m_{in} \\ \vdots & \cdots & \vdots & \cdots & \vdots \\ m_{n1} & \cdots & m_{nj} & \cdots & m_{nn} \end{bmatrix} \tag{1}$$

*Step 2: Obtain the initial direct influence matrix $\mathbf{P} = \left[p_{ij}\right]_{n \times n}$, which is the multiplication of **M** and v.*

$$P = v \times M \tag{2}$$

$$v = \min \left[\frac{1}{\max\limits_{i} \sum_{j}^{n} |d_{ij}|}, \ \frac{1}{\max\limits_{j} \sum_{i}^{n} |d_{ij}|} \right] \tag{3}$$

Step 3: Calculate the total influence matrix T with Equation (4). The element t_{ij} indicates the indirect effects that criteria i has on criteria j. Thus,

$$T = P + P^2 + \cdots + P^K = P(I - P)^{-1} \tag{4}$$

where $T = \left[t_{ij} \right]_{n \times n}$ and I is an identity matrix.

Step 4: Derive each column sum (c_j) and row sum (r_i) from matrix T as follows:

$$c_i = (c_j)_{n \times 1} = (c_j)'_{1 \times n} = \left[\sum_{i=1}^{n} t_{ij} \right]' \tag{5}$$

$$r_i = (r_i)_{n \times 1} = \left[\sum_{j=1}^{n} t_{ij} \right] \tag{6}$$

The element c_j in vector $\mathbf{c}$ denotes that the total effects received by criterion j received, from the other criteria. Similarly, r_i represents the direct and indirect effects of factor i on the other criteria.

Step 5: Derive matrix T_C based on the criteria and T_D based on the dimensions. Matrix T could be differentiated into T_C based on the criterion and T_D based on the dimensions. Matrix T_D is found by averaging the degree of criterion influence in each dimension.

$$T_C = \begin{array}{c} D_1 \\ D_2 \\ \vdots \\ D_n \end{array} \begin{array}{c} c_{11} \\ c_{12} \\ c_{1m_1} \\ c_{21} \\ c_{22} \\ c_{2m_2} \\ \vdots \\ c_{n1} \\ c_{n2} \\ c_{nm_n} \end{array} \begin{bmatrix} T_c^{11} & T_c^{12} & \cdots & T_c^{1n} \\ T_c^{21} & T_c^{22} & \cdots & T_c^{2n} \\ \vdots & \vdots & \cdots & \vdots \\ T_c^{n1} & T_c^{n2} & \cdots & T_c^{nn} \end{bmatrix} \tag{7}$$

Step 6: Get the INRM. Thus, $r_i + c_i$ reflects the strength of the influences given and received on factor i, while $r_i - c_i$ shows the net effect of factor i on the other factors. Clearly, if $r_i - c_i$ is positive, factor i is a causal component, and if $r_i - c_i$ is negative, then factor i is an affected component. As a result, the influence relationship map (INRM) can be finished by mapping the data set $(r_i + c_i, r_i - c_i)$.

3.2. Obtain the ANP Weights Based on the DEMATEL Technique

The DEMATEL method is used to establish the interrelationship between factors used to construct an INRM. The method is summarized as follows. ANP is the method customarily used to construct an unweighted supermatrix for distributing the factor weights. In contrast to the AHP technique, the ANP considers the interdependency and relationship between factors or criteria. However, there are three main problems in the original ANP method. First, we have to assume the relationship structure of the evaluation system before using ANP. Second, it is difficult to understand the ANP questionnaire due to its complexity [39], the process of pairwise comparison is time-consuming and it is not easy to obtain consistent results. Third, the assumption that each cluster will have same equal weight seems irrational because of the different degrees of influence among the dimensions or clusters [16]. Luckily, these three shortcomings can be solved by using the DEMATEL-based ANP method. The degrees of influence of each dimension can be obtained by DEMATEL, then the ANP process can be applied to normalize the unweighted supermatrix. The details are as follows:

Step 7: Obtain the unweighted supermatrix. We get a matrix T_C^δ by normalizing T_C.

$$T_c^\delta = \begin{bmatrix} T_c^{\delta 11} & T_c^{\delta 12} & \cdots & T_c^{\delta 1n} \\ T_c^{\delta 21} & T_c^{\delta 22} & \cdots & T_c^{\delta 2n} \\ \vdots & \vdots & \cdots & \vdots \\ T_c^{\delta n1} & T_c^{\delta n2} & \cdots & T_c^{\delta nn} \end{bmatrix}$$

(8)

For example, T_C^{pq}, which is a submatrix of T_C can be normalized to $T_C^{\delta pq}$, as follows:

$$T_c^{pq} = \begin{bmatrix} t_{11}^{pq} & \cdots & t_{1j}^{pq} & \cdots & t_{1m_q}^{pq} \\ \vdots & & \vdots & & \vdots \\ t_{i1}^{pq} & \cdots & t_{ij}^{pq} & \cdots & t_{im_q}^{pq} \\ \vdots & & \vdots & & \vdots \\ t_{m_p1}^{pq} & \cdots & t_{m_pj}^{pq} & \cdots & t_{m_pm_q}^{pq} \end{bmatrix} \rightarrow t_i^{pq} = \sum_{j=1}^{m_q} t_{ij}^{pq}$$

(9)

where $i = 1, 2, \cdots, m_p$.

$$T_C^{\delta pq} = \begin{bmatrix} t_{11}^{pq}/t_1^{pq} & \cdots & t_{1j}^{pq}/t_1^{pq} & \cdots & t_{1m_q}^{pq}/t_1^{pq} \\ \vdots & & \vdots & & \vdots \\ t_{i1}^{pq}/t_i^{pq} & \cdots & t_{ij}^{pq}/t_i^{pq} & \cdots & t_{im_q}^{pq}/t_i^{pq} \\ \vdots & & \vdots & & \vdots \\ t_{m_p1}^{pq}/t_{m_p}^{pq} & \cdots & t_{m_pj}^{pq}/t_{m_p}^{pq} & \cdots & t_{m_pm_q}^{pq}/t_{m_p}^{pq} \end{bmatrix} = \begin{bmatrix} t_{11}^{\delta pq} & \cdots & t_{1j}^{\delta pq} & \cdots & t_{1m_q}^{\delta pq} \\ \vdots & & \vdots & & \vdots \\ t_{i1}^{\delta pq} & \cdots & t_{ij}^{\delta pq} & \cdots & t_{im_q}^{\delta pq} \\ \vdots & & \vdots & & \vdots \\ t_{m_p1}^{\delta pq} & \cdots & t_{m_pj}^{\delta pq} & \cdots & t_{m_pm_q}^{\delta pq} \end{bmatrix}$$

(10)

The unweighted matrix supermatrix W_{pq} is transposed from matrix $T_C^{\delta pq}$ as follows:

$$t_D^{nm} = \frac{\sum\limits_{i=1}^{i_n} \sum\limits_{j=1}^{j_m} t_{ij}}{i_n \times j_m}$$

(11)

$$W_{pq} = \begin{bmatrix} t_{11}^{pq} & \cdots & t_{i1}^{pq} & \cdots & t_{m_p1}^{pq} \\ \vdots & & \vdots & & \vdots \\ t_{1j}^{pq} & \cdots & t_{ij}^{pq} & \cdots & t_{m_pj}^{pq} \\ \vdots & & \vdots & & \vdots \\ t_{1m_q}^{pq} & \cdots & t_{im_q}^{pq} & \cdots & t_{m_pm_q}^{pq} \end{bmatrix} = \left(T_c^{\delta pq} \right)'$$

(12)

As shown in Equation (11), D_i denotes the ith dimension; c_{ij} denotes the jth criteria in the ith dimension.

Step 8: Derive the weighted supermatrix. Referring to step 5, we can get matrix T_D by averaging the degree of the criterion influence in each dimension, which is derived by

$$t_D^{nm} = \frac{\sum\limits_{i=1}^{i_n} \sum\limits_{j=1}^{j_m} t_{ij}}{i_n \times j_m} \tag{13}$$

where i_n is the number of criteria in dimension n; and j_m is the number of criteria in the dimension m.

$$T_D = \begin{bmatrix} t_D^{11} & t_D^{12} & \cdots & t_D^{1n} \\ t_D^{21} & t_D^{22} & \cdots & t_D^{2n} \\ \vdots & \vdots & & \vdots \\ t_D^{n1} & t_D^{n2} & \cdots & t_D^{nn} \end{bmatrix} \rightarrow \delta_2 = \sum_{j=1}^{n} t_D^{\delta 2j} \tag{14}$$

$$T_D^\delta = \begin{bmatrix} \frac{t_D^{11}}{\delta_1} & \frac{t_D^{12}}{\delta_1} & \cdots & \frac{t_D^{1n}}{\delta_1} \\ \frac{t_D^{21}}{\delta_2} & \frac{t_D^{22}}{\delta_2} & \cdots & \frac{t_D^{2n}}{\delta_2} \\ \vdots & \vdots & \vdots & \vdots \\ \frac{t_D^{n1}}{\delta_n} & \frac{t_D^{n2}}{\delta_n} & \frac{t_D^{n3}}{\delta_n} & \frac{t_D^{nn}}{\delta_1} \end{bmatrix} = \begin{bmatrix} t_D^{\delta 11} & t_D^{\delta 12} & \cdots & t_D^{\delta 1n} \\ t_D^{\delta 21} & t_D^{\delta 22} & \cdots & t_D^{\delta 2n} \\ \vdots & \vdots & & \vdots \\ t_D^{\delta n1} & t_D^{\delta n2} & \cdots & t_D^{\delta nn} \end{bmatrix} \tag{15}$$

Then, we can obtain the weighted supermatrix W_δ by multiplying the unweighted supermatrix W with T_D^δ as follows:

$$T_D^\delta = \begin{bmatrix} \frac{t_D^{11}}{\delta_1} & \frac{t_D^{12}}{\delta_1} & \cdots & \frac{t_D^{1n}}{\delta_1} \\ \frac{t_D^{21}}{\delta_2} & \frac{t_D^{22}}{\delta_2} & \cdots & \frac{t_D^{2n}}{\delta_2} \\ \vdots & \vdots & \vdots & \vdots \\ \frac{t_D^{n1}}{\delta_n} & \frac{t_D^{n2}}{\delta_n} & \frac{t_D^{n3}}{\delta_n} & \frac{t_D^{nn}}{\delta_1} \end{bmatrix} = \begin{bmatrix} t_D^{\delta 11} & t_D^{\delta 12} & \cdots & t_D^{\delta 1n} \\ t_D^{\delta 21} & t_D^{\delta 22} & \cdots & t_D^{\delta 2n} \\ \vdots & \vdots & & \vdots \\ t_D^{\delta n1} & t_D^{\delta n2} & \cdots & t_D^{\delta nn} \end{bmatrix} \tag{16}$$

Step 9: Calculate the DANP weights. Limit the weighted supermatrix W_δ by Equation (17) until the supermatrix has converged and become stable. The DANP weights can then be found by

$$W_W = \lim_{\lambda \to \infty} (W)^\lambda \tag{17}$$

4. Establishment of a Green Building Weighting System for China

We apply the proposed method that combines DEMATEL and ANP to determine the structure among the factors, and to survey the weights of each criterion and dimension. A city level green building assessment applied in Xiamen, which is located in southeastern China, is illustrated as an empirical example. Xiamen is a city with an external population of more than 65%, and many new buildings have been built in recent years. This city is famous for its good ecological environment and its higher requirements for green buildings. It can represent the average level of green building requirements in China.

4.1. Identification Dimensions and Criteria for Green Building Assessment

Many international rating systems share common indicators for the evaluation of building sustainability, including the quality of the indoor environment, energy and water efficiency, use of green material, innovativeness, site ecology and waste and pollution solutions [1,4,5,11,23,26,29,40–42]. Government, citizens and other stakeholders monitoring and participating in the operation management are also considered important factors for green building evaluation. Table 1 lists the main dimensions and criteria of the evaluation system described above.

First, we construct an evaluation system with 7 dimensions and 30 criteria that were obtained based from a review of the existing literature and international green building rating systems. To simplify some criteria and reflect the special characteristics of the Chinese environments, we designed a questionnaire and requested 10 experts to respond to the questions asked with linguistic variables, which ranged from very unimportant (0,0,1) to very important (9,10,10) (Table 2). Then, after discussion with the 10 experts, we extracted the essential criteria, as those having of a mean of 4.5 points or above and the results are displayed in Table 3. One dimension and 8 criteria were deleted after the initial survey.

Table 1. Green building evaluation system.

Dimension	Criteria	Source
Management	Sustainable procurement Stakeholder participation Ease of maintenance Management system	[4]; [34]; [16]; [26];
Indoor environment	Indoor air quality Natural ventilation Smoking ban Indoor thermal control Lighting zones and controls	[38]; [4]; [1]; [26]; [9];
Materials	Sourcing of materials Recycled materials Renewable materials Hard landscaping	[1]; [11]; [29]; [34]; [4];
Energy and Water efficiency	Reduction of CO_2 Drying spacing Renewable energy Energy delivery performance Water saving system Water-saving facilities Water monitoring	[22]; [23]; [9]; [41] [34]; [4];
Innovation	Smart technology application Culture of innovation Support of national economy	[26]; [4] [34]; [4];
Site ecology	Public transportation Maximum car parking Outdoor environment Cyclist landscaping	[38]; [4];
Waste and Pollution solutions	Garbage classification and sorting Drainage system Flood risk	[11]; [34]; [11];

Table 2. Linguistic variables for ranking criteria.

Linguistic Variables	Very Unimportant	Unimportant	Slightly Unimportant	Fair	Slightly Important	Important	Very Important
Triangular fuzzy set	(0,0,1)	(0,1,3)	(1,3,5)	(3,5,7)	(5,7,9)	(7,9,10)	(9,10,10)

Table 3. Extracted dimensions and criteria.

Dimension	Criteria
Management (D_1)	Sustainable procurement (C_{11}) Stakeholder participation (C_{12}) Ease of maintenance (C_{13}) Management system (C_{14})
Indoor environment (D_2)	Natural ventilation (C_{21}) Indoor thermal (C_{22}) Smoking ban (C_{23}) Indoor air quality (C_{24})
Material (D_3)	Sourcing of materials (C_{31}) Recycled materials (C_{32}) Renewable materials (C_{33})
Energy and Water efficiency (D_4)	Reduction of CO_2 (C_{41}) Renewable energy (C_{42}) Energy delivery performance (C_{43}) Water saving system (C_{44})
Site ecology (D_5)	Public transportation (C_{51}) Maximum car parking (C_{52}) Outdoor environment (C_{53}) Cyclist landscaping (C_{54})
Innovation (D_6)	Smart technology application (C_{61}) Culture of innovation (C_{62}) Support of national economy (C_{63})

4.2. DANP Method for Measuring the Relationship between Dimensions and Criteria

To measure the initial direct influence matrix, we designed a questionnaire with the aim of obtaining the degree of influence between any two indicators according to Table 3. During the survey, experts were asked to respond to a question by making pairwise comparisons of the degrees of influence between the criteria. As seen in Table A1 in Appendix A, a 22 × 22 average initial direct influence matrix was obtained by averaging the experts' questionnaire responses. The consistency gaps of the 10 questionnaires are 4.86%, which is smaller than 5%, and the confidence level is 95.14%, which is little more than 95%. Although the 10 experts cannot represent all relative stakeholders, the result shows a good consistency that can reflect parts of real situations.

The normalized directed-relation matrix can be obtained by Equations (2) and (3). After that, the total-influence matrix T (Table A2) is calculated by Equation (4). The total influence matrix of the dimensions can be calculated by averaging the influence matrix T within each dimension, as seen in Table A3. Table A4 reflects the sum of the influences given and received among criteria and dimensions and is obtained by implementing Equations (5) and (6) of step 4.

As we can see in Table A4, the largest $(r_i - c_i)$ value (0.13) is for innovation (D_6), so this is the most influential dimension. It means that innovation is the key to the development of a green building. Furthermore, innovation technology will have a deep influence on the other dimensions. In addition, management (D_1) has the maximum value $(r_i + c_i)$ (1.49), which means that it has the largest total influence degree within dimensions. As we know, green building operation, use of sustainable materials, improving energy efficiency, water saving, indoor environment and so on are also closely related to management in China. Therefore, management affects the other dimensions, like indoor environment, material, energy and water efficiency and site ecology, and it is also affected by innovation. The influence value, which is greater than the average of all criteria in the total influence matrix (0.09) is set as the threshold to determine the directions of influence between dimensions and criteria. As we can see in Figure 1, the influence network-relationship map (INRM) of the six dimensions and their subsystems can be plotted according to Tables A2 and A4.

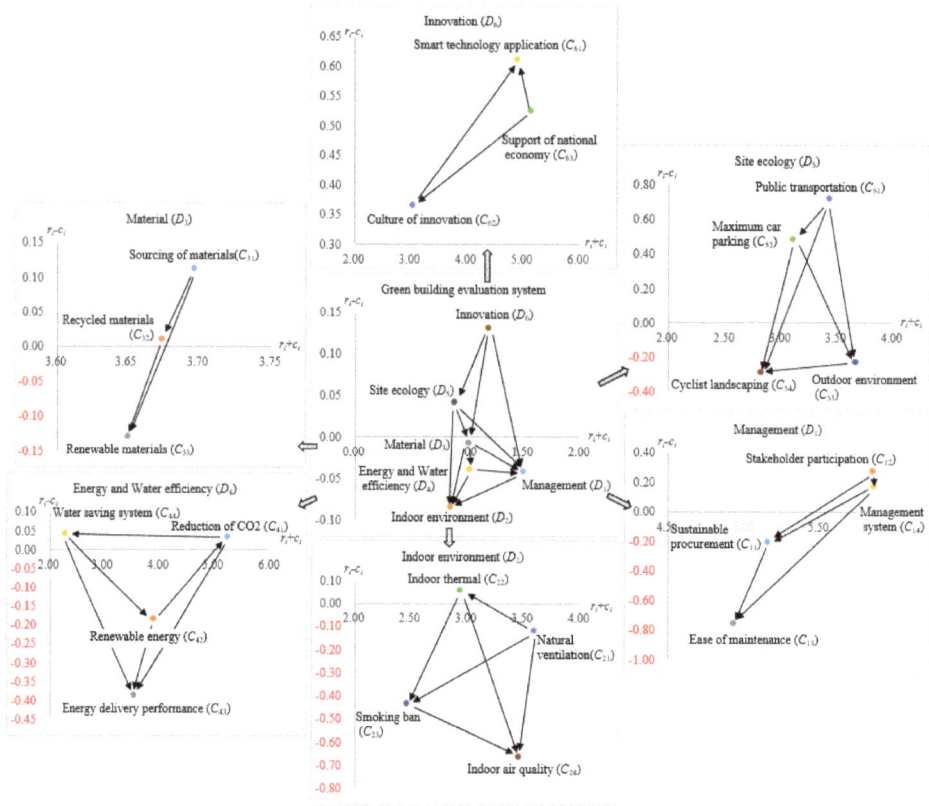

Figure 1. Influential network relation map (INRM).

5. Discussion

While most studies applied AHP or ANP to investigate the indicators of green buildings, this study used DANP to derive the importance of indicators. DANP adopts the results of DEMATEL and concepts of ANP to calculate the weights of criteria. The influential degrees among criteria (T_c) and dimensions (T_D) are obtained from DEMATEL, where we normalized T_c by considering the influential strength of each criterion within its dimension. Thus, the weighted supermatrix (step 12 in Section 3) takes into account the proportion of each criterion within its dimension and the degrees of influence between each dimension and other dimensions. As a result, we get the weight of the sub-factor first, then we obtain the weight of each dimension by summing the weights of criteria that belong to the same dimension. This process can avoid the time-consuming pairwise comparisons in the original ANP and obtain consistent results.

As illustrated in Figure 1, in dimension D_1, stakeholder participation (C_{12}) influences sustainable procurement (C_{11}), ease of maintenance (C_{13}) and management system (C_{14}). This means that residents pursuing a greener environment, active governmental promotion of green building development, the development of green building materials and technology by engineers and participation by other stakeholders all have positive effects on green building sustainable development. In terms of the indoor environment (D_2), the relationships between the four criteria are relatively weak, but natural ventilation (C_{21}), indoor thermal environment (C_{22}) and the banning of smoking (C_{23}) all affect indoor air quality (C_{24}). Indoor air quality is a very important criterion for green building assessment, being of the most concern to people and it is easily affected by the other elements. In the material dimension (D_3), all three criteria influence each other. The sourcing of building materials is gradually getting more and more attention, with recycled and renewable materials becoming more and more attractive, a trend expected to continue into the future. Carbon dioxide emissions can be reduced by the use of renewable energy and an increase in energy transfer efficiency. Hence, experts consider renewable energy (C_{42}) and energy delivery performance (C_{43}) to have a strong influence on the reduction of CO_2 (C_{41}). With the increased use of private vehicles, the demand for parking spaces has become an urgent problem. With the increase in popularity of the sharing economy and the enhancement of green awareness, people also expect to have access to bicycle systems (C_{54}) and public transportation (C_{51}) near these buildings. In dimension D_6, smart technology (C_{61}), a culture of innovation (C_{62}) and support of the national economy (C_{63}) all affect each other.

The assessment system for a green building is mapped in Figure 1 with innovation (D_6) on the top and management (D_1) on the right. This means that D_6 is the main factor that influences the other dimensions. At the same time, management is a complex factor, which affects the other dimension, and it is also affected by all other dimensions except for the indoor environment (D_2). Thus, we must consider innovation and management seriously if we want to extend and develop green building assessment systems effectively in China. The results can also be plotted, as shown in Figure 2, where the innovation is a major cause and management is located in the center of the system. "Energy and water efficiency" and "indoor environment" are affected by the other dimensions. Furthermore, the influence weight can be calculated by the DANP method (Table 4). In Table 4, management (D_1) and innovation (D_6) are the two most significant dimensions, sharing an importance of 40.91% in the green building assessment system. This also exactly confirms the findings of the INRM.

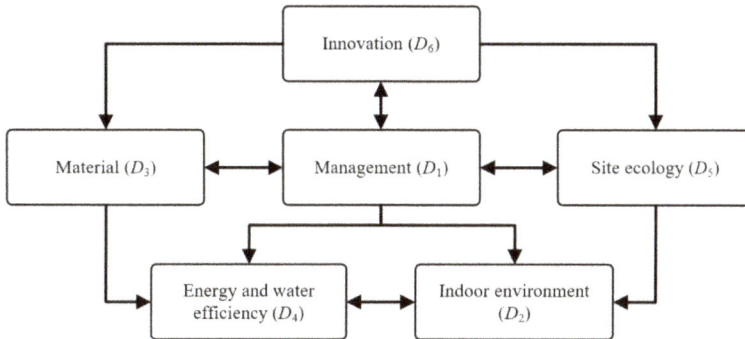

Figure 2. Major influence route.

From Table 4, management (D_1) has the largest weight, followed by innovation (D_6), indoor environment (D_2), materials (D_3), energy and water efficiency (D_4) and site ecology. In terms of the criteria, supporting the national economy (C_{63}) (6.74%), management system (C_{14}) (6.18%), and application of smart technology (C_{61}) (6.16%) are the three most significant criteria, followed by ease of maintenance (C_{13}), stakeholder participation (C_{14}), and sustainable procurement (C_{11}). On the other hand, water saving systems (C_{44}) (2.39%), maximum car parking (C_{52}) (2.76%) and access to public transportation (C_{51}) (2.89%) are the least important criteria in the system. These results verify the results of the DEMATEL analysis again where management (D_1) and innovation (D_6) are the top two priorities in the green building evaluation system, because the six most significant criteria are related to management and innovation.

Table 4. Influence weights of the green building assessment system.

	Local Weight	Ranking		Local Weight	Ranking	Global Weight	Ranking
D_1	0.2374	1	C_{11}	0.242	4	0.0573	6
			C_{12}	0.251	3	0.0596	5
			C_{13}	0.253	2	0.0601	4
			C_{14}	0.254	1	0.0604	2
D_2	0.1468	5	C_{21}	0.275	2	0.0404	15
			C_{22}	0.211	4	0.0310	19
			C_{23}	0.212	3	0.0311	18
			C_{24}	0.302	1	0.0443	11
D_3	0.1583	4	C_{31}	0.326	3	0.0516	10
			C_{32}	0.332	2	0.0525	9
			C_{33}	0.342	1	0.0542	8
D_4	0.1656	2	C_{41}	0.338	1	0.0560	7
			C_{42}	0.266	2	0.0440	12
			C_{43}	0.254	3	0.0421	13
			C_{44}	0.142	4	0.0235	22
D_5	0.1284	6	C_{51}	0.221	3	0.0283	20
			C_{52}	0.211	4	0.0271	21
			C_{53}	0.319	1	0.0410	14
			C_{54}	0.249	2	0.0320	17
D_6	0.1636	3	C_{61}	0.368	2	0.0603	3
			C_{62}	0.229	3	0.0374	16
			C_{63}	0.403	1	0.0659	1

Although management has the largest degree of total influence, its net influence value is negative, which means that the quality of the management is related to other dimensions, especially innovation (D_6). Therefore, technical and cultural innovation, as well as the support of the national economy, are important for the improvement of management quality. Indoor environment (D_2), materials (D_3) and energy and water (D_4) have almost the same weights, and they are affected by management and innovation. Although site ecology (D_5) accounts for the smallest weight, we cannot ignore its contribution for the assessment system. The increasing popularity of the sharing economy in work and in daily life have a positive impact on green building.

6. Conclusions

Six dimensions and 22 criteria were selected based on the literature review and expert questionnaire responses were used to build a green building assessment system for China. These dimensions include management, indoor environment, materials, energy and water, site ecology and innovation. One should be conscious of the fact that these dimensions and criteria are quite similar to those of international tools such as BREEAM and LEED, but their priorities are somewhat different, due to consideration of the Chinese context.

Differing from previous studies, this paper applied a hybrid MCDM model to construct the green building assessment system. First, the DEMATEL method was used to find the complex relationship among dimensions and criteria. Then, the INRM was constructed as shown in Figure 1. The INRM can reflect the causal relationship among the dimensions and criteria. After that, the DEMATEL-based ANP method was applied to calculate the influence weights of the criteria.

The results indicate that management and innovation are the two most important factors for assessing green buildings practices in China, making up about 41% of the total evaluation weight. This includes the three highest weighted criteria, support of the national economy, management system and smart technology innovation, which represent 19% of the total assessment criteria weights. Unlike the results from past studies, the weights of energy and water resources, indoor environment and other indicators are not very high, but this does not mean that these indicators are not important. In fact, the weight distribution of the evaluation index system in this study is balanced, and the

Sustainability **2018**, *10*, 1173

difference between the highest weight and the minimum is not very large, just 4.3%. This means that the development of green buildings in China is at a preliminary stage, and there is still a certain gap between developed countries and China. It is urgent to promote the development and popularization of green buildings through management and innovation.

This paper offers several contributions to the literature. First, most previous studies ignored the relationship between criteria. In this study, a hybrid model with DANP model was used to analyze the complex assessment system, and a clear map can be obtained showing the causal relationships among dimensions and criteria. The key dimensions and criteria are found to help the Chinese government and the construction industry to realize how to implement green buildings in China. Second, this hybrid model can reduce the complexity of the ANP method while taking into account the extent of the impact between dimensions and criteria.

There are some limitations on the application of the DANP model. We just interviewed 10 experts and got the data. Although the average gap-ratio in consistency is smaller than 5%, we cannot conclude that they can represent all stakeholders' consensus. The experts' opinions differ somewhat from each other owing to their different backgrounds. We used the fuzzy logic method to remedy this problem. Perhaps other methods, such as the Delphi or Grey relational analysis, can be used to solve this problem. The empirical data are limited to Xiamen city in China, so research findings may vary for other areas and countries. With regards to future research, some green building cases can be collected for performance evaluation by the application of TOPSIS and other models based on the DANP model in this paper. At the same time, the proposed model can also be used for handling similar decision making problems in other industries.

Acknowledgments: The authors are extremely grateful to Sustainability Journal editorial team's valuable comments on improving the quality of this article. This research was supported by the Social Science Fund of Fujian Province under Grant FJ2016B101, China.

Author Contributions: Qi-Gan Shao analyzed the data, literature review, and article writing. James J. H. Liou and Sung-Shun Weng dealt with the research design, and article writing. Yen-Ching Chuang, completed the article writing and formatting. Finally, James J. H. Liou and Sung-Shun Weng revised the paper.

Conflicts of Interest: The authors declare no conflict of interest.

Appendix A Results in Detail

This study applied the DANP model to build an influential network relationship map (INRM) between assessment systems and to derive the 22 criterion weights. As mentioned in Section 4, we collected 10 experts' questionnaires and the results to measure the total influence matrix T. Table A1 shows the 22 × 22 average initial direct influence matrix that was obtained by averaging the experts' questionnaire responses.

Table A2 shows the total influential relationship among 22 criteria. Table A3 illustrates the relationship among the six dimensions. Table A4 reflects the sum of influences received or given from the degree of the influence of each dimension and criterion within its dimension. The results indicate that innovation (D_6) is the main cause in the green building assessment systems because it has largest net influence ($r_i - c_i$).

The weights of criteria were obtained from the DANP process. Table A5 illustrates the unweighted supermatrix, which was obtained by transposing the normalized influence matrix Tc based on Equations (8)–(12). The weighted supermatrix W based on Equations (13)–(16) is shown in Table A6.

Table A1. Initial direct influence matrix.

	C_{11}	C_{12}	C_{13}	C_{14}	C_{21}	C_{22}	C_{23}	C_{24}	C_{31}	C_{32}	C_{33}	C_{41}	C_{42}	C_{43}	C_{44}	C_{51}	C_{52}	C_{53}	C_{54}	C_{61}	C_{62}	C_{63}
C_{11}	0.00	1.80	3.60	2.80	1.00	1.00	0.00	1.20	3.00	3.20	3.20	2.20	2.40	1.80	1.80	0.20	0.20	1.20	0.80	2.80	0.60	2.00
C_{12}	2.40	0.00	3.00	3.60	1.20	0.80	3.00	1.80	2.80	2.40	2.40	1.80	2.40	1.60	2.20	1.80	2.00	3.00	2.60	2.40	2.80	2.00
C_{13}	3.20	1.80	0.00	3.20	1.80	1.40	0.80	1.60	1.00	1.40	1.40	1.00	1.20	1.00	1.00	1.80	1.20	1.60	1.00	1.80	0.80	1.00
C_{14}	3.20	3.80	3.20	0.00	1.20	1.20	3.00	1.40	1.40	1.60	1.60	1.80	1.60	2.20	2.00	2.80	3.00	2.60	2.60	2.80	2.40	1.40
C_{21}	0.20	1.00	1.80	1.00	0.00	3.40	2.20	3.40	1.40	0.60	0.60	3.20	2.80	2.60	0.00	0.00	0.00	0.60	0.00	1.00	0.00	1.40
C_{22}	0.20	1.00	1.60	1.00	2.80	0.00	0.80	3.40	0.40	0.60	0.60	1.20	0.60	0.60	0.00	0.00	0.00	0.60	0.00	0.40	0.00	0.00
C_{23}	0.20	1.40	1.20	1.60	2.20	1.20	0.00	0.00	0.00	0.40	0.40	1.00	1.40	1.20	0.00	0.00	0.00	1.00	0.00	0.40	0.00	1.60
C_{24}	0.80	2.00	2.00	1.40	2.60	0.60	2.60	0.00	0.80	3.40	3.40	1.00	1.40	0.60	0.00	0.20	0.00	0.60	0.00	0.40	0.20	0.00
C_{31}	2.20	2.80	1.80	2.00	1.80	1.20	0.00	2.60	0.00	0.00	0.00	1.20	1.00	1.20	0.00	0.20	0.00	0.80	0.00	0.20	0.40	1.60
C_{32}	3.40	2.00	1.60	1.60	0.80	0.40	0.40	0.80	2.20	3.40	3.00	3.00	1.60	0.80	0.00	0.20	0.00	0.80	0.00	0.80	0.40	2.20
C_{33}	3.40	2.00	1.80	1.60	1.00	0.40	0.40	1.00	2.20	1.80	1.60	2.60	1.40	0.80	0.00	0.00	0.00	0.40	0.00	1.20	0.40	2.20
C_{41}	2.60	2.20	1.60	2.00	3.20	1.80	2.20	2.20	1.80	1.80	1.20	0.00	2.20	2.20	0.80	1.80	1.60	2.20	1.80	2.20	0.80	2.20
C_{42}	2.80	1.80	1.40	1.40	1.60	1.00	0.60	1.40	0.80	1.40	1.20	3.20	0.00	1.60	0.20	0.40	0.40	1.00	0.60	1.40	0.40	1.80
C_{43}	2.00	1.40	1.80	1.60	1.20	1.60	0.60	1.40	0.80	0.80	0.80	1.40	0.60	0.00	0.40	0.20	0.20	0.60	0.20	1.00	0.60	2.20
C_{44}	2.40	1.40	1.80	1.60	0.00	0.00	0.00	0.00	0.20	0.00	0.00	2.20	0.60	0.80	0.00	0.00	0.20	2.40	2.20	1.20	0.60	2.40
C_{51}	1.80	2.60	2.60	3.00	0.60	0.60	0.00	0.60	0.40	0.40	0.40	2.20	0.60	0.60	0.20	0.00	2.80	2.40	2.20	2.00	1.00	2.40
C_{52}	1.40	2.20	2.20	2.80	0.60	0.60	0.40	1.00	0.40	0.40	0.40	1.20	0.60	0.60	0.60	2.00	0.00	0.00	2.60	2.00	0.80	1.00
C_{53}	0.80	2.60	2.20	2.20	0.80	0.80	0.20	1.20	0.60	0.80	0.40	1.60	0.60	1.00	0.60	1.40	1.20	0.00	0.00	1.20	1.00	1.40
C_{54}	1.20	2.20	1.20	1.80	0.20	0.20	0.00	0.20	0.20	0.20	0.20	1.60	0.20	0.40	0.00	1.20	1.60	1.20	0.00	0.00	0.80	1.40
C_{61}	1.60	1.80	3.20	3.40	1.60	2.00	1.20	2.00	1.60	1.00	1.60	2.20	2.20	2.80	3.00	2.40	2.20	1.60	1.60	0.00	2.00	1.80
C_{62}	1.60	2.40	1.80	2.40	0.60	0.60	1.20	0.80	0.60	0.80	0.80	1.00	0.80	0.80	0.80	0.80	0.60	1.00	1.00	2.00	0.00	2.00
C_{63}	2.80	3.00	2.00	1.80	1.00	0.80	0.80	1.20	2.80	2.40	2.40	2.40	2.00	2.00	1.60	1.80	1.60	2.40	2.60	3.00	2.80	0.00

Note: The average gap-ratio in consensus (%) $= \frac{1}{m(m-1)}\sum_{i=1}^{m}\sum_{j=1}^{m}\left(\frac{|d_{ij}^{s} - d_{ij}^{s-1}|}{d_{ij}^{s}}\right) * 100\% = 4.86\% < 5\%$, where m is the number of criteria ($m = 22$), s is the sample of 10 experts ($s = 10$) whose practical experience and significant confidence reach 95.14% (more than 95%).

Table A2. Total-influence matrix of criteria.

	C_{11}	C_{12}	C_{13}	C_{14}	C_{21}	C_{22}	C_{23}	C_{24}	C_{31}	C_{32}	C_{33}	C_{41}	C_{42}	C_{43}	C_{44}	C_{51}	C_{52}	C_{53}	C_{54}	C_{61}	C_{62}	C_{63}
C_{11}	0.11	0.15	0.18	0.17	0.09	0.08	0.06	0.10	0.14	0.14	0.15	0.15	0.13	0.12	0.08	0.06	0.06	0.10	0.08	0.14	0.07	0.13
C_{12}	0.18	0.14	0.20	0.21	0.11	0.08	0.13	0.13	0.14	0.14	0.14	0.16	0.14	0.13	0.10	0.10	0.11	0.16	0.13	0.15	0.12	0.15
C_{13}	0.15	0.13	0.10	0.16	0.10	0.08	0.07	0.10	0.08	0.09	0.09	0.11	0.09	0.09	0.06	0.08	0.07	0.10	0.07	0.11	0.06	0.10
C_{14}	0.19	0.21	0.20	0.14	0.11	0.09	0.13	0.13	0.11	0.12	0.12	0.16	0.13	0.14	0.10	0.12	0.13	0.15	0.13	0.16	0.11	0.14
C_{21}	0.08	0.09	0.11	0.09	0.06	0.11	0.09	0.13	0.08	0.06	0.06	0.14	0.12	0.11	0.03	0.03	0.03	0.06	0.04	0.08	0.03	0.09
C_{22}	0.07	0.08	0.10	0.08	0.11	0.04	0.06	0.10	0.05	0.06	0.06	0.12	0.11	0.10	0.03	0.03	0.03	0.05	0.03	0.07	0.04	0.04
C_{23}	0.05	0.07	0.07	0.08	0.08	0.06	0.03	0.11	0.03	0.04	0.04	0.07	0.05	0.05	0.03	0.03	0.02	0.04	0.02	0.04	0.02	0.08
C_{24}	0.07	0.10	0.10	0.09	0.10	0.06	0.09	0.05	0.06	0.05	0.05	0.08	0.08	0.07	0.03	0.02	0.03	0.06	0.03	0.05	0.03	0.08
C_{31}	0.13	0.14	0.12	0.13	0.09	0.06	0.05	0.12	0.06	0.05	0.05	0.11	0.09	0.09	0.03	0.03	0.04	0.07	0.04	0.07	0.03	0.12
C_{32}	0.15	0.12	0.12	0.12	0.07	0.05	0.05	0.08	0.11	0.06	0.13	0.14	0.10	0.08	0.02	0.04	0.04	0.07	0.04	0.09	0.05	0.11
C_{33}	0.16	0.12	0.15	0.11	0.07	0.05	0.05	0.08	0.10	0.10	0.12	0.13	0.13	0.07	0.03	0.04	0.04	0.06	0.04	0.13	0.05	0.14
C_{41}	0.14	0.16	0.11	0.16	0.14	0.10	0.11	0.13	0.11	0.11	0.06	0.13	0.06	0.13	0.03	0.09	0.09	0.13	0.10	0.09	0.07	0.11
C_{42}	0.11	0.12	0.11	0.11	0.09	0.07	0.06	0.09	0.07	0.09	0.11	0.11	0.08	0.09	0.06	0.05	0.05	0.08	0.06	0.08	0.05	0.09
C_{43}	0.10	0.10	0.09	0.10	0.07	0.07	0.05	0.08	0.06	0.06	0.08	0.14	0.05	0.05	0.04	0.04	0.04	0.07	0.05	0.07	0.04	0.09
C_{44}	0.13	0.08	0.09	0.09	0.03	0.03	0.03	0.04	0.04	0.04	0.07	0.11	0.08	0.05	0.03	0.03	0.03	0.05	0.04	0.08	0.04	0.12
C_{51}	0.10	0.15	0.15	0.16	0.07	0.06	0.05	0.08	0.07	0.07	0.07	0.08	0.05	0.08	0.05	0.05	0.11	0.12	0.11	0.12	0.07	0.09
C_{52}	0.09	0.13	0.13	0.14	0.06	0.05	0.05	0.08	0.06	0.06	0.06	0.13	0.07	0.07	0.05	0.09	0.04	0.11	0.10	0.11	0.06	0.09
C_{53}	0.08	0.13	0.12	0.12	0.07	0.06	0.05	0.08	0.06	0.07	0.06	0.10	0.07	0.07	0.03	0.07	0.07	0.06	0.10	0.09	0.06	0.09
C_{54}	0.15	0.10	0.08	0.10	0.04	0.03	0.03	0.04	0.04	0.04	0.04	0.08	0.05	0.05	0.11	0.06	0.10	0.12	0.04	0.07	0.05	0.07
C_{61}	0.11	0.16	0.19	0.19	0.11	0.10	0.09	0.13	0.11	0.10	0.11	0.16	0.13	0.14	0.05	0.11	0.07	0.12	0.10	0.09	0.10	0.14
C_{62}	0.11	0.13	0.12	0.13	0.06	0.05	0.07	0.07	0.06	0.07	0.07	0.09	0.07	0.07	0.05	0.06	0.05	0.08	0.07	0.10	0.04	0.10
C_{63}	0.18	0.19	0.17	0.17	0.10	0.08	0.08	0.11	0.14	0.13	0.14	0.16	0.13	0.13	0.09	0.10	0.09	0.14	0.12	0.16	0.12	0.10

Table A3. Total-influence matrix of dimensions.

	D_1	D_2	D_3	D_4	D_5	D_6
D_1	0.16	0.10	0.12	0.12	0.10	0.12
D_2	0.08	0.08	0.05	0.07	0.04	0.06
D_3	0.13	0.07	0.10	0.08	0.05	0.08
D_4	0.12	0.07	0.07	0.08	0.06	0.08
D_5	0.12	0.06	0.06	0.07	0.08	0.08
D_6	0.16	0.09	0.10	0.11	0.09	0.11

Table A4. Sum of influences given and received on criteria and dimensions.

Dimensions	r_i	c_i	$r_i + c_i$	$r_i - c_i$	Criteria	r_i	c_i	$r_i + c_i$	$r_i - c_i$
D_1	0.73	0.77	1.49	−0.04	C_{11}	2.47	2.68	5.15	−0.20
					C_{12}	3.06	2.79	5.86	0.27
					C_{13}	2.09	2.84	4.93	−0.75
					C_{14}	3.01	2.85	5.86	0.17
D_2	0.38	0.46	0.85	−0.08	C_{21}	1.74	1.86	3.60	−0.12
					C_{22}	1.50	1.44	2.94	0.06
					C_{23}	1.02	1.45	2.47	−0.43
					C_{24}	1.40	2.06	3.46	−0.66
D_3	0.50	0.51	1.01	−0.01	C_{31}	1.90	1.79	3.70	0.11
					C_{32}	1.84	1.83	3.67	0.01
					C_{33}	1.76	1.89	3.65	−0.13
D_4	0.49	0.53	1.02	−0.04	C_{41}	2.64	2.60	5.24	0.03
					C_{42}	1.86	2.04	3.90	−0.18
					C_{43}	1.58	1.97	3.55	−0.39
					C_{44}	1.16	1.12	2.28	0.04
D_5	0.46	0.42	0.88	0.04	C_{51}	2.07	1.35	3.43	0.72
					C_{52}	1.79	1.31	3.11	0.48
					C_{53}	1.72	1.95	3.67	−0.23
					C_{54}	1.26	1.55	2.82	−0.29
D_6	0.66	0.53	1.19	0.13	C_{61}	2.75	2.14	4.90	0.61
					C_{62}	1.70	1.34	3.04	0.36
					C_{63}	2.83	2.31	5.15	0.52

Table A5. The un-weighted supermatrix W.

	C_{11}	C_{12}	C_{13}	C_{14}	C_{21}	C_{22}	C_{23}	C_{24}	C_{31}	C_{32}	C_{33}	C_{41}	C_{42}	C_{43}	C_{44}	C_{51}	C_{52}	C_{53}	C_{54}	C_{61}	C_{62}	C_{63}
C_{11}	0.18	0.24	0.30	0.28	0.28	0.23	0.17	0.31	0.32	0.34	0.34	0.31	0.27	0.24	0.17	0.20	0.19	0.35	0.26	0.41	0.19	0.39
C_{12}	0.25	0.19	0.27	0.29	0.24	0.18	0.28	0.29	0.34	0.33	0.33	0.30	0.27	0.24	0.19	0.21	0.21	0.31	0.26	0.36	0.29	0.35
C_{13}	0.29	0.24	0.18	0.30	0.29	0.23	0.19	0.29	0.31	0.34	0.35	0.31	0.27	0.25	0.17	0.26	0.22	0.30	0.23	0.41	0.23	0.36
C_{14}	0.26	0.28	0.27	0.19	0.24	0.20	0.28	0.27	0.32	0.34	0.34	0.30	0.25	0.26	0.19	0.24	0.24	0.28	0.25	0.39	0.28	0.33
C_{21}	0.20	0.25	0.30	0.25	0.15	0.29	0.23	0.34	0.38	0.31	0.31	0.35	0.30	0.28	0.07	0.20	0.19	0.38	0.23	0.38	0.17	0.45
C_{22}	0.20	0.25	0.29	0.29	0.36	0.13	0.19	0.33	0.30	0.35	0.35	0.33	0.31	0.28	0.10	0.21	0.20	0.36	0.23	0.34	0.20	0.46
C_{23}	0.17	0.27	0.27	0.24	0.33	0.17	0.11	0.41	0.26	0.37	0.37	0.38	0.31	0.26	0.10	0.19	0.19	0.41	0.21	0.42	0.20	0.37
C_{24}	0.20	0.27	0.28	0.24	0.31	0.20	0.30	0.17	0.36	0.32	0.32	0.31	0.31	0.28	0.10	0.20	0.18	0.41	0.22	0.32	0.21	0.48
C_{31}	0.25	0.24	0.24	0.23	0.30	0.18	0.15	0.37	0.19	0.40	0.41	0.33	0.27	0.28	0.11	0.22	0.19	0.38	0.23	0.29	0.21	0.50
C_{32}	0.30	0.24	0.23	0.25	0.29	0.20	0.20	0.31	0.36	0.21	0.43	0.40	0.28	0.22	0.10	0.21	0.19	0.37	0.24	0.33	0.20	0.47
C_{33}	0.26	0.25	0.24	0.23	0.30	0.21	0.22	0.28	0.33	0.34	0.23	0.39	0.30	0.23	0.12	0.23	0.21	0.35	0.25	0.35	0.19	0.45
C_{41}	0.29	0.23	0.23	0.25	0.30	0.22	0.19	0.30	0.33	0.35	0.34	0.25	0.19	0.30	0.14	0.21	0.20	0.31	0.24	0.37	0.19	0.41
C_{42}	0.26	0.22	0.26	0.23	0.27	0.26	0.18	0.29	0.33	0.33	0.33	0.42	0.28	0.18	0.12	0.22	0.19	0.34	0.24	0.35	0.20	0.45
C_{43}	0.28	0.22	0.25	0.24	0.27	0.22	0.22	0.30	0.35	0.32	0.34	0.40	0.25	0.26	0.14	0.22	0.21	0.31	0.24	0.35	0.21	0.40
C_{44}	0.22	0.25	0.26	0.27	0.28	0.23	0.19	0.30	0.33	0.33	0.34	0.37	0.24	0.23	0.12	0.13	0.28	0.32	0.27	0.38	0.23	0.34
C_{51}	0.21	0.28	0.26	0.28	0.26	0.22	0.21	0.32	0.33	0.33	0.34	0.34	0.25	0.24	0.17	0.25	0.13	0.19	0.34	0.42	0.23	0.38
C_{52}	0.19	0.28	0.26	0.27	0.27	0.23	0.19	0.32	0.33	0.35	0.34	0.34	0.23	0.26	0.16	0.24	0.22	0.32	0.16	0.36	0.25	0.39
C_{53}	0.22	0.23	0.23	0.27	0.26	0.22	0.21	0.30	0.33	0.33	0.34	0.41	0.22	0.24	0.13	0.25	0.29	0.30	0.23	0.28	0.30	0.41
C_{54}	0.22	0.26	0.27	0.28	0.25	0.24	0.20	0.30	0.34	0.31	0.35	0.29	0.24	0.26	0.21	0.25	0.24	0.30	0.23	0.28	0.30	0.41
C_{61}	0.26	0.27	0.24	0.24	0.27	0.20	0.26	0.29	0.31	0.34	0.35	0.32	0.25	0.25	0.18	0.25	0.21	0.27	0.26	0.42	0.16	0.42
C_{62}	0.22	0.26	0.24	0.27	0.25	0.20	0.26	0.30	0.34	0.33	0.35	0.32	0.26	0.25	0.18	0.23	0.21	0.30	0.26	0.42	0.31	0.42
C_{63}	0.26	0.27	0.24	0.24	0.27	0.21	0.21	0.30	0.34	0.33	0.33	0.32	0.26	0.25	0.17	0.22	0.21	0.30	0.27	0.42	0.31	0.27

Table A6. The weighted supermatrix W_W.

	C_{11}	C_{12}	C_{13}	C_{14}	C_{21}	C_{22}	C_{23}	C_{24}	C_{31}	C_{32}	C_{33}	C_{41}	C_{42}	C_{43}	C_{44}	C_{51}	C_{52}	C_{53}	C_{54}	C_{61}	C_{62}	C_{63}
C_{11}	0.06	0.06	0.06	0.06	0.06	0.06	0.06	0.06	0.06	0.06	0.06	0.06	0.06	0.06	0.06	0.06	0.06	0.06	0.06	0.06	0.06	0.06
C_{12}	0.06	0.06	0.06	0.06	0.07	0.07	0.07	0.07	0.06	0.06	0.06	0.06	0.06	0.06	0.06	0.06	0.06	0.06	0.06	0.06	0.06	0.06
C_{13}	0.06	0.06	0.06	0.06	0.07	0.07	0.07	0.07	0.06	0.06	0.06	0.06	0.06	0.06	0.06	0.06	0.06	0.06	0.06	0.06	0.06	0.06
C_{14}	0.06	0.06	0.06	0.06	0.07	0.07	0.07	0.07	0.06	0.06	0.06	0.06	0.06	0.06	0.06	0.06	0.06	0.06	0.06	0.06	0.06	0.06
C_{21}	0.04	0.04	0.04	0.04	0.05	0.05	0.05	0.05	0.04	0.04	0.04	0.04	0.04	0.04	0.04	0.04	0.04	0.04	0.04	0.04	0.04	0.04
C_{22}	0.04	0.04	0.04	0.04	0.04	0.04	0.04	0.04	0.04	0.04	0.04	0.04	0.04	0.04	0.04	0.04	0.04	0.04	0.04	0.04	0.04	0.04
C_{23}	0.05	0.05	0.05	0.05	0.05	0.05	0.05	0.05	0.05	0.05	0.05	0.05	0.05	0.05	0.05	0.05	0.05	0.05	0.05	0.05	0.05	0.05
C_{24}	0.05	0.05	0.05	0.05	0.06	0.06	0.06	0.06	0.05	0.05	0.05	0.05	0.05	0.05	0.05	0.05	0.05	0.05	0.05	0.05	0.05	0.05
C_{31}	0.05	0.05	0.05	0.05	0.06	0.06	0.06	0.06	0.05	0.05	0.05	0.05	0.05	0.05	0.05	0.05	0.05	0.05	0.05	0.05	0.05	0.05
C_{32}	0.06	0.06	0.06	0.06	0.06	0.06	0.06	0.06	0.06	0.06	0.06	0.06	0.06	0.06	0.06	0.06	0.06	0.06	0.06	0.06	0.06	0.06
C_{33}	0.05	0.05	0.05	0.05	0.06	0.06	0.06	0.06	0.05	0.05	0.05	0.05	0.05	0.05	0.05	0.05	0.05	0.05	0.05	0.05	0.05	0.05
C_{41}	0.06	0.06	0.06	0.06	0.06	0.06	0.06	0.06	0.06	0.06	0.06	0.06	0.06	0.06	0.06	0.06	0.06	0.06	0.06	0.06	0.06	0.06
C_{42}	0.05	0.05	0.05	0.05	0.05	0.05	0.05	0.05	0.05	0.05	0.05	0.05	0.05	0.05	0.05	0.05	0.05	0.05	0.05	0.05	0.05	0.05
C_{43}	0.04	0.04	0.04	0.04	0.04	0.04	0.04	0.04	0.04	0.04	0.04	0.04	0.04	0.04	0.04	0.04	0.04	0.04	0.04	0.04	0.04	0.04
C_{44}	0.02	0.02	0.02	0.02	0.03	0.03	0.03	0.03	0.02	0.02	0.02	0.02	0.02	0.02	0.02	0.02	0.02	0.02	0.02	0.02	0.02	0.02
C_{51}	0.03	0.03	0.03	0.03	0.03	0.03	0.03	0.03	0.03	0.03	0.03	0.03	0.03	0.03	0.03	0.03	0.03	0.03	0.03	0.03	0.03	0.03
C_{52}	0.03	0.03	0.03	0.03	0.03	0.03	0.03	0.03	0.03	0.03	0.03	0.03	0.03	0.03	0.03	0.04	0.04	0.04	0.04	0.03	0.03	0.03
C_{53}	0.04	0.04	0.04	0.04	0.05	0.05	0.05	0.05	0.04	0.04	0.04	0.04	0.04	0.04	0.04	0.03	0.04	0.03	0.04	0.04	0.04	0.04
C_{54}	0.03	0.03	0.03	0.03	0.03	0.03	0.05	0.03	0.04	0.04	0.04	0.03	0.03	0.04	0.04	0.04	0.03	0.04	0.03	0.04	0.03	0.03
C_{61}	0.04	0.04	0.04	0.04	0.05	0.05	0.07	0.05	0.04	0.04	0.04	0.04	0.04	0.06	0.06	0.06	0.06	0.06	0.04	0.04	0.06	0.06
C_{62}	0.04	0.04	0.03	0.04	0.03	0.04	0.04	0.04	0.03	0.03	0.04	0.03	0.03	0.04	0.04	0.04	0.03	0.04	0.04	0.04	0.04	0.04
C_{63}	0.07	0.07	0.07	0.07	0.07	0.07	0.07	0.07	0.07	0.07	0.07	0.07	0.07	0.07	0.07	0.07	0.07	0.07	0.07	0.07	0.07	0.07

References

1. Shad, R.; Khorrami, M.; Ghaemi, M. Developing an Iranian green building assessment tool using decision making methods and geographical information system: Case study in Mashhad city. *Renew. Sustain. Energy Rev.* **2017**, *67*, 324–340. [CrossRef]

2. Dixit, M.K.; Culp, C.H.; Fernández-Solís, J.L. System boundary for embodied energy in buildings: A conceptual model for definition. *Renew. Sustain. Energy Rev.* **2013**, *21*, 153–164. [CrossRef]

3. Yeheyis, M.; Hewage, K.; Alam, M.S.; Eskicioglu, C.; Sadiq, R. An overview of construction and demolition waste management in Canada: A lifecycle analysis approach to sustainability. *Clean Technol. Environ. Policy* **2012**, *15*, 81–91. [CrossRef]

4. Al-Jebouri, M.F.A.; Saleh, M.S.; Raman, S.N.; Rahmat, R.A.A.B.O.K.; Shaaban, A.K. Toward a national sustainable building assessment system in Oman: Assessment categories and their performance indicators. *Sustain. Cities Soc.* **2017**, *31*, 122–135. [CrossRef]

5. Ali, H.H.; Al Nsairat, S.F. Developing a green building assessment tool for developing countries: Case of Jordan. *Build. Environ.* **2009**, *44*, 1053–1064. [CrossRef]

6. Ye, L.; Cheng, Z.; Wang, Q.; Lin, H.; Lin, C.; Liu, B. Developments of Green Building Standards in China. *Renew. Energy* **2015**, *73*, 115–122. [CrossRef]

7. Balaban, O.; Puppim de Oliveira, J.A. Sustainable buildings for healthier cities: Assessing the co-benefits of green buildings in Japan. *J. Clean Prod.* **2017**, *163*, 68–78. [CrossRef]

8. Wong, S.C.; Abe, N. Stakeholders' perspectives of a building environmental assessment method: The case of CASBEE. *Build. Environ.* **2014**, *82*, 502–516. [CrossRef]

9. Dwaikat, L.; Ali, K. Measuring the Actual Energy Cost Performance of Green Buildings: A Test of the Earned Value Management Approach. *Energies* **2016**, *9*, 188. [CrossRef]

10. Kang, H.J. Development of a systematic model for an assessment tool for sustainable buildings based on a structural framework. *Energy Build.* **2015**, *104*, 287–301. [CrossRef]

11. Darko, A.; Chan, A.P.C.; Owusu-Manu, D.G.; Ameyaw, E.E. Drivers for implementing green building technologies: An international survey of experts. *J. Clean Prod.* **2017**, *145*, 386–394. [CrossRef]

12. Chenari, B.; Dias Carrilho, J.; Gameiro da Silva, M. Towards sustainable, energy-efficient and healthy ventilation strategies in buildings: A review. *Renew. Sustain. Energy Rev.* **2016**, *59*, 1426–1447. [CrossRef]

13. Hsu, C.C.; Liou, J.J.H.; Chuang, Y.C. Integrating DANP and modified grey relation theory for the selection of an outsourcing provider. *Expert Syst. Appl.* **2013**, *40*, 2297–2304. [CrossRef]

14. Huang, C.N.; Liou, J.J.H.; Chuang, Y.C. A method for exploring the interdependencies and importance of critical infrastructures. *Knowl.-Based Syst.* **2014**, *55*, 66–74. [CrossRef]

15. Hung, Y.H.; Huang, T.L.; Hsieh, J.C.; Tsuei, H.J.; Cheng, C.C.; Tzeng, G.H. Online reputation management for improving marketing by using a hybrid MCDM model. *Knowl.-Based Syst.* **2012**, *35*, 87–93. [CrossRef]

16. Liou, J.J.H. Building an effective system for carbon reduction management. *J. Clean Prod.* **2015**, *103*, 353–361. [CrossRef]

17. Chen, S.H.; Lin, W.T. Analyzing determinants for promoting emerging technology through intermediaries by using a DANP-based MCDA framework. *Technol. Forecast. Soc.* **2017**, *9*, 1–17. [CrossRef]

18. Liou, J.J.H.; Hsu, C.C.; Chen, Y.S. Improving transportation service quality based on information fusion. *Transp. Res. Part A-Policy Pract.* **2014**, *67*, 225–239. [CrossRef]

19. Bozovic-Stamenovic, R.; Kishnani, N.; Tan, B.K.; Prasad, D.; Faizal, F. Assessment of awareness of Green Mark (GM) rating tool by occupants of GM buildings and general public. *Energy Build.* **2016**, *115*, 55–62. [CrossRef]

20. Chen, X.; Yang, H.; Lu, L. A comprehensive review on passive design approaches in green building rating tools. *Renew. Sustain. Energy Rev.* **2015**, *50*, 1425–1436. [CrossRef]

21. Li, Y.Y.; Chen, P.H.; Chew, D.A.S.; Teo, C.C. Exploration of critical resources and capabilities of design firms for delivering green building projects: Empirical studies in Singapore. *Habitat Int.* **2014**, *41*, 229–235. [CrossRef]

22. Alyami, S.H.; Rezgui, Y. Sustainable building assessment tool development approach. *Sustain. Cities Soc.* **2012**, *5*, 52–62. [CrossRef]

23. Lee, W.L. A comprehensive review of metrics of building environmental assessment schemes. *Energy Build.* **2013**, *62*, 403–413. [CrossRef]

24. Guo, Q.; Wu, Y.; Ding, Y.; Feng, W.; Zhu, N. Measures to enforce mandatory civil building energy efficiency codes in China. *J. Clean Prod.* **2016**, *119*, 152–166. [CrossRef]
25. Hong, T.; Li, C.; Yan, D. Updates to the China Design Standard for Energy Efficiency in public buildings. *Energy Policy* **2015**, *87*, 187–198. [CrossRef]
26. Yu, W.; Li, B.; Yang, X.; Wang, Q. A development of a rating method and weighting system for green store buildings in China. *Renew. Energy* **2015**, *73*, 123–129. [CrossRef]
27. Si, J.; Marjanovic-Halburd, L.; Nasiri, F.; Bell, S. Assessment of building-integrated green technologies: A review and case study on applications of Multi-Criteria Decision Making (MCDM) method. *Sustain. Cities Soc.* **2016**, *27*, 106–115. [CrossRef]
28. Banani, R.; Vahdati, M.M.; Shahrestani, M.; Clements-Croome, D. The development of building assessment criteria framework for sustainable non-residential buildings in Saudi Arabia. *Sustain. Cities Soc.* **2016**, *26*, 289–305. [CrossRef]
29. Liu, Z.W.; Gao, P.Z.; Kang, Y.W. Low-Carbon Building Assessment and Model Construction. *Appl. Mech. Mater.* **2013**, *361–363*, 903–907. [CrossRef]
30. Nilashi, M.; Zakaria, R.; Ibrahim, O.; Majid, M.Z.A.; Mohamad Zin, R.; Chugtai, M.W.; Aminu Yakubu, D. A knowledge-based expert system for assessing the performance level of green buildings. *Knowl.-Based Syst.* **2015**, *86*, 194–209. [CrossRef]
31. Sabaghi, M.; Mascle, C.; Baptiste, P.; Rostamzadeh, R. Sustainability assessment using fuzzy-inference technique (SAFT): A methodology toward green products. *Expert Syst. Appl.* **2016**, *56*, 69–79. [CrossRef]
32. Chen, F.H.; Hsu, T.S.; Tzeng, G.H. A balanced scorecard approach to establish a performance evaluation and relationship model for hot spring hotels based on a hybrid MCDM model combining DEMATEL and ANP. *Int. J. Hosp. Manag.* **2011**, *30*, 908–932. [CrossRef]
33. Varmazyar, M.; Dehghanbaghi, M.; Afkhami, M. A novel hybrid MCDM model for performance evaluation of research and technology organizations based on BSC approach. *Evaluation Program Plan.* **2016**, *58*, 125–140. [CrossRef] [PubMed]
34. BRE. *BREEAM New Construction: Non-Domestic Buildings (Technical Manual SD5073–2.0:2011)*; Building Research Establishment Ltd.: Watford, UK, 2011.
35. Supeekit, T.; Somboonwiwat, T.; Kritchanchai, D. DEMATEL-modified ANP to evaluate internal hospital supply chain performance. *Comput. Ind. Eng.* **2016**, *102*, 318–330. [CrossRef]
36. Uygun, Ö.; Kaçamak, H.; Kahraman, Ü.A. An integrated DEMATEL and Fuzzy ANP techniques for evaluation and selection of outsourcing provider for a telecommunication company. *Comput. Ind. Eng.* **2015**, *86*, 137–146. [CrossRef]
37. Ou Yang, Y.P.; Shieh, H.M.; Tzeng, G.H. A VIKOR technique based on DEMATEL and ANP for information security risk control assessment. *Inf. Sci.* **2013**, *232*, 482–500. [CrossRef]
38. BRE. *Background to the Green Guide to Specification: Methodology and Sponsors*; Building Research Establishment Ltd.: Watford, UK, 2015; Available online: http://www.bre.co.uk/greenguide/page.jsp?id=2069 (accessed on 6 April 2018).
39. Chen, I.S. A combined MCDM model based on DEMATEL and ANP for the selection of airline service quality improvement criteria: A study based on the Taiwanese airline industry. *J. Air Transp. Manag.* **2016**, *57*, 7–18. [CrossRef]
40. Darko, A.; Chan, A.P.C.; Ameyaw, E.E.; He, B.J.; Olanipekun, A.O. Examining issues influencing green building technologies adoption: The United States green building experts' perspectives. *Energy Build.* **2017**, *144*, 320–332. [CrossRef]
41. Arslan, F. The Role of Green Buildings in Sustainable Production: Example of Inci Aku Industrial Battery Factory, Turkey. *Dokuz Eylül Üniversitesi Sosyal Bilimler Enstitüsü Dergisi* **2017**, *19*, 119–145. [CrossRef]
42. Chiu, W.Y.; Tzeng, G.H.; Li, H.L. A new hybrid MCDM model combining DANP with VIKOR to improve e-store business. *Knowl.-Based Syst.* **2013**, *37*, 48–61. [CrossRef]

sustainability

MDPI

Article

Construction Projects Assessment Based on the Sustainable Development Criteria by an Integrated Fuzzy AHP and Improved GRA Model

Seyed Morteza Hatefi [1] and Jolanta Tamošaitienė [2,*]

[1] Department of Civil Engineering, Faculty of Engineering, Shahrekord University, P.O. Box 115, 64165478 Shahrekord, Iran, smhatefi@alumni.ut.ac.ir

[2] Civil Engineering Faculty, Vilnius Gediminas Technical University, Saulėtekio al. 11, LT 2040 Vilnius, Lithuania; jolanta.tamosaitiene@vgtu.lt

* Correspondence: jolanta.tamosaitiene@vgtu.lt

Received: 21 February 2018; Accepted: 23 March 2018; Published: 27 March 2018

Abstract: Due to the increasing population and earth pollution, managing construction and infrastructure projects with less damage to the environment and less pollution is very important. Sustainable development aims at reducing damage to the environment, making projects economical, and increasing comfort and social justice. This study proposes fuzzy analytic hierarchy process (AHP) and improved grey relational analysis (GRA) to assess construction projects based on the sustainable development criteria. For doing so, sustainable development criteria are first identified in economic, social, and environmental dimensions using literature review, and are then customized for urban construction projects using experts' opinions. After designing questionnaires and collecting data, fuzzy AHP is used for determining the importance of sustainable development criteria and their subcriteria. Then, improved GRA is employed for assessing six recreational, commercial, and official centers in Isfahan regarding the weights of criteria and subcriteria. The proposed fuzzy AHP-improved GRA help us to prioritize construction projects based on the sustainable development criteria. The results of applying fuzzy AHP show that the weights of economic, social, and environmental criteria are equal to 0.330, 0.321, and 0.349, respectively, which are close to each other. This means that the importance of all three aspects of sustainability is almost equal to each other. Furthermore, "Having profits for the society", "Increasing social justice", and "Adherence to environmental policies" are identified as the most important indicators of sustainable development in terms of economic, social, and environmental aspects, respectively. Finally, the results of employing improved GRA determine Negin Chaharbagh recreational and commercial complex as the best project.

Keywords: sustainability; project evaluation; construction projects; fuzzy theory; fuzzy multi-attribute decision-making methods

1. Introduction

The term "sustainable development" was first introduced in 1980 in the reports of International Union for Conservation of Nature (IUCN). This union used this term in a report named "the strategy for conservation of nature" to describe the condition that development can be useful for nature, not harmful. Sustainable development is, in fact, the balance between development and environment [1] with four aspects of sustainability in nature, politics, society, and economy. Not only does sustainable development focus on the environment, but it also pays attention to social and economical aspects. Sustainable development connects society, economy, and environment [2]. One important international event in this aspect is the international meeting for sustainable development in which participants agreed on points for sustainable development:

(1) Reducing, by half, the number of people who do not have access to water until 2015,
(2) Minimizing chemical materials harmful for human health and nature until 2000,
(3) Reducing, by half, the speed of decrease in maritime stores and bringing maritime resources to a sustainable level until 2015,
(4) Reducing the trend of nature destruction until 2010,
(5) Increasing sustainability in using renewable energy,
(6) Plans for preparing a ten year program about sustainable development.

The reasons behind the importance and the need for studying and reviewing sustainable development are air pollution (carbon dioxide and other gases) and its negative effects, greenhouse gases, ozone destruction, and acid rain, which result in incurable diseases [3]. From many decades ago, some documents read the need for economical and environmental policies at the global level, and predicted an increase in the above-mentioned problems. If there is an increase in population rate, air pollution, and fossil resources exploitation in industrial countries, these will cause a sudden and uncontrollable decrease in resources in the next years, and will change environmental and economical sustainability in a way that leads the international balance to severe shortages and the loss of opportunities for realizing human potentials [4,5]. Sustainability has penetrated many areas, especially the construction industry, due to its significant environmental impact. The green building is one of the most important issues of sustainable development. There are many definitions of green construction, in which the common elements of all definitions include the perspective of a life cycle, sustainable environment, health issues, and its effects on society. In simple terms, green building can be considered as building on the basis of environmental considerations, which seeks to reduce pollution during construction, operation, and destruction, such as noise pollution, water, dust, and greenhouse gases, as well as the use of renewable energy and minimizing non-renewable energy. Therefore, sustainability in the construction industry has important implications for economic and environmental sustainability. With the rapid growth of this industry in the coming years, and the need for materials as well as waste generation, the use of sustainable materials and technologies in this industry will be a major contributor to sustainable development. Regarding these issues, the evaluation of construction projects based on the dimensions of sustainable development is very essential [6]. In an article named "designing sustainable construction", it is stated that extensive and multi-fields efforts in mechanics, electronic and electrical devices, communication, acoustics, architecture, and structural engineering are made all over the world for designing sustainable constructions. This is cooperation among owners, suppliers, contractors, and users, and extensive studies have been done on sustainable constructions since many years ago. The results show that saving energy and water and making constructions compatible with the environment significantly prevents carbon dioxide emission in the atmosphere. This study presents strategies for designing sustainable constructions, using the integration of smart structure technology and a combination of semi-active vibration control that lead into lighter and more efficient constructions [6]. In one study, authors identified key criteria of construction, and using fuzzy analytic hierarchy process assessed smart constructions. This study introduced key criteria related to smart sustainable constructions in the domains of environment, society, economy, and technical factors. Then, based on the group analytic hierarchy process, assessed the sustainability of smart constructions [7]. In one study, global warming was referred to as a big challenge for human survival. Carbon dioxide density increase helps global warming. Due to increase in constructions and change in climate, end to fossil fuels, restriction in area and resources, geothermal energy can be used for sustainable development. This study presents methods for employing such energy in infrastructure construction, based on the history and classification of criteria of thermal energy performance which is presented in the form of different designs, applications, setting, and climate table [8]. There are different studies on assessing construction projects in the literature review. Mousavi et al. [9] used a model based on fuzzy logic and fuzzy TOPSIS for determining the performance of construction projects. Functional criteria for assessing projects in this study are operational criteria, financial criteria, technological criteria, and criteria related to state requirements. In another study, Vahdani et al. [10] presented a hybrid model based on support vector machine for assessing and selecting the best construction project. The study utilized technological criteria, budget amount, state requirements, and resource restrictions for determining the performance of construction projects. Some other studies utilized risk factors for assessing construction projects [11,12]. Tailan et al. [13] utilized five risk criteria and evaluated 30 construction projects based on fuzzy analytic hierarchy process model and fuzzy TOPSIS model. Time risk, expense risk, quality risk, and risk related to environmental sustainability were utilized for assessing construction projects. Islam et al. [14] used risk factors to assess construction projects based on fuzzy multi-criteria decision-making methods. One recent study has assessed a tunneling project based on five risk factors namely time risk, expense risk, quality risk, safety risk, and environment sustainability risk [15,16]. According to the literature review, most studies utilize risk factors for assessing construction projects, while this study uses sustainable development criteria for construction projects evaluation. Regarding the method of construction in our country and utilizing modern construction materials, this industry suffers from different problems among which inadequate resistance against earthquake, short life of the building, erosion due to climate, excessive waste construction while execution, inefficiency of recycling construction materials, and etc., and utilizing unskilled labor force result in unendurable buildings and structures. Moreover, no attention is paid to environment in the domains of destruction, co-existence, and utilization of recycled materials. Due to the importance of this issue, construction projects are assessed via fuzzy analytic hierarchy process and improved grey relational analysis based [17] on sustainable development criteria.

2. The Proposed Model

This study utilized fuzzy analytic hierarchy process and improved grey relational analysis for assessing six commercial, tourism, and recreational projects in Isfahan megacity. Isfahan is a historical city at the center of Iran. Its center is Isfahan and is the 3rd most-populated city after Tehran and Mashhad, and known as "Nesfe Jahan" in Iranian culture. It is famous for its Islamic architecture and beautiful boulevards, covered bridges, beautiful tunnels, palaces, mosques, and unique minarets (Figure 1).

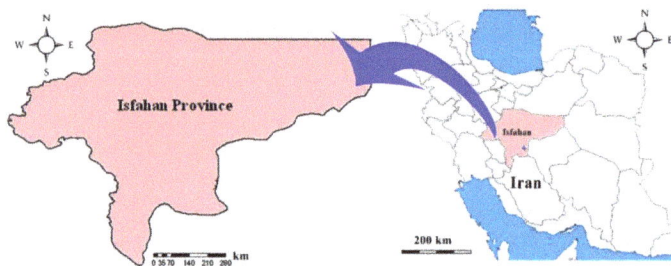

Figure 1. The study area.

The proposed model of fuzzy AHP and improved GRA for assessing construction projects in Isfahan megacity include the steps shown graphically in Figure 2. In the proposed model, criteria of sustainable development are identified in three aspects of economy, society, and environment using literature and experts' ideas. Then two questionnaires related to fuzzy AHP and improved GRA were designed. After distributing the questionnaires and collecting experts' ideas, fuzzy AHP was used for determining the weights of sustainable development criteria, and finally, improved GRA is applied for determining the final weights and ranks of construction projects in Isfahan megacity.

2.1. Fuzzy AHP

Fuzzy AHP, which was first introduced by Tomas Saaty in 1980 [18], is one of the most comprehensive systems for multi-criteria decision-making. This technique makes possible formulating the problem hierarchically, and can consider different quantity and quality criteria in the problem. This process integrates different options in decision-making and can analyze the sensitivity of criteria and subcriteria. Pairwise comparisons facilitate judgements and computations. It shows the compatibility and incompatibility of the decisions which is a benefit of multi-criteria decision-making. It contains a strong theoretical basis and is based on axiomatic principles [19]. Although experts utilize their abilities and aptitude for doing comparisons in AHP, we should consider that traditional AHP cannot completely reflect human thoughts. Fuzzy numbers are more compatible with phrases and ambiguities, so, it is better to utilize them in decisions in the real world. The following part shows the steps of implementing fuzzy AHP introduced by Chang [20].

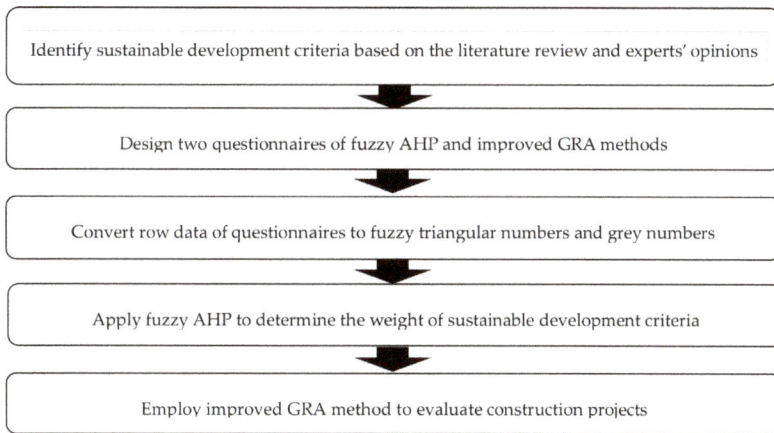

Figure 2. The proposed model.

2.1.1. Defining Relevant FUZZY Numbers for Determining the Importance of Criteria

At first, the convenient fuzzy numbers were defined for doing pairwise comparisons. Table 1 is used for preprocessing data and converting row data of questionnaires to corresponding fuzzy numbers.

Table 1. Linguistic terms and the corresponding triangular fuzzy numbers [21].

Linguistic Term	Fuzzy Number	Reverse Fuzzy Number
Equally Preferred	(1, 1, 1)	(1, 1, 1)
Moderately Preferred	(2, 3, 4)	$\left(\frac{1}{4}, \frac{1}{3}, \frac{1}{2}\right)$
Strongly Preferred	(4, 5, 6)	$\left(\frac{1}{6}, \frac{1}{5}, \frac{1}{4}\right)$
Very Strongly Preferred	(6, 7, 8)	$\left(\frac{1}{8}, \frac{1}{7}, \frac{1}{6}\right)$
Extremely Preferred	(9, 9, 9)	$\left(\frac{1}{9}, \frac{1}{9}, \frac{1}{9}\right)$

2.1.2. Forming Fuzzy Pairwise Comparison Matrix

In this stage, the matrix for fuzzy pairwise comparison matrix is provided. If there is more than one expert, geometric average is used for aggregating experts' opinions. In this stage, the aggregated fuzzy decision matrix is constructed based on the geometric average of fuzzy decision matrixes obtained by all experts.

2.1.3. Calculating the Relative Weight for Each Row of Fuzzy Pairwise Comparison Matrix ($\tilde{S}_i$)

If fuzzy numbers are triangular, they are shown as (l_i, m_i, u_i), then $\tilde{S}_i$ is calculated in the following way.

$$\tilde{S}_i = \sum_{j=1}^{m} M_{gi}^j \times \left[\sum_{i=1}^{n} \sum_{j=1}^{m} M_{gi}^j \right]^{-1} \tag{1}$$

$$\sum_{j=1}^{m} M_{gi}^j = \left(\sum_{j=1}^{m} l_j, \sum_{j=1}^{m} m_j, \sum_{j=1}^{m} u_j \right) \tag{2}$$

$$\sum_{i=1}^{n} \sum_{j=1}^{m} M_{gi}^j = \left(\sum_{i=1}^{n} l_i, \sum_{i=1}^{n} m_i, \sum_{i=1}^{n} u_i \right) \tag{3}$$

$$\left[\sum_{i=1}^{n} \sum_{j=1}^{m} M_{gi}^j \right]^{-1} = \left(\frac{1}{\sum_{i=1}^{n} u_i}, \frac{1}{\sum_{i=1}^{n} m_i}, \frac{1}{\sum_{i=1}^{n} l_i} \right) \tag{4}$$

2.1.4. Calculating the Bigness Degree of $\tilde{S}_i$

If $\tilde{M}_1 = (l_1, m_1, u_1)$ and $\tilde{M}_2 = (l_2, m_2, u_2)$ are two triangular fuzzy numbers, then the degree of possibility of $\tilde{M}_2 \geq \tilde{M}_1$ is calculated as

$$V(M_2 > M_1) = hgr(M_2 > M_1) = \mu_{M_2}(d) = \begin{cases} 1 & if\ m_2 \geq m_1 \\ 0 & if\ l_1 \geq u_2 \\ \frac{l_1 - u_2}{(m_2 - u_2) - (m_1 - l_1)} & otherwise \end{cases} \tag{5}$$

2.1.5. Calculating the Weights by the Possibility Degree

Possibility degree of a convex fuzzy number is more than that of k convex fuzzy numbers can be calculated according to the following formulation. This concept can be stated in the following way.

$$V(\tilde{M} > \tilde{M}_1, \tilde{M}_2, \ldots, \tilde{M}_k) = V(\tilde{M} > \tilde{M}_1) and\ V(\tilde{M} > \tilde{M}_2) \ldots and\ V(\tilde{M} > \tilde{M}_k)$$
$$= \min V(\tilde{M} > \tilde{M}_k),\ i = 1, 2, \ldots, k \tag{6}$$

Based on the above concept, the weight vector is calculated as follows:

$$d_i'(A_i) = \min V(S_i > S_k),\ k = 1, 2, \ldots, n,\ k \neq i$$
$$W' = (d'(A_1), d'(A_2), \ldots, d'(A_n))^T \tag{7}$$

2.1.6. Normalizing the Weight Vector

After normalizing W' vector, the normalized weight vector (W) whose elements are non-fuzzy numbers is defined as

$$W = (d(A_1), d(A_2), \ldots, d(A_n))^T \tag{8}$$

2.2. Improved Grey Relational Analysis

Deng [22–24] presented the first article on grey system theory named controlling grey systems, and then introduced grey system theory. Generally, its idea is that by concentrating on minor or restricted details about a decision, a general picture of it will be imagined. So, this method deals with defective, weak, and uncertain issues. This method can have satisfactory results with a little information and a lot of change in criteria. Grey theory refers to an effective mathematical model for solving uncertain and vague problems by combining with a MCDM method [25,26]. When the units are assessed by different criteria, the effects of some will be ignored. This happens especially when performance criteria are large in their amount, and when the aims and instructions of these criteria are different; then the results of analysis will be incorrect. Suppose that $\otimes G = [\underline{G}, \overline{G}]$ is a grey number, with $\underline{G}$ as the lower bound and $\overline{G}$ as the upper bound of the grey number. Furthermore, suppose that we have a decision-making matrix with all its entries as grey numbers with m units of decisions (alternatives) and n criteria shown in the following way:

$$\otimes G = \begin{bmatrix} \otimes G_{11} & \otimes G_{12} & \cdots & \otimes G_{1n} \\ \otimes G_{21} & \otimes G_{22} & \cdots & \otimes G_{2n} \\ \vdots & \vdots & \ddots & \vdots \\ \otimes G_{m1} & \otimes G_{m2} & \cdots & \otimes G_{mn} \end{bmatrix} \tag{9}$$

In the above matrix, $\otimes G_{ij}$ is the j-th criterion for the i-th choice. The algorithm of improved grey relational analysis is introduced by Hashemi et al. [17], and has the following steps.

2.2.1. Preparing a Decision Matrix and Normalized Decision Matrix

Interval grey numbers are used for assessing alternatives based on criteria. For doing so, a questionnaire is designed and sent to respondents. Then, after gathering row data of questionnaires, they are converted to the corresponding interval grey numbers according to Table 2.

Table 2. Linguistic terms and correspondent interval grey numbers.

	Very Low	Low	Medium	High	High
$\otimes G$	$[0, 1]$	$[1, 4]$	$[4, 6]$	$[6, 9]$	$[9, 10]$

After preparing the decision matrix, a normalized decision matrix is prepared based on the following relations.

$$\otimes y_{ij} = \frac{\otimes G_{ij}}{Max\{\overline{G}_{ij},\ i = 1,2,\ldots,m\}} \quad for\ i = 1,2,\ldots,m,\ j = 1,2,\ldots,n \tag{10}$$

$$\otimes y_{ij} = \frac{Min\{\underline{G}_{ij},\ i = 1,2,\ldots,m\}}{\otimes G_{ij}} \quad for\ i = 1,2,\ldots,m,\ j = 1,2,\ldots,n \tag{11}$$

Relation (10) is used for benefit criteria, and Relation (11) is used for cost criteria. The benefit and cost criteria are identified based on the experts' ideas and the literature review.

2.2.2. Defining Reference Alternative

Reference alternative are taken from normalized decision matrix according to the following equation:

$$y^0 = \{y_1^0,\ y_2^0,\ \ldots,\ y_n^0\}$$
$$\otimes y_j^0 = \left(max_{i=1}^m \underline{y}_{ij},\ max_{i=1}^m \overline{y}_{ij}\right)\ for\ j = 1,2,\ldots,n \tag{12}$$

In the above relation, y_j^0 is the reference amount for the j-th criterion and y_{ij} are the entries of the normalized matrix.

2.2.3. Calculating the Difference between Alternatives and Reference Alternative

In this stage, the difference between each alternative and reference alternative is obtained, and then the difference matrix is constructed as follows:

$$\otimes \Delta = \begin{bmatrix} \otimes \Delta_{11} & \otimes \Delta_{12} & \cdots & \otimes \Delta_{1n} \\ \otimes \Delta_{21} & \otimes \Delta_{22} & \cdots & \otimes \Delta_{2n} \\ \vdots & \vdots & \ddots & \vdots \\ \otimes \Delta_{m1} & \otimes \Delta_{m2} & \cdots & \otimes \Delta_{mn} \end{bmatrix}$$

(13)

$$\otimes \Delta_{ij} = \left[y_j^0 - \overline{y}_{ij}, y_j^0 - \underline{y}_{ij} \right] \ for \ i = 1, \ldots, m; \ j = 1, \ldots, m$$

2.2.4. Calculating Grey Relational Coefficient

Grey relational coefficient is defined for determining how to make the alternatives closer to the reference alternative. Equation (14) is utilized for calculating grey relational coefficient.

$$\otimes \gamma_{ij} = [\underline{\gamma}_{ij}, \overline{\gamma}_{ij}]$$

$$\underline{\gamma}_{ij} = \frac{\min_{i=1}^m \min_{j=1}^n \Delta_{ij} + \rho \max_{i=1}^m \max_{j=1}^n \overline{\Delta}_{ij}}{\Delta_{ij} + \rho \max_{i=1}^m \max_{j=1}^n \overline{\Delta}_{ij}}$$

$$\overline{\gamma}_{ij} = \frac{\min_{i=1}^m \min_{j=1}^n \Delta_{ij} + \rho \max_{i=1}^m \max_{j=1}^n \overline{\Delta}_{ij}}{\Delta_{ij} + \rho \max_{i=1}^m \max_{j=1}^n \overline{\Delta}_{ij}}$$

(14)

where $\otimes \gamma_{ij}$ is grey relational coefficient ρ is distinguishing coefficient, taking the value of 0.5 in this study.

2.2.5. Calculating Grey Relational Degree for Each Alternative

After calculating grey relational coefficient, grey relational degree for each alternative is calculated in the following way:

$$\otimes \Gamma_i = [\underline{\Gamma}_i, \overline{\Gamma}_i] = \left[\sum_{j=1}^n \underline{\gamma}_{ij} \cdot w_j, \sum_{j=1}^n \overline{\gamma}_{ij} \cdot w_j \right] \ for \ i = 1, 2, \ldots, m$$

(15)

2.2.6. Whitening the Grey Relational Degree for Each Alternative

The grey relational degrees for each alternative can be whitened by using Equation (16). The alternatives are prioritized based on the rule that the bigger the whitened relational degree, the better the corresponding alternative.

$$\Gamma_i = \frac{\underline{\Gamma}_i + \overline{\Gamma}_i}{2}$$

(16)

3. Results and Discussion

3.1. Case Study

Fuzzy AHP and improved GRA are applied to assess six recreational, tourism, and commercial projects in Isfahan megacity based on the dimensions of sustainable development. In this research, only the investment projects that are under construction or ready for exploitation are considered. These projects are diverse, and just recreational and tourism projects are considered. These projects form decision alternatives and are Fadak center recreational, tourism, and commercial complex (P_1), Anoshirvan recreational, commercial, and official complex (P_2), Goldasteh garden recreational, service, and commercial complex (P_3), Negin Chaharbagh recreational and commercial complex (P_4),

Sepahan recreational and commercial complex (P_5), and ShahrRoyaha recreational and commercial complex (P_6). Criteria of sustainable development in three aspects of economy, society, and environment should be first identified for assessing the above projects. Table 3 shows criteria of sustainable development, which are extracted from literature review and experts' opinions.

Table 3. Criteria and subcriteria of sustainable development in construction projects.

Main Criteria	Subcriteria	Notation
Economical (C_1)	Technology promotion after project completion	C_{11}
	Absorbing foreign capital	C_{12}
	Paying attention to society and market needs	C_{13}
	Having profits for the society	C_{14}
	Cost effectiveness/Economic profit	C_{15}
	Positive impact on the region's economy	C_{16}
	Information on risks and financial risks	C_{17}
	Construction and architecture requirement for making the projects economical	C_{18}
Social (C_2)	Welfare and economical growth	C_{21}
	Society development and renovation	C_{22}
	Creating equal job opportunities	C_{23}
	Paying attention to cultural heritage	C_{24}
	Paying attention to the beauty and customs of the area	C_{25}
	Public participation and project control	C_{26}
	No risk to people in the region	C_{27}
	Increasing social justice	C_{28}
Environmental (C_3)	Use of eco-friendly technologies	C_{31}
	Waste management system	C_{32}
	Energy efficient management	C_{33}
	No greenhouse gas emissions	C_{34}
	Protection of plants and animals	C_{35}
	Optical and visual non-pollution	C_{36}
	Preventing nature destruction	C_{37}
	Using thermal, wind, and sun energy	C_{38}
	Non-pollution of surface water and underground water	C_{39}
	Adherence to environmental policies	C_{310}

This study collects the intended information based on library research, questionnaire, field research, and experts' ideas. The statistical population is a set of units that are common in one or more than one attribute. It is obvious that we should take into account the judgements of experts and theorists in the domain of project management, civil engineering (construction management, civil, metro), architecture (sustainable architecture), and urban management and planning, to have an exact and comprehensive response. An attempt is made to identify these experts and connect to them. For this purpose, 20 experts were identified to complete the questionnaires. These experts had enough information about the construction projects being investigated, and met one of the following requirements: having at least one professional degree in project management, having significant compilations in one domain of sustainable development and project management, postgraduate students at industrial engineering, project management, or one of the fields of engineering, and having professional resume in researching and running and managing infrastructure projects. The study reached the following results by extracting data from questionnaires and analyzing them using fuzzy AHP and improved GRA.

3.2. The Results of Applying Fuzzy AHP

Computational results of fuzzy AHP method are as follows:

- Pairwise comparison of main criteria with respect to goal and determining the local weights of main criteria.
- Pairwise comparison of subcriteria with respect to their criteria and determining the local weights of subcriteria.
- Multiplying the local weights of subcriteria to the local weight of the related criterion for determining the final weight of subcriteria

The fuzzy rating reported in Table 1 was used to obtain pairwise comparisons and construct pairwise comparison matrix. Aggregated fuzzy pairwise matrix is calculated based on the geometric mean of the ideas of 20 experts for the main criteria of sustainable development, and is reported in Table 4. For calculating local weights related to main criteria of sustainable development, Equations (1)–(8) should be applied on integrated fuzzy pairwise comparison matrix. For example, fuzzy expansion $\tilde{S}_i$ is calculated for each main criterion. According to Equations (1)–(4), fuzzy expansion of each main criterion is calculated in the following way:

$$\left[\sum_{i=1}^{n} \sum_{j=1}^{m} M_{gi}^j \right]^{-1} = (0.081, 0.092, 0.104)$$

$$\tilde{S}(C_1) = (3.400, 3.630, 3.890) \times (0.081, 0.092, 0.104) = (0.277, 0.333, 0.403)$$

$$\tilde{S}(C_2) = (3.040, 3.560, 4.130) \times (0.081, 0.092, 0.104) = (0.248, 0.326, 0.428)$$

$$\tilde{S}(C_3) = (3.210, 3.720, 4.250) \times (0.081, 0.092, 0.104) = (0.262, 0.341, 0.440)$$

Table 4. Pairwise comparison matrix of main criteria.

	Economical Criterion (C_1)	Social Criterion (C_2)	Environmental Criterion (C_3)
Economical criterion (C_1)	(1, 1, 1)	(0.39, 0.48, 0.61)	(0.33, 0.39, 0.47)
Social criterion (C_2)	(1.64, 2.08, 2.54)	(1, 1, 1)	(0.40, 0.48, 0.59)
Environmental criterion (C_3)	(2.14, 2.56, 3.03)	(1.69, 2.08, 2.48)	(1, 1, 1)

After calculating $\tilde{S}_i$ for main criteria, the bigness degree of $\tilde{S}_i$ is calculated based on Equation (5) for main criteria. The results are reported in Table 5, in which element of row i and column j shows $V(S_i > S_j)$. Finally, the non-normalized and normalized weights of main criteria can be extracted from Equations (6)–(8). The normalized and non-normalized weights of the main criteria are as follows:

$$W' = (0.945, 0.919, 1.000)^T$$
$$W = (0.330, 0.321, 0.349)^T$$

Table 5. The bigness degree of $\tilde{S}_i$ in relation to other criteria $V(S_i > S_j)$.

	$j = 1$	$j = 2$	$j = 3$
$i = 1$	1	1	0.945
$i = 2$	0.959	1	0.919
$i = 3$	1	1	1

As the results show, environmental dimension with the weight of 0.349 has the most importance among the sustainable development dimensions in construction projects in Isfahan. The second and the third ranks belong to economic and social themes, respectively. Similarly, local weights of subcriteria of sustainable development can be calculated. Table 6 shows the local weights of sustainable development criteria, local weights of subcriteria, and the final weights of subcriteria. It should be mentioned that final weights of each subcriterion are obtained by multiplying the local weight of the respective main criterion to the local weight of that subcriterion. Based on the results reported in Table 6, it can be concluded that among subcriteria of sustainable development in construction projects, "adherence to environmental policies" (C_{310}), "having profits for the society" (C_{14}), and "paying attention to society and market needs" (C_{13}) are the most important subcriteria.

Table 6. The weights of criteria and subcriteria of sustainable development in infrastructure projects.

Main Criteria	Local Weight	Subcriteria	Local Weight	Final Weight
C_1	0.330	C_{11}	0.065	0.021
		C_{12}	0.057	0.019
		C_{13}	0.172	0.057
		C_{14}	0.174	0.057
		C_{15}	0.126	0.042
		C_{16}	0.147	0.049
		C_{17}	0.131	0.043
		C_{18}	0.129	0.043
C_2	0.321	C_{21}	0.114	0.037
		C^{22}	0.105	0.034
		C_{23}	0.091	0.029
		C_{24}	0.131	0.042
		C_{25}	0.089	0.029
		C_{26}	0.095	0.030
		C_{27}	0.111	0.036
		C_{28}	0.164	0.053
C_3	0.349	C_{31}	0.093	0.032
		C_{32}	0.057	0.020
		C_{33}	0.070	0.024
		C_{34}	0.089	0.031
		C_{35}	0.095	0.033
		C_{36}	0.133	0.046
		C_{37}	0.093	0.032
		C_{38}	0.107	0.037
		C_{39}	0.118	0.041
		C_{310}	0.166	0.058

3.3. The Results of Applying Improved GRA

For implementing the improved GRA, the ideas of 20 experts about assessing alternatives with respect to subcriteria will be collected based on a questionnaire. To do this, experts state their ideas about satisfying each subcriterion in each alternative based on linguistic terms reported in Table 2. After collecting the row data in the form of linguistic terms, they are converted to grey numbers according to Table 2. Then, the grey relational decision matrix is obtained based on the arithmetic mean of grey numbers taken from 20 experts' opinions. Table 7 shows grey decision matrix in which elements are in the form of interval grey numbers, and present the scores of alternatives with respect to subcriteria.

Table 7. Grey relational decision matrix.

	P_1	P_2	P_3	P_4	P_5	P_6
C_{11}	[5.35, 7.35]	[4.88, 6.88]	[5.61, 6.78]	[5.77, 7.19]	[3.94, 5.17]	[4.23, 5.28]
C_{12}	[3.24, 5.24]	[4.97, 6.55]	[1.64, 2.75]	[6.01, 7.60]	[6.41, 7.94]	[4.79, 6.18]
C_{13}	[3.69,5.18]	[5.41, 7.02]	[3.79, 5.04]	[6.02, 7.41]	[1.19, 2.83]	[3.76, 5.19]
C_{14}	[5.66,6.67]	[5.52, 6.58]	[5.06, 6.13]	[6.54, 8.43]	[5.75, 6.96]	[4.38, 5.93]
C_{15}	[1.76,3.21]	[5.03, 6.36]	[5.06, 6.14]	[6.39, 7.46]	[1.96, 3.50]	[4.15, 6.08]
C_{16}	[4.76,6.17]	[5.84, 6.84]	[4.79, 6.60]	[5.56, 7.41]	[1.70, 3.06]	[4.51, 6.34]
C_{17}	[2.65,4.01]	[4.78, 6.67]	[5.09, 6.91]	[6.94, 7.97]	[5.21, 6.59]	[3.85, 5.64]
C_{18}	[1.16,2.53]	[5.94, 7.38]	[1.69, 3.21]	[6.52, 8.05]	[6.30, 7.37]	[5.18, 6.32]
C_{21}	[2.73,4.22]	[6.20, 7.38]	[6.26, 7.61]	[6.80, 8.16]	[1.71, 2.92]	[5.13, 6.16]
C_{22}	[5.62,6.64]	[5.37, 7.11]	[1.11, 2.41]	[5.35, 7.33]	[1.27, 2.87]	[4.12, 5.67]
C_{23}	[2.19,3.48]	[5.11, 6.65]	[5.83, 7.09]	[5.94, 7.30]	[0.76, 2.52]	[4.61, 6.35]
C_{24}	[2.17,3.26]	[5.30, 6.55]	[0.91, 2.51]	[6.29, 7.60]	[2.12, 3.21]	[3.49, 5.00]
C_{25}	[6.18,7.65]	[5.23, 6.30]	[2.71, 4.44]	[6.28, 7.68]	[5.21, 6.82]	[4.70, 5.96]
C_{26}	[6.38,7.40]	[4.24, 6.24]	[4.85, 6.16]	[5.66, 7.14]	[4.78, 6.54]	[4.10, 5.23]
C_{27}	[6.46,7.93]	[5.95, 7.24]	[3.86, 5.38]	[6.35, 8.02]	[2.65, 4.37]	[3.71, 5.40]
C_{28}	[4.54,5.85]	[5.23, 6.51]	[5.09, 6.20]	[6.34, 8.31]	[2.57, 3.89]	[5.22, 6.33]
C_{31}	[3.61,4.84]	[5.14, 6.74]	[0.87, 2.65]	[6.06, 7.69]	[1.13, 2.58]	[3.99, 5.35]
C_{32}	[6.84,7.87]	[4.47, 6.25]	[2.33, 4.16]	[5.68, 7.13]	[1.17, 2.89]	[4.41, 6.05]
C_{33}	[5.81,7.07]	[5.01, 6.26]	[3.70, 5.57]	[5.96, 7.05]	[4.91, 6.36]	[4.11, 5.51]
C_{34}	[3.95, 5.29]	[4.94, 6.25]	[2.36, 3.41]	[6.11, 7.64]	[2.25, 3.90]	[4.39, 6.15]
C_{35}	[4.63, 6.40]	[4.97, 6.67]	[5.82, 7.80]	[7.05, 8.28]	[2.51, 3.67]	[4.34, 5.68]
C_{36}	[4.28, 5.30]	[4.54, 6.38]	[0.74, 2.03]	[6.38, 8.33]	[3.78, 5.62]	[3.93, 5.71]
C_{37}	[6.68, 8.43]	[5.06, 6.32]	[4.79, 6.10]	[6.87, 8.32]	[1.80, 2.84]	[4.84, 6.37]
C_{38}	[5.49, 7.44]	[4.11, 6.03]	[0.17, 2.08]	[5.47, 7.20]	[1.07, 2.73]	[4.72, 6.12]
C_{39}	[6.19, 7.58]	[4.47, 6.23]	[2.72, 4.26]	[5.85, 7.36]	[2.80, 4.31]	[4.23, 5.65]
C_{310}	[2.17, 3.95]	[4.68, 6.35]	[3.75, 4.97]	[7.15, 8.41]	[2.73, 4.61]	[5.17, 6.21]

In the next step of applying improved GRA, the grey relational decision matrix is normalized by using Equations (10) and (11). Then, reference alternative is determined by Equation (12) and its distance with the alternatives is calculated using Equation (12). Grey relational coefficient is obtained using Equation (14). Equation (15) is utilized to provide the grey relational degree for each alternative, which is reported in the second column of Table 8. It should be mentioned that in Equation (15), w_j shows the weight of j-th subcriterion, which is the result of applying fuzzy AHP. Finally, the grey relational degree for each alternative is whitened by using Equation (16). The whitened relational degree is reported in the third column of Table 8. The last column of Table 8 shows the priority of construction projects. As the results show, Negin Chaharbagh recreational and commercial complex (P_4) is the best project based on sustainable development dimensions. The grey relational degree for this project is [0.190, 0.279], and its whitened grey relational degree is equal to 0.234, which is the largest value among whitened degree values. After that, Anooshirvan recreational, commercial, and official complex (P_2) and Fadak center recreational, tourism, and commercial complex (P_1) have the largest values of the whitened relational degree. Therefore, they have the second and third ranks based on sustainable development concept among six construction projects, respectively.

Table 8. The results of improved grey relational analysis for assessing construction projects.

Construction Project	Grey Relational Coefficient ($\otimes\Gamma_i$)	Whitened Grey Relational Degree (Γ_i)	Rank
P_1	[0.160, 0.220]	0.190	3
P_2	[0.170, 0.237]	0.203	2
P_3	[0.150, 0.199]	0.174	5
P_4	[0.190, 0.279]	0.234	1
P_5	[0.142, 0.187]	0.165	6
P_6	[0.158, 0.214]	0.186	4

3.4. Summary of Findings

In this paper, 26 indicators of sustainability in the economic, environmental, and social fields were utilized to assess six construction projects in Isfahan, Iran. The importance and weight of each aspect of sustainability criteria was calculated using the fuzzy AHP method. The weight of economic, social, and environmental criteria has been equal to 0.330, 0.321, and 0.349, respectively. These results indicate that environmental criteria are more important than the economic and social criteria. However, the results are very close to each other, which indicate that the importance of all three aspects of sustainability is almost equal to each other. In summary, there are two methods for demonstrating the effects and relations among aspects of sustainability. In the first method, as shown in Figure 3, there is an overlap among the aspects of sustainability. In the second method, depicted in Figure 4, the economic dimension is located within the social dimension. The social dimension is also embedded in the environmental dimension. Figure 4 focuses on the central role of environment that plays in human society, and in turn, in the economy.

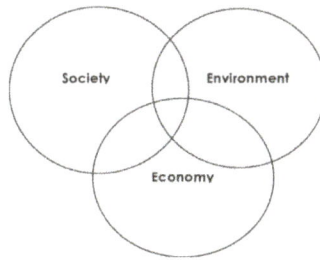

Figure 3. Overlapping aspects of sustainable development.

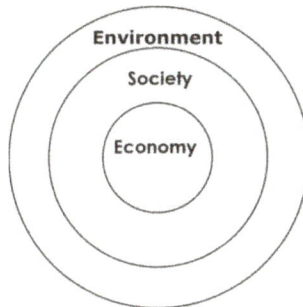

Figure 4. Concentric circles visualization of sustainable development.

According to the weight of subcriteria, "adherence to environmental policies" (C_{310}), "having profits for the society" (C_{14}), and "paying attention to society and market needs" (C_{13}) are the most important sustainability subcriteria. Furthermore, "having profits for the society" (C_{14}), "Increasing social justice" (C_{28}), and "adherence to environmental policies" (C_{310}) with the weight scores of 0.057, 0.053, and 0.058, are the most important criteria in the economic, social, and environmental aspects, respectively. The results of improved GRA are applied to prioritize the construction projects based on their whitened grey relational degrees. According to these results, the priority of construction projects is as: $P_4 > P_2 > P_1 > P_6 > P_3 > P_5$.

4. Conclusions

Due to the importance of sustainable development in construction, the study assesses construction projects based on sustainable development indices using fuzzy AHP and improved GRA. For doing so, 8 subcriteria in the economical dimension, 8 subcriteria in the social dimension, and 10 subcriteria in the environmental dimension were extracted from literature review, and customized by experts' opinions. After that, two questionnaires related to fuzzy AHP and improved GRA were designed in order to collect data in the form of linguistic terms. After distributing questionnaires and collecting data, linguistic terms were converted to the corresponding fuzzy and grey ratings. Fuzzy AHP was used for determining the importance of criteria of sustainable development and their subcriteria. The results of applying fuzzy AHP shows that among sustainable development dimensions, environmental criteria are more important than other dimensions. Also, among sustainable development subcriteria, "adherence to environmental policies", "having profits for the society", and "paying attention to society and market needs" are the most important. Finally, considering the weights acquired through fuzzy AHP method, improved GRA was used for assessing and prioritizing six construction projects in Isfahan based on sustainable development indices. The results of employing this method revealed that the best construction project regarding the sustainable development indices is Negin ChaharBagh recreational and commercial complex.

Author Contributions: The individual contribution and responsibilities of the authors were as follows: Seyed Morteza Hatefi and Jolanta Tamošaitienė—Developing a new quantitative; Seyed Morteza Hatefi and Jolanta Tamošaitienė-Providing an appropriate method; Seyed Morteza Hatefi—Implementation of the proposed model in a real case study; Seyed Morteza Hatefi—Designed the research, methodology, performed the development of the paper; Seyed Morteza Hatefi—Collected and analyzed the data and the obtained results; Jolanta Tamošaitienė—Provided extensive advice throughout the study; Jolanta Tamošaitienė—Regarding the research design, revised the manuscript; Seyed Morteza Hatefi and Jolanta Tamošaitienė—Methodology, findings. All the authors have read and approved the final manuscript.

Conflicts of Interest: The authors declare no conflict of interest.

References

1. Tabassi, A.A.; Roufechaei, K.M.; Ramli, M.; Bakar, A.H.A.; Ismail, R.; Pakir, A.H.K. Leadership competences of sustainable construction project managers. *J. Clean. Prod.* **2016**, *124*, 339–349. [CrossRef]
2. Hatefi, S.M. Strategic planning of urban transportation system based on sustainable development dimensions using an integrated SWOT and fuzzy COPRAS approach. *Glob. J. Environ. Sci. Manag.* **2018**, *4*, 99–112.
3. Mohammadizadeh, M.J.; Karbassi, A.R.; Nabi Bidhendi, G.R.; Abbaspour, M. Integrated environmental management model of air pollution control by hybrid model of DPSIR and FAHP. *Glob. J. Environ. Sci. Manag.* **2016**, *2*, 381–388.
4. Haghshenas, H.; Vaziri, M.; Gholamialam, A. Evaluation of sustainable policy in urban transportation using system dynamics and world cities data: A case study in Isfahan. *Cities* **2015**, *45*, 104–115. [CrossRef]
5. Pazouki, M.; Jozi, S.A.; Ziari, Y.A. Strategic management in urban environment using SWOT and QSPM model. *Glob. J. Environ. Sci. Manag.* **2017**, *3*, 207–216.
6. Wang, N.; Adeli, H. Sustainable building design. *J. Civ. Eng. Manag.* **2014**, *20*, 1–10. [CrossRef]
7. Alwaer, H.; Clements-Croome, D.J. Key performance indicators (KPIs) and priority setting in using the multi-attribute approach for assessing sustainable intelligent buildings. *Build. Environ.* **2010**, *45*, 799–807. [CrossRef]
8. Alkaff, S.A.; Sim, S.C.; Ervina Efzan, M.N. A review of underground building towards thermal energy efficiency and sustainable development. *Renew. Sustain. Energy Rev.* **2016**, *60*, 692–713. [CrossRef]
9. Mousavi, S.M.; Gitinavard, H.; Vahdani, B. Evaluating construction projects by a new group decision-making model based on intuitionistic fuzzy logic concepts. *Int. J. Eng.* **2015**, *28*, 1312–1319.
10. Vahdani, B.; Mousavi, S.M.; Hashemi, H.; Mousakhani, M.; Ebrahimnejad, S. A new hybrid model based on least squares support vector machine for project selection problem in construction industry. *Arab. J. Sci. Eng.* **2014**, *39*, 4301–4314. [CrossRef]

11. Zavadskas, E.K.; Turskis, Z.; Tamošaitiene, J. Risk assessment of construction projects. *J. Civ. Eng. Manag.* **2010**, *16*, 33–46. [CrossRef]

12. Chatterjee, K.; Zavadskas, E.K.; Tamošaitienė, J.; Adhikary, K.; Kar, S. A hybrid MCDM technique for risk management in construction projects. *Symmetry* **2018**, *10*, 46. [CrossRef]

13. Taylan, O.; Bafail, A.O.; Abdulaal, R.M.S.; Kabli, M.R. Construction projects selection and risk assessment by fuzzy AHP and fuzzy TOPSIS methodologies. *Appl. Soft Comput.* **2014**, *17*, 105–116. [CrossRef]

14. Islam, M.S.; Nepal, M.P.; Skitmore, M.; Attarzadeh, M. Current research trends and application areas of fuzzy and hybrid methods to the risk assessment of construction projects. *Adv. Eng. Inform.* **2017**, *33*, 112–131. [CrossRef]

15. Yazdani-Chamzini, A. Proposing a new methodology based on fuzzy logic for tunnelling risk assessment. *J. Civ. Eng. Manag.* **2014**, *20*, 82–94. [CrossRef]

16. Iqbal, S.; Choudhry, R.; Holschemacher, K.; Ali, A.; Tamošaitienė, J. Risk management in construction projects. *Technol. Econ. Dev. Econ.* **2015**, *21*, 65–78. [CrossRef]

17. Hashemi, S.H.; Karimi, A.; Tavana, M. An integrated green supplier selection approach with analytic network process and improved Grey relational analysis. *Int. J. Prod. Econ.* **2015**, *159*, 178–191. [CrossRef]

18. Saaty, T. *Analytic Hierarchy Process: Planning, Priority Setting*; RWS: Chalfont St Peter, UK, 1980.

19. Saaty, T. *Theory and Applications of the Analytic Network Process: Decision Making With Benefits, Opportunities, Costs, and Risks*; RWS Publications: Chalfont St Peter, UK, 2005.

20. Chang, D.Y. Applications of the extent analysis method on fuzzy AHP. *Eur. J. Oper. Res.* **1996**, *95*, 649–655. [CrossRef]

21. Kahraman, C. Fuzzy Multi-Criteria Decision Making: Theory and Applications with Recent Developments. Available online: http://www.springer.com/gp/book/9780387768120 (accessed on 20 March 2018).

22. Deng, J.L. Control problems of grey system. *Syst. Control Lett.* **1982**, *1*, 288–294.

23. Deng, J.L. Introduction to Grey system theory. *J. Grey Syst.* **1988**, *1*, 1–24.

24. Deng, J.L. Properties of relational space for grey systems. In *Essential Topics on Grey System—Theory and Applications*; China Ocean Press: Beijing, China, 1988; pp. 1–13.

25. Zavadskas, E.K.; Kaklauskas, A.; Turskis, Z.; Tamošaitienė, J. Selection of the effective dwelling house walls by applying attributes values determined at intervals. *J. Civ. Eng. Manag.* **2008**, *14*, 85–93. [CrossRef]

26. Zavadskas, E.K.; Kaklauskas, A.; Turskis, Z.; Tamošaitienė, J. Multi-attribute decision-making model by applying grey numbers. *Informatica* **2009**, *20*, 305–320.

sustainability

MDPI

Case Report

The Accelerated Window Work Method Using Vertical Formwork for Tall Residential Building Construction

Taehoon Kim [1], Hyunsu Lim [2,*], Chang-Won Kim [2], Dongmin Lee [1], Hunhee Cho [1] and Kyung-In Kang [1]

[1] School of Civil, Environmental and Architectural Engineering, Korea University, Anam-Dong, Seongbuk-Gu, Seoul 02841, Korea; kimth0930@hanmail.net (T.K.); ldm1230@korea.ac.kr (D.L.); hhcho@korea.ac.kr (H.C.); kikang@korea.ac.kr (K.-I.K.)

[2] BK21 Innovative Leaders for Creating Future Value in Civil Engineering, Korea University, Anam-Dong, Seongbuk-Gu, Seoul 02841, Korea; wraith304@korea.ac.kr

* Correspondence: iroze00@korea.ac.kr; Tel.: +82-2-921-5920

Received: 29 January 2018; Accepted: 7 February 2018; Published: 9 February 2018

Abstract: In tall residential building construction, there is a process gap between the window work and the structural work. This process gap extends the total period of the project and increases its cost. In addition, as this process gap increases external exposure to noise and dust, it negatively affects the environment of a site and often causes civil complaints. This paper introduces a new window work process called the accelerated window work (AWW) method, which minimizes the process gap and can reduce construction cost and duration and the number of civil complaints. We provide technical details and management elements of the AWW method with a case study that demonstrates the reductions in construction costs and duration compared with the conventional method. This work contributes to the body of knowledge in window work in tall buildings by introducing and validating a new window work method and process. The proposed method will be useful for practitioners who are under short-term constraints.

Keywords: accelerated window work method; vertical formwork; tall residential building construction; window work

1. Introduction

In tall residential building construction, there is a process gap between the window work and the structural work [1,2]. For tall building structural work, large formwork with several working platforms is generally installed outside the building [3]. These platforms limit the operation height of the construction hoists that lift window components. In addition, because a residential building has many walls and small windows compared with office buildings, there is a relatively large difference in productivity between the structural work and window work, which increases the process gap between these types of work. Because of these spatial constraints and productivity differences, window construction begins after several floors of structural work.

This process gap extends the duration of the project and increases the cost of the entire construction [4]. As the window work is on the critical path of the process, the entire duration of construction is delayed if window work is delayed. In addition, because the construction section between the structural work and the window work is open externally, safety facilities such as safety rails and nets must be installed in this section. These additional safety facilities increase material and labor costs by repetitive installation during the construction. A larger gap between the two processes increases the effect on the duration and cost of construction. In addition, this process gap negatively affects the environment of a site and often causes civil complaints because it increases external exposure to noise and dust. In particular, dust caused by construction-related activities is a major source of particulate matter (PM) in the urban atmosphere [5], which has been associated with higher rates of morbidity and mortality in urban areas [6,7].

To solve this problem, various attempts have been made to reduce this process gap [2,8]. One method is to raise the maximum floor of hoist operation by modifying the platform of the vertical formwork above where the hoist is installed. With this method, window construction can start one or two floors earlier. Another method is to reduce the unit of construction for window work. This method can reduce the gap from the structural work by performing window installation more frequently. However, these methods cannot fundamentally eliminate the gaps.

Previous studies on window works have focused on preventing delays to the period of window work by management methods, or increasing efficiency by automated construction [9–13]. However, these studies have not solved the process gaps. Kang et al. [14] proposed the concept of using an auto climbing system independently to perform external insulation and finishing work. However, this method was costly because of the additional system, which must operate five or more floors below the formwork floor. Lim et al. [15] proposed a window proximity-tracking method to reduce the process gap. However, they proposed only the concept and did not perform economic analysis based on an actual application in a construction site.

This study introduces the accelerated window work (AWW) method using vertical forms for tall residential building construction, which can reduce the vertical constraint between structural work and window work. We describe the advantages of the AWW method compared with conventional construction methods in terms of the construction period and the costs through a case study of a tall residential building. The proposed method will help construction managers to ensure cost-effective alternatives in tall building constructions with limited duration.

2. The Conventional Window Work Method for Tall Residential Building Construction

Figure 1 shows the conventional process of window work and structural work in tall residential building construction.

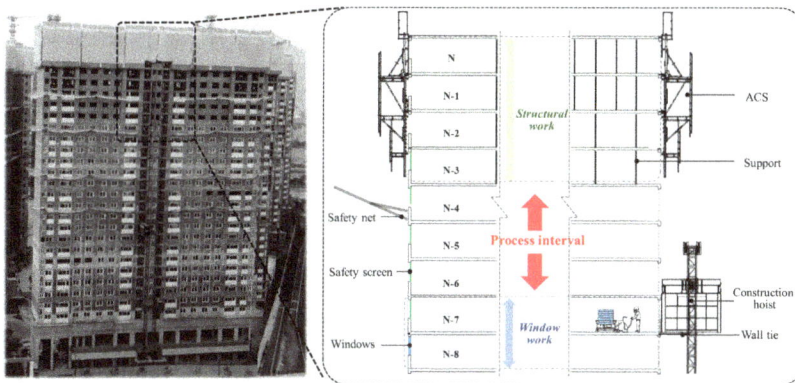

Figure 1. Schematic diagram of conventional processes for the structural work and window work.

From the N-th floor, which is the structural work floor, to the (N − 2)-th floor, curing is performed, and supports are dismantled on the (N − 3)-th floor. The automatic climbing system (ACS) is installed from the N-th floor to the (N − 2)-th floor. From the (N − 4)-th to the (N − 6)-th floor, where windows cannot be lifted into place and installed, temporary safety facilities, such as safety nets and guardrails, are installed to prevent accidents involving falling workers and dropping objects, while dust filter nets or similar resources are installed to prevent air pollution around the site via dust scattering. Windows can be lifted and installed on the (N − 7)-th floor.

The window work in tall residential building constructions is generally performed with an interval of five or more floors. Because the productivity of window work is higher than that of structural work, window work teams only work when there is a sufficient quantity of window work available.

After structural work on all floors is completed, the caulking work for external windows is performed outside by using a suspended working platform or gondola with a wire rope that is secured to the top of the building. This work method potentially involves falls by workers because of the unstable working environment.

As a result, the conventional window work method has a vertical constructability constraint of six or more floors between structural work and window work. This increases total construction duration because of the late start time of the window work and increases costs directly due to the safety facilities required as well as indirectly due to the extended construction period.

3. The AWW Method Using a Vertical Formwork for Tall Residential Building Construction

In the AWW method, window work is carried out on the lower work plates of the ACS while structural work is performed on the upper work plates, and window work is conducted in the same work cycle as structural work (Figure 2). This method removes the vertical constructability constraint between the two processes. It allows for an earlier start of finishing work and reduces the total period of construction by an amount equal to the decrease of the process gap between structure work and window work. In this method, an additional level of work plate is added to the ACS, and window work is performed on this level. As in the traditional method, the structural work is performed from the N-th to the (N − 2)-th floor, and the support is dismantled on the (N − 3)-th floor. Windows are lifted to the (N − 2)-th floor by construction hoists to complete window work on the (N − 2)-th and (N − 3)-th floors. To apply the AWW method, the following core technologies and management efforts are required: (1) a vertically extended ACS, (2) construction hoist operation, and (3) window work cycle management.

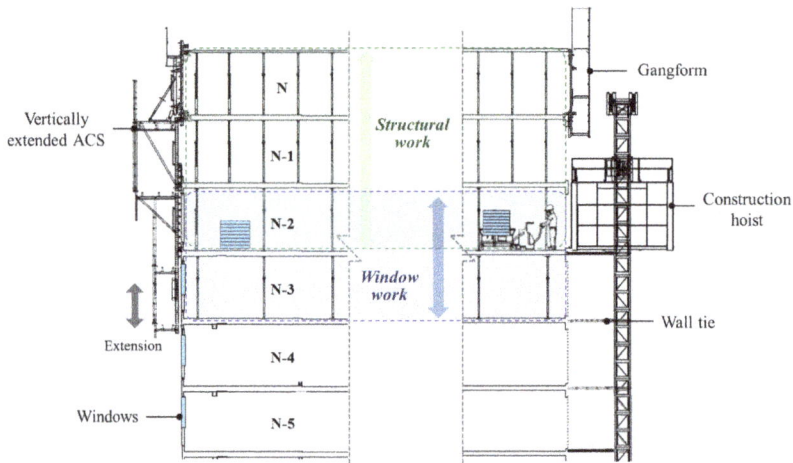

Figure 2. Schematic diagram of the AWW method.

3.1. Vertically Extended ACS

To perform window work in the ACS, the additional work plate should be installed on the bottom of the ACS, which requires some simple modifications. First, the vertically extended ACS requires structural strengthening because of the increase in self-weight and workloads. Second, the height of the work plate between the (N − 2)-th and (N − 3)-th floors is adjusted to allow workers to perform window work outside. Finally, pollution prevention plates are installed on the work plate below the structure work floor, which prevents windows from being polluted by concrete paste from the upper floor.

3.2. Construction Hoist Operation

To lift window materials to the (N − 2)-th floor, some of the construction hoists should be able to access this floor. Some modifications of the construction hoist operation method are required. First, the extension cycle of the hoistway is shortened. In the conventional method, the hoistway is extended at intervals of three floors considering the finishing work cycle. However, in the AWW method, the hoistway should be extended at intervals of one floor to lift windows to each floor. Second, a gangform that has no work plates should be installed above the construction hoist to avoid spatial interference with the construction hoist. The formwork system is lifted by a tower crane. This enables the construction hoist to access the (N − 2)-th floor for window work.

3.3. Window Work Cycle Management

In the AWW method, the window work and the structural work are performed on the same ACS that climbs with the cycle of structural work, typically 4–7 days per floor. The window work on a floor should therefore be completed within the cycle of structural work. The window work crew would work with a lower work quantity but a shorter cycle. In the conventional method, a window work crew comes on site when windows are required for five or more floors; in the AWW method, they must install windows on one floor within the cycle of structural work. This requires some changes in window work cycle management.

First, it is important to ensure the minimum quantity of work for the window work crew. This can be achieved by appropriately zoning the workspace. Second, the construction manager should ensure an optimal stock of windows on each floor so as not to delay the window work. If they are not hoisted in advance and the window work is delayed, the structural work in the critical path would be delayed, which extends the total construction period.

As noted, the AWW method solves the vertical constructability constraint of the conventional method. In the AWW method, the window work is performed on the (N − 3)-th floor, while structural work proceeds on the N-th floor, which allows finishing work to start four or more floors earlier than the conventional method. Moreover, there are no openings that have the potential for fall accidents because the windows are installed in the ACS on every floor. A workspace enclosed by windows improves work safety and prevents noise and dust from being released outside. Furthermore, additional safety facilities and temporary facilities for openings are no longer required, thereby decreasing costs.

4. Case Study

4.1. Case Description

The AWW method was applied to two mixed-use residential buildings: a 36-story building (Building A) and a 33-story building (Building B) located in Gongneung-dong, Seoul, South Korea. The construction period was about three years and one month. Figure 3 shows the layout plan and cross-section of this project. Commercial facilities are located from the first floor to the third floor, and residential facilities from the fourth floor to the top floor. Fire escape floors are located on the 31st and 28th floors.

The ACS used for external wall formwork was installed on the outside of the buildings, and aluminum forms were used for the internal formwork. Plastic window frames were used, and there were 34 windows per floor for Building A, taking up an area of 110 m². For lifting materials, one construction hoist in each building and one tower crane for the site were installed.

In the scheduling management of this project, the start time of finishing work was a critical factor because the total construction period was short, and the focus was placed on finishing work because of the characteristics of tall residential buildings. In addition, the start of structural work was delayed by about a month compared with the planned schedule. For this reason, the construction managers decided to apply the AWW method from the time the ACS was installed.

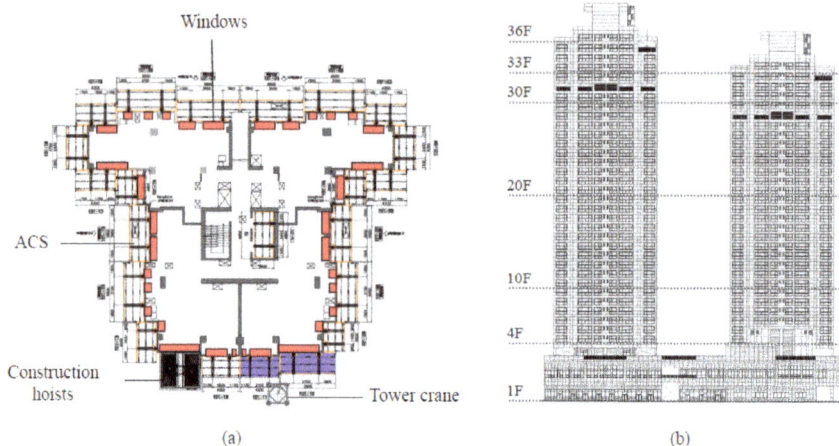

Figure 3. Case project: (**a**) layout plan of equipment; (**b**) cross section of the case project.

4.2. AWW Method Application

Figure 4 shows Building A, subject to the AWW method. The method was used for 20 units of the ACS where the walk plate was extended to the (N − 3)-th floor. The added work plate was made of aluminum bars, walk plates, and a handrail. Pollution prevention plates were installed on the (N − 1)-th and (N − 2)-th floors to prevent installed windows from being polluted by concrete paste (Figure 4d).

Two units of the ACS located above the construction hoist were installed with only gangform without work plates (Figure 4c). The gangforms were raised by the tower crane. The cycle of extending the hoistway was the same as that for raising the ACS at an interval of one floor. Wall ties fixing the hoistway to the wall were installed at intervals of two floors.

Figure 5 shows the process of the AWW method. Structural work for each floor was performed over six working days. Formwork, rebar work, and concrete placement were conducted on the N-th floor while curing took place on the (N − 1)-th floor. On the (N − 2)-th floor, 50% of the supports and forms were dismantled. On the (N − 3)-th floor, the remaining supports were dismantled and lifted to the N-th floor. Structural work was carried out in the same way as the conventional method, and only forms were removed on the (N − 2)-th floor in advance for window work.

The window work for each floor was also conducted within the same six working days as the structural work. This work consisted of installing, filling, and caulking window frames and installing glass. First, window frames were lifted and installed on the (N − 2)-th floor (Figure 4a). Then, filling, caulking, and glass installation were performed on the (N − 3)-th floor (Figure 4b). A window crew performed the installing, filling, and caulking of the window frames and installed the glass. To ensure work continuity, a crew performed four processes for 4 days in the building, and after the crew finished window work on Building A, they immediately began working on Building B. No safety facilities for openings were installed because the entire workspace was enclosed by the ACS and the windows.

(a) Preparation of windows

(b) Window installation

(c) Gangform above construction hoist

(d) Pollution prevention plate

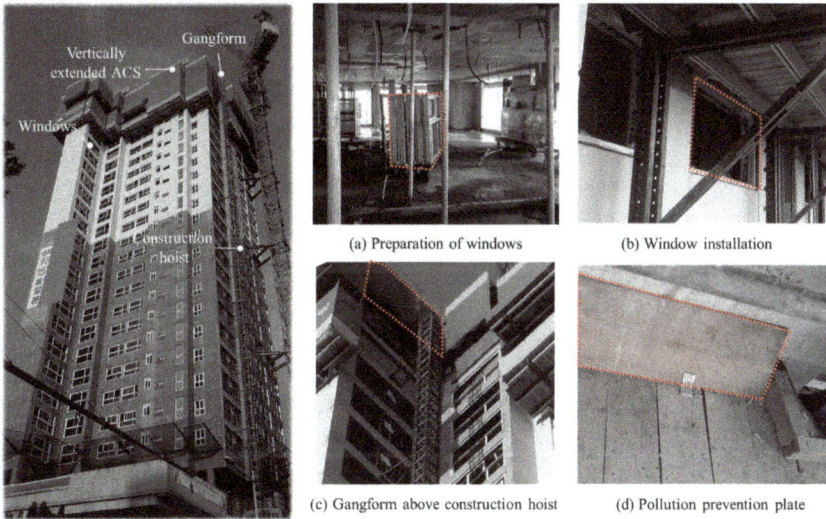

Figure 4. The AWW method in the case project.

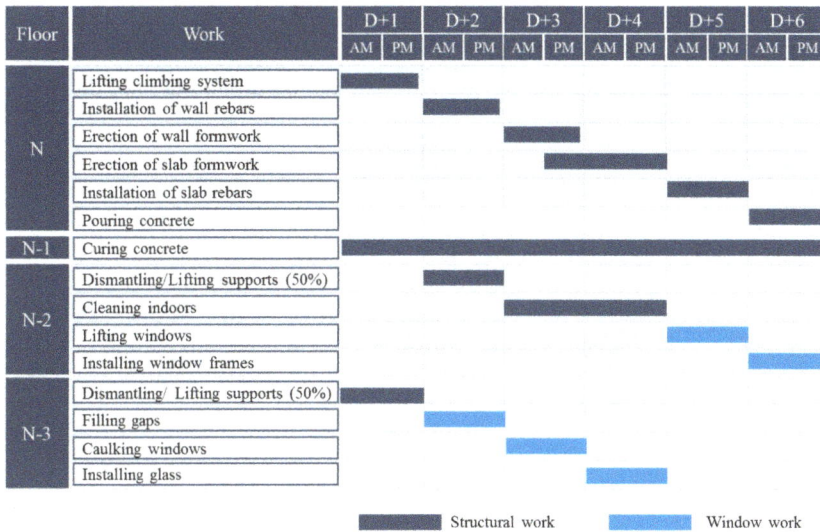

Floor	Work	D+1 AM	D+1 PM	D+2 AM	D+2 PM	D+3 AM	D+3 PM	D+4 AM	D+4 PM	D+5 AM	D+5 PM	D+6 AM	D+6 PM
N	Lifting climbing system	▉	▉										
	Installation of wall rebars			▉	▉								
	Erection of wall formwork					▉	▉						
	Erection of slab formwork						▉	▉					
	Installation of slab rebars									▉	▉		
	Pouring concrete											▉	▉
N-1	Curing concrete	▉	▉	▉	▉	▉	▉	▉	▉	▉	▉	▉	▉
N-2	Dismantling/Lifting supports (50%)			▉	▉								
	Cleaning indoors					▉	▉	▉	▉				
	Lifting windows									▇	▇		
	Installing window frames											▇	▇
N-3	Dismantling/ Lifting supports (50%)	▉	▉										
	Filling gaps			▇	▇								
	Caulking windows					▇	▇						
	Installing glass							▇	▇				

▉ Structural work   ▇ Window work

Figure 5. Construction process of the AWW method.

4.3. Results

Figure 6 compares the conventional window work method and the AWW method in terms of the duration of construction. In total, there were 24 more working days of window work with the AWW method than with the conventional method. This was because window work using the AWW method has a shorter cycle than that of the conventional method. However, the start time of window work began earlier using the AWW method compared with the conventional method because window work began as soon as the ACS was installed, while window work using the conventional method does not start until structural work reaches the 8th floor. Moreover, window work using the AWW method was completed 48 days earlier than it is using the conventional method because the delay caused by dismantling the ACS after the curing of the top floor structure is eliminated. Furthermore, internal finishing works requiring curing, such as masonry and plasterwork, were completed earlier than with the conventional method because of the speeding-up of curing by the windows, which were installed in advance and therefore blocked the outside cold air in the winter. Accordingly, the duration of internal finishing works was shorter by 4 days compared with the conventional method. Consequently, finishing works were conducted earlier by reducing the duration of curing and performing window work in advance, thereby decreasing the entire period of construction by 52 days.

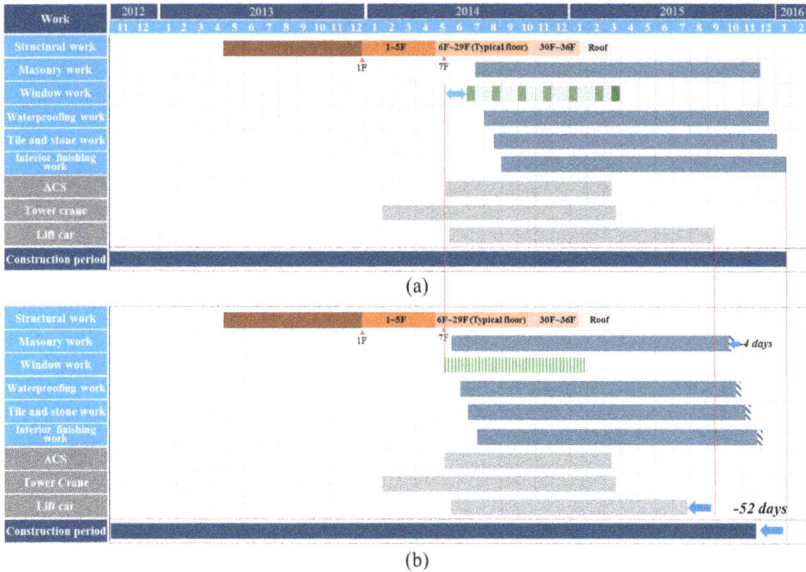

(a)

(b)

Figure 6. Comparison of project schedules: (**a**) conventional method; (**b**) AWW method.

Table 1 shows the cost variations by applying the AWW method compared with the conventional method. Costs increased in three way. One was the increase in labor cost for window work. First, the quantity of window work per floor in the project was approximately 60% of the quantity of window work that could be performed by a crew in a day. This increased the total number of working days for window work by 24 days because of the decreased productivity of window work. The increase in cost was $25,455. Second, the cost of construction hoist operation increased. Because the number of required extensions of the hoistway and wall ties increased by 1.5 times, the labor cost and the material cost increased by $1364 and $455, respectively. Finally, the equipment rental cost for adding a level of work plate to the ACS increased. The ACS was extended vertically by a total of 18 units, increasing the cost by $24,545.

However, other costs were reduced. First, the equipment rental cost for the suspended working platform used for external caulking work was eliminated because the caulking work was performed on the ACS. The cost of renting the suspended working platforms for 14 days was reduced by $4073. Second, the material costs of safety facilities decreased because openings were closed by windows. Safety nets and guardrails to prevent accidents involving falling workers and dropping objects were not required; neither were dust filter nets to prevent air pollution around the site by dust scattering. Costs therefore decreased by $99,179. Finally, the equipment rental cost for construction hoists was reduced by the decrease in the overall duration of construction. The cost of renting construction hoists decreased by $31,491 because the construction period was reduced by 52 days.

In sum, the reduction in costs of $134,743 was greater than the increase of $51,819; overall costs were therefore reduced by $82,924.

Table 1. Estimated cost savings by applying the AWW method compared with the conventional method.

Items			Cost Change ($)			
			Labor	Materials	Equipment	Total
Cost increases	Window work	Installation labor	25,455			+25,455
		Construction hoist	1364	455		+1819
	Structural work	ACS		24,545		+24,545
	Subtotal (b)		26,819	25,000	–	+51,819
Cost reductions	Window work	Suspended platform			4073	−4073
	Safety work	Safety nets	26,771	23,184		−49,955
		Safety handrails	16,426	3230		−19,656
		Safety screens	29,229	339		−29,568
	Finishing work	Tower crane			28,364	−28,364
		Construction hoist			3127	−3127
	Subtotal (a)		72,426	26,753	35,564	−134,743
	Total Reduction (a−b)		**45,607**	**1753**	**35,564**	**−82,924**

5. Discussion

The AWW method reduced duration and cost by eliminating process gaps between structural and window processes. The main reduction was achieved by continuously installing windows directly after structural work. The AWW method also provided a stable environment that was closed with windows for finishing work, thereby reducing the risk of delays by rework. The cost of window work increased, but overall costs of construction were reduced by eliminating additional safety facilities. Because the cost of safety facilities increases as the numbers of gaps and floors increase, the AWW method used in large buildings is expected to offer greater overall cost savings.

The application of the AWW method also contributed to improved work safety and fewer civil complaints. In the case project, no dropping or falling accidents occurred during the construction period, even though no external safety facilities were installed. In addition, even though the case project was performed in an urban setting, the number of civil complaints about dust and noise because of the structural work decreased by approximately 70% compared with the sites of other buildings of about the same size as that of this case study. This prevention of dust diffusion to the outside reduces air pollution in the vicinity of the construction site. This is because windows installed earlier eliminated unintentional falling and dropping, and effectively prevented dust and noise from being released to the outside. Furthermore, the stable work environment using the ACS reduced the number of window caulking flaws, thereby improving the quality of the window work.

Despite these advantages, several factors should be considered to ensure that the AWW method is applied successfully. First, sufficient windows for one cycle of window work are required to minimize productivity loss. If the quantity of windows for one floor is too small, window work at an interval of two floors or zoning the workspace should be considered. Second, structural work can be delayed if any problem occurs during window work, given that both are conducted at the same time. Thus, efforts should be made to reduce the risk of delaying construction by regularly maintaining the stocks of windows and adjusting the relevant processes through weekly meetings.

6. Conclusions

This study introduces a new window work method for tall building construction called the accelerated window work (AWW) method, which employs vertical formwork for tall residential building construction. We described the method with technical details and management elements and show that this method can reduce the duration of construction by starting and finishing window work earlier. The results of the case study showed that (1), by decreasing the process gap between structural

Sustainability **2018**, *10*, 456

and window work, the proposed method can reduce the construction duration and (2), by eliminating the need for safety facilities for openings, the AWW method can significantly reduce costs. Moreover, the case study verified the effectiveness of the AWW method in constructing tall buildings in urban areas by increasing safety performance and reducing the noise and released dust, which helped reduce civil complaints. This work contributes to the body of knowledge about window work in tall buildings by introducing and validating a new window work method. It is expected that this new method will be useful for practitioners under schedule pressures and for those specializing in building construction in urban areas where many civil complaints and claims are expected. In future work, we plan to investigate the zoning method to minimize the loss of productivity in the AWW method. Also required is further research on curtain wall work, which is on the critical path for initiating internal finishing works including window work.

Acknowledgments: This research was supported by a grant (18AUDPB106327-04) from the Architecture & Urban Development Research Program funded by the Ministry of Land, Infrastructure and Transport of the Korean Government.

Author Contributions: Taehoon Kim and Hyunsu Lim developed the research ideas and completed the writing; Chang-Won Kim and Dongmin Lee collected and analyzed data; Hunhee Cho and Kyung-In Kang designed and organized the overall research flow.

Conflicts of Interest: The authors declare no conflict of interest.

References

1. Thabe, W.Y.; Beliveau, Y.J. HVLS: Horizontal and Vertical Logic Scheduling for multistory projects. *J. Const. Eng. Manag.* **1994**, *120*, 875–892. [CrossRef]
2. Lee, D.; Lim, H.; Cho, H.; Kang, K.I. The Study on integrated ACS and lift car system for early beginning of windows work in tall building construction. In Proceedings of the Korea Institute of Building Construction Spring Conference, Korea, 29 May 2014; Korea Institute Of Building Construction: Seoul, Korea, 2014; Volume 14, pp. 72–73.
3. Kim, T.H.; Shin, Y.S.; Lee, U.K.; Kang, K.I. Decision support model using a decision tree for formwork selection in tall building construction. *J. Arch. Inst. Korea* **2007**, *23*, 177–184.
4. Bogus, S.M.; Diekmann, J.E.; Molenaar, K.R.; Harper, C.; Patil, S.; Lee, J.S. Simulation of overlapping design activities in concurrent engineering. *J. Constr. Eng. Manag.* **2011**, *137*, 950–957. [CrossRef]
5. Chuersuwan, N.; Nimrat, S.; Lekphet, S.; Kerdkumrai, T. Levels and major sources of PM2. 5 and PM10 in Bangkok Metropolitan Region. *Environ. Int.* **2008**, *34*, 671–677. [CrossRef] [PubMed]
6. Nel, A. Air pollution-related illness: Effects of particles. *Science* **2005**, *308*, 804–806. [CrossRef] [PubMed]
7. Sioutas, C.; Delfino, R.J.; Singh, M. Exposure assessment for atmospheric ultrafine particles (UFPs) and implications in epidemiologic research. *Environ. Health Perspect.* **2005**, *113*, 947–955. [CrossRef] [PubMed]
8. Si, D.; Lee, Y. *The Construction Project of the Sharp Lakepark in Chungna*; Korea Institute of Building Construction: Seoul, Korea, 2012; pp. 44–49.
9. Chung, J.; Lee, S.H.; Yi, B.J.; Kim, W.K. Implementation of a foldable 3-DOF master device to a glass window panel fitting task. *Autom. Constr.* **2010**, *19*, 855–866. [CrossRef]
10. Lee, Y.M.; Kim, Y.S. A study for major delay risk factors in curtain wall work of high-rise building using FMEA. *J. Arch. Inst. Korea* **2011**, *27*, 189–196.
11. Moon, S.D.; Ock, J.H. A Study on the construction management by design and construction phase for economic improvement of the curtain wall construction. *J. Arch. Inst. Korea* **2014**, *30*, 53–62. [CrossRef]
12. Vähä, P.; Heikkilä, T.; Kilpeläinen, P.; Järviluoma, M.; Gambao, E. Extending automation of building construction—Survey on potential sensor technologies and robotic applications. *Autom. Constr.* **2013**, *36*, 168–178. [CrossRef]
13. Yu, S.N.; Lee, S.Y.; Han, C.S.; Lee, K.Y.; Lee, S.H. Development of the curtain wall installation robot: Performance and efficiency tests at a construction site. *Auton. Robots* **2007**, *22*, 281–291. [CrossRef]

14. Kang, H.M.; Yi, J.S. A suggestion on the design-build integrated management for revitalization of exterior insulation and finishing system (EIFS)—Based on the Case Study of Apartment Housing. *J. Arch. Inst. Korea* **2012**, *28*, 157–166.
15. Lim, H.; Kim, T.; Kim, C.W.; Cho, H.; Kang, K.I. Window Proximity Tracking Method for Tall Building Construction. *Procedia Eng.* **2017**, *196*, 76–81. [CrossRef]

sustainability

MDPI

Article

Towards Sustainable Renovation: Key Performance Indicators for Quality Monitoring

Tatjana Vilutiene [1,*] and Česlovas Ignatavičius [2]

[1]   Department of Construction Management and Real Estate, Vilnius Gediminas Technical University,
     Sauletekio Ave. 11, Vilnius LT-10223, Lithuania
[2]   The Lithuanian Expanded Polystyrene Association (LEPA), Vilniaus Str. 31, Vilnius 01402, Lithuania;
     info@epsa.lt
*   Correspondence: tatjana.vilutiene@vgtu.lt; Tel.: +370-5-274-5233

Received: 1 May 2018; Accepted: 29 May 2018; Published: 1 June 2018

Abstract: The aim of this study is to propose the rational quality monitoring of the renovation process with methodology for data collection and analysis. The presented approach is based on a complex system of criteria that enables the comprehensive evaluation of the quality of the renovation process. Methodology that is developed for the rational quality monitoring of the renovation process can be used for long term monitoring activities to ensure that the system is up to date, while reflecting the concerns of the key stakeholders and the transfer of requirements. The main emphasis lies on the identification of the rapidly changing environment (regulations, technologies, needs and expectations of building owners, etc.). Quality monitoring will also serve as an analytical framework to analyze the effects of renovation and to identify what measures shall be undertaken to ensure that the renovation delivers the most positive results. This paper presents the case study analysis of renovated multi-family apartment buildings, the existing indicators of the renovation monitoring scheme and the structure of the proposed monitoring system of the renovation processes, key indicators, the main components of the system, and their links.

Keywords: civil engineering; sustainability; sustainable renovation; renovation measures; monitoring system; key performance indicators; multi-family buildings

---

1. Introduction

A pressing task that the world is facing today is the sustainable development of cities that is addressed through the constructive interaction of environmental, economic, and social factors. Sustainability priorities focus on the problems of energy efficiency, environment protection, and other issues appearing throughout all of the building life cycle, and deal with various interest groups with different targets. Sustainable development is becoming more relevant, with recent studies showing that more and more research is being published that relates to sustainability across the whole building life-cycle [1]. Sustainable building design [2], integrated planning for sustainable building [3], climate change challenges [4], solutions to achieve Nearly Zero-Energy Buildings [5], building-integrated green technologies [6], and the assessment of building refurbishment measures [7–9] are among the most topical areas of recent research.

EU members were ruled by The Energy Performance of Buildings Directive (EPBD) [10] to reduce energy consumption in the building sector. The existing buildings represent the biggest potential for energy savings, and therefore they are crucial in achieving the objective of reducing environmental impacts by about 95% by 2050 compared to 1990. Thus, the rate of building renovations needs to be increased. Although the recast of the EPBD 2010/31/EU [10] focuses on new buildings, the refurbishment of existing buildings requires investigation because of the huge potential for energy savings. The Energy Efficiency Directive 2012/27/EU [11] especially encourages thorough renovations.

It covers residential and non-residential buildings and requires the establishment of long-term strategies for the renovation of the national building stock. These requirements were transferred to the national legislative documents. So, the technical regulation of construction [12] in Lithuania sets the requirements for the thermal insulation of residential building envelopes according to the exact energy class. In accordance with the requirements for energy performance [12], the energy efficiency of new buildings constructed after 1 January 2016 should not be lower than A class, and after 1 January 2021, the energy efficiency class of newly constructed buildings should not be lower than A++. Less strict requirements have been set for the energy efficiency class of renovated buildings. Class C of energy efficiency is desirable in many renovation cases so far.

Energy efficiency potential could be identified by considering different factors which influence energy consumption, i.e., climate parameters, solutions for the building envelope, energy systems, the mode of building operation, and the behavior of occupants [13].

Interest groups, or stakeholders, who participate in the renovation process have different influences on the process and the results of the renovation. Figure 1 depicts the main key stakeholders on different levels participating in the renovation process. The research of Risholtn and Berker [14] revealed that private building owners are a key group that are urging to increase the dwelling energy efficiency rates. The analysis of the effect of domestic occupancy profiles on energy performance revealed that the type and size of household influences the energy demand [15]. If building owners were conscious consumers or had specific knowledge, they would succeed in saving energy. Therefore, relevant and reliable advice is crucial and can get homeowners to realize energy savings. However, recent feedback from implemented renovation projects is insufficient for the provision of reliable advices.

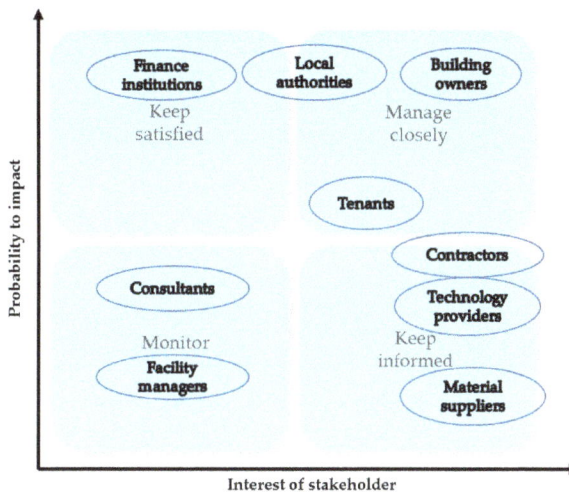

Figure 1. Mapping of key stakeholders of the renovation.

The aim of this study is to propose rational quality monitoring of the renovation process, while taking into account the needs and level of involvement of different stakeholders. The presented approach is based on a complex system of criteria that enables comprehensive evaluation of the quality of the renovation process. Methodology that is developed for the rational quality monitoring of the renovation process can be used for long term monitoring activities to ensure that the system is up to date, while reflecting changes in technologies, the concerns of stakeholders, and the increase of requirements. The main emphasis lies on the identification of the rapidly changing environment (regulations, technologies, needs and expectations of building owners, etc.). Quality monitoring will also serve as an analytical framework to analyze the effects of renovation, and as a means to assist in finding out the measures that shall be undertaken to ensure that the renovation delivers the most positive results. If correctly applied, the developed methodology will help to ensure that the renovation process is directed towards sustainability.

The analysis made by the authors revealed that there is the possibility to collect and analyze renovation related data from different perspectives. This is useful for different stakeholders of the renovation process and would obviously help to facilitate decision-making processes. The study also revealed that in order to allow the systematic data collection, analysis, and transfer, it is necessary to formulate the requirements for the submission of data. The methodological approach proposed in this study can be used for the generation of specific procedures, and it can contribute to the transparency of data collection, analysis, and transfer. The proposed methodology is versatile and could be adapted for other cases by changing the set of criteria so that they respond to the different environment.

The paper is organized as follows: the second section describes the methodology used, the data sources, and presents a workflow diagram of the proposed approach. The third section presents the results of the data analysis of the renovated case buildings. The fourth section presents the specific approach for gathering the information on all stages of the renovation process, and describes the set of monitoring indicators including their codes, links with the processes of the renovation scheme and methods for capturing, measurement, and the calculation of every indicator.

2. Materials and Methods

The approach is based on a literature analysis of the generally applied indicators for the monitoring of the renovation process, the analysis of existing monitoring scheme and applied indicators for performance measurement, interviewing specialists of the Housing Energy Efficiency Agency (HEEA), desk research of documentation, and the analysis of data on renovated case buildings (see Figure 2). The analysis resulted in the generalization of findings, conclusions, and suggestions for the improvement of the monitoring scheme for renovations. A sample of 74 buildings located in different regions of Lithuania was taken to analyze the energy performance indicators before and after renovation, the applied renovation measures, and energy renovation scenarios.

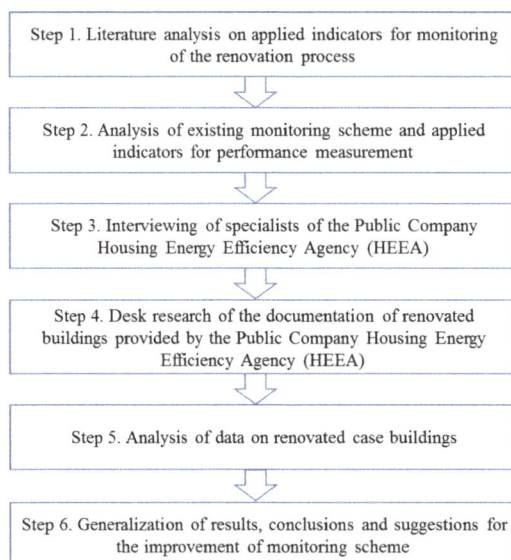

Step 1. Literature analysis on applied indicators for monitoring of the renovation process
Step 2. Analysis of existing monitoring scheme and applied indicators for performance measurement
Step 3. Interviewing of specialists of the Public Company Housing Energy Efficiency Agency (HEEA)
Step 4. Desk research of the documentation of renovated buildings provided by the Public Company Housing Energy Efficiency Agency (HEEA)
Step 5. Analysis of data on renovated case buildings
Step 6. Generalization of results, conclusions and suggestions for the improvement of monitoring scheme

Figure 2. Steps of the research methodology.

3. Results of the Renovation Data Analysis

This section provides a concise and precise description of the results of the analysis of data of the renovated case buildings, the interpretation, and preliminary conclusions.

3.1. Description of Case Buildings

The sample size for the analysis compounded of 74 multi-apartment buildings that had received funding to renovate from the Multi-apartment buildings renovation (modernization) program that was implemented under the leadership of the Public Company Housing Energy Efficiency Agency (HEEA) [16]. The sample buildings were renovated in the period of 2014–2016. Buildings renovated in 2017 were not included in the sample due to the necessity to collect data on the actual values of thermal energy consumption for heating when the renovation was finished. These values were observed during the heating period of 2016–2017. Most of the renovated buildings were located in Vilnius County and were constructed in the period of 1961–1990 (see Figures 3 and 4). This could be explained by the fact that until 1992, in Lithuania there were no strict requirements for the heat preservation in buildings, and therefore multi-apartment buildings constructed before this year were poorly insulated or were not insulated. The rise of requirements for the energy performance of buildings has been gradually tightened since 1992, thus fostering the endeavors to rise the energy performance of buildings. This led to the application of energy efficiency measures including better thermal insulation of envelopes. Therefore, the Multi-apartment buildings renovation (modernization) program only provided funding for buildings constructed before 1993. In buildings constructed before this year, prevail non-insulated precast reinforced concrete and masonry structures, and such construction materials consequently make up the largest share in the analyzed sample (see Figure 5).

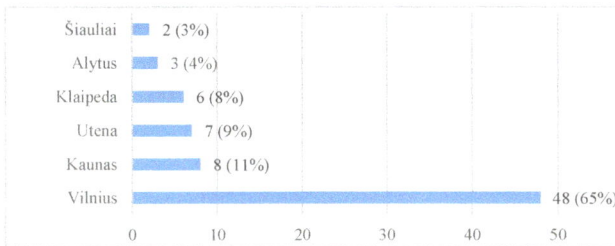

Figure 3. The distribution of case buildings by county (sample size n = 74).

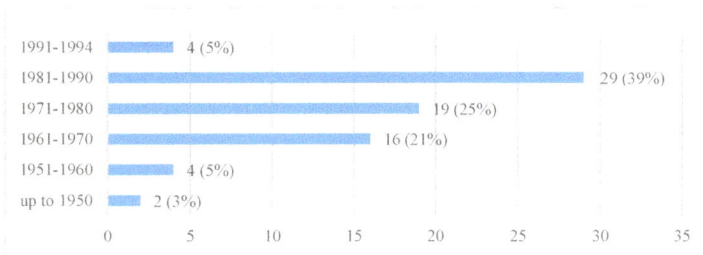

Figure 4. The distribution of case buildings by periods of construction (sample size n = 74).

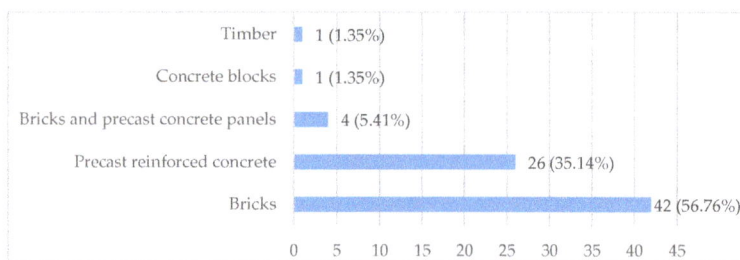

Figure 5. The distribution of case buildings by the type of construction materials of the bearing walls (sample size n = 74).

3.2. Description of the Renovation Scenarios

Data of the main renovation scenarios applied in the analyzed case buildings is depicted in Table 1. For each scenario, the table provides the number of cases in the sample and the values of thermal energy consumption based on the data of the energy certificates that were issued before and after the renovation. About 84% of the renovated buildings achieved a C class of energy efficiency (see Figure 6).

Table 1. Renovation scenarios in the analyzed case buildings.

Renovation Scenarios	Change of Energy Class	Number of Cases in the Sample	Thermal Energy Consumption before the Renovation (Max-Min), kWh/m²	Thermal Energy Consumption after the Renovation (Max-Min), kWh/m²	Thermal Energy Consumption before the Renovation (Average), kWh/m²	Thermal Energy Consumption after the Renovation (Average), kWh/m²	Thermal Energy Consumption Reduction (Average), %
Scenario 1	D → B	7	322.68–205.48	89.43–59.84	242.56	75.02	69.07
Scenario 2	D → C	23	322.43–175.10	159.41–70.99	237.64	104.01	56.23
Scenario 3	E → B	4	374.34–248.89	146.68–85.08	301.95	108.91	63.93
Scenario 4	E → C	37	553.73–216.84	199.4–68.38	309.68	104.86	66.14
Scenario 5	E → D	1	-	-	400.00	170.90	57.28
Scenario 6	F → C	1	-	-	806.25	215.86	73.23
Scenario 7	G → C	1	-	-	1095.47	316.79	71.08

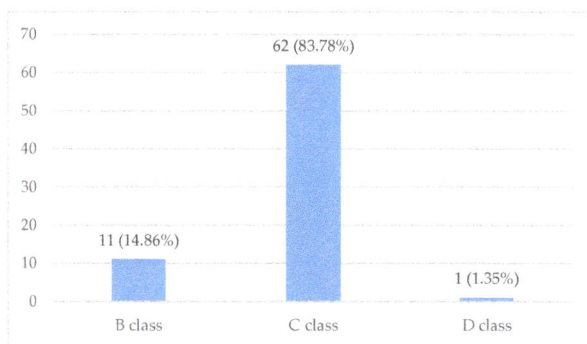

Figure 6. Number of certificates by achieved energy class (sample size n = 74).

3.3. Results of the Renovation in Figures

The results of the comparison of the calculated and actual measured values of the thermal energy consumption for heating in the sample of the analyzed buildings is depicted in Table 2. The figures of the actual calculated thermal energy consumption for heating (kWh/m²) were taken from the energy certificates that were issued after the modernization. The figures of the actual measured thermal energy consumption for heating (kWh/m^2) were provided by the exact district heating supplier. During the research, this information was provided only for half of the analyzed sample, i.e. for 35 buildings of the 74 that were analyzed. The objects' data in the table is sorted by the calculated values of the percentage decrease (+) or increase (−) in the consumption of thermal energy for heating. In most cases (about 83% of the sample's buildings), the difference between the calculated and the actual measured values show a decrease of the actual measured thermal energy consumption. Only 6 from 35 cases (about 17% of the sample's buildings) show an increase of actual measured thermal energy consumption (see Figure 7).

Table 2. The difference between the calculated and the actual measured values of thermal energy consumption for heating (sample size n = 35).

Renovated Building	Type of Envelope Insulation	Actual Calculated Thermal Energy Consumption for Heating, kWh/m^2	Actual Measured Thermal Energy Consumption for Heating, kWh/m^2	Difference between the Calculated and the Actual Measured Values, kWh/m^2	Percentage Decrease (+)/Increase (−) in Consumption Thermal Energy for Heating, %
Biliūno 28	Mineral wool	61.13	99.02	−37.89	−61.98
Liudiškių 31C	Mineral wool	41.57	46.63	−5.06	−12.17
Kosmonautų 11	Mineral wool	68.38	73.09	−4.71	−6.89
Smilties pylymo 3	Mineral wool	57.39	62.90	−5.51	−5.93
Panerių 15	Mineral wool	46.14	47.99	−1.85	−4.01
Vilniaus 29	Mineral wool	64.89	67.70	−2.81	−2.82
Šviesos 5	EPS	68.93	67.51	1.42	2.06
Šviesos 3	EPS	68.98	64.05	4.93	7.15
Chemikų 55	Mineral wool	62.33	57.55	4.78	7.67
Lozoraičio 3	Mineral wool	48.57	38.80	9.77	9.64
Baltijos 65	Mineral wool	47.57	33.60	13.97	12.43
Mickevičiaus 1A	EPS	87.60	76.60	11.00	12.56
Mickevičiaus 1	EPS	76.79	65.22	11.57	15.07
Mickevičiaus 16	EPS	83.75	69.84	13.91	16.61
Kretingos 27	EPS	97.34	80.79	16.55	17.00
Liudiškių 31B	Mineral wool	53.57	43.58	9.99	18.65
Liudiškių 31A	Mineral wool	52.73	42.03	10.70	20.29
Trakų 18	EPS	85.44	67.69	17.75	20.77
Mickevičiaus 7	EPS	85.18	61.40	23.78	27.92
Saulės 17	EPS	85.25	55.52	29.73	34.87
Trakų 29	EPS	85.39	55.30	30.09	35.24
Trakų 11	EPS	74.19	47.97	26.22	35.34
Taikos 97	Mineral wool	56.47	19.67	36.80	37.55
Trakų 27	EPS	90.83	55.86	34.97	38.50
Trakų 2	EPS	66.08	40.22	25.86	39.13
Trakų 4	EPS	80.51	47.95	32.56	40.44
Atgimimo 6	Mineral wool	94.00	53.94	40.06	42.62
Lozoraičio 24	Mineral wool	192.00	101.74	90.26	47.01
Vilniaus 26	EPS	109.11	55.16	53.95	49.45
Trakų 25	EPS	82.76	41.11	41.65	50.33
Taikos 103	Mineral wool	73.59	36.32	37.27	50.65
Riomerio 6	Mineral wool	90.09	41.15	48.94	54.32
Kęstučio 6	Mineral wool	93.17	34.25	58.92	63.24
Liepų 20	EPS	87.65	28.90	58.75	67.03
Kęstučio 2	Mineral wool	98.21	26.17	72.04	73.35

The energy consumption reduction measures for the renovation are based on the passive and active measures. The insulation of the envelope is most often applied as a passive measure, and the instalment of new efficient equipment is applied as an active measure (see Table 3). The Programme for the renovation/upgrading of multi-apartment buildings [16] (hereinafter–Programme) aims to encourage the apartment owners to comprehensively upgrade multi-apartment buildings and residential districts in order to achieve higher living standards, rational use of energy resources, and compensation for the reduced budgetary expenditure on heating costs. Therefore, the Programme provides the support for both energy efficiency improvement measures and to other building renovation measures, including (i) major repairs or reconstruction of the heating systems and hot and cold water supply systems; (ii) the replacement of windows and exterior doors; (iii) roof thermal insulation, including the installation of new sloping roofs (excluding the construction of attic premises); (iv) glassing of balconies (loggia) under a unified project; (v) thermal insulation of exterior walls; (vi) thermal insulation of cellar ceilings; (vii) thermal insulation of basement walls; (viii) the installation of the equipment for alternative energy sources (sun, wind, etc.); (ix) major repairs and the replacement of elevators; (x) replacement or repair of the buildings' systems (wastewater systems, electrical and fire prevention installations, and drinking water pipes).

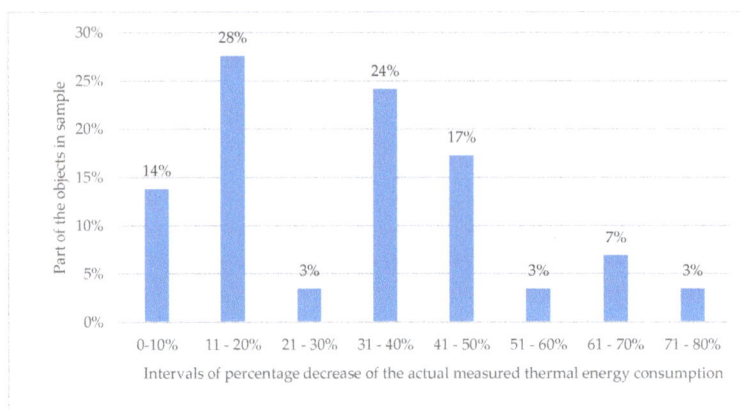

Figure 7. The distribution of buildings by percentage decrease of the actual measured thermal energy consumption (sample size n = 35).

The analysis of the renovated buildings sample shows that the most popular renovation measures are: modernization of the heating system; modernization of the ventilation system; modernization of the hot water system; the replacement of windows in apartments and other premises; the change of windows and doors in other common areas; glazing of balconies; exterior wall insulation; and roof insulation (see Table 3). These measures were applied in most renovation scenarios of the case buildings.

Table 3. Measures applied in the main renovation scenarios.

Applied Measures *	Renovation Scenarios						
	Scenario 1	Scenario 2	Scenario 3	Scenario 4	Scenario 5	Scenario 6	Scenario 7
	D → B	D → C	E → B	E → C	E → D	F → C	G → C
	Number of Case Buildings Were Measures Applied						
	7	23	4	37	1	1	1
	Energy efficiency improvement measures						
HS	+	+	+	+	+	+	+
VS	+	+	+	+	+	+	+
RES		+		+			
HWS	+	+	+	+	+		
AWR	+	+	+	+	+	+	+
CWR	+	+	+	+	+	+	+
BG	+	+	+	+	+		+
WI	+	+	+	+		+	+
RI	+	+	+	+	+	+	+
BSI	+		+	+			
EU			+	+			
	Other (non-energy saving measures)						
WWS	+	+	+	+			
DWS	+	+		+			
DRS	+						
RWS		+		+			
EWS		+	+	+			+

* HS—modernization of heating system; VS—modernization of ventilation system; RES—installation of renewable energy sources (solar, wind, etc.); HWS—modernization of the hot water system; AWR—replacement of windows in apartments and other premises; CWR—change of windows and doors in other common areas; BG—glazing of balconies; WI—exterior wall insulation; RI—roof insulation; BSI—basement slab insulation; EU—replacement of old elevators to higher energy class elevators; WWS—modernization of the waste water drainage system; DWS—modernization of the drinking water supply system; DRS—drainage system instalment; RWS—instalment of a rainwater drainage system; EWS—upgrade of electrical wiring system.

3.3.1. Energy Saving Measures

This section presents the distribution of active and passive energy-saving measures in the case buildings. Main energy-saving measures include the modernization of heating systems; the modernization of ventilation systems; the replacement of windows in apartments and other premises; the change of windows and doors in other common areas; glazing of balconies; exterior wall insulation; and roof insulation (see Figure 8).

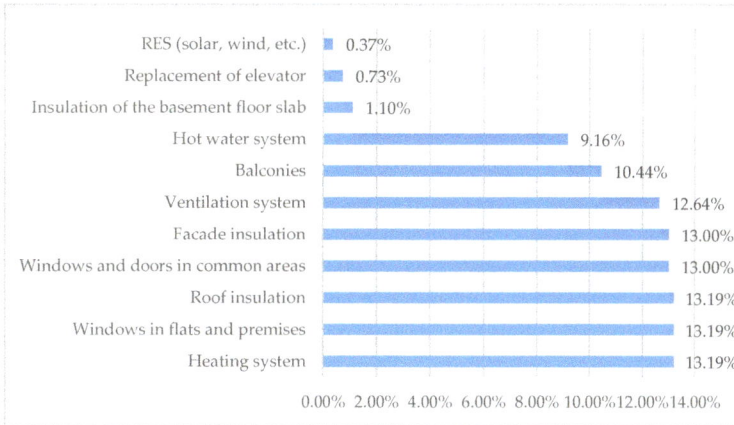

Figure 8. Distribution of applied energy saving measures in the case buildings (sample size n = 74).

The modernization of ventilation systems mainly covers the repair of existing natural ventilation systems (in 73% of all cases). It was only in 24% of the sample buildings that the mechanical insulation with heat recovery was installed (see Figure 9).

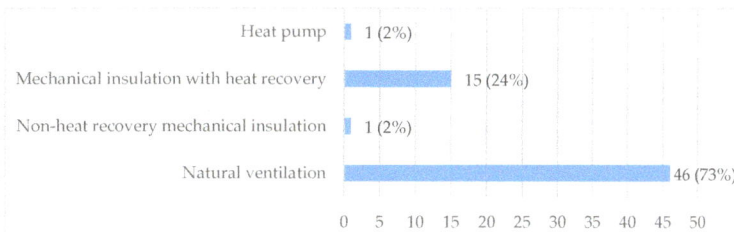

Figure 9. Installed ventilation systems by type (sample size n = 74).

3.3.2. Non-Energy Saving Measures

The most commonly used other (non-energy saving) measures are the replacement of the drinking water system and the installation of a rainwater system (in about 60% of all renovation cases). The replacement of domestic waste water systems, the repair of electric wiring systems, and the repair of drainage is less common (about 40% of all renovation cases) (see Figure 10).

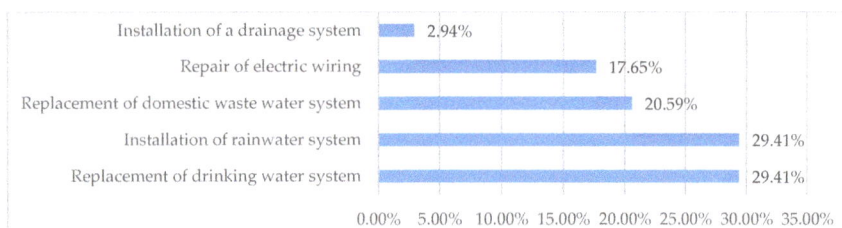

Figure 10. The distribution of the applied non-energy saving measures in the case buildings (sample size n = 74).

3.3.3. Façade Insulation Materials

The type and material of insulation for the facade was specified by HEEA specialists. The data provided revealed that expanded polystyrene (EPS) foam and mineral wool are the most popular types of insulation materials. EPS was used almost twice as often as mineral wool (see Figure 11). In most cases, the façade insulation works included the insulation of a basement wall, the repair of defects, and the repair of pathways adherent to the façade surface.

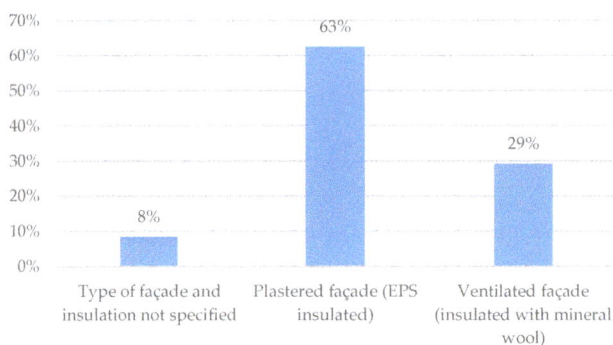

Figure 11. The distribution of case buildings by the type of façade and insulation (sample size n = 48).

Table 4. Economic estimates of reduced heat losses through walls after renovation.

Economic Estimates of Reduced Heat Losses through Walls	Type of Façade	
	EPS Insulated Plastered Façade	Ventilated Façade Insulated with Mineral Wool
Investments (interval values), Eur/m^2	42.24–128.19	57.47–227.04
Investments (average values), Eur/m^2	76.19	104.54
The estimated reduction of heat losses through the walls of the building (interval values), kWh/m^2/a	40.30–120.61	35.22–181.84
The estimated reduction of heat losses through the walls of the building (average values), kWh/m^2/a	89.19	82.63
The share of investments for the reduction of heat losses through the walls of the building (interval values), Eur/kWh/m^2/a	256.02–4280.31	233.96–6962.53
The share of investments for the reduction of heat losses through the walls of the building (average values), Eur/kWh/m^2/a	1178.76	2531.05

The analysis revealed that plastered facade insulated with EPS required less investments than ventilated facade insulated with mineral wool, while the estimated reduction of heat losses through the walls of the building is similar (respectively, 89.19 and 82.63 kWh/m^2/a) for both types of façades (see Table 4). In some cases, the investments depicted in the table include the cost of repairs of defects detected on exterior walls, the cost of basement wall insulation, and the cost of repairs of pathways adherent to the façade surface. The cost of these improvements was the cause of the large interval of costs.

3.3.4. Observations of the Existing Monitoring Scheme

This section provides statistics on the indicators of the existing monitoring carried out by the Public Company Housing Energy Efficiency Agency (HEEA) of the Lithuanian Ministry of the Environment acting as the main managing authority for the programme. The existing monitoring scheme collects information on annual thermal energy savings (GWh/year) in all country renovated buildings (see Figure 12), the reduction of CO_2 emissions (thsd. t/year) that resulted from the renovation (see Figure 13), and calculates the distribution of thermal energy savings in all country renovated buildings by the source of funding (see Figure 14). This is obviously insufficient information to support the decisions of renovation interest groups.

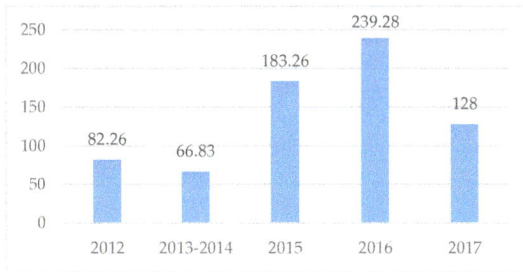

Figure 12. The estimated annual (2012–2017) thermal energy savings in all country renovated buildings (GWh/year). Data source: www.betalt.lt.

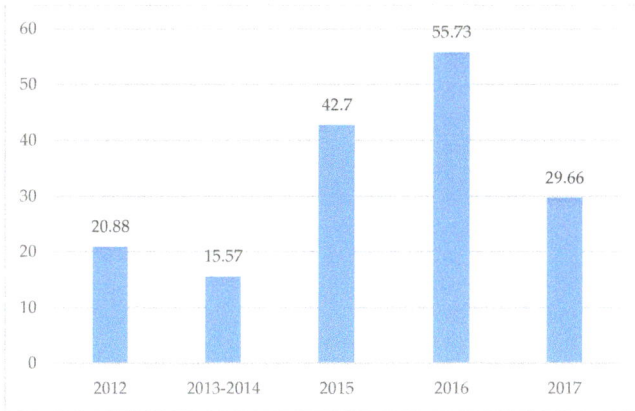

Figure 13. The estimated annual (2013–2017) reduction of CO_2 emissions (thsd. t/year). Data source: www.betalt.lt.

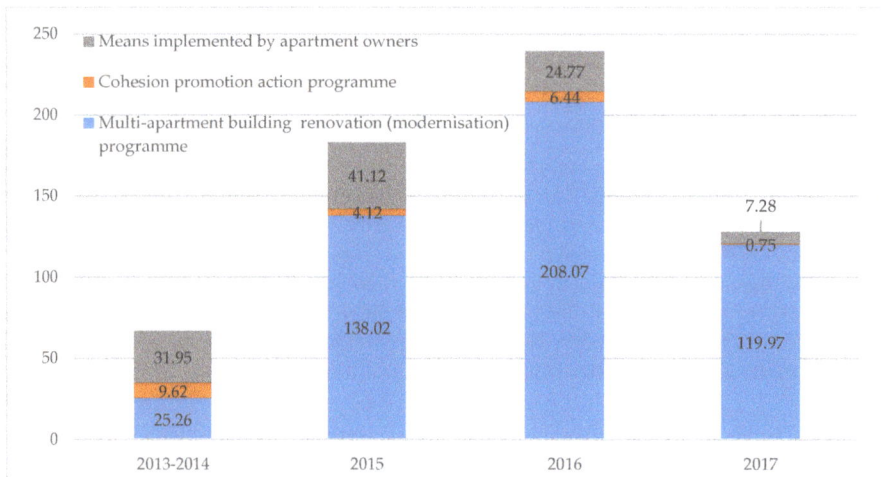

Figure 14. The estimated annual (2013–2017) thermal energy savings in all country renovated buildings classified by source of funding (GWh/year). Data source: www.betalt.lt.

3.4. Main Findings and Suggestions

The existing monitoring scheme collects insufficient information on the renovation process, and this therefore reduces the ability to rely on objective and reliable information when deciding on renovation scenarios. The analysis made by the authors revealed that there is the possibility to collect and analyze renovation related data from different perspectives that is useful for different stakeholders of the renovation process, and this would obviously help to facilitate decision-making processes.

Currently, the process of information collection and its transfer to different interest groups is mainly manual, the data is not integrated and is collected in different sources, and in some cases, access is closed. In order to facilitate the flow of information, the data needs to be digitalized.

In order to digitize the information that is related to the renovation process and the results, it is necessary to formulate the requirements for the submission of data in order to allow for systematic data collection, analysis, and transfer.

4. Suggested Key Performance Indicators for the Monitoring of the Renovation Process

Based on the findings of the analysis, two groups of indicators suggested for the quality monitoring of the renovation process:

IR1 The indicators assessing the requirements for renovation.
IR2 The indicators assessing the results of the renovation process.

Information for quality indicators comes from:

(1) Register of applicants:

- IR1.1. The number of applications.
- IR1.2. The number of approved applications.
- IR1.3. The number of changed investment plans.

(2) The Investment plans and Energy certificates:

- IR1.4. The planned thermal energy savings (calculations made in Investment plans).
- IR1.5. The planned reduction of CO_2 emissions (calculations made in Investment plans).

- IR1.6. The existent energy class of the building (Energy certificates).
- IR1.7.–IR 1.10. Number of target A, B, C, D energy class certificates (calculations made in Investment plans).

(3) Actual data:

- IR2.1. Reduction of thermal energy consumption (the calculations made are based on the actual thermal energy consumptions).
- IR2.2. Actual thermal energy savings (data on the actual thermal energy consumptions).
- IR2.3. Actual reduction of CO_2 emissions (calculations based on the actual thermal energy consumptions).
- IR2.3.–IR 2.7. Number of achieved A, B, C, D energy class certificates (data from energy certificates).

(4) Other primary and secondary sources of information (Surveys; Expert reports; Feasibility studies; Certification reports).

Table 5 presents a detailed description of all the indicators proposed for quality monitoring of the renovation process, including calculation methods, data sources, and data provider.

Table 5. Key performance indicators for the quality monitoring of the renovation process.

Code of Indicator	Monitoring Indicator	Measuring Unit	Calculation Method	Data Sources	The Moment of Attainment *	Data Provider
1	2	3	4	5	6	7
			IR1 The indicators assessing the requirements for renovation			
IR1.1.	Number of applications	Quantity	Data transfer Assignment to groups Aggregation	Submitted applications Register of applications	The application is submitted	Housing Energy Efficiency Agency
IR1.2.	Number of approved applications	Quantity	Data transfer Assignment to groups Aggregation	Approved applications Register of applications	The application is approved	Housing Energy Efficiency Agency
IR1.3.	Number of changed investment plans	Quantity	Data transfer	Approved applications	The final revision of the investment plan is approved	Housing Energy Efficiency Agency
IR1.4.	Planned thermal energy savings (all buildings)	GWh/year	Data transfer	Investment plan	The final revision of the investment plan is approved	Housing Energy Efficiency Agency
IR1.5.	Planned reduction of CO_2 emissions	t/year	Data transfer	Investment plan	The final revision of the investment plan is approved	Housing Energy Efficiency Agency
IR1.6.	Existent energy class of building	Code of energy class	Aggregation	Investment plan Energy certificate issued before the renovation	The energy certificate is issued	Housing Energy Efficiency Agency
IR1.7.	Number of target A energy class certificates	Code of energy class	Aggregation	Investment plan	The final revision of the investment plan is approved	Housing Energy Efficiency Agency
IR1.8.	Number of target B energy class certificates	Code of energy class	Aggregation	Investment plan	The final revision of the investment plan is approved	Housing Energy Efficiency Agency
IR1.9.	Number of target C energy class certificates	Code of energy class	Aggregation	Investment plan	The final revision of the investment plan is approved	Housing Energy Efficiency Agency
IR1.10.	Number of target D energy class certificates	Code of energy class	Aggregation	Investment plan	The final revision of the investment plan is approved	Housing Energy Efficiency Agency
			IR2 The indicators assessing the results of renovation process			
IR2.1.	Reduction of thermal energy consumption (single buildings)	%	Calculated using the following formula: $t_1 = \frac{t_b - t_a}{t_b} \times 100\%$ (C_a—actual thermal energy consumption after renovation; C_b—thermal energy consumption before renovation)	Investment plan/actual measurements	The yearly data of actual thermal energy consumption is provided	Housing Energy Efficiency Agency Thermal energy provider
IR2.2.	Actual thermal energy savings (all buildings)	GWh/year	Data transfer Aggregation of actual thermal energy savings of single buildings	Investment plan/actual measurements	The yearly data of actual thermal energy consumption is provided	Housing Energy Efficiency Agency Thermal energy provider
IR2.3.	Actual reduction of CO_2 emissions	t/year	Recalculated according to actual thermal energy savings	Investment plan	The yearly data of actual thermal energy consumption is provided	Housing Energy Efficiency Agency

Table 5. *Cont.*

Code of Indicator	Monitoring Indicator	Measuring Unit	Calculation Method	Data Sources	The Moment of Attainment *	Data Provider
1	2	3	4	5	6	7
IR2.4.	Number of achieved A energy class certificates	Code of energy class	Data transfer	Investment plan Energy certificate issued after the renovation	The energy certificate is issued	Housing Energy Efficiency Agency
IR2.5.	Number of achieved B energy class certificates	Code of energy class	Data transfer	Investment plan Energy certificate issued after the renovation	The energy certificate is issued	Housing Energy Efficiency Agency
IR2.6.	Number of achieved C energy class certificates	Code of energy class	Data transfer	Investment plan Energy certificate issued after the renovation	The energy certificate is issued	Housing Energy Efficiency Agency
IR2.7.	Number of achieved D energy class certificates	Code of energy class	Data transfer	Investment plan Energy certificate issued after the renovation	The energy certificate is issued	Housing Energy Efficiency Agency

* Column explains the condition when the indicator is considered to be attained.

5. Conclusions

The existing monitoring scheme collects insufficient information on the renovation process, and this therefore reduces the ability to rely on objective and reliable information when deciding on renovation scenarios. The current paper presents a case study analysis of renovated multi-family apartments. The analysis made by the authors revealed that there is the possibility to collect and analyze renovation related data from different perspectives that is useful for different stakeholders of the renovation process, and this would obviously help to facilitate decision-making processes.

The study proposes methodology for the rational quality monitoring of the renovation process. The presented approach is based on a complex system of criteria that enables the comprehensive evaluation of the quality of the renovation process. The methodology that has been developed for rational quality monitoring of the renovation process can be used for long term monitoring activities to ensure that the system is up to date, while reflecting the concerns of key stakeholders and the increase of requirements.

Quality monitoring could also serve as an analytical framework to analyze the effects of renovation, and may assist in finding out the measures that shall be undertaken to ensure that the renovation process delivers the most positive results. The proposed approach is based on a complex system of criteria that reflects the needs and expectations of the different interest groups of renovation and ensures the reliability of analysis.

There are no strict limitations for the use of the presented approach. The advantage of the proposed approach is that the proposed methodology is versatile and could be adapted for other cases by changing the set of criteria so that is responds to the different environment. Nevertheless, the correctness of application depends on the type and the accuracy of the initial data, the experience level of experts, and the decision-makers.

The methodological approach proposed in this study can be used for the generation of specific procedures, and the authors believe that new methodology could be the baseline for future research and that it can contribute by ensuring the transparency of data collection, analysis, and transfer. Currently, the process of information collection and transfer to different interest groups is mainly manual, data is not integrated, is stored in different sources, and in some cases, access is closed. In order to facilitate the flow of information, the data needs to be digitalized. In order to digitize the information related to the renovation process and the results, it is necessary to formulate the requirements for the submission of data to ensure systematic data collection, analysis, and transfer.

Author Contributions: All authors contributed equally to this work.

Funding: This research received no external funding.

Acknowledgments: The authors are grateful to the Public Company Housing Energy Efficiency Agency (HEEA) for providing access to the documentation of the renovated case study buildings. Also, the authors would like to acknowledge the Public Company, the Lithuanian District Heating Association (LDHA) for providing data on the actual measured thermal energy consumption of some of the renovated case study buildings.

Conflicts of Interest: The authors declare no conflict of interest.

References and Note

1. Zavadskas, E.K.; Antucheviciene, J.; Vilutiene, T.; Adeli, H. Sustainable decision-making in civil engineering, construction and building technology. *Sustainability* **2018**, *10*, 14. [CrossRef]
2. Wang, N.M.; Adeli, H. Sustainable Building Design. *J. Civ. Eng. Manag.* **2014**, *20*, 1–10. [CrossRef]
3. Mikaelsson, L.A.; Larsson, J. Integrated Planning for Sustainable Building—Production an Evolution over Three Decades. *J. Civ. Eng. Manag.* **2017**, *23*, 319–326. [CrossRef]
4. Akbari, H.; Cartalis, C.; Kolokotsa, D.; Muscio, A.; Pisello, A.L.; Rossi, F.; Santamouris, M.; Synnefa, A.; Wong, N.H.; Zinzi, M. Local Climate Change and Urban Heat Island Mitigation Techniques—The State of the Art. *J. Civ. Eng. Manag.* **2016**, *22*, 1–16. [CrossRef]

5. Zavadskas, E.K.; Antucheviciene, J.; Kalibatas, D.; Kalibatiene, D. Achieving Nearly Zero-Energy Buildings by applying multi-attribute assessment. *Energy Build.* **2017**, *143*, 162–172. [CrossRef]
6. Si, J.; Marjanovic-Halburd, L.; Nasiri, F.; Bell, S. Assessment of building-integrated green technologies: A review and case study on applications of Multi-Criteria Decision Making (MCDM) method. *Sustain. Cities Soc.* **2016**, *27*, 106–115. [CrossRef]
7. Vilches, A.; Garcia-Martinez, A.; Sanchez-Montañes, B. Life cycle assessment (LCA) of building refurbishment: A literature review. *Energy Build.* **2017**, *135*, 286–301. [CrossRef]
8. Sierra-Pérez, J.; Boschmonart-Rives, J.; Gabarrell, X. Environmental assessment of façade-building systems and thermal insulation materials for different climatic conditions. *J. Clean. Prod.* **2016**, *113*, 102–113. [CrossRef]
9. Yang, J.; Chong, A.; Santamouris, M.; Kolokotsa, D.; Lee, S.E.; Tham, K.W.; Sekhar, C.; Cheong, D.K.W. Energy utilizability concept as a retrofitting solution selection criterion for buildings. *J. Civ. Eng. Manag.* **2017**, *23*, 541–552. [CrossRef]
10. European Parliament and the Council. Directive 2010/31/EU on the energy performance of buildings. *Off. J. Eur. Union* **2010**. [CrossRef]
11. European Parliament and the Council of the European Union. Directive 2012/27/EU on energy efficiency, amending Directives 2009/125/EC and 2010/30/EU and repealing Directives 2004/8/EC and 2006/32/EC. *Off. J. Eur. Union* **2012**, *55*, 1–57.
12. Building Technical Regulation *STR 2.01.02:2016, Design and Certification of Energy Performance of Buildings*; adopted on 11 November 2016 by Order No. D1-754 of the Minister of Environment of the Republic of Lithuania. (In Lithuanian)
13. Džiugaitė-Tumėnienė, R.; Jankauskas, V.; Motuzienė, V. Energy balance of a low energy house. *J. Civ. Eng. Manag.* **2012**, *18*, 369–377. [CrossRef]
14. Risholtn, B.; Berker, T. Success for energy efficient renovation of dwellings—Learning from private homeowners. *Energy Policy* **2013**, *61*, 1022–1030. [CrossRef]
15. Martinaitis, V.; Zavadskas, E.K.; Motuzienė, V.; Vilutienė, T. Importance of occupancy information when simulating energy demand of energy efficient house: A case study. *Energy Build.* **2015**, *101*, 64–75. [CrossRef]
16. Programme for the Renovation/Upgrading of Multi-Apartment Buildings—Lithuania. Available online: http://www.buildup.eu/en/explore/links/programme-renovationupgrading-multi-apartment-buildings-lithuania (accessed on 10 April 2018).

sustainability

MDPI

Article

Development of the CO₂ Emission Evaluation Tool for the Life Cycle Assessment of Concrete

Taehyoung Kim [1], Sanghyo Lee [2,*], Chang U. Chae [1], Hyoungjae Jang [2] and Kanghee Lee [3]

[1] Building and Urban Research Institute, Korea Institute of Civil Engineering and Building Technology, Daehwa-dong 283, Goyandae-Ro, Ilsanseo-Gu, Goyang-Si, Gyeonggi-Do 10223, Korea; kimtaehyoung@kict.re.kr (T.K.); cuchae@kict.re.kr (C.U.C.)

[2] School of Architecture & Architectural Engineering, Hanyang University, Sa 3-dong, Sangrok-Gu, Ansan-Si 04763, Korea;duethj@hanyang.ac.kr

[3] Department of Architectural Engineering, Andong National University, 1375, Gyeongdong-Ro, Andong-Si 36729, Korea; leekh@andong.ac.kr

* Correspondence: mir0903@hanyang.ac.kr; Tel.: +82-31-436-8182; Fax: +82-31-436-8100

Received: 13 October 2017; Accepted: 15 November 2017; Published: 17 November 2017

Abstract: With the goal of reducing greenhouse gas (GHG) emissions by 26.9% below business-as-usual by 2020, the construction industry is recognized as an environmentally harmful industry because of the large quantity of consumption and waste with which it is associated, and the industry has therefore been requested to become more environmentally friendly. Concrete, a common construction material, is known to emit large amounts of environmentally hazardous waste during the processes related to its production, construction, maintenance, and demolition. To aid the concrete industry's efforts to reduce its GHG emissions, this study developed a software program that can assess GHG emissions incurred over the life cycle of a concrete product, and a case study was conducted to determine the impact of the proposed concrete assessment program on a construction project.

Keywords: concrete; greenhouse gas (GHG) emissions; life cycle assessment (LCA)

1. Introduction

As part of a policy system aiming to reduce GHG emissions in the industrial sector through quantitative assessment, the carbon labeling system was started in Europe and has spread and developed globally according to the situation in each country [1]. Through the implementation of a carbon labeling certification system in South Korea, various construction materials have been certified, and the use of such materials will increase in the future. Among construction materials, concrete is recognized as a material that emits a large quantity of greenhouse gases (GHG) across its lifecycle, from production, through construction, maintenance, and the dismantling and disposal processes, and environmental impact assessments of the material must therefore include diverse studies. Concrete emits a large amount of GHG from the production of its raw materials, including cement, aggregate, and admixture, to transportation and concrete production. However, research assessing the environmental impact of the concrete industry is at an early stage, and insufficient data is available to establish policies and support technologies that would increase the sustainability of the concrete industry. In the COOL software [2] that is currently used for carbon labeling certification, a life cycle index (LCI) database (DB) [3] has not yet been constructed for the admixture that is one of the concrete components, and thus the overseas LCI DB [4] must be directly examined and entered. In addition, although diversified impact assessment analysis techniques must be employed to effectively reduce GHG emissions over the concrete life cycle, the currently available assessment technologies are insufficient. Therefore, the purpose of this study was to develop a software program that will assess the GHG emissions associated with concrete in a manner that is consistent with the assessment method of the carbon labeling certification program, and in which data can easily be inputted, stored, calculated, and managed. For this purpose, the life cycle assessment (LCA) guideline ISO 14044 [5] was applied, and product category rules (PCRs) that are prescribed by the carbon labeling certification program to calculate a product's life cycle GHG emission information were analyzed. The process flow and data collection characteristics of actual concrete production were analyzed to derive the GHG emission assessment factors for each stage [6,7]. A software program that can conduct inventory and comparative analyses of the assessment results by impact category was developed, and its reliability and applicability were examined through case analysis.

This program assesses greenhouse gas emissions using the amount of raw materials and energy inputted into concrete production in terms of LCA.

Also, this program has built raw material (cement, aggregate, admixture, etc.) and energy (electricity, diesel, kerosene, etc.) LCI (Life cycle Inventory) DB for users to assess the greenhouse gas emissions of concrete more easily.

This program is a concrete-specialized program, which is used to assess greenhouse gas emissions among LCA assessment programs for building materials that have been developed in Korea so far.

2. Previous Studies

The assessment factors and technologies used in existing domestic and overseas LCA software programs were analyzed. Representative domestic LCA software programs include TOTAL and COOL, which commonly perform assessments based on data from the national LCI DB information network. TOTAL [8] was developed by the Ministry of Environment (ME) in South Korea and can perform general LCA. Because this software supports the data format required by the life cycle environmental impact assessment certification system, as well as the domestic environmental labeling system, TOTAL can perform LCAs optimized for the appropriate environment labeling system. COOL [9] was developed by the Korean Environmental Industry & Technology Institute to provide convenience to companies applying for carbon labeling certification. This software includes a carbon labeling emission factor that is automatically displayed according to the material, and can be used to estimate the carbon emissions of a product.

Representative overseas LCA software programs include BEES, ATHENA, GaBi, ENVEST2, BECOST, and LISA, which commonly support LCAs for all industrial products. BEES [10] is currently the leading LCA tool in the United States (US). It was developed by the US National Institute of Standards and Technology in 1994 to assist with the selection of environmentally friendly construction materials. The program estimates the energy amount using an industry-related analysis of the construction industry, with a DB constructed using the integration method for construction materials including paper, pig iron, copper, and plywood. This technique was designed to consider both environmental and economic efficiencies, and its purpose was to develop tools with which building owners and the government can determine appropriate investment priorities. The program includes greenhouse effect, acidification, fossil fuel reduction, natural resources depletion, and air pollution standards as assessment items. Another program, ATHENA [11], was developed in Canada, and it provides detailed LCI DBs. Detailed LCI calculation results are summarized in graphic or tabular format. The program assesses the acquisition of ecological resources and intrinsic energy inputs by type, potential greenhouse effect, measurements of water and air toxicity, and solid waste. GaBi [12] was developed in Europe to manage sustainability through LCAs of organizations, plants, processes, or products. In addition to LCAs, GaBi can also assess life cycle engineering, environmental design, and energy efficiency, and materials and production processes are distinguished in detail. The LCA can be performed after the "GaBi DB manager" selects values for normalization and weighting, and the sensitivity analyses of various variables can be conducted.

Of particular relevance to this paper, Eco-Quanum [13] is the first building LCA software that assesses environmental impacts based on the energy consumption in a building. ENVEST2 [14] is another LCA software program, developed by Building Research Establishment, Ltd. (UK) for buildings. ENVEST2 supports the building assessment items stipulated by BREEAM, an eco-friendly building certification system used in the UK. ENVEST2 provides environmental and economic efficiency information through an environmental impact average value called Eco Point. In addition, ENVEST2 can be easily used at the initial stages of planning and designing a building. BECOST [15] is a building LCA software program developed by the VTT Technical Research Center of Finland, Ltd. BECOST uses environmental impact data derived from the LCA of a building, including design, construction, maintenance, and deconstruction, and the program is used to develop reduction technologies. LISA [16] is an LCA-building software program developed by Newcastle University and BHP Institute. LISA offers an impressive environmental factor analysis capability because of its built-in DB in various materials. In addition, because LISA's interface consists of simple input and output formats, it can be easily used by the assessor.

Upon analyzing domestic and overseas LCA software programs, it was found that they have similar stepwise assessment methods, with material input by process flow and DB application, whereas they were found to be different in terms of data input, calculation system, and result analysis. In particular, a common problem was the lack of a standard data format that can be implemented to systematically manage the assessment factor DB. Therefore, even if the purposes of the software programs are the same, the data collected by each user and the assessment methods applied will be different. Therefore, the assessment results are not only inconsistent but are also limited to one-time assessments, and the different systems impose a limit on the ability to objectively compare and analyze their results. For these reasons, data entry and allocation technologies are required to support the universal use of concrete-specific LCA, and LCA software is required that simplifies the assessment process, including the establishment of a concrete LCA process and the automated determination of the input data items.

Marinković, S. reviews the research by Panesar et al. The research was an analysis of the impact of the selection of functional unit (FU) on the life cycle assessment of green concretes. The research presented the influence of six different FUs (which included volume, compressive strength, durability of concrete, binder intensity, and a combination of these) on several impact categories [17].

Mohammadi, J. et al. investigated the environmental impacts associated with the manufacture of fourteen benchmark concrete products in Australia. This research provides datasets for reference cementitious construction materials (including concrete, mortar, grout, and render), which aid the construction industry in evaluating its environmental impact [18].

Yazdanbakhsh, A. et al. compared the environmental impacts of using coarse natural aggregate (NA) and coarse recycled aggregate (RA) to produce concrete in the New York City area, by means of a unique LCA framework that incorporates comprehensive regional data [19].

Vieira, D. R. analyzed a literature review conducted to present the state-of-the-art of LCA methods applied to the manufacturing of common and ecological concrete. Concepts and tools are discussed [20].

This research presented relevant landmarks on sustainable development and materials efficiency on the assessment of the environmental impact of construction products. An overview on the European Construction Products Regulation (CPR) is given followed by an outline of the book [21].

3. GHG Emissions LCA of Concrete

3.1. Overview

To develop assessment software for concrete-specific life cycle GHG emissions, the LCA guideline ISO 14044 was applied to concrete.

3.2. Aim and Scope

(1) Function and Function Unit Setting

A general type of processed concrete was selected as the target product of the LCA in this study, and among the various functions of concrete, the formation of concrete structures and concrete products was selected as the main function on which the LCA was based. As shown in Table 1, the function unit was set at 1 m^3 of concrete production, a value that is convenient for data management and utilization, based on the aforementioned main function.

Table 1. Function and function unit.

	Content
Name	concrete
Function	concrete structures formation
Function unit	1 m^3 production of concrete

(2) System Boundary Setting

The production stage (cradle to gate) shown in Figure 1 was selected as the LCA range for this study. The concrete production stage was subdivided into raw material acquisition, transportation, and manufacturing stages. The GHG emissions contribution of each stage to the atmosphere was assessed [17].

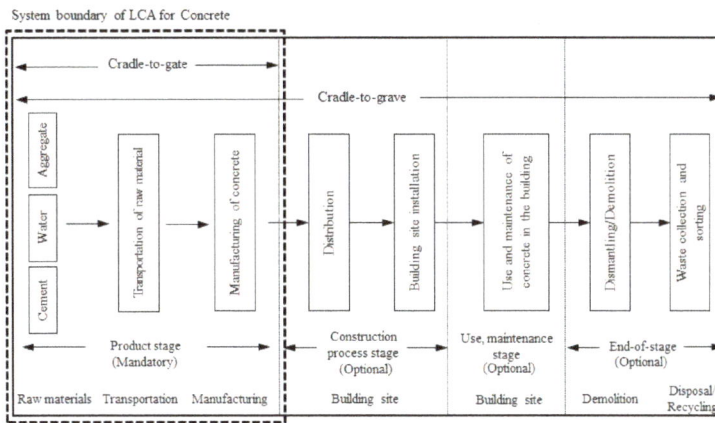

Figure 1. System boundary of concrete life cycle assessment (LCA).

3.3. Inventory Analysis

The input and output factors, including energy, raw materials, products, and waste, were analyzed for the LCA range of concrete. The LCI DB for each of the materials and energy sources inputted to the concrete production was investigated, as shown in Table 2. The LCI DBs for the input materials and energy sources that were used in this GHG emissions LCA were obtained from by the Korea Ministry of Land, Infrastructure, and Transport, the Ministry of Knowledge Economy, and the Ministry of Environment in South Korea. Because LCI DBs differ by country, DBs provided by the relevant country must be used.

However, because the LCI data for recycled aggregate, Ground Granulated Blast Furnace Slag (*GGBS*), fly ash, and admixtures are not yet available in Korea, this study used foreign LCI data of Ecoinvent database.

Especially, LCI database of Ground Granulated Blast Furnace slag (GGBS) applies the process of industrial by-product recycling. Thus, it only assesses the environmental impact of processes after the blast furnace slag is discharged as a form of a by-product. The database is derived from the amount of energy used in cooling, crushing, and handling, after the slag is discharged from the blast furnace of a steel mill.

Swiss ecoinvent database is a reliable database utilized in life cycle assessments conducted in various fields in Korea. In the future, if a Korean database were developed, comparative analysis could be conducted with the ecoinvent database.

Table 2. Categories of life cycle index (LCI) database (DB) name and source.

	Division	Reference	Nation
	Cement	National LCI	Korea
	Coarse aggregate	National LCI	Korea
	Fine aggregate	National LCI	Korea
	recycled fine aggregate	Ecoinvent	Switzerland
Raw material	recycled coarse aggregate	Ecoinvent	Switzerland
	Blast furnace slag	Ecoinvent	Switzerland
	Fly ash	Ecoinvent	Switzerland
	Water	National LCI	Korea
	Chemical admixture	Ecoinvent	Switzerland
	Electric	National LCI	Korea
Energy	Diesel	National LCI	Korea
	Kerosene	National LCI	Korea
Transportation	Truck	National LCI	Korea
	Train	National LCI	Korea

3.4. Impact Assessment

3.4.1. Raw Materials Stage

The GHG emissions of the concrete raw materials stage were calculated as the sum of the amount of each constituent material (kg) included in the 1-m^3 concrete production multiplied by the corresponding GHG emission per unit, as follows [18]:

$$GHG_M = \sum(M(i) \times Unit\ MCO_{2eq}) \qquad (1)$$

where GHG_M represents the GHG emissions during the production of concrete raw materials [kg-CO$_2$eq/m^3]; M(i) is the amount of each raw material in 1 m^3 of concrete production [kg], where i = 1, 2, 3 and 4, representing cement, aggregate, admixture, and water, respectively; and Unit MCO$_{2eq}$ is the GHG emission of each raw material per unit [kg-CO$_2$eq/kg].

3.4.2. Transportation Stage

In general, cement is transported by railway from a production plant to a shipment base near the demand site. The cement is then transported to a ready-mixed concrete plant using a bulk cement truck. To estimate GHG emissions during the transportation stage, the number of transportation equipment components required to transport each concrete constituent material was calculated using the total quantity of each material and the loading capacity of the relevant transportation means. Considering the calculated number of transportation equipment components, transportation distance, and fuel efficiency, GHG emissions during the transportation stage were assessed. The equation for calculating GHG emissions during the transportation stage is as shown in Equation (2) [19].

$$GHG_T = \sum(M(i)/L_t) \times (d/e) \times Unit\ TCO_{2eq} \qquad (2)$$

where GHG_T represents the GHG emissions during the transportation of concrete raw materials [kg-CO$_2$eq/m^3]; M(i) is the amount of each raw material in 1 m^3 of concrete production [ton], where i = 1, 2 and 3, representing cement, aggregate, and admixture, respectively; L_t is the loading capacity of the transportation equipment component for the relevant material [ton]; d is the transportation distance [km]; e is the fuel efficiency [km/L]; and Unit TCO$_{2eq}$ is the GHG emission of the transportation equipment per unit [kg-CO$_2$eq/L].

3.4.3. Manufacturing Stage

In this stage, GHG emissions can be derived using the energy consumption of the concrete manufacturing facilities as shown in Figure 2. The amount of energy (power and/or oil) used in the concrete batch plant must be derived. In this study, energy ratios were calculated by applying the differences in the operation time of the mixing equipment corresponding to different concrete strengths. In addition, because by-product and waste emissions data are annually collected, the values assigned according to the ratios of the production by concrete strength to the total concrete production were derived as shown in Figure 3. The equation for calculating GHG emissions during the manufacturing stage is as shown in Equation (3) [20].

$$GHG_F = \sum[(E(i)/R) \times \text{Unit } FCO_{2eq}] \tag{3}$$

where GHG_F represents the GHG emissions during the unit concrete manufacturing stage [kg-CO$_2$ eq/m^3]; R is the annual concrete production [m^3/year]; E(i) is the annual consumption of each energy source [unit/year], where i = 1, 2, and 3, representing power, oil, and water; and Unit FCO_{2eq} is the GHG emission of each energy source per unit [kg-CO$_2$eq/unit].

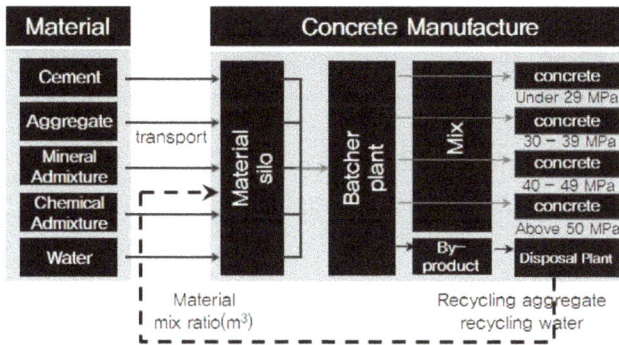

Figure 2. By-product data flow diagram in concrete manufacturing stage.

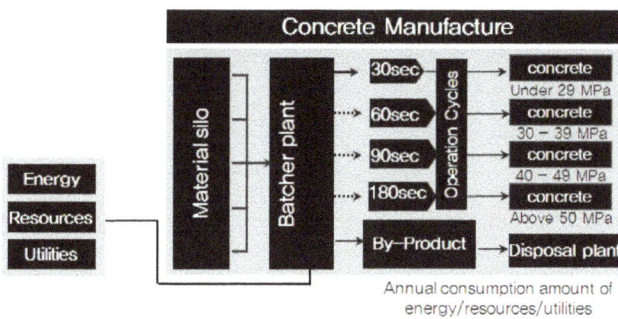

Figure 3. Energy data flow diagram in concrete manufacturing stage.

4. Assessment Software Development

4.1. Overview

The concrete GHG emissions LCA software was developed based on the previously constructed assessment factor database and data allocation system. The assessment process consists of an information input process and an assessment analysis process. The information input process refers to the process of collecting and inputting concrete data, and inputs the project overview, combined materials, and manufacturing process. The assessment analysis process refers to the process of calculating the LCA and GHG emissions using the previous input information [21,22].

4.2. Concrete GHG Emissions LCA Software

As shown in Figure 4, the concrete GHG emissions LCA software is executed in the following order: project overview input, combined material information input, manufacturing process information input, LCA stage, and project completion stage. The data allocation system refers to a system that properly distributes collected data for each stage and inputs the emission data. The assessment automation system refers to a system in which the procedures for the concrete GHG emissions LCA are automatically performed in the software's assessment module by setting basic initial values. In the result analysis, the effects of GHG, resources, energy, and economic efficiency are analyzed based on the assessment results derived through inventory analysis and impact assessment, considering the special characteristics of each stage of the life cycle as shown in Table 3. The program's management technology stores the database that is used for the assessment and the assessment results as data in the material list format, so that projects can be compared with one another.

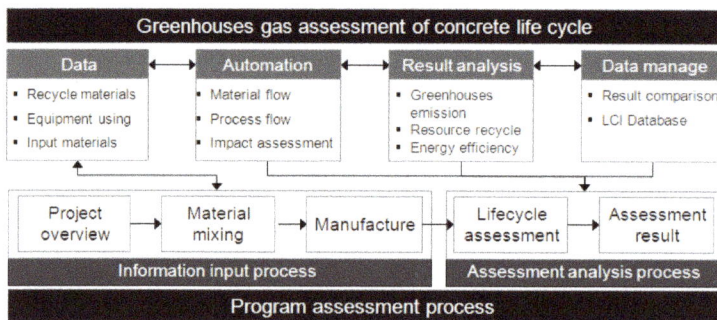

Figure 4. Linkages of greenhouse gas (GHG) assessment techniques for concrete.

Table 3. LCA process for concrete GHG emissions.

	Detailed Assessment Items
Project overview	- basic information: manufacturer, product specifications
Input of the combined material information	- material information: type of material, mix proportion, area of production, transport mode - supplier information: material suppliers, location - transport details: material transport, transportation routes/destinations
Input of manufacturing process information	- annual input material date: materials, energy, fuel, renewable energy - annual output material date: concrete, by-product/waste, recycling materials

Table 3. *Cont.*

	Detailed Assessment Items
LCA	- system boundaries, functional units - environmental impact categories: global warming - material flow analysis and review - applying assignment, material balance review - derived materials and resources list - applying emission factors of list items - emission factors (classification, characterization) - GHG emissions, resource/energy usage
Assessment result	- LCA results of GHG - project assessment database storage

4.2.1. Project Overview Input Stage

In the project overview input stage, the assessor, assessment purpose, and assessment scope are set, and basic information for assessing the GHG emissions of the project are entered. The basic information includes the manufacturer's name and address, material specifications, and application period (winter or standard period) of the concrete [23,24].

4.2.2. Combined Material Information Input Stage

In the combined material information input stage, the types and quantities of cement, water, aggregate, and admixture are directly entered based on the mixing design of 1 m^3 of concrete. In addition, the location information or transportation distance of each material company can be directly entered or automatically calculated, and the information on the transportation equipment used for each material is entered when the national LCI DB item is selected.

4.2.3. Manufacturing Process Information Input Stage

The manufacturing process information input stage establishes the data flow of all inputs and outputs during concrete production by applying a data allocation system. A period for collecting the data on various specifications of the concrete produced in the manufacturing process is set, and the data on the input and output materials collected during the period are entered. The allocation system is applied to the input data so that the data for the LCA can be created.

4.2.4. LCA Performance Stage

In the LCA performance stage, the LCA of concrete products is performed using the previously entered combined material information and manufacturing process information, GHG emissions are estimated, and the assessment results are compared and analyzed. The LCA was designed to perform goal and scope definition, life cycle inventory analysis, life cycle impact assessment, and life cycle result interpretation processes, and default values were established for each process so that the stepwise data flow could be consistently entered. In this way, the input data and the emissions data are allocated to conform to the concrete product function unit, and GHG emissions are calculated using the national LCI DBs. The LCA performance stage is divided into a process flow chart creation stage, a material list creation stage, an impact assessment stage, and a result analysis stage [25,26].

(1) Process Flow Chart Creation

The input data and the emissions data entered in the information input process are automatically placed in each unit process and in function unit levels corresponding to 1 m^3 of concrete production.

Thus, a process flow chart is created for concrete production. The types and quantities of material flows can be identified for each unit process, which include raw materials, resources and energy, renewable energy, products, by-products, and waste.

(2) Material List Creation

The material list is divided into data categories, material names, numerical values, and units. The data categories refer to the data types of the corresponding materials, and the flows and the purposes of the materials can be confirmed. In addition, a mass balance check is possible by comparing the weights of the items in the material list.

(3) Impact Assessment and Result Analysis

As shown in Figure 5D,E, GHG emissions can be estimated based on the material list. Detailed analysis functions include the material reuse rate and the energy reduction rate. The material reuse rate is a value in an index form obtained by analyzing the amount of raw material input and the recycled amount of the recovered material. The energy reduction rate is a value in an index form representing fuel consumption and renewable energy production in each process [27].

(**A**) Main page

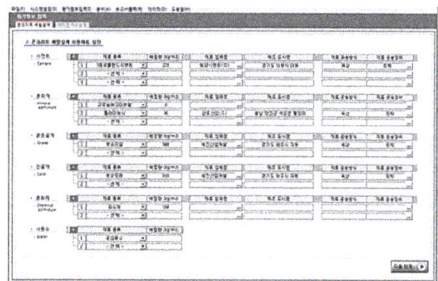

(**B**) Raw material information input page

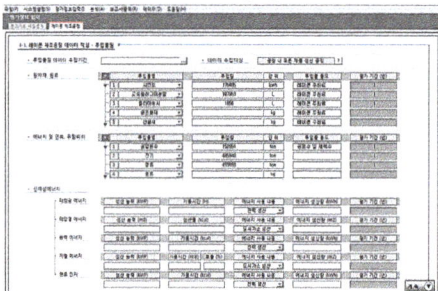

(**C**) Manufacturing information input page

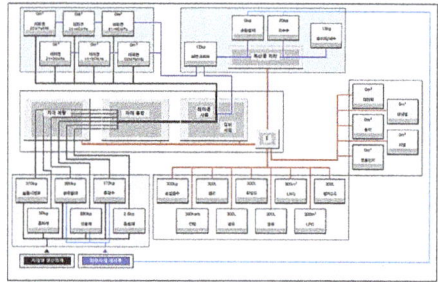

(**D**) Flow chart of LCA for concrete

Figure 5. *Cont.*

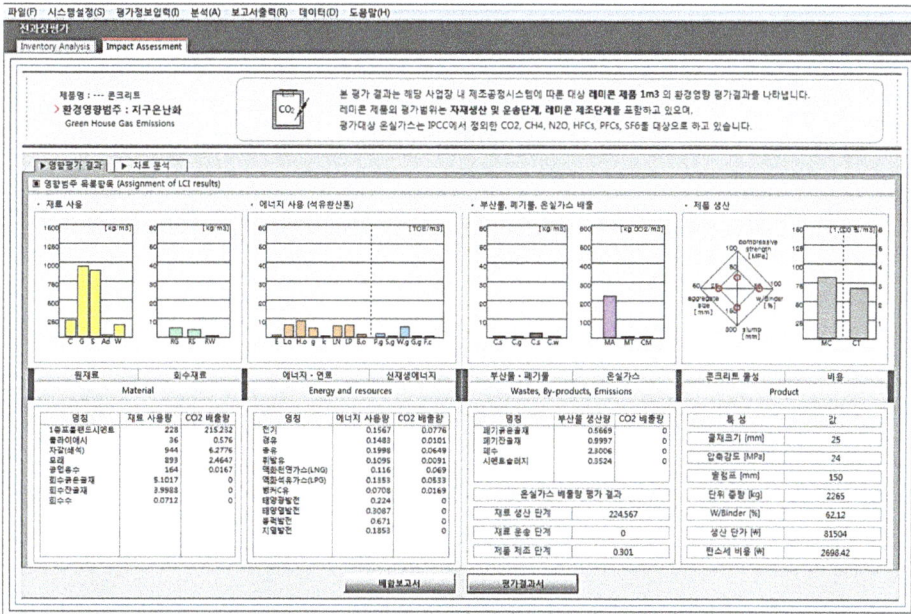

(**E**) Impact assessment result page

Figure 5. GHG emissions LCA program for concrete.

5. Case Analysis

5.1. Overview

To verify the reliability of the software developed in this study, a case analysis was conducted. The concrete procured at a domestic construction site was selected as the analysis target. In addition, the results of the proposed software were compared with those obtained using the COOL software to verify reliability.

5.2. Analysis Target

The concrete procured at a construction site in Seongnam City, Gyeonggi-do, was selected as the analysis target. As shown in Tables 4 and 5, the mix report, energy consumption, and production information were collected for the procured concrete. For the raw material consumption, the mix design for each concrete strength was applied. For the energy consumption, the annual consumption of the concrete manufacturer was examined and applied. For the concrete production, production by concrete strength was examined and applied. Waste generation was set at 2% of material input based on waste disposal regulations for the ready-mixed concrete industry, and the recovery rate of the aggregate was set at 100%. For the water recovery and reuse rate, the use details of the mix report were applied. For the transportation of concrete combined materials, the origins of the materials listed in the mix report were examined, and the relevant distances were calculated [28,29].

Table 4. Information on analysis target.

Division		Contents
Energy and resource consumption (per year)	electronic (kwh)	1,708,481
	diesel (L)	842,582
	kerosene (L)	57,475
	water (L)	93,274
Production (per year)	1–20 MPa	68,962
	21–30 MPa	490,041
	31–40 MPa	20,542
By-products emissions and disposal	by-products	All material inputs 2% (assumed)
	recycling rate (aggregate)	100% recovery of aggregate remaining (assumed)
	recycling rate (water)	60%
Materials/products during transportation	utility	Truck (limited to the same fuel vehicles)
	Distance (km)	Specific materials transport route

Table 5. Concrete mix design for target building.

	Mix Design of Concrete (kg/m^3)							
	OPC	FA	G	S	R/S	AE	W	R/W
18 MPa	236	35	958	454	455	1.22	97	64
24 MPa	336	0	936	453	443	3.05	131	0
30 MPa	334	59	954	393	394	2.95	103	68

OPC: ordinary portland cement, AE: chemical admixture, FA: fly ash, G: coarse aggregate, S: fine aggregate, R/S: recycled aggregate, R/W: recycled water, W: water.

5.3. Assessment Results

The LCAs of GHG emissions of the concrete products used at the target construction site, which had strengths of 18, 24, and 30 MPa, showed emissions of 267.5 kg-CO$_2$ eq/m^3, 364.9 kg-CO$_2$ eq/m^3, and 365.4 kg-CO$_2$ eq/m^3, respectively as shown in Figure 6. As the strength increased, the GHG emissions tended to rise. In addition, GHG emitted from the production of the concrete's raw materials (including cement and aggregate) represented the majority of the total GHG emissions as shown in Table 6.

This is related with the sealing process in which the highest amount of energy input and greenhouse gas are emitted during the cement production. To increase the temperature of rotary kiln 1000~1450 °C for producing clinker, carbon dioxide (CO$_2$) and methane (CH$_4$) were emitted due to the input of fuels such as B.C. oil, waste tire, waste plastics, bituminous coal, etc. Comparing the above results with those obtained using the COOL software, the error rate was found to be 1–2%. Based on this result, the GHG emissions LCA produced by the developed software, which is specialized for concrete and easy to use, was considered to be suitable for analyses, without using the existing software (COOL) that requires more expertise.

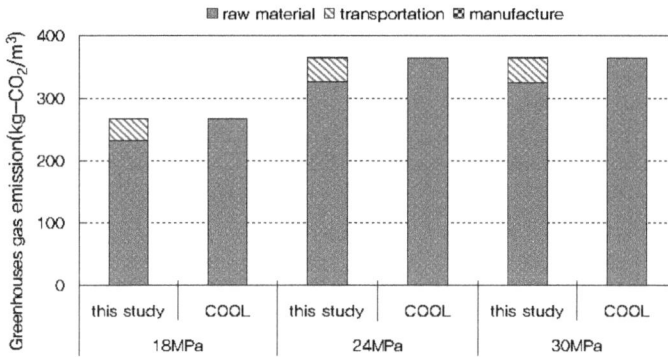

Figure 6. Assessment results and comparison of GHG emissions of different concrete strengths.

Table 6. Assessment results of GHG emissions of different concrete strengths.

Division		Greenhouse Gas Emissions [kg-CO$_2$ eq/m^3]	
Concrete Standard	Method \\ Stage	This Study	COOL Program
case 1 (18 MPa)	raw material	232.18	266.62
	transportation	34.61	
	manufacturing	0.67	0.12
	total	267.46	266.74
case 2 (24 MPa)	raw material	325.74	364.15
	transportation	38.57	
	manufacturing	0.67	0.12
	total	364.99	364.27
case 3 (30 MPa)	raw material	324.81	363.85
	transportation	39.21	
	manufacturing	1.34	0.12
	total	365.36	363.97

6. Conclusions

This study was conducted to develop concrete-specific GHG emissions LCA software, with the following conclusions.

Because the developed software has built-in data allocation technology, LCA automation technology, and result analysis and management technology, the software showed improved usability for GHG emissions LCAs of concrete compared to existing software programs.

Using the developed software, GHG emissions of concrete with 18, 24, and 30 MPa strengths were determined to be 267.5 kg-CO$_2$ eq/m^3, 364.9 kg-CO$_2$ eq/m^3, and 365.4 kg-CO$_2$ eq/m^3, respectively. These results were similar to those obtained using the existing COOL software: 266.7 kg-CO$_2$ eq/m^3, 364.3 kg-CO$_2$ eq/m^3, and 363.9 kg-CO$_2$ eq/m^3, respectively.

The reliability of the developed software was verified by comparing the case analysis assessment results of the proposed software with those of the existing software.

The case analysis indicated that the raw material stage accounted for more than 90% of the total GHG emissions, and the emissions during the transportation and manufacturing stages were found to be insignificant. Considering that the basic unit of greenhouse gas is 944 kg-CO$_2$ eq/ton from ordinary portland cement (OPC) among the raw materials of concrete, it is a greater influence than aggregate and admixture. Therefore, in order to decrease the greenhouse gas emissions of concrete, up to 60~80% of cement mixture amount (kg/m^3) is replaced with Furnace Slag Powder and Fly Ash. Currently, research is being conducted on this, which is likely to improve durability and be more economical because the basic unit of greenhouse gas emissions of GGBS (41.8 kg-CO_2 eq/ton) and F/A (15.3 kg-CO_2 eq/ton) is smaller than OPC.

The limitations of this study are applied to domestic and overseas LCI DB in the developed program. Because LCI DBs differ by country, DBs provided by the relevant country must be used. However, because the LCI DB for the admixture is not established in South Korea, the overseas ecoinvent DB was applied.

Further studies on the assessment of various environmental impact categories are required, including the database supplementation of the developed software and various case analysis tests. In addition to CO_2, substances such as NOx and SO_2 are also emitted during the concrete production process, and these also affect the acidification and eutrophication of the ecosystem, as well as global warming.

Acknowledgments: This research was supported by Basic Science Research Program through the National Research Foundation of Korea (NRF) funded by the Ministry of Education (NRF-2015R1D1A1A01059286).

Author Contributions: All the authors contributed substantially to all aspects of this article.

Conflicts of Interest: The authors declare no conflict of interest.

References

1. Leggett, J. A guide to the Kyoto protocol a treaty with potentially vital strategic implications for the renewables industry. *Renew. Sustain. Energy Rev.* **2008**, *12*, 345–351. [CrossRef]
2. National Life Cycle Index Database Information Network. Available online: http://www.edp.or.kr (accessed on 15 October 2017).
3. Ministry of Land, Transport and Maritime Affairs of the Korean Government. *National Database for Environmental Information of Building Products*; Ministry of Land, Transport and Maritime Affairs of the Korean Government: Sejong City, Korea, 2008.
4. Ecoinvent, Swiss Centre for Life Cycle Inventories. Available online: http://www.ecoinvent.org/database (accessed on 15 October 2017).
5. ISO 14044. *Life Cycle Assessment (Requirements and Guidelines)*; International Organization for Standardization: Geneva, Switzerland, 2006.
6. ISO/DIS 13315-2. *Environmental Management for Concrete and Concrete Structures-Part 2: System Boundary and Inventory Data*; International Organization for Standardization: Geneva, Switzerland, 2014.
7. Korea Environmental Industry & Technology Institute. *Carbon Labeling Guidelines*; Korea Environmental Industry & Technology Institute: Seoul, Korea, 2009.
8. Korea Environmental Industry & Technology Institute. Environmental Declaration of Product. Available online: http://www.edp.or.kr/edp/edp_intro.asp (accessed on 15 October 2017).
9. Korea National Cleaner Production Center. Environmental Declaration of Product. Available online: http://www.kncpc.or.kr/main/main.asp (accessed on 15 October 2017).
10. European thematic network on practical recommendations for sustainable construction (PRESCO WP2). *Inter-Comparison and Benchmarking of LCA-Based Environmental Assessment and Design Tool, Final Report*; PRESCO: Paris, France, 2005.
11. Available online: http://www.athenasmi.org (accessed on 15 October 2017).
12. Available online: http://www.gabi-software.com/ (accessed on 15 October 2017).
13. Available online: http://ecoquantum.com.au/ (accessed on 15 October 2017).
14. Available online: https://envest2.bre.co.uk/ (accessed on 15 October 2017).
15. Available online: http://www.vtt.fi/vtt/index.jsp (accessed on 15 October 2017).
16. Available online: http://www.lisa.au.com/ (accessed on 15 October 2017).
17. Marinković, S. On the selection of the functional unit in LCA of structural concrete. *Int. J. Life Cycle Assess.* **2017**, *22*, 1634–1636. [CrossRef]

18. Mohammadi, J.; South, W. Life cycle assessment (LCA) of benchmark concrete products in Australia. *Int. J. Life Cycle Assess.* **2017**, *22*, 1588–1608. [CrossRef]
19. Yazdanbakhsh, A.; Bank, L.C.; Baez, T.; Wernick, I. Comparative LCA of concrete with natural and recycled coarse aggregate in the New York City area. *Int. J. Life Cycle Assess.* **2017**, 1–11. [CrossRef]
20. Vieira, D.R.; Calmon, J.L.; Coelho, F.Z. Life cycle assessment (LCA) applied to the manufacturing of common and ecological concrete: A review. *Constr. Build. Mater.* **2016**, *124*, 656–666. [CrossRef]
21. Pacheco-Torgal, F.; Cabeza, L.F.; Labrincha, J.; De Magalhaes, A.G. *Eco-Efficient Construction and Building Materials: Life Cycle Assessment (LCA), Eco-Labelling and Case Studies*; Elsevier: Cambridge, MA, USA, 2014.
22. ISO 21930. *Environmental Declaration of Building Products*; International Organization for Standardization: Geneva, Switzerland, 2007.
23. Jung, J.S.; Lee, J.S.; An, Y.J.; Lee, K.H.; Bae, K.S.; Jun, M.H. Analysis of Emission of Carbon Dioxide from Recycling of Waste Concrete. *Arch. Inst. Korea* **2008**, *24*, 109–116.
24. Van den Heede, P.; De Belie, N. Environmental impact and life cycle assessment (LCA) of traditional and "green" concretes: Literature review and theoretical calculations. *Cem. Concr. Compos.* **2012**, *34*, 431–442. [CrossRef]
25. Kim, T.H.; Tae, S.H. Proposal of Environmental Impact Assessment Method for Concrete in South Korea: An Application in LCA (Life Cycle Assessment). *Int. J. Environ. Res. Public Health* **2016**, *13*, 1074. [CrossRef] [PubMed]
26. Jesus, G.M.; Justo, G.N. Assessment of the decrease of CO_2 emissions in the construction field through the selection of materials: Practical case study of three houses of low environmental impact. *Build. Environ.* **2006**, *41*, 902–909.
27. Kim, T.H.; Tae, S.H.; Chae, C.U. Analysis of Environmental Impact for Concrete Using LCA by Varying the Recycling Components, the Compressive Strength and the Admixture Material Mixing. *Sustainability* **2016**, *8*, 389. [CrossRef]
28. Kim, S.H.; Chae, C.W. *LCA Study on Concrete and Environmental Evaluation Method by Concrete Strength*; Korea Concrete Institute: Seoul, Korea, 2012.
29. Korea Concrete Institute. *Environmental Impact and CO_2 Assessment of Concrete Structures*; Korea Concrete Institute: Seoul, Korea, 2013.

sustainability

MDPI

Article

Emission of Volatile Organic Compounds (VOCs) from Dispersion and Cementitious Waterproofing Products

Mateusz Kozicki *, Michał Piasecki, Anna Goljan, Halina Deptuła and Adam Niesłochowski

Department of Thermal Physics, Acoustic and Environment, Building Research Institute, 00-611 Warsaw, Poland; m.piasecki@itb.pl (M.P.); a.goljan@itb.pl (A.G.); h.deptula@itb.pl (H.D.); a.nieslochowski@itb.pl (A.N.)
* Correspondence: m.kozicki@itb.pl; Tel.: +48-22-5796-187

Received: 30 April 2018; Accepted: 20 June 2018; Published: 26 June 2018

Abstract: Many different methods and indicators are commonly used for the assessment of indoor air quality (IAQ). One of them is pollution source control; among the sources, building materials are of special concern. This study presents a source characterization of waterproofing products used mainly in non-industrial buildings. The authors have attempted to fill some research gaps by determining emission factors for waterproofing materials. The work contains a summary of the volatile organic compounds (VOCs) emitted from dispersion and cementitious liquid-applied water-impermeable products. VOC emissions were determined in a 100-L stainless steel ventilated emission test chamber. Air samples were collected by an active method on Tenax TA®, while VOCs were analyzed using a TD-GC/MS method. Identified VOCs were also expressed as the total volatile organic compounds (TVOCs) and converted into area-specific emission rates q_A. The results for different groups of identified compounds (alcohols, benzene derivatives, aldehydes, ketones, ethers and esters) were compared. It was found that VOC emissions clearly decreased with time during each experiment, which lasted 28 days. It is further noted that different types of products were characterized by the emission of specific groups of compounds that were not emitted by other types of products. An essential factor in the elimination and minimization of the occurrence of sources of indoor air pollution is the appropriate selection of finishing materials, which should be characterized by as low as possible emission of VOCs. The results presented in this work can lead to practical applications in the selection of low-emission products for certified green buildings.

Keywords: VOC; indoor air quality; dispersion waterproofing products; cementitious waterproofing products; emission test chamber

---

1. Introduction

1.1. Sustainable Construction Materials

The analysis of the existing trend for the market of eco-construction materials indicates that the number of products with certain eco and environmental characteristics on the market is evidently growing. Investors and developers are willing to buy eco-construction products if they offer the same technical performance with the same price in comparison to the "standard products". The growing eco-awareness regarding adverse impact of buildings on the environment and human health in Europe is slowly becoming a decisive criterion in decision-making process of market stake holders, especially large contractors and investors. The accessibility of information concerning environmental aspects of construction products and elements is considered as a one of the driving forces of green market development in some leading EU countries (UK, Holland, France, Germany, and Sweden).

The worldwide trends show the parameters of well-being of building occupants from the very beginning should be included in design activities and further steps of construction process.

All commercial environmental assessment systems (Building Research Establishment Environmental Assessment Method (BREEAM), Leadership in Energy and Environmental Design (LEED) or German Sustainable Building Council (DGNB)) include the issue of low-emitting materials selection in the context of minimizing emissions and impact on users [1–3]. The results of studies on volatile organic compound concentrations may influence the results of certification of new office buildings.

The building design stage should include the development of an IAQ plan, which aims to include specifications, installations and activities that reduce IAQ. The plan must also consider the following issues: elimination and control of pollution sources; procedures for air exchange in the facility after completion of finishing works; conducting air quality tests by an independent laboratory; and procedures for maintaining proper air quality during use. An important element is the correct design and use of the building's ventilation system in accordance with the requirements of the relevant standards (EN 13779: 2007) [4], proper location of ventilation inlets and outlets, and proper design of the location of building objects in relation to external sources of pollution.

1.2. VOC Emission from Building Products

VOCs include various organic chemicals that have a high vapor pressure at room temperature and boiling points in an interval of approximately 50–100 °C to 240–260 °C [5]. According to the United States Environmental Protection Agency (EPA), the average level of VOCs in homes is five times higher than outdoors. For several hours after activities such as paint stripping, levels may be 100–1000 times background outdoor levels [6].

Two different techniques are used for the isolation and preconcentration of VOCs from air samples: a passive method employing a homemade passive sampler and a dynamic method based on a sorption tube [7]. Passive samplers fall into one of two categories: passive diffusion samplers, in which the transport of analytes takes place by way of free diffusion of the analyte through a stagnant gas layer, and passive samplers, in which the transport of analytes takes place by way of permeation of the analyte through a semipermeable membrane [8]. The dynamic method involves passing through the air sampler on a solid sorbent located in a glass or stainless steel tube. There are two methods of dynamic sampling with a sorbent tube—active and passive sampling. During active sampling, a calibrated pump is attached to the tube to draw air through the tube at a constant rate. Active sampling is used for short-term monitoring. In turn, passive sampling is designed for long-term monitoring and it works by molecular diffusion process. VOC samples are thermally desorbed from the adsorptive trap and concentrated by two-stage cryo-focusing process. Detection and identification are carried out by gas chromatography (GC) or gas chromatography mass spectrometry (GC/MS).

In recent years, an increasingly common alternative method of VOCs acquisition is based on sensors that change chemical information from the environment into analytically useful signals [9]. There are various types of commercially available chemical sensors: infrared radiation absorption (NDIR, Nondispersive Infrared Sensors) and thermal sensors (Pellistores) for evaluation of explosion risk and electrochemical (EC, Electrochemical), photoionization (PID, Photoionization Sensor) and semiconductor (MOS, Metal Oxide Semiconductor Sensors) for detection of toxic compounds [9,10]. Recent literature reviews [9–12] indicate advantages of the chemical sensors include mainly economic aspects such as simple design or low cost of manufacturing. However, these papers describe also limitations of sensors metrological parameters, such as poor sensitivity and selectivity, narrow measurement range or too high limit of detection.

VOCs are emitted by a wide array of products numbering in the thousands. Building materials play an important role in determining indoor environmental quality due to their large surface areas and permanent exposure to indoor air. Building materials release a wide variety of volatile organic compounds that are a main source of indoor air pollutants, worsening the air quality. They are absorbed into the body through respiratory system and the skin, causing headaches or eye and respiratory irritation. It is recognized that such contamination is responsible for long-term health effects associated with the development of allergy or asthma and discomfort among occupants [13–18].

The situation in which building occupants experience acute health and comfort effects that appear to be linked to the time spent in a building, but no specific illness or cause can be identified, is called "sick building syndrome" (SBS) [19,20]. Wargocki et al. assessed the perceived air quality and SBS symptoms while performing simulated office work in an existing office in which the air pollution level was modified by introducing or removing a pollution source [21,22].

A comprehensive review of emission rates from different building materials was described by Yu et al. [23]. The work compares the results of emission rates from liquid and paste products and indicates their average levels, e.g., wall primer (6 µg/m²·h), floor waxes (80,000 µg/m²·h) and wall and floor adhesives (220,000–271,000 µg/m²·h) [23]. Wilke et al. (2004) studied VOC emissions from low-emitting adhesives. The TVOC emission rates from tested adhesives ranged 900–10,000 µg/m²·h after 24 h. The emission rates decreased to below 200 µg/m²·h in 28 days for all adhesives. Acetic acid, 2-ethylhexanol and phenoxy propanol were the most abundant compounds [24].

For indoor air quality assessment using the IAQ_{index}, it is suggested to specify the dominant air VOC contaminants [25]. Some building materials are high VOC emitters, such as solvent paints and adhesives and the coatings and coverings on walls, ceilings and floors [26,27]. Paints which are widely used in residential indoor environmental might contribute significantly to poor IAQ, especially as they are applied to large surfaces. Many research projects have been undertaken to determine emissions of VOCs from freshly painted surface and their impact on IAQ [28]. Chang et al. investigated commercially available low-VOC latex paints and performed test emission inside 53-l stainless steel chambers [29]. Schieweck and Bock [30] studied the content of semi-volatile organic compounds (SVOCs) and the emission potential from paints by means of solvent extraction and chamber emission tests. Few studies have been performed so far concerning low VOC-emitting coatings [31–33] and no data have been reported on VOC emissions from waterproofing products. In the contemporary literature, no detailed data on this subject have been published.

Extensive research has been conducted on VOC emissions from building materials, including emission chamber studies [34,35] and modeling to predict gaseous emissions. Modeling methods are generally based on theory and fluid dynamics, and many kinds of analytical or numerical models have been proposed for characterizing VOC emissions from wet and dry materials [33,36]. Guo et al. [37] developed a classical vapor pressure and boundary layer model to predict VOC emissions, and Li et al. [38] improved the model by considering VOC emissions from paint applied to an absorptive substrate (mass transfer-based model).

Some European countries introduced rating systems for building products based on volatile pollutant emission levels. They developed numerous voluntary ecolabels, such as EMICODE®, The Blue Angel® or Finnish M1®, and obligatory rating systems, such as German AgBB and French VOC Regulations, with their emission classes (from A+ to C) based on emission testing. An example of a procedure for liquid building materials (adhesives, coatings or products used for installation) and decorative products (wallpaper or paint) is the guideline proposed by the French Agency for Food, Environmental and Occupational Health & Safety (Protocole AFFSET). The document contains a list of VOCs that may be emitted by these products together with the lower concentrations of interest (LCIs) that must be followed [39]. These regulations can increase the number of low-emitting products on the market.

1.3. Waterproofing Products

According to Standard EN 14891: 2017 [40], liquid-applied water-impermeable products are classified into three types: dispersion (DM, dispersion liquid-applied water-impermeable products), cementitious (CM, cementitious liquid-applied water-impermeable products) and reaction resin based (RM, reaction resin liquid-applied water-impermeable products). The cementitious products presented in this publication are divided into two subgroups: products mixed with water (RCM, rigid and waterproof) and polymer-modified products (FCM, flexible and waterproof). A summary of the parameters and typical applications for the tested types of waterproofing products is given in Table 1.

Waterproofing products based on reaction resins are applied mainly outdoors and/or in industrial buildings; hence, few data are presented in this study on VOC emissions indoors.

The waterproofing products have been described in various data sheets issued by the ZDB (Federation of the German Construction Industry), BEB (German Screed and Coverings Association) and DIBt (German Institute for Construction Technology). Since 2012, composite waterproofing has been classified into five exposure classes [41]: A0, moderate exposure to water without hydrostatic pressure; A, high exposure to water without hydrostatic pressure; B0, moderate exposure to water without hydrostatic pressure; B, high exposure to permanent internal water pressure; and C, high exposure to water without hydrostatic pressure with additional chemical action (used only for reaction resin products). These classes must be specified by the designer in function of the projected loads and moisture exposure.

DM are made on modified synthetic resin. After applying to the surface, they form a moisture-impermeable dense coating. They are commonly used for seamless and jointless sealing under ceramic coverings, gypsum plasterboard, cement-lime plasters and screeds or concrete surfaces in damp and wet areas, e.g., in private bathrooms with bathtubs, showers, and toilets and kitchens. After drying, waterproofed foils are a ready-to-use substratum for all finishing materials, such as glazed tiles, terracotta, wood floors and panels, and synthetic floor coverings. They are designed for places temporarily exposed to the impact of flowing water. Settlement and edge joints must be secured with sealing tape, and wall junctions and floor drains with sealing collars for walls/floors. DM are particularly convenient for use in tight or poorly accessible places.

Table 1. A summary of the parameters and typical applications for dispersion and cementitious types of waterproofing products.

Type of Waterproofing	Typical Applications	Consistency	Chemical Composition	Moisture Exposure Class
DM	for indoor use; for damp and wet areas, e.g., in private bathrooms with bathtubs, showers, toilets and kitchens	emulsion (semi-liquid damp waterproofing foil)	modified synthetic resin dispersion mixed with fillers and modifying agents	A A0
RCM	for indoor and outdoor use; balconies, patios, in showers, washrooms, toilets; swimming pools and water tanks	powder mix	cement, calcium carbonate, crystalline silica	A A0 B B0
FCM	for indoor and outdoor use; balconies, plaza decks, weathered roofs or topped roofs, mechanical and equipment rooms; water, wastewater, seawater and marine aquarium tanks and other reinforced concrete structures	emulsion + powder mix	dry mix: cement, mineral fillers and modifying agents + emulsion: water dispersion of plastic	

Execution of waterproofing jointless sealing is also possible by using FCM. These products are manufactured mortars, whose tightness provides appropriately selected hydrophobic additives. RCM are two-component products—their preparation requires the addition of an adequate amount of batched water to the cementitious component. FCM are two-component products composed of a dry mix of cement, mineral fillers and modifying agents and aqueous dispersions of plastics mixed in proportions recommended by the manufacturers.

2. Materials and Methods

Hydroisolations were obtained directly from domestic manufacturers as freshly made products. Nine different commercially available waterproofing products were evaluated by estimating their VOC emissions. These products were also admitted to trading on the European Union market. This study presents the results of research carried out in the period of 2015–2017.

The waterproofing products were tested under the same test conditions to ensure comparable and reproducible results. According to EN 16516: 2017 [42] and EN ISO 16000-9: 2006 [43], emission rates on Day 3 and Day 28 of a chamber test of waterproofing products should be reported. The VOC emission experiment for these products was carried out in an emission test chamber with a capacity of 0.1 m³. The size of the test chamber was appropriate for the intended use of building products with regard to the reference room described in accordance with Harmonized Standard of EN 16516: 2017 [42] and European Standard EN ISO 16000-9: 2006 [43]. The reference room serves as a convention for calculating and reporting product emissions reflecting the real room conditions.

2.1. Sample Preparation

The layer number and coating thickness depended on the data recommended by the manufacturer. In the case of the tested products, two layers were used, and the coating thickness was between 1 mm and 4 mm depending on the product. Minimum dry coat thickness was verified by on-site coat thickness measurements on the cured membrane using caliper rule. Then, each sample was placed onto a fiber-cement board (Figure 1) with defined dimensions appropriate to the size of the considered test chamber and the product loading factor L (20 cm × 20 cm for floors—L = 0.4; 20 cm × 70 cm for walls and floors—L = 1.4; 30 cm × 80 cm for all surfaces—L = 1.8). The loading factor is defined as the ratio of the area of the tested material in the reference room to the volume of the reference room (m²/m³). Each product was applied to the top side of a board by a paintbrush. After a drying period of 24 h under room conditions, the fiber-cement boards covered with the water-impermeable products were placed into an emission chamber to obtain the emissions from the effective area of the examined product. The consumption values are specified by the manufacturer in the product sheets. The emission area, consumption and loading factor for each tested waterproofing product are summarized in Table 2. The sample preparation procedure is in accordance with Harmonized Standard of EN 16516: 2017 [42] and European Standard EN ISO 16000-11: 2006 [44].

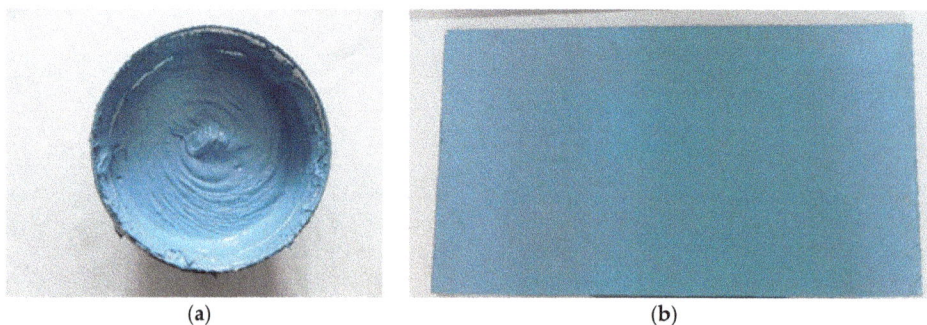

(a) (b)

Figure 1. Picture of DM product: in the form provided by the manufacturer (**a**); and as sample No. 2 which placed on top of fiber-cement board about 20 cm × 70 cm dimensions (**b**) (photo taken by an author).

Table 2. Sample parameters for the dispersion and cementitious types of waterproofing products tested in the study.

Sample Number	Type of Waterproofing	Sample Parameters		
		Emission Area (m^2)	Consumption (kg/m^2)	Loading Factor (m^2/m^3)
1		0.14	0.8	1.4
2	DM	0.14	1.1	1.4
3		0.18	1.5	1.8
4		0.09	3.0	0.4
5	RCM	0.14	2.8	1.4
6		0.18	4.0	1.8
7		0.14	3.4	1.4
8	FCM	0.14	3.5	1.4
9		0.14	3.0	1.8

2.2. Experimental System

The experimental system contained a stainless steel chamber, a clean air generation and humidification system, an air mixing system and a monitoring system (Figure 2). The mixing devices were made of low-emitting and low-adsorbing materials. The test chamber was connected to an electronic mass flow controller, which controlled the air flow, test time and air change rate. A small fan was running continuously to ensure good mixing of the air inside the chamber. The air velocity above the test samples falls within the range 0.1 m/s to 0.3 m/s. The cover from the emission chamber was filled with sealing material to avoid uncontrolled air exchange with external air. The volume of the chamber was 0.1 m^3. During the experimental period, the air exchange rate (n) was 0.5 h^{-1}, and the conditions inside the chamber were controlled by an air conditioning unit at $23 \pm 1\,°C$ and $50 \pm 5\%$ relative humidity.

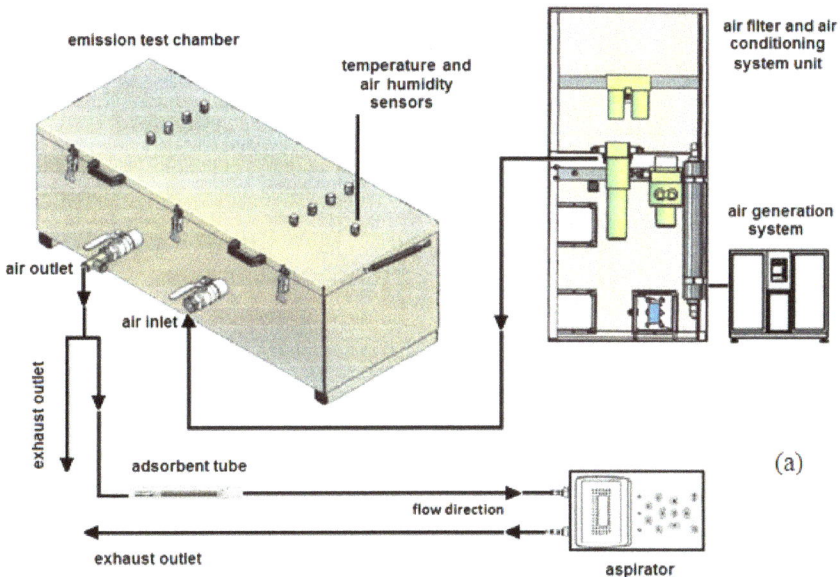

Figure 2. *Cont.*

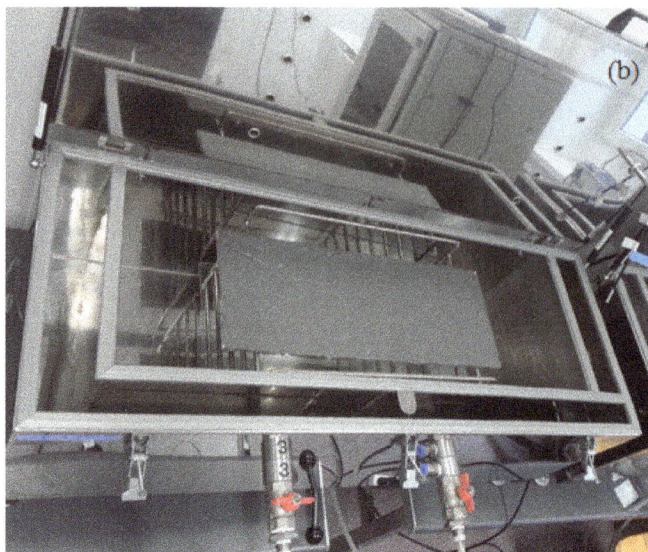

Figure 2. Schematic of the experimental system (**a**) and photograph of the emission test chamber with an examined sample (**b**) photo taken by an author.

2.3. VOC Identification

Air samples were adsorbed onto adsorption tubes filled with resin material. Tenax TA® tubes were chosen because of their efficiency to trap low concentrations of organics from air samples, chemical inertness, thermal stability and good storage stability.

Volatile organic compounds were thermally desorbed using a thermal desorption cold trap injector on a Perkin Elmer Turbo Matrix 350 system (Akron, OH, USA). The tubes were heated at 300 °C for 10 min under a helium flow (30 mL/min), and the substances were cryo-focused at −13 °C. The volatile compounds were injected into a GC capillary column by heating the cold trap to 280 °C for 3 min. The splitless injection mode was applied.

The separation and analysis of volatile compounds was achieved using a gas chromatograph equipped with a mass spectrometer (Perkin Elmer Clarus 500). The following GC oven temperature program was applied: initial temperature of 40 °C for 3 min, 10 °C per min to 300 °C, and final temperature of 300 °C for 1 min. The GC-MS transfer line temperature was held at 250 °C, and source temperatures was kept at 180 °C. Chromatographic separation was performed on a Restek RXI-5 capillary column (30 m, 0.25 mm ID, 1.0 μm df). The MS analysis was carried out over a scan range of 20–600 m/z with an ionization energy of 70 eV in electron ionization mode. The limit of quantification was <1 μg·m^{-3}.

The volatile compounds were identified by comparing the retention times of chromatographic peaks with the retention times of reference compounds and by searching the NIST 2011 mass spectral database. All volatile compounds with mass spectra with match factors $p \geq 90\%$ were considered identified.

Quantitative analysis of volatile organic compounds was performed with the external standard method that uses the relationship between peak areas (or height) and analyte concentration. The calibration curves were obtained based on six distinct known concentration levels. Standard solutions were prepared by injecting them into a tube in the inert gas flow; the spiked adsorbent tubes were then thermally desorbed in the same conditions as the samples.

3. Results and Discussion

To estimate the concentrations of identified VOCs, compound-specific response factors determined by certificated standard solution were used. Identified compounds were quantified using their individual response factors when the reference compound was available. In order cases, quantification was estimated by referring to a substance with similar chemical structure. For compounds that were not identified, the total area under the chromatographic curve was converted on a molecular mass basis to a concentration using the toluene equivalent [45].

To compare different products with similar target VOC profiles coming from different emission areas, an area-specific emission rate q_A [μg/m²·h] (also called an emission factor) was used. The relationship between the VOC concentration in the outlet air from the emission test chamber (c), the area-specific emission rate (q_A) and the area-specific air flow rate (q) of the emission test chamber can be expressed as a dependency:

$$q_A = c \times q = c \times (n/L) \tag{1}$$

where L is the product loading factor (m²/m³) (see Table 2) and n is the air change rate (h^{-1}), which is equal to 0.5 h^{-1} for the chamber used in the experiments.

For each product (1–9), double sampling was performed. The compound concentrations determined from duplicate air samples did not deviate more than ±15% from their mean. Table 3 presents the average concentrations (μg/m³) of identified VOCs after 3- and 28-day emission chamber experiments. These results ranged from 2 to 2646 μg/m³ and from 2 to 357 μg/m³ after 3 and 28 days, respectively. For RCM (Products 4–6), the VOCs identified were determined at the lowest level of concentration among the studied product groups, with a maximum of 36 μg/m³ (2-butoxyethanol) and 10 μg/m³ (benzoic acid) after 3 and 28 days, respectively. DM (Products 1–3) reached values of 382 μg/m³ (n-butanol) at 3 days and 74 μg/m³ (n-butanol) at 28 days of VOC emissions. The highest VOC concentrations were identified for FCM (Products 7–9), with a maximum of 2646 μg/m³ (n-butanol) after 3 days and 357 μg/m³ (2-propyl-1-pentanol) after 28 days. The number of identified compounds ranged between 1 (Product No. 1 for RCM) and 27 (Product No. 4 for FCM).

Table 3. Concentrations (μg/m³) and area-specific emission rates (μg/m²·h) of identified VOCs for three types of liquid-applied water-impermeable products (average values after 3- and 28-day emission test chamber experiments).

Identified VOC	Average Value after 3 Days		Average Value after 28 Days	
	Concentration [μg/m³]	Area-Specific Emission Rate [μg/m²·h]	Concentration [μg/m³]	Area-Specific Emission Rate [μg/m²·h]
DM				
product 1				
n-Butanol	11 ± 2	4 ± 1	3 ± 1	1 ± 1
Styrene	17 ± 3	6 ± 1	4 ± 1	1 ± 1
(1-Methylethyl)benzene	7 ± 1	2 ± 1	<2	<1
Propylbenzene	4 ± 1	1 ± 1	3 ± 1	1 ± 1
Benzaldehyde	19 ± 3	7 ± 1	<2	<1
2-Ethyl-1-hexanol	64 ± 12	23 ± 4	32 ± 6	11 ± 2
2-Ethylhexyl acetate	50 ± 9	18 ± 3	28 ± 5	10 ± 2
2-Ethylhexyl acrylate	9 ± 2	3 ± 1	6 ± 1	2 ± 1
Octyl propionate	11 ± 2	4 ± 1	6 ± 1	2 ± 1
TVOC/Sum	211 ± 38	75 ± 14	78 ± 14	28 ± 5

Table 3. *Cont.*

Identified VOC	Average Value after 3 Days		Average Value after 28 Days	
	Concentration [μg/m³]	Area-Specific Emission Rate [μg/m²·h]	Concentration [μg/m³]	Area-Specific Emission Rate [μg/m²·h]
DM				
product 2				
n-Butanol	382 ± 69	136 ± 24	74 ± 14	26 ± 5
Butyl ether	47 ± 8	17 ± 3	7 ± 1	2 ± 1
Butyl acrylate	47 ± 8	17 ± 3	7 ± 1	2 ± 1
Butyl propionate	6 ± 1	2 ± 1	<2	<1
2-Ethyltoluene	7 ± 1	2 ± 1	<2	<1
Propylbenzene	4 ± 1	1 ± 1	<2	<1
Butyl butyrate	4 ± 1	1 ± 1	<2	<1
1-(2-Butoxyethoxy)ethanol	37 ± 7	13 ± 2	42 ± 8	15 ± 3
Others	18 ± 3	6 ± 1	18 ± 3	6 ± 1
TVOC/Sum	383 ± 69	137 ± 25	120 ± 22	43 ± 8
product 3				
n-Butanol	109 ± 20	30 ± 5	8 ± 1	2 ± 1
1,2-Propanediol	62 ± 11	17 ± 3	14 ± 3	4 ± 1
Cyclohexanol	99 ± 18	2 ± 1	53 ± 10	15 ± 3
Styrene	216 ± 39	60 ± 11	20 ± 4	6 ± 1
TVOC/Sum	486 ± 88	135 ± 24	95 ± 17	26 ± 5
RCM				
product 4				
2-Ethyl-1-hexanol	8 ± 2	10 ± 2	2 ± 1	2 ± 1
TVOC/Sum	8 ± 2	10 ± 2	2 ± 1	2 ± 1
product 5				
2-Methyl-1-propanol	20 ± 4	7 ± 1	<2	<1
2-Methylcyclopentanol	2 ± 1	1 ± 1	<2	<1
Benzoic acid	12 ± 2	4 ± 1	10 ± 2	4 ± 1
Others	17 ± 3	6 ± 1	2 ± 1	1 ± 1
TVOC/Sum	51 ± 9	18 ± 3	12 ± 2	4 ± 1
product 6				
n-Butanol	16 ± 3	4 ± 1	5 ± 1	1 ± 1
1-Methoxy-2-propanol	6 ± 1	2 ± 1	2 ± 1	1 ± 1
1-Pentanol	5 ± 1	1 ± 1	2 ± 1	1 ± 1
Hexanal	8 ± 2	2 ± 1	2 ± 1	1 ± 1
2-Butoxyethanol	36 ± 6	10 ± 2	6 ± 1	2 ± 1
TVOC/Sum	71 ± 13	20 ± 4	17 ± 3	5 ± 1
FCM				
product 7				
2-Methyl-1-propanol	76 ± 14	27 ± 5	11 ± 2	4 ± 1
n-Butanol	1382 ± 249	494 ± 89	138 ± 25	49 ± 9

Table 3. *Cont.*

Identified VOC	Average Value after 3 Days		Average Value after 28 Days	
	Concentration [μg/m³]	Area-Specific Emission Rate [μg/m²·h]	Concentration [μg/m³]	Area-Specific Emission Rate [μg/m²·h]
Ethylene glycol	39 ± 7	14 ± 3	10 ± 2	4 ± 1
3-Methyl-2-heptene	6 ± 1	2 ± 1	<2	<1
Hexanal	12 ± 2	4 ± 1	<2	<1
Butyl acetate	140 ± 25	50 ± 9	4 ± 1	1 ± 1
2-Methyl-1-pentanol	4 ± 1	1 ± 1	<2	<1
Ethylbenzene	49 ± 9	17 ± 3	<2	<1
o,m-Xylene	146 ± 26	52 ± 9	5 ± 1	2 ± 1
Butyl ether	227 ± 41	81 ± 15	2 ± 1	1 ± 1
p-Xylene	68 ± 12	24 ± 4	<2	<1
Butyl propionate	113 ± 20	40 ± 7	<2	<1
3-Methyl-4-heptanone	44 ± 8	16 ± 3	<2	<1
3-Methyl-2-heptanone	8 ± 2	3 ± 1	<2	<1
Allylbenzene	7 ± 1	2 ± 1	<2	<1
Propylbenzene	37 ± 7	13 ± 2	<2	<1
Benzaldehyde	42 ± 8	15 ± 2	2 ± 1	1 ± 1
3-Ethyltoluene	12 ± 2	4 ± 1	<2	<1
Butyl butyrate	47 ± 8	17 ± 3	3 ± 1	1 ± 1
3-Carene	5 ± 1	12 ± 2	<2	<1
2-Propyl-1-pentanol	475 ± 86	170 ± 30	357 ± 64	127 ± 23
Benzyl alcohol	55 ± 10	20 ± 4	35 ± 6	12 ± 2
Dibutoxymethane	9 ± 2	3 ± 1	<2	<1
2-Ethylhexyl acetate	115 ± 21	41 ± 7	85 ± 15	30 ± 5
Butyl 2-ethylhexyl ether	6 ± 1	2 ± 1	4 ± 1	1 ± 1
2-Ethylhexyl acrylate	11 ± 2	4 ± 1	5 ± 1	2 ± 1
2-Ethylhexyl butyrate	4 ± 1	1 ± 1	2 ± 1	1 ± 1
TVOC/Sum	2647 ± 476	945 ± 170	614 ± 111	219 ± 39
product 8				
n-Butanol	2524 ± 454	901 ± 162	230 ± 41	82 ± 15
Butyl ether	578 ± 104	206 ± 37	60 ± 11	21 ± 4
1-Methoxy-2-propanol	128 ± 23	46 ± 8	6 ± 1	2 ± 1
1,2-Propanediol	114 ± 21	41 ± 7	21 ± 4	7 ± 1
Butyl acetate	146 ± 26	52 ± 9	2 ± 1	1 ± 1
Ethylbenzene	104 ± 19	37 ± 7	2 ± 1	1 ± 1
Butyl acrylate	125 ± 23	45 ± 8	2 ± 1	1 ± 1
Butyl propionate	178 ± 32	64 ± 12	7 ± 1	2 ± 1
Butyl butyrate	98 ± 18	35 ± 6	7 ± 1	2 ± 1
Benzaldehyde	52 ± 27	19 ± 3	6 ± 1	2 ± 1
Others	66 ± 12	24 ± 4	10 ± 2	4 ± 1
TVOC/Sum	4146 ± 750	1481 ± 267	345 ± 62	123 ± 22

Table 3. *Cont.*

Identified VOC	Average Value after 3 Days		Average Value after 28 Days	
	Concentration [µg/m³]	Area-Specific Emission Rate [µg/m²·h]	Concentration [µg/m³]	Area-Specific Emission Rate [µg/m²·h]
	FCM			
	product 9			
n-Butanol	2646 ± 476	735 ± 132	173 ± 31	48 ± 9
Butyl acetate	82 ± 15	23 ± 4	3 ± 1	1 ± 1
Ethylbenzene	62 ± 11	17 ± 3	3 ± 1	1 ± 1
Xylene	28 ± 5	8 ± 2	3 ± 1	1 ± 1
Propylbenzene	70 ± 13	19 ± 3	5 ± 1	1 ± 1
4-Heptanone	27 ± 5	7 ± 1	<2	<1
3-Methyl-4-heptanone	44 ± 8	12 ± 2	<2	<1
Butyl propionate	92 ± 17	26 ± 5	8 ± 2	2 ± 1
Butyl butyrate	31 ± 6	9 ± 2	<2	<1
Benzaldehyde	96 ± 17	27 ± 5	7 ± 1	2 ± 1
Isobutyl ether	178 ± 32	49 ± 9	3 ± 1	1 ± 1
TVOC/Sum	3229 ± 580	897 ± 161	205 ± 37	57 ± 10

Table 3 is supplemented by uncertainty for particular compounds. The expanded uncertainty was calculated using a factor of k = 2, which corresponds to the level of confidence of approximately 95%. Uncertainty was determined based on available data including: data on the accuracy of the measurement system used and experimentally obtained data on repeatability.

The only compound emitted from DM was styrene (Product Nos. 1 and 3). DM are modified synthetic resin dispersions often consisting of styrene and/or acrylate copolymers. Acrylates and acetates were also identified from products in the two-component polymer-modified cementitious group. Butyl acetate is more commonly used for industrial coatings, but ethyl acetate is preferable when faster evaporation is needed [23]. In turn, ketones were present only in FCM emissions (Product Nos. 7 and 9). These compounds are added in small amounts to increase the solubility of difficult-to-dissolve resins in industrial products [23].

The compounds not assigned to the groups of alcohols, benzene derivatives, aldehydes and ketones, or ethers and esters were identified as 3-methyl-2-heptene or 3-carene (Product No. 7).

In addition, Table 3 lists the calculated total volatile organic compound (TVOC) concentrations and area-specific emission rates q_A [µg/m²·h]. TVOC was calculated by summing identified and unidentified compounds eluting between n-hexane and n-hexadecane using response factor for toluene. It was found that the amount of TVOCs emitted from each type of waterproofing product after 3 days was approximately 10 times more than the emission after 28 days.

Figure 3 indicates that the area-specific emission rate values for FCM (violet on the chart) were considerably higher (up to 1481 µg/m²·h after 3 days and 219 µg/m²·h after 28 days) than those of the other groups on both days tested. Indirect q_A values were determined for DM (green on the chart). The lowest q_A values (down to 10 µg/m²·h after 3 days and 3 µg/m²·h after 28 days) were identified for RCM (blue on the chart).

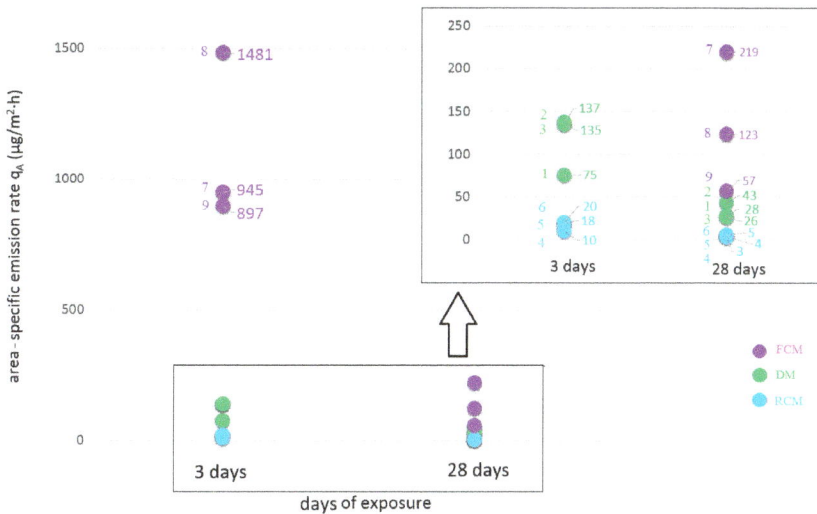

Figure 3. Comparison of the area-specific emission rate q_A [$\mu g/m^2 \cdot h$] after 3 and 28 days of the chamber experiment with dispersion and cementitious types of waterproofing products. On the left side of each point is located the product number, while on the right is the proper q_A value [mg/m^3].

The area-specific emission rates of the identified VOCs are summarized in view of the most numerous group of compounds, including alcohols, benzene derivatives, aldehydes and ketones, and ethers and esters. The results are tabulated in Table 4, and a distribution chart is presented in Figure 4. These data show that RCM did not emit benzene derivatives or ethers and esters. Two other groups (FCM and DM) were diverse in terms of the compounds emitted from them.

The data analysis shows that alcohols were the dominant group of emitted compounds. They constituted 69% of the total emissions after 3 days and 76% after 28 days. The most frequently occurring compound was butyl alcohol, which could be used as a solvent. The group with the second highest emissions was ethers and esters, composing 20% of the total emissions after 3 days and 17% after 28 days. Other groups, i.e., benzene derivatives and aldehydes and ketones, constituted a low percentage of the total emissions in all cases.

Table 4. The area-specific emission rates q_A [$\mu g/m^2 \cdot h$] of the four groups of compounds identified from the different types of liquid-applied water-impermeable products after 3 and 28 days in the chamber test emission experiment.

Group of Compounds	Area-Specific Emission Rate q_A [$\mu g/m^2 \cdot h$]					
	DM					
	Product 1		Product 2		Product 3	
	3 Days	28 Days	3 Days	28 Days	3 Days	28 Days
Σ alcohols	27	13	150	41	75	21
Σ benzene derivatives	10	3	4	1	60	56
Σ aldehydes and ketones	7	1	-	-	-	-
Σ ethers and esters	25	14	37	6	-	-

Table 4. *Cont.*

Group of Compounds	Area-Specific Emission Rate q_A [$\mu g/m^2 \cdot h$]					
	RCM					
	Product 4		Product 5		Product 6	
	3 days	28 days	3 days	28 days	3 Days	28 Days
Σ alcohols	10	3	7	1	18	4
Σ benzene derivatives	-	-	-	-	-	-
Σ aldehydes and ketones	-	-	15	2	2	1
Σ ethers and esters	-	-	-	-	-	-
	FCM					
	Product 7		Product 8		Product 9	
	3 Days	28 Days	3 Days	28 Days	3 Days	28 Days
Σ alcohols	725	198	988	92	735	48
Σ benzene derivatives	129	6	37	1	44	3
Σ aldehydes and ketones	50	9	19	2	46	3
Σ ethers and esters	238	38	402	28	106	4

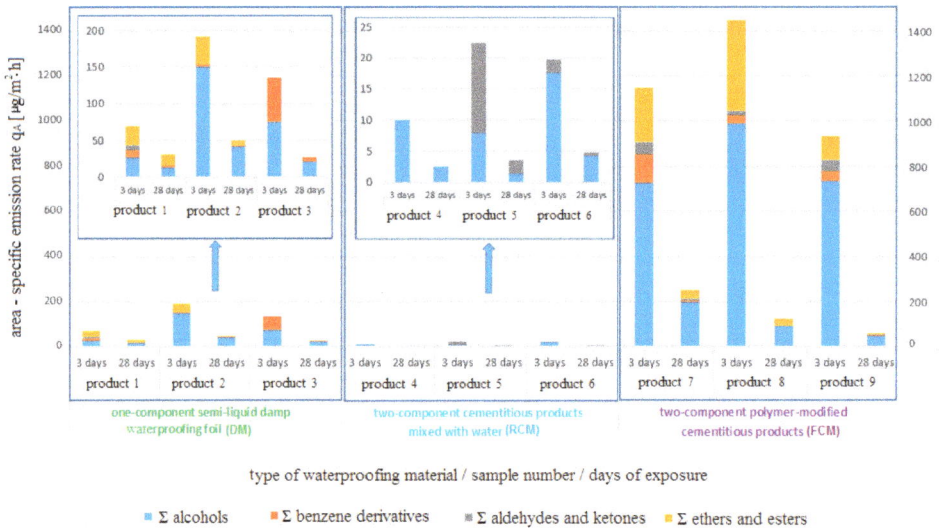

Figure 4. Distribution of main groups of compounds emitted from the tested waterproofing products in terms of the area-specific emission rate q_A [$\mu g/m^2 \cdot h$].

4. Conclusions

Emission chamber experiments revealed differences in the emission rates and groups of released compounds between dispersion and cementitious waterproofing products. The area-specific emission rates from the examined waterproofing products after the first 3 days was 10 times higher than the emission after 28 days, which suggests that interior spaces should be ventilated before their uses. Consumers need emission information to assist in making purchasing decisions of waterproofing products.

The starting area-specific emission rates (after three days) reached values between 10 $\mu g/m^2 \cdot h$ and 1481 $\mu g/m^2 \cdot h$. After 28 days, the area-specific emission rates were under 219 $\mu g/m^2 \cdot h$ for all examined products. Among them, the highest emission proved to be from FCM, while the lowest emission was observed from RCM. Furthermore, no benzene derivatives, ethers or esters were identified from the RCM group. The dominant group of detected compounds were alcohols in each case. n-Butanol, found in most of the tested products, was the most frequently detected chemical. Styrene, acrylates and acetates were identified from DM and FCM. All tested waterproofing products did not contain volatile carcinogens classified in categories 1A and 1B (from UE Regulation (EC) No 1272/2008, Annex VI Part 3).

The area-specific emission rate was used for waterproofing product characterization and categorization. The tested waterproofing products have been proven to belong to the low-emitting group. Among these products, the group with the lowest emission consisted of RCM, while FCM had the highest emission and the most diverse composition. Referring to Standard EN 15251: 2007, products from the DM and RCM groups can be included in the *very low* VOC-emitting class, while those from the FCM group belong to the *low* VOC-emitting class. The verification of low-emitting materials and products is important because the use of low-emitting components will result in low-emitting complete building structures. VOC-emitting source material characterization is important for manufacturers, contractors and building designers to provide a healthy and comfortable environment for building occupants. Moreover, the producers of waterproofing materials require VOC content information to be able to improve quality of their product. In addition, this characterization allows these individuals to select low-emission products for certified green buildings.

Author Contributions: M.K. analyzed the results and wrote the paper. The experimental plan was formulated under the supervision of M.P. The experiments were performed by A.G., H.D. and A.N. All authors have read and approved the final manuscript.

Conflicts of Interest: The authors declare no conflict of interest.

References

1. Cole, R.J.; Valdebenito, M.J. The importation of building environmental certification systems: International usages of BREEAM and LEED. *Build. Res. Inf.* **2013**, *41*, 662–676. [CrossRef]
2. Bernardi, E.; Carlucci, S.; Cornaro, C.; Bohne, R.A. An Analysis of the Most Adopted Rating Systems for Assessing the Environmental Impact of Buildings. *Sustainability* **2017**, *9*, 1226. [CrossRef]
3. Park, J.; Yoon, J.; Kim, K.-H. Critical Review of the Material Criteria of Building Sustainability Assessment Tools. *Sustainability* **2017**, *9*, 186. [CrossRef]
4. Standard PN-EN 13779: 2007 Ventilation for Non-Residential Buildings. Performance Requirements for Ventilation and Room-Conditioning Systems. Available online: https://wiedza.pkn.pl/web/guest/wyszuk iwarka-norm?p_auth=j8sXFgvO&p_p_id=searchstandards_WAR_p4scustomerpknzwnelsearchstandard sportlet&p_p_lifecycle=1&p_p_state=normal&p_p_mode=view&p_p_col_id=column-1&p_p_col_count= 1&_searchstandards_WAR_p4scustomerpknzwnelsearchstandardsportlet_standardNumber=PN-EN+13 779%3A2007E&_searchstandards_WAR_p4scustomerpknzwnelsearchstandardsportlet_javax.portlet.actio n=showStandardDetailsAction (accessed on 18 June 2018).
5. Dürkop, J.; Horn, W.; Englert, N.; Plehn, W. *Building Products: Determining and Avoiding Pollutants and Odours*; Umweltbundesamt: Dessau, Germany, 2007.
6. Hess-Kosa, K. *Indoor Air Quality: Sampling Methodologies*; CRC Press: Reno, NV, USA, 2001.
7. Zabiegała, B.; Przyjazny, A.; Namieśnik, J. Passive dosimetry as an alternative technique to dynamic enrichment of organic pollutants of indoor air. *J. Environ. Pathol. Toxicol. Oncol.* **1999**, *18*, 47–59. [PubMed]
8. Zabiegała, B.; Sărbu, C.; Urbanowicz, M.; Namieśnik, J. A Comparative Study of the Performance of Passive Samplers. *J. Air Waste Manag. Assoc.* **2011**, *61*, 260–268. [CrossRef] [PubMed]
9. Szulczyński, B.; Gębicki, J. Currently Commercially Available Chemical Sensors Employed for Detection of Volatile Organic Compounds in Outdoor and Indoor Air. *Environments* **2017**, *4*, 21. [CrossRef]
10. Spinelle, L.; Gerboles, M.; Kok, G.; Persijn, S.; Sauerwald, T. Review of Portable and Low-Cost Sensors for the Ambient Air Monitoring of Benzene and Other Volatile Organic Compounds. *Sensors* **2017**, *17*, 1520. [CrossRef] [PubMed]

11. Ahmad, M.W.; Mourshed, M.; Mundow, D.; Sisinni, M.; Rezgui, Y. Building energy metering and environmental monitoring—A state-of-the-art review and directions for future research. *Energy Build.* **2016**, *120*, 85–102. [CrossRef]

12. Eusebio, L.; Derudi, M.; Capelli, L.; Nano, G.; Sironi, S. Assessment of the Indoor Odour Impact in a Naturally Ventilated Room. *Sensors* **2017**, *17*, 778. [CrossRef] [PubMed]

13. Gostner, J.M.; Zeisler, J.; Alam, M.T.; Gruber, P.; Fuchs, D.; Becker, K.; Neubert, K.; Kleinhappl, M.; Martini, S.; Überall, F. Cellular reactions to long-term volatile organic compound (VOC) exposures. *Sci. Rep.* **2016**, *6*, 37842. [CrossRef] [PubMed]

14. Bernstein, J.A.; Alexis, N.; Bacchus, H.; Bernstein, I.L.; Fritz, P.; Horner, E.; Li, N.; Mason, S.; Nel, A.; Oullette, J.; et al. The health effects of non-industrial indoor air pollution. *J. Allergy Clin. Immunol.* **2008**, *121*, 585–591. [CrossRef] [PubMed]

15. Batterman, S.; Su, F.C.; Li, S.; Mukherjee, B.; Jia, C. Personal exposure to mixtures of volatile organic compounds: Modeling and further analysis of the RIOPA data. *Res. Rep. Health Eff. Inst.* **2014**, *181*, 3–63.

16. WHO/IARC International Agency for Research on Cancer (IARC). *IARC Monographs on the Evaluation of Carcinogenic Risks to Humans*; WHO/IARC International Agency for Research on Cancer (IARC): Lyon, France, 2006.

17. Wolkoff, P. Impact of air velocity, temperature, humidity, and air on long-term VOC emissions from building products. *Atmos. Environ.* **1998**, *32*, 2659–2668. [CrossRef]

18. Wolkoff, P.; Wilkins, C.K.; Clausen, P.A.; Nielsen, G.D. Organic compounds in office environments—Sensory irritation, odor, measurements and the role of reactive chemistry. *Indoor Air* **2006**, *16*, 7–19. [CrossRef] [PubMed]

19. United States Environmental Protection Agency. *Indoor Air Facts No. 4 (Revised): Sick Building Syndrome*; United States Environmental Protection Agency: Washington, DC, USA, 1991.

20. Kostyrko, K.; Wargocki, P. *Measurements of Odours and Perceived Indoor Air Quality in Buildings*; Scientific Studies Research Building Institute: Warsaw, Poland, 2012.

21. Bakó-Biró, Z.; Wargocki, P.; Weschler, C.J.; Fanger, P.O. Effects of pollution from personal computers on perceived air quality, SBS symptoms and productivity in offices. *Indoor Air* **2004**, *14*, 178–187. [CrossRef] [PubMed]

22. Wargocki, P.; Wyon, D.P.; Baik, Y.K.; Clausen, G.; Fanger, P.O. Perceived air quality, sick building syndrome (SBS) symptoms and productivity in an office with two different pollution loads. *Indoor Air* **1999**, *9*, 165–179. [CrossRef] [PubMed]

23. Yu, C.; Crump, D. A review of the emission of VOCs from polymeric materials used in buildings. *Build. Environ.* **1998**, *33*, 357–374. [CrossRef]

24. Wilke, O.; Jann, O.; Brödner, D. VOC- and SVOC-emissions from adhesives, floor coverings and complete floor structures. *Indoor Air* **2004**, *14*, 98–107. [CrossRef] [PubMed]

25. Piasecki, M.; Kostyrko, K.; Pykacz, S. Indoor environmental quality assessment: Part 1: Choice of the indoor environmental quality sub-component models. *J. Build. Phys.* **2017**, *41*, 264–289. [CrossRef]

26. Wang, H.L.; Nie, L.; Li, J.; Wang, Y.F.; Wang, G.; Wang, J.H.; Hao, Z.P. Characterization and assessment of volatile organic compounds (VOCs) emissions from typical industries. *Chin. Sci. Bull.* **2013**, *58*, 724–730. [CrossRef]

27. Lim, J.; Kim, S.; Kim, A.; Lee, W.; Han, J.; Cha, J.S. Behavior of VOCs and carbonyl compounds emission from different types of wallpapers in Korea. *Int. J. Environ. Res. Public Health* **2014**, *11*, 4326–4339. [CrossRef] [PubMed]

28. Sparks, L.E.; Guo, Z.; Chang, J.C.; Tichenor, B.A. Volatile Organic Compound Emissions from Latex Paint—Part 1. Chamber Experiments and Source Model Development. *Indoor Air* **1999**, *9*, 10–17. [CrossRef] [PubMed]

29. Chang, J.C.; Fortmann, R.; Roache, N.; Lao, H.C. Evaluation of low-VOC latex paints. *Indoor Air* **1999**, *9*, 253–258. [CrossRef] [PubMed]

30. Schiewecka, A.; Bock, M.C. Emissions from low-VOC and zero-VOC paints—Valuable alternatives to conventional formulations also for use in sensitive environments? *Build. Environ.* **2015**, *85*, 243–252. [CrossRef]

31. Guo, Z.; Chang, J.C.S.; Sparks, L.E.; Fortmann, R.C. Estimation of the rate of VOC emissions from solvent-based indoor coating materials based on product formulation. *Atmos Environ.* **1999**, *33*, 1205–1215. [CrossRef]

32. Kwok, N.H.; Lee, S.C.; Guo, H.; Hung, W.T. Substrate effects on VOC emissions from an interior finishing varnish. *Build. Environ.* **2003**, *38*, 1019–1026. [CrossRef]

33. Xiong, J.; Wang, L.; Bai, Y.; Zhang, Y. Measuring the characteristic parameters of VOC emission from paints. *Build. Environ.* **2013**, *66*, 65–71. [CrossRef]

34. Chang, J.C.S.; Tichenor, B.A.; Guo, Z.; Krebs, K.A. Substrate Effects on VOC Emissions from a Latex Paint. *Indoor Air* **1997**, *7*, 223–286. [CrossRef]

35. Yang, X.; Chen, Q.; Zeng, J.; Zhang, J.S.; Shaw, C.Y. Effects of environmental and test conditions on VOC emissions from "wet" coating materials. *Indoor Air* **2001**, *11*, 270–278. [CrossRef] [PubMed]

36. Li, M. Diffusion-controlled emissions of volatile organic compounds (VOCs): Short-, mid, and long-term emission profiles. *Int. J. Heat Mass Transf.* **2013**, *62*, 295–302. [CrossRef]

37. Guo, Z.; Tichenor, B.A. Fundamental mass transfer models applied to evaluating the emissions of vapor-phase organics from interior architectural coatings. In Proceedings of the 1992 EPA/AWMA Symposium, Durham, NC, USA, 4–9 May 1992.

38. Li, F.; Niu, J.; Zhang, L. A physically-based model for prediction of VOCs emissions from paint applied to an absorptive substrate. *Build. Environ.* **2006**, *41*, 1317–1325. [CrossRef]

39. AFSSET. *Procédure De Qualification Des Émissions De Composes Organiques Volatils Par Les Matériaux De Construction Et Produits De Decoration*; French Agency for Environmental and Occupational Health Safety: Maisons-Alfort, France, 2009.

40. Standard EN 14891: 2017 Liquid Applied Water Impermeable Products for Use Beneath Ceramic Tiling Bonded with Adhesives—Requirements, Test Methods, Assessment and Verification of Constancy of Performance, Classification and Marking. Available online: https://standards.cen.eu/dyn/www/f?p=204:110:0::::FSP_PROJECT,FSP_ORG_ID:59329,6050&cs=1915F706706485DF975E96506D90A0C62 (accessed on 18 June 2018).

41. ZDB Merkblatt; Verbundabdichtungen. Hinweise fur Die Ausfiihrung von Flussig zu Verarbeitenden Verbundabdichtungen mit Bekleidungen und Belagen aus Fliesen und Platten fur den Innen- und AuBenbereich. 2012. Available online: https://www.baufachmedien.de/hinweise-fuer-die-ausfuehrung-von-fluessig-zu-verarbeitenden-verbundabdichtungen-fuer-den-innen-und-auss.html (accessed on 18 June 2018).

42. Harmonized Standard EN 16516: 2017 Construction Products: Assessment of Release of Dangerous Substances—Determination of Emissions into Indoor Air. Available online: https://standards.cen.eu/dyn/www/f?p=204:110:0::::FSP_PROJECT,FSP_ORG_ID:41188,510793&cs=1216364B7D630CA10EDC767A41CE75645 (accessed on 18 June 2018).

43. International Organization for Standardization. *Standard ISO 16000-9 Indoor Air—Part 9: Determination of the Emission of Volatile Organic Compounds from Building Products and Furnishing—Emission Test Chamber Method*; International Organization for Standardization: Geneva, Switzerland, 2006.

44. International Organization for Standardization. *Standard ISO 16000-11 Indoor Air—Part 11: Determination of the Emission of Volatile Organic Compounds from Building Products and Furnishing—Sampling, Storage of Samples and Preparation of Test Specimens*; International Organization for Standardization: Geneva, Switzerland, 2006.

45. ECA-IAQ (European Collaborative Action "Indoor Air Quality and its Impact on Man"). *Total Volatile Organic Compounds (TVOC) in Indoor Quality Investigations*; Report No. 19; Office for Official Publications of the European Community: Luxembourg, 1997.

![sustainability logo] *sustainability*

MDPI

Article

Classification of Economic Regions with Regards to Selected Factors Characterizing the Construction Industry

Bożena Hoła and Tomasz Nowobilski *

Department of Construction Technology and Management, Faculty of Civil Engineering,
Wroclaw University of Science and Technology, 50-370 Wrocław, Poland; bozena.hola@pwr.edu.pl
* Correspondence: tomasz.nowobilski@pwr.edu.pl; Tel.: +48-71-320-39-71

Received: 27 April 2018; Accepted: 16 May 2018; Published: 18 May 2018

Abstract: This article presents the methodology for classifying economic regions with regards to selected factors that characterize a region, such as: the economic structure of the region and share of individual sectors in the economy; employment; the dynamics of the development of individual sectors expressed as an increase or decrease in production value; population density, and the level of occupational safety. Cluster analysis, which is a method of multidimensional statistical analysis available in Statistica software, was used to solve the task. The proposed methodology was used to group Polish voivodeships with regards to the speed of economic development and occupational safety in the construction industry. Data published by the Central Statistical Office was used for this purpose, such as the value of construction and assembly production, the number of people employed in the construction industry, the population of an individual region, and the number of people injured in occupational accidents.

Keywords: economic regions; regional classification; classification methodology; construction industry; cluster analysis; accidents in construction

---

1. Introduction

On the basis of published indicators of economic development [1,2], it can be stated that individual regions of the world are economically developed to a different extent. An inherent attribute of every economic activity is the phenomenon of the accident rate, and one of the most accident-prone sections of the economy is the construction industry [3,4]. When analyzing construction statistics, it can be noticed that the values of indicators that characterize the construction industry in developed countries are definitely higher than in developing countries, while the values of indicators characterizing occupational safety in developed countries are significantly lower than those in developing countries. The same differentiation can be observed between the internal economic regions of countries [5].

In scientific and engineering research, the problem of identifying objects with similar characteristics is very common. When carrying out such research, it is essential to properly classify objects that are described by many features into appropriate groups. Based on an analysis of the subject literature, it should be stated that the problem of classifying objects that are described by specific features often occurs in various areas of scientific research. Authors of such works use various methods of mathematical statistics, and various supporting tools with regards to this problem. Based on our own research, it was found that no attempt has been made to classify economic regions regarding the prevalence of accidents in the construction industry.

It was also found that the use of different mathematical tools to solve the same task very often leads to different final results. These observations gave an impetus to undertake research that aimed to fill the identified research gap by:

1. proposing our own universal methodology for the classification of economic regions that are characterized by different factors
2. applying the developed methodology for the classification of Polish regions with regards to selected factors that characterize the construction industry within the aspect of occupational safety and economic development.

A method of multidimensional statistical analysis—cluster analysis, which is available in Statistica software—was used to solve the task. It involves segmenting the data set into subsets, in order to distinguish homogeneous objects in an analyzed set [6,7].

The article was organized in the following way: Section 2 presents a review of literature related to the topic of the article, and also the justification of undertaking the research topic; Section 3 presents and discusses the proposed methodology for the classification of economic regions with regards to selected factors that characterize the construction industry; Section 4 contains an example illustrating the application of the developed methodology for the classification of Polish voivodeships; and finally, Section 5 contains conclusions from the authors' research.

2. Literature Review—Application of Data Classification Methods in Scientific Research

Conducting scientific research requires the processing and analyzing of large amounts of data. In some scientific disciplines, this may complicate or make it impossible to properly investigate phenomena, which may result in situations in which information relevant to researchers remains invisible. In this case, it is necessary to properly organize the observed data into structures, or to group it into categories [6,7]. Methods of multidimensional statistical analysis, including cluster analysis, are helpful for solving such research problems. The concept of cluster analysis was introduced in [8], and covers various classification algorithms for data exploration. The final effect of the calculations carried out using the above algorithms is the allocation of the analyzed data to appropriate groups, in which the individual elements show mutual similarities. It gives an opportunity to capture structures that in the real world create the analyzed data, and also to reduce them to a level that allows them to be properly analyzed.

Data grouping methods are used in many fields of science, including medicine, social sciences, agricultural and technical sciences, as well as economics. In the publication of Hartigan [9], many examples of the use of taxononic methods can be found, and for their fundamental application the author considers the development of the classification of animals and plants that had already been done in the times of Aristotle, and then later by Linnaeus.

In addition, the author gives a number of examples of the use of classification methods in: archaeology, anthropology, phytosociology, psychology, and psychiatry, as well as in other fields. In economic sciences, classification methods are used, for, among others: determining the market structure, determining a product's position on the market, identifying test markets, market segmentation, classifying sectors due to financial conditions, classification of the labour market, and spatio-temporal analyses [10,11].

One of the most often-used methods of data grouping is cluster analysis. For example, in [12], cluster analysis was used to segment the real estate market. The calculations were based on a non-hierarchical method, and Euclidean distance was used to determine the similarity between objects. In turn, in [13], the real estate investment environment was divided into 5 levels in 17 cities in China using cluster analysis. The classification was based on 25 indicators that influence construction investments and characterize: the economic environment, the market environment, the infrastructure environment, and the social environment of an investment. The analysis of the results indicated significant regional differences in the real estate investment environment. The results of the conducted analysis were the basis for the development of strategies and remedial measures that aimed at equalizing the existing differences.

In [14,15], various methods used in the classification of objects were widely described. In [14], the hierarchical Ward method, the non-hierarchical method of K-means and fuzzy clustering were used to identify groups of objects on the real estate market. In turn, in [15], cluster analysis was used to analyze the real estate market in combination with factor analyses. The advantages and disadvantages of the used methods were demonstrated.

References [16,17] show that cluster analysis is a good tool for grouping areas and regions of a given country. For example, in [16], this method was used to develop groups of Polish voivodeships that are characterized with a similar level of development in terms of transport infrastructure. Statistical data obtained from the Central Statistical Office was used in the analysis. Another example of the grouping of voivodeships in Poland is [17], in which voivodeships were classified based on the similarity of the state of higher education.

Grouping methods are also used in issues related to occupational safety in the construction industry. In [18], the authors carried out an analysis of accidents in the construction industry in Hong Kong using Principal component analysis and cluster analysis. The results of the research allowed the most probable accident situations to be identified. As a result of the conducted calculations, it was found that the largest number of accidents occurs in the private sector. The tools that were used, and the results obtained using them, are a good starting point for the analysis of accidents in the construction industry around the world. A similar study can be found in [19], in which an analysis of accidents related to electrical and mechanical work was carried out. In these studies, a two-stage cluster analysis was used. The obtained results allowed the knowledge gap regarding the causes of accidents related to electrical and mechanical works to be filled, as well as preventive measures to be defined and promoted.

Defining accident patterns at work in the Polish construction industry was the subject of [20]. The author carried out an analysis of the features of a construction site in order to determine the most probable circumstances of the occurrence of accidents. The research was carried out on the basis of data obtained from the register of the Regional Labour Inspectorate in Cracow.

The review of the literature shows that the problem of classifying objects to groups of similar objects occurs in many areas of science. Invaluable tools in such studies are methods of multidimensional statistical analyses, including cluster analysis. Based on the review of the literature, no scientific studies were found regarding the classification of economic regions with regards to selected features that characterize the construction industry. Such classification may be necessary for conducting research in which knowledge from similar sources is required. Due to the above, the authors of the article attempted to develop a universal methodology for classifying economic regions. This classification is based upon selected factors that characterize the construction industry in a region.

3. Proposed Research Methodology

The subject of the research is the classification of the set of economic regions V with regards to selected factors that characterize the construction industry regarding aspects of economic development and occupational safety.

$$V = \{v; v = 1, \ldots, V\} \tag{1}$$

Based on a literature review, previous research on factors causing hazards in the construction industry [21] and the authors' own experience, it was stated that the basic factors that characterize economic regions are: the economic structure of a region, thus the share of individual sectors of the economy in the entire economy; the dynamics of development of individual sectors of the economy expressed as an increase or decrease in production value; employment in individual sectors of the economy; and, population density in a region and also the level of occupational safety in sectors of the economy.

Therefore, each economic region can be described by the vector of factors F_v:

$$F_v = [F_1, \ldots, F_n, \ldots, F_N]; \qquad v = 1, \ldots, V, \tag{2}$$

where:

F_n—the factor taken for analysis ($n = 1, 2, 3, \ldots, N$).

The set of economic regions V is characterized by the following matrix of factors:

$$F_V = \begin{bmatrix} F_{1,1} & \cdots & F_{n,1} & \cdots & F_{N,1} \\ F_{1,2} & \cdots & F_{n,2} & \cdots & F_{N,2} \\ \vdots & \vdots & \vdots & \vdots & \vdots \\ F_{1,v} & \cdots & F_{n,v} & \cdots & F_{N,v} \\ \vdots & \vdots & \vdots & \vdots & \vdots \\ F_{1,V} & \cdots & F_{n,V} & \cdots & F_{N,V} \end{bmatrix} \tag{3}$$

where:

v—the economic region ($v = 1, 2, 3, \ldots, V$),

n—the factor taken for analysis ($n = 1, 2, 3, \ldots, N$).

In mathematical analyses, knowledge about numerical values of indicators that describe particular factors is important. Thus, the set of economic regions V is characterized by a two-dimensional matrix of indicators, which takes the following form:

$$I_V = \begin{bmatrix} I_{1,1} & \cdots & I_{n,1} & \cdots & I_{N,1} \\ I_{1,2} & \cdots & I_{n,2} & \cdots & I_{N,2} \\ \vdots & \vdots & \vdots & \vdots & \vdots \\ I_{1,v} & \cdots & I_{n,v} & \cdots & I_{N,v} \\ \vdots & \vdots & \vdots & \vdots & \vdots \\ I_{1,V} & \cdots & I_{n,V} & \cdots & I_{N,V} \end{bmatrix} \tag{4}$$

The values of indicators adopted for calculations often differ by a measuring unit or scale, which may negatively affect the grouping [6]. In order to deal with this, all numerical data should be subjected to standardization; the choice of the appropriate standardization formula depends on the type of data [22]. In the proposed methodology, a standardization of variables was adopted as one of the standardization methods, according to the following formula:

$$P_{n,v} = \frac{I_{n,v} - \overline{I_n}}{\sigma} \tag{5}$$

where:

$P_{n,v}$—the values of indicators after standardization,

$I_{n,v}$—the values of unstandardized indicators,

$\overline{I_n}$—the average value in the analysed set of objects,

σ—the standard deviation of the value of I_n indicator.

The effect of standardization is the creation of a set of P_V parameters that describe the analyzed set of economic regions F_V, which are written in the form of a two-dimensional matrix. In this matrix, each row contains values of all the parameters that are related to one region, whereas each column contains the data of one parameter for all the regions. This matrix is described by the following formula:

$$P_V = \begin{bmatrix} P_{1,1} & \cdots & P_{n,1} & \cdots & P_{N,1} \\ P_{1,2} & \cdots & P_{n,2} & \cdots & P_{N,2} \\ \vdots & \vdots & \vdots & \vdots & \vdots \\ P_{1,v} & \cdots & P_{n,v} & \cdots & P_{N,v} \\ \vdots & \vdots & \vdots & \vdots & \vdots \\ P_{1,V} & \cdots & P_{n,V} & \cdots & P_{N,V} \end{bmatrix} \tag{6}$$

The calculated parameters form the basis for grouping economic regions with the use of cluster analysis. Cluster analysis is one of the methods of data exploration, the main idea of which is to group the analyzed objects in such a way that, in a given group, there are objects which are deemed to be "similar" to each other, and also "dissimilar" to objects from other groups [6,7]. The criterion for assessing the affiliation of an object to a given group is the measure of similarity. The function, which is inverse to the measure of "similarity", and thus the function of "dissimilarity" of objects that is a measure of the distance between them, is used for practical considerations. This means that if the distance between object O_a and O_b is greater than the distance between objects O_a and O_c, and therefore:

$$d(O_a, O_b) > d(O_a, O_c); \ where: \ a \neq b \neq c; \ a, b, c \in \{v\} \tag{7}$$

then object O_a is more "dissimilar" to object O_b than to object O_c. Consequently, this leads to a situation where objects O_a and O_c can create a cluster because they are more "similar" to each other. Various distance measures that are used in cluster analysis. Geometric distance in the multidimensional space—the Euclidean distance [6], was used to solve the discussed issue. The general formula of the Euclidean distance takes the following form:

$$d(O_a, O_b) = \sqrt{\sum_n (P_{n,a} - P_{n, \ b})^2} \tag{8}$$

where:

O_a, O_b—the assessed objects, i.e., economic regions a and b, where $a \neq b$, and $a, b \in \{v\}$,

$P_{n,a}$, $P_{n,b}$—the designated parameter values P_n for economic regions a and b, where $a \neq b$, and $a, b \in \{v\}$.

In the analyzed task, objects are the individual economic regions v. For grouping objects, the use of hierarchical and agglomeration grouping techniques is proposed. The agglomeration technique, which is the most often used in research [6], involves the gradual connection of objects, which together constitute separate clusters, into new clusters until all objects form one cluster. Each connection of two clusters is called a step. An important issue when determining the appropriate distance between clusters, apart from the choice of the above-mentioned distance measure, is to determine the method of merging objects. Different methods of merging objects were analyzed, including methods in which the distance between specific locations of clusters is determined (e.g., between a given object or center of gravity of a cluster), and also methods that use variance analysis—e.g., the Ward's method [23]. Based on the conducted analyses, it was found that the most unambiguous results are achieved using the Ward method, the main advantage of which is the grouping of objects in a way that allows clusters with a similar number of objects to be formed. This eliminates the so-called "chain-linking" effect, and the newly created clusters are characterized by the smallest possible diversity between their individual elements. In the developed methodology, the method of merging objects using the Ward method was adopted.

The result of hierarchical cluster analysis is a tree-shaped graph—the so-called dendrogram. It shows in which step the objects connect with each other. However, it does not give an unambiguous answer for the correct number of clusters. This number depends on the place where the branches of the tree are cut off on the chart. Due to this, an important issue is to correctly determine the place of the cut-off. According to [6], there is no objective rule of how to do it. There are only supportive methods, such as the method of graphical dendrogram analysis that involves the examination of the distance between successive bonds, the method using the Grabinski meter [24] or the Mojen rule [25]. The developed methodology involves the method of graphical dendrogram analysis.

After selecting the appropriate place of cutting off the branches on the dendrogram, the clusters should be identified, and on their basis, the final classification of the analyzed economic regions and their assessment and ranking should be made. The developed methodology for this classification is shown in Figure 1.

Classification of the set of economic regions v with regards to selected criteria

$$v = \{\, v;\ v = 1, ..., V \,\}$$

STAGE 1: Identification and selection of factors that characterize the analysed economic regions

Identification of factors that characterize economic regions with regards to social, economic or other aspects

Literature review

Selection of F_v factors for the analysis
$$F_v = [F_1, ..., F_v, ..., F_V]\,; \quad v = 1, ..., V$$

STAGE 2: Acquiring calculation data

Published statistical data

Acquiring data on indicators $I_{n,v}$ that describe *factors* $F_{n,v}$ in individual regions, where n - the factor $(n = 1, 2, 3, ..., N)$ and v - the economic region $(v = 1, 2, 3, ..., V)$

Development of a matrix of I_v indicators that characterize the analysed economic regions

$$I_V = \begin{bmatrix} I_{1,1} & \cdots & I_{n,1} & \cdots & I_{N,1} \\ I_{1,2} & \cdots & I_{n,2} & \cdots & I_{N,2} \\ \vdots & \vdots & \vdots & \vdots & \vdots \\ I_{1,v} & \cdots & I_{n,v} & \cdots & I_{N,v} \\ \vdots & \vdots & \vdots & \vdots & \vdots \\ I_{1,V} & \cdots & I_{n,V} & \cdots & I_{N,V} \end{bmatrix}$$

STAGE 3: Normalization of the data that constitutes the basis for the classification

Transformation of indicators $I_{n,v}$ into parameters $P_{n,v}$

Data normalization

Development of a matrix of P_V parameters that characterize the analysed economic regions

$$P_V = \begin{bmatrix} P_{1,1} & \cdots & P_{n,1} & \cdots & P_{N,1} \\ P_{1,2} & \cdots & P_{n,2} & \cdots & P_{N,2} \\ \vdots & \vdots & \vdots & \vdots & \vdots \\ P_{1,v} & \cdots & P_{n,v} & \cdots & P_{N,v} \\ \vdots & \vdots & \vdots & \vdots & \vdots \\ P_{1,V} & \cdots & P_{n,V} & \cdots & P_{N,V} \end{bmatrix}$$

STAGE 4: Classification of economic regions

Cluster analysis

Assumptions:
- measure of similarity - Euclidean distance,
- grouping method - hierarchical,
- grouping technique - agglomeration,
- object merging method - the Ward method.

Statistica package

Performing calculations

Analysis of results

CLASSIFICATION OF ECONOMIC REGIONS

Identified clusters

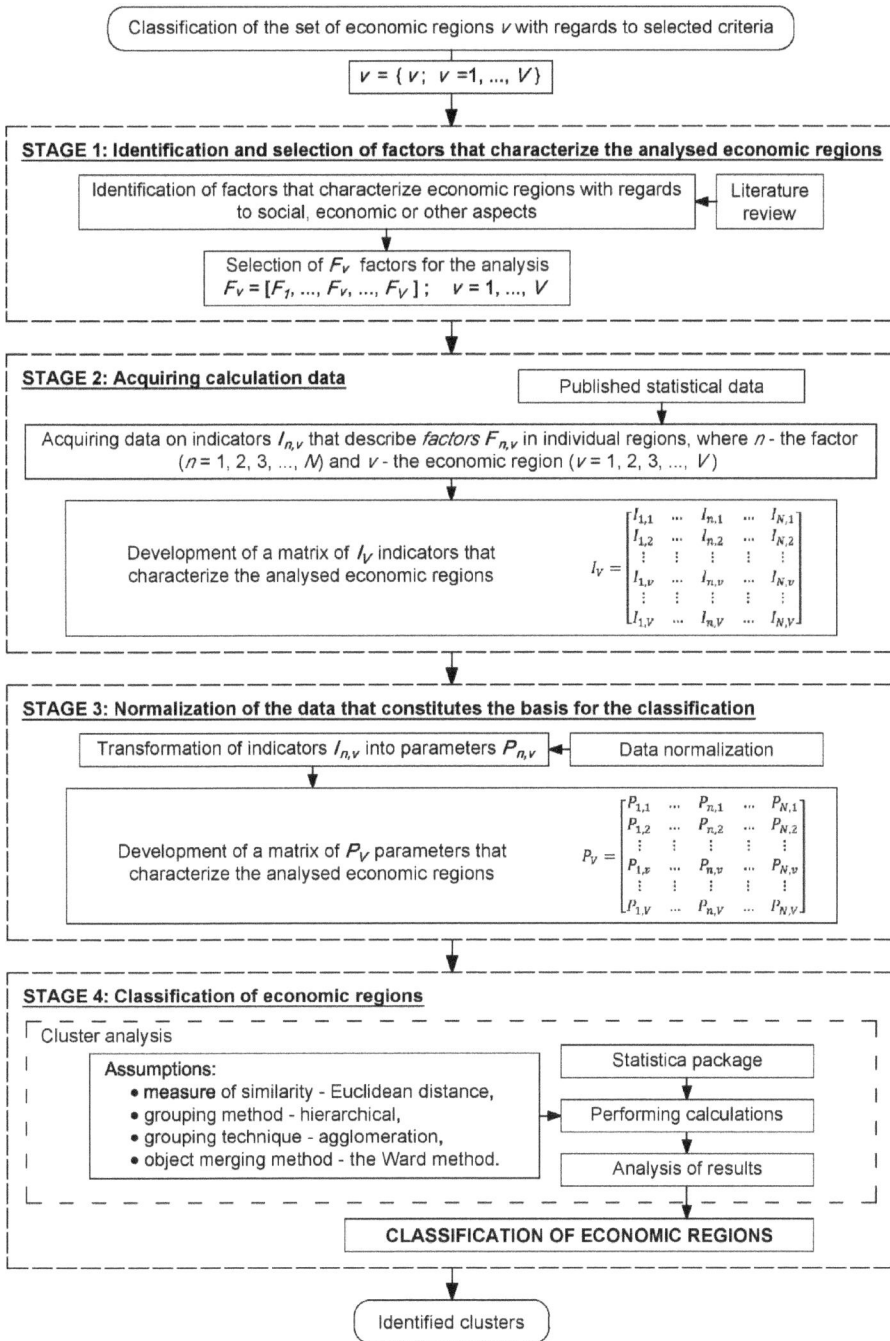

Figure 1. The developed methodology for classifying economic regions.

The proposed methodology was developed on the basis of previous experience and analyses carried out by the authors. The final elements used for its construction, such as the data standardization method, Euclidean distance, the grouping of objects using the hierarchical method, and also the binding of objects using the Ward method, resulted from the analysis of the results of many previous attempts that aimed to develop a universal and accurate methodology. The authors of the article believe that the proposed methodology is a good and well thought-out tool for grouping economic regions, and can be used to classify whole countries, regions of a given country and other territorial units, as well as economic facilities, e.g., enterprises. The grouping can be conducted with regards to various classification criteria, including criteria related to occupational safety in the construction industry. The necessary condition for an appropriate calculation is a substantive identification of factors that characterize regions or economic objects within the analyzed aspect, and also the obtaining of reliable numerical data on them. The results of the conducted classification may otherwise be subject to error.

4. Application of the Proposed Methodology Using the Classification of Polish Voivodeships as an Example

The proposed methodology was used to group Polish voivodeships. The classification was based on data published by the Central Statistical Office [5]. The following indicators were adopted to describe the voivodeships: the value of construction and assembly production ($I_{1,y}$), the number of people employed in the construction industry ($I_{2,y}$), the population of a given region ($I_{3,y}$), and the number of people injured in occupational accidents ($I_{4,y}$). The calculations include data for the period from 2008 to 2016 that was obtained for 16 Polish voivodeships. Therefore, the number of analyzed indicators for individual voivodeships amounted to 36; some of them are shown in Table 1.

Table 1. Values of selected indicators.

	Economic Indicators											
v	Value of Construction and Assembly Production $I_{1,v,y}$			Number of People Employed in the Construction Industry $I_{2,v,y}$			Population of a Given Region $I_{3,v,y}$			Number of People Injured in Occupational Accidents $I_{4,v,y}$		
	2008 $y=1$	...	2016 $y=9$	2008 $y=1$	...	2016 $y=9$	2008 $y=1$	...	2016 $y=9$	2008 $y=1$	...	2016 $y=9$
1	14,568.1	...	13,721.2	68,874	...	65,039	2,877,059	...	2,903,710	892	...	413
2	5288.9	...	6216.6	43,952	...	42,918	2,067,918	...	2,083,927	607	...	269
...	...	...	...	...	...	...	...	...	...	...	...	...
16	7325.1	...	9104.0	39,288	...	35,483	1,692,957	...	1,708,174	362	...	144

y—year ($y=1, \ldots, 9$), v—voivodeship ($v=1, \ldots, 16$).

The obtained values of indicators were standardized. The effect of this action was the creation of parameters $P_{n,v,y}$, constituting a set of normalized indicators which were the basis for classification. Some of the data obtained after standardization is presented in Table 2.

After conducting the calculations using Statistica software, the dendrogram shown in Figure 2 was obtained. When analyzing the dendrogram, it can be noted that some voivodeships very quickly create clear clusters. It was proposed to cut off the branch of the dendrogram at the place marked with a dashed line in Figure 2. With such a cut, four groups of voivodeships were obtained, in which each voivodeship has a similar level of the development of the construction industry and occupational safety. The identified groups of voivodeships are given in Table 3.

Table 2. Values of selected parameters.

v	Parameters											
	Value of Construction and Assembly Production $P_{1,v,y}$			Number of People Employed in the Construction Industry $P_{2,v,y}$			Population of a Given Region $P_{3,v,y}$			Number of People Injured in Occupational Accidents $P_{4,v,y}$		
	2008 $y=1$	$\cdots$	2016 $y=9$	2008 $y=1$	$\cdots$	2016 $y=9$	2008 $y=1$	$\cdots$	2016 $y=9$	2008 $y=1$	$\cdots$	2016 $y=9$
1	0.734	...	0.394	0.481	...	0.291	0.397	...	0.395	0.534	...	0.288
2	−0.537	...	−0.504	−0.264	...	−0.323	−0.254	...	−0.251	−0.110	...	−0.294
...	...	...	...	...	...	...	...	...	...	...	...	...
16	−0.258	...	−0.158	−0.403	...	−0.530	−0.555	...	−0.546	−0.664	...	−0.798

y—year ($y = 1, \ldots, 9$), v—voivodeship ($v = 1, \ldots, 16$).

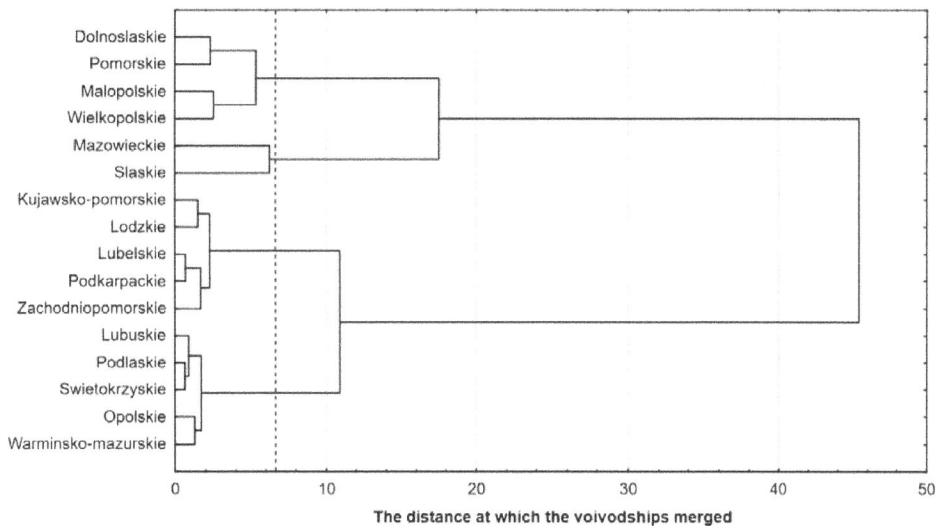

Figure 2. Dendrogram—a graph showing the connection of individual voivodeships in the subsequent calculation steps.

Table 3. The obtained groups of voivodeships, which are characterized by a similar speed of construction industry development and a similar level of occupational safety.

Cluster	Voivodeships	The Distance at Which the Voivodeships Merged
I	Dolnoslaskie, Pomorskie, Malopolskie, Wielkopolskie	5.38
II	Mazowieckie, Slaskie	6.25
III	Kujawsko-Pomorskie, Lodzkie, Lubelskie, Podkarpackie, Zachodniopomorskie	2.26
IV	Lubuskie, Podlaskie, Swietokrzyskie, Opolskie, Warminsko-Mazurskie	1.73

Based on the analysis of the obtained results, four clusters were distinguished, which include 2 to 5 voivodeships. Cluster I consists of the voivodeships Dolnoslaskie, Pomorskie, Malopolskie and Wielkopolskie; cluster II is formed from the voivodeships Mazowieckie and Slaskie; cluster III consists of the voivodeships Kujawsko-Pomorskie, Lodzkie, Lubelskie, Podkarpackie and Zachodniopomorskie; while cluster IV includes the voivodeships Lubuskie, Podlaskie, Swietokrzyskie, Opolskie and Warminsko-Mazurskie.

A very fast creation of connections between voivodeships in clusters III and IV was observed. These clusters include the voivodeships most similar to each other in terms of construction and assembly production, the number of people injured in occupational accidents, the number of people employed in the construction industry, and the population living in the voivodeship. The merging distance in the case of cluster III is equal to 2.26, while in the case of cluster IV is equal to 1.73

The most different voivodeships from all of the others are Mazowieckie and Slaskie. They form one cluster; however, the distance at which the merging between them occurred is equal to 6.25, and is more than twice as high as in the case of clusters III and IV. In turn, a comparable level of similarity between pairs of voivodeships can be observed in cluster I, namely: Dolnoslaskie and Pomorskie, and also Malopolskie and Wielkopolskie. The merging distance in both pairs is equal to 2.31 and 2.53, respectively. However, the merging distance of these two pairs is equal to 5.38.

5. Conclusions

The following conclusions were made on the basis of our research and analyses:

1. The proposed methodology for classifying objects is of universal character and can be used to group countries, regions of a given country and other territorial units, as well as economic facilities, e.g., enterprises. The grouping can be conducted with regards to various classification criteria, including criteria related to occupational safety in the construction industry. The necessary condition for an appropriate calculation is a substantive identification of factors that characterize regions or economic objects within the analyzed aspect, and also the obtaining of reliable numerical data on them.
2. The basis of the proposed methodology is a method of multidimensional analysis of statistical data, namely cluster analysis. Specific solutions include data standardization, a measure of similarity in the form of Euclidean distance, grouping objects using the hierarchical method, and binding objects using the Ward method.
3. The developed methodology was used to classify Polish voivodeships with regards to factors that characterize the rate of economic development in the construction industry and the level of occupational safety. The conducted calculations and analysis of the results allowed the following conclusions to be formulated:

    -   The qualitative and quantitative structure of statistical data, which was the basis for the classification of voivodeships, allowed four distinct clusters consisting from two to five voivodeships to be separated. Voivodeships included in a cluster are characterized with a similar level of occupational safety in the construction industry.
    -   The isolated clusters are characterized by different levels of similarity, which is confirmed by the values of the merging distance measure for individual clusters. Cluster ranking with regards to the similarity of the voivodeships that form clusters is as follows:

        1. cluster IV consists of the voivodeships Lubuskie, Podlaskie, Swietokrzyskie, Opolskie and Warminsko-Mazurskie,
        2. cluster III consists of the voivodeships Kujawsko-Pomorskie, Lodzkie, Lubelskie, Podkarpackie and Zachodniopomorskie,
        3. cluster I consists of the voivodeships Dolnoslaskie, Pomorskie, Malopolskie and Wielkopolskie,
        4. cluster II consists of the voivodeships Mazowieckie and Slaskie.

    -   The very big similarity between voivodeships located in clusters III and IV means that voivodeships included in these clusters are characterized by a similar level of construction and assembly production value, occupational safety, the number of people employed in the construction industry, and the number of people living in the voivodeship.

- The Mazowieckie and Slaskie are atypical voivodeships. They are the most different when compared with the others. Although they form one cluster, the distance at which the merging between them occurs is relatively large when compared to the merging distance in the other clusters.

4. The proposed methodology can be applied in both the area of scientific research and engineering practice. The results of tests and analyses obtained using this methodology can be the basis for classifying and comparing objects and determining their rankings. The correct classification of objects (which are described by many factors) into groups can be important in determining the characteristics of a given community, making an assessment, or looking for dependencies that apply to this community. The practical aspect of the proposed methodology is connected to the possibility of formulating conclusions, which could be important at a higher management level.

5. In the research conducted by the authors, information about voivodeships belonging to the same cluster will be used to possess statistical data from these voivodeships, which in turn will be used for the construction of multifactorial linear regression models for predicting indicators describing the level of occupational safety in the construction industry in the group of voivodeships. Mathematical models, which were developed in this way, will be more accurate when compared to the general model that was built for the whole of Poland. The creation of separate mathematical models for individual voivodeships is impossible, due to the insufficient amount of reliable statistical data that can be used to construct them.

Author Contributions: Formal analysis, T.N.; Methodology, B.H. and T.N.; Project administration, B.H.; Software, T.N.; Supervision, B.H.

Funding: This research was funded by NCBiR within the framework of the Programme for Applied Research grant number PBS3/A2/19/2015.

Acknowledgments: The article is the result of the implementation by the authors of research project No. 244388 "Model of the assessment of risk of the occurrence of building catastrophes, accidents and dangerous events at workplaces with the use of scaffolding".

Conflicts of Interest: The authors declare no conflict of interest.

References

1. Central Statistical Office in Katowice. *Indicators of the Sustainable Development of Poland 2015*; Central Statistical Office in Katowice, Silesian Center for Regional Research: Katowice, Poland, 2015; ISBN 978-83-89641-54-0.
2. Eurostat. *Eurostat Regional Yearbook 2017 Edition*; Publications office of the European Union: Luxembourg, 2017; ISBN 978-92-79-71616-4.
3. International Labour Organization. Safety and Health at Work: A Vision for Sustainable Prevention. In Proceedings of the 20th World Congress on Safety and Health at Work 2014, Global Forum for Prevention, Frankfurt, Germany, 24–27 August 2014.
4. Hoła, B.; Szóstak, M. Analysis of the state of the accident rate in the construction industry in European Union countries. *Arch. Civil Eng.* **2015**, *61*, 19–34. [CrossRef]
5. Central Statistical Office. *Statistical Yearbook 2008–2016*; Department of Statistical Publishing: Warsaw, Poland, 2009–2017.
6. Stanisz, A. *A Simple Statistic Course with the Use of STATISTICA PL on the Basis of Medical Examples, Volume 3: Multivariate Analysis*; StatSoft: Cracow, Poland, 2007; ISBN 978-83-88724-19-0.
7. Wierzchoń, S.; Kłopotek, M. *Algorithms of Cluster Analysis*; Publishing House of WNT: Warsaw, Poland, 2015; ISBN 978-83-7926-261-8.
8. Tryon, R.C. *Cluster Analysis: Correlation Profile and Orthometric (Factor) Analysis for the Isolation of Unities in Mind and Personality*; Edwards Brothers: Ann Arbor, MI, USA, 1939.
9. Hartigan, J.A. *Clustering Algorithms*; Wiley: New York, NY, USA, 1975.
10. Tkaczynski, A. Segmentation Using Two-Step Cluster Analysis. In *Segmentation in Social Marketing*; Springer: Singapore, 2017; pp. 109–125.

11. Markowska, M.; Sokołowski, A.; Strahl, D.; Sobolewski, M. Dynamic classification of regions of the European Union at the NUTS 2 level with regards to sensitivity to the economic crisis in the area of the labour market. *HSS* **2015**, *22*, 37–50.

12. Napoli, G.; Giuffrida, S.; Valenti, A. Forms and functions of the real estate market of Palermo (Italy). Science and knowledge in the cluster analysis approach. *Green Energy Technol.* **2017**, 191–202. [CrossRef]

13. Wang, W.; Yang, J.; Gong, X. The Regional Real Estate Investment Environment Research Based on Prime Component Analysis: The Case of Shandong. In *ICCREM 2017: Real Estate and Urbanization—Proceedings of the International Conference on Construction and Real Estate Management*; ASCE: Reston, VA, US, 2017; pp. 217–224.

14. Gabrielli, L.; Giuffrida, S.; Trovato, M.R. Gaps and overlaps of urban housing sub-market: Hard clustering and fuzzy clustering approaches. *Green Energy Technol.* **2017**, 203–219. [CrossRef]

15. Wang, N.; Li, H.-M.; Tan, X.; Zhong, X.-R. Development evaluation of real estate industry in China's major cities based on factor and cluster analysis. Xi'an Jianzhu Keji Daxue Xuebao. *J. Xi'an Univ. Archit. Technol.* **2010**, *42*, 590–594.

16. Migała-Warchoł, A.; Sobolewski, M. Evaluation of voivodeships diversification in Poland according to transport infrastructure indications. *Quant. Methods Econ.* **2013**, *14*, 89–98.

17. Zalewska, E. Application of cluster analysis and methods of linear ordering in the assessment of Polish higher education. *Sci. Works Wroclaw Univ. Econ.* **2017**, *469*, 234–242.

18. Chiang, Y.-H.; Wong, F.K.-W.; Liang, S. Fatal Construction Accidents in Hong Kong. *J. Constr. Eng. Manag.* **2018**, *144*. [CrossRef]

19. Wong, F.K.; Chan, A.P.; Wong, A.K.; Hon, C.K.; Choi, T.N. Accidents of Electrical and Mechanical Works for Public Sector Projects in Hong Kong. *Int. J. Environ. Res. Public Health* **2018**, *15*, 485. [CrossRef] [PubMed]

20. Drozd, W. Identification and profiling the pattern of construction accidents with the cluster analysis. *Mater. Bud.* **2017**, 101–102. [CrossRef]

21. Hoła, B.; Nowobilski, T.; Szer, I.; Szer, J. Identification of factors affecting the accident rate in the construction industry. *Procedia Eng.* **2017**, *11*. [CrossRef]

22. Jarocka, M. Selection of the normalization formula in the comparative analysis of multi-feature objects. *Econ. Manag.* **2015**, 113–126. [CrossRef]

23. Ward, J.H., Jr. Hierarchical Grouping to Optimize an Objective Function. *J. Am. Stat. Assoc.* **1963**, *58*, 236–244. [CrossRef]

24. Grabiński, T. *Methods of Taxononometry*; Cracow University of Economics: Cracow, Poland, 2007.

25. Mojena, R. Hierarchical grouping methods and stopping rulet: An evaluation. *Comput. J.* **1977**, *20*, 359–363. [CrossRef]

sustainability

MDPI

Article

Sustainable Construction Industry in Cambodia: Awareness, Drivers and Barriers

Serdar Durdyev [1,*], Edmundas Kazimieras Zavadskas [2], Derek Thurnell [1], Audrius Banaitis [2] and Ali Ihtiyar [3]

[1] Department of Engineering and Architectural Studies, Ara Institute of Canterbury, 130 Madras Street, Christchurch 8140, New Zealand; Derek.Thurnell@ara.ac.nz

[2] Department of Construction Management and Real Estate, Vilnius Gediminas Technical University, Sauletekio al. 11, LT-10223 Vilnius, Lithuania; edmundas.zavadskas@vgtu.lt (E.K.Z.); audrius.banaitis@vgtu.lt (A.B.)

[3] Department of Business Administration, Zaman University, No: 8, St: 315, Boeng Kak 1, Tuol Kouk, 12151 Phnom Penh, Cambodia; aihtiyar@gmail.com

* Correspondence: durdyevs@ara.ac.nz

Received: 23 January 2018; Accepted: 1 February 2018; Published: 2 February 2018

Abstract: Although sustainability is of utmost importance, anecdotal evidence suggests that the concept is not adequately implemented in many developing countries. This paper investigates industry stakeholders' awareness of the current state of, factors driving, and barriers hindering the adoption of sustainable construction (SC) in Cambodia. Using an empirical questionnaire survey targeting local construction professionals, respondents were invited to rate their level of awareness, knowledge and understanding of SC, as well as to rate the level of importance of 31 drivers and 10 barriers identified from the seminal literature. The data set was subjected to the relative importance index method. The results suggest that the industry-wide adoption of SC practices is poor, which is believed to be due to a lack of awareness and knowledge, and reluctance to adopt new sustainable technologies. Furthermore, more efforts must be put into the selection of more durable materials for the extension of buildings' lives and to minimize material consumption, as well as to develop energy-efficient buildings with minimal environmental impact and a healthy indoor environment, so that the ability of future generations to meet their own needs will not be compromised. The outcomes of this study have enriched knowledge about the current state of, drivers of, and barriers to sustainable construction in a typical developing economy. Although the outcomes of this study were a short scoping exercise, it has formed a significant base for future SC work within Cambodia.

Keywords: sustainability; construction industry; Cambodia; drivers; barriers

1. Introduction

The construction industry makes a very significant contribution to the sustainable development of the overall economy by achieving basic objectives of development including employment creation, re-distribution, and the generation of output and income [1]. Furthermore, the construction industry helps by taking a significant role in satisfying basic social and physical needs, including the provision of infrastructure, the production of accommodation, and of consumer goods [2]. It has a highly visible output and stimulates a sizeable amount of economic growth through inter-sectoral linkages between construction and other sectors, giving it a powerful role in economic development [3]. The construction sector evidently has a strong relationship with sustainable development on the 'triple bottom line' of economic growth, environmental impact and social progress [4]. Therefore, the sector cannot be examined by only considering its impact on socio-economic development, but neither should its environmental and social impact be disregarded [5]. The detrimental effect of the sector on the environment can be explained by the production of 35–40% of CO_2 emissions, 40% of raw materials and 25% of timber consumption, 40% of solid waste production, and it consumes 40% of total energy production and 16% of water usage worldwide [6,7]. Moreover, a sustainable construction industry plays a significant role in balancing human needs through the provision of a safe, healthy and physiologically comfortable building environment, which eventually will enhance human satisfaction and productivity [4].

According to Agenda 21 of the Sustainable Construction Industry in Developing Countries (SCIDC), sustainable construction is perceived to be a holistic and integrative concept striving to restore harmony and balance between the environment, economy and society [8]. The SCDIC definition implies that, initially, approaches to sustainable construction were more concerned with technical issues (resource efficiency and reducing the environmental impacts of construction) rather than the economic and social aspects of sustainability [9]. However, when it comes to developing countries, the adoption of even very basic aspects of sustainability are still in their infancy. For instance, developed countries have been focusing on efficient resource utilization and the reduction of their environmental impact, whereas developing countries have a lower degree of achievement (i.e., the non-existence of building energy codes), which makes the process of adoption challenging.

Drivers for a more sustainable construction industry and barriers to its implementation have received broad attention by researchers worldwide [10–12]. While some studies have presented models or frameworks towards the implementation of sustainability principles [4,13–18], other studies have proposed the benefits/drivers and constraints to a sustainable construction industry [5,6,12,19–21]. Egan [21] identified five key drivers, which are: project team; a quality driven agenda; process integration; leaders' commitment; commitment to people; and a focus on the client. He further quantified and presented seven targets for enhancing levels of sustainability. Also, Vanegas and Pearce [13] presented a framework to increase the sustainability of the construction industry and a road map for industry stakeholders to move toward sustainable infrastructure and facilities. Furthermore, the study identified drivers for change to achieve sustainability of the built environment, which are resource degradation and depletion, and stressed the noticeable influences of the construction industry environment on human health. In another study via case studies, Berardi [22] found that the stakeholders with power, who are uninterested, are the main barrier to the adoption of sustainable practices in construction projects. Lack of information and communication were found to be the reasons for stakeholders being reluctant to adoption. The study further recommends improving the relationship among the construction stakeholders. Albino and Berardi [23] studied three case studies and presented the significance of the integration between construction stakeholders, where a supplier's engagement and design team with special experience in sustainable practices are found to be most significant. From the construction project developers' perspective, Zainul-Abidin [11] investigated the level of knowledge and awareness about sustainable construction in Malaysia. The vast majority of developers were reluctant to implement sustainable construction concepts due to their lack of knowledge and cost concerns, and even amongst large project developers, few were adopting the concepts. In another study, Shen et al. [10] evaluated the feasibility reports of 87 projects in China with regard to their sustainability performance, environmental, economic and social aspects, in particular. The findings mainly revealed that the socio-economic aspect has received the least attention. Survey results of construction practitioners in Chile revealed that larger companies along with those involved in infrastructure projects were more aware of environmental aspects of sustainability, while those in the building sector had the lowest level of awareness [6]. Furthermore, a common problem among stakeholders was found to be a lack of knowledge about sustainability, where, surprisingly, developers were perceived to be the least knowledgeable in both sectors. Chew [24] provides a state-of-the-art review of the strategic thrusts promoted and implemented in Singapore, such as government support, legislation on research and development support, as well as the development of demolition protocols and a sustainable-construction capability development fund.

1.1. Problem Statement

A review of the relevant literature reveals that the status of the industry (in regards to sustainability) in developing countries is not promising, which is apparent from the low level of awareness and the lack of knowledge of its stakeholders. In fact, there is a similar picture in developed countries, as the focus in those countries is mainly on the economic aspects of sustainability [6]. The review also found that the main factors hindering implementation of sustainability in the construction industry are lack of education and training on sustainable construction (SC), technologies, capacities and, more importantly, policies for the development and successful implementation of sustainability practices. The conventional wisdom is that there is a long way to go to achieve a sustainable construction industry, and this process needs an input from all industry stakeholders; however, it is important to get a clear picture of the current situation as a starting point [8,11].

Despite the similarity in findings of previous studies worldwide, the situation in each country, due to its sui generis socio-economic-politic context, requires a particular diagnosis. The intention of this survey-based research is to investigate industry professionals' awareness of the current state of, and their experience-based feedback on the factors driving and hindering, sustainable construction in Cambodia. It is hoped that the outcomes of this study enrich knowledge about the current state of, drivers of, and barriers to sustainable construction in Cambodia and form a significant base for future SC work within the country.

1.2. Construction Industry in Cambodia

Cambodia is one of the developing countries located in South-East Asia, and has been a member of the Association of Southeast Asian Nations (ASEAN) since 1999 [25]. With an annual contribution of 9–10% to GDP, the construction industry is one of the main pillars of the Cambodian economy [25]. Moreover, the industry contributes to the country's employment creation, which accounts for 200,000 jobs per day [26], and the strengthening of allied sectors due to its strong linkages. Due to these achievements of the construction industry, it seems likely that the industry will continue to remain promising, and a driver of the country's economic development and appetite for overseas investors, particularly from the ASEAN region and China [27]. For instance, in comparison with 2015, the sector experienced a marked acceleration in the value of construction project approvals in 2016, which increased by 260% [28]. However, notwithstanding the construction industry's significant economic contribution, sustainability issues in the sector have attracted very little attention from industry practitioners, as the vast majority of construction projects are profit-indexed [29]. Although there are individual LEED (Leadership in Engineering and Environmental Design) awarded construction projects in the country (i.e., Vattanac Capital tower), the level of awareness of, and knowledge about SC is low, which affects the delivery of sustainable projects [29]. Therefore, this study aims to provide a snapshot view of sustainability awareness and knowledge in the Cambodian construction context. Moreover, it is hoped that the quantified drivers and recommendations for overcoming barriers from the viewpoint of industry stakeholders will be useful for the successful implementation of sustainability issues.

2. Literature Review

Sustainability has become such a buzzword, which was defined simply by Pearce [30] as lasting or in perpetuity. In the current body of knowledge, the perception of sustainability is that it only concerns environmental protection [31]. However, other elements of sustainability, which are economic and social, cannot be disregarded and, therefore, in order to keep a balance between the three elements the concept of sustainability should be considered as a holistic and integrative approach [8]. As the construction industry provides various physical facilities (e.g., dams, roads, bridges, residential and commercial buildings, factories, recreational facilities amongst others), these have an impact on society, environment and economy. Therefore, the balance between the three elements of sustainability plays a significant role in the construction industry compared to other industries, and it is strongly recommended that the industry's success must be considered based on the triple-bottom-line, rather than traditionally used measures focusing on time, cost and quality [32].

Various sustainability-assessment tools and methods have been introduced, particularly by advanced countries and a few developing ones, to assess the sustainability performance of buildings. For instance, in the UK, the Building Research Establishment Environmental Assessment Method (BREEAM) was the first method of rating, assessing and certifying the sustainability of buildings, based on an overall score of pass, good, very good, excellent and outstanding [5]. The Leadership in Energy and Environmental Design (LEED) assessment system, which was developed by the United States Green Building Council (USGBC), rates the sustainability of green buildings according to their design, construction and operation [33,34]. In Hong Kong, Building Environmental Assessment Method (BEAM) Plus is one of the green-building certifications. Compatible with internationally-renowned green-building certifications such as LEED and BREEAM, the original intent of BEAM Plus was to be a voluntary scheme for the assessment of buildings' environmental performance [35]. The International Initiative for a Sustainable Built Environment (iiSBE) developed the SBTool assessment method to rate the sustainability performance of buildings, which focuses on the environmental, economic as well as social aspects of sustainability [36]. Despite the significant impact of the construction industry on economic and social development, the extant literature and the existing sustainability assessment tools suggest that the bulk of the studies and methods on the subject focus on the environmental aspect of construction sustainability [5].

To find ways to implement sustainable construction, firstly it is important to identify possible drivers and constraining factors, so that frontline industry professionals (government authorities, contractors, project managers) can subsequently act upon them effectively. Identification of the factors (both negative and positive) influencing SC is crucial [37], and will facilitate the adoption of those that have a positive effect and in the elimination or control of those that have a negative effect. Thus, Table 1 depicts a snapshot of the studies conducted in different countries focused particularly on sustainability awareness, and the drivers/opportunities to achieve, and barriers hindering, implementation of SC practices. Additionally, it provides a robust backing from the literature to achieve the aim of this study, which is the quantification of those identified drivers and barriers, as well as measuring the awareness of sustainable construction among industry practitioners in Cambodia. This is particularly the case due to the fact that there is a dearth of notable research conducted on the subject in the Cambodian construction context; therefore, an international context is required to identify and adopt the drivers, and remove the factors that hinder implementation of sustainable construction.

Table 1. Overview of the literature on the drivers of, and barriers to, sustainable construction (SC).

Location	Reference	Findings
Korea	Whang and Kim [5]	Environmental: energy efficiency, indoor environmental quality, waste management, management, ecological environment. Economic: life-cycle cost, knowledge management, value for money in delivery, retention of skilled labour, innovation. Social: well-being, community, education/training, service quality, health and safety.
Canada	Ruparathna and Hewage [38]	Barriers: lack of consideration of sustainability criteria in the evaluation of bids, unavailability of standard methods for procurement, lack of knowledge of local conditions, lack of explicit statutory requirements that cover sustainable procurement.
Malaysia	Shafii et al. [9]	Barriers: lack of awareness on sustainable building, lack of training and education, the higher cost of sustainable building options, procurement issues, regulatory barriers, lack of professional capabilities/designers, disincentive factors for local material production, lack of case studies/examples.

Table 1. Overview of the literature on the drivers of, and barriers to, sustainable construction (SC).

Location	Reference	Findings
Chile	Serpell et al. [6]	Drivers: regulations, company awareness, corporate image, client demand, cost reduction, market differentiation, suppliers. Barriers: lack of financial incentives, designers work alone, economic needs of higher priority, environmental costs not included in the cost structure, governmental bureaucracy, lack of knowledge on sustainable technologies, lack of environmental concern, affordability.
Greece	Manoliadis et al. [39]	Drivers: energy conservation, waste reduction, indoor environmental quality, environmentally-friendly energy technologies, resource conservation, incentive programmes, performance-based standards, land-use regulations and urban-planning policies, education and training, re-engineering the design process, sustainable construction materials, new cost metrics based on economic and ecological value systems, new kinds of partnerships and project stakeholders, product innovation and/or certification, recognition of commercial buildings as productivity assets.
US	Ahn et al. [37]	Drivers: energy conservation, improving indoor environmental quality, environmental/resource conservation, and waste reduction. Barriers: first cost premium of the project, long pay-back periods from sustainable practices, tendency to maintain current practices, and limited knowledge and skills of subcontractors.
International experts' survey	Darko et al. [40]	Drivers: greater energy-efficiency of buildings, reduction in the environmental impact of buildings, greater water-efficiency of buildings, enhancement of occupants' health and comfort and satisfaction, good company image/reputation or marketing strategy, better indoor environmental quality.
Finland	Hakkinen and Belloni [41]	Drivers: development of the awareness of clients about the benefits of sustainable building, the development and adoption of methods for sustainable building requirement management, the mobilization of sustainable building tools, the development of designers' competence and team-working, and the development of new concepts and services. Barriers: steering mechanisms, economics, a lack of client understanding, process (procurement and tendering, timing, cooperation and networking), and underpinning knowledge (knowledge and common language, the availability of methods and tools, innovation).
Hong Kong	Lam et al. [42]	Factors: green technology and techniques, reliability and quality of specification, leadership and responsibility, stakeholder involvement, guide and benchmarking systems.
UK	Pitt et al. [43]	Drivers: client awareness, building regulations, client demand, financial incentives, investment, labelling/measurement, planning policy, taxes/levies. Barriers: affordability, building regulations, lack of client awareness, lack of business case understanding, lack of client demand, lack of proven alternative technologies, lack of one single labelling/measurement standard, planning policy.

3. Research Method

This study adopts a mixed-method approach, which includes an in-depth analysis of previous studies on the subject (a qualitative approach) for the identification of the drivers and barriers [6]; a descriptive survey method (Appendix A) for the data collection (a combined qualitative and quantitative approach) from industry stakeholders in Cambodia [44]; and statistical methods that are consistent with the nature of this study, which were used to quantify the collected data in order to prioritise the drivers and barriers in terms of their levels of impact [2].

Content analysis via a systematic literature review was conducted to initially identify and list the general drivers and barriers of SC in the Cambodian context. Table 1 depicts an overview of the literature on the subject worldwide. In addition, in order to assure consistency with the aim of this paper, a descriptive survey method comprising three stages was adopted to investigate a new country-specific set of constructs (drivers and barriers) [45]: the interview, pilot study and open-ended questionnaire.

The target population for the study comprised government representatives, project managers, architects, engineers, contractors and subcontractors in the Cambodian construction sector, as well as academics. The sampling frames of data collection comprised the Ministry of Land Management, Urban Planning and Construction (MLMUPC), the Board of Engineers Cambodia (BEC), Board of Architects Cambodia (BAC), Cambodia Constructors Association (CCA) and the Housing Development Association of Cambodia (HDAC), as they represent the bulk of the actors responsible for a sustainable construction industry. Initially, a convenience sample of two contractors, two architects, one project management consultant and two government representatives agreed to devote their time to participating in face-to-face interviews in order to explore drivers and barriers of SC in the Cambodian context, and to validate the relevance of the factors mainly sourced via the aforementioned content analysis. An open-ended questionnaire was then designed (based on the new constructs identified during the interviews). The questionnaire was divided into three parts. In the first part, the respondents were asked to provide details of their respective demographic backgrounds (i.e., position, length of experience, professional affiliation); the second part focused on measuring their level of awareness, knowledge, and understanding about SC; and the third part sought feedback from industry stakeholders on the drivers and barriers of SC, which were subsequently ranked according to their relevant importance index (RII).

Prior to the administration of the open-ended questionnaire survey, the relevance and clarity of these open-ended questions were pre-tested [46]. Only Cronbach's alpha at a minimum of 0.6 was accepted for this pilot study, following Hinton et al.s' recommendation [47]. A score of 0.875 was achieved for this study, allowing the collection of data via open-ended questionnaire survey. Consequently, during the open-ended questionnaire survey, the respondents were given an opportunity to include any further constraint(s) in the open-ended section of the questionnaire. Emails bearing a link to the online survey were circulated to potential study participants encouraging them to respond before the cut-off date set for receiving responses. Finally, the results were analysed and interpreted accordingly in order to achieve the aim of this paper.

4. Data Analysis

Data preparation is of crucial importance before starting the data analysis stage. This study adopted a Likert scale rating of influence level from 1 to 5 where 1 stands for "Very low" and 5 for "Very high" influence of the drivers and barriers collected via the web-based questionnaire survey. Once the data were prepared in a suitable form for the statistical package used, the next step of the research process adopted in this study was the analysis of the data. This study aimed to prioritise the identified drivers and barriers of SC in Cambodia according to their importance, using the RII method [48,49]. For each driver and barrier, the RII was calculated using Equation (1):

$$ RII = \frac{\sum W}{(A \times N)} \tag{1} $$

where: W = weighting of each driver or barrier given by respondents; A = highest weight, which is 5 for this study; N = total number of respondents. Calculated RIIs range in value from 0 to 1 (0 not inclusive), indicating that the higher the RII, the more important was the driver or barrier.

Finally, for reliability analysis, which is particularly required when the items are used to form a scale (i.e., Likert scale), Cronbach's alpha (Cα) test was used to measure the internal consistency of the set of items forming the scale, which varies from 1.00 (the maximum value of reliability coefficient) to 0.6 (the minimum value of reliability coefficient) [50,51].

5. Results and Discussion

5.1. Survey Results

5.1.1. Demographic Background

The research was conducted between June 2017 and November 2017. Invitations (135 distributed) to participate in the questionnaire survey were sent to potential respondents. By the cut-off date set for the survey, it had been completed by 104 respondents representing companies operating in the architecture, engineering and construction (AEC) sector in Cambodia at various stages of project delivery, as well as government representatives. Table 2 tabulates the participant demographics for this study, where responses from contractors, architects and engineers accounted for 37%, 27% and 20%, respectively. Responses from project consultants and government authorities accounted for 12% and 5%, respectively. Preliminary analysis of the participant demographics showed that the respondents occupied high-ranking positions such as director or executive director, project managers and supervisors with an aggregated average of 6 to 18 years of work experience in the construction industry in Cambodia. The responses were therefore from those with higher status who are responsible for the implementation of sustainable practices in the construction industry.

Table 2. Demographic background of the respondents (*n* = 104).

Respondent	# of Responses	%	Length of Experience *	Affiliation
Contractor	38	37%	11	CCA
Architect	28	27%	7	BAC
Engineer	21	20%	6	BEC
Project consultants	12	12%	13	BEC
Government authority	5	5%	18	

* Average.

5.1.2. Awareness, Knowledge and Understanding of Sustainable Construction

Zainul-Abidin [11] claims that the adoption of SC practices starts with awareness, and that after awareness, knowledge can be a good catalyst in achieving adoption. Therefore, the first objective of this study was to gain an insight into the level of awareness and knowledge about the concept of SC. Respondents were asked to rate their level of awareness (from 0% = not at all aware to 100% = extremely aware) and knowledge (from 0% = no knowledge to 100% = extremely knowledgeable) on SC. It can be seen in Figure 1 that the awareness level of sustainable construction among the respondents ranges between 23% and 5%, which is relatively low. Similarly, the level of knowledge of sustainable construction among the stakeholders is not also promising, which ranges from 27% to 52%. This finding is not surprising as the level of implementation of SC practices in Cambodia overlaps with the level of awareness and knowledge. In his study, Brown [52] identified that a lack of awareness and knowledge on sustainability is one of the factors hindering the implementation of SC practices in the construction industry. It is possible the reason behind the lack of awareness and knowledge could be the lack of concrete, explicit regulations [38] or lack of expressed interest from clients [37]. Furthermore, this finding from the survey is justified by a review of the literature on the subject, which suggests that the low level of knowledge and awareness on the concept of SC is not very different in other developing countries [6].

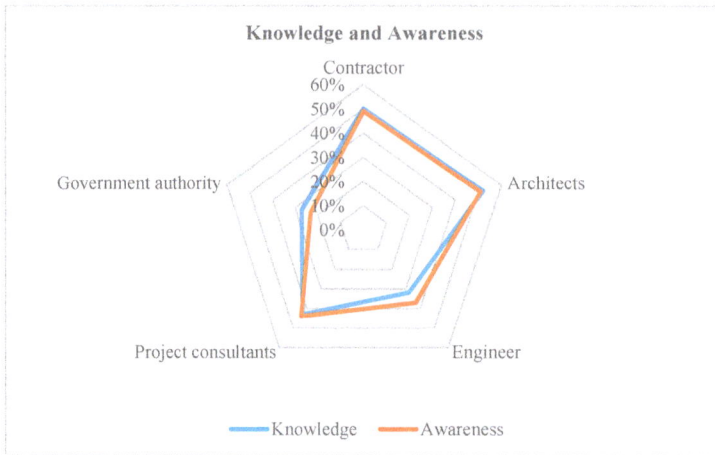

Figure 1. Respondents' levels of awareness and knowledge of SC.

As part of the survey, the respondents were additionally asked to select statement that best fit with their understanding about SC. The statements were written based on the three pillars of sustainable construction: environment; economic, and social issues [4], which are depicted by Figure 2. Statements *'to protect the environment'* and *'efficient use of natural resources'* received the highest percentage of 81% and 72%, respectively, which are both related to the environmental aspect of sustainability. This finding is not surprising, as it is self-evident that the word sustainability has been promoted as environmental awareness or protection of our environment [53]. While the statements related to the social aspect of sustainability, which are *'life quality'* and *'progress in social life'* received moderate percentages of 67% and 51%, respectively, issues related to the economic aspect, which are *'maintains economic growth'* and *'profit-generation without compromising the future needs'* received the lowest percentages of 49% and 43%, respectively. The low percentages of the economic-related issues reveal that the respondents do not know the economic consequences of a sustainable construction industry, which is unsurprising as sustainability has not been discussed at the sectoral level that much [30], particularly in developing countries [6].

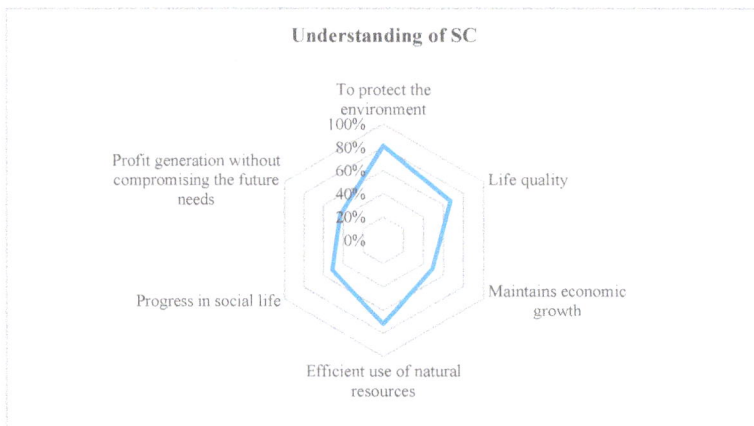

Figure 2. Perceived understanding of SC.

5.2. Drivers of Sustainable Construction

The second objective of this study was to rank the factors, which are believed to be the drivers of the implementation of sustainable construction in Cambodia. The RII method was utilized to rank the drivers of SC under each sustainability pillar (environmental, economic and social), which were identified through a rigorous review of the relevant literature, together with interviews with construction professionals in Cambodia. The following sub-sections present results of the data analysis and discussion on the outcomes of the analytical techniques adopted in this study. However, only the five most significant ranked drivers are considered, as suggested by Durdyev and Ismail [2], thus are further discussed in the following sub-sections.

5.2.1. Environmental Drivers

Table 3 shows the RII of 10 factors, which were identified through a review of the literature, to be environmental drivers of sustainable construction in Cambodia. In addition, SPSS-based results of Cα (0.816) show internal consistency between the items within the category [54]. The results clearly show that the respondents perceive the environmental drivers are significant, as the lowest mean score is slightly over 3.50. This finding aligns with the previous discussed results of the respondents' understanding of SC, where the environment-related statements received the highest percentage. Moreover, it has been proven that the construction industry is responsible for a large proportion of carbon-dioxide emissions worldwide, as well as the consumption of energy reserves and natural resources [55], which has an extremely detrimental impact on the environment. Another significant issue worthy of note is that there is not much difference between the perceived ratings of the different environmental drivers, which shows the significance of each individual item under this category.

With an RII of 4.10, the most important environmental driver of SC is perceived to be '*resource conservation*'. This result is justified, as the annual resource consumption of the construction industry worldwide accounts for 40% of the gravel, sand and stone, 25% of the timber and 16% of the water used [56], which makes the construction industry very resource intensive. This finding is consistent with previous findings of Akadiri et al. [57], who described it as achieving more with less, and identified the driver as one the objectives to best achieve the concepts of sustainability in the building sector. Halliday [58] states that the use of stocks of certain resources (material, energy, land and water), which are frequently consumed by the construction industry, should be treated cautiously as their availability is becoming extremely limited. Substitution of those materials with, preferably, renewable ones was recommended, or at least with less-scarce resources. Hence, the use of local supplies and materials, particularly those with low embodied energy [57] to minimise the dependency on imported materials, design for construction, and lifecycle design [59] to minimise material wastage, are some of the initiatives pursued in other countries to create ecologically sustainable buildings, which could be implemented in the Cambodian context.

'*Greater energy efficiency*' ranked as the second most important environmental driver (RII = 3.96) for SC in Cambodia, which is not surprising, as the construction industry is the most dominant energy consumer [60]. Energy consumption by the construction industry in Organization for Economic Co-operation and Development (OECD) countries, which is about 25–40% of total energy consumed, is clear evidence of the dominancy of the sector in this regard [61] and justifies the significant ranking of the driver. Despite its rapid population growth and dynamic economic development, Cambodia is experiencing macro-economic instability, particularly in the monetary domain. The net effect of these factors is due to the rapid growth in demand for energy in Cambodia, but the improvements necessary to compensate for the growing demand do not seem to be occurring at the desired pace [62]. However, meeting energy demand in the country is of strategic importance, and so promotion of energy efficiency and conservation in Cambodia via the implementation of SC practices is vital, particularly to achieve reductions in electricity demand in the construction sector. Consideration of the optimum insulation for buildings in Cambodia is recommended, as almost none of them are insulated and, therefore, energy demand is assumed to be higher than for insulated buildings. Despite its initial cost, insulation materials are proven to be a significant cost-saving solution by reducing the energy demand of commercial and residential buildings. Furthermore, the implementation of energy-efficiency technologies and energy generation from natural (renewable) resources is recommended. Ultimately, it is hoped that conservation of energy will reduce Cambodia's dependency on imported energy.

Table 3. Relative importance index (RII) of environmental drivers of sustainable construction.

Environmental Drivers (Cα = 0.816)	RII	Rank	Reference
Resource conservation	4.1	1	Dimoudi and Tompa [56]
Greater energy efficiency	3.96	2	Darko et al. [63]
Reduces the environmental impact of buildings	3.85	3	Sev [4]
Improving indoor environmental quality	3.83	4	Darko et al. [40]
Waste reduction	3.83	5	Wong et al. [64]
Land-use regulations and urban planning policies	3.71	6	Ahn et al. [37]
Sustainable construction materials	3.68	7	Sev [4]
Ecological environment	3.62	8	Akadiri et al. [57]
Greater water-efficiency of buildings	3.59	9	Ahn et al. [37]
Environmentally-friendly energy technologies	3.51	10	Chan et al. [65]

Besides being the most energy-consuming end-user, buildings are the major source of both direct and indirect carbon emissions, which has a detrimental impact on our living environment [4]. This justifies the third significant driver *'reduces the environmental impact of buildings'* rated by the respondents. As previously mentioned, the construction industry is one of the most significant contributors to the detriment of the environment, producing large amounts of CO_2 emissions and solid waste, 40% of raw materials, and 25% of timber consumption, and consuming 40% of total energy production and 16% of water usage worldwide [6]. Therefore, the adoption of green building practices or technologies (one of the drivers perceived to be significant with an RII = 3.51) in the construction industry has become strategically necessary to reduce carbon emissions and energy shortages, particularly in developing countries [65]. This driver was followed by *'improving indoor environmental quality'* which was perceived by respondents to be the joint fifth most significant driver. This result aligns with the findings of Ahn et al. [37], where both drivers have been ranked within the top 5 drivers of SC. They claimed that indoor quality of buildings is significant in providing healthy conditions and ensuring the occupants' well-being, as in our modern society, we spend more than 70% of our time indoors [4]. This finding further aligns with the rankings of the drivers under the social category of SC (Table 5), where *'well-being'* and *'health and safety'* are perceived to be the most significant drivers. Receiving the same rating from the respondents, *'waste reduction'* was believed to be driven by the successful implementation of SC practices. This is not surprising, as SC practices promote the use of pre-fabricated building components and adopt modular coordination techniques in construction projects, which avoids measuring and cutting on site [4]. Other factors, which are *'land use regulations and urban planning policies'* [37], *'sustainable construction materials'* [4], *'ecological environment'* [57], and *'greater water-efficiency of buildings'* [37], were perceived to be important in driving change in the construction industry towards the implementation of sustainable practices.

5.2.2. Economic Drivers

As previously mentioned, construction practitioners' perceptions of SC is mainly towards environmental issues. Although the adoption of SC practices is believed to have high initial cost [37], their benefits can be seen in the long-term, for instance the reduced cost of operation and maintenance during the facility's lifecycle [57]. Therefore, the economic aspect of SC practices cannot be disregarded. On this issue, the respondents were asked to rate the economic drivers of SC, and their respective RIIs are tabulated in Table 4. The SPPS-based result of Cα of 0.761 confirms the internal consistency among the economic drivers of SC [51].

Table 4. RII of economic drivers of sustainable construction.

Economic Drivers (Cα = 0.761)	RII	Rank	Reference
Value for money in the delivery	3.58	1	Whang and Kim [5]
Retention of skilled labour	3.51	2	Whang and Kim [5]
Lifecycle cost	3.31	3	Sev [4]
Construction cost efficiency	3.28	4	Akadiri et al. [57]
Quality management for durability	3.27	5	Sev [4]
Innovation/R & D	3.23	6	Chan et al. [65]
Profitability/Productivity	3.23	7	Nahmens and Ikuma [66]
Image and reputation	3.22	8	Serpell et al. [6]
Commercial viability	3.17	9	Whang and Kim [5]
Cost of operation and maintenance	2.82	10	Ahn et al. [37]
Knowledge management	2.71	11	Whang and Kim [5]
Competitive industry	2.65	12	Wong et al. [64]
Support of local economy	2.54	13	Pearce [30]

The relatively lower RII values of the economic drivers compared to the environmental drivers supports the previous finding of this study, where the respondents demonstrated their tendency to have a higher level of understanding towards the environmental drivers of SC. *'Value for money in the delivery'* and *'retention of skilled labour'* are pioneering drivers that received the two highest ratings from the survey participants, which is consistent with the findings of Whang and Kim [5]. The results reveal that the respondents' expectations of, and focus on SC practices is short-term, as the majority of long-term drivers received lower ratings. The higher ratings of *'construction cost efficiency'* and *'quality management for durability'* drivers compared to *'profitability/productivity'* reinforces the short-term expectations of the respondents from SC. It is believed that SC practices will enhance construction productivity on a long-term basis [66], and enhanced productivity drives faster delivery of construction projects with lower cost and higher quality [45,54]. Another finding worthy of note is that drivers such as *'competitive industry'* and *'support of local economy'* received relatively lower ratings, which suggests that the feedback was biased towards the opinions of the professionals at the project level. Finally, despite being one of its long-term benefits [37], *'cost of operation and maintenance'* ranked among the lowest drivers, which reaffirms the recognition of SC practices among the industry practitioners.

5.2.3. Social Drivers

The respondents were asked to rate the drivers under the social pillar of SC, which are presented in Table 5. The Cα of 0.71 further confirms the internal consistency between the social pillars of SC [51]. Compared to the other pillars of SC, social drivers received the lowest ratings on average, which clearly shows the respondents' lack of understanding and knowledge about the correlation between the social drivers and the concept of SC, or the social benefits of SC implementation. This finding is also consistent with the findings depicted by Figures 1 and 2.

Table 5. RII of social drivers of sustainable construction.

Social Drivers (Cα = 0.710)	RII	Rank	Reference
Well-being	3.5	1	Akadiri et al. [57]
Health and Safety	3.48	2	Sev [4]
Culture/heritage	3.35	3	Whang and Kim [5]
Employment	2.74	4	Ofori [67]
Partnership working	2.46	5	Ahn et al. [37]
Service quality	2.46	6	Zainul-Abidin [11]
Security	2.41	7	Whang and Kim [5]
Community	2.32	8	Ahn et al. [37]

Unfortunately, in current building design practices the social aspects of sustainability have been disregarded, or in other words buildings are not designed for humans [57], which affects *'well-being'* (ranked as the highest driver of SC), *'health and safety'* (the second highest driver of SC), physiological comfort and satisfaction, and consequently the productivity of an occupant. In this regard, the selected construction material or preferred design can be energy efficient, cost effective and perform well; however, it should not be considered as a sustainable product or design if it does not provide sufficient comfort for, and improve productivity of the occupant [11]. The driver *'Culture/heritage'* ranked as the third most significant in this category of SC, which confirms the findings of Sev [4]. Although cultural resources are important in transmitting knowledge to the next generation, as far as the authors are aware, in the majority of buildings around the country this issue has been disregarded. Clear evidence of this is that overseas construction investors bring their own designs without taking into account the values of the Khmer culture; however, it is believed that a building design contrary to the urban context will eventually damage the existing cityscape and skyline [4]. An in-depth review of the literature on the subject reveals that only studies from the architectural point of view have considered the cultural aspect to be a driver of SC, while from the contractors' point of view it was disregarded. This clearly indicates that there is disagreement among industry stakeholders on the drivers of SC. Thus, sustainable building design is expected to provide healthy, safe and comfortable indoor conditions while protecting the nature and the cultural heritage of a country. While the perceived less significant social drivers of SC were *'employment'*, *'partnership working'*, *'service quality'*, *'security'*, and *'community'*, this does not necessarily mean that they can be disregarded. However, this finding shows the misunderstanding of the long-term impact of SC concepts among the respondents. For instance, it has been asserted that successful implementation of SC practices will generate employment growth within a country [67].

5.3. Barriers to Sustainable Construction

Despite the increasing demand for more sustainable construction practices, factors hindering its implementation still exist, which warrants further exploration. Research on the barriers to the implementation of SC practices has gained worldwide attention [37,55,65,68]; however, there is a dearth of studies on the subject from developing countries in the relevant literature [69]. Therefore, this study has identified and ranked the barriers to SC according to their importance level as perceived by industry professionals in Cambodia. Table 6 shows the barriers to SC identified from the literature and their respective rankings. The results reveal that all the barriers are perceived to be significant for successful adoption of SC practices, while the Cα of 0.825 confirms the internal consistency between the items within the category [51].

'The higher cost of sustainable building option' barrier was perceived to be the most significant among other barriers to SC in Cambodia. This result reaffirms an element of bias in the perceptions of industry practitioners in SC practices, despite clear evidence that it is possible to procure sustainable buildings without significantly higher initial costs, which is in fact consistent with the outcomes of many studies worldwide [9,37,70–72]. The ranking of the barrier *'long pay-back periods from sustainable practices'* reveals the reluctance of construction companies to invest in environmentally, economically and socially sustainable designs/projects in Cambodia. Furthermore, this result reaffirms the short-term expectations of industry practitioners previously discussed in the economic driver sub-section of this paper.

Table 6. Barriers to implementation of sustainable-construction practices.

Barriers (Cα = 0.825)	RII	Rank	Reference
The higher cost of sustainable-building option	4.17	1	Qian et al. [68]
Lack of government incentives	4.14	2	Chan et al. [65]
Economic needs of higher priority	4.05	3	Serpell et al. [6]
Lack of statutory requirements that cover sustainable procurement	4.00	4	de Souza Dutra et al. [73]
Lack of professional capabilities/designers	3.97	5	Chan et al. [65]
Lack of client demand	3.97	6	Swarup et al. [74]
Lack of training and education	3.88	7	AlSanad [55]
Long pay-back periods from sustainable practices	3.77	8	Ahn et al. [37]
Tendency to maintain current practices	3.73	9	AlSanad [55]
Lack of environmental concern	3.61	10	Serpell et al. [6]

'*Lack of government incentives*' was rated the second most significant barrier, which is justified as it is believed that government's role in promotion of SC practices is unquestionably important [65]. For the adoption of SC practices it is important to promote among, and motivate the industry stakeholders, particularly in Cambodia where SC is in its infancy. To the authors' knowledge there are no government incentives that would trigger the adoption of sustainable-building practices, which perhaps justifies the perception of the respondents towards this barrier. Further, the fourth ranking of the hindering factor of SC, '*lack of statutory requirements that cover sustainable procurement*' reinforces the previous finding of the study, which is in agreement with the outcomes reported by de Souza Dutra et al. [73].

The ranking of third most significant barrier for '*economics needs of higher priority*' [6], implies that sustainability among industry practitioners is not seen as a priority, particularly the social and environmental pillars. This finding, therefore, correlates with the barrier '*tendency to maintain current practices*' [37], which inhibits the implementation of SC practices in Cambodia. The reason for this finding seems to be due to the lack of awareness and knowledge about the SC concept; indeed, it would seem that the current practices of project procurement employed in Cambodia satisfy industry stakeholders.

'*Lack of professional capabilities/designers*' is perceived to be the joint 6th most significant barrier to the adoption of SC practices. This result aligns with the finding of AlSanad [55] and Chan et al. [65] that a lack of skilled professionals limits the implementation of SC practices. This perception is valid, as the majority of construction professionals are unfamiliar/inexperienced with SC practices, and those (very few) who are familiar with SC are either from neighbouring countries, or local professionals educated overseas. It is clear that the country is lacking in the expertise necessary for the uptake of SC, and tertiary education currently remains focused on conventional methods of construction, rather than providing knowledge of SC practices for the future. '*Lack of client demand*' ranked as the other joint 6th most significant barrier to be overcome for successful adoption of SC practices in Cambodia. The relative importance of this barrier aligns with the outcomes reported by Korkmaz et al. [75]. Also, Serpell et al. [6] reported that client demand has a significant influence on encouraging practitioners to implement SC practices on their projects. Furthermore, Swarup et al. [74] identified that clients' strong and continuous commitment towards sustainable practices has an important impact on the achievement of sustainability goals. Although '*lack of training and education*' was ranked 7th among the barriers to SC, the general consensus [6,11] is on the criticality of this barrier in overcoming the aforementioned hindering factors, and in further improving the understanding of '*environmental concern*' among industry practitioners at various levels of responsibilities. The government's support via educational institutions would be beneficial for further improvement of the awareness and knowledge of sustainable-construction practice among future construction professionals.

6. Conclusions

The popularity of sustainable practices has grown over the last decade; adoption levels are increasing and seem set to continue as long as the barriers to its adoption are overcome. Successful implementation of sustainability principles worldwide have demonstrated the significance of the concept for the reduction of the construction industry's detrimental impact on our environment and social life, as well as sustaining continuous economic contribution. However, notwithstanding its contribution to socio-economic development, unfortunately the implementation of sustainable construction practices in Cambodia does not look promising, which would undoubtedly seem to be the only solution to current and future problems concerning the environment, economy and social life. Although known for its benefits, such as energy and resource conservation, cost efficiency, well-being, consistent quality and health and safety, the adoption of SC at the level it deserves is not easy due to industry- or country-specific barriers. Moreover, rather than focusing on general benefits or barriers, priority must be given to those which fit well with local conditions or are of utmost significance. To this end, being the first work undertaken in the Cambodian construction context, the authors' intention for this survey-based study was to investigate industry practitioners' awareness, knowledge and understanding to develop a picture of the current state of SC, as this is believed to be the stepping stone for successful SC adoption. Furthermore, significance levels of the drivers of, and barriers to, SC were quantified, to encourage widespread implementation of SC practices and to discover strategies to overcome those barriers to successful adoption of the concept in Cambodia.

A questionnaire survey was designed to ask industry stakeholders' viewpoints regarding SC, and to prioritise 31 driving and 10 hindering factors of the adoption of SC principles, which are reported in the relevant literature. Only 104 usable responses (of 135 distributed) were received from construction professionals in Cambodia by the cut-off date set for the survey. Collected data from the respondents using linguistic terms as very low, low, moderate, high and very high for the drivers and barriers, were expressed numerically using a Likert scale. The relative importance index method was further used to quantify the significance level of each identified driver and barrier.

The findings suggest that the level of industry-wide adoption of SC practices is poor, which is believed to be due to the legislative or stakeholder-related factors, namely lack of legislative requirements, lack of education focusing on sustainability, lack of awareness and knowledge, and reluctance to adopt new sustainable technologies. Thus, the first implication of this study for decision makers is to enable a legal and policy environment that will facilitate the achievement of sustainable development through sustainable-construction practices, where government plays significant role. Along with effective enforcement of these policies, it is essential to improve the knowledge and understanding about sustainable practices among the industry stakeholders. Early stage and long-term sustainability awareness and knowledge could be achieved through regular training, seminars and workshops, and more importantly incorporating the concept of sustainability into the higher education curriculum. Moreover, incentives to promote sustainability among practitioners and investors would be crucial, particularly financial incentives for the companies willing to adopt sustainable practices in their projects.

The results of statistical analysis showed that the top 5 environmental drivers were also pioneers among the other pillars of SC in Cambodia, which are resource conservation, greater energy efficiency, reduces the environmental impact of buildings, improving indoor environmental quality, and waste reduction. This result indicates that more effort must be invested in the selection of more durable materials for the extension of buildings' lives and for minimal material consumption. Furthermore, it is essential to develop energy-efficient buildings with minimal environmental impact and a healthy indoor environment; hence, the ability of future generations to meet their own needs will not be compromised.

Despite the factors driving the implementation of SC practices, barriers to SC practices still exist and have been evaluated in this study. The top 5 rated barriers were the higher cost of the sustainable-building option, lack of government incentives, economic needs of higher priority, lack of statutory requirements that cover sustainable procurement and lack of professional capabilities/designers. Although it was not listed as one of the barriers (to avoid duplication), the low level of awareness should be considered as the leading barrier to the adoption of SC practices, particularly in countries where SC adoption is in its infancy. However, it is worth pointing out that the top barriers are due to the economic concerns of the stakeholders, which reinforces their lack of knowledge about the concept. Moreover, this study reaffirms that the high cost of sustainable practice is perceived to be the most hindering factor worldwide regardless of the socio-economic status of the country.

The findings of this study have enriched knowledge about the current state of, drivers of, and barriers to sustainable construction in a typical developing economy setting. Although the outcomes of this study were a short scoping exercise, it has formed a significant foundation for future SC work within Cambodia. Due to the lack of available information, this study did not provide data on the extent of SC practices in Cambodia. However, the scope of the findings is sufficient to draw a picture of the lack of SC adoption and implementation in the country and report that the drivers are strong enough to motivate policy makers and industry stakeholders to address the barriers and invest in overcoming them. Efforts to overcome these barriers would provide impetus for significant SC adoption and implementation of its benefits, and consequently promote the adoption of SC practices in Cambodia.

Although the study objectives were achieved, further investigation based on a larger sample size of practising professionals is recommended. It is further recommended that feedback from construction clients is obtained in future research, as these clients are key influencers of decisions and outcomes in the project-delivery process.

Acknowledgments: The support of the Department of Engineering and Architectural Studies, Ara Institute of Canterbury, is acknowledged. Appreciation is given to the key professionals from the construction industry in Cambodia for providing their opinions, support and assistance.

Author Contributions: Serdar Durdyev, Derek Thurnell and Ali Ihtiyar together designed the research and wrote the paper, Edmundas Kazimieras Zavadskas and Audrius Banaitis provided extensive advice throughout the study regarding the abstract, introduction, literature review, research methodology, data analysis, results and discussion, and conclusions of the manuscript. The discussion was a team task. All authors have read and approved the final manuscript.

Conflicts of Interest: The authors declare no conflict of interest.

Appendix A. The Questionnaire Survey Sample

Part II

Using the 5-point rating scales provided, please rate the level of importance of each item based on your experience? It will be appreciated if you could also suggest other items that have not been included in the list which may drive the sustainable-construction industry in Cambodia.

Table A1. Level of importance: 5 = Very High; 4 = High; 3 = Moderate; 2 = Low; 1 = Very Low.

Drivers of Sustainable-Construction Industry	Level of Agreement					No Idea
	5	4	3	2	1	
Environmental						
1 Resource conservation						
2 Greater energy efficiency						
3 Reduces the environmental impact of buildings						
4 Improving indoor environmental quality						
5 Waste reduction						
6 Land-use regulations and urban planning policies						
7 Sustainable-construction materials						
8 Ecological environment						
9 Greater water-efficiency of buildings						
10 Environmentally-friendly energy technologies						
Other environmental drivers? Please list and assess:						

References

1. Durdyev, S.; Ismail, S. The build-operate-transfer model as an infrastructure privatisation strategy for Turkmenistan. *Util. Policy* **2017**, *48*, 195–200. [CrossRef]
2. Durdyev, S.; Ismail, S. On-site construction productivity in Malaysian infrastructure projects. *Struct. Surv.* **2016**, *34*, 446–462. [CrossRef]
3. Durdyev, S.; Ismail, S. Role of the construction sector in the economic development of Turkmenistan. *EEST Part A Energy Sci. Res.* **2012**, *29*, 883–890.
4. Sev, A. How can the construction industry contribute to sustainable development? A conceptual framework. *Sustain. Dev.* **2009**, *17*, 161–173. [CrossRef]
5. Whang, S.W.; Kim, S. Balanced sustainable implementation in the construction industry: The perspective of Korean contractors. *Energy Build.* **2015**, *96*, 76–85. [CrossRef]
6. Serpell, A.; Kort, J.; Vera, S. Awareness, actions, drivers and barriers of sustainable construction in Chile. *Technol. Econ. Dev. Econ.* **2013**, *19*, 272–288. [CrossRef]
7. Energy Information Administration (EIA). *Annual Energy Review 2011*; Energy Information Administration: Washington, DC, USA, 2012.
8. Du Plessis, C. *Agenda 21 for Sustainable Construction in Developing Countries*; CSIR Report BOU/E0204; CSIR, UNEP-IET C: Pretoria, South Africa, 2002.
9. Shafii, F.; Ali, Z.A.; Othman, M.Z. Achieving sustainable construction in the developing countries of Southeast Asia. In Proceedings of the 6th Asia-Pacific Structural Engineering and Construction Conference, Kuala Lumpur, Malaysia, 5–6 September 2006.
10. Shen, L.Y.; Hao, J.L.; Tam, V.W.Y.; Yao, H. A checklist for assessing sustainability performance of construction projects. *J. Civ. Eng. Manag.* **2007**, *13*, 273–281. [CrossRef]
11. Zainul-Abidin, N. Investigating the awareness and application of sustainable construction concept by Malaysian developers. *Habitat Int.* **2010**, *34*, 421–426. [CrossRef]
12. Holloway, S.; Parrish, K. The contractor's role in the sustainable construction industry. *Proc. Inst. Civ. Eng. Eng. Sustain.* **2015**, *168*, 53–60. [CrossRef]
13. Vanegas, J.A.; Pearce, A.R. Drivers for change: An organizational perspective on sustainable construction. In Proceedings of the Construction Congress VI, Orlando, FL, USA, 20–22 February 2000.
14. Khaksar, E.; Abbasnejad, T.; Esmaeili, A.; Tamošaitienė, J. The effect of green supply chain management practices on environmental performance and competitive advantage: A case study of the cement industry. *Technol. Econ. Dev. Econ.* **2016**, *22*, 293–308. [CrossRef]
15. Işik, Z.; Aladağ, H. A fuzzy AHP model to assess sustainable performance of the construction industry from urban regeneration perspective. *J. Civ. Eng. Manag.* **2017**, *23*, 499–509. [CrossRef]
16. Oliveira, R.A.F.; Lopes, J.; Sousa, H.; Abreu, M.I. A system for the management of old building retrofit projects in historical centres: The case of Portugal. *Int. J. Strateg. Prop. Manag.* **2017**, *21*, 199–211. [CrossRef]
17. Turskis, Z.; Morkunaite, Z.; Kutut, V. A hybrid multiple criteria evaluation method of ranking of cultural heritage structures for renovation projects. *Int. J. Strateg. Prop. Manag.* **2017**, *21*, 318–329. [CrossRef]
18. Yazdani, M.; Chatterjee, P.; Zavadskas, E.; Hashemkhani Zolfani, S. Integrated QFD-MCDM framework for green supplier selection. *J. Clean. Prod.* **2017**, *142*, 3728–3740. [CrossRef]
19. Schmidt, J.-S.; Osebold, R. Environmental management systems as a driver for sustainability: State of implementation, benefits and barriers in German construction companies. *J. Civ. Eng. Manag.* **2017**, *23*, 150–162. [CrossRef]
20. Koo, C.; Hong, T. Development of a dynamic incentive and penalty program for improving the energy performance of existing buildings. *Technol. Econ. Dev. Econ.* **2018**, *24*, 295–317. [CrossRef]
21. Egan, J. Rethinking Construction, The Report of the Construction Task Force. Available online: http://constructingexcellence.org.uk/wp-content/uploads/2014/10/rethinking_construction_report.pdf (accessed on 31 January 2018).
22. Berardi, U. Stakeholders' influence on the adoption of energy-saving technologies in Italian homes. *Energy Policy* **2013**, *60*, 520–530. [CrossRef]
23. Albino, V.; Berardi, U. Green buildings and organizational changes in Italian case studies. *Bus. Strategy Environ.* **2012**, *21*, 387–400. [CrossRef]

24. Chew, K.C. Singapore's strategies towards sustainable construction. *IES J. Part A Civ. Struct. Eng.* **2010**, *3*, 196–202. [CrossRef]

25. Durdyev, S.; Omarov, M.; Ismail, S. SWOT Analysis of the Cambodian construction industry within the ASEAN economic community. In Proceedings of the 28th International Business Information Management Association Conference, Seville, Spain, 9–10 November 2016.

26. Sum, M. Strong Employment in the Construction Sector. *Khmer Times.* 2017. Available online: http://www.khmertimeskh.com/news/37571/strong-employment-in-the-construction-sector/ (accessed on 4 October 2017).

27. Yonn, R. The effects of Cambodia economy on ASEAN economic moving forward. *Manag. Econ. Ind. Organ.* **2017**, *1*, 1–16.

28. Hawkins, H.; Sek, O. Construction Investment Skyrockets in 2016. *The Cambodia Daily.* 2017. Available online: https://www.cambodiadaily.com/business/construction-investment-skyrockets-in-2016-123668/ (accessed on 12 October 2017).

29. Cambodia Constructors Association (CCA). The Evolution of Cambodia's Construction Industry. Construction & Property, 2015. Available online: http://www.construction-property.com/read-more-101 (accessed on 5 October 2017).

30. Pearce, D. Is the construction sector sustainable? Definitions and reflections. *Build. Res. Inf.* **2006**, *34*, 201–207. [CrossRef]

31. Burgan, B.A.; Sansom, M.R. Sustainable steel construction. *J. Constr. Steel Res.* **2006**, *62*, 1178–1183. [CrossRef]

32. Huovila, P.; Koskela, L. Contribution of the principles of lean construction to meet the challenges of sustainable development. In Proceedings of the 6th Annual Conference of the International Group for Lean Construction, Guarujá, Brazil, 13–15 August 1998.

33. Lee, W.L.; Burnett, J. Benchmarking energy use assessment of HK-BEAM, BREEAM and LEED. *Build. Environ.* **2008**, *43*, 1882–1891. [CrossRef]

34. Akcay, E.C.; Arditi, D. Desired points at minimum cost in the "Optimize Energy Performance" credit of leed certification. *J. Civ. Eng. Manag.* **2017**, *23*, 796–805. [CrossRef]

35. Hui, E.C.M.; Tse, C.; Yu, K. The effect of BEAM Plus certification on property price in Hong Kong. *Int. J. Strateg. Prop. Manag.* **2017**, *21*, 384–400. [CrossRef]

36. Larsson, N. The International Initiative for a Sustainable Built Environment. Available online: http://www.iisbe.org/sbmethod (accessed on 31 January 2018).

37. Ahn, Y.H.; Pearce, A.R.; Wang, Y.; Wang, G. Drivers and barriers of sustainable design and construction: The perception of green building experience. *Int. J. Sustain. Build. Technol. Urban Dev.* **2013**, *4*, 35–45. [CrossRef]

38. Ruparathna, R.; Hewage, K. Sustainable procurement in the Canadian construction industry: Current practices, drivers and opportunities. *J. Clean. Prod.* **2015**, *109*, 305–314. [CrossRef]

39. Manoliadis, O.; Tsolas, I.; Nakou, A. Sustainable construction and drivers of change in Greece: A Delphi study. *Constr. Manag. Econ.* **2006**, *24*, 113–120. [CrossRef]

40. Darko, A.; Chan, A.P.C.; Owusu-Manu, D.; Ameyaw, E.E. Drivers for implementing green building technologies: An international survey of experts. *J. Clean. Prod.* **2017**, *145*, 386–394. [CrossRef]

41. Häkkinen, T.; Belloni, K. Barriers and drivers for sustainable building. *Build. Res. Inf.* **2011**, *39*, 239–255. [CrossRef]

42. Lam, P.T.I.; Chan, E.H.W.; Poon, C.S.; Chau, C.K.; Chun, K.P. Factors affecting the implementation of green specifications in construction. *J. Environ. Manag.* **2010**, *91*, 654–661. [CrossRef] [PubMed]

43. Pitt, M.; Tucker, M.; Riley, M.; Longden, J. Towards sustainable construction: Promotion and best practices. *Constr. Innov.* **2009**, *9*, 201–224. [CrossRef]

44. Durdyev, S.; Mbachu, J. On-site labour productivity of New Zealand construction industry: Key constraints and improvement measures. *Constr. Econ. Build.* **2011**, *11*, 18–33. [CrossRef]

45. Durdyev, S.; Mbachu, J. Key constraints to labour productivity in residential building projects: Evidence from Cambodia. *Int. J. Constr. Manag.* **2018**, in press. [CrossRef]

46. Tabachnick, B.G.; Fidell, L.S. Principal components and factor analysis. In *Using Multivariate Statistics*, 5th ed.; Tabachnick, B.G., Ed.; Pearson/Allyn & Bacon: Boston, MA, USA, 2007; pp. 582–633.

47. Hinton, P.R.; Brownlow, C.; McMurray, I.; Cozens, B. *SPSS Explained*; Routledge: New York, NY, USA, 2004.

48. Lessing, B.; Thurnell, D.; Durdyev, S. Main factors causing delays in large construction projects: Evidence from New Zealand. *J. Manag. Econ. Ind. Organ.* **2017**, *1*, 63–82.

49. Gudienė, N.; Banaitis, A.; Banaitienė, N. Evaluation of critical success factors for construction projects – an empirical study in Lithuania. *Int. J. Strateg. Prop. Manag.* **2013**, *17*, 21–31. [CrossRef]

50. Durdyev, S.; Sherif, M.; Lay, M.L.; Ismail, S. Key factors affecting construction safety performance in developing countries: Evidence from Cambodia. *Constr. Econ. Build.* **2017**, *17*, 48–65. [CrossRef]

51. Durdyev, S.; Omarov, M.; Ismail, S. Causes of delay in residential construction projects in Cambodia. *Cogent Eng.* **2017**, *4*. [CrossRef]

52. Brown, K.A. *Incorporating Green-Building Design Principles into Campus Facilities Planning: Obstacles and Opportunities*; Ohio University: Athens, OH, USA, 2006.

53. Mardani, A.; Streimikiene, D.; Zavadskas, E.K.; Cavallaro, F.; Nilashi, M.; Jusoh, A.; Zare, H. Application of Structural Equation Modeling (SEM) to solve environmental sustainability problems: A comprehensive review and meta-analysis. *Sustainability* **2017**, *9*, 1–65. [CrossRef]

54. Durdyev, S.; Ismail, S.; Kandymov, N. Structural equation model of the factors affecting construction labor productivity. *J. Constr. Eng. Manag.* **2018**, *144*, 1–11. [CrossRef]

55. AlSanad, S. Awareness, drivers, actions, and barriers of sustainable construction in Kuwait. *Procedia Eng.* **2015**, *118*, 969–983. [CrossRef]

56. Dimoudi, A.; Tompa, C. Energy and environmental indicators related to construction of office buildings. *Resour. Conserv. Recycl.* **2008**, *53*, 86–95. [CrossRef]

57. Akadiri, P.O.; Chinyio, E.A.; Olomolaiye, P.O. Design of a sustainable building: A conceptual framework for implementing sustainability in the building sector. *Buildings* **2012**, *2*, 126–152. [CrossRef]

58. Halliday, S. *Sustainable Construction*; Butterworth Heinemann: London, UK, 2008.

59. Graham, P. *Building Ecology—First Principles for a Sustainable Built Environment*; Blackwell, Publishing: Oxford, UK, 2003.

60. Schimschar, S.; Blok, K.; Boermans, T.; Hermelink, A. Germany's path towards nearly zero-energy buildings—Enabling the greenhouse gas mitigation potential in the building stock. *Energy Policy* **2011**, *39*, 3346–3360. [CrossRef]

61. Asif, M.; Muneer, T.; Kelley, R. Life cycle assessment: A case study of a dwelling home in Scotland. *Build. Environ.* **2007**, *42*, 1391–1394. [CrossRef]

62. Economic Research Institute for ASEAN and East Asia. Cambodia National Energy Statistics 2016. 2016. Available online: http://www.eria.org/RPR_FY2015_08.pdf (accessed on 7 September 2017).

63. Darko, A.; Zhang, C.; Chan, A.P.C. Drivers for green building: A review of empirical studies. *Habitat Int.* **2017**, *60*, 34–49. [CrossRef]

64. Wong, J.M.W.; Ng, S.T.; Chan, A.P.C. Strategic planning for the sustainable development of the construction industry in Hong Kong. *Habitat Int.* **2010**, *34*, 256–263. [CrossRef]

65. Chan, A.P.C.; Darko, A.; Olanipekun, A.O.; Ameyaw, E.E. Critical barriers to green building technologies adoption in developing countries: The case of Ghana. *J. Clean. Prod.* **2018**, *172*, 1067–1079. [CrossRef]

66. Nahmens, I.; Ikuma, L.H. Effects of lean construction on sustainability of modular homebuilding. *J. Archit. Eng.* **2012**, *18*, 155–163. [CrossRef]

67. Ofori, G. Sustainable construction: Principles and a framework for attainment—Comment. *Constr. Manag. Econ.* **1998**, *16*, 141–145. [CrossRef]

68. Qian, Q.K.; Chan, E.H.W.; Khalid, A.G. Challenges in delivering green building projects: Unearthing the transaction costs (TCs). *Sustainability* **2015**, *7*, 3615–3636. [CrossRef]

69. Darko, A.; Chan, A.P.C. Critical analysis of green building research trend in construction journals. *Habitat Int.* **2016**, *57*, 53–63. [CrossRef]

70. Kats, G. *The Costs and Financial Benefits of Green Buildings, A report to California's Sustainable Building Task Force*; Massachusetts Technology Collaborative: Sacramento, CA, USA, 2003.

71. Enshassi, A.; Mayer, E. Barriers to the application of sustainable construction concepts in Palestine. In Proceedings of the 2005 World Sustainable Building Conference, Tokyo, Japan, 27–29 September 2005.

72. Ametepey, O.; Aigbavboa, C.; Ansah, K. Barriers to successful implementation of sustainable construction in the Ghanaian construction industry. *Procedia Manuf.* **2015**, *3*, 1682–1689. [CrossRef]

73. de Souza Dutra, C.T.; Rohan, U.; Branco, R.R.; Chinelli, C.K.; de Araujo, A.J.V.B.; Soares, C.A.P. Barriers and challenges to the sustainability requirements implementation in public procurement of engineering works and services. *Open J. Civ. Eng.* **2017**, *7*, 1–13. [CrossRef]
74. Swarup, L.; Korkmaz, S.; Riley, D. Project delivery metrics for sustainable, high-performance buildings. *J. Constr. Eng. Manag.* **2011**, *137*. [CrossRef]
75. Korkmaz, S.; Riley, D.; Horman, M. Piloting evaluation metrics for sustainable, high-performance building project delivery. *J. Constr. Eng. Manag.* **2010**, *136*, 877–885. [CrossRef]

sustainability

MDPI

Article

Promoting Sustainability through Investment in Building Information Modeling (BIM) Technologies: A Design Company Perspective

Marius Reizgevičius [1,2], **Leonas Ustinovičius** [3,*], **Diana Cibulskienė** [2], **Vladislavas Kutut** [1] **and Lukasz Nazarko** [3]

[1] Department of Construction Management and Real Estate, Vilnius Gediminas Technical University; 11 Saulėtekio al., 10223 Vilnius, Lithuania; mariusreizgevicius@gmail.com (M.R.); vladislavas.kutut@vgtu.lt (V.K.)

[2] Siauliai University, 88 Vilnius Street, 76285 Šiauliai, Lithuania; cibulskiene@yahoo.de

[3] Faculty of Engineering Management, Bialystok University of Technology, 45A Wiejska Street, 15-351 Bialystok, Poland; l.nazarko@pb.edu.pl

* Correspondence: l.uscinowicz@pb.edu.pl; Tel.: +370-5-2745-233

Received: 3 January 2018; Accepted: 14 February 2018; Published: 26 February 2018

Abstract: The aim of this article is to enhance the understanding of how design companies perceive the benefits of Building Information Modeling (BIM) technologies application. BIM is recognized in the literature as a (potentially) powerful driver leading the construction sector towards sustainability. However, for design companies, the choice to invest in BIM technologies is basically an economic one. Specifically, a design company assesses economic benefits and efficiency improvements thanks to the application of BIM technologies. The article discusses the return on investments (ROI) in BIM technologies and reviews ROI calculation methodologies proposed by other authors. In order to evaluate BIM return on investment correctly practical ROI calculations are carried out. Appropriate methods, together with the relevant variables for ROI calculation, are developed. The study allows for adjusting the calculation method making it more accurate and understandable using the Autodesk Revit based ROI calculation of the first year.

Keywords: building information modeling; BIM; sustainability; sustainable design; return on investment; ROI; efficiency; BIM evolution

1. Introduction

Building Information Modeling (BIM) technology has radically altered the organization of construction operations. Building information models and digital construction allow for assessing building structure and to identify its advantages and disadvantage, taking into account technological, economic, and environmental aspects. It is widely believed that the popularization of BIM should have a positive impact on the sustainability of the whole construction sector [1]. Representatives of the construction industry base their decisions to invest in BIM on its expected impact on the construction project performance. Therefore, the evaluation of quantitative and qualitative benefits of BIM is necessary. In this paper, the analysis of the concept of Return on Investment (ROI) and of ROI calculation methodologies are carried out. Significance of the criteria of the ROI is assessed. Data on cost reduction thanks to BIM implementation is scarce, with design data confidentiality being one of the primary reasons. Published studies are often criticized for scope, accuracy, calculations, objectivity, and general bias (in cases when studies are published by companies selling BIM-related software and services) [2].

There are very few frameworks and examples of ROI calculations relation to design processes. The results generated by the most common ROI tool, Autodesk Revit evoke certain doubts (analyzed later in the paper), therefore, the authors have decided to develop an alternative method of ROI calculation that would be suitable not only to market giants, but also to small and middle design studios.

The goal of this paper is to analyze the benefits of BIM technologies application in design companies. Specifically, the application of ROI approach to measuring BIM efficiency is studied based on the current body knowledge. An applicable ROI calculation model is also proposed. Regarding the research methods, the authors first performed the analysis of the scientific literature on the ROI concept and on measuring BIM efficiency [2–9]. Then, the logical construction approach is used to propose a specific ROI calculation method. Finally, the application of the method is presented based on the data from a Lithuanian design company.

2. Building Information Modeling and Sustainability

Building Information Modeling may be described as a process of: (a) elaborating an integrated and holistic building creation strategy encompassing design, construction, and life-cycle management based on modelling and computer simulation [10]; (b) creation and utilization of system of integrated graphical data management and information flow in connection with the description of construction process; and, (c) turning single contractors into teams that work as decentralized units that tackle complex problems and integrate separate tasks into coherent processes. In consequence, an increase in the efficiency and the lowering of costs of various operations throughout the entire building lifecycle is expected [11].

Conventional three-dimensional (3D) BIM is being transformed into a four-(4D) [12], five-(5D), six-(6D), or even seven-(7D), and, in the future, eight-dimensional (8D) versions on the basis of the application of PLM (Product Lifecycle Management) to construction [13,14]. This solution has been named BLM (Building Lifecycle Management) or unified project management [15]. This trend is a logical consequence of the use of large amounts of information available in smart 3D building models [16]. Synthesis of aspects integrated in the BIM framework is presented in Figure 1.

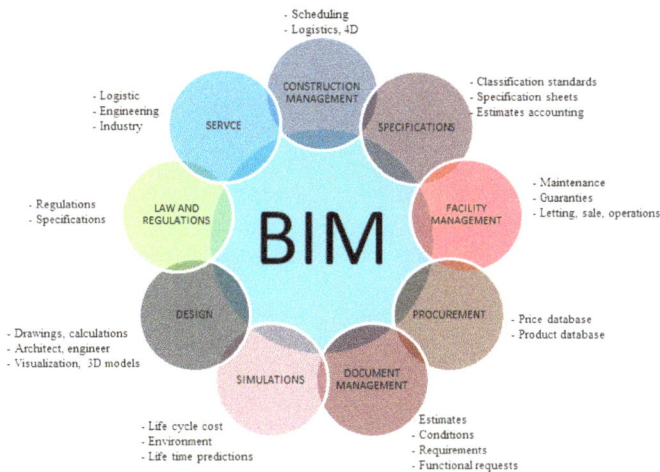

Figure 1. Building Information Modeling (BIM) lifecycle view [17].

The BIM dimensions beyond the third one may be described as [17,18]:

- 4D—virtual model of the built structure with construction plans and work progress control capability; with additional possibility to prospectively visualize a virtually constructed building in any moment in time;

- 5D—cost data is fed into a 3D model coupled with the construction schedule. Benefits of the fifth dimension of BIM may consist in the higher precision and predictability of changes occurring in the project together with a more reliable cost analysis of different construction scenarios.
- 6D—introduction of sustainable development principle into the investment process with an emphasis on energy efficiency. The sixth dimension of BIM allows for obtaining information about the building's projected energy consumption at a very early (concept) stage.
- 7D—integration of the Facility Management concept into BIM. It allows tracking of the status of given building components, their specifications and guarantee periods. The seventh dimension of BIM encompasses the management of the full life cycle of a building from the concept to the demolition.
- 8D—supplementing the model with security and healthcare information. This dimension focuses on three tasks: identification of threats resulting from chosen design and construction solutions, indication of alternatives to the most risky solutions, signaling the need to control specific risks on the construction site.

From the position of sustainability paradigm, the evolution of BIM promises many possible ways in which designing, constructing, maintaining, and decommissioning buildings may be done in a more sustainable manner. Most commonly, the positive impact of BIM implementation on the sustainability of the construction sector is expressed through the concept of sustainable (green) design [19], which may be defined as creating and operating a healthy built environment based on resource efficiency and ecological design [20]. Additionally, wider spread of BIM technologies is expected to have a positive impact not only on environmental aspects of sustainability, but also on economic and social ones. It may be stated that BIM promotes sustainability in its three classical dimensions:

- Environmental sustainability—more environmentally conscious decisions throughout the whole life cycle of a building. Thanks to the BIM's capacity to store, process, and share all kinds of building-related information it is possible to minimize environmental impact of the asset in relation to: energy and water consumption, used materials, waste management, carbon footprint etc.
- Economic sustainability—assuring economic viability, increasing productivity, and reducing waste. Most authors concentrate on the economic benefits that BIM poses for investors in terms of early detection of potential clashes, better engineering decisions, more efficient logistics, precise *ex-ante* calculation of costs throughout the whole life cycle of an asset, etc. This paper studies the question of economic viability of implementing BIM technology in design companies since this topic is currently insufficiently covered in the literature.
- Social sustainability—contributing to creating healthy and livable communities by providing tools to improve building operation in such aspects as: waste management, indoor air quality, noise pollution, safety at the construction site, more precise, and less disturbing operations on municipal infrastructure, onerousness of maintenance activities (thanks to better timing and synchronization). Moreover, sustainability-related information continuously gathered and analyzed thanks to BIM may be turned into an instrument of social engagement by involving occupants in setting common sustainability goals for their buildings, monitoring the progress, and celebrating achievements. Such use of BIM does not only contribute to "greening" of the built environment, but also encourages interaction among occupants and strengthens social bonds.

By modelling, managing, and predicting such aspects of construction as waste management, indoor air quality, protection during construction, erosion, and sedimentation control, etc. All of that is related to gathering, storing and processing large amounts of specific data that needs to be treated in an integrated way just the way BIM does it [20]. In other words, BIM's contribution to sustainable design could be summarized in five points [21]: better informed decision-making, better analysis, easier access to information, better communication between stakeholders, and simpler certification. It is therefore in the public interest that as many designers as possible adopt BIM technologies in their practice.

3. Benefits of Implementing BIM in A Design Company

Yan and Damian [22] carried out a study to assess building information model implementation advantages and barriers. The aim was to assess the ratio of BIM in architecture, engineering, construction, management companies of the United Kingdom, the United States, and other countries [23–26]. Companies that were not using BIM were questioned as well. The questionnaires of BIM implementation barriers were sent to the Architecture, Engineering and Construction (AEC) industry practitioners and academics. USA is the undisputed leader in BIM use. Besides USA, only few UK companies use BIM technology; feasibility studies are still in process. However, local government stated that by 2016 all public sector construction projects should be carried out using 2nd BIM competence level [27,28].

21 respondents (out of 67 questioned) were from UK, 23 from USA, and 23 from other countries [22,29]. 25% of respondents in the US and UK noted the biggest advantage of BIM to be short designing time. 25% of respondents in the United Kingdom assumed it to reduce costs and human resources. More than 10% of respondents of both the countries believed that BIM technology advantages were: creativity, sustainability, and increased quality. About 40% of US respondents and 20% of UK respondents believed that their firms would have to use a lot of time and human resources for training. It was emphasized that decisions taken by organizations were based on business prospects (profit). AEC industry is hesitant to invest in BIM because of the lack of information about BIM financial benefits. In addition, social and personal work habits of architects and designers need to be assessed. Designers are usually satisfied with their work quality and any changes are not easily acceptable. Other barriers were also mentioned: currently used technologies (software, skills) are sufficient; workers refuse to be trained, BIM is not suitable for all projects; expensive training and uncertain copyright ownership.

According to McGraw Hill Construction BIM studies two thirds of users foresee positive ROI on their investments. Almost half of the users (48%) get the average ROI indicator. Users that appreciate BIM identified key factors affecting the use of BIM software [3]:

(1) improved project results (less requests for information (RFI) from customers),
(2) more efficient cooperation using 3D visualizations,
(3) personnel productivity improvement,
(4) the positive impact on winning new contracts,
(5) BIM lifecycle value, and
(6) initial staff training costs.

The authors have conducted supplementary survey among design companies to determine the most important qualitative benefits of implementing BIM. Respondents were asked to indicate which charactersitcs of BIM would be most important in the decision to start using BIM software. The following features were listed:

(1) user-friendliness,
(2) easy sharing of information in one central file,
(3) ability to perform automatic analyses (lighting, building's location in relation to cardinal directions etc.),
(4) transparency of projects,
(5) detailed visualisations,
(6) better conditions for designing prefabricated elements,
(7) precise calculation of the needed amount of work,
(8) more possibiities to innovate, and
(9) avoidance of errors.

According to McGraw-Hill Construction studies [5], 27% of the potential BIM users in Western Europe are not interested in BIM software due to the economic factors. Western European countries carry out much lower volume projects; existing projects are often reconstructed/modernized causing less favorable environment for BIM. According to the survey [5] 70% professional users use BIM for more than 60% of their projects. Meanwhile, 46% of new BIM users building information model use it in about 15% of their projects. Taking all Western European BIM users into account, 59% use BIM software in more than 30% of their projects. In conclusion, those users who have comprehended the essence of BIM have effectively integrated it into their internal design processes [30].

Finnish national authorities, local governments, and representatives of industry begin to appreciate opportunities that follow BIM implementation [27]. The guide of general requirements of BIM, published in 2012, highlighted the increasing need to develop general rules and standardized BIM. Initial mandates and requirements were only for building designs, but in 2015, general requirements for infrastructure projects were issued [27,31]. In many countries around the world one observes the promotion of BIM use at governmental levels [32].

BIM Institute of Canada suggested that BIM designing should be binding in the country's public sector. The authors [33] state the inability of the responsible organizations to standardize and patent BIM operation to be the dominant factor for BIM installation in public sectors. Standards and protocols of common language, as well as software packages that can interact with each other, are necessary in order to make the information freely available to all construction market participants.

BIM benefits within the building lifecycle are identified by processes according to "Digital Construction" guidelines for 2014–2020 [34] by Lithuanian construction organizations: improved new spatial planning and design, fewer mistakes, and more rational decisions. Design process is of higher quality and more rational, there is less need for amendments in the future [35,36]. BIM model visualizes design solutions of all design parts, for the client to gain more understanding of the results, eliminating mistakes caused by the overlapping parts of the project [37–40]. According to the theory [41], different BIM use scenarios show BIM network to be effective in reducing the number of mistakes. Mistakes resistance mechanism: permanent and direct communication (in real time), accident detection, automated scrutiny of standards/codes, design versatility, and continuous learning are compelling reasons reducing the number of design defects. According to the authors, further research is necessary to verify the theoretical studies carried out by other authors.

There is no widespread method of BIM return on investments (ROI) calculation. Most of the users regard ROI as time, money, and BIM deployment effort return [6]. Calculated ROI rate is an estimated payback of BIM software, but it is not a return on investment in a specific project. Negative or equal ROI is noticeable in smaller organizations during the first year of primary BIM mastering stage, as it is harder for the organization to absorb the initial costs because of the software price, training, and business development using BIM. Main beneficiaries of the projects that use BIM design are designers and contractors [42]. Usually contractors reach positive ROI faster than designers because they get more financial benefits—less duplication of work, increased profits. The highest ROI is achieved by users with a longer experience in BIM use, strong skills, years of experience, a high level mastering BIM [6]. According to Poirier [43], investment in new equipment and technology may improve productivity but productivity indicators may actually decrease if the price of new equipment is greater than relative savings as a result of labor costs and profit results. BIM implementation creates a challenge that can be met only if BIM technologies are profitable for a company and the profit potentially outweighs the implementation costs. In the next part of the paper, ROI in BIM technologies are calculated as compared with "Autodesk Revit" ROI calculations.

4. Research on Economic Effects of BIM-Based Design

According to the review of latest literature there are several methods that could be used to analyze the investment in BIM technologies. The best known is the return on investment analysis (Return on Investment—ROI), "Prioritaring Efter NyytoGrunder" (PENG) model and total economic impact (TEI) model. ROI analysis of few projects is performed by the Center for Integrated Facility Engineering (CIFE) [1]. ROI of the first year is also calculated by Autodesk Revit company. PENG model is used in the Swedish construction industry to calculate savings using BIM. TEI model, used by CIFE, is to evaluate the impacts of the Virtual Design and Construction (VDC) and to calculate the payback time. Azhar's [44] research indicates that BIM technology return on investment (ROI) may be much higher than traditional construction projects without BIM showing potential economic benefits [45].

Research of Al-Zwainy at el. [46] show that BIM return on investment can be analyzed in various aspects. CIFE performs ROI analysis in Holder Construction Company. It makes solid ROI calculations of 10 projects of BIM technologies and assesses the savings. Comparative studies of two options are often performed: two similar projects are estimated (one designed using BIM, the second one designed with traditional methods). Autodesk Revit has developed the first year of ROI (return on investment) calculation model evaluating software and labor costs, loss and growth of productivity, as well as training time. CIFE uses the total economic impact model (TEI) to find out the benefits of digital design. This method is based more on BIM payback time [3].

Return on investment in BIM was estimated in the United Kingdom, France, and Germany [6]. 71% of UK consumers receive a positive return on investment (37% of users get 25% or bigger return on investment) and 13% of consumers get negative ROI. Out of the three studied countries, the percentage of consumers with negative ROI is the highest in the UK. In France, the positive ROI is observed by 82% of the users and the negative ROI ratio only by 5% of the users. Positive ROI is obtained by 67% of consumers in Germany (the lowest result out of the three studied countries) and 9% register negative ROI. 55% of the companies that do not use BIM software state that the first reason for not using BIM is no such requirement from the clients. They also add time requirements to evaluate the appropriate software and the scale of investments as the reasons. Small businesses in Western Europe are in no hurry to apply BIM technologies because BIM is less effective for smaller projects. Based on the last AEC field studies BIM also brings new challenges [47]. Very often, models become so large and complex that they exceed the capacity of the computers, making it problematic to track the models in real time.

Study of Giel and Issa [48] suggest that return on investment in BIM-supported projects may be much higher than the return from the traditional investments. Works [49–52] describe the benefits from BIM implementation in the qualitative terms. Some papers based on statistics on establish ROI in quantitative terms [45,48]. These studies conclude that short-term and small contracts may benefit from BIM in some qualitative aspects, but monetary savings related to BIM implementation were relatively low. The analysis of ROI also indicated that the design costs would rise if they were to operate with BIM [45]. It is an understandable result of more workload imposed on designer(s) by BIM in the initial project phases. It is the owner (investor) and contractor who may count on the largest gains from implementing BIM. This observation is confirmed by the case studies presented in [53]. The study [54] proposes a structured method for analyzing the BIM ROI based on the avoidance costs of rework due to design errors.

Method of calculating ROI proposed by Autodesk evaluates not only the cost of the system, but also the productivity changes [55]. A sharp decline in productivity is noticeable when the new system is acquired, as users have to learn to use new program. After training and the time needed to master the program, the curve of productivity noticeable goes up (Figure 2).

ROI in BIM technologies could not be measured from design to construction completion or demolition. Construction companies that are unwilling or unable to create their own design departments are often BIM initiators by requiring it from external designers that they work with. Meanwhile, a designer looks for the profitability of the offered services. Designer's work starts from the design task and the preparation of a commercial offer, and ends with a construction permit and project acceptance certificate. Construction contractor usually appears only when the construction permit is ready. The construction contractor, as well as the designer, are often chosen in the basis of the lowest price criterion. Assessing ROI, the return curve may be significantly distorted. Taking the period from the design to construction completion is one way to calculate the return. Designer does not care for the end of construction works if it has no author supervision of the execution or technical supervision contract, so ROI should not be measured the same for design and construction companies. BIM design should not be forced on designers—it has to be implemented on a voluntary basis and is consistent with the designer's whole work system.

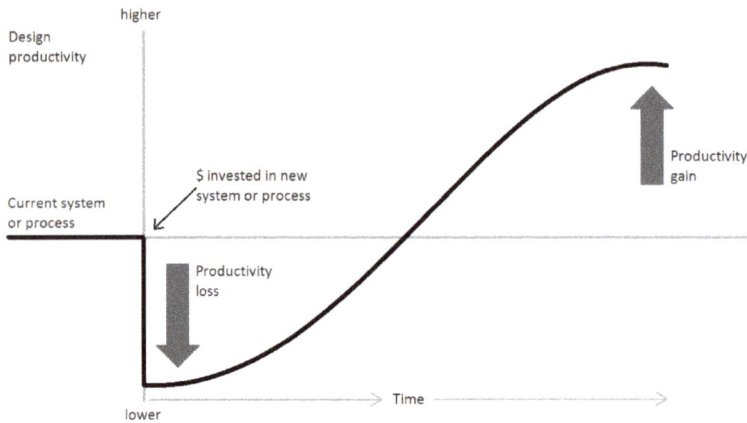

Figure 2. Design productivity after BIM system implementation [56].

In the process of assessing the benefits of BIM, it is important to mention the wider communication and cooperation opportunities of the staff and of individual members of the project team [57]. BIM establishes closer cooperation between members of the project team. BIM-integrated interactive possibilities are much more favorable to the customer information. Database information of building information model may be freely accessible via the internet and shared using cloud technology between members of the design team, customers, suppliers, producers, etc. [57].

Figure 3 illustrates the links between the customer (builder), construction contractor, and the designer (design company or architect) [58]. Such lines of communication are described by the standard design company model. Contractor search starts when the project is ready. It is noted that instructions from the customer are directly transmitted to the construction contractor that coordinates construction issues with subcontractors, suppliers, and other participants in the construction works. In this configuration, the designer is left aside and only cooperates with the client who implements designer's instructions [59]. This model provides a hierarchical management system where separate participants of the construction project are in no contact; there is no cooperation between different parties. Most design companies work according to this model, therefore a fair assessment of ROI is debatable. Contractor, designer, customer—different parties are mostly concerned about their own, narrowly understood interest.

Autodesk Revit calculates ROI, i.e., the indicator characterizing the return on investment required for BIM technologies installation, in the following manner [55]:

$$ROI = \frac{Earnings}{Investment or\ Cost}. \tag{1}$$

The first year, ROI calculation is the following:

$$First\ year\ ROI = \frac{(B - (\frac{B}{1+E})) \cdot (12 - C)}{A + (B \cdot C \cdot D)}. \tag{2}$$

ROI formula variables:

A- cost of hardware and software (EUR)

B-   monthly labor cost (EUR)

C-   training time (months)

D-   relative productivity loss during training in relation to the productivity before the training (percentage)

E-   relative productivity gain after training in relation to the productivity during the training (percentage)

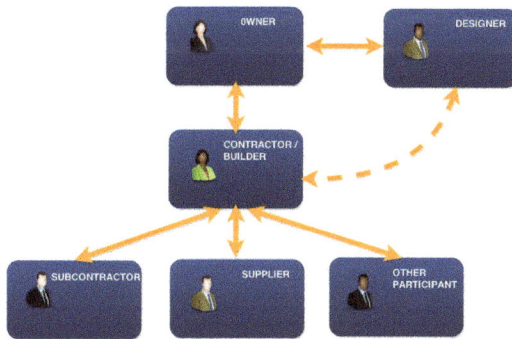

Figure 3. Standard cooperation model between construction parties.

The question is whether ROI calculation method by Autodesk assesses all the relevant variables. When calculating the return on investment the following elements should be considered:

1.   The software and hardware
2.   Training
3.   Development processes
4.   Interoperability solutions
5.   3D library development
6.   Consultant services

The authors suggest a different methodology of ROI calculation ("ROI-DC") which takes into account more variables. The proposal of first year ROI verbal formula expression is the following:

$$ROI\ DC\ (first\ year) = \frac{(Gain\ after\ Investment - Investment\ Cost)}{Investment\ Cost}. \tag{3}$$

As the next step the following ROI formula is proposed:

$$First\ year\ ROI\ DC = \frac{((12 - C) \cdot (1 + E) \cdot (B \cdot F \cdot G) - (A + (B \cdot C \cdot D))}{A + (B \cdot C \cdot D)}. \tag{4}$$

1.   Increase (Gain) after investment = (12 months—training time (months)) × relative productivity gain 12 months after training in relation to the productivity during the training (times) × monthly labor costs (euro) × direct design work (times) × company works specificity (times).
2.   Investment Price (Cost)-software price + (training time (months)) × monthly labor costs (euro) × relative productivity loss during training in relation to the productivity before the training (times).

Visualization, productivity increase, process improvement (less duplication of work and fewer coordination problems), competitive advantage, improvement of communication, and collaboration [5] are important issues to consider while assessing the return on investment in BIM. In order to assess the benefits of BIM and to assess the return on investment it is necessary to take into account the following factors: what savings the new system will make, what influence it will have on the company's share value and profitability, how much time it will take to train new users [48].

Some of the variables in the formula have bigger weight than others. ROI calculation methodology by Autodesk [55] shows that the productivity index (deriving from the productivity gains and losses) has the greatest impact on ROI. The results also show that the most productive were the users, who have been trained from the scratch and had no contact with the program before. Productivity of the consumers who purchased the program and independently studied it does not equal the one achieved by the fully trained users. Staff with no training were the least productive. The data shows that the critical training time is 1–2 months. If the trainings start within two months after purchasing the program, bad habits develop and incorrect program use jeopardizes the productivity [55].

Figure 4 indicates the areas of ROI measurement. The fact that ROI cannot be measured in only one of those areas is underscored. The figure shows that the building information model is most beneficial to the customer through the entire lifetime of the construction. The customer has the longest payback time. Meanwhile, for the designer/architect who uses BIM model only for the designing time, the payback time for the particular project is much shorter. BIM payback time for contractors is slightly longer so they feel bigger benefits of BIM. Other construction participants, such as suppliers, subcontractors, etc., are involved in the project only partially, and therefore ROI time for them is even shorter.

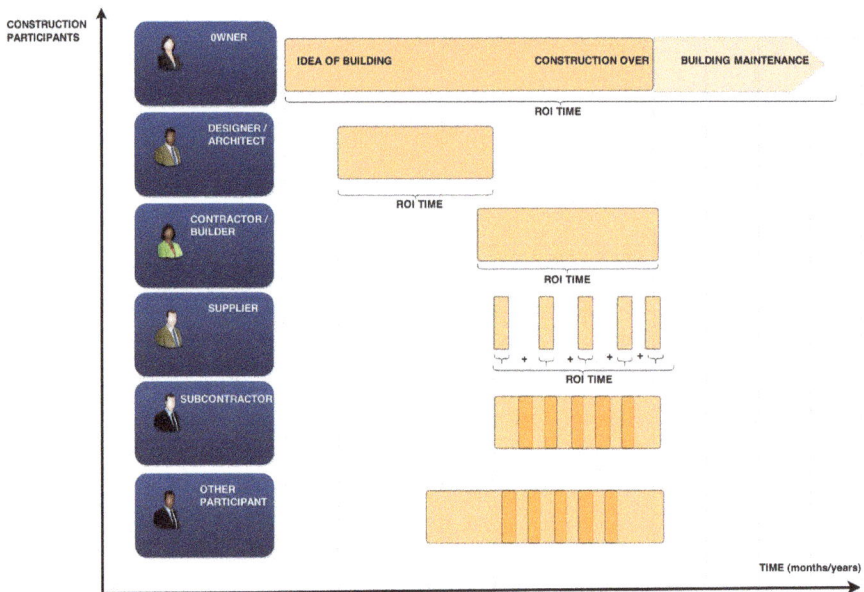

Figure 4. Return on Investment (ROI) assessed areas [50].

5. Evaluation of ROI in BIM Software

Different actors in the construction project have different needs that may be satisfied by BIM (Figure 5).

The analysis of existing research and calculations of BIM technologies return on investment by various authors and companies have shown that the advantages of high ROI in BIM technology has not persuaded many design firms and that they are in no hurry to implement the building information model software. A number of reasons have led to this situation. In particular, ROI calculations are carried out for large projects where BIM is used not only in design, but also in construction and management. High ROI ratio seems inadequate when calculation parameters exceed 100% of the first year of BIM software installation. Such indicators seem inadequate and they describe only construction industry giants—design centers, consortia, companies engaged in both the design, and construction activities. A medium or small design company only engaged in design will certainly not experience such a high first year return on investment (ROI ≥ 100%). The existing projects will suffer first of all because of staff training, large investments requiring a loan that will effect in paying back the interest afterwards.

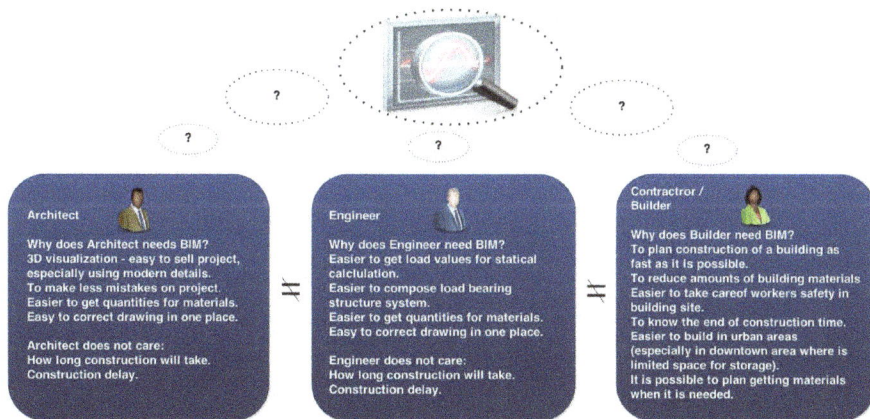

Figure 5. Different needs of participants in a construction project [59].

In order to evaluate the return on investment in BIM correctly designers of companies that installed BIM software were interviewed. The results are shown in Table 1.

Table 1. Results of the survey of designers who have installed BIM software.

Expected productivity loss after starting to use BIM software	34%
Time needed to restore efficiency to the previous level	2.3 month
Maximum productivity gain after staff training	31%
Time period for the productivity to reach the maximum level	4.8 month

ROI calculation methodologies that are presented by other authors show that there is a number of criteria (variables) to be assessed in order to correctly calculate the return on investment. Known BIM return on investment calculations show that not all of the relevant variables are evaluated. In addition, it needs to be pointed out that ROI rate of first year BIM exceeding 100% should be seen as inadequate. The same numerical values of ROI indicators are not obtained after replicating the calculations of the model, which raises the questions whether the calculations were correct, whether they reflect the reality, and whether there are any hidden variables in the formula. Since the detailed calculation process is not revealed the ROI calculator subtleties are impossible to clarify. In addition, it is necessary to consider the fact that every country has different working conditions, wages, and different possibilities for big and small companies. The adjusted ROI calculation formula is presented in this paper based on the existing studies of BIM return on investment.

Design company analysis is performed in order to assess the benefits of BIM. A small design company operating for more than two years is assessed together with its architects, engineers, and energy performance certification experts. The evaluation of various software configurations is aimed at finding a combination of programs or a program that will suit both the architect and the engineer work. Research may also be useful for new or recently started companies.

Three design software distributing companies were interviewed. There are two alternatives on which BIM introduction for a company is based: Autodesk and Archicad. In order to get more comprehensive results an alternative is introduced—ZWCAD + Professional software that does not have a BIM design possibility. It is important to mention that the ZWCAD design program is an alternative to AutoCAD. This software does not require additional money for qualification and training. Autodesk company, as the dominant player in the field of design software, sells its products grouped into packages. Autodesk offered package includes the following programs: AutoCAD, AutoCAD Architecture, AutoCAD MEP, AutoCAD Structure Detailing, Autodesk Showcase, Autodesk Revit, Autodesk 3ds Max Design, Autodesk Raster Design, Autodesk ReCap, Autodesk Navisworks Manage, Autodesk Inventor, Autodesk InfraWorks, Autodesk Robot structure, and Analysis Professional (design calculation software). ArchiCAD software package also includes software add-ons: BIM Server, BIMx, and Eco Evaluation (power rating). It is noted that the large sets of programs offered are in the excess in respect to micro or small companies with few employees that do not engage in big projects. In the best case, the company will use 2–3 programs from the package. Nevertheless, the design company is the limited on the grounds that the purchase price for the entire software package or a program will remain the same.

In case of an enterprise that is engaged only in a fragment of the whole construction project, like designing a building, BIM benefits are difficult to assess. The method of ROI calculation that is presented in this paper is tailored to address the design phase with its specificity. The following issues in a design company implementing BIM technology have been identified:

1. purchasing software with required functionalities,
2. expectation of improved results after BIM implementation,
3. slowing down of the design process directly after BIM deployment, and
4. training time after the deployment of BIM tools,

   Determining a company's specificity.

All of these issues are reflected in the coefficients that need to be determined before ROI in BIM in a design company. Autodesk Revit offers a similar calculation method, however, the use of the formula visible in the provided Excel file does not give the same results as Autodesk Revit. That suggests that there might be some hidden coefficients built into the Autodesk Revit formula.

The coefficients of ROI-DC methodology have been established as a result of interviews with project staff in design companies, which have implemented and use BIM design. F and G coefficients (direct design work, company's work specificity) are to be calculated individually for each design enterprise planning to invest in BIM technologies.

Table 2 contains three alternatives of software that are analyzed to choose the optimal version for a micro company. The table summarizes data for two positions (architect and designer). It is noted that the proposed duration of the training is relatively short—24 to 40 academic hours (0.6–1 working week). In general, this is believed to be a marketing tool allowing for the user to understand that the management of the program is easy enough to learn in such a short time. It is necessary to assess the size and number of projects, the need of BIM for projects, computer settings in order to choose the right alternative for optimal work. If the capacity of a computer is insufficient it needs to be improved or new ones have to be purchased to meet the needs of more demanding programs.

Table 2. Evaluation of software configuration alternatives in a micro enterprise design company.

Alternative	Cost of the Program, € (with Taxes)	Computer Settings	Cost of the Annual Subscription, €	Suggested Training Time, Hours	Cost of the Training for 1 Person, €
I (architect work place)	4485.64	64-bit operating system 4GB RAM internal memory Processor: dual-core	588.74	40/1 workplace	173.77 (4.34 €/hour)
II	562.0	1GB RAM internal memory Processor: Intel®Pentium 4 1.5 Ghz or equvalent AMD®	242.0	Unnecessary	Unnecessary
III (architect + engineer workplace)	23,788.6 (8349 + 15,439.60)	64-bit operating system 4GB RAM internal memory Single- or Multi-Core Intel®Pentium®, Xeon®, or i-Series processor or AMD®	3218,.6 (1089.0 + 2129.6)	24/1 workplace	1250 (52.08 €/hour)

New variables and the calculation model is offered by the authors to assess the return on investment. *"Autodesk Revit"* ROI and the proposed ROI-DC (Return on Investment—Design Company) calculation models are compared. Obtained results and variables are presented for comparison. It is important to note that both models calculate the first year return on investment. Table 3 shows the ROI assessing variables.

Table 3. ROI variables.

Variable Symbol	ROI "*Autodesk Revit*"	Numeric Value	ROI "*ROI-DC*"	Numeric Value
A	Cost of software, $	6000 $	Cost of software [1], €	4485,64; 562 € *; 15,439.60 €
B	Monthly labor cost, $	4200 $	Monthly labor cost [2], €	(713.9)
C	Training time, months	3 months	Training time [3], months	2 months
D	Relative productivity loss during training, %	50%	Relative productivity loss during training [4], %	0.34
E	Relative productivity gain after training, %	25%	Relative productivity gain after training [5], times	0.31
F	Used in Revit ROI calculations, but direct position was not found	82%	Direct design work [6], times	0.82
NEW VARIABLES				
G	-	-	Company's work specificity [7], times	0.465
	First Year ROI, %	61%	**First Year ROI, %**	xxx

*—Program does not support BIM. [1]—Optimal software is chosen on the basis of its subscription costs. [2]—The average salary of 713.9 EUR as of September of 2015 according to the Lithuanian Department of Statistics (K2—average gross of all positions of 2015). [3]—2 months of training is set by Autodesk Revit studies as a minimum training period to acquire necessary skills. Although the program distributors offer shorter training terms of 0.6–1 week (this is only the introduction training term that is not enough to master the knowledge). Therefore, the training terms are submitted not as a training time but as the time required to achieve the same level of productivity as in the use of the original software. [4]—Designers of 10 companies with the installed BIM design software were questioned. [5]—Designers of 10 companies with the installed BIM design software were questioned. [6]—It is important to mention that this indicator of ROI calculations by Autodesk Revit in the present case does not describe specific work hours—time is required not only to carry out the design work but for printing, interest in the legislative framework, documentation preparation. According to ROI by Autodesk Revit 36% of time is spent to design, 46% for documentation (respectively 52 and 68 hours per month), and total of (36 + 46 = 82%) of the time for general project work. As it is appropriate to evaluate this factor it is taken in the same form for calculations. [7]—The opportunity of the relevant micro company to use BIM software for ongoing projects. 46.5% of the ongoing projects needed BIM design. The company has its own business strategy to achieve the best results. Only the company itself is aware of the kind of projects they work on the most often and what type of projects need BIM. Common companies in Lithuania are engaged in more than one activity, i.e., they design not only new buildings, but also carry out projects of reconstruction, repair and detailed plans. In some projects BIM design cannot be used and will not bring the expected benefits.

It is important to note that other indicators could be assessed in the calculation of return on investment in BIM design:

1. number of employees for one project,
2. time consumption for the project,
3. the profitability of the project, and
4. turnover rates, etc. of the company.

For the first comparison of ROI-DC model the same measures are used by Autodesk Revit experts assessing new factor G (company specificity) in addition. The results are shown in Table 4.

One can observe a considerable difference in the way ROI indicators are calculated. Return on investment calculation by Autodesk Revit shows the 32% of the first year costs return despite the fact that the monthly salary is six times lower than the price of the software. Logically, the rate of return seems unlikely disregarding the two variables of F and G. Employee of an average salary of 714 euro per month working with BIM program would earn only 8568 euro per year and this amount is slightly above the design program cost. If in addition estimating to working with the loss of 34% in the first two to three months (i.e., 0.34 times less work done), while the remaining nine to seven months with efficiency increased by 31% the results of return would still be inadequate. In other words, the results are too optimistic.

Table 4. Comparison of Autodesk Revit and ROI-DC calculations.

Variable Symbol	ROI Autodesk Revit	Numeric Value	ROI "ROI-DC"	Numeric Value
A	Cost of software, €	4485.64 €	Cost of software, €	4485.64 €
B	Monthly labor cost, €	714 €	Monthly labor cost, €	714 €
C	Training time, months	2.3 months	Training time, months	2.3 months
D	Productivity lost during training, %	34%	Productivity lost during training, %	0.34
E	Productivity gain after training, %	0.31%	Productivity gain after training, times	0.31
		New Variables		
F	Used in Revit ROI calculations, but direct position was not found	82%	Direct design work, times	0.82
G	-	-	Company's work specificity, times	0.465
	First Year ROI, %	32%	**First Year ROI, %**	−31.4

Estimated ROI ratio shows that the design software will have no return in the first year. The resulting rate of return is negative. The difference between the required time and cost of carrying out the work is substantial. In this case, the situation of a Lithuanian company in the present market situation is assessed. The average wage of professionals working in the Lithuanian labor market is taken into account and the price of software offered in the local market is used for the calculations.

No categorical stance can be taken that there is no financial gain from installing BIM software. One of the most influential factors in ROI is wages. Wages rise while leaving other variables at fixed values would give the positive return values. Relevant calculations for a five year ROI are presented in Table 5.

Table 5. ROI in BIM in five years.

Variable Symbol	Numeric Value	Numeric Value
A	4485.64 €	4485.64 €
B	1375 €	1375 €
C	2.3 months	2.3 months
D	34%	0.34
E	31%	0.31
F	82%	0.82
G	-	0.465
	ROI Autodesk Revit 57%	**ROI "ROI-DC"** 20%

The model suggests that the forced installation of BIM would result in nearly doubling the prices of design services (714 € → 1375 €), as companies will have to pay higher salaries (thus higher earnings for the company will remain). For a simple comparison, it can be mentioned that a company X is currently offering an individual house project up to 200 m² (project parts: general, architectural, construction, site plan) for 1500 €. The company would carry out the same project for 3000 € in the situation of the imposed installation of BIM, and that is only in case when a company chooses one of the cheapest BIM design programs.

The performed calculation is made excluding the new G factor avoiding the doubt of ROI-DC calculations. These results are compared with the ROI results obtained from Autodesk Revit. B coefficient is the average wage of the relevant market. Calculation results are presented in Table 6.

In conclusion, the first year return of BIM installation is unnoticeable assessing G variable at level of current market wages and BIM software installation prices in Lithuania. It is important to mention that the second year gives better results because the employees will not lose time for trainings, work will be more efficient.

Table 6. Comparison of Autodesk Revit and ROI-DC calculations excluding the G factor.

Variable Symbol	Numeric Value	Numeric Value
A	4485.64 €	4485.64 €
B	714 €	714 €
C	2.3 months	2.3 months
D	34%	0.34
E	31%	0.31
F	82%	0.82
G	-	not evaluated
ROI *Autodesk Revit* 17%	**ROI "ROI-DC"** −22%	

Getting back to the purpose of the study—the most acceptable design software option for the company X is looked for. The task is to find an alternative design program for a micro company. Three options available at the market taken into account. Received commercial offers are presented in the Table 7.

Table 7. Software evaluation alternatives for a micro enterprise in design services.

Alternative	General Program (Package) Price (With Taxes)	Cost of One Workplace, €		BIM Design Function
		Architect	Engineer-Constructor	
I	4485.64		4485.64	+
II	562		562	−
III	20,570	8349	15,439.60	+

All three options are now compared using ROI-DC assessment methodology. With the help of a decision-making method the investment in design to the program will be made (Table 8).

Table 8. Software evaluation alternatives of a micro enterprise in design services for an architect.

Variable Symbol	Numeric Value Alternative I	Numeric Value Alternative II	Numeric Value Alternative III
A	4485.64 €	562 €	8349 €
B	714 €	714 €	714 €
C	2.3 months	-	2.3 months
D	0.34	-	0.34
E	0.31	-	0.31
F	0.82	0.82	0.82
G	0.465	0.465	0.465
First Year ROI, "ROI-DC"	−31.4%	481.3%	−61.2%

Architect workplace is necessary for BIM design, so only two alternatives with BIM design capabilities are compared. The salary increase is calculated for the first year return to reach at least 20% (Table 9).

Table 9. Software alternatives of design services for an architect (only those with BIM design opportunities).

Variable Symbol	Numeric Value Alternative I	Numeric Value Alternative III
A	4485.64 €	8349 €
B	1380 €	2560 €
C	2.3 months	2.3 months
D	0.34	0.34
E	0.31	0.31
F	0.82	0.82
G	0.465	0.465
First year ROI, "ROI-DC"	20%	20%
Rise in price of design services, times	1.93	3.59

The obtained results show that the choice of the less expensive BIM design software would raise the price of the services of the company about two times (714 € → 1380 €). Meanwhile, services that are choosing a more expensive design program can price up about four times (714 € → 2560 €). Another calculation considers BIM design programs targeted at a design engineer. The results are shown in Table 10.

Table 10. Software alternatives for an engineer-constructor.

Variable Symbol	Numeric Value Alternative I	Numeric Value Alternative II	Numeric Value Alternative III
A	4485.64 €	562 €	15,439.60 €
B	714 €	714 €	714 €
C	2.3 months	-	2.3 months
D	0.34	-	0.34
E	0.31	-	0.31
F	0.82	0.82	0.82
G	0.465	0.465	0.465
First year ROI, "ROI-DC"	−31.4%	481.3%	−78.4%

When comparing design programs for engineers it is important to mention that a third alternative has the computer calculation function that facilitates the work of an engineer. Those two alternatives give an engineer the possibility to design or use BIM system for structures building modeling. Constructor work program needs additional evaluation of return to determine how much work he performs using BIM program, and what sort of building projects an engineer cannot fulfill without BIM computer calculation programs. It is therefore considered that the G factor to be higher for the engineer designing special structures where special structures take the biggest part of the design.

In another option the calculation of engineer's salary at the maximum workload all year round (i.e., working without down-time). The engineer is supposed to develop only special structures requiring BIM design and construction calculations. In this case, the G factor set at level 1 and only alternative III of the software configuration used (the only that has BIM structural design and calculation functions). The desired payback time of five years is taken, i.e., the first year ROI is at least 20 percent (Table 11).

Table 11. Software alternatives for an engineer-constructor (only those with BIM design opportunities).

Variable Symbol	Numerical Value Alternative III
A	15,439.60 €
B	2200 €
C	2.3 months
D	0.34
E	0.31
F	0.82
G	0.9
1 year "ROI-DC"	20%

This example shows that the company could think about BIM design program investment only if it is able to pay 2200 € per month to the engineer.

Figure 6 shows the ROI (%) dependence on the wage (€/month). The linear relationship is evident. Software configurations with the BIM design option are compared. The comparison reveals that the less expensive alternative I ROI is negative at −31.41%, while the more expensive alternative III scores −61.16% (assuming the average wage as reported by the Lithuanian Department of Statistics). Cheaper program one year ROI is possible at 2740 €/month wages, meanwhile, the more expensive program—at 5100 €/month. It shows that the turning point (ROI = 0%) choosing alternative I is at 1104 €/month wage, meanwhile, the turning point (ROI = 0%) of the more expensive alternative is at 2055 €/month.

Figure 6. ROI dependence on monthly labor cost.

The obtained results of the ROI assessments of BIM programs reflect the current construction market situation where the salaries of specialists do not follow the market reality. ROI in BIM programs is not just equal to zero, but is negative. It is revealing that the dependence is not completely linear. Raising the salary of the employee by 1.55 times (714€ → 1104€) at the alternative I and by 2.88 times (714 € → 2055 €) at the alternative III, ROI = 0% is obtained. This means no first year ROI but it is likely that in the same situation of a company next year ROI will be positive.

For the possibility of the first year ROI = 100% of BIM programs the salary of the alternative I should be 2740 €/month and 5100 €/month salary for the alternative III.

Figure 7 shows ROI (%) dependence on the gross wage (€/month). The relationship is very strong ($R^2 = 0.9893$ and $R^2 = 0.9873$).

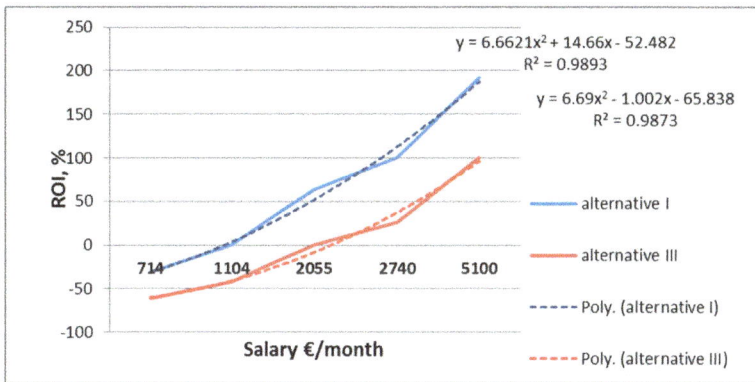

Figure 7. Strength evaluation of statistical relationship.

Strong statistical relationship established by assessing the link between ROI and average wage is described, respectively, by the exponential regression model. Results of both I and II alternatives lead to the conclusion that—in the state of low average wages—a higher ROI is possible only with necessary higher wage growth. With high wages, similar wage increase will affect ROI growth.

6. Conclusions

The presented study is based not only on theoretical considerations and examples present in scientific literature, but also on the interviews with project staff engaged in BIM-based design in 10 enterprises. ROI-DC method is a conceptual computational model in which the Return on Investment formula is adjusted to the design process in construction projects. It is important to remember that the calculation of the ROI-DC coefficients should be carried out on the case-by-case basis.

The developed ROI-DC calculation method evaluates additional variables (direct design work and company's specificity), influencing the return on investment in BIM technologies. The calculations have shown that it takes time for the ROI in BIM technology to achieve positive figures. The following conclusions may be drawn:

1. The calculations have shown that the ROI of the 4500 € program is possible in five years if the service prices rise about two times (employee salary growth: 714 € → 1375 €).
2. Estimated BIM software installation for the architect workplace may provide the first year ROI = 20% under the following conditions. Company's service prices should rise about two times (714 € → 1380 €) in case of the less expensive design program and up to four times (714 € → 2560 €) when the more expensive design program is chosen.
3. Estimated BIM software installation for the engineer constructor workplace may provide the first year ROI = 20% under the following conditions. It is worth for a company to invest in BIM design program, only when company is able to pay 2200 € monthly salary to an engineer (respectively, service prices will increase three times: 714 € → 2200 €).
4. ROI (%) dependence on the gross wage (€/month) is almost linear. Alternatives of programs with the BIM design program possibility are compared. Less expensive program would yield ROI within one year only with the 2740€/month employee's salary; the more expensive program—with the 5100 €/month salary.

The relationship between the ROI (%) and the gross wage (€/month) is very strong (R^2 = 0.9893 and R^2 = 0.9873). There is a strong statistical dependence. With high wages increase ROI is accordingly higher.

In conclusion, the proposed ROI-DC method may be seen as a practical management instrument [60], facilitating economically rational decisions in the sector, thus increasing its productivity [60]. The results of the calculations indicate that BIM adoption may suffer a significant slowdown for purely economic reasons. Therefore, if one counts on a quicker and more substantial contribution of BIM to the sustainability of the construction sector, isomorphic pressures from policy circles, regulatory bodies, and industry associations to adopt BIM on the project level are needed [61]. Local authorities in many countries are committed to BIM and require that all new public projects be carried out with use of BIM. It is done with the assumption that when project owners see the benefits of BIM-based asset management BIM would be free-willingly implemented by private actors from the earliest project stages.

Policy interventions might be combined with various financial and non-financial incentives to adopt BIM by designers who would otherwise choose not to do it for economic reasons. Moreover, high positive sustainability-related externalities that are offered by BIM justify a concerted effort of public and private stakeholders to apply BIM more extensively to construction of objects other that buildings (roads, bridges, engineering objects, line infrastructure) [62].

Acknowledgments: Part of the research for this paper has been conducted in the framework of projects no. S/WZ/1/2014 and S/WZ/4/2015 financed from the funds of the Ministry of Science and Higher Education of Poland.

Author Contributions: All the authors contribute equally to this paper.

Conflicts of Interest: The authors declare no conflict of interest.

References

1. Wong, K.-D.; Fan, Q. Building information modelling (BIM) for sustainable building design. *Facilities* **2013**, *31*, 138–157. [CrossRef]
2. Giel, B.K.; Issa, R.; Olbina, S. Return on investment analysis of building information modeling in construction. In Proceedings of the International Conference on Computing in Civil and Building Engineering; Tizani, W., Ed.; University of Florida: Gainesville, FL, USA, 2010. Available online: http://www.engineering.nottingham.ac.uk/icccbe/proceedings/pdf/pf77.pdf (accessed on 8 February 2018).
3. Sen, S. The Impact of BIM/VDC on ROI: Developing a Financial Model for Savings and ROI Calculation of Construction Projects. Master's Thesis, Royal Institute of Technology, Stockholm, Sweden, 2012.
4. *The Business Value of BIM for Construction in Global Market*; Smart Market Report; McGraw Hill Construction: Bedford, MA, USA, 2008.
5. *The Business Value of BIM for Construction in Global Market*; Smart Market Report; McGraw Hill Construction: Bedford, MA, USA, 2009.
6. *The Business Value of BIM for Construction in Global Market*; Smart Market Report; McGraw Hill Construction: Bedford, MA, USA, 2010.
7. *The Business Value of BIM for Construction in Global Markets*; Smart Market Report; McGraw Hill Construction: Bedford, MA, USA, 2014.
8. Santos, R.; Costa, A.A.; Grilo, A. Bibliometric analysis and review of Building Information Modelling literature published between 2005 and 2015. *Autom. Construct.* **2017**, *80*, 118–136. [CrossRef]
9. Hu, Z.-Z.; Zhang, J.-P.; Yu, F.-Q.; Tian, P.-L.; Xiang, X.-S. Construction and facility management of large MEP projects using a multi-Scale building information model. *Adv. Eng. Softw.* **2016**, *100*, 215–230. [CrossRef]
10. Miettinen, R.; Paavola, S. Beyond the BIM utopia: Approaches to the development and implementation of building information modeling. *Autom. Construct.* **2014**, *43*, 84–91. [CrossRef]
11. Love Peter, E.D.; Matthews, J.; Simpson, I.; Hill, A.; Olatunji, O.A. A benefits realization management building information modeling framework for asset owners. *Autom. Construct.* **2014**, *37*, 1–10. [CrossRef]
12. Kim, H.; Anderson, K.; Lee, S.; Hildreth, J. Generating construction schedules through automatic data extraction using open BIM (building information modeling) technology. *Autom. Construct.* **2013**, *35*, 285–295. [CrossRef]

13. Migilinskas, D.; Ustinovichius, L. Computer-aided modelling, evaluation and management of construction project according PLM concept. *Lect. Notes Computer Sci.* **2006**, *4101*, 242–250. [CrossRef]
14. Popov, V.; Juocevicius, V.; Migilinskas, D.; Ustinovichius, L.; Mikalauskas, S. The use of a virtual building design and construction model for developing an effective project concept in 5D environment. *Autom. Construct.* **2010**, *19*, 357–367. [CrossRef]
15. Future directions for IFC-based interoperability. Available online: http://www.itcon.org/2003/17 (accessed on 8 February 2018).
16. Ding, L.; Zhou, Y.; Akinci, B. Building information modeling (BIM) application framework: The process of expanding from 3D to computable nD. *Autom. Construct.* **2014**, *46*, 82–93. [CrossRef]
17. Ustinovičius, L.; Rasiulis, R.; Nazarko, L.; Vilutienė, T.; Reizgevicius, M. Innovative research projects in the field of building lifecycle management. *Procedia Eng.* **2015**, *122*, 166–171. [CrossRef]
18. Ustinovičius, L.; Walasek, D.; Rasiulis, R.; Cepurnaite, J. Wdrażanie technologii informacyjnych w budownictwie—Praktyczne studium przypadku. *Econ. Manag.* **2015**, *1*, 290–310. [CrossRef]
19. Azhar, S.; Carlton, W.A.; Olsen, D.; Ahmad, I. Building information modeling for sustainable design and LEED®rating analysis. *Autom. Construct.* **2011**, *20*, 217–224. [CrossRef]
20. Kibert, C.J. *Sustainable Construction: Green Building Design and Delivery*, 4th ed.; John Wiley & Sons, Inc.: Hoboken, NJ, USA, 2016; p. 23.
21. BIM and Sustainable Design. Available online: http://buildipedia.com/aec-pros/design-news/bim-and-sustainable-design (accessed on 27 December 2017).
22. Yan, H.; Damian, P. Benefits and Barriers of Building Information Modelling. In Proceedings of the 12th International Conference on Computing in Civil and Building Engineering, Beijing, China, 16–18 October 2008.
23. Aladag, H.; Demirdögen, G.; Isık, Z. Building information modeling (BIM) use in Turkish construction industry. *Procedia Eng.* **2016**, *161*, 174–179. [CrossRef]
24. Jia, J.; Sun, J.; Wang, Z.; Xu, T. The construction of BIM application value system for residential buildings' design stage in China based on traditional DBB mode. *Procedia Eng.* **2017**, *180*, 851–858. [CrossRef]
25. Ma, S.; Li, C.; Hu, D.; Zhao, Q.; Zhang, X. The application of BIM technology in construction management. *J. Residuals Sci. Technol.* **2016**, *13*. [CrossRef]
26. Zhang, J.; Seet, B.-C.; Lie, T.T. Building information modelling for smart built environments. *Buildings* **2015**, *5*, 100–115. [CrossRef]
27. Tulenheimo, R. Challenges of implementing new technologies in the world of BIM: Case study from construction engineering industry in Finland. *Procedia Econ. Finance* **2015**, *21*, 469–477. [CrossRef]
28. Lin, Y.-C.; Lee, H.-Y.; Yang, I.-T. Developing as-built BIM model process management system for general contractors: A case study. *J. Civil Eng. Manag.* **2016**, *22*, 608–621. [CrossRef]
29. Zuo, J.; Zillante, G.; Xia, B.; Chan, A.; Zhao, Z. How Australian construction contractors responded to the economic downturn. *Int. J. Strat. Property Manag.* **2015**, *19*, 245–259. [CrossRef]
30. Ning, G.; Junnan, L.; Yansong, D.; Zhifeng, Q.; Qingshan, J.; Weihua, G.; Geert, D. BIM-based PV system optimization and deployment. *Energy Build.* **2017**, *150*, 13–22. [CrossRef]
31. Bradley, A.; Li, H.; Lark, R.; Dunn, S. BIM for infrastructure: An overall review and constructor perspective. *Autom. Construct.* **2016**, *71*, 139–152. [CrossRef]
32. Petri, I.; Beach, T.; Rana, O.F.; Rezgui, Y. Coordinating multi-site construction projects using federated clouds. *Autom. Construct.* **2017**, *83*, 273–284. [CrossRef]
33. Porwal, A.; Hewage, K.N. Building information modeling (BIM) partnering framework for public construction projects. *Autom. Construct.* **2013**, *31*, 204–214. [CrossRef]
34. Guidelines for Digital Construction in Lithuania 2014–2020. Available online: http://www.skaitmeninestatyba.lt/files/On_development_of_BIM_and_Digital_Construction_Lithuania.pdf (accessed on 23 March 2014).
35. Čereška, A.; Zavadskas, E.K.; Cavallaro, F.; Podvezko, V.; Tetsman, I.; Grinbergienė, I. Sustainable assessment of aerosol pollution decrease applying multiple attribute decision-making methods. *Sustainability* **2016**, *8*, 586. [CrossRef]
36. Zolfani, S.H.; Maknoon, R.; Zavadskas, E.K. Multiple attribute decision making (MADM) based scenarios. *Int. J. Strat. Property Manag.* **2016**, *20*, 101–111. [CrossRef]

37. Książek, M.V.; Nowak, P.O.; Kivrak, S.; Rosłon, J.H.; Ustinovichius, L. Computer-aided decision-making in construction project development. *J. Civil Eng. Manag.* **2015**, *21*, 248–259. [CrossRef]

38. Bucoń, R.; Sobotka, A. Decision-making model for choosing residential building repair variants. *J. Civil Eng. Manag.* **2015**, *21*, 893–901. [CrossRef]

39. Suder, A.; Kahraman, C. Multicriteria analysis of technological innovation investments using fuzzy sets. *Technol. Econ. Dev. Econ.* **2016**, *22*, 235–253. [CrossRef]

40. Cid-López, A.; Hornos, M.J.; Carrasco, R.A.; Herrera-Viedma, E. A hybrid model for decision-making in the information and communications technology sector. *Technol. Econ. Dev. Econ.* **2015**, *21*, 720–737. [CrossRef]

41. Hattab, M.; Hamzeh, F. Using social network theory and simulation to compare traditional versus BIM–lean practice for design error management. *Autom. Construct.* **2015**, *52*, 59–69. [CrossRef]

42. Cao, D.; Wang, G.; Li, H.; Skitmore, M.; Huang, T.; Zhang, W. Practices and effectiveness of building information modelling in construction projects in China. *Autom. Construct.* **2015**, *49*, 113–122. [CrossRef]

43. Poirier, E.A.; Staub-French, S.; Forgues, D. Measuring the impact of BIM on labor productivity in a small specialty. *Autom. Construct.* **2015**, *58*, 74–84. [CrossRef]

44. Azhar, S. Building information modeling (BIM): Trends, benefits, risks, and challenges for the AEC industry. *Leader. Manag. Eng.* **2011**, *11*, 241–252. [CrossRef]

45. Walasek, D.; Barszcz, A. Analysis of the adoption rate of building information modeling [BIM] and its return on investment [ROI]. *Procedia Eng.* **2017**, *172*, 1227–1234. [CrossRef]

46. Al-Zwainy, F.M.S.; Mohammed, I.A.; Al-Shaikhli, K.A.K. Diagnostic and assessment benefits and barriers of BIM in construction project management. *Civ. Eng. J. Tehran* **2017**, *3*, 63–77.

47. Johansson, M.; Roupé, M.; Bosch-Sijtsema, P. Real-time visualization of building information models (BIM). *Autom. Construct.* **2015**, *54*, 69–82. [CrossRef]

48. Giel, B.; Issa, R. Return on investment analysis of using building information modeling in construction. *J. Comput. Civ. Eng.* **2013**, *27*, 511–521. [CrossRef]

49. Love, P.; Simpson, I.; Hill, A.; Standing, C. From justification to evaluation: Building information modeling for asset owners. *Autom. Construct.* **2013**, *35*, 208–216. [CrossRef]

50. Jin, R.; Hancock, C.M.; Tang, L.; Wanatowski, D. BIM investment, returns, and risks in China's AEC industries. *J. Const. Eng. Manag.* **2017**, *143*. [CrossRef]

51. Gu, Y.; Storey, V.C.; Woo, C.C. Conceptual modeling for financial investment with text mining. *Lect. Notes Comput. Sci.* **2015**, *9381*, 528–535. [CrossRef]

52. Won, J.; Lee, G. How to tell if a BIM project is successful: A goal-driven approach. *Autom. Construct.* **2016**, *69*, 34–43. [CrossRef]

53. Barlish, K.; Sullivan, K. How to measure the benefits of BIM: A case study approach. *Autom. Construct.* **2012**, *24*, 149–159. [CrossRef]

54. Lee, G.; Park, H.K.; Won, J. D-3 city project: Economic impact of BIM-assisted design validation. *Autom. Construct.* **2012**, *22*, 577–586. [CrossRef]

55. Calculating BIM's Return on Investment. Available online: http://www.cadalyst.com/aec/calculating-bim039s-return-investment-2858 (accessed on 20 February 2018).

56. Return on Investment with Autodesk Revit. Autodesk Building Solutions White Paper. Available online: http://usa.autodesk.com/revit/white-papers/ (accessed on 15 April 2015).

57. Chi, H.-L.; Wang, X.; Jiao, Y. BIM-enabled structural design: Impacts and future developments in structural modelling, analysis and optimisation processes. *Arch. Comput. Methods Eng.* **2015**, *22*, 135–151. [CrossRef]

58. Ashcraft, H.; Shelden, R.D. BIM Implementation Strategies, Hanson Bridgett, Gehry Technologies, 2015. Available online: http://www.nibs.org/?page=bsa_proceedings (accessed on 10 March 2015).

59. Reizgevičius, M.; Reizgevičiūtė, L.; Ustinovičius, L. The need of BIM technologies implementation to design companies. *Econ. Manag.* **2015**, *7*, 45–53. [CrossRef]

60. Ejdys, J.; Matuszak-Flejszman, A. New management systems as an instrument of implementation sustainable development concept at organizational level. *Technol. Econ. Dev. Econ.* **2010**, *16*, 202–218. [CrossRef]

61. Nazarko, J.; Chodakowska, E. Measuring productivity of construction industry in Europe with data envelopment analysis. *Procedia Eng.* **2015**, *122*, 204–212. [CrossRef]

62. Radziszewski, P.; Nazarko, J.; Vilutiene, T.; Dębkowska, K.; Ejdys, J.; Gudanowska, A.; Halicka, K.; Kilon, J.; Kononiuk, A.; Kowalski, J.K. Future trends in road technologies development in the context of environmental protection. *Baltic J. Road Bridge Eng.* **2016**, *11*, 160–168. [CrossRef]

![sustainability logo] *sustainability*

MDPI

Article

An Algorithm for Modelling the Impact of the Judicial Conflict-Resolution Process on Construction Investment

Andrej Bugajev *,† and Olga R. Šostak †

Faculty of Fundamental Sciences, Vilnius Gediminas Technical University, Sauletekio ave. 11, LT-10223 Vilnius, Lithuania; olgaolgaregina@gmail.com
* Correspondence: zvex77777@gmail.com; Tel.: +370-677-10-123
† These authors contributed equally to this work.

Received: 27 November 2017; Accepted: 9 January 2018; Published: 12 January 2018

Abstract: In this article, the modelling of the judicial conflict-resolution process is considered from a construction investor's point of view. Such modelling is important for improving the risk management for construction investors and supporting sustainable city development by supporting the development of rules regulating the construction process. Thus, this raises the problem of evaluation of different decisions and selection of the optimal one followed by distribution extraction. First, the example of such a process is analysed and schematically represented. Then, it is formalised as a graph, which is described in the form of a decision graph with cycles. We use some natural problem properties and provide the algorithm to convert this graph into a tree. Then, we propose the algorithm to evaluate profits for different scenarios with estimation of time, which is done by integration of an average daily costs function. Afterwards, the optimisation problem is solved and the optimal investor strategy is obtained—this allows one to extract the construction project profit distribution, which can be used for further analysis by standard risk (and other important information)-evaluation techniques. The overall algorithm complexity is analysed, the computational experiment is performed and conclusions are formulated.

Keywords: construction investment; algorithms on graphs; decision graphs; greedy algorithm; dynamic programming; judicial conflict resolution; sustainable city development; stakeholder theory

---

1. Introduction

In this article, it is considered how society rights violations may affect the implementation of a construction investment project. The infringement of society rights can be harmful to investors, since the judicial process can greatly increase the costs of the project and reduce the profit or even make it negative. On the one hand, it is important to avoid conflicts that can greatly increase the costs of the construction investment project or even force it to stop. On the other hand, in the case when the judicial process is avoided, some slight violations may greatly reduce the construction project's costs or increase its value, leading to increased profit for construction investors. Thus, investors should make some decisions carefully during the preparation of the project. For a project's successful execution, it is crucial to evaluate and manage risk factors in cases of different decisions. The correct decision strategy for a conflict-resolution process would minimise risks or costs. More generally, it is needed to choose the ratio between risks and costs. Usually, investors underestimate the risks, and thus the correct strategy selection made by construction investors leads one to avoid conflicts, which leads to avoidance of project failures and makes a great deal for sustainable city development. However, even if the optimal decision strategy is known, it is important to evaluate the project, since at the project initialization stage, such an evaluation helps one to select the most suitable project from all possible alternatives or to refuse to start any projects at all. Nowadays, planning and management technologies can greatly improve the quality of investor decisions. Integration of such technologies into project implementation is very significant for investors.

Investment efficiency plays a critical role in investment prioritisation, and therefore, it has become an important topic in recent research [1,2]; the irreversibility of potential negative outcomes of projects was studied [3], and the optimal project portfolio creation was studied [4]. There exist methods for supporting investments and resource allocation in risky environments (see [5]). The investment-supporting methods are important in the construction field; for example, a contractor's selection can be supported by special methods [6,7]. Special problems are formulated and solved for assessment of unfinished construction projects [8]. The problem of optimal project selection can be very complex due to the risks associated with uncertainties in the projects, which should be properly addressed [9]. Some risks are highly related to conflicts between decision makers; for example, a recent study addressed the conflict between a construction company and selected suppliers [10]. However, there are no studies dedicated to evaluate the project from the perspective of risks that come from the judicial process that arise from some slight law violations that were made purposely. In this paper, we develop an algorithm that is aimed to help investors to manage the risks that are connected to the judicial conflict-resolution process, that is, an algorithm for modelling the impact of the judicial conflict-resolution process on construction investment.

In this paper, we concentrate on a specific case when the problem is defined by a graph with cycles, and some natural properties of the problem can be used to obtain a tree. Another important result of this work is formal representation of such a process with estimation of time and decision cycles. More specifically, the investment into construction of buildings is considered. In our case, decisions that should be made are dependent on the past (i.e., after some cycles it depends on which state it was in before), so initially it is not a Markov decision process [11]. However, during the execution of the proposed algorithm, it eventually becomes a Markov decision process, and, after the optimal strategy is selected, the process becomes a Markov chain. When the problem is formulated in the form of the Markov chain, there are many ways to solve it [12]. However, in this paper we consider this problem as a general problem on graphs. Since the Markov decision process structure is described by a tree, the solution of the considered problem is easy to find—we provide algorithms to solve it.

The practical importance of the analysis of the investment in building construction processes is well described in [13]. This forces companies to build on the limits of construction law restrictions and sometimes they even slightly violate the law. Some non-critical construction law violations can be widely found in construction projects that are already successfully finished [14]. In this situation, we are inclined to blame the drawbacks of laws that regulate urban planning and protection of visual identity (investors cannot always be expected to abandon their self-centred ends for the sake of urban values, etc.). This is largely influenced by confusing, non-effective systems for the coordination with government institutions and the public. The regulation of construction is confusing, and the builders breach the introduced requirements. Inappropriate distribution of functions among government institutions and private subjects raises a lot of problems. One of the outcomes of inappropriate legal regulation is violation of the society's rights (i.e., the parties that are not directly related to the investment construction process, owners of neighbouring plots, users, communities of residential districts, etc.). Even if the construction is done without any violations of the procedures, negative impact on the surrounding environment may exist. This can be described as social cost in construction projects, and a state-of-the-art overview of social costs in the construction industry is presented in [15].

It may seem that a smart decision to make the law violations protects the interests of investors only. However, the investors may violate the law purposely in some cases only, and in other cases the modelling of processes would objectively indicate that breaking the law is not profitable—this would contribute to sustainable city development. More specifically, it is profitable to violate the law in the case when the risks of the conflict between the construction company and the society are small compared to the rewards. Obviously, if the potential profit is big, the investors would make decisions that lead to a high probability of the conflict. In this case, such an investment strategy stands against the higher degree of sustainability of city development. Thus, the proposed algorithm can be used to identify the cases when the behaviour of the investors may contribute to unsustainable development. Additional analysis of such cases would lead one to identify the rules that do not regulate the construction process strictly enough. An example of such a detailed analysis of territorial planning procedures is provided in [16], where a detailed description of Lithuanian laws and its regulating system can be found. The application of this research could contribute to the sustainable city development in two ways: by stopping the investors from taking risks that will probably lead to the conflict with society (leading to loses), and by indicating the drawbacks of the regulatory system when the law violations (i.e., risks) are profitable. In the first case, such modelling would prevent the investors from taking socially destructive actions (i.e., violation of the law). In the second case, it would work in two ways: it would encourage the investors to violate the law, and indicate that the regulation procedures must be modified to make the expected penalties large enough (in terms of probability of identifying the violations by inspecting institutes and the size of fines) to make it not profitable to violate the law. Thus, we assume that such modelling could be used (along with the investors) by institutions that control the development of territorial planning procedures. It is especially important in the case when the investor chooses to violate some critical laws, such as ones that are related to toxic waste [17]. The distinction between more and less important (in some sense) laws is out of the scope of the current research—it must be analysed in each case separately. However, the application of this research can indicate in which cases such analysis must be performed.

The modelling of such a process was already performed in [13,14], and was done using recurrence equations that were solved using the dynamic programming concept. However, the proposed method is limited to the decision tree, thus it does not support cycles. Moreover, it solves the fixed optimisation problem, which gives a limited amount of information. For example, it can give the average profit value, but cannot give any information on the risks of losing a large amount of invested money. These results were extended in [18,19], where the conceptual model was applied to describe the behaviour of two conflicting systems (the investor and community) in different states and times. It revealed the importance of duration of different events during a judicial conflict-resolution process and showed how to estimate it and include in the model. As a consequence, it became possible to estimate cycles in the graph that describe the process—this can be an important part of project evaluation. However, such modelling of judicial process resolution with estimation of time and cycles was not performed before, and this is what this research is dedicated to.

The main aim of this research is to perform the evaluation of the judicial conflict-resolution process's impact on investment with estimation of time and decision cycles. More specifically, the authors develop an algorithm that allows one to extract the project's profit distribution semi-automatically. The probability of occurrence of different events in different states must be evaluated by experts or by other means, as well as the whole set of events and the set of dependencies between these events—we assume that this data is known; for considered cases this data is partially taken from other publications. As for parameter estimation from empirical data—sometimes it is even possible to develop the patterns describing judicial decisions; in a recent research a large collection of judicial cases was used to measure the probabilities quite precisely [20]. It is important to note that the sensitivity analysis is out of the scope of this research; for example, probability values can be perturbed in order to gather more general results that are robust to errors of estimation of probability values in a similar way to what was done by other authors [21,22]. Some recent studies propose a comprehensive sensitivity analysis to identify the uncertain parameters that significantly influence the decision-making process in investments, and quantify their degree of influence [23].

The research topic deals with strategic management; it is strongly related to agency [24] and stakeholder [25] theories. According to the agency theory, the investor can be represented as a principal, and the construction company as an agent. One of problems being addressed in the agency theory is different risk toleration by a principal and an agent. The construction company (compared to the investors) may be less sensitive to a project failure, because the financial losses are not direct—they depend on contract clauses, the impact on reputation of the company [26], and so on. As for the investor—he loses the invested money directly as the project fails and stops. Thus, we assume that the construction company and the investors have different attitudes towards the risk of project failure. The proposed modelling supports the interests of the investors; these interests overlap with interests of society, and contribute to sustainable city development. In other words, this research contributes to the increase of investment protection; it can also affect corporate social responsibility for construction companies. The impact of investors' protection on the company earnings management was analysed in recent research [27]. The stakeholder theory considers the will of an organisation to be directly affected by interests of different stakeholders. According to that theory, the society and the investors are external stakeholders of the organisation (construction company). In this case, the investors are shareholding stakeholders. In the stakeholder theory, the business decisions are usually optimised from the point of view of a business organisation (construction company)—optimisation is done taking into account the interests of all stakeholders. In this paper, we concentrate on the optimisation of decision strategy according to the interests of the investors. Thus, this research contributes to the stakeholder theory by investigating the risks taken by the investors. It is important to note that this does not mean that the investors are aware of the risks and will act according to them—the risks for the investors can be partially neglected by the construction company; however, according to stakeholder theory, the construction company should care for the interests of the investors. In recent work, there was performed an experimental scenario study in which investment behaviour was analysed in situations when management had to make trade-offs between shareholders' and non-shareholding stakeholders' interests [28]. In a similar way, the modelling proposed in this paper can potentially be applied to model the behaviour of the investors and their reactions to construction company decisions.

This research is dedicated to show how to model judicial processes and their impact; that is, it is focused on algorithms for computer simulation. This paper makes the following contributions. (1) We investigate the modelling of the investors' behaviour that impacts the sustainability of the city development. The developed algorithm can be applied as a tool in other research in order to identify the drawbacks of the regulatory system and to evaluate impacts of regulation changes. Assuming that the investors take the risks that are too big, such modelling can help the investors to avoid socially destructive actions—law violations that would lead to judicial conflicts. If the risks are recommended to be taken, it would indicate that more investigation of some cases may be needed to evaluate the negative consequences of such violations on the sustainability. (2) We propose a modelling approach to evaluate the impact of the judicial conflict-resolution process on the investment. The model was based on another work [13], where the modelling was very limited to a specific case, which lacked the flexibility (without time or cycle estimation) and formalised algorithmic representation. (3) It provides the algorithm that lets one estimate the time and cycles (cyclic repeat of the events), giving a big variety of applications for similar process-modelling problems. Moreover, we propose a simple algorithm to calculate cost and time parameters for different scenarios (in [13], it was assumed to be known without any formal calculation procedure). (4) It provides the algorithms to simulate the impact of the judicial conflict-resolution process on the investment, when the costs of failure depend on time. More specifically, it shows how to extract the profit distribution from the model. This simulation result lets one evaluate different projects and compare them. (5) It shows how to model the Markov decision process (that is a part of the considered problem) using the algorithms on a tree that are represented from the perspective of the general theory of algorithms, independently from approaches that are usually used in these cases. The algorithms are represented in a simple-to-understand form as a recursion. Greedy algorithm and dynamic programming techniques are used.

This paper is organised as follows. In Section 2, we present an example of the process of judicial conflict resolution. In accordance with this example, we identify aspects that are important for the model, and in Section 3 propose the data structure that is suitable to describe the information that is needed for the algorithms. In Section 3, we propose a solution in the form of a set of algorithms. In Section 4, we apply these algorithms to the example from Section 2. The conclusions are formulated in Section 5.

2. Example and Problem Formulation

In this section we consider an example of the process of judicial conflict resolution that was already described before in [13,14]. This process is related to an investment project of a residential building that is located in Vilnius (Lithuania) at the address Seskines 45C [29]. In this section we will present only some details of this process. There are three sides involved in a conflict—the investor, the society and the court. We will refer to these sides by abbreviations I_{k_1} (investor), S_{k_2} (society) and C_{k_3} (court), where k_1, k_2, k_3 are the identification numbers of different events in which decisions are made by different sides. The whole process is represented by the decision graph (Figure 1), in which decision nodes are I_{k_1}, chance nodes are S_{k_2} and C_{k_3}, and the end nodes are R_{k_4}. R_{k_4} represent the judicial conflict-resolution scenarios.

The description of graph edges is presented in Table 1 with time values, costs and chance node probabilities or decision variable abbreviations. All money costs are measured in TEUR (thousands euros), time is measured in days. Some data was originally provided in Lithuanian litas—we converted it to euros by dividing by a factor 3.4528, and for simplicity performed some rounding off. For decision nodes the variables are denoted X_{ij}, which can be 1 or 0; i is the index of the investor node I_i in which the decision is made; j is the number of decisions in this case. This means that for each i, values X_{ij} are equal to 1, and remaining values are equal to 0.

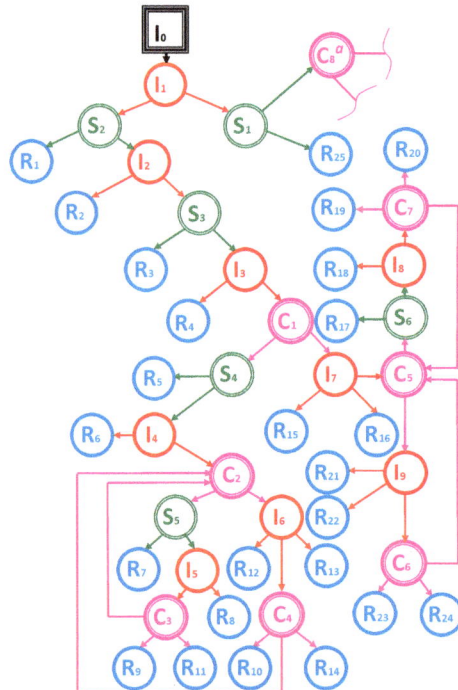

Figure 1. The graph for the considered example. C_8^a copies the node C_1 and all nodes following from it.

Table 1. The description of the conflict participants' behaviour, and the losses due to the judicial process. Costs are presented in TEUR (thousands euros); I, S, C denotes the Investor, the Society and the Court, respectively. R denotes the Result nodes (the ends of scenarios).

Edge	Actions (Decisions) Description	t (Days)	p	Losses (TEUR)
(I_0, I_1)	Project begins	0	$x_{0,1}$	0
(I_1, S_2)	I informs S (according to law)	180	$x_{1,1}$	0
(I_1, S_1)	I informs S (violates the law)	630	$x_{1,2}$	0
(S_2, R_1)	S does not submit its suggestions	30	0.8	0
(S_2, I_2)	S submits its suggestions	14	0.2	0
(I_2, R_2)	I accepts suggestions	14	$x_{2,1}$	1000
(I_2, S_3)	I rejects suggestions	14	$x_{2,2}$	0
(S_3, I_3)	S applies to an advanced hearing institution	7	0.7	0
(S_3, R_3)	S does not apply to an advanced hearing institution	30	0.3	0
(I_3, R_4)	I makes a peace treaty with S	14	$x_{3,1}$	150
(I_3, C_1)	I does not make a peace treaty with S	21	$x_{3,2}$	10
(C_1, S_4)	I loses	30	0.4	0
(C_1, I_7)	I wins	30	0.6	0
(S_1, C_8)	S applies to an advanced hearing institution	28	0.1	0
(S_1, R_{25})	S does not apply to an advanced hearing institution	30	0.9	0
(S_4, R_5)	S does not apply to a court of first instance	14	0.25	-10
(S_4, I_4)	S applies to a court of first instance	7	0.75	0
(I_4, C_2)	I does not make a peace treaty with S	7	$x_{4,2}$	20
(I_4, R_6)	I makes a peace treaty with S	7	$x_{4,1}$	100
(C_2, S_5)	I wins	90	0.5	0
(C_2, I_6)	I loses	90	0.5	0
(S_5, R_7)	S does not apply to a court of appeal	14	0.15	-30
(S_5, I_5)	S applies to a court of appeal	7	0.85	0
(I_5, C_3)	I does not make a peace treaty with S	7	$x_{5,2}$	10
(I_5, R_8)	I makes peace treaty with S	7	$x_{5,1}$	100
(C_3, R_9)	I wins	120	0.35	-40
(C_3, R_{11})	I loses	120	0.35	0
(C_3, C_2)	C returns a lawsuit to a court of first instance	120	0.3	0
(I_6, R_{13})	I does not apply to a court of appeal	14	$x_{6,1}$	0
(I_6, C_4)	I applies to a court of appeal	14	$x_{6,2}$	10
(I_6, R_{12})	I makes a peace treaty with S	14	$x_{6,3}$	200
(C_4, R_{14})	I loses	120	0.35	0
(C_4, C_2)	C returns a lawsuit to a court of first instance	120	0.3	0
(I_7, R_{16})	I does not apply to a court of appeal	14	$x_{7,1}$	0
(I_7, C_5)	I applies to a court of first instance	14	$x_{7,2}$	20
(I_7, R_{15})	I makes a peace treaty with S	14	$x_{7,3}$	150
(C_5, S_6)	I wins	90	0.5	0
(C_5, I_9)	I loses	90	0.5	0
(S_6, R_{17})	S does not apply to a court of appeal	14	0.15	-30
(S_6, I_8)	S applies to a court of appeal	7	0.85	0
(I_8, C_7)	I does not make a peace treaty with S	7	$x_{8,2}$	10
(I_8, R_{18})	I makes a peace treaty with S	7	$x_{8,1}$	100
(C_7, R_{20})	I wins	120	0.35	-40
(C_7, R_{19})	I loses	120	0.35	0
(C_7, C_5)	C returns a lawsuit to a court of first instance	120	0.3	0
(I_9, R_{22})	I does not apply to a court of appeal	14	$x_{9,1}$	0
(I_9, C_6)	I applies to a court of appeal	14	$x_{9,2}$	10
(I_9, R_{21})	I makes a peace treaty with S	14	$x_{9,3}$	200
(C_6, R_{24})	I wins	120	0.35	-40
(C_6, R_{23})	I loses	120	0.35	0
(C_6, C_5)	C returns a lawsuit to a court of first instance	120	0.3	0
(C_4, R_{10})	I wins	120	0.35	-40

The description of different ending scenarios is provided in Table 2. However, it is unclear what are the actual profits (positive or negative) in different scenario cases—it depends on the graph (Figure 1).

Table 2. The description of possible conflict endings.

End Nodes	Description	Success
R_1, R_3, R_{25}	successful completion of the judicial conflict	yes
R_2	project solution is changed, costs are increased	yes
$R_4, R_6, R_8, R_{12}, R_{15}, R_{18}, R_{21}$	peace treaty is signed	yes
$R_5, R_7, R_9, R_{10}, R_{17}, R_{20}, R_{24}$	investor wins the judicial conflict	yes
$R_{11}, R_{13}, R_{14}, R_{16}, R_{19}, R_{22}, R_{23}$	project failure	no

To estimate costs, we use data from documents that are related to this investment project. Reference [30] is used to estimate the building price for sale by accumulating the quadrature of apartments (which is equal to 5998.8 m^2) and multiplying it by its market value in 2007 taken from [31], which is mentioned to be from 1.39 to 2.17 TEUR, and for simplicity we assume that it is 1.5 TEUR per square metre. Thus, the estimation of the price the building was sold for gives 8998.2 TEUR. The building costs data is taken from [32] and is equal to 5508.6 TEUR. We note that we do not expect to estimate the needed data precisely—we consider this example for tests and demonstration purposes only, as our goal is to provide the algorithm for those who have access to this type of data directly. The building price and its distribution in time is important in order to estimate the amount of possible losses in the case when the investor loses the judicial process and fails to complete the project. Unfortunately, we do not have the exact data of the project's cost distribution in time. However, we have the information on costs of the project that were paid before the judicial process was started—these costs were equal to 236.04 TEUR [33]. We use this number as project costs during the first phase of a project, which lasted for 613 days. The rest of $5508.573 - 236.04032$ we distribute as follows. We assume that the project costs in TEUR per day are higher at the beginning of the building process by 50% compared to the last phase of the building process due to additional costs of expensive machinery (cranes, etc.). The calculated data is used to compute average daily costs for different time intervals.

In Table 3, we provide the interval lengths and average daily costs measured in TEUR per day for different project phases. Phase 1 is the phase before the actual building process has begun. Phases 2 and 3 describe the actual construction process; we separate the process into these phases taking into account that the project costs per day are higher at the beginning of the building process. Phase 4 describes the phase where the judicial process can be continued, however, the house construction is finished. The length of the first phase l_1 is taken from [19], the total cost is taken from [33] (as was mentioned before). The average daily costs parameter for phase 1, c_1, is calculated by dividing the total costs by l_1. The lengths for last two phases l_2, l_3 are taken as 90 and 478 days (the estimation for total length $l_2 + l_3$ is taken from [32]) and the average daily costs c_2, c_3 are derived from the requirements $l_1 c_1 + l_2 c_2 + l_3 c_3 = 5508.573$ and $c_2 = 1.5 c_3$. The average daily costs are presented in Table 3. From this data we construct the average costs function $F(t)$ (Figure 2):

$$F(x) = \begin{cases} 0.3851, & 0 \le t \le 613 \\ 12.9018, & 613 < t \le 703 \\ 8.6012, & 1181 < t \le 1668 \\ 0, & 1668 < t. \end{cases} \quad (1)$$

As was mentioned before, our goal is to provide an algorithm to extract the project profit distribution.

Table 3. The description of different project phases.

The Phase	Begin Date	End Date	Duration (Days)	Average Daily Costs (TEUR/Day)
1	1 October 2003	5 June 2005	613	0.3851
2	6 June 2005	4 September 2005	90	12.9018
3	5 September 2005	27 December 2006	478	8.6012
4	28 December 2006	28 April 2008	487	0

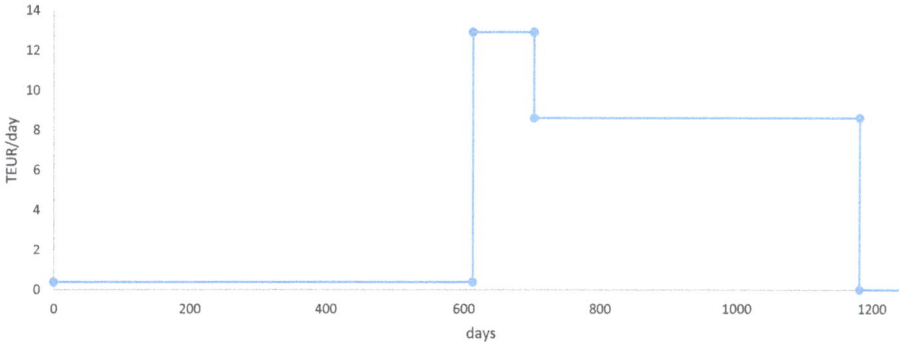

Figure 2. Average daily costs.

First of all, there is a degree of freedom for the investor, since in some situations he can make decisions on his own—as was mentioned before, there exist decision nodes I_k. Each decision node has a set of undefined probabilities $X_k = \{x_{k,j}, j = 1, \ldots, m_k\}$. We define the investment strategy as a set of all graph decisions $X = \{X_k, k = 1, \ldots, M\}$. The investment strategy must be chosen before the project evaluation is performed. There are different ways to choose the optimal strategy for the considered project, and one of the simplest ways to do that is to solve the optimisation problem

$$\max_X P_0(X), \tag{2}$$

where $P_0(X)$ is a function that computes the expected profit value for root node with number 0, and in the general case for each node n the expectation value is defined by a classical probabilistic expectation value formula

$$P_n(X) = \sum_{i \in children(n)} P_i(X) \cdot p_{n,i}, \tag{3}$$

where $children(n)$ returns the list of children indexes for the node with index n, $p_{n,i}$ is the probability for the node with number i to follow after the node with number n. However, there are no restrictions for the P_n computational algorithm (algorithmic form is presented in Section 3), for example, if for risk evaluation purposes it is needed to avoid big loses, additional weights or nonlinear functions may be used. Moreover, it can be integrated into a business-intelligence system, followed by prediction of business-intelligence system effectiveness (see [34]) in order to select the appropriate technique. The Bellman [35] principle gives the rule of optimal strategy, which states that in the case when the graph is a tree, (2) is fulfilled when X is chosen such as

$$\max_X P_n(X) \ \forall n, \tag{4}$$

which means that we can apply the dynamic programming principle and begin the computation from the bottom of the tree, where the values P_n are known. However, in our case we cannot apply the dynamic programming principle directly, because we have cycles in a graph (Figure 1). So firstly, we must convert the graph into a tree with some assumptions. We should note that our graph can be interpreted as a tree with some exceptions—edges that form cycles. More specifically: (C_3, C_2), (C_4, C_2), (C_6, C_5), (C_7, C_5). The main assumption is based on the real meaning of these edges—these edges mean the process is returning to the first-instance court. In our graph, it was assumed that the first-instance court event can be repeated indefinitely; however, in reality, a court of appeal usually avoids repeating this event many times. Thus, naturally, we assume the maximum repeat count for this event to be 2. Note that we do not provide any proof for the optimal maximum of the event repeat count to be 2, since our algorithm supports any positive integer value. This assumption lets us convert the considered graph into a tree, since we can recursively make copies of the subgraph from the nodes the process returns to.

P_n for end nodes (leaves) can be calculated by accumulating the project costs and times from the top of the tree to the bottom, and computing the profit at the end nodes using a pre-order tree-traversal algorithm idea, the exact algorithm of which will be presented in Section 4. Time must be converted to costs at success and failure nodes in different ways. In the case of success, all costs of the whole building project must be measured and subtracted from sale income (as we mentioned, we consider it to be 8998.2 TEUR). For failure nodes, the building costs must be measured up to the moment of failure.

To choose the optimal strategy, we use (2) and (3). After this, we can evaluate the project by extracting the scenario profit distribution from the tree.

3. The Data Structure and the Algorithms

Firstly, we declare the data structure that will be sufficient to store all data needed. Edges in Figure 1 represent decisions, and nodes represent events that occur after decisions are made. Assuming that there are no cycles in a graph (it is a tree), we let each node correspond to the directional edge that points to that node (i.e., end node of the edge). The data of all edges can be stored in nodes, so we can represent the whole process as a tree.

Each event has the price value and the time value. In our approach, the price value represents the price of that event, that is, the actual price needed to perform the decision. However, the time value represents the time needed to pass before the event starts. This allows us to store all information from the considered example into a decision tree.

Thus, the basic element of our data structure (a tree) is a node with connections to parent and children nodes (Figure 3).

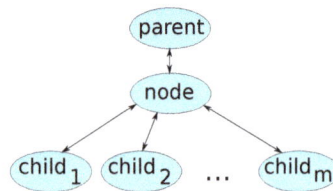

Figure 3. The relations between a node and its parent and child nodes.

A more detailed description is presented in Algorithm 1, where the symbol "//" denotes comments that describe the corresponding fields. Field *type* describes the type of a node. For our purposes, it is enough to define these types:

- 2—unsuccessful scenario end node,
- 1—investor decision node,
- 0—other nodes.

Note that types 1 and 2 are mutually exclusive, so we can use one field to describe this information. Field *cycle* is the pointer that forms a cycle. We assume that if this pointer is not equal to *NULL*, then this node describes the edge to return to one of the previous events; we refer to it as *cycle node*. Here, we present some notes for such nodes:

- The node is fictive, that is, it must be added to the list of nodes to describe the additional edge (since in a graph there is more than one edge pointing to the same node).
- The node has no children—it is reserved to describe the parameters of an edge forming a cycle.
- After cycles are converted to the extended tree, such nodes must be deleted if they are leaves of the tree; the probabilities must be recalculated with the same proportions but without cycle nodes.

Algorithm 1: Data structure

---

struct {
    int type; //node type
    float p; //the probability for this node to be selected by parent
    float price; //the price of event
    float time; //the time before event starts
    node* parent; //the pointer to parent node
    vector< node* > children; //the list of children
    node* cycle; //the pointer to the ancestor that forms a cycle
    float priceTotal; //accumulated price
    float timeTotal; //accumulated time
    float value; //node expected profit value
} *node;*

Note, that in all presented functions it is assumed that the arguments are passed by reference, i.e., the changes of arguments are seen outside of these functions. We propose the algorithm that consists of these steps:

1. Recursively make copies of the subgraph from nodes the process is returning to (i.e., looking for the pointer *cycle*), extend the tree by adding additional nodes. For this purpose we define a function *Expand* in Algorithm 2.

Algorithm 2: Cyclic expansion algorithm

Function *Expand(node)*
    if *node.cycle* $!=$ *NULL* **then**
        if $depth < depth_{max}$ **then**
            $newNode = CopyOfSubtree(node.cycle)$
            $newNode.p = 1$
            $newNode.parent = node$
            $node.children.add(newNode)$
            $depth = depth + 1$
            $Expand(newNode)$
            $depth = depth - 1$
        else
            // deletes unneeded fictive node and recalculates the probabilities
            // of parent's children nodes leaving the same proportions
            $smartDelete(node)$
        end
    else
        for *each node t in node.children* **do**
            $Expand(t)$
        end
    end
end

---

2. Calculate total times (field *timeTotal*) and costs (field *priceTotal*) up to the moment when events are finished—this is implemented in the function *CalcPars* in Algorithm 3.
3. Evaluate end-node scenarios (calculate field *value*) and select optimal strategy—function *CalcValues* in Algorithm 4.

4. Create profit (field *value*) distribution by calculating different scenario probabilities. A simple implementation is provided in Algorithm 5.

Algorithm 3: Costs parameters calculation algorithm

Function *CalcPars(node)*
    if *node.parent = NULL* **then**
        node.timeTotal = node.time
        node.priceTotal = node.price
    else
        node.timeTotal = node.time + node.parent.timeTotal
        node.priceTotal = node.price + node.parent.priceTotal
    end
    for *each node t in node.children* **do**
        CalcPars(t)
    end
end

Algorithm 4: Scenario profit evaluation and strategy selection algorithm

---

Function *CalcValues(node)*
    for *each node t in node.children* **do**
        CalcValues(*t*)
    end
    if *node.children = NULL* **then**
        Estimate(*node*)
    else
        if *node.type* = 1 **then**
            for *each node t in node.children* **do**
                t.p = 0
            end
            best = t that optimise $\max\limits_{t \in node.children}$ *t.value*
            best.p = 1
        end
        n = node.children.size
        $node.value = \sum\limits_{i=1}^{n} node.children[i].p \cdot node.children[i].value$
    end
end

Algorithm 5: Probabilities calculation algorithm with distribution extraction

---

Function *CalcProbs(node, distribution)*
    if *node.parent = NULL* **then**
        node.probTotal = 1
    else
        node.probTotal = node.p ∗ *node.parent.probTotal*
    end
    if *node.children.size()* > 0 **then**
        for *each node t in node.children* **do**
            CalcProbs(*t*)
        end
    else
        distribution.add(node.value, node.probTotal)
    end
end

In Algorithm 2, *CopyOfSubtree(node)*, there is the procedure of making a copy of a tree with the root *node* and returning the root of this copy (see Algorithm 6). The Algorithm 2 uses a pre-order tree-traversal principle, and therefore, it travels through a tree, dynamically extending it. Since *CopyOfSubtree* must make a copy of original data before extension, it uses the field *cycle* as a barrier to ignore the extended part of a tree during algorithm execution.

Algorithm 6: Subtree-copying algorithm

Function *CopyOfSubtree(node)*
    newNode(node) // copies the node
    newNode.children.clear()
    if *node.cycle* = *NULL* **then**
        for *each node t in node.children* **do**
            newNode.children.add(CopyOfSubtree(t))
        end
        for *each node t in newNode.children* **do**
            t.parent =(newNode)
        end
    end
    return *newNode*
end

Algorithm 3 calculates time and total costs accumulated until the according node's event is finished. Here, the principle of pre-order tree-traversal is used as well, and this greedy-algorithm strategy guarantees that each node will have the sum of parameters of their ancestors added to the values of this node. Thus, at the end of this algorithm, leaves will have total time and cost parameters for different process scenarios.

After total time and cost parameters are obtained, the values of leaves (which represent end-scenario project profits) can be calculated directly using function *Estimate* that is provided in Algorithm 7. However, values for the rest of the nodes must be calculated taking into account values of probabilities. The function *CalcValues*, which is presented in Algorithm 4, calculates values for all nodes and selects the optimal strategy. In Algorithm 4, the post-order tree-traversal idea is applied, that is, we use the dynamic programming principle with Formula (3). Note that Formula (3) calculates the classical expectation value, however, it can be easily changed to any other value-estimation procedure—it is especially useful when the user wants to add additional penalties for risk-management purposes. This value is used to select the optimal strategy, that is, it chooses the one with maximal value. Moreover, the last procedure *CalcProbs* calculates the distribution of end-scenario profits. As can be seen from Algorithm 5, it is implemented using the pre-order tree-traversal principle to accumulate end-node probabilities and profit values; these probabilities are added to the distribution. *distribution.add* represents collecting values and probabilities for distribution, that is, it accumulates (sums up) probabilities for same values.

Algorithm 7: Estimator

Function *Estimate(node)*
    if *node.type* = 2 **then**

$$node.value = -\left(node.priceTotal + \int_{0}^{node.timeTotal} F(t)\, dt \right)$$

    else

$$node.value = P - \left(node.priceTotal + \int_{0}^{t_{max}} F(t)\, dt \right)$$

    end
end

Let m be the maximum repeat counts for the first-instance court event, which is usually small (we have assumed it to be 2 before). At the step of graph expansion, depending on a graph, $\mathcal{O}(n\,a^m)$ operations must be performed and the extended graph gets $\bar{n} = \mathcal{O}(n\,a^m)$ edges, where $a \geq 1$ is the maximum number of edges pointing to a root of a subtree leading to cyclic repeats. The rest of the procedures at other steps are based on recursive tree traversal that have the complexity order $\mathcal{O}(\bar{n})$. Thus, the overall algorithm complexity is $\mathcal{O}(n\,a^m)$. In the next section we present the example formulas with more exact numbers.

4. Computational Experiment

In this section we apply the proposed project-evaluation algorithm to the example considered in Section 2. Firstly, we apply the data structure proposed in Algorithm 1 to the proposed example. The graph (Figure 1) has 83 nodes (excluding the beginning node I_0), however, as was mentioned before, we describe it as a tree with some exceptions that form cycles, more specifically: (C_3, C_2), (C_4, C_2), (C_6, C_5), (C_7, C_5) and the other four edges that are in the subtree with root C_8^a. Thus, in total there are eight cycles in a graph, as was mentioned before in Section 3; to describe them, an additional eight nodes are used.

Now we consider the consequence of applying Algorithm 2 to our example. First of all, the number of nodes dramatically increases, because in this case each cyclic return during Algorithm 2 approximately doubles the number of cycles for future returns. That is, if we let each cycle repeat two times, we get $\bar{n} = 91 + 4 \times (14 \times 4) + 2 \times ((14 + 2) \times 4) = 91 + 2^2 \times 56 + 2^1 \times 64$. Here we get 91 by adding 8 cycle nodes to 83 initial nodes, the total number of nodes in all cycles is equal to 56 and we add 8 (resulting in 64) cycle nodes for all extension iterations except for the last one. In the general case (with any maximum depth), we can calculate the number of nodes with the formula

$$\bar{n}(m) = \begin{cases} 91 + 56 \cdot 2^m + \sum_{i=1}^{m-1} 64 \cdot 2^i = \mathcal{O}(2^m), & m \geq 1 \\ 83, & m = 0 \end{cases} \tag{5}$$

where m is the maximum number of repeats for cycles, that is, with a large m the number of nodes doubles as m increases by 1. However, as we already mentioned before, m is small in the considered case, and therefore, this technique is suitable. The graphical illustrations for a graph without cycles and with cycles, and $m = 2$, are presented in Figure 4.

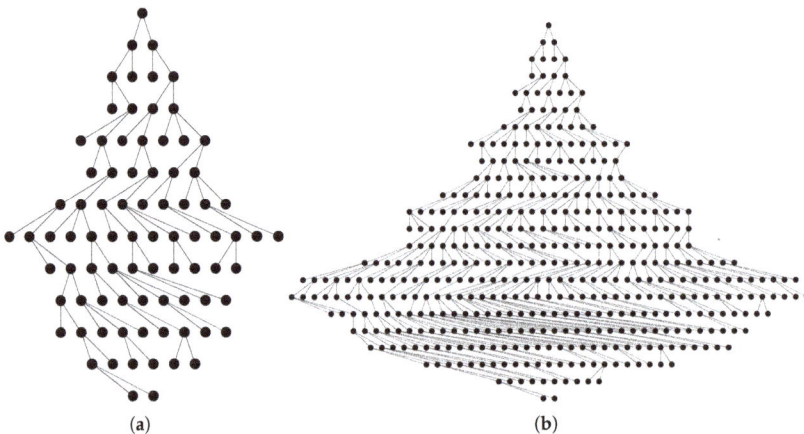

(a)   (b)

Figure 4. The impact of the Algorithm 2 on graph. (a) The graph without cycles, $m = 0$, $\bar{n} = n = 83$; (b) the graph with $m = 2$ ($\bar{n} = 443$).

Next, we apply Algorithms 3–5 to obtain the profit distribution that is provided in Table 4.

Table 4. The profit distribution.

X	−4831.46	−2794.58	−417.96	3329.63	3389.63	3439.63	3489.63
P(X)	0.0000305	0.0000770	0.0002775	0.06	0.00227966	0.00593781	0.931398

As we see from Table 4, only seven different values can occur. Negative values represent project failures, however, some of them are considerably smaller than others; for example, we see that with probability 0.0002775, the profit is −417.96, which is not as bad as the value of −4831.46. This difference is caused by different time intervals before the failure occurs, that is, in the beginning of the project the costs spent on the building process are small. Thus, this information can be very important for risk management. The project profit expected value is equal to 3477.68, however, it is not informative for risk evaluation. The decisions that were made in decision nodes are stored in a graph—the chosen edges have probabilities equal to 1, and the remaining ones have probabilities equal to 0. In total, 63 decisions were made by the program. Since we have no notations for the subtree from node C_8^a (as well as all nodes that were generated by the cycle-expansion algorithm), and some decisions are different than in the subtree from node C_1, we do not provide the set of decisions; we consider this as non-informative in the current research.

We see some negative values in Table 4, which means that in the considered case, the expected profit value is bigger in the case of the law violation. That is, small probabilities of getting negative incomes lower the expected profit by a smaller value than the possible project modification does (i.e., building according to an alternative project without the law violations). It is an open question whether such violation makes a negative impact on sustainability of city development. However, the bigger profit for construction investors can be followed by further investments resulting in a more sustainable city growth. The more detailed study of advantages and disadvantages of such a case in terms of sustainable city development is out of the scope of this research. Such a study would rely on more detailed information about violations, which can be found in other papers for Lithuanian cases [17].

We note that in the general case, after cyclic repeats of events, the decisions can become different. This means that if we did not expand these cycles, the strategies would be selected (by some means) and fixed—which could be not optimal. On the other hand, the number of cycles could be unlimited, for example, if we apply some implementation of a Monte Carlo method. However, as was mentioned before, it is reasonable to limit the number of cyclic repeats due to specificity of the real process (the initial assumption of unlimited repeats of events with constant probabilities, and its usage in a graph, is quite artificial). Thus, we conclude that our proposed algorithm fits the considered problem very well.

5. Conclusions

In this paper, the real-life example of the process of judicial conflict resolution was considered. It was found that the time estimation is important when the costs of failures strongly depend on time. It is obvious that in the considered case, the failure at the beginning stages of building construction is much less costly than the failure at the later construction stages. Thus, for modelling of the impact of such judicial processes on investment, the evaluation of time is critical. The proposed approach allows us to evaluate the project with an estimation of time. Identifying big risks is important to avoid project failures, and it also lets investors select the appropriate projects for investment.

It is normal for judicial processes to return back to the previous stages (e.g., a court of appeal returns to a court of first-instance), increasing the time of the judicial process, and therefore, it is important to evaluate such cyclic event repeats. However, in practice, the number of repeats for such events is limited, and therefore, we used that idea directly in the proposed cyclic-expansion algorithm to convert a graph into a tree by limiting the number of cyclic event repeats. Firstly, we apply the cyclic-expansion algorithm. Secondly, we apply the algorithm to calculate the parameters for end nodes by recursively accumulating time and costs through a tree (using pre-order tree traversal). Thirdly, we use an algorithm that is based on a post-order tree-traversal idea to evaluate all nodes and select the optimal strategy. Thereafter, the simple pre-order tree-traversal algorithm is applied to accumulate the probabilities and compute the distribution of scenario values.

In the general case, after cyclic repeats of events, the decisions can become different. Thus, the tree expansion is important, because it allows repeated decision nodes to have different values. The initial assumption of unlimited repeats of events with constant probabilities, and its usage in a graph, is quite artificial. Thus, it is optimal to limit the number of cyclic repeats due to specificity of the real process; this leads to the proposed cyclic-expansion approach directly.

After extracting the profit distribution of the project, it is impossible to estimate the project uniquely—it strongly depends on investors' demands (i.e., risk versus profit evaluation). The analysis of the risk evaluation from the profit distribution is out of the scope of this research, and therefore, we consider the profit distribution as a final result of application of the proposed algorithm.

It is important to note that the algorithm does not identify the importance of the law violations in terms of sustainability. However, it could indicate in which cases the law violations are profitable for construction investors, that is, which cases need an additional investigation. In practice, it can be used in two ways. (1) By investors in a decision-support system to avoid risks of conflicts with society. (2) By institutions that control the development of territorial-planning procedures to identify and eliminate the weakness in law and regulations systems. If such kinds of tools were used in both suggested cases simultaneously, this would greatly contribute to a sustainable city development, because it would lead to much more predictable and sustainable behaviour of construction-process participants. In a similar way, the algorithm can be used by a scientific community to model how changes of territorial-planning procedures and laws impact the behaviour of construction-process participants. Thus, the proposed algorithm can support the development of rules regulating the construction process.

Author Contributions: Olga R. Šostak have formulated the problem, collected and analysed the data; Andrej Bugajev have created appropriate algorithms, implemented it in C++ and wrote pseudocode. Both authors have contributed to performing experiments, analysis of results and writing the paper equally.

Conflicts of Interest: The authors declare no conflict of interest.

References

1. Davidov, S.; Pantoš, M. Stochastic assessment of investment efficiency in a power system. *Energy* **2017**, *119*, 1047–1056.
2. Li, C. Empirical Study on the Impact Factors of the Investment Efficiency of Urbanization Construction in Guizhou Province, China. *DEStech Trans. Econ. Manag.* **2016**, doi:10.12783/dtem/icem2016/4112.
3. Focacci, A. Managing project investments irreversibility by accounting relations. *Int. J. Proj. Manag.* **2017**, *35*, 955–963.
4. Shariatmadari, M.; Nahavandi, N.; Zegordi, S.H.; Sobhiyah, M.H. Integrated resource management for simultaneous project selection and scheduling. *Comput. Ind. Eng.* **2017**, *109*, 39–47.
5. Marugán, A.P.; Márquez, F.P.G.; Lev, B. Optimal decision-making via binary decision diagrams for investments under a risky environment. *Int. J. Prod. Res.* **2017**, *55*, 5271–5286.
6. Zavadskas, E.K.; Turskis, Z.; Antucheviciene, J. Selecting a contractor by using a novel method for multiple attribute analysis: Weighted Aggregated Sum Product Assessment with grey values (WASPAS-G). *Stud. Inform. Control* **2015**, *24*, 141–150.
7. Trinkūnienė, E.; Podvezko, V.; Zavadskas, E.K.; Jokšienė, I.; Vinogradova, I.; Trinkūnas, V. Evaluation of quality assurance in contractor contracts by multi-attribute decision-making methods. *Econ. Res. Ekon. Istraž.* **2017**, *30*, 1152–1180.
8. Lazauskas, M.; Kutut, V.; Zavadskas, E.K. Multicriteria assessment of unfinished construction projects. *Gradevinar* **2015**, *67*, 319–328.

9.   Suh, S.; Suh, W.; Kim, J.I. Risk analysis model for regional railroad investment. *Eng. Comput.* **2017**, *34*, 164–173.

10. Xu, J.; Zhao, S. Noncooperative Game-Based Equilibrium Strategy to Address the Conflict between a Construction Company and Selected Suppliers. *J. Constr. Eng. Manag.* **2017**, *143*, 04017051.

11. Bellman, R. A Markovian Decision Process. *Indiana Univ. Math. J.* **1957**, *6*, 679–684.

12. Richey, M. The evolution of Markov chain Monte Carlo methods. *Am. Math. Mon.* **2010**, *117*, 383–413.

13. Šostak, O.R.; Vakrinienė, S. Mathematical modelling of dispute proceedings between investors and third parties on allegedly violated third-party rights. *J. Civ. Eng. Manag.* **2011**, *17*, 126–136.

14. Šostak, O.R. Planning Development of Construction by Taking Into Account Interests of the Third Parties (Doctoral Dissertation, in Lithuanian). Ph.D. Thesis, Vilnius Gediminas Technical University, Vilnius, Lithuania, 2011.

15. Celik, T.; Kamali, S.; Arayici, Y. Social cost in construction projects. *Environ. Impact Assess. Rev.* **2017**, *64*, 77–86.

16. Mitkus, S.; Šostak, O.R. Modelling the process for defence of third party rights infringed while implementing construction investment projects. *Technol. Econ. Dev. Econ.* **2008**, *14*, 208–223.

17. Mitkus, S.; Šostak, O.R. Preservation of healthy and harmonious residential and work environment during urban development. *Int. J. Strateg. Prop. Manag.* **2009**, *13*, 339–357.

18. Šostak, O.R.; Makutėnienė, D. Timely determining and preventing conflict situations between investors and third parties: Some observations from Lithuania. *Int. J. Strateg. Prop. Manag.* **2013**, *17*, 390–404.

19. Šostak, O.R.; Makutėnienė, D. Modelling a dispute hearing between an investor and the public concerned in administrative courts of the republic of Lithuania. *Technol. Econ. Dev. Econ.* **2013**, *19*, 489–509.

20. Zhou, W.Z.; Peng, Y.; Bao, H.J. Regular pattern of judicial decision on land acquisition and resettlement: An investigation on Zhejiang's 901 administrative litigation cases. *Habitat Int.* **2017**, *63*, 79–88.

21. Manasse, P.; Savona, R.; Vezzoli, M. Danger Zones for Banking Crises in Emerging Markets. *Int. J. Financ. Econ.* **2016**, *21*, 360–381.

22. Stević, Ž.; Pamučar, D.; Vasiljević, M.; Stojić, G.; Korica, S. Novel Integrated Multi-Criteria Model for Supplier Selection: Case Study Construction Company. *Symmetry* **2017**, *9*, 279.

23. Santos, S.F.; Fitiwi, D.Z.; Bizuayehu, A.W.; Shafie-Khah, M.; Asensio, M.; Contreras, J.; Cabrita, C.M.P.; Catalão, J.P.S. Impacts of Operational Variability and Uncertainty on Distributed Generation Investment Planning: A Comprehensive Sensitivity Analysis. *IEEE Trans. Sustain. Energy* **2017**, *8*, 855–869.

24. Eisenhardt, K.M. Agency theory: An assessment and review. *Acad. Manag. Rev.* **1989**, *14*, 57–74.

25. Mitroff, I.I. *Stakeholders of the Organizational Mind*; Jossey-Bass: San Francisco, CA, USA, 1983.

26. Gras-Gil, E.; Manzano, M.P.; Fernandez, J.H. Investigating the relationship between corporate social responsibility and earnings management: Evidence from Spain. *Brq-Bus. Res. Q.* **2016**, *19*, 289–299.

27. Martinez-Ferrero, J.; Gallego-Alvarez, I.; Garcia-Sanchez, I.M. A Bidirectional Analysis of Earnings Management and Corporate Social Responsibility: The Moderating Effect of Stakeholder and Investor Protection. *Aust. Account. Rev.* **2015**, *25*, 359–371.

28. Schwarzmüller, T.; Brosi, P.; Stelkens, V.; Spörrle, M.; Welpe, I.M. Investors' reactions to companies' stakeholder management: the crucial role of assumed costs and perceived sustainability. *Bus. Res.* **2017**, *10*, 79–96.

29. GooGle. Object Location on GooGle Maps. 2017. Available online: https://www.google.lt/maps/place/%C5%A0e%C5%A1kin%C4%97s+g.+45C,+Vilnius+07158/@54.7184175,25.2479231,17z/data=!4m5!3m4!1s0x46dd915d44d79665:0x5a00399e0ea3ff3b!8m2!3d54.7184175!4d25.2501118?hl=en (accessed on 11 January 2018).

30. Lithuanian State Enterprise Centre of Registers. On the Data About the Real Estate; Document Number (7.2./1147)s-1689; Lithuania, 2007. Available online: http://techmat.vgtu.lt/~ab/docs/EnterpriseRegisters.pdf (accessed on 11 January 2018). (In Lithuanian)

31. Www.inreal.lt.   Real Estate Market Overview for 2007 Year. 2008.   Available online: http://www.inreal.lt/media/editor/inreal/rinkos-apzvalgos/2007_metine_apzvalga.pdf (accessed on 11 January 2018). (In Lithuanian)

32. Vilnius County Head Administration. The Copy of the Act of Building Usage Suitability. Document Number (100)11.55-134. Lithuania, 2007.   Available online: http://techmat.vgtu.lt/~ab/docs/HeadAdministrationDoc.pdf (accessed on 11 January 2018). (In Lithuanian)

33. 690th Dwelling House Construction Association, (Lithuania, Company Code 125112766). Reference Number 20: About Funds Used for Object Financing. 2005. Available online: http://techmat.vgtu.lt/~ab/docs/Bendrija.jpg (accessed on 11 January 2018). (In Lithuanian)

34. Weng, S.S.; Yang, M.H.; Koo, T.L.; Hsiao, P.I. Modeling the prediction of business intelligence system effectiveness. *SpringerPlus* **2016**, *5*, 737.

35. Bellman, R.E.; Dreyfus, S.E. *Applied Dynamic Programming*; Princeton University Press: Princeton, NJ, USA, 2015.

![sustainability logo] *sustainability*

MDPI

Article

Cost Calculation of Construction Projects Including Sustainability Factors Using the Case Based Reasoning (CBR) Method

Agnieszka Leśniak and Krzysztof Zima *

Institute of Management in Construction, Faculty of Civil Engineering, Cracow University of Technology, 31-155 Krakow, Poland; alesniak@L3.pk.edu.pl
* Correspondence: kzima@L3.pk.edu.pl; Tel.: +48-628-23-54

Received: 23 April 2018; Accepted: 11 May 2018; Published: 17 May 2018

Abstract: The idea of sustainable development and the resulting environmentally friendly attitudes are increasingly used in construction projects. Designing in accordance with the principles of sustainable development has an impact on the costs of construction works. The authors of this paper proposed an approach to estimate the costs of sports field construction using the Case Based Reasoning method. In their analysis, they distinguished 16 factors that affect the cost of a construction project and are possible to already be described at an early stage of its preparation. The original elements of the work include: consideration of such environmental factors as the environmental impact of the building, materials used, the impact of the facility on the surroundings affecting the amount of implementation costs and development of own database containing 143 construction projects that are related to sports fields. In order to calculate the similarity of cases, different calculation formulas were applied depending on the type of data (quantitative, qualitative, uncertain, no data). The obtained results confirmed that the CBR method based on historical data and using criteria related to sustainable development may be useful in cost estimation in the initial phase of a construction project. Its application to the calculation of the costs that are related to the implementation of sports fields generates an error of 14%, which is a very good result for initial calculations. In the short run, such factors as the impact of the object and the type of materials that are used from the perspective of their influence on the environment may be decisive as far as the costs determined in the life cycle of the building are concerned, as well as the lowest costs of the building construction ensuring the appropriate quality and respect for the environment.

Keywords: Case Based Reasoning; construction; cost estimation; sports field; sustainability

---

1. Introduction

The implementation processes of sustainable development in the area of construction are of great importance. Construction processes have an important role in creation of built environment and their impacts have to be measured as construction contributes to air pollution, land use and contamination, usage of resources, water and materials depletion, water pollution, impacts on human health, and climate change [1]. The erection of a construction object is usually associated with the development of the biologically active surface, and subsequently, with putting a burden on the environment. The severity of these kinds of impact on the environment varies depending on the stage of the life cycle of the building. The analysis of the object vs. natural environment relation allows for distinguishing four basic stages of impact that are linked to the following processes: extraction of raw materials, production of materials, construction of the facility, operation of the facility, and its demolition [2]. As [3] proved, the results of developing sustainable architecture are based on changing the function of a building from a linear approach to a closed circulation plan. The linear approach treats the building as the "place of processing natural resources into waste" (for instance, water is transformed into sewage, energy into heat losses, building materials into waste). The other approach, namely in the closed circulation plan, a building can change from a consumer of energy and other resources into a virtually self-sufficient unit (through energy recovery or the re-use of water or waste). Designed buildings are characterized by diversified energy demand, which depends on many factors; for example, the material and construction solutions applied, the type of object, heating system, and its efficiency. According to the Energy Performance of Buildings Directive [4], all new buildings must be nearly zero-energy buildings by the end of 2020 and all new public buildings by 2018. In general, in terms of sustainable development and sustainable construction, refurbishment of buildings is preferred to new construction, because this helps to save energy and building materials in construction phase, also reduces generation of waste and other emissions [5]. A building object should be designed, constructed, operated, and demolished in accordance with the requirements of sustainable development. The construction materials used to build objects greatly influence the impact of the building on the natural environment, which was noticed in many works, as in [6–10]. The choice of building materials with the appropriate sustainability criteria is not straightforward. What is relatively easy to determine are such objective factors as cost constraints and design considerations, yet other, often subjective, factors have an impact on the selection, which influences the achievement of sustainability goals. In [11], one can find an optimization model for sustainable materials selection, while in [12], a model using a multi-criteria analysis enabling the selection of sustainable materials. Researchers [13] noticed that due to possible harmful effects of construction equipment on the environment, their evaluation with sustainability considerations can be considered as a helpful activity to move toward the sustainability in construction. The paper [14] presents a review of the literature on the sustainable built environment, which was made on the basis of the articles that were published between 1998 and 2015. The authors believe that the welfare of the whole society depends on the sustainability of the built environment [14].

2. Literature Review

The construction objects that are built today should show adequate durability, affect the environment in a harmless way, be economical in the consumption of materials and energy, and take into account the consequences of failure from the point of view of human life and health [15]. Choosing the right construction and material solutions, as well as the technologies that are applied in the first stage of the investment process influences the costs of its implementation and subsequent operation. All of the construction projects are risky, and different techniques and tools are proposed for assessment of risk in Construction project [16,17]. In [18], authors noticed that the fluctuation of material prices is one of the risk factor leading to cost overruns problems. In construction, reliable estimation of costs is important for both the investor who finances the investment and the contractor who has to estimate the costs and achieve a satisfactory level of profit.

To determine the price of construction works, the direct costs that are connected with realization of the works, overhead costs and profit must be consider. The contractors can calculate costs of works using unit price. This method requires a lot of experience from the contractor. The unit price must include all of the mentioned cost's elements. Traditional method of cost estimating is based on calculation of costs elements of separately: the direct cost (labour, materials, and equipment), indirect cost, and profit [19]. This traditional approach is accurate but time consuming, and therefore new methods are still being sought by means of new mathematical tools that can support the effectiveness of the calculation. Studies that are worth considering include ones that are employing artificial neural networks [20–22], linear regression [23], fuzzy sets [19], and support vector machines [24]. Researchers [25] proposed a hybrid model where multivariate regression method and the artificial neural network (ANN) method have been combined to provide a cost estimate model.

One of the methods that is proposed in the literature, which can be used in estimating costs in construction, is CBR (Case Based Reasoning) [26]. Acquiring knowledge as a result of researching the correctness of data, assimilating or formulating new concepts that are based on examples from the past, (CBR—Case Based Reasoning) in contrast to relying on individual experiences, can accelerate the process of estimating costs in construction. CBR can be defined as systems that solve new problems by adapting the results that were used to solve old issues [27]. In [28], the authors compare the system of inference from cases to the black box. The input data describe the problem; the output data form a solution to the problem, while the memory of past cases and the box contain a reasoning mechanism. In the case-based inferencing, the basic source of knowledge is a database that is containing not rules but a set of cases from the problems encountered and resolved. New problems are solved by searching for the most similar cases and their possible adaptation. The CBR models used for cost estimates may be based on both quantitative and qualitative data [29]. In [30,31], it was noticed that, although some information is not specified, CBR models for long periods of use maintain quality and ability to solve problems and work better than other models. The use of CBR to estimate construction costs is based on searching for similar investments already completed. This provides a simple way to measure construction costs, given that, according to most studies, there are non-linear relationships between cost and factors that affect it [32–35]. An interesting example of the use of CBR in the cost estimation process can be a model using the AHP method to determine the weights of criteria, proposed by [26] or the CBR model using genetic algorithms to estimate the construction costs [36] or unit cost of residential construction projects [37]. Models are also being created that predict both construction time and cost at an early stage of a construction project [38]. Ryu et al. [39] proposed the CONPLA-CBR tool that generates master schedules at the preconstruction stage.

The paper presents the concept of supporting the estimation of construction costs in the initial investment phase based on Case Based Reasoning (the CBR method). The process of information management and the use of historical data differ from the models that have been proposed so far. Typically, global assessment of construction projects or facilities based on a few or several criteria describing the construction project in a general way, such as the usable area, cubic volume of the building, height, complexity, or location. Such criteria often do not take into account differences in the type of materials used and the details of solutions or environmental impact. In cost calculations, fuzzy logic is rarely used when there is uncertain or imprecise information for the problem of cost calculation for the implementation of construction works.

In the proposed algorithm, among the proposed 16 explanatory variables, it is suggested to take into account the features related to the environmental impact of the building, its influence on the surroundings and the parameters of construction materials in relation to the idea of sustainable development. Four different calculation formulas are also proposed for quantitative and qualitative data, for cases without data or ones with uncertain or inaccurate data. It is worth noting that there are no proposals for methods and tools supporting the cost estimation of such sports facilities as football pitches, treadmills, or skate parks.

3. Cost Calculations Method Based on Case Based Reasoning with Sustainability Criterion

The database CSDB (Cost Solution DataBase) that was developed to support index cost estimation with the CBR method includes cases containing four groups of information: information about geometry of building object—GEO, solution of problem (unit cost of object)—SoP, description of the construction—DoC, data for adaptation process—Ad. The database CSDB was defined by the following formula:

$$CSDB = U_{i=1}^{n} Case_i \{GEO_i, SoP_i, DoC_i, Ad\} \qquad (1)$$

where:

$Case_i$—*i*-th case from the database,
GEO_i—graphical representation of the construction elements for the *i*-th case,
SoP_i—solution of problem,
DoC_i—description of construction,
Ad—data for adaptation process such date, localization of construction, and

n—number of old cases in the database.

GEO is information about the graphic appearance of the construction object or element in the form of quantitative data. For example, it can be information about the volume of an object, surface, or dimensions, such as length, width, and height. Solution of problem SoP contains information about a unit price of construction object referenced to, for example, surface area or volume of object. Description of construction DoC contains all the necessary information about factors (including sustainability factors) influencing to the cost of the project in the form of qualitative data.

The presented CSBD database was created in order to support the process of estimating costs in the early phase of the investment, including quantitative variables (defining the bill of quantities) and qualitative variables. The novel element of supporting the process of cost estimation is one that is involving the variables affecting sustainable development. The authors of the paper note the increasing influence of factors that are related to sustainable development and their impact on the investment price. The assessment of sustainable development consists of three basic elements: impact on the environment (consumption of natural resources, consumption of energy resources, and emissions of pollutants into the environment), comfort, and quality of life, as well as cost (the amount of funds allocated to achieve the expected value in use and reduction of the impact on the environment). In terms of costs, this means constructing a building at the lowest possible cost while obtaining adequate quality and respect for the environment, as well as the low costs of its operation.

Figure 1 presents the algorithm for the cost estimation method based on the Case Based reasoning method, taking into account the above assumptions.

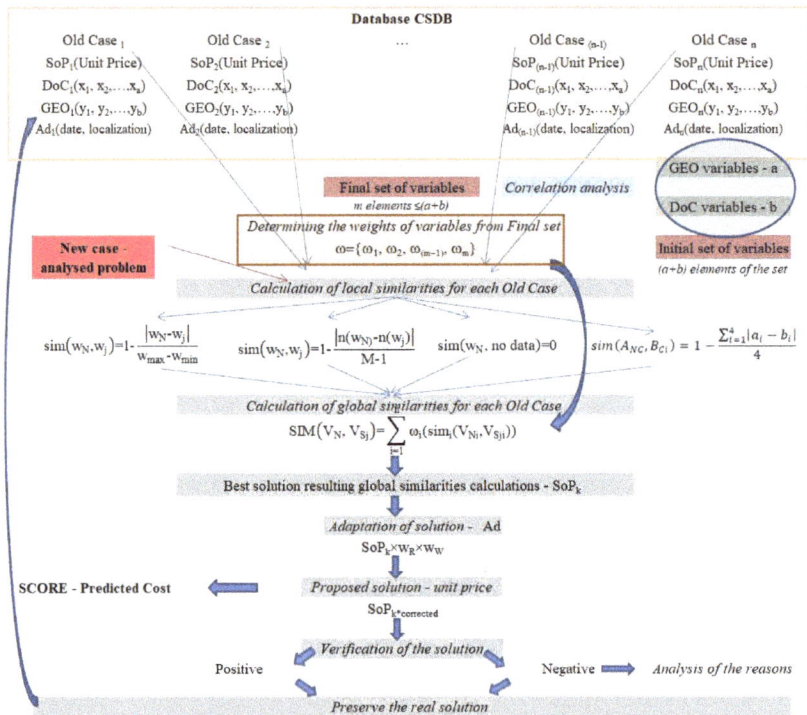

Figure 1. Algorithm used in the method proposed.

The first step in the method that is proposed is to define a preliminary set of variables, divided into GEO geometry variables, and variables describing the DoC structure. Using the case database collected, one can perform a correlation analysis in step 2 to eliminate the variables that have no impact or have a small impact on the unit price of construction works. The result of the correlation analysis is to create the final set of variables (Final set). The third step of the analysis is the assessment of the validity of variables, which can be performed by means of expert assessment; alternatively, it may result from the determined strength of correlation between the individual variables and the price.

The fourth step of the method begins the proper analysis based on the inference from cases. The new case providing a problem to analyse is compared in pairs with all the old cases from the database. The purpose of the comparison is to find the most similar cases, and, by analogy, to determine the unit price of the construction works for the new case.

The similarity of cases can be calculated using different formulas depending on the type of explanatory variables. The calculation formulas applied in the algorithm are presented below:

- For quantitative explanatory variables (as per [40]):

$$sim(w_N, w_j) = 1 - \frac{|w_N - w_j|}{w_{max} - w_{min}}$$ (2)

where:

w_N—value of the explanatory variable for the new case,
w_j—value of the explanatory variable for the *j*-th old case, and
w_{max}, w_{min}—minimum and maximum values for all the old cases included in the database.

- For qualitative explanatory variables:

$$sim(w_N, w_j) = 1 - \frac{|n(w_N) - n(w_j)|}{M - 1}$$ (3)

where:

$n(w_N)$, $n(w_j)$—place in an ordered array of values $n(w) = 1, 2, \ldots, n$,
M—number of values.

- For uncertain or inaccurate variables:

$$sim(A_{NC}, B_{Ci}) = 1 - \frac{\sum_{i=1}^{4} |a_i - b_i|}{4}$$ (4)

where:

$sim(A_{NC}, B_{Ci})$—similarity between fuzzy numbers,
A_{NC}—fuzzy number for the new case,
B_{Ci}—fuzzy number for the old case taken from the database, and
a_i, b_i—characteristic points for fuzzy numbers $A_{NC} = (a_1, a_2, a_3, a_4)$ and $B_{Ci} = (b_1, b_2, b_3, b_4)$.

The formula for uncertain or inaccurate variables is given as an example and concerns a situation in which the shape of the membership function is trapezoidal. It is defined for a situation in which two trapezoidal membership functions assume values $A_{NC} = (a_1, a_2, a_3, a_4)$ and $B_{Ci} = (b_1, b_2, b_3, b_4)$. The characteristic points allow for describing the limit values for the shape of the membership function that accepts the values 0 and 1. This allows for describing the fuzzy number using the four real numbers, which allows for the quick execution of actions using only these characteristic point values. When adopting other forms of membership functions, other calculation formulas should be used.

- In the absence of data for an explanatory variable for a new case, an old case or both:

$$sim(w_N, w_S) = 0$$ (5)

where:

w_N—an explanatory variable for the new case and
w_S—an explanatory variable for the old case.

The formulas given above are applicable when comparing local similarity, defined as a pair comparison of similarities for subsequent explanatory variables.

The next, fifth calculation step is to calculate the global similarity, which is the weighted sum of the local similarities of all the explanatory variables that are used in analysis and collected in the final set of variables. In order to calculate global similarity, the following formula was used:

$$SIM(V_N, V_{Sj}) = \sum_{i=1}^{n} \omega_i (sim_i(V_{Ni}, V_{Sji})) \qquad (6)$$

where:

ω_i—weight of the *i*-th explanatory variable,
$SIM(V_N, V_{Sj})$—global similarity between the old V_j and the new case V_N, and
$sim_i(V_{Ni}, V_{Sji})$—local similarity for the *i*-th explanatory variable between the old V_j and the new case V_N.

It is necessary to determine a few basic assumptions in the course of the analysis that uses the CBR method:

1. old cases with the highest global similarity $SIM(V_N, V_j)$ are selected. The minimum number of cases entering the selected set of cases is 3. This assumption is to limit the possibility of choosing an accidental solution—if three solutions are chosen, then the possible extreme solution is rejected to limit the possibility of overestimating or underestimating the price of works;
2. the minimum value of global similarity for cases included in the set of solutions must be greater than 70%;
3. the resulting value of similarities is given as a percentage and is a natural number; and,
4. cases from the set of selected cases are rejected as extreme when the difference between the selected cases is greater than 50%.

The next step is to adapt the calculated unit price for the new case. The value correction concerns a situation in which there is a difference between the duration of the construction works and the location between the new case and the selected old cases. To do the adaptation, two coefficients were applied: regional coefficient determined in the case of a difference in the location of a construction project, (introduces price correction due to the difference in the local prices of services and materials) and inflation rate (introduces a price correction due to changes in prices of services and materials over time). Corrections are made individually for each of the unit prices that were selected in the course of the analysis, resulting from the selection of the old case as the most similar and meeting the conditions that are given in the assumptions.

The unit price of the new case, resulting from several finally selected old cases after adaptation, is calculated as the weighted average of the weights of individual old cases (where the basis for the calculation of the weight is the degree of the similarity to the new case) and the unit prices of old cases. The proposed unit price is accepted as the final price of the new case and in the next step it is multiplied by the amount of works from the bill of quantities. After the settlement of the construction works, the price result obtained is verified by the actual costs. The verified result is saved in the database for its subsequent use.

4. An Example of Supporting Cost Calculation with the Use of the CBR Method, Taking into Account the Factors of Sustainable Construction

The calculation example concerns the estimation of the costs of the implementation of sports fields in the early phase of the investment, and is thus based on the investment concept. The cost analysis is performed in accordance with the algorithm provided above, which is based on the CBR method. The purpose of the exemplary cost analysis is to determine the price for the implementation of a new sports facility, which is a sports field that is based on historical data from the investments already completed.

In the first step, a preliminary set of criteria was compiled. The set of criteria was created on the basis of literature studies and the analysis of advertisements that were placed in Poland on the public procurement website. The analysis included the selection of information describing the subject of the contract.

On the basis of literature studies [22,41–43] and the analysis of announcements in public procurement, 16 variables explaining sports fields were distinguished:

1. Quantitative variables (type of information—GEO)
   intended use of the field (five types of fields)
   surface area of the field (variables range: 275–8714 m^2)
   surface area of the access paths and routes (variables range: 0–1753 m^2)
   green surface area (variables range: 0–6017 m^2)
   surface area of the ball containment netting (variables range: 0–2212 m^2)
   fence length (variables range: 0–602.5 m)
2. Qualitative variables (type of information—DoC)
   type of the material for sports surface (six types of materials)
   type of the material for access routes (five types of materials)
   type of the fence (five types of fences)
   type of sports equipment—handball (yes or no)
   type of sports equipment—volleyball (yes or no)
   type of sports equipment—basketball (yes or no)
   type of sports equipment—football (yes or no)
   type of sports equipment—tennis (yes or no)
   impact of the construction on the environment (rating 1–5)
   impact on the surroundings (rating 1–5)

The presented example includes the analysis of the costs of sports fields, both single- and multifunctional. The information contained in the database comes from the investments executed in 2014–2016 and includes 143 building projects concerning the construction of sports fields. The data comes from advertisements and building cost estimates prepared for tender proceedings for the implementation of sports fields in Poland. The data from cost estimates and the public procurement description were placed in the database that was created. The data includes all of the variables highlighted in the course of the analysis of the initial set of criteria: GEO and DoC, and the information about the SoP solution and data allowing for the later adaptation of the Ad solution (tender date and location of a construction project). The method of recording information in subsequent records in the database is described below.

Case$_n$ {GEO = (intended use of the field, surface area of the field, surface area of the access paths and routes, green surface area, surface area of the ball containment netting, fence length); DoC = (type of the material for sports surface, type of the material for access paths, type of the fence, type of sports equipment—handball, type of sports equipment—volleyball, type of sports equipment—basketball, type of sports equipment—football, type of sports equipment—tennis, impact of the construction on the environment, impact on the surroundings); SoP = (unit price of the field surface area); and, Ad = (location, date of the bid)}

The GEO information contains basic parameters concerning the size of individual construction works. On the other hand, the DoC information contain a description of the scope of works and the types of solutions used in the implementation of individual elements of the sports field and its surroundings, and the characteristics of the undertaking related to the idea of sustainable development.

The features that are related to the impact of a building object on the environment (sustainability factors) included in the initial set of variables involve in point 2. Qualitative variables (type of information—DoC) such variables:

- impact of the construction on the environment (for instance, energy demand, use of renewable energy, efficiency of energy systems);
- materials (for instance, materials with low environmental impact, materials with low risk of health hazard, recyclable materials): type of the material for sports surface and type of the material for access routes; and,
- impact on the surroundings (such as air pollution, noise, vibration, wind effects and shading of the area, the effect of the thermal island)

The determination of the correlation strength was made on the basis of the Guilford scale:

$r = 0$ no correlation,
$0 < r < 0.1$ barely perceptible correlation,
$0.1 < r < 0.3$ poor correlation,
$0.3 < r < 0.5$ average correlation,
$0.5 < r < 0.7$ high correlation,
$0.7 < r < 0.9$ very high correlation,
$0.9 < r < 1$ almost full correlation, and
$r = 1$ full correlation.

Table 1 depicts the results of the Pearson and Spearman correlation analysis. For quantitative variables, the analysis involved Pearson's correlation; while for quality variables, it used Spearman's correlation. In order to create the final set of explanatory variables, variables that showed no correlation or very low correlation were rejected in the course of the analysis. The final set of criteria with designated weights is shown in Table 2. The weights were determined based on the correlation strength of individual variables with the unit price of a construction work.

Table 1. The correlation coefficients of explanatory variables from initial set.

Variables	Correlation Coefficient:
Surface area of the fields	−0.192
Surface area of the access paths	0.232
Green surface area	0.079
Fence length	0.249
Surface area of the ball containment netting	0.056
Intended use	−0.472
Material for sport surface	−0.299
Material for access paths	−0.283
Type of sports equipment—Handball	0.244
Type of sports equipment—Basketball	0.477
Type of sports equipment—Volleyball	0.363
Type of sports equipment—Football	−0.289
Type of sports equipment—Tennis	−0.071
Fence type	−0.045
Impact of the construction on the environment	0.095
Impact on the surroundings	−0.640

Table 2. The final set of explanatory variables with their weights.

Variables	Weights ω_i
Surface area of the fields	6.2%
Surface area of the access paths	7.5%
Fence length	8.0%
Intended use	15.2%
Material for sport surface	9.7%
Material for access paths	9.1%
Type of sports equipment—Handball	7.9%
Type of sports equipment—Basketball	15.4%
Type of sports equipment—Volleyball	11.7%
Type of sports equipment—Football	9.3%

The final set of variables consists of two explanatory variables that are related to the idea of sustainable development and included in the Materials (materials for sports surfaces and materials for access). The total validity of these variables is 18.7%, thus they are quite a significant element affecting the unit price of a construction project, although not being decisive. Due to the low correlation with the unit price, the other variables that are related to sustainable construction were rejected, that is, the impact of the building on the environment and the impact on the surroundings, which is two of the four factors that were originally considered.

After creating the final set of variables and determining their validity, local similarities were calculated for the subsequent explanatory variables between the New Case and all the Old Cases from the database. Local Formulas (2)–(5) were used to calculate local similarities. An example of calculating a local similarity for a quantitative variable of an access surface, between the New Case and the Old Case 7 using the Formula (2) has been shown below:

$$\text{sim}(w_N, w_j) = 1 - \frac{|458 - 295|}{5569 - 0} = 1 - 0.029 = 0.971 \cong 97\% \tag{7}$$

where:

w_N equals 458 and determines the size of the access surface for the New Case—458 m^2,

w_j equals 295, which means the access surface for Old Case 7 equal 295 m^2,

w_{max} equals 5569, which means the maximum access area for the whole set of cases is equal 5569 m^2, and

w_{min} is equal 0, which means the minimum access space for the whole set of cases is equal 0 m^2 (some orders for the implementation of sports fields did not include the implementation of access routes).

On the other hand, the local similarity for the quality variable: material for sports surfaces, between the New Case and the Old Case 7 using the formula (3) is shown below:

$$\text{sim}(w_N, w_j) = 1 - \frac{|5 - 5|}{6 - 1} = 1 - 0 = 1 = 100\% \tag{8}$$

where:

$n(w_N)$ is equal 5 and determines the value given to the material which is the polyurethane from which the New Case surface is planned to be made,

$n(w_j)$ is equal 5, which means the value given to the material which is the polyurethane from which the Old Case 7 surface was made, and

$M = 6$, which results from the specification of six types of materials used for sports surfaces found in the database (natural grass; surface from natural dried wood chips—technologically softened along fibers with 5–50 mm fraction; artificial grass; brick flour; polyurethane surface; asphalt surface).

A list of calculated local similarities and global similarity computed in accordance with the formula (6) taking into account the weights given in Table 2 between the New Case and the selected Old Cases is shown in Table 3.

Table 3. The values of the local similarities and the global similarities for the chosen cases.

Variables	Local Similarities		
	Case 7	Case 8	Case 9
Surface area of the fields	94%	88%	94%
Surface area of the access paths	97%	87%	92%
Fence length	86%	79%	79%
Intended use	86%	0%	100%
Material for sport surface	100%	0%	100%
Material for access paths	100%	100%	0%
Type of sports equipment—Handball	100%	100%	100%
Type of sports equipment—Basketball	100%	0%	100%
Type of sports equipment—Volleyball	100%	0%	100%
Type of sports equipment—Football	100%	100%	100%
Global similarities	**98%**	**45%**	**88%**

The highest similarity of 98% occurred for three cases: Case 7—SIM($V_{\text{Test case 1}}$, $V_{\text{Case 7}}$); Case 23—SIM($V_{\text{Test case 1}}$, $V_{\text{Case 23}}$); and, Case 83—SIM($V_{\text{Test case 1}}$, $V_{\text{Case 83}}$).

For the selected cases in the analysis of three cases with the highest similarity rate, the adaptation of unit prices of these old cases was made due to the time difference in the calculation and the difference in the prices of services and materials in different regions of Poland. The inflation factor considering the time difference was calculated on the basis of the price index "The Sekocenbud forecasting and indexation bulletin" [44]. The regional coefficient, in turn, was calculated based on the newsletter "The Sekocenbud regional price bulletin" [45]. The adjusted unit prices are shown in Table 4.

Table 4. Adjusted unit prices of selected cases.

Case	Case 7	Case 23	Case 83
Unit price	80.25 €	84.23 €	79.34 €
Regional factor	1.028	0.960	1.028
Indexation factor	100.1%	101.3%	101.3%
Adjusted price	82.58 €	81.91 €	82.62 €

For example, the price after the adaptation of Case 7 has been calculated as 80.5 € × 1.028 × 1.001, where the regional factor is—1.028 and the indexation factor is equal to (1 + 0.01%) = 100.1%. The final price for test case 1 is an arithmetic mean of the unit prices of case 7, 23, 83, and equals:

$$\frac{(C_{jA}^{\text{Case 7}} + C_{jA}^{\text{Case 23}} C_{jA}^{\text{Case 83}})}{3} = \frac{(82.58 + 81.91 + 82.62)}{3} = 82.37 \ \text{€}/\text{m}^2 \tag{9}$$

After calculating the unit price after adaptation for the New Case equal 82.37 €/m² it is necessary to specify the value of the works. For this purpose, the unit price was multiplied by the number of the works representing the construction works consisting in the performance of the sports field, namely the area of the pitch.

The value of the works for the New Case is thus finally: 82.37 [€/m²] × 1200 [m²] = 98 844.00 €.

5. Discussion

An analysis of the error that was generated by the forecasts for the costs of sports field construction using the presented method was made. Testing was performed on selected 15 cases for which the costs of building a construction project were known. Mean Absolute Estimate Error (MAEE) for 10 test cases was 14%. The error range for individual cases is 2–34%, except that only for three test cases it exceeded 20%. In five cases, the error was smaller than 20%. Figure 2 shows the distribution of the MAEE error in the following ranges: up to 10%, 10–20%, and above 20%.

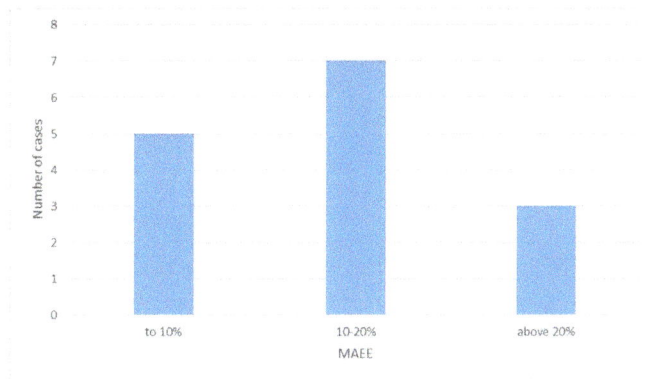

Figure 2. The distribution of the Mean Absolute Estimate Error (MAEE) error with the cost calculation broken down into percentage ranges.

In accordance with the PMI 2008 [46] guidelines, the error of 14% meets the requirements of an acceptable cost calculation error made in the early stages of construction projects in the range of −30% to +50%.

Also, the requirements of the American Association of Cost Engineers (AACE)—error size of −10% to +15%—and the requirements of the Construction Industry Institute—error size of −30% to +50% (for: [47]) are met.

It is worth noting that, in similar studies, only the use of artificial neural networks in supporting costs in such an early investment phase gives similar good results [48,49]. However, even in these cases, the maximum calculation error exceeds 50%, and in the presented method it was 34%.

It seems that both the use of the CBR method in supported cost calculation and the use of variables based on the idea of sustainable development can bring good results. According to the authors, along with the development of the idea of sustainable development of such variables affecting price, there will be increasingly more research on the impact of these factors on costs.

For the CBR method, further research must address the impact of changing calculation formulas that estimate local and global similarities on the MAEE error. The search for formulas reflecting the specificity of data used during calculations may bring even better results. In-depth research into the adaptation process of selected solutions should also be performed.

6. Conclusions

Supporting cost calculations at an early stage of a construction project is a vital problem. Despite the fact that cost-support models that are based on numerous mathematical methods have already been presented in the world literature many times, only a few generate a calculation error below 20%. It seems that only these models that are based on historical data, artificial neural networks and case based reasoning methods reach an acceptable level of error. The vast majority of methods do not take into account the variables that are based on the idea of sustainable development.

The authors believe that the proposed method using Case Based Reasoning has the potential and can be useful for practice. It generated an Mean Absolute Estimate Error of 14%. It should be noticed that only for 3 test cases the MAE Error exceeded 20%. To present performing of the model an example of calculation the unit price for the New Case was successfully conducted. The value of the works for the New Case finally equalled 98 844.00 €.

The original element of the of cost estimation process using the proposed model is involving the factors affecting sustainable development. The authors note the increasing influence of factors that are related to sustainable development on the investment price. Environmental impact, comfort, and quality of life, in the near future can be decisive in terms of costs determined in the lifecycle of a building, taking into account the construction of a building object as the cheapest cost when obtaining adequate quality and respect for the environment, while obtaining low operating costs.

Author Contributions: The individual contribution and responsibilities of the authors were as follows: K.Z. designed the research main idea and collected the data. Authors together analyzed the data and the obtained results. A.L. provided extensive advice throughout the study results and methodology. K.Z. and A.L. wrote the paper. All the authors have read and approved the final manuscript.

Funding: This research received no external funding.

Conflicts of Interest: The authors declare no conflict of interest.

References and Notes

1. Zavadskas, E.K.; Vilutienė, T.; Tamošaitienė, J. Harmonization of cyclical construction processes: A systematic review. *Procedia Eng.* **2017**, *208*, 190–202. [CrossRef]
2. Golański, M. Wybór materiałów budowlanych w kontekście efektywności energetycznej i wpływu środowiskowego. *Budownictwo i Inżynieria Środowiska* **2012**, *3*, 39–53.
3. Bonenberg, W.; Kapliński, O. The Arcithect and the Paradigms of Sustainable Development: A Review of Dilemmas. *Sustainability* **2018**, *10*, 100. [CrossRef]
4. European Union (EU). *Directive 2010/31/EU of 19 May 2010 on the Energy Performance of Buildings*; European Union: Brussels, Belgium, 2010.
5. Zavadskas, E.K.; Antucheviciene, J.; Kalibatas, D.; Kalibatiene, D. Achieving Nearly Zero-Energy Buildings by applying multi-attribute assessment. *Energy Build.* **2017**, *143*, 162–172. [CrossRef]
6. Calkins, M. *Materials for Sustainable Sites: A Complete Guide to the Evaluation, Selection, and Use of Sustainable Construction Materials*; John Wiley & Sons: New York, NY, USA, 2008.
7. Thormark, C. The effect of material choice on the total energy need and recycling potential of a building. *Build. Environ.* **2006**, *41*, 1019–1026. [CrossRef]
8. Leśniak, A.; Zima, K. Comparison of traditional and ecological wall systems using the AHP method. In *International Multidisciplinary Scientific GeoConference Surveying Geology and Mining Ecology Management*; SGEM: Albena, Bulgaria, 2015; Volume 3, pp. 157–164.
9. Drozd, W.; Leśniak, A.; Zaworski, S. Construction Time of Three Wall Types Made of Locally Sourced Materials: A Comparative Study. *Adv. Mater. Sci. Eng.* **2018**, 1–8. [CrossRef]
10. Švajlenka, J.; Kozlovská, M. Houses Based on Wood as an Ecological and Sustainable Housing Alternative—Case Study. *Sustainability* **2018**, *10*, 1502. [CrossRef]
11. Florez, L.; Castro-Lacouture, D. Optimization model for sustainable materials selection using objective and subjective factors. *Mater. Des.* **2013**, *46*, 310–321. [CrossRef]
12. Akadiri, P.O.; Olomolaiye, P.O.; Chinyio, E.A. Multi-criteria evaluation model for the selection of sustainable materials for building projects. *Autom. Constr.* **2013**, *30*, 113–125. [CrossRef]
13. Ghorabaee, M.K.; Amiri, M.; Zavadskas, E.K.; Antucheviciene, J. A new hybrid fuzzy MCDM approach for evaluation of construction equipment with sustainability considerations. *Arch. Civ. Mech. Eng.* **2018**, *18*, 32–49. [CrossRef]
14. Ubarte, I.; Kaplinski, O. Review of the sustainable built environment in 1998–2015. *Eng. Struct. Technol.* **2016**, *8*, 41–51. [CrossRef]
15. Runkiewicz, L. Realizacja obiektów budowlanych zgodnie z zasadami zrównoważonego rozwoju. *Przegląd Budowlany* **2010**, *81*, 17–23.
16. Burtonshaw-Gunn, S.A. *Risk and Financial Management in Construction*; Routledge: Abington-on-Thames, UK, 2017.
17. Chatterjee, K.; Zavadskas, E.K.; Tamošaitienė, J.; Adhikary, K.; Kar, S. A hybrid MCDM technique for risk management in construction projects. *Symmetry* **2018**, *10*, 46. [CrossRef]

18. Odimabo, O.O.; Oduoza, C.; Suresh, S. Methodology for Project Risk Assessment of Building Construction Projects Using Bayesian Belief Networks. *Int. J. Constr. Eng. Manag.* **2017**, *6*, 221–234. [CrossRef]
19. Plebankiewicz, E.; Leśniak, A. Overhead costs and profit calculation by Polish contractors. *Technol. Econ. Dev. Econ.* **2013**, *19*, 141–161. [CrossRef]
20. Juszczyk, M. Application of PCA-based data compression in the ANN-supported conceptual cost estimation of residential buildings. *AIP Conf. Proc.* **2016**, *1738*, 200007. [CrossRef]
21. Leśniak, A.; Juszczyk, M. Prediction of site overhead costs with the use of artificial neural network based model. *Arch. Civ. Mech. Eng.* **2018**, *18*, 973–982. [CrossRef]
22. Juszczyk, M.; Leśniak, A.; Zima, K. ANN Based Approach for Estimation of Construction Costs of Sports Fields. *Complexity* **2018**, 7952434. [CrossRef]
23. Sonmez, R. Parametric range estimating of building costs using regression models and bootstrap. *J. Constr. Eng. Manag.* **2008**, *134*, 1011–1016. [CrossRef]
24. Kim, G.H.; Shin, J.M.; Kim, S.; Shin, Y. Comparison of school building construction costs estimation methods using regression analysis, neural network, and support vector machine. *J. Build. Constr. Plan. Res.* **2013**, *1*, 1–7. [CrossRef]
25. Yazdani-Chamzini, A.; Zavadskas, E.K.; Antucheviciene, J.; Bausys, R. A Model for Shovel Capital Cost Estimation, Using a Hybrid Model of Multivariate Regression and Neural Networks. *Symmetry* **2017**, *9*, 298. [CrossRef]
26. An, S.-H.; Kim, G.; Kang, K. A Case-based reasoning cost estimating model using experience by analytic hierarchy process. *Build. Environ.* **2007**, *42*, 2573–2579. [CrossRef]
27. Riesbeck, C.K.; Schank, R.C. *Inside Case-Based Reasoning*; Lawrence Erlbaum Associates: Hillsdale, NJ, USA, 1989.
28. Pal Sankar, K.; Shiu Simon, C.K. *Foundations of Soft Case-Based Reasoning*; John Wiley & Sons, Inc.: Hoboken, NJ, USA, 2004.
29. Mendes, E.; Mosley, N.; Counsell, S. The application of casebased reasoning to early web project cost estimation. In *Proceedings—IEEE Computer Society's International Computer Software and Applications Conference*; Institute of Electrical and Electronics Engineers Computer Society: Oxford, UK, 2002.
30. Kim, G.-H.; An, S.-H.; Kang, K.-I. Comparison of construction cost estimating models based on regression analysis, neural networks, and case-based reasoning. *Build. Environ.* **2004**, *39*, 1235–1242. [CrossRef]
31. Duverlie, P.; Castelain, J.M. Cost estimation during design step: Parametric method versus case based reasoning method. *Int. J. Adv. Manuf. Technol.* **1999**, *15*, 895–906. [CrossRef]
32. Chou, J.-S.; Peng, M.; Persad, K.; O'Connor, J. Quantity-based approach to preliminary cost estimates for highway projects. *Transp. Res. Rec.* **2006**, *1946*, 22–30. [CrossRef]
33. Emsley, M.W.; Lowe, D.J.; Duff, A.R.; Harding, A.; Hickson, A. Data modelling and the application of a neural network approach to the prediction of total construction costs. *Constr. Manag. Econ.* **2002**, *20*, 465. [CrossRef]
34. Lowe, D.J.; Emsley, M.W.; Harding, A. Predicting construction cost using multiple regression techniques. *J. Constr. Eng. Manag.* **2006**, *132*, 750–758. [CrossRef]
35. Marir, F.; Wang, F.; Ouazzane, K. A case-based expert system for estimating the cost of refurbishing construction buildings. In Proceedings of the 4th International Conference on Enterprise Information Systems, Ciudad Real, Spain, 3–6 April 2002; Volume 1, pp. 391–398.
36. Ji, S.-H.; Park, M.; Lee, H.-S. Cost estimation model for building projects using case-based reasoning. *Can. J. Civ. Eng.* **2011**, *38*, 570–581. [CrossRef]
37. Dogan, S.Z.; Arditi, D.; Günaydın, H.M. Determining attribute weights in a CBR model for early cost prediction of structural system. *J. Constr. Eng. Manag.* **2006**, *132*, 1092–1098. [CrossRef]
38. Koo, C.; Hong, T.; Hyun, C.; Koo, K.A. CBR-based hybrid model for predicting a construction duration and cost based on project characteristics in multi-family housing projects. *Can. J. Civ. Eng.* **2010**, *37*, 739–752. [CrossRef]
39. Ryu, H.G.; Lee, H.S.; Park, M. Construction planning method using case based reasoning (CONPLA-CBR). *J. Comput. Civ. Eng.* **2007**, *21*, 410–422. [CrossRef]
40. Traczyk, W. *Inżynieria Wiedzy/Knowledge Engineering*; Akademicka Oficyna Wydawnicza EXIT: Warszawa, Poland, 2010.

41. Zima, K. Cost Estimating of Football Pitches Construction in the Concept Phase Using Case Based Reasoning. In Proceedings of the International Conference on Economics and Management Engineering (ICEME 2017), Wuhan, China, 24–26 March 2017; DEStech Transactions on Economics, Business and Management; 2017; pp. 24–26.

42. Polish Green Building Council Homepage. Available online: https://plgbc.org.pl/od-czego-zaczac/ (accessed on 3 March 2018).

43. Public Procurement Control Department Homepage. Available online: https://www.uzp.gov.pl/ (accessed on 6 March 2018).

44. Sekocenbud. Forecasting and Indexation Bulletin—ZWW, 2014–2017; PROMOCJA Sp. z o.o.: Warszawa, Poland.

45. Sekocenbud. Regional Price Bulletin—BCR, 2014–2017; PROMOCJA Sp. z o.o.: Warszawa, Poland.

46. PMI. *A Guide to the Project Management Body of Knowledge*; Project Management Institute: Newtown Square, PA, USA, 2008.

47. Kim, S.; Shim, J.H. Combining case-based reasoning with genetic algorithm optimization for preliminary cost estimation in construction industry. *Can. J. Civ. Eng.* **2014**, *41*, 65–73. [CrossRef]

48. Gunaydin, H.; Dogan, S. A neural network approach for early cost estimation of structural systems of buildings. *Int. J. Proj. Manag.* **2004**, *22*, 595–602. [CrossRef]

49. Juszczyk, M. Application of committees of neural networks for conceptual cost estimation of residential buildings. *AIP Conf. Proc.* **2015**, *1648*, 600008. [CrossRef]

sustainability

MDPI

Article

Cost Based Value Stream Mapping as a Sustainable Construction Tool for Underground Pipeline Construction Projects

Murat Gunduz * and Ayman Fahmi Naser

Department of Civil Engineering, Qatar University, P.O. Box: 2713 Doha, Qatar; an1401569@student.qu.edu.qa
* Correspondence: mgunduz@qu.edu.qa

Received: 9 November 2017; Accepted: 23 November 2017; Published: 27 November 2017

Abstract: This paper deals with application of Value Stream Mapping (VSM) as a sustainable construction tool on a real construction project of installation of underground pipelines. VSM was adapted to reduce the high percentage of non-value-added activities and time wastes during each construction stage and the paper searched for an effective way to consider the cost for studied construction of underground pipeline. This paper is unique in its way that it adopts cost implementation of VSM to improve the productivity in underground pipeline projects. The data was observed and collected from site during construction, indicating the cycle time, value added and non-value added of each construction stage. The current state was built based on these details. This was an eye-opening exercise and a process management tool as a trigger for improvement. After the current state assessment, a future state is attempted by Value Stream Mapping tool balancing the resources using a Line of Balance (LOB) technique. Moreover, a sustainable cost estimation model was developed during current state and future state to calculate the cost of underground pipeline construction. The result shows a cost reduction of 20.8% between current and future states. This reflects the importance of the cost based Value Stream Mapping in construction as a sustainable measurement tool. This new tool could be utilized in construction industry to add the sustainability and effective cost management.

Keywords: Value Stream Mapping (VSM); Line of Balance (LOB); sustainable construction; underground pipeline project

---

1. Introduction

The construction industry is increasingly moving towards the adoption of sustainable strategies and increased efficiency targets [1]. Value Stream Mapping (VSM) is a new phrase that originates from Toyota's material and information flow diagrams and was designed to help Toyota's suppliers learn the Toyota Production System [2]. VSM is a lean management system to assess the current state and a designed future state for the series of activities from beginning to hand-over. VSM is not just only a tool and limited to identify wastes in a system, but it is used to analyze and assist in designing processes, tracing material flow, and documenting information flow of a given product family. VSM adapts symbols to represent a clear and visual process from the customer's requirements to the final accomplishment.

Despite the efforts for sustainability studies in building and infrastructure construction, the sustainability issues in industrial construction remain understudied [3]. The application of VSM in real construction industry has not received enough attention by researchers due to the difficulty in implementation of VSM in a real construction activity. In this paper, construction of an underground pipeline project was practically assessed by VSM as a sustainable construction method. In this context, cost based Value Stream Mapping is developed as a measurement tool to build an optimized future state with less cost. Cost based Value Stream Mapping considers the customer value as the guide which basically eliminates the time wastes to get more production with high quality. Thus, the integration of VSM and cost methods were used together to develop a systematic framework. The contribution of this paper to existing knowledge can be stated as introduction of a cost based VSM methodology in underground pipeline projects. The cost based VSM technique was utilized on a real case study to evaluate the difference between traditional cost and VSM improved cost. This paper attempts to identify, develop and apply a concept of cost based VSM in construction sector, which can serve as a new way forward for future cost estimation.

2. Literature Review

2.1. Value Stream Mapping Principle

Value Stream Mapping (VSM) technique was introduced in second half of twentieth century by Toyota Motor Company which became one of the important lean management methods. Value Stream Mapping (VSM) has become a popular method for lean thinking and implementation in recent years [4]. Basically, it is graphically presentation production processes starting by using a raw material as an input to deliver the customer's product as an output. Martin et al. [5] defined VSM as "the sequence of activities an organization undertakes to deliver on a customer request".

The VSM is strictly divided into three types of operations that are undertaken: (1) Non-value adding; (2) Necessary but non-value adding; and (3) Value-adding. The first is pure waste with unnecessary actions that should be completely eliminated. The second one involves actions that are necessary but might be wasteful. VSM was not limited only to identifying waste in a system, but could also be used to analyze and assist in designing processes, tracing material flow, and documenting information flow of a given product or product family [2]. Wastes originates mostly from schedule deviation and can be generated in stationary or nonstationary productions, Figure 1 illustrates the type of wastes and provides example in each case.

VSM uses symbols to represent a clear and visual process from the customer's requirement to the final accomplishment. Drawing a value stream map is the result of implementing a VSM tool.

In summary, VSM provides a visual, full-cycle macro view of how work progresses from a customer request to the final fulfilment of that request. The mapping process deepens the understanding of work systems that deliver value to customers and reflect the work flow from a customer's perspective. As a result, the process of Value Stream Mapping provides effective ways to establish strategic directions for better decision making and work design.

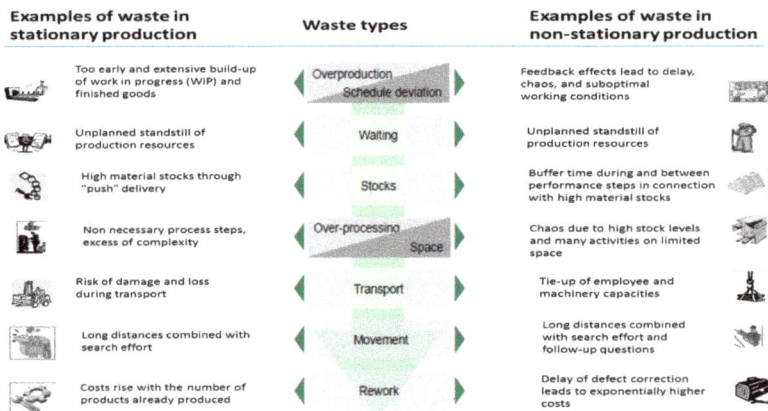

Examples of waste in stationary production	Waste types	Examples of waste in non-stationary production
Too early and extensive build-up of work in progress (WiP) and finished goods	Overproduction / Schedule deviation	Feedback effects lead to delay, chaos, and suboptimal working conditions
Unplanned standstill of production resources	Waiting	Unplanned standstill of production resources
High material stocks through "push" delivery	Stocks	Buffer time during and between performance steps in connection with high material stocks
Non necessary process steps, excess of complexity	Over-processing / Space	Chaos due to high stock levels and many activities on limited space
Risk of damage and loss during transport	Transport	Tie-up of employee and machinery capacities
Long distances combined with search effort	Movement	Long distances combined with search effort and follow-up questions
Costs rise with the number of products already produced	Rework	Delay of defect correction leads to exponentially higher costs

Figure 1. Classification of wastes in stationary and non-stationary productions.

Schmidtke et al. [6] developed an enhanced VSM method, which utilizes discrete event simulation (DES). The method featured a feasibility and trade-off analysis which is incorporated into the VSM procedure. Sabaghi et al. [7] focused on three lean manufacturing techniques, which are Kanban production system, setup time reduction, and total productive maintenance (TPM) in a plastic fabrication industry. Suarez-Barraza et al. [8] describe the implementation of a tool called Supply Chain Value Stream Mapping (SCVSM) in order to thoroughly understand competitive priorities of volume and delivery (On-time Delivery (OTD)) for any supply chain in organizations. Throughout the case study by Yuvamitra et al. [9], implementing changes in both the information flow system and the material flow system would save an estimated 75% for the manufacturing time of the rope.

2.2. Value Stream Mapping and Sustainability in Construction

Value Stream Mapping has been used in many industries. Tyagi et al. [10] used Value Stream Mapping (VSM) to explore the wastes, inefficiencies, non-valued added steps in a single, definable process out of complete product development process. Wang et al. [11] proposed a supplier selection framework tailored to effective information integration for supply chain management. Jeong et al. [12] examined critical factors affecting productivity at the operational level, and then forecast the productivity dynamics. By integrating BIM with construction operation simulations, they were able to develop reliable construction plans that adapted to project changes. Lee et al. [13] developed a hybrid estimation model to predict the quantity and cost of waste in the early stage of construction. Jia et al. [14] proposed a Therblig-embedded Value Stream Mapping (TVSM) method to improve the energy wastes.

Construction projects are complex systems that inherently contain complex interface problems [15]. In spite of being successfully applied into different industries, the application of Value Stream Mapping in construction is limited. There are only few studies on VSM in construction, and they refer more to construction supplies rather than to the process itself. Pasqualini et al. [16] described the modified application of VSM in a Brazilian construction company; some adaptations were made in each stage of VSM. Another example is a case study of housing construction. Yu et al. [17] worked on a case study of the standard wood platform-frame structure, the case of 400 houses construction was regarded as repetitive works. These two cases demonstrate that VSM is a tool used to identify the sources of environmental and production waste, quantify them, and suggest reduction strategies.

Shou et al. [18] presented a literature review of the Critical Success Factors in the implementation of VSM across five different sectors: manufacturing, healthcare, construction, product development, and service. The review covers the peer-reviewed journal articles on VSM in Scopus from 1999 to 2015. The findings of this study provide a good basis for industry practitioners to effectively implement VSM in construction industry.

Pasqualini et al. [19] presented a methodology for modeling multi-product manufacturing systems with dynamic material, energy and information flows with the aim to generate economic and environmental value stream maps (E^2VSM). The proposed methodology is validated with an industrial case not in construction industries.

Dinesh et al. [20] described in a case study a production of a heavy-duty electrical transformer with some approximations and simplifications with a VSM application.

Rosenbaum et al. [21] utilized a case study of the VSM application as a green-lean approach in the construction of a hospital in order to improve its environmental and production performance during the structural concrete work stage. The paper presented the implementation process of the value stream for future state by aggregate tools to enhance productivity improvement.

Mok et al. [22] showed that VSM and simulation tools when combined together form a strong integration tool for improving the work productivity. This paper deals and covers complex environments of projects such as introduction of pipe spools (pipe with elbows or tees or flanges, etc.), which was not covered in Reference [22]. Moreover, this paper differs from this study by evaluating the effectiveness of integration of VSM, Line of Balance (LOB) and cost based VSM as a tool to improve the value added. After the assessment of the current state, the future state of VSM is attempted by using LOB for resource levelling which has a major contribution in the cost reduction. The application of LOB technique and cost based VSM are totally new contribution to the literature in these types of projects. In this paper, the application of VSM tools in the complex environment conclude that the micro-concepts, takt time, supermarket to facilitate pull, continuous flow and space maker was properly utilized to facilitate VSM applications, one often requires adjustments, averaging and approximation in terms data collection, linearity in flow and timings. These new tools are contribution to current body of knowledge. This paper developed a better understanding of VSM and how it can be modified, with LOB method and how the manpower aspect is considered in this research. In the contrary, the manpower aspect was not considered in Reference [22]. Moreover, it introduced a developed concept to calculate the cost of Value Stream Mapping on a weekly basis. This technique can be utilized to understand the construction progress status and the future expectation. Moreover, it can be used to calculate the cost of similar projects for tendering purposes.

The above thorough literature in construction shows the importance of implementation of VSM and sustainability in construction as a lean management tool to reduce the time wastes and increase the value added in the processes. As currently understood, VSM needs to be linked with the cost to a better optimal production rate. This study will focus on this perspective.

2.3. Line of Balance in Lean Construction

The literature reviewed papers emphasize on VSM philosophy, identifying activities that develop value and in line with concept of reduction of time waste. Therefore, LOB is studied in order to identify its potential relationship as a resource levelling tool to the principles of VSM in construction. According to Pinheiro et al. (2009), repeating units of work might cause learning effects, what leads to the reduction in activities duration. The smaller the repeating unit, the greater the learning effect, what reduces also global lead time.

This study utilized LOB tool as a planning device that gives support to the VSM concept by proper utilizing the resources to calculate the cost based VSM.

3. Methodology

There is limited research that utilized a VSM to calculate the cost for current and future states in the construction industry. Such studies were concentrated and focused on manufacturing domain only. In this study, the concept of cost reduction was addressed in construction by using a Value Stream Mapping. More specifically, the Value Stream Mapping was applied on a real underground pipeline project. The construction processes were optimized based on the cost according to Value Stream Mapping technique. This was achieved by comparing a cost of current state with a cost of future state, which attempts to evaluate the objective of Value Stream Mapping in construction. The same can be set as a model for proper utilization and estimation to similar repetitive projects in future. Cost items include labor cost, materials cost, production support cost, equipment cost, operation support cost, facilities and maintenance cost, all other value stream cost, as shown in Figure 2.

The very first step in this study is to form a comprehensive current state for the construction process. This is a vital step as it lays the ground for a full understanding the processes in construction and find out the time wastes. By the Value Stream Mapping concept, the wastes were measured and identified in the current state. Then, the future state was formulated accordingly to improve the productivity and reduce the time wastes. Work plan and required manpower was set according to feedback from current state to optimize the process and ensure wastes are minimized in combination with Line of Balance technique (LOB).

Figure 2. The elements for cost based Value Stream Mapping (VSM).

The basic concepts of LOB have been applied in the construction industry as a resource driven scheduling method. Line of Balance (LOB) is a method of showing the repetitive work that may exist in a project as a single line on a graph. Recently LOB has been associated with lean construction applications, especially as a tool for tactical planning of works. In this study, LOB will be introduced along with the VSM to reduce the lead time. The smaller the repeating unit, the greater the learning effect, what reduces also global lead time as shown in Figure 3.

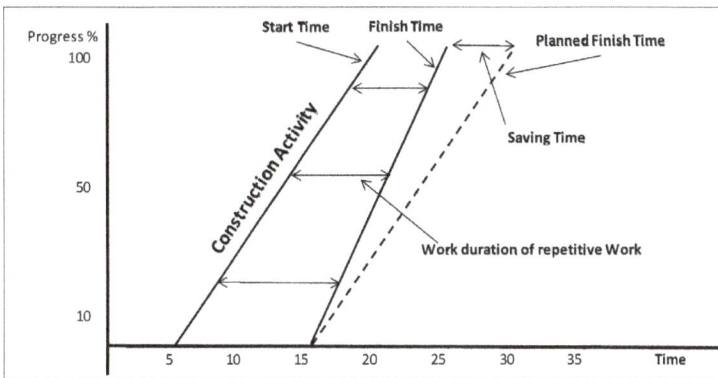

Figure 3. Improvement of the total lead time for repetitive work.

The cost based of Value Stream Mapping is calculated weekly and it takes into account of all costs related to value stream. The cost calculation was done for current and future states and then a comparison between them has been evaluated for estimation and bidding purpose for similar future projects.

4. Data Characteristics

In this study, an underground pipeline project was selected to apply VSM. The main purpose of selection is the effectiveness in data collection and repetitiveness of the activities. As per project contract, the scope of work was construction of a pipeline (36-inch diameter) with a network of 36.5 km. The first step in the Value Stream Mapping technique is to develop a process flow diagram to provide a clear picture about every stage for the construction of underground pipeline. Figure 4 shows the work flow diagram for the Value Stream Mapping.

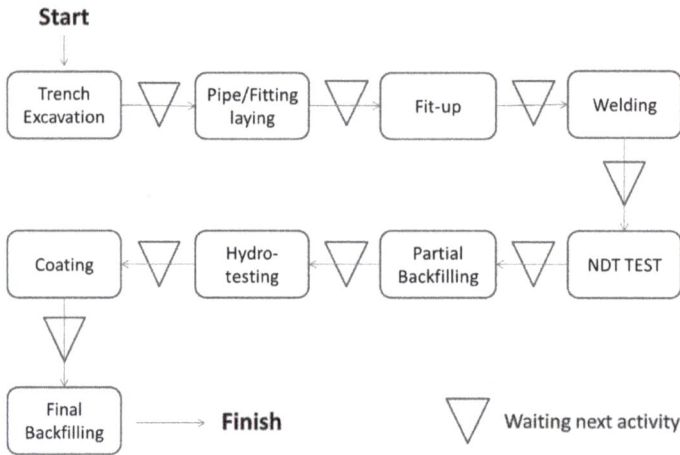

Figure 4. Work flow diagram.

Monitoring and recording with stop-watch was conducted to record the actual cycle time for each construction stage accurately as presented in the below Table 1.

Table 1. Cycle time for each construction stage.

Construction Activity	Cycle Time (Minutes)
Trench Excavation	193.0
Laying Pipes & Fittings	130.0
Fit-up	128.0
Welding	600.0
NDT Test	320.0
Partial Backfilling	48.0
Hydro-testing	181.0
Painting	179.0
Final Backfilling	601.0

With collected data, the Value Stream Mapping for construction of six pipes can be built based on VSM basic concept. The Value Stream Mapping is the most efficient and functional tool which furnishes the current state of the construction stages to find out the wastes. It envisions the value-added time (VAT) and Non-value-added time (NVAT) for each construction stage and introduces the actual stage for the entire process to evaluate the current total lead time. From the collected data, the following is the current state of construction of pipeline as shown in Figure 5. Meaning of Samples and Icons for current and future states of Value Stream Mapping is available in Appendix A.

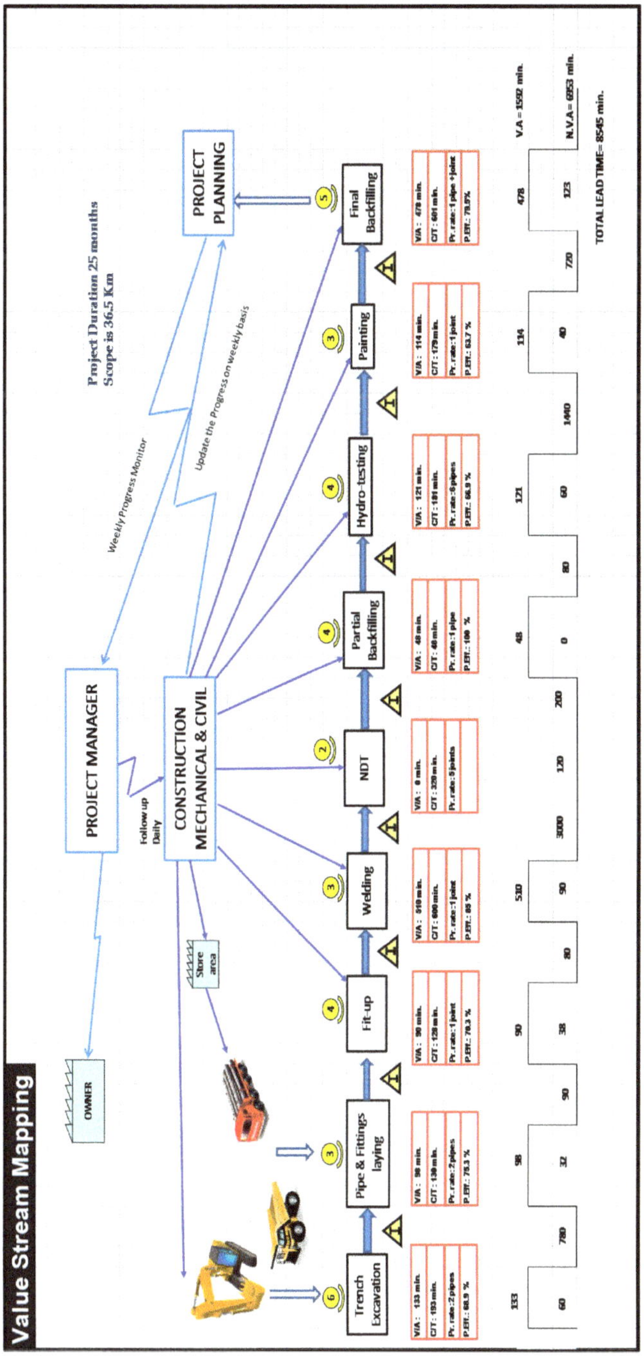

Figure 5. Current state of VSM for construction of underground pipeline.

5. Data Analysis for Current State of VSM

The duration was observed for each individual activity. The actual time spent during physical and productive construction was recorded as VAT and the time consumed during non-productive construction such as waiting, idle, rest and standby times were recorded as NVAT.

The percentage of VAT to the total lead time was 18.6%. The waste percentage is quite high compared to the total lead time in the construction process. Therefore, the total lead time has to be reduced by enhancing the progress and eliminating the non-value-added activity.

The customer would specify the quantity of the required products on daily, weekly, monthly or yearly basis based on the placed purchase order. The takt time is defined as the rate at which a pipe joint must be completed to complete the construction of underground pipeline to satisfy the client requirements.

This difference implied in a new form of calculating takt time for construction, obtained by means of the division of the effectively available time for each stage by amount of joint to be executed. As result, takt time of the construction will indicate the time in which one pipe joint should be completed, stipulated in the contract planning base line.

For aforementioned case study at the beginning, the project scope is to install 36,500 m of underground pipeline; the pipe size is 36-inch in diameter and shall be completed within 25 months. Based on the above information and as contingency plan, an internal schedule has been built and the project shall be completed with 23 months consequently. According to project scope 3343 joints must to be completed.

$$\text{Targeted productivity rate} = \frac{\text{Number of joints}}{\text{Available time}}$$

$$\text{Available time} = 23 \text{ months} \times 26 \text{ days} = 598 \text{ days.}$$

$$\text{Number of joints} = 3343$$

$$\text{Targeted productivity rate} = \frac{3343}{598} = 5.6 \text{ joint/day.}$$

As a result, the takt time in construction is defined by a required time to produce one joint as it is stipulated in the contract.

$$\text{Takt time} = \frac{\text{Available working time per day}}{\text{productivity rate per day}} = \frac{600 \text{ min}}{5.6} = 107 \text{ min.}$$

(This is the takt time required to complete installation of one pipe and one joint.)

The cycle time for all entire construction activities is higher than the takt time. The welding and final backfilling processes are the bottleneck of the entire process; this means there will be waiting time to complete these activities and expecting idle manpower. Thus, the construction activities have to be synchronized to achieve the rhythm of takt time and complete within contract duration.

5.1. Cost Based Current State of VSM

The value stream cost is basically calculated weekly and takes into account of all costs in the VSM. As mentioned previously all costs related to VSM is considered as a direct cost. Any cost outside the Value Stream Mapping is not included in the costing of Value Stream Mapping. Weekly cost of VSM is listed in detail in Table 2. The total cost is $19,477 which will be compared with a cost of future state Value Stream Mapping later to investigate and evaluate the difference between these two costs.

Table 2. Cost of current state activities.

Current State of VSM	Employee Cost ($)	Production Support Cost ($)	Maintenance Cost ($)	Machine/Equipment Cost ($)	Other Costs ($)	Total Cost ($)
Trench Excavation	1306	250	550	1096		3202
Pipe & fittings laying	583	286	389	603		1861
Fit-up	755	250				1005
Welding	617	500	150	274	300	1841
NDT		20		350		370
Partial Backfilling	1062		150	822		2034
Hydro-testing	754	145	130	250		1279
Painting	617	250	95			962
Final Backfilling	822	250	255	1096		2423
	4500 (Construction Managers and Site Engineers)					4500
Total						*19,477*

Note: Employee cost: is the basic salary including the employee benefit. Production support cost: Expenses not directly associated with construction activities such as design, engineering, and procurement. Maintenance cost: Cost to keep equipment or machine in good working condition. Machine/Equipment cost: In this research, rental and operating cost.

5.2. Suggested Improvement and Future State of VSM

The following suggested improvements were developed according to the construction status in guidance with the implementation of the modified basic concept of Value Stream Mapping.

(a)   Pull system and synchronize first-in/first-out flow:

Trench excavation, partial and final backfilling: the process cycle time for each activity is higher than takt time. The required improvement is to reduce the non-value added especially the waiting time which is required to prepare the trench, bedding and inspection prior to place and laying the pipes. The excavation is the first construction activity; thus, it would be better to assign an excavation team to start the excavation activity one hour in advance which would provide enough time for the next activity to start earlier. Moreover, it was proposed to introduce a continuous flow and develop an open work front by eliminating the partial backfilling activity. The agreement between the main contractor and consultant was developed to proceed with painting and coating after completion of NDT (Non-destructive test). Then, the pipeline will be backfilled finally to be ready for final testing. In Value Stream Mapping, it is called first-in/first-out or pull system, where a downstream pulls the upstream activity to improve the work flow and improve the utilization of manpower and equipment. Thus, 24 hours was given to backfilling team to complete the final backfilling after completion of five joints. The application of continuous flow, the non-value-added time for trench excavation, partial backfilling and final backfilling has been reduced.

(b)   Combined and restructured the work activities (fit-up and welding activities):

The time required to complete the installation of one joint is 107 min. Welding activity has high cycle time, therefore and in coordination with the engineering team, a technique called Weld Map Drawing was initiated. Basically, welding map is an isometric drawing that shows the location for all pipe joints in the project, as shown in Figure 6.

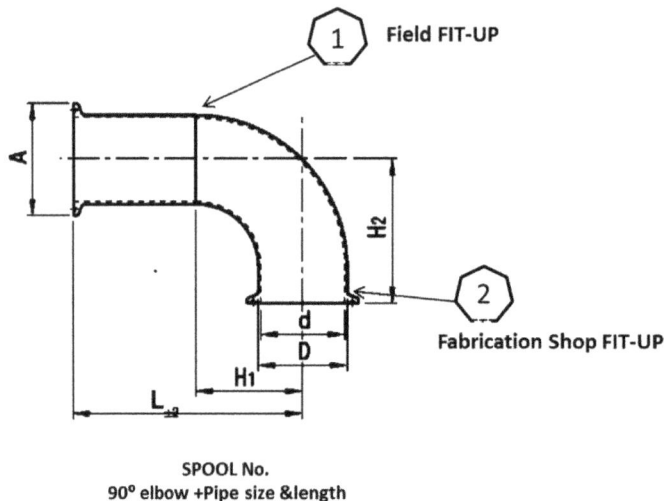

Figure 6. Sample of weld map drawing.

Normally, the field engineering section in coordination with construction field would study and investigate thoroughly in detail each joint and specify the joint status with respect to field weld or manufacturer joint weld named as spools. After finalizing the list, the spools would be submitted by construction team to manufacturer, this would allow construction and spool manufacturer to track and expedite the spool fabrication till completion date. By weld map drawing and spool concept, the welding cycle and lead times will be reduced accordingly. To speed up the welding process, the construction team may ask the engineering to generate a list of tack welding spools.

In this study, the joints that can be fabricated in the shop by manufacturer are 1171 joints and the construction team suggested to reduce the idle time of the workers by merging the group of fit-up and welding teams for identical tasks. The pipefitters were subject to tack welding training for two weeks period which enhanced the productivity of welding process. The implication for the above strategy was significant. The observation for welding activity cycle time is reduced and total lead time is reduced as well.

(c)   Production leveling by pacemaker/supermarket:

Welding and NDT activities: The concept of Supermarket Pull System between welding process and NDT process was introduced. Basically, the supermarket pull system is a controlled inventory of joints that is subject to testing schedule in due course according to site situation. Therefore, the joints were accumulated to perform NDT test in one shot. Moreover, the NDT was schedule to be during the night to perform more tests without any interruption. According to project baseline.

Schedule, the takt time was schedule to produce 5 joints per day to meet the project completion date. The completion date was scheduled two months ahead, as a safety factor. The proposal was introduced to prepare 5 joints for test which facilitate an open front for other activities and optimize the resources effectively.

The concept of pacemaker loop encompasses the flow of information and material between the mainly construction and testing. The downstream loop is impacted by the upstream loop and this can't be done without proper scheduling and continuous development.

(d)   Work restructuring to improve the construction reliability:

Hydro-testing was scheduled after completing the NDT test, painting and final backfilling, which was in coordination and approval of project resident engineer. To ensure a proper stability and anchorage for the network integrity, the final backfilling has taken place prior to proceed with hydro-test. This was suggested to apply the concept of Supermarket Pull system and to ensure a continuous flow. This can be achieved by batch of Kanban (It is a Japanese term that gives authorization and instructions for the production or withdrawal conveyance of items in a pull system) by preparing a minimum of 5 completed welded joints for NDT test and build a ready pipeline for hydro-test according to site condition. By this improvement hydro-testing activity were improved significantly.

(e)   Painting:

The cycle time of the painting process is higher than takt time, thus, the wastes and waiting time has to be minimized. Due to the super market concept which was introduced and developed between welding and NDT activities, the continuous flow is generated to catch-up the quantity of joints generated after hydro-testing.

(f)   Manpower leveling and LOB (Line of Balance):

Integration of LOB with VSM played a vital role to level the required manpower. In this study, the resource leveling was achieved by identifying the number of crews required to complete the project on the time. The study used velocity rating diagram to find required resources for each stage so each work stage can be done by synchronized crew and continue without interruption. According to LOB technique, the following information was applied:

-       Vertical axis plots cumulative progress of number of joints completed in the project. Horizontal axis plots time and sloping lines represent rate of production i.e., number of joints per day.

- To complete the project and minimize the waste and lead time, the crews are distributed as following:

  (1) Two crews for excavation and final backfilling. Each crew consists three civil workers under one supervisor.
  (2) One crew for pipes and fittings laying (three pipe fitters)
  (3) One crew for fit-up and welding (four welders and three helpers)
  (4) One crew for NDT testing (two NDT technicians)
  (5) One crew for painting (three painters)
  (6) One crew for hydro-testing (four mechanical/pipe fitters).

The repetitive activities are distributed uniformly to ensure a proper utilization of manpower and tools, as in Figure 7.

5.3. Proposed Future State of VSM

The previously mentioned improvements were implemented in the current state to achieve the future state as illustrated below in the Figure 8.

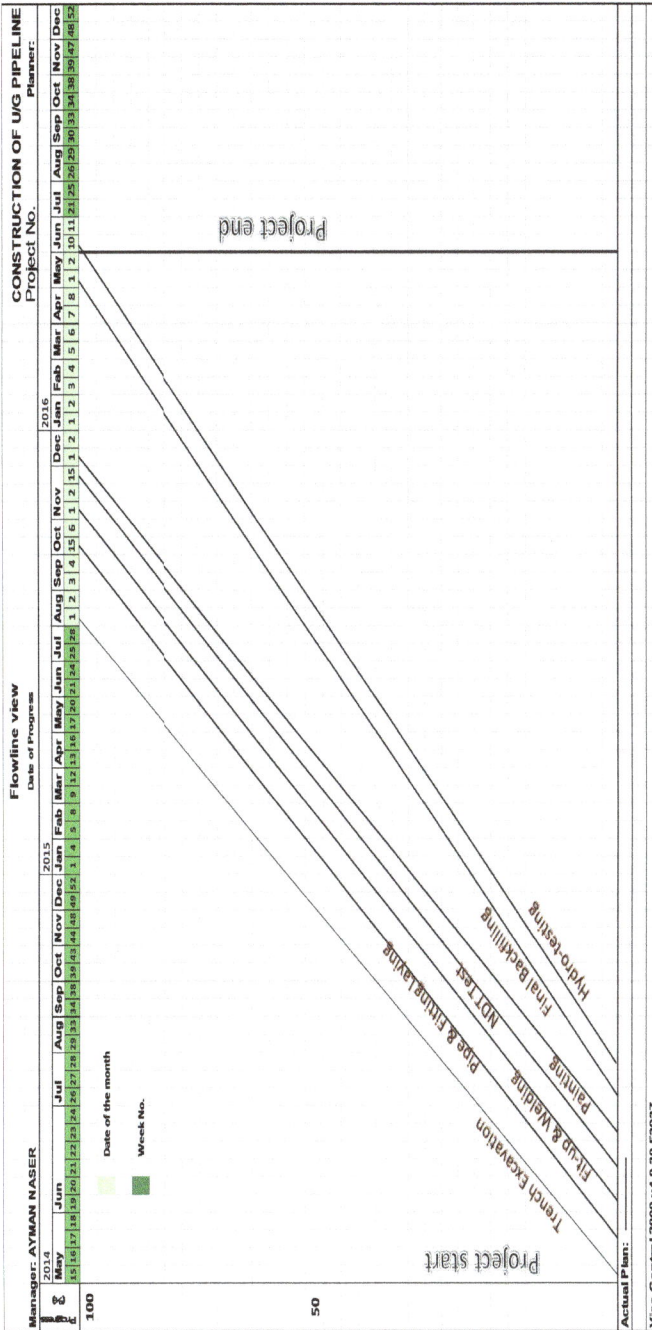

Figure 7. Line of Balance (LOB) for resource levelling.

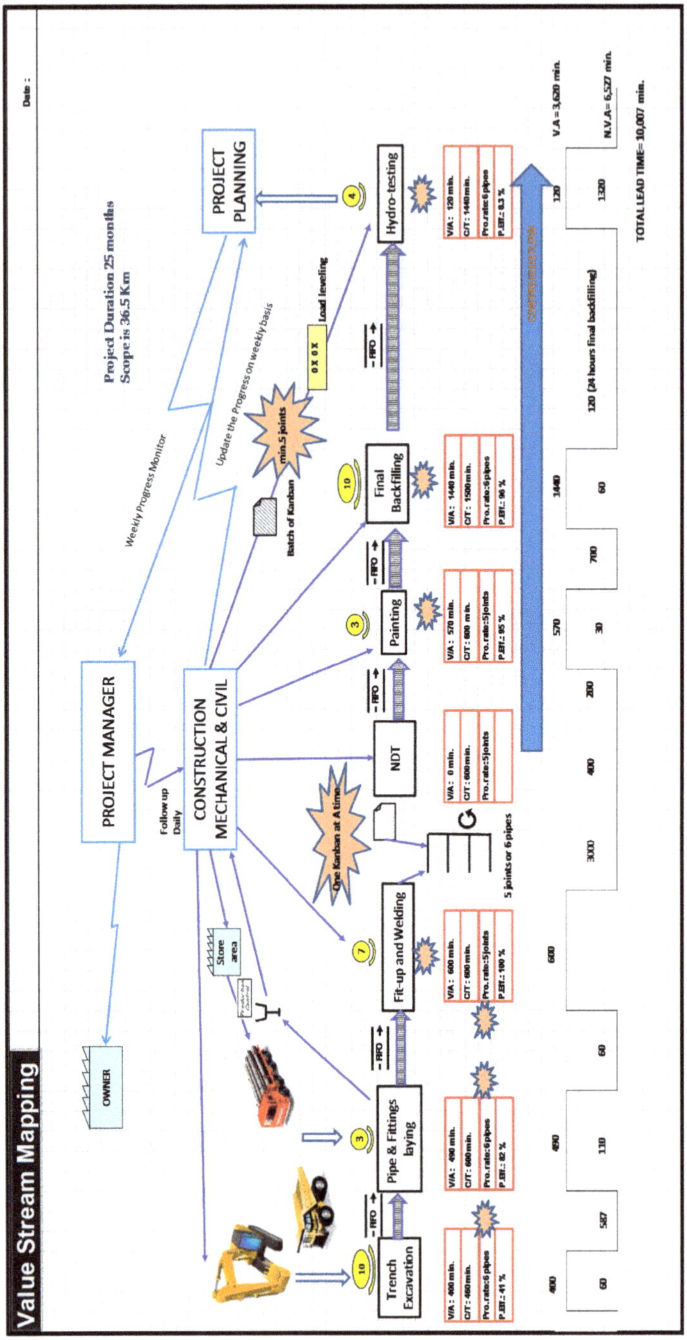

Figure 8. Future state of VSM for construction of underground pipeline.

Future state of VSM shows a significant improvement in the construction process. It has been observed with a reduction of total lead time, reduction in the time wastes (NVA) and increase in value added time (VA). Table 3 shows the achieved improvements:

- The lead time was calculated in current state 8545 min based on construction of one joint. While, the total lead time in the future state is 5922 min. There is an improvement of 30.7%.
- The value-added time in current state is 1592 min, and it is improved by 43.3% as 903 min.
- With application of VSM, the non-value-added time is reduced by 27.8%. The NVA was reduced from 5019 min to 6953 min.
- The comparison of production capacity for current and future states is based on one joint and one pipe for current and future states of VSM.

Table 3. Production capacity comparison between current and future states of VSM.

Construction Activity	Current State	Future State	Improvement
Lead Time (min)	8545	5922	30.7%
Value Added (min)	1592	903	43.3%
Non-Value added (min)	6953	5019	27.8%
Production capacity	1 joint/1 pipe	1 joint/1 pipe	

As a result, future state map has to be validated on continuous improvement basis. Takt time for the construction can be calculated based on the customer's requirement and according to signed contract between the parties. Thus, each process possesses different takt time to be executed and calculated accordingly. Therefore, takt time shall be developed by establishing a pull system, introduce a levelling system, and establish a pacemaker loop for the future state of VSM. The takt time for supply and pacemaker loops was developed to meet the contract completion date. The time for supply loop is 2307 min and the required time to complete the pacemaker loop is 4700 min with a control inventory of supermarket time of 3000 min. The construction capacity was adjusted to construct five joints and six pipes. Thus, the takt time for each process loop was balanced to meet the project completion date accordingly.

5.4. Cost Based Future State of VSM

Cost of future state was calculated to evaluate the importance of total cost reduction by Value Stream Mapping. After suggestion improvements and the formulation of future state of Value Stream Mapping, the cost has been calculated and summarized in Table 4 of the weekly period as previously implemented in the current state.

In comparison between the cost based for current and future states of Value Stream Mapping, the reduction in cost is 20.8%. It shows the importance of utilizing the Value Stream Mapping in construction as a management tool. Moreover, this technique provides an accurate cost calculation method that can be used for tendering purposes.

Table 4. Cost of Future state activities.

Future State of VSM	Employee Cost ($)	Production Support Cost ($)	Maintenance Cost ($)	Machine/ Equipment Cost ($)	Other Costs ($)	Total Cost ($)
Trench Excavation	1424.7	250	550	310.5		2535.2
Pipe & fittings laying	602.7	286	389	125.4		1403.1
Fit-up						
Welding	1095.9	500	150	82.2	300	2128.1
NDT		20		350		370
Partial Backfilling						
Hydro-testing	754	145	130	250		1279
Painting	617	250	95			962
Final Backfilling	1424.7	250	255	310.5		2240.2
	4500 (Construction Managers and Site Engineers)					4500
Total						*15,417.6*

6. Discussion of Results

The objective of this paper is to enhance and develop the utilization of VSM as a lean management tool in construction industry and emphasis the VSM benefits by adapting of a cost based VSM.

The paper deals with the possibility of implementing the design method of Value Stream Mapping (VSM) supported by Line of Balance (LOB) on a construction project focused on the installation of underground pipelines with a perspective of cost. More specifically, the cost implementation of the proposed method enables to reduce the project by comparing its current and future costs and improving its productivity.

Alvandi et al. [19] validated an industrial manufacturing case study to present a methodology for modeling an environmental Value Stream Maps without consideration of cost and resource levelling.

On other hand, Dinesh et al. [20] utilized a manufacturing plant with some approximations and simplifications in VSM application. While in this paper, construction environments are systematically studied for the manpower and cost with Value Stream Mapping as a lean tool.

The research by Jeong et al. [12] presented a Building Information Modeling (BIM)-integrated simulation framework for predicting productivity dynamics at the construction planning phase only. While in this paper, the concept is applicable for all phases of planning, bidding and execution.

Mok et al. [22] represent a flow diagram to depict the VSM tools. While in this paper, the application of VSM tools in the complex environment such as the mico-concepts, takt time, supermarket to facilitate pull, continuous flow and spacemaker were properly utilized to facilitate VSM applications. In addition, it evaluates the effectiveness for integration of VSM, LOB and cost as a tool to improve the value added and reduce the time wastes in construction activities.

The proposed method in this research introduces a significant contribution deals with resource levelling and cost estimation can be utilized in the entire project phases. This research has focused on developing a stable production flow rather than eliminating individual waste. VSM, a powerful lean construction tool, was used to analyze the construction process and restructure the production system.

Several papers described the achieving constant process improvement in manufacturing/plant production is by Value Stream Mapping only. By being the part of construction industry with the ability to bring in cost and productivity advantages, a Value Stream Mapping stands as a more attractive alternative when compared with the other conventional management tools. To capitalize on such benefits, the followings were observed as remarkable improvements:

After analysing the application of Value Stream Mapping in construction industry and demonstrate the physical application by using a real case study, the total VAT was improved by 43.3% with respect to cycle time While the NVAT was reduced by 27.8%.

After introduction of the LOB with VSM, the total lead time was reduced by 30.7% and required manpower was reduced by 12.5%. The major contribution of integration of LOB tool with VSM was to optimize the resources utilization and reduce the total lead time.

The cost is reduced by 20.8% in comparison between two VSM states.

7. Conclusions and Recommendations

This paper developed a better understanding of VSM and how it can be modified, with LOB to be applied in the construction industry. Moreover, it is introduced a developed concept to calculate the cost of Value Stream Mapping on a weekly basis. This technique can be utilized to understand the construction progress status and the future expectation. Moreover, it can be used to calculate the cost of similar projects for tendering purposes.

This paper attempts to address the application of Value Stream Mapping to a real-life problem of an underground pipeline project. It analyzed the improvement of VAT and reduction of NVAT compared with the cycle time. It studied the reduction opportunity in total lead time. LOB tool with VSM as a new tool provided a powerful attempt for proper resources utilization and monitoring to improve the VAT and decrease the total lead time.

Although several papers addressed using VSM as a lean tool to reduce waste and add value during the construction process, no detailed and unified VSM instructions exist concerning how to implement it in construction to evaluate the improvements and calculate the cost reduction. This paper tried to cover this gap. Suggested improvements will result in achieving the production and capacity demand to meet project completion deadline. Future research can be extended and focused further on the following: Applying Value Stream Mapping for other repetitive and non-repetitive construction activities; integrating and combining VSM with other tactics and lean management tools to improve the benefits of Value Stream Mapping in the construction industry; incorporating and evaluating the cost based VSM for bidding and estimation purpose; introduction of a simulation based VSM software that can be used to study the construction time and cost in advance for future construction.

This study has some limitations as it deals with a specific construction company and different environments, such as—but not limited to—a project, owner, project specifications, main contractor, operations and other environment-based constraints.

Another limitation is the process of integration the cost-based of Value Stream Mapping concept in the budget and estimation calculation as a vital tool in the construction companies for biding purpose. This can be developed for future studies.

Moreover, no simulation modeling was used to develop different scenario requirements to compare current state with different scenarios for future state and optimize the cost. This can be expanded for future research.

To extend the body of knowledge, the paper findings can be attempted for other areas such as construction of high-rise building. Similarly, the construction challenges for green building can also attended under the umbrella of Value Stream Mapping.

Author Contributions: Both authors were involved in the development of this paper. Murat Gunduz and Ayman Fahmi Naser organized the paper together. Both authors worked together during the development of graduate study outputs of Ayman Fahmi Naser into this paper.

Conflicts of Interest: The authors declare no conflict of interest.

Appendix A

Table A1. CValue Stream Mapping Symbols & Icons.

Symbol/Icon	Meaning and Description
	Customer/Supplier Icon: represents the Supplier when in the upper left, customer when in the upper right, the usual end point for material
	Manpower Icon: represents skilled manpower. Shows the number of manpower required to process the VSM family at a particular workstation
Activity Name	Dedicated Activity flow Icon: a process, operation, machine or department, through which material flows. Represents one department with a continuous, internal fixed flow.
V/A : Value Added time in minutes / C/T : Cycle time in munites / Pr. rate: Production rate per pipe / P.Eff.: Production Efficiency (%)	Data Box Icon: it goes under other icons that have significant information/data required for analysing and observing the system.
VA: Value Added in minutes / NVA: Non-Value Added in minutes / Waiting time in minutes	Timeline Icon: shows value added times (Cycle Times) and non-value added (wait) times. Use this to calculate Lead Time and Total Cycle Time.
	Inventory Icons: show inventory and waiting time between two activities/processes.
	Electronic Info Icon: represents electronic flow such as electronic data interchange, the Internet, indicate the frequency of information/data interchange.
	Manual Information Icon: A straight, thin arrow shows general flow of information from memos, reports, or conversation. Frequency and other notes may be relevant.
FIFO	FIFO Lane Icon: First-In-First-Out inventory. Use this icon when processes are connected with a FIFO system that limits input.

Table A1. *Cont.*

Symbol/Icon	Meaning and Description
	Push Arrow Icon: represents the "pushing" of material from one activity to the next activity.
	Kaizen Burst Icon: used to highlight improvement needs and plan kaizen workshops at specific activities that are critical to achieving the Future State Map of the value stream.
Production Control	Kanban Post Icon: a location where kanban signals reside for pickup. Often used by a central production scheduling or control production department.
	Withdrawal Kanban Icon: represents a device that instructs a material handler to transfer parts from a supermarket to the receiving process. He goes to the supermarket and withdraws the necessary items.
	Supermarket Icon: an inventory "supermarket" (kanban stockpoint) with a "Pull" icon that indicates physical removal.
	Production Kanban Icon: triggers production of a predefined number of parts. Signals a supplying process to provide parts to a downstream process.
0X 0X	Load Leveling Icon: a tool to batch kanbans in order to level the production volume and mix over a period of time.

References

1. Carvalho, A.C.V.; Granja, A.D.; Silva, V.G. A Systematic Literature Review on Integrative Lean and Sustainability Synergies over a Building's Lifecycle. *Sustainability* **2017**, *9*, 1156. [CrossRef]
2. Rother, M.; Shook, J.; Womack, J.; Jones, D. *Learning to See: Value Stream Mapping to Add Value and Eliminate MUDA*, 3rd ed.; Lean Enterprise Institute: Cambridge, MA, USA, 2003.
3. Yun, S.; Jung, W. Benchmarking Sustainability Practices Use throughout Industrial Construction Project Delivery. *Sustainability* **2017**, *9*, 1007. [CrossRef]
4. Wenchi, S.; Jun, W.; Peng, W.; Xiangyu, W.; Heap-Yih, C. A cross-sector review on the use of Value Stream Mapping. *Int. J. Prod. Res.* **2017**, *55*. [CrossRef]
5. Martin, K.; Osterling, M. *Value Stream Mapping: How to Visualize Work and Align Leadership for Organizational Transformation*; McGraw-Hill: New York, NY, USA, 2013.
6. Schmidtke, D.; Heiser, U.; Hinrichsen, O. A Simulation-enhanced Value Stream Mapping Approach for Optimisation of Complex Production Environments. *Int. J. Prod. Res.* **2014**, *52*, 6146–6160. [CrossRef]
7. Sabaghi, M.; Rostamzadeh, R.; Mascle, C. Kanban and value stream mapping analysis in lean manufacturing philosophy via simulation: A plastic fabrication (case study). *Int. J. Serv. Oper. Manag.* **2015**, *20*, 118–140. [CrossRef]
8. Suarez-Barraza, M.F.; Miguel-Davila, J.; Vasquez-García, C.F. Supply chain value stream mapping: A new tool of operation management. *Int. J. Q. Reliab. Manag.* **2016**, *33*, 518–534. [CrossRef]
9. Yuvamitra, K.; Lee, J.; Dong, K. Value Stream Mapping of Rope Manufacturing: A Case Study. *Int. J. Sustain. Eng.* **2017**, *2017*. [CrossRef]
10. Tyagi, S.; Choudhary, A.; Cai, X.; Yang, K. Value Stream Mapping to reduce the lead-time of a product development process. *Int. J. Prod. Econ.* **2015**, *160*, 202–212. [CrossRef]

Sustainability **2017**, *9*, 2184

11. Wang, T.; Zhang, Q.; Chong, H.; Wang, X. Integrated Supplier Selection Framework in a Resilient Construction Supply Chain: An Approach via Analytic Hierarchy Process (AHP) and Grey Relational Analysis (GRA). *Sustainability* **2017**, *9*, 289. [CrossRef]

12. Jeong, W.; Chang, S.; Son, J.; Yi, J. BIM-Integrated Construction Operation Simulation for Just-In-Time Production Management. *Sustainability* **2016**, *8*, 1106. [CrossRef]

13. Lee, D.; Kim, S.; Kim, S. Development of Hybrid Model for Estimating Construction Waste for Multifamily Residential Buildings Using Artificial Neural Networks and Ant Colony Optimization. *Sustainability* **2016**, *8*, 870. [CrossRef]

14. Jia, S.; Yuan, Q.; Lv, J.; Liu, Y.; Ren, D.; Zhang, Z. Therblig-embedded Value Stream Mapping method for lean energy machining. *Energy* **2017**, *138*, 1081–1098. [CrossRef]

15. Chien-Liang, L.; Chen-Huu, J. Exploring Interface Problems in Taiwan's Construction Projects Using Structural Equation Modeling. *Sustainability* **2017**, *9*, 822. [CrossRef]

16. Pasqualini, F.; Zawislak, P.A. Value Stream Mapping in construction: A case study in a Brazilian construction company. In Proceedings of the 13th Annual Conference of the International Group for Lean Construction, Sydney, Australia, 19–21 July 2005; pp. 117–125.

17. Yu, H.; Tweed, T.; Al-Hussein, M.; Nasseri, R. Development of lean model for house construction using Value Stream Mapping. *J. Constr. Eng. Manag.* **2009**, *135*, 782–790. [CrossRef]

18. Shou, W.; Wang, J.; Chong, H.Y.; Wang, X. Examining the Critical Success Factors in the Adoption of Value Stream Mapping. In Proceedings of the 24th Annual Conference of the International Group for Lean Construction (IGLC), Boston, MA, USA, 20–22 July 2016; pp. 93–102.

19. Alvandi, S.; Li, W.; Schönemann, M.; Kara, S.; Herrmann, C. Economic and Environmental Value Stream Map (E2VSM) Simulation for Multiproduct Manufacturing Systems. *Int. J. Sustain. Eng.* **2016**, *9*, 354–362. [CrossRef]

20. Dinesh, S.; Nitin, S.; Pratik, D. Application of Value Stream Mapping (VSM) for lean and cycle time reduction in complex production environments: A case study. *Prod. Plan. Control* **2017**, *28*, 398–419. [CrossRef]

21. Rosenbaum, S.; Toledo, M.; González, V. Improving environmental and production performance in construction projects using Value-Stream Mapping: Case study. *J. Constr. Manag.* **2014**, *140*. [CrossRef]

22. Mok, K.L.; Sin, E.; Han, S.H.; Jang, W. Value stream mapping and simulation as integrated lean approach tool for improving productivity in the installation of natural gas pipes. In Proceedings of the Construction, Building and Real Estate Research Conference of the Royal Institution of Chartered Surveyors (COBRA 2010), Paris, France, 2–3 September 2010.

![sustainability logo] *sustainability*

MDPI

Article

The Impact of Aircraft Noise on Housing Prices in Poznan

Radoslaw Trojanek [1,*], **Justyna Tanas** [2], **Saulius Raslanas** [3] **and Audrius Banaitis** [3]

1   Department of Microeconomics, Poznan University of Economics and Business, Al. Niepodleglosci 10, 61-875 Poznan, Poland
2   Department of Organization and Management Theory, Poznan University of Economics and Business, Al. Niepodleglosci 10, 61-875 Poznan, Poland; justyna.tanas@ue.poznan.pl
3   Department of Construction Management and Real Estate, Vilnius Gediminas Technical University, Sauletekio al. 11, LT-10223 Vilnius, Lithuania; saulius.raslanas@vgtu.lt (S.R.); audrius.banaitis@vgtu.lt (A.B.)
*   Correspondence: r.trojanek@ue.poznan.pl; Tel.: +48-668116511

Received: 27 October 2017; Accepted: 10 November 2017; Published: 13 November 2017

Abstract: In the paper, we analyzed the impact of aircraft noise on housing prices. We used a dataset containing geo-coded transactions for 1328 apartments and 438 single-family houses in the years 2010 to 2015 in Poznan. In this research, the hedonic method was used in OLS (ordinary least squares), WLS (weighted least squares), SAR (spatial autoregressive model) and SEM (spatial error model) models. We found strong evidence that aircraft noise is negatively linked with housing prices, which is in line with previous studies in other parts of the world. In our research, we managed to distinguish the influence of aircraft noise on different types of housing. The noise depreciation index value we found in our study was 0.87% in the case of single-family houses, and 0.57% regarding apartments. One of the reasons for the difference in the level of impact of aircraft noise may be the fact that the buyers of apartments may be less sensitive to aircraft noise than the buyers of single-family houses.

Keywords: impact of aircraft noise; housing prices; hedonic methods; airport

1. Introduction

Noise coming from aviation and its supporting operations is a crucial issue at airports across the world. The aviation industry has come a long way in efficiency and sustainability thanks to improvements in operations and technology [1]. It must be stated that huge improvements in technology have been made, so the level of noise coming from a single airplane is much lower than a few decades ago. Sustainable development in different aspects [2], as well as of air transport through the reduction of aircraft noise pollution at airports is promoted by the EU Environmental Noise Directive [3] and the associated Balanced Approach Regulation [4].

In recent years, air transport has grown in significance. In the pre-accession of Poland to the European Union period, in the years 1989–2004, air transport was developing very slowly [5]. After accession, post-socialist countries eliminated the barriers to entering their aviation markets [6,7]. Moreover, market liberalization resulted in new EU member countries being penetrated by low-cost carriers, which introduced new routes to destinations mainly in Western Europe [8].

Apart from the undoubted benefits of the sustainable development of society, this form of transport also generates some broadly defined costs (social and economic). There is no doubt that an increase in the level of aircraft noise is and will be an increasingly serious problem for people living in the vicinity of airports (both large international airports and less important local ones). This is connected with the development of regional airports and the intensification of air traffic in their area, but, most importantly, with the growing number of international flights. Three factors influence noise burden: the number of flights, the level of noise emitted by each airplane, and the time of flight.

Other factors that may have an impact include flight paths and procedures, the distribution of flights in flight paths, or the use of runways. The characteristic features of aircraft noise are the fact that it occurs instantly, quickly obtains its maximum level, and then rapidly decreases. Consequently, many inhabitants of the areas surrounding airports complain about the level of noise, although the results of aircraft noise measurements show that it does not contribute to permissible noise levels being exceeded significantly. Given the above, it seems necessary to examine the consequences of the vicinity of an airport.

An overview of the studies of the negative influence of aircraft noise allows us to distinguish its most important spheres [9]:

- Physical and mental health of people influenced by an airport (numerous studies show that exposure to aircraft noise destabilises one's mental condition, causes anxiety, increases aggression and excitability, raises blood pressure, disturbs heart and breath rhythms, reduces brain efficiency, and is the cause of an increased number of heart attacks and coronary diseases, as well as contributing to hearing deficiency or loss and speech disorders [10];
- Sleep quality (which is directly affected by aircraft noise at night and indirectly influenced by noise during the day) [11,12];
- Work efficiency (noise sensitivity increases the probability of disturbances in the execution of tasks and reduces work efficiency);
- Learning at schools (recent research shows a relationship between noise and children's ability to learn and absorb information) [13];
- Voice communication (both indoors and outdoors; this may involve interfering a conversation, watching television, or listening to the radio);
- Using park and leisure areas (research shows that users find noise a very important factor influencing the quality of rest) [14];
- Air traffic noise has the most negative effect on housing prices. Meanwhile, road and train noises have similar but smaller effects [15];
- The market value of residential properties (almost all studies confirm the negative impact of noise on the market value of properties located near the airport).

In the paper, the influence of aircraft noise on the last of the above spheres will be discussed. Real properties are a specific good, which is a result of their physical, economic, institutional–legal and environmental features. The specificity of the real estate market is determined by the unique attributes of a property [16]. Structural and locational attributes could have been considered by house buyers as a vital factor in property transactions [17]. Having examined these features, it is justifiable to say that the market value of a property is influenced not only by its direct characteristics (such as the size and shape of a plot, the age of a building, construction type, technical condition [18]), but also factors that involve its broadly defined surroundings [19]. Studies of the determinants of housing prices in developed markets often take into account environmental components [20–24]. These factors may be divided into two groups according to the kind of impact: positive influence (e.g., the vicinity of green areas, bodies of water) and negative influence (e.g., noise, air pollution) [21]. The indoor environment of each building depends on some criteria, like temperature, humidity, noise, etc. [25,26].

The remainder of this paper is structured as follows. The next section presents a review of the literature on the relationship between aircraft noise and real estate prices in the areas surrounding an airport assessment. Section three presents study area, data collection and variables, and the methodological background of the hedonic models applied. Section four presents and analyzes the results obtained. The last section presents concluding remarks.

2. Literature Review

The externalities resulting from airport operation, particularly aircraft noise, represent social costs, which may be identified as a change in the value of properties located in the area affected by airport activities. The most frequently used methods of noise cost estimation included are based on revealed preferences. Revealed preferences are consumer choices, and they are analyzed with the use of historical data on property sales. Of all the models based on revealed preferences, the hedonic price model is the most frequently used method for analyzing the influence of airport operation on the property market.

In order to determine the annoyance costs related to noise, noise depreciation indices (NDIs) are used. NDIs are defined as the percent increase in the loss of property market values caused by a unit increase in noise exposure and are identified with the use of hedonic price methods. By now, there are approximately 50 HP studies for airports in Canada and the US, and probably an equal number of non-North American airports [27]. The aircraft noise literature has been previously reviewed by Nelson [27], Schipper et al. [28], Bateman et al. [29] and Wadud [30]. A summary of these studies is presented in Table 1.

Table 1. Summary of recent reviews of literature on aircraft NDIs (noise depreciation indices).

Author(s)	NDI	Research Period	Study Area	Subject Scope
Nelson [27]	0.28–1.49%	1969–1993	USA (17), Canada (6)	23 airports, 33 NDI
Schipper et al. [28]	0.1–3.57%	1967–1996	USA (21), Canada (5), Australia (2), UK (2)	19 studies, 30 NDI
Bateman et al. [29]	0.29–2.3%	1960–1996	USA (20), UK (5), Canada (3), Australia (2)	30 studies
Wadud [30]	0–2.3%	1970–2007	USA (35), Canada (8), Australia (8), the UK (8), the Netherlands (1), France (1), Switzerland (3) and Norway (1)	65 studies

Source: own research.

These NDI estimates indicate that housing prices react differently across countries. This variation may be the result of different noise metrics (Noise Exposure Forecast (NEF), Noise Number Index (NNI), Australian Noise Exposure Forecast (ANEF), day–night sound Level (Leq, Ldn)), or different airport scales or different urban spatial structure. Otherwise different functional forms of models (linear, log-linear) used also account for a considerable part of the variation in these NDI estimates [30]. Moreover, some researchers argue that NDI and wealth are positively correlated. Wadud [30] carried out meta-regression analysis and concluded that the NDI tends to be higher in developed countries. Figure 1 summarizes the NDI estimates through a frequency distribution based on 79 studies carried out form 1970 till 2016 all over the world.

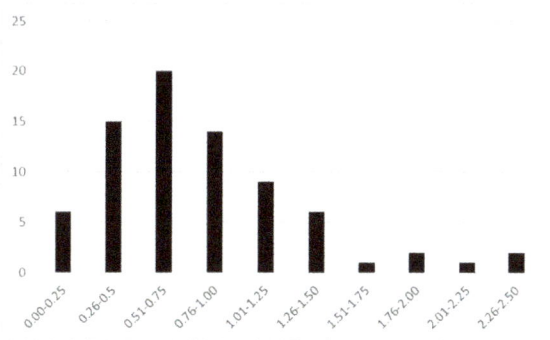

Figure 1. Frequency distribution of NDIs (79 studies from 1970 to 2016). Source: based on Wadud [27] and own research.

Taking into account some recent studies, most of them were carried out in Europe (Table 2). There are a few new analyses regarding European case studies [31–44]. All of these European analyses of the relationship between aircraft noise and real estate prices in the areas surrounding an airport confirmed a negative influence of aircraft noise on the market value of properties. The NDI ranges from 0.5% to 1.7% per decibel. However, difficulties may arise when comparing the obtained results since different noise indicators, thresholds, types of property and sources of data were used in these studies.

Table 2. Summary of recent reviews of literature on aircraft NDIs.

Id	Author(s)	Location	Noise Measure	Threshold dB	NDI	Research
1	Nguy et al. [31]	Beijing, China	–	–	1.05%	130 observations; sales; apartments; 2006–2012
2	Baranzini and Ramirez [32]	Switzerland, Geneva	Ldn	50 dB	1.17%	13,034 observations; rents; apartments; 2003
3	Salvi [33]	Switzerland, Zurich	Leq16	50 dB	0.97%	3737 observations; sales; single-family houses; 1995–2005
4	Dekkers and van der Straaten [34]	Netherlands, Amsterdam	Lden	45 dB	0.77%	66,636 observations; sales; different types of properties; 1999–2003
5	Brandt and Maenning [35]	Germany, Hamburg	Lden	62 dB	1.29%	4832 observations; for sale; apartments; 2002–2008
6	Thanos et al. [36]	Greece, Athens	Lden	55 dB	0.49%	1613 observations; sales; different types of properties; 1995–2001
7	Püschel and Evangelinos [37]	Germany, Düsseldorf	Lden	55 dB	1.04%	1370 observations; for sale; apartments; November 2009
8	Huderek-Glapska and Trojanek [38]	Poland, Warsaw	Laeq (The study is based on the Limited Use Area. The LUA is based on the actual noise indicators (L_{AeqD} and L_{AeqN}) around Warsaw airport and include the aircraft movements over the next five years.)	55 dB	~0.2%	130324 observations; for sale; apartments; 2007–2011
9	Trojanek [39]	Poland, Warsaw	Laeq Laeq (The study is based on the Limited Use Area. The LUA is based on the actual noise indicators (L_{AeqD} and L_{AeqN}) around Warsaw airport and include the aircraft movements over the next five years.)	55 dB	~0.8%	5290 observations; sale; apartments; 2010
10	Winke [40]	Germany, Frankfurt	Lden	55 dB	1.70%	19148 observations; for sale; apartments; 2006–2014
11	Chalermpong [41]	Thailand, Bangkok	NEF	30 dB	2.12%	384 observations; sales; new homes; 2002–2008
12	Boes and Nuesch [42]	Switzerland, Zurich	Leq16	30 dB–50 dB	0.5%	19,721 observations; for sale; rents; 2001–2006
13	Ahlfeldt and Maenning [43]	Germany, Berlin	Lden	45 dB	0.5–0.6%	31,289 observations; sales; different types of houses; 2000–2007
14	Lavandier et al. [44]	France, Paris	Lden	50 dB	Mean value 1.08%	19,891 observations; sales; single-family houses; 2002–2008 (except 2007)
					Mean value 1.51%	23,264 observations; sales; apartments; 2002–2008 (except 2007)

Source: own research.

3. Methods and Data

3.1. Study Area

Poznan is located in Central-Western Poland, in the central part of Wielkopolskie Province. It is the fifth largest city in Poland by population (541.6 thousand inhabitants) and the eight largest by size

(262 sq km). There are two airports within the administrative borders of the city of Poznan: Poznan Lawica International Airport and Poznan Krzesiny military airport, part of NATO structures.

Henryk Wieniawski Poznan Lawica International Airport (code IATA: POZ, code ICAO: EPPO), an international airport and one of the oldest airports in Poland, is located seven kilometres west of the centre of Poznan. As of 2015, it was the seventh largest Polish airport regarding the number of passengers carried and the number of airport operations (see Figure 2). Two modern passenger terminals ensure the capacity of 1900 passengers arriving and 1100 passengers departing per hour. In the years 2011–2013, owing to European funds, Lawica Airport was extended, and now it has a complex of passenger terminals which can handle up to 3.5 m passengers yearly.

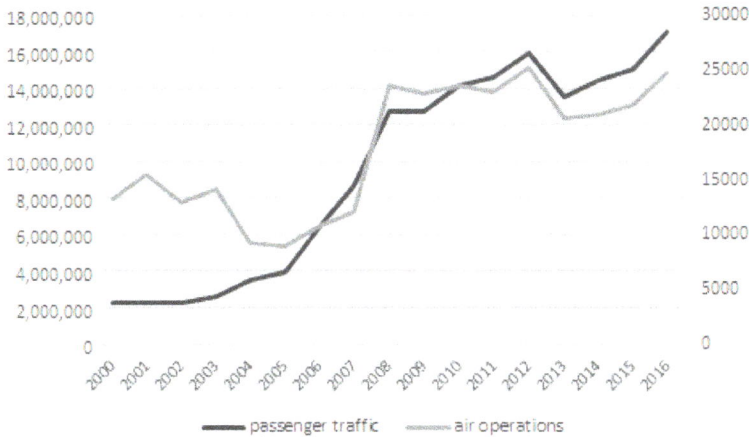

Figure 2. Some passenger traffic and air operations at Lawica Airport. Source: Poznan Airport.

Henryk Wieniawski Poznan Lawica International Airport operates regular flight connections to more than 30 airports. In recent years, it has handled approximately 1.5 million passengers a year (in the years 2010–2016).

In the case of Krzesiny, the 31st Air Base, it is an air force base located in South-East Poznan. The 31st Air Base is an air force unit for military operations conducted as part of the national defence system and a NATO very high readiness joint task force. In 2001, it was modernised so that it could handle F–16 aircraft. The grounds of the base have a rectangular shape. In the years 2001–2002, it was thoroughly modernized. It was actually entirely re-built. Now, it can handle practically all aircraft that are operated.

In comparison to 2011, in 2013 the total number of operations fell by 20.6% (from 6481 to 5143), with the number of night operations on almost the same level as in 2011 (see Figure 3). Such a big drop in the number of air operations significantly contributes to the improvement of the acoustic climate in the vicinity of the Poznań-Krzesiny airport. At the military airport in Poznan, supersonic multirole fighter aircraft F-16 Block 52+ are presently used. The 31st Air Base has 32 such planes.

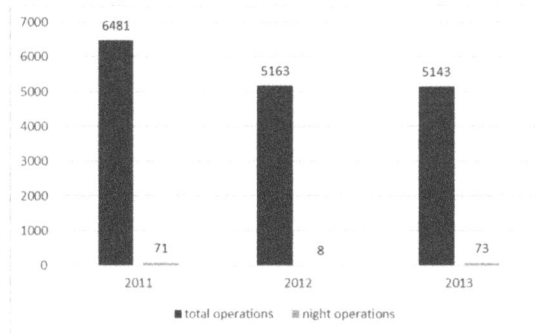

Figure 3. A number of air operations at Krzesiny Military Airport. Source: Environmental protection program.

3.2. Data Collection and Variables

The research is based on transaction data (Figure 4 shows the map of Poznan, noise counters produced by two airports and property sales). The data on apartment transactions conducted from the first quarter of 2010 to the fourth quarter of 2015 was obtained from the Board of Geodesy and Municipal Cadastre in Poznan. The obtained data referred to the transactions concerning all kinds of properties, both residential and non-residential (e.g., commercial properties or garages). In the process of data cleansing, purchases of more than one residential unit and non-free market transactions (e.g., debt collector sales) were removed. The data included in notarial contracts concerning apartments contain the following information: the transaction date, the price, the area of an apartment, the floor on which an apartment is located, and the area of any auxiliary premises (e.g., a garage/parking spot in an indoor parking lot or a cellar/residents' cupboard). Such a set of factors may bias the results of the research, as notarial contracts do not include information on strong pricing components such as, for example, the construction technology. Then, thanks to the cadastre data, a great deal of information on the height of buildings and the year of construction was added. Using the street view application on maps.google.com, the missing data concerning the height, year of construction, and, first of all, technology was provided. Then, with the use of Googlemaps API (application program interface), the addresses were geocoded (addresses of transactions).

The data included in notarial contracts concerning single-family houses contain the following information: the transaction date, the price, the area of plot, gross covered area. Then, thanks to the cadastre data, a great deal of information on the number of floors and the year of construction was added. Using the street view application on maps.google.com, the missing data concerning the number of overground and underground floors and the year of construction was provided. Moreover, we specified the type of roof (flat, sloping), the type of building (detached, semi-detached, terraced), the type of garage (whether an integral part of a house or detached), and, first of all, the technical condition of a building on the basis of external elements (based on historical photos). There is no doubt that floor space is an important factor regarding. Unfortunately, it is included in less than 10% of the observations in notarial deeds. That is why we decided to establish the area of a building taking into account the built-up area, number of aboveground floors, correction factor, the existence of a garage and the type of roof.

Spatial analyses were performed using the QGIS software. Those transactions, which took place within the area affected by aircraft noise (treatment group) and within 1.0 km from this zone (control group), were analyzed. In this research, apartments built before 1950 were excluded because such apartments are located mainly in the city center. Moreover, in the case of such apartments, the technical condition of a building is a significant determinant of their value (our dataset does not include this factor), which might affect the results obtained.

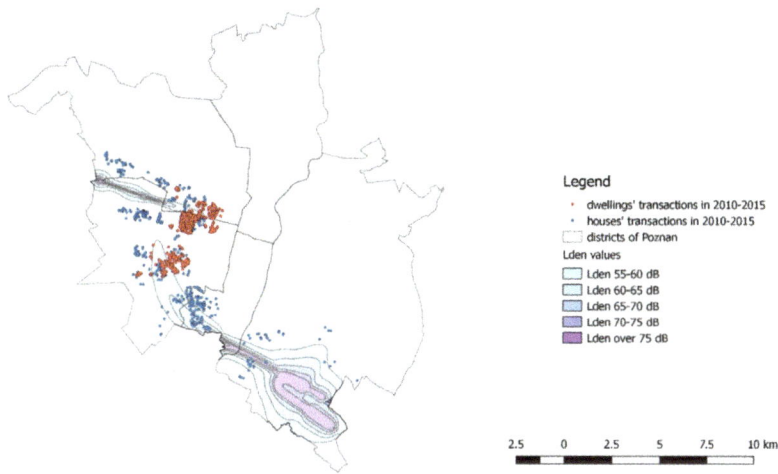

Figure 4. Aircraft noise boundaries and property transactions included in the analysis. Source: Based on the Board of Geodesy and Municipal Cadastre in Poznan and own research.

Tables 3 and 4 summarize the descriptive statistics (mean and standard deviation) of the variables used in the study. Based on a distance of 1.0 km from the aircraft noise, we sorted the housing transactions into a treatment group consisting of properties located in the noise zone (single-family houses, 107 observations, and apartments, 158 observations) and a control group with properties located outside (1 km buffer, e.g., [45,46]) the aircraft noise zone (single-family houses, 331 observations, and apartments, 1170 observations). We used the transaction prices in the logarithm term as the response variable in our models.

Table 3. Descriptive statistics of single-family house transactions.

	Control Group		Treatment Group	
	Mean	Standard Deviation	Mean	Standard Deviation
y2010	0.10	0.30	0.16	0.37
y2011	0.16	0.37	0.16	0.38
y2012	0.17	0.37	0.14	0.35
y2013	0.19	0.39	0.15	0.36
y2014	0.17	0.38	0.20	0.40
y2015	0.21	0.41	0.19	0.39
Area	157.08	52.71	155.86	58.55
Transaction price (in PLN)	579,935.68	212,738.75	540,635.91	213,968.57
Age/10	2.72	1.72	2.98	1.63
q1	0.11	0.31	0.10	0.31
q2	0.15	0.36	0.20	0.40
q3	0.29	0.45	0.35	0.48
q4	0.24	0.43	0.22	0.42
q5	0.21	0.41	0.13	0.34
Underfloor	0.45	0.50	0.50	0.50
Garage	0.76	0.43	0.78	0.42
Areaplot	453.19	253.80	443.53	270.23
PU	0.13	0.34	0.18	0.38
Distance to CC	6.12	1.31	6.16	0.70
Airnoise5560	0	0	0.59	0.49
Airnoise6065	0	0	0.24	0.43
Airnoise6570	0	0	0.17	0.37
No of observations	331		107	

Source: own research.

The information on aircraft noise zones was taken from an acoustic map from 2012. Directive 2002/49/EC of the European Parliament requires the carrying out of a long-term policy of environmental protection against noise in the European Union countries. Its realization is based on the estimation of the long-term noise indicators Lden and Ln in the areas under protection. The threshold value used in this study was 55 dB. In order to establish both airports' noise influence on the acoustic map of Poznań, the following data was used: the acoustic characteristics of the aircraft used, arrival and departure routes, glide paths, take-off and landing profiles, and the distribution of the intensity of flights during daytime, in the evenings and at night.

Table 4. Descriptive statistics of apartment transactions.

	Control Group		Treatment Group	
	Mean	Standard Deviation	Mean	Standard Deviation
y2010	0.18	0.38	0.11	0.31
y2011	0.17	0.37	0.09	0.29
y2012	0.15	0.36	0.16	0.37
y2013	0.16	0.37	0.18	0.38
y2014	0.18	0.39	0.18	0.38
y2015	0.16	0.37	0.38	0.45
Area	44.34	13.26	48.41	13.47
Transaction price (in PLN)	226,411.72	75,402.95	241,305.46	72,835.37
Age	0.57	0.50	0.29	0.46
Basement	0.60	0.49	0.68	0.47
Floor1	0.13	0.33	0.23	0.42
Floor2	0.51	0.50	0.34	0.48
Floor3	0.36	0.48	0.43	0.50
Distance to CC	4.32	1.03	5.24	0.90
Airnoise5560	0	0	1	0
Height2	0.31	0.46	0.22	0.42
Technology2	0.52	0.50	0.55	0.50
No of observations	1170		158	

Source: own research.

Within the reach of aircraft noise Lden (Lawica and Krzesiny airports combined), there were 2137 inhabitants in the range from 65 to 75 dB. The number of inhabitants exposed to Lden noise at a level of 55–65 dB is about 26,000.

3.3. Hedonic Price Models

Mathematical statistics methods are broadly applied to analyze the pricing of real estate [47–54]. The most commonly applied methods of housing evaluation are divided into two groups: traditional and advanced methods. The advanced methods include techniques such as hedonic price modelling (HPM), artificial neural networks (ANN), case-based reasoning, and spatial analysis methods. The HPM is an ideal analytical tool to analyze a non-homogeneous commodity regarding its attributes.

The first documented use of hedonic regression dates back to 1922, when G. A. Hass developed the farmland price model [55]. The first researcher to use the hedonic method to analyze the real estate market was probably Ridker, who aimed to identify the influence of pollution reduction on house prices [56]. The theoretical framework of the hedonic method was developed by Lancaster [57] and Rosen [58].

The idea of the hedonic method lies in the assumption that the price of heterogeneous goods may be characterised by their attributes. In other words, this method allows us to estimate the value of the particular attributes of a given product. The price of a given good is the response variable, while its quantitative and qualitative attributes are the explanatory variables. The equation may be written as follows (1):

$$P = \beta_0 + \sum_{i=1}^{K} \beta_i X_i + u, \tag{1}$$

where P is the price of a good, β is the regression coefficient, X is an attribute of a good (value driver), u is a random error.

One of the key issues in hedonic methods is the choice of the form of the regression function. The log-linear (natural logarithm) form of the regression function is most frequently used for studying changes in the real estate market in empirical research. As housing is a heterogeneous good, it is difficult to indicate a full list of crucial attributes. The heterogeneity of real estate hinders the measurement of price impacts. Taking into account Malpezzi [59] and Crompton's [60] suggestions, six major categories of characteristics of housing may be distinguished: (1) structural attributes, (2) neighborhood related services and features, (3) location and accessibility, (4) environmental attributes, (5) community attributes and (6) time-related features. In our study, we examine the implicit value of the aircraft noise. We hypothesize that transaction price is a function of structural features, locational attributes, time and aircraft noise. The basic hedonic function of price (y) can be stated as:

$$\ln(P) = f\ (structural\ atributes,\ location,\ aircraft\ noise,\ time) \tag{2}$$

In this research, we used several variants of hedonic regression, namely standard ordinary least squares (OLS), robust weighted least squares (WLS) and spatial models. According to WLS, the estimation was made with the following steps: an OLS regression was run, then the logs of the squares of residuals become the dependent variable in an auxiliary regression, on the right-hand side of which are the original independent variables plus their squares. The fitted values from the auxiliary regression were then used to construct a weighted series, and the original model was re-estimated using weighted least squares [61]. In recent years, a growing concern has risen regarding the spatial dependence found in most house price data. Spatial dependence intuition was presented by Tobler [62], who concluded that there is a reason to believe that things that are near will be more related than distant ones. As one of the most important features of the housing market is the importance of location, the hypothesis of the spatial dependence of house prices seems plausible. The spatial-lag model is based on the assumption that the spatially weighted average of housing prices in a neighbourhood affects the price of each house (indirect effects) in addition to the features of housing and neighbourhood characteristics (direct effects) [63]. The ordinary least squares (OLS) hedonic estimates are not biased, but estimation efficiency may be lowered by spatial dependence. The obtained results can be biased, especially regarding their statistical significance [64,65]. The model that deals with this interpretation of spatial dependence is called the spatial error model (SEM). In contrast, the spatial error model does not include indirect effects, but it assumes that there may be one or more omitted variable in the hedonic price equation and that the omitted variable(s) vary spatially [63]. Due to this spatial pattern in the omitted variables, the error term of the hedonic price equation tends to be spatially autocorrelated. The econometric model dealing with this kind of spatial dependence is called the spatial lag model, or spatial autoregressive (SAR) model. In the spatial lag model, spatial dependence is assumed to be present in the additional explanatory variable.

4. Results

Among the apartment characteristics checked for in the research were the following: year of transaction, area of the apartment, age, construction technology, floor, the height of the building, basement, distance to city center and finally range of aircraft noise. In the case of single-family houses, we used: year of the transaction, the area of the house, age of the building, underground floor, quality of the building, basement, garage, type of plot ownership, distance to city center and range of aircraft noise. The choice of qualitative and quantitative data was limited by the availability of information in the database. Table 5 presents the variables used in the study in case of single-family houses and Table 6 regarding apartments.

Table 5. Qualitative and quantitative variables applied in the models in case of single-family houses.

Variable	Symbol	Description
Price	Price	Price for property (in PLN)
Year	y2010, y2011, y2012, y2013, y2014, y2015	6 time dummy variables used in the global model. If the apartment was sold in a given year, it takes the value 1; otherwise it takes 0
Area	Area	Area of building = built-up area x number of overground floors (type of roof taken into account) × 0.8–20 m² (if there is a garage in the building)
Age	Age	Age of the building divided by 10
Quality	q1—new building to finish q2—the building is in bad condition q3—the building is in average condition q4—the building is in good condition q5—the building is in very good condition	5 dummy variables. If the apartment is located on a given floor, it takes the value 1; otherwise it takes 0
Underground floor	Underfloor	If there is underground floor than 1, if not 0
Garage	Garage	If there is garage than 1, if not then 0
Area of plot	Plotarea	Area of plot
PU	PU	0—ownership of the plot 1—perpetual usufruct
Dcc	Dcc	Distance to city centre
Airnoise	Airnoise	1—Lden 55–60 dB 2—Lden 60–65 dB 3—Lden 65–70 dB

Source: own elaboration.

Table 6. Qualitative and quantitative variables applied in the models in case of apartments.

Variable	Symbol	Description
Price	Price	Price of an apartment (in PLN)
Year	y2010, y2011, y2012, y2013, y2014, y2015	6 time dummy variables used in the global model. If the apartment was sold in a given year, it takes the value 1; otherwise it takes 0
Area	Area	Area of apartment m²
Construction technology	Technology1—if the apartment is located in a building made with a prefabricated technology Technology2—if the apartment is located in a building made with a traditional technology	2 dummy variables. If the apartment is located in a building made with given technology, it takes the value 1; otherwise it takes 0
Age	Age	Age of the building
Floor	Floor1—ground and top floor Floor2—intermediate floors Floor3—first and second floor	3 dummy variables. If the apartment is located on a given floor, it takes the value 1; otherwise it takes 0
Height	Height1—buildings up to 4 floors Height2—buildings above 5 floors	2 dummy variables. If the building has given height it takes the value 1; otherwise it takes 0
Airnoise	Airnoise	If an apartment is located in aircraft noises in the range of 55–60 dB (Lden) then it takes value 1, otherwise it takes 0 (value under 55 dB)
Dcc	Dcc	Distance to city center in km
Basement	Basement	If an apartment has a basement then it takes value 1, otherwise it takes 0

Source: own elaboration.

To address the research questions, hedonic regression equations using ordinary least squares and spatial models were estimated. The dependent variable was the natural log of a sales price. Gretl and Geodaspace software were used to estimate the parameters of functions.

Houses are heterogeneous in nature. This heterogeneity may be the reason for heteroscedasticity in the residuals of the estimation of the function. Indeed, we found heteroscedasticity in the case of apartments (there was no problem in case of single-family homes, according to Breusch–Pagan and Koenker–Basset tests). That is why we used OLS with heteroscedasticity correction (WLS). Moreover, we tested for the presence of multicollinearity as it leads to unstable coefficients and inflated standard errors. The variance inflation factors (VIFs) was used to detect multicollinearity. The VIF values in the model do not exceed 4.7 in case of single-family houses and 2.8 in case of apartments, which is in line with the most conservative rules of thumb that the mean of the VIFs should not be considerably higher than 10. Tests for the normality of residuals are presented in Table 7.

Table 7. Test for normality of residuals (Ordinary Least Squares models).

Name of the Test	Single-Family Houses	Apartments
Doornik-Hansen	2.34913, with *p*-value 0.308954	3.33477, with *p*-value 0.18874
Shapiro-Wilk W	0.995164, with *p*-value 0.189877	0.997589, with *p*-value 0.0494359
Lilliefors	0.0234289, with *p*-value ~= 0.81	0.0314622, with *p*-value ~= 0
Jarque-Bera	2.6041, with *p*-value 0.271974	3.27364, with *p*-value 0.194597

Source: own elaboration.

In order to test for the presence of spatial effects in the data, we calculated spatial weights between observations [66]. Based on the geographical coordinates, we created (438×438 for sing-family houses and 1328×1328 for apartments) spatial weight matrixes based on the distance between them; a 200 m for apartments and 400 m for single-family houses threshold distance d was assumed. We tested different threshold distances and decided to use these as they had the highest value of I-Moran statistics. Following the arguments of Anselin [67], tests for the presence of spatial effects were carried out (both spatial autocorrelation and spatial lag dependence). To conclude, we found strong evidence of spatial dependence in the form of spatial autocorrelation and spatial lag.

The estimation results for single-family houses are presented in Table 8 and for apartments in Table 9.

The estimated models were well-fitted in case of apartments, as they explained about 82% of the price variations. As far as for the single-family houses, the models explained from 65–66%, depending on the model. Almost all of the variables applied in the models turned out to be statistically relevant, and the expected coefficient signs were correct. The results of the spatial models (in the case of single-family houses and apartments) suggest that spatial effects were present in the data, in the form of unobserved variables and significant spatial processes (in the case of single-family houses).

Table 8. Estimation results (dependent variable is a natural logarithm of single-family house sale price).

	OLS		WLS		SAR		SEM	
Variable	Coefficient	Probability	Coefficient	Probability	Coefficient	Probability	Coefficient	Probability
Constant	13.5745	0.0000	13.5696	0.0000	13.2867	0.0000	13.6318	0.0000
y2011	−0.1062	0.0029	−0.0968	0.0069	−0.1024	0.0028	−0.1110	0.0006
y2012	−0.1817	0.0000	−0.1736	0.0000	−0.1788	0.0000	−0.1768	0.0000
y2013	−0.2629	0.0000	−0.2487	0.0000	−0.2604	0.0000	−0.2817	0.0000
y2014	−0.2672	0.0000	−0.2514	0.0000	−0.2647	0.0000	−0.2742	0.0000
y2015	−0.3336	0.0000	−0.3247	0.0000	−0.3300	0.0000	−0.3334	0.0000
Age	−0.0148	0.0626	−0.0128	0.0750	−0.0145	0.0589	−0.0205	0.0085
Area	0.0020	0.0000	0.0021	0.0000	0.0020	0.0000	0.0018	0.0000
Areaplot	0.0001	0.0001	0.0002	0.0001	0.0002	0.0000	0.0002	0.0000
Underfloor	0.0865	0.0001	0.0774	0.0002	0.0882	0.0000	0.0726	0.0003
Garage	0.0443	0.0502	0.0552	0.0171	0.0448	0.0396	0.0374	0.0751
q2	−0.3008	0.0000	−0.3398	0.0000	−0.3024	0.0000	−0.2626	0.0000
q3	−0.2049	0.0000	−0.2337	0.0000	−0.2060	0.0000	−0.1696	0.0000
q4	−0.1024	0.0115	−0.1498	0.0002	−0.1044	0.0074	−0.0685	0.0816
q5	0.1078	0.0068	0.0561	0.1528	0.1008	0.0085	0.1272	0.0008
PU	−0.0478	0.0806	−0.0565	0.0216	−0.0479	0.0692	−0.0499	0.0552
dcc	−0.0720	0.0000	−0.0729	0.0000	−0.0675	0.0000	−0.0001	0.0000
Airnoise	−0.0458	0.0001	−0.0417	0.0001	−0.0440	0.0001	−0.0447	0.0061
W_lnprice					0.0188	0.0039		
Lambda							0.40073	0.0000
R-squared	0.6582		0.6617					
Pseudo R-squared					0.6675		0.6565	
N	438		438		438		438	

Source: own elaboration.

Table 9. Estimation results (dependent variable is a natural logarithm of apartment sale price).

Variable	OLS Coefficient	Probability	WLS Coefficient	Probability	SAR Coefficient	Probability	SEM Coefficient	Probability
Constant	11.6100	0.0000	11.5313	0.0000	11.5641	0.0000	11.6653	0.0000
y2011	−0.0052	0.6653	−0.0139	0.2697	−0.0051	0.6710	−0.0086	0.4505
y2012	−0.0577	0.0000	−0.0576	0.0000	−0.0576	0.0000	−0.0595	0.0000
y2013	−0.0687	0.0000	−0.0743	0.0000	−0.0685	0.0000	−0.0621	0.0000
y2014	−0.0273	0.0216	−0.0328	0.0085	−0.0271	0.0216	−0.0291	0.0101
y2015	0.0058	0.6291	−0.0080	0.4847	0.0062	0.6009	0.0022	0.8489
Dcc	−0.0334	0.0000	−0.0338	0.0000	−0.0333	0.0000	−0.0387	0.0000
Area	0.0282	0.0000	0.0322	0.0000	0.0282	0.0000	0.0282	0.0000
Area^2	−0.0001	0.0000	−0.0002	0.0000	−0.0001	0.0000	−0.0001	0.0000
Basement	0.0288	0.0128	0.0262	0.0343	0.0284	0.0136	0.0136	0.2916
Age	−0.0058	0.0000	−0.0056	0.0000	−0.0058	0.0000	−0.0063	0.0000
Floor2	0.0132	0.2382	0.0073	0.4731	0.0134	0.2267	0.0174	0.0972
Floor3	0.0328	0.0028	0.0287	0.0045	0.0330	0.0024	0.0310	0.0024
Height2	−0.0402	0.0000	−0.0378	0.0000	−0.0398	0.0000	−0.0279	0.0143
Technology	0.0495	0.0000	0.0463	0.0002	0.0493	0.0000	0.0313	0.0271
Airnoise	−0.0252	0.0255	−0.0263	0.0028	−0.0251	0.0250	−0.0290	0.0661
W_lnprice					0.0037	0.5140		
Lambda							0.5739	0.0000
R-squared	0.8234		0.8221					
Pseudo R-squared					0.8236		0.8213	
N	1328		1328		1328		1328	

Source: own elaboration.

We observed that within the period under study (2010–2015), time had a significant impact on transaction prices. It is worth mentioning that housing prices in the biggest cities in Poland increased by about 100% between 2006 and 2007 [68,69]. At the end of 2007, the subsequent decreasing phase in the property price cycle began, resulting from this abnormal price increase and the beginning of financial crisis [70]. It is worth noticing that in the case of single-family houses, this downturn was higher than considering apartments in the analyzed locations.

Taking into account the perspective of this paper's objectives, the statistical relevance of the air-noise variable is important. The application of the log-linear model enabled the percentage difference in the price of the single-family house/apartment with similar characteristics located within aircraft noise zones and the 1.0 km buffer zone to be identified. The value of the air-noise coefficient in the SEM model (the regression coefficients obtained in models were similar, however for the interpretation we used the SEM model as it is most robust) reached the value of −0.0447 (Table 8), which indicates that a single-family house located in area affected by the aircraft noise was about 4.59% cheaper (in the area with aircraft noise level values of Lden 55–60 dB), 9.18% cheaper (in the area with aircraft noise level values of Lden 60–65 dB) and 13.77% cheaper (in the area with aircraft noise level values of Lden 65–70 dB) than a single-family house (with the same characteristics) located in the 1.0 km buffer zone (with aircraft noise under 55 dB) in the years 2010–2015. In case of apartments, this decrease was smaller- it was about 2.86% (Lden 55–60 dB).

5. Conclusions

To our best of our knowledge, this study is one of the first studies to address the effects of aircraft noise (measured with Lden) on real estate value in the post-socialist urban context. While the problem has been addressed in many articles, most of them focused on areas in the USA, Canada and Western European countries. Moreover, we managed to distinguish the influence of aircraft noise on different types of housing. In earlier studies, mainly one type of housing (for example apartments) was the basis of the analysis. It was difficult, actually impossible, to compare NDIs for different types of housing, taking into account differences in the location of airports, residential markets, various periods, measures of noise, model specifications.

This article aimed to identify the impact of aircraft noise created by airports on apartment and single-family house prices in Poznan. In this research, the hedonic method in OLS, WLS, SEM and SAR models were used. The application of the log-linear model allowed the identification of the percentage difference in the price of an apartment with similar features located within the noise zones and outside. In order to compare the obtained results with previous studies, we estimated the NDI values. The NDI value we found in our study was 0.87 in the case of single-family houses, which means there is a 0.87% value discount per decibel (Lden noise indicator). Regarding apartments, the NDI was a 0.57% decrease of value per 1 dB of aircraft noise. The reason for the difference in the level of impact of aircraft noise may be the fact that the buyers of such apartments may be less sensitive to aircraft noise than the owners of single-family houses, who, because of higher noise levels cannot fully take advantage of the benefits related to a single family home (e.g., limited enjoyment of their garden).

Our investigation showed that the sensitivity of buyers differs depending on housing type. Although the results are reasonable, our proposal is not without limitations, which are mainly related to the possible change of acoustic climates and the limitations of the dataset regarding the variables describing properties. Our study was made on based on the acoustic map from the year 2012, so some limitation may arise, as the acoustic climate of the city may have changed. However, we were not able to overcome this issue as the map is created every five years. On the other hand, taking into account the number of flights (a rather stable number), it may be assumed that aircraft noise did not change significantly. As far as the dataset is concerned, gathering the data for a housing market analysis is always a huge challenge. In the case of our research, we used different sources of information, however, we were not able to control directly for the quality of an apartment or a single-family house (inside). Moreover, the impact of aircraft noise in other cities in Poland may be different as the influence of wealth effects is regionally distributed.

In this regard, future research should be carried out in other cities in Poland so that it would be possible to compare NDI values from different airports. In order to increase the comparability of future research, they should be based on the same assumptions (variables describing the properties, time scope, methods used). It could provide a chance to examine the impact of regional wealth effects on NDI values.

Author Contributions: Radoslaw Trojanek and Justyna Tanas together designed the research and wrote the paper, Saulius Raslanas and Audrius Banaitis provided extensive advice throughout the study regarding the abstract, introduction, literature review, research methodology and data, and results of the manuscript. The discussion was a team task. All authors have read and approved the final manuscript.

Conflicts of Interest: The authors declare no conflict of interest.

References

1.	Hind, P. *The Sustainability of UK Aviation: Trends in the Mitigation of Noise and Emissions*; Independent Transport Commission: London, UK, 2016.
2.	Manzhynski, S.; Siniak, N.; Źróbek-Różańska, A.; Źróbek, S. Sustainability performance in the Baltic Sea Region. *Land Use Policy* **2016**, *57*, 489–498. [CrossRef]
3.	EC Directive 2002/49/EC of the European parliament and the Council of 25 June 2002 relating to the assessment and management of environmental noise. *Off. J. Eur. Communities* **2002**, *189*, 12–25. Available online: http://eur-lex.europa.eu/legal-content/EN/TXT/?uri=celex:32002L0049 (accessed on 27 October 2017).
4.	EC Regulation (EU). No 540/2014 of the European Parliament and of the Council of 16 April 2014 on the sound level of motor vehicles and of replacement silencing systems, and amending Directive 2007/46/EC and repealing Directive 70/157/EEC. *Off. J. Eur. Communities* **2014**, *173*, 65–78.
5.	Jankiewicz, J.; Huderek-Glapska, S. The air transport market in Central and Eastern Europe after a decade of liberalisation—Different paths of growth. *J. Transp. Geogr.* **2016**, *50*, 45–56. [CrossRef]
6.	Augustyniak, W. Efficiency change in regional airports during market liberalization. *Econ. Sociol.* **2014**, *7*, 85–93. [CrossRef] [PubMed]
7.	Olipra, Ł.; Augustyniak, W. Analysis of business traffic at Wroclaw Airport—Implications for economic development of the city and the region. *J. Int. Stud.* **2015**, *8*, 175–190. [CrossRef]
8.	Dobruszkes, F. New Europe, new low-cost air services. *J. Transp. Geogr.* **2009**, *17*, 423–432. [CrossRef]

9.   National Academies of Sciences, Engineering, and Medicine. *Effects of Aircraft Noise: Research Update on Select Topics*; Transportation Research Board: Washington, DC, USA, 2008; ISBN 978-0-309-09806-9.

10. Swift, H. *A Review of the Literature Related to Potential Health Effects of Aircraft Noise*; Purdue University: West Lafayette, IN, USA, 2010.

11. Basner, M.; Samel, A.; Isermann, U. Aircraft noise effects on sleep: Application of the results of a large polysomnographic field study. *J. Acoust. Soc. Am.* **2006**, *119*, 2772–2784. [CrossRef] [PubMed]

12. Holt, J.B.; Zhang, X.; Sizov, N.; Croft, J.B. Airport noise and self-reported sleep insufficiency, United States, 2008 and 2009. *Prev. Chronic Dis.* **2015**, *12*, E49. [CrossRef] [PubMed]

13. Clark, C.; Martin, R.; van Kempen, E.; Alfred, T.; Head, J.; Davies, H.W.; Haines, M.M.; Lopez Barrio, I.; Matheson, M.; Stansfeld, S.A. Exposure-Effect Relations between Aircraft and Road Traffic Noise Exposure at School and Reading Comprehension: The RANCH Project. *Am. J. Epidemiol.* **2005**, *163*, 27–37. [CrossRef] [PubMed]

14. Rapoza, A.S.; Fleming, G.G.; Lee, C.S.Y.; Roof, C.J. *Study of Visitor Response to Air Tour and Other Aircraft Noise in National Parks*; Volpe National Transportation Systems Center: Cambridge, MA, USA, 2005.

15. Kopsch, F. The cost of aircraft noise? Does it differ from road noise? A meta-analysis. *J. Air Transp. Manag.* **2016**, *57*, 138–142. [CrossRef]

16. Renigier-Bilozor, M.; Wisniewski, R.; Bilozor, A. Rating attributes toolkit for the residential property market. *Int. J. Strateg. Prop. Manag.* **2017**, *21*, 307–317. [CrossRef]

17. Kam, K.J.; Chuah, S.Y.; Lim, T.S.; Ang, F.L. Modelling of property market: The structural and locational attributes towards Malaysian properties. *Pac. Rim Prop. Res. J.* **2016**, *22*, 203–216. [CrossRef]

18. Raslanas, S. Research of market value of multistory housing in Vilnius. *Technol. Econ. Dev. Econ.* **2004**, *10*, 167–173. [CrossRef]

19. Trojanek, R.; Gluszak, M. Spatial and time effect of subway on property prices. *J. Hous. Built Environ.* **2017**, 1–26. [CrossRef]

20. Kaklauskas, A.; Zavadskas, E.K.; Raslanas, S. Modelling of real estate sector: The case for Lithuania. *Transform. Bus. Econ.* **2009**, *8*, 101–120.

21. Zavadskas, E.; Kaklauskas, A.; Maciunas, E.; Vainiunas, P.; Marsalka, A. Real estate's market value and a pollution and health effects analysis decision support system. In Proceedings of the 4th International Conference on Cooperative Design, Visualization, and Engineering, Shanghai, China, 16–20 September 2007; p. 191.

22. Raslanas, S.; Tupenaite, L. Peculiarities of private houses valuation by sales comparison approach. *Technol. Econ. Dev. Econ.* **2005**, *11*, 233–241. [CrossRef]

23. Źróbek, S.; Trojanek, M.; Źróbek-Sokolnik, A.; Trojanek, R. The influence of environmental factors on property buyers' choice of residential location in Poland. *J. Int. Stud.* **2015**, *8*, 164–174. [CrossRef]

24. McCord, J.; McCord, M.; McCluskey, W.; Davis, P.T.; McIlhatton, D.; Haran, M. Effect of public green space on residential property values in Belfast metropolitan area. *J. Financ. Manag. Prop. Constr.* **2014**, *19*, 117–137. [CrossRef]

25. Raslanas, S.; Kliukas, R.; Stasiukynas, A. Sustainability assessment for recreational buildings. *Civ. Eng. Environ. Syst.* **2016**, *33*, 286–312. [CrossRef]

26. Le Boennec, R.; Salladarré, F. The impact of air pollution and noise on the real estate market. The case of the 2013 European Green Capital: Nantes, France. *Ecol. Econ.* **2017**, *138*, 82–89. [CrossRef]

27. Nelson, J.P. Meta-Analysis of Airport Noise and Hedonic Property Values: Problems and Prospects. *J. Transp. Econ. Policy* **2004**, *38*, 1–28.

28. Schipper, Y.; Nijkamp, P.; Rietveld, P. Why do aircraft noise value estimates differ? A meta-analysis. *J. Air Transp. Manag.* **1998**, *4*, 117–124. [CrossRef]

29. Bateman, I.; Day, B.; Lake, I. The Effect of Road Traffic on Residential Property Values: A Literature Review and Hedonic Pricing Study. Available online: http://www.gov.scot/Publications/2001/07/9535/File-1 (accessed on 27 October 2017).

30. Wadud, Z. Using meta-regression to determine Noise Depreciation Indices for Asian airports. *Asian Geogr.* **2013**, *30*, 127–141. [CrossRef]

31. Nguy, A.; Sun, C.; Zheng, S. Airport noise and residential property values: Evidence from Beijing. In *Proceedings of the 17th International Symposium on Advancement of Construction Management and Real Estate*; Springer: Berlin/Heidelberg, Germany, 2014; Volume 1, pp. 473–481.

32. Baranzini, A.; Ramirez, J.V. Paying for Quietness: The Impact of Noise on Geneva Rents. *Urban Stud.* **2005**, *42*, 633–646. [CrossRef]

33. Salvi, M. Spatial Estimation of the Impact of Airport Noise on Residential Housing Prices. *Swiss J. Econ. Stat.* **2008**, *144*, 577–606. [CrossRef]

34. Dekkers, J.E.C.; van der Straaten, J.W. Monetary valuation of aircraft noise: A hedonic analysis around Amsterdam airport. *Ecol. Econ.* **2009**, *68*, 2850–2858. [CrossRef]

35. Brandt, S.; Maennig, W. Road noise exposure and residential property prices: Evidence from Hamburg. *Transp. Res. Part D Transp. Environ.* **2011**, *16*, 23–30. [CrossRef]

36. Thanos, S.; Wardman, M.; Bristow, A.L. Valuing Aircraft Noise: Stated Choice Experiments Reflecting Inter-Temporal Noise Changes from Airport Relocation. *Environ. Resour. Econ.* **2011**, *50*, 559–583. [CrossRef]

37. Püschel, R.; Evangelinos, C. Evaluating noise annoyance cost recovery at Düsseldorf International Airport. *Transp. Res. Part D Transp. Environ.* **2012**, *17*, 598–604. [CrossRef]

38. Huderek-Glapska, S.; Trojanek, R. The impact of aircraft noise on house prices. *Int. J. Acad. Res. Int. J. Acad. Res. Part B* **2013**, *5*, 397–408. [CrossRef]

39. Trojanek, R. The impact of aircraft noise on the value of dwellings—The case of Warsaw Chopin airport in Poland. *J. Int. Stud.* **2014**, *7*. [CrossRef] [PubMed]

40. Winke, T. The impact of aircraft noise on apartment prices: A differences-in-differences hedonic approach for Frankfurt, Germany. *J. Econ. Geogr.* **2016**, *101*. [CrossRef]

41. Chalermpong, S. Impact of airport noise on property values. Case of Suvarnabhumi International Airport, Bangkok, Thailand. *Transp. Res. Rec. J. Transp. Res. Board* **2010**, *2177*, 8–16. [CrossRef]

42. Boes, S.; Nüesch, S. Quasi-experimental evidence on the effect of aircraft noise on apartment rents. *J. Urban Econ.* **2011**, *69*, 196–204. [CrossRef]

43. Ahlfeldt, G.M.; Maennig, W. External productivity and utility effects of city airports. *Reg. Stud.* **2013**, *47*, 508–529. [CrossRef]

44. Lavandier, C.; Sedoarisoa, N.; Desponds, D.; Dalmas, L. A new indicator to measure the noise impact around airports: The Real Estate Tolerance Level (RETL)—Case study around Charles de Gaulle Airport. *Appl. Acoust.* **2016**, *110*, 207–217. [CrossRef]

45. Cohen, J.P.; Coughlin, C.C. Changing noise levels and housing prices near the Atlanta airport. *Growth Chang.* **2009**, *40*, 287–313. [CrossRef]

46. Cohen, J.P.; Coughlin, C.C. Spatial hedonic models of airport noise, proximity, an housing prices. *J. Reg. Sci.* **2008**, *48*, 859–878. [CrossRef]

47. Raslanas, S.; Tupenaite, L.; Steinbergas, T. Research on the prices of flats in the south east London and Vilnius. *Int. J. Strateg. Prop. Manag.* **2006**, *10*, 51–63. [CrossRef]

48. Deaconu, A.; Lazar, D.; Buiga, A.; Fatacean, G. Marginal prices of improvements made to blocks of flats: Empirical evidence from Romania. *Int. J. Strateg. Prop. Manag.* **2016**, *20*, 156–171. [CrossRef]

49. Taltavull de La Paz, P.; López, E.; Juárez, F. Ripple effect on housing prices. Evidence from tourist markets in Alicante, Spain. *Int. J. Strateg. Prop. Manag.* **2017**, *21*, 1–14. [CrossRef]

50. Lee, C.-C.; Lee, C.-C.; Chiang, S.-H. Ripple effect and regional house prices dynamics in China. *Int. J. Strateg. Prop. Manag.* **2016**, *20*, 397–408. [CrossRef]

51. Chen, J.-H.; Ong, C.F.; Zheng, L.; Hsu, S.-C. Forecasting spatial dynamics of the housing market using Support Vector Machine. *Int. J. Strateg. Prop. Manag.* **2017**, *21*, 273–283. [CrossRef]

52. Yang, H.; Song, J.; Choi, M. Measuring the Externality Effects of Commercial Land Use on Residential Land Value: A Case Study of Seoul. *Sustainability* **2016**, *8*, 432. [CrossRef]

53. Del Giudice, V.; De Paola, P.; Manganelli, B.; Forte, F. The monetary valuation of environmental externalities through the analysis of real estate prices. *Sustainability* **2017**, *9*, 229. [CrossRef]

54. Jayantha, W.M.; Lau, J.M. Buyers' property asset purchase decisions: An empirical study on the high-end residential property market in Hong Kong. *Int. J. Strateg. Prop. Manag.* **2016**, *20*, 1–16. [CrossRef]

55. Colwell, P.F.; Dilmore, G. Who Was First? An Examination of an Early Hedonic Study. *Land Econ.* **1999**, *75*, 620–626. [CrossRef]

56. Coulson, E. *Monograph on Hedonic Estimation and Housing Markets*; Penn State University: State College, PA, USA, 2008.

57. Lancaster, K.J. A new approach to consumer theory. *J. Political Econ.* **1966**, *74*, 132. [CrossRef]

58. Rosen, S. Hedonic Prices and Implicit Markets: Product Differentiation in Pure Competition. *J. Political Econ.* **1974**, *82*, 34–55. [CrossRef]

59. Malpezzi, S. Hedonic Pricing Models: A Selective and Applied Review. In *Housing Economics and Public Policy*; Blackwell Science Ltd.: Oxford, UK, 2008; pp. 67–89, ISBN 9780470690680.

60. Crompton, J.L. The impact of parks on property values: A review of the empirical evidence. *J. Leis. Res.* **2001**, *33*, 1–31. [CrossRef]

61. Cottrell, A. *Gretl Manual: Gnu Regression, Econometrics and Time-Series Library*; Wake Forest University: Winston-Salem, NC, USA, 2005.

62. Tobler, W.R. A Computer Movie Simulating Urban Growth in the Detroit Region. *Econ. Geogr.* **1970**, *46*, 234–240. [CrossRef]

63. Kim, C.W.; Phipps, T.T.; Anselin, L. Measuring the benefits of air quality improvement: A spatial hedonic approach. *J. Environ. Econ. Manag.* **2003**, *45*, 24–39. [CrossRef]

64. Anselin, L.; Rey, S. Properties of Tests for Spatial Dependence in Linear Regression Models. *Geogr. Anal.* **1991**, *23*, 112–131. [CrossRef]

65. Anselin, L. GIS Research Infrastructure for Spatial Analysis of Real Estate Markets. *J. Hous. Res.* **1998**, *9*, 113–133. [CrossRef]

66. Wilhelmsson, M. Spatial Models in Real Estate Economics. *Hous. Theory Soc.* **2002**, *19*, 92–101. [CrossRef]

67. Anselin, L. Exploring Spatial Data with GeoDa: A Workbook. *Geography* **2005**, *244*. Available online: http://www.csiss.org/ (accessed on 2 October 2017).

68. Trojanek, R. An analysis of changes in dwelling prices in the biggest cities of Poland in 2008–2012 conducted with the application of the hedonic method. *Actual Probl. Econ.* **2012**, *7*, 5–14.

69. Trojanek, M.; Trojanek, R. Profitability of Investing in Residential Units: The Case of Real Estate Market in Poland in the Period from 1997 to 2011. *Actual Probl. Econ.* **2012**, *2*, 73–83.

70. Trojanek, R. Dwelling's price fluctuations and the business cycle. *Econ. Sociol.* **2010**, *3*. [CrossRef] [PubMed]

![sustainability logo] *sustainability*

MDPI

Article

A New Group Decision Model Based on Grey-Intuitionistic Fuzzy-ELECTRE and VIKOR for Contractor Assessment Problem

Hassan Hashemi [1], Seyed Meysam Mousavi [2], Edmundas Kazimieras Zavadskas [3,*],
Alireza Chalekaee [3,4] and Zenonas Turskis [3]

[1]   Young Researchers and Elite Club, South Tehran Branch, Islamic Azad University, Tehran 1584743311, Iran; hashemi.h@live.com
[2]   Department of Industrial Engineering, Faculty of Engineering, Shahed University, Tehran 3319118651, Iran; sm.mousavi@shahed.ac.ir
[3]   Institute of Sustainable Construction, Faculty of Civil Engineering, Vilnius Gediminas Technical University, Saulėtekio ave. 11, LT-10223 Vilnius, Lithuania; zenonas.turskis@vgtu.lt
[4]   School of Civil Engineering, Iran University of Science and Technology, Tehran 1684613114, Iran; a_chalekaee@civileng.iust.ac.ir
*   Correspondence: edmundas.zavadskas@vgtu.lt; Tel.: +370-5274-4910

Received: 18 April 2018; Accepted: 16 May 2018; Published: 18 May 2018

Abstract: This study introduces a new decision model with multi-criteria analysis by a group of decision makers (DMs) with intuitionistic fuzzy sets (IFSs). The presented model depends on a new integration of IFSs theory, ELECTRE and VIKOR along with grey relational analysis (GRA). To portray uncertain real-life situations and take account of complex decision problem, multi-criteria group decision-making (MCGDM) model by totally unknown importance are introduced with IF-setting. Hence, a weighting method depended on Entropy and IFSs, is developed to present the weights of DMs and evaluation factors. A new ranking approach is provided for prioritizing the alternatives. To indicate the applicability of the presented new decision model, an industrial application for assessing contractors in the construction industry is given and discussed from the recent literature.

Keywords: multi-criteria group decision-making; ELECTRE; VIKOR; IFSs; GRA; contractor assessment problem

---

1. Introduction

The contractor selection process (CSP) includes five main stages in practice as follows [1]:

- Project packaging;
- Invitation;
- Prequalification;
- Shortlisting;
- Bid evaluation.

Multi-criteria decision-making (MCDM) approach can be suitable in solving complex problems, such as executing system selection and contractor evaluation [2–4]. The CSP can be taken in the MCDM framework, considering an overall strategy [5–7]. Different criteria must be considered along with the interest of a group of experts or decision makers (DMs) [8]. For the CSP, analytical methods have not properly improved, despite a high increase in the multifaceted nature of projects along with a relative increase in candidate forms of executing systems for the projects. Hence, these decision tools and methods should be highlighted and employed [9].

The outranking methods, as an uncommon category of MCDM methods, meet the particular requirements of the soft decisions which properly handle the real-decision situations e.g., [10–13]. To start with outranking method, ELECTRE (ELimination and Choice Expressing the Reality) was created by Roy [14]. Some outranking approaches, based on ELECTRE as well-known model, were reported in recent years i.e., [15,16]. Hashemi et al. [17] utilized the interval-valued intuitionistic fuzzy (IVF)-ELECTRE III as a suitable choice, keeping in mind the end goal to illuminate an investment project selection problem. Azadnia et al. [18] used the fuzzy C-Means (FCM) clustering regarded as a data-mining approach to categorize suppliers, and ELECTRE has been utilized to rank the suppliers. Sevkli [19] compared and contrasted crisp and fuzzy ELECTRE approaches for the supplier evaluation in an industry case. Teixeira de Almeida [20] proposed a model that integrated ELECTRE and utility function regarding to outsourcing contracts appraisement. Montazer et al. [21] developed a fuzzy expert system that was utilized to assist firms with fuzzy ELECTRE III. Marzouk [22] regarded MCDM approach with ELECTRE III for the CSP. You et al. [23] extended MCDM approach based on ELECTRE III and best-worst techniques with the multiplicative preference relations and intuitionistic fuzzy sets (IFSs).

Classical MCDM methods assume that the ratings of alternatives and the evaluation factors' relative importance regarded ascertain numbers, but in real engineering applications and management situations, these assumptions are not practical [24]. Therefore, various types of membership functions by concentrating on ambiguous components have been applied in solving engineering and management problems [25–30]. The IFSs propose a generalization of fuzzy sets theory [31,32]. Recently, this theory regards the explicit presentation and expression with both likes and dislikes. Various scientists have displayed new approaches and methodologies to adapt to the fuzzy MCDM (FMCDM) issues with taking IFSs. Chen [33] developed an IFS-approach to the problem solving, by utilizing decision tree induction. Ye [34] regarded decision problems by unknown information on weights of criteria with Entropy and IFSs. Li et al. [35] provided a linear programming approach for handling the MCDM with DMs and IFSs. Liu [36] extended power-average operator with IFSs for dealing with the MCDM. Fouladgar et al. [37] proposed a model to regard a specific end goal to figure the importance weights of assessment components and to rank feasible projects, respectively. Hashemi et al. [38] developed a compromise ratio approach with IFSs theory to water resources area. Zhao et al. [39] reported an IFS-VIKOR method to handle the supplier selection. Hosseinzadeh et al. [40] designed an MCDM model with a combination of IFS, grey relational analysis (GRA) and TOPSIS method to select the best precursor. Zavadskas et al. [41] extended the MULTIMOORA approach with IVIFSs for analyzing real-world civil engineering problems. Keshavaraz Ghorabaee et al. [42] reported the compromise solution by T2FSs for project selection problem.

This study designs a new multi-criteria group decision-making (MCGDM) model in light of novel hybrid approaches of the GRA, IF-ELECTRE and VIKOR along with multi-criteria analysis. In the IF-ELECTRE method, the calculation process of concordance dominance (CD) and discordance dominance (DD) matrixes are in light of the idea that the potential candidate or alternative ought to have the most limited distance from the positive ideal solution (PIS) and farthest distance from the negative ideal solution (NIS). Further, a weighting approach is regarded and extended in view of a generalized version of the Entropy and IFSs to determine weights of both DMs and the criteria. Finally, in view of the idea of the VIKOR method, a new index is introduced for appraising the alternatives.

The rest of this study is arranged as follows. An overview of IFSs is reported in Section 2. A decision model is illustrated in Section 3. A real application example is presented for the contractor selection problem according to the literature in Section 4 to indicate the steps of the model. In the final section, conclusions will be given.

2. Preliminaries

Atanassov [31] developed traditional fuzzy set to the IFS with regard to a hesitation degree. An IF is defined as:

$$I = \{\chi, \mu_I(\chi), \upsilon_I(\chi) | \chi \in X\}, \tag{1}$$

which is described with a membership function μ_I and a non-membership function υ_I, where

$$\mu_I : \chi \to [0,1], \chi \in X \to \mu_A(\chi) \in [0,1], \tag{2}$$

$$\upsilon_I : \chi \to [0,1], \chi \in X \to \upsilon_A(\chi) \in [0,1] \tag{3}$$

with the condition

$$0 \le \mu_I(\chi) + \upsilon_I(\chi) \le 1 \; for \; all \; \chi \in X. \tag{4}$$

The third parameter of IFS is $\pi_I(\chi)$, regarded as the intuitionistic fuzzy index as below [43]:

$$\pi_I(\chi) = 1 - \mu_I(\chi) - \upsilon_I(\chi). \tag{5}$$

and

$$0 \le \pi_I(\chi) \le 1. \tag{6}$$

Definition 1 [31,44]. Let I and I' be two IFSs, then

$$I \oplus I' = \{\chi, \mu_I(\chi) + \upsilon_{I'}(\chi) - \mu_I(\chi).\mu_{I'}(\chi), \upsilon_I(\chi).\upsilon_{I'}(\chi) | \chi \in X\}, \tag{7}$$

$$I \otimes I' = \{\chi, \mu_I(\chi).\mu_{I'}(\chi), \upsilon_I(\chi) + \upsilon_{I'}(\chi) - \upsilon_I(\chi).\upsilon_{I'}(\chi) | \chi \in X\}. \tag{8}$$

From these Equations, the following relations are obtained:

$$nI = \{\chi, (1 - (1 - \mu_I(\chi))^n, (\upsilon_I(\chi))^n | \chi \in X\}, n \ge 0, \tag{9}$$

$$I^n = \{\chi, (\mu_I(\chi))^n, (1 - (1 - \upsilon_I(\chi))^n, | \chi \in X\}, n \ge 0. \tag{10}$$

Definitionn 2 [45,46]. Let I be an IFS. Then the score function S and the accuracy function H may be represented as below:

$$S(I) = \mu_I - \upsilon_I, \tag{11}$$

and

$$H(I) = \mu_I - \upsilon_I, \tag{12}$$

respectively. Clearly $S(I) \in [-1,1]$ and $H(I) \in [0,1]$ for any IFS I.

Definitionn 3 [47]. Intuitionistic fuzzy weighted geometric with respect to a weighting vector ω, $IFWG_\omega$ is characterized as

$$IFWG_\omega(I_1, I_2, \ldots, I_n) = \prod_{j=1}^{n} I_j^{\omega_j} = \langle \prod_{j=1}^{n}(\mu_{I_j})^{\omega_j}, 1 - \prod_{j=1}^{n}(1 - \upsilon_{I_j})^{\omega_j}\rangle, \tag{13}$$

where $\omega_k \in [0,1]$, and $\sum_{j=1}^{n} \omega_j = 1, (j = 1, 2, \ldots, n)$.

Definitionn 4 [48]. Distance between two IFSs I and I' can be characterized as takes after:

$$D(I, I') = \sqrt{\frac{1}{2n} \sum_{j=1}^{n} \left[(\mu_I(x_j) - \mu_{I'}(x_j))^2 + (v_I(x_j) - v_{I'}(x_j))^2 + (\pi_I(x_j) - \pi_{I'}(x_j))^2 \right]} \quad (14)$$

3. Proposed Uncertain Group Decision Model

For the MCGDM problem with IF uncertainty, let $CA = \{CA_1, CA_2, \ldots, CA_m\}$ be a set of m candidates or alternatives, and $CR = \{CR_1, CR_2, \ldots, CR_n\}$ be the set of n conflicting criteria, and let $DM = \{DM_1, DM_2, \ldots, DM_t\}$ be a set of t DMs. Let $X^{(e)} = \left(\tilde{x}_{ij}^{(e)} \right)_{m \times n}$ be an IF-decision matrix, where $\tilde{x}_{ij}^{(e)} = \left(\mu_{ij}^{(e)}, v_{ij}^{(e)}, \pi_{ij}^{(e)} \right)$ is a criterion value provided by eth DM, denoted by an IFN, for the alternative CA_i versus the criterion CR_j.

The process of the proposed group decision model based on GRA, IF-ELECTRE and VIKOR methods are provided as below.

3.1. Determine the DMs' Importance, Criteria' Weights and Aggregated IFS Decision Matrix

There are various tools and approaches to regard the criteria' weights. This study adopts information regarding Entropy method to provide criteria' weights. The Entropy was one of the ideas in thermodynamics originally by Shannon [49]. The steps of determining the DMs' importance and criteria' weights by Entropy method are reported as below:

(1) To denote the DMs' importance from the IF-decision matrix, the method of Entropy weights [50] is given by:

$$\lambda_{ij}^{(e)} = \frac{1 - J_{ij}^{(e)}}{t - \sum_{k=1}^{t} J_{ij}^{(e)}}, \quad (15)$$

where $\lambda_{ij}^{(e)} \in [0,1]$, $\sum_{e=1}^{t} \lambda_{ij}^{(e)} = 1$, $i = 1, 2, \ldots, m$, $j = 1, 2, \ldots, n$, $e = 1, 2, \ldots, t$ and $J_{ij}^{(e)}$ is computed by:

$$J_{ij}^{(e)} = \frac{1}{\sqrt{2} - 1} \times \left\{ \sin \frac{\pi \left(1 + \mu_{ij}^{(e)} - v_{ij}^{(e)} \right)}{4} + \sin \frac{\pi \left(1 - \mu_{ij}^{(e)} + v_{ij}^{(e)} \right)}{4} - 1 \right\}, \quad (16)$$

where $0 \leq J_{ij}^{(e)} \leq 1$, $i = 1, 2, \ldots, m$, $j = 1, 2, \ldots, n$ and $e = 1, 2, \ldots, t$.

(2) After weights' values for the DMs are obtained, the evaluating values described by different DMs are aggregated regarding the IFWG operator by:

$$\tilde{r}_{ij} = \left(\tilde{x}_{ij}^{(1)} \right)^{\lambda_{ij}^{(1)}} \otimes \left(\tilde{x}_{ij}^{(2)} \right)^{\lambda_{ij}^{(2)}} \otimes \cdots \otimes \left(\tilde{x}_{ij}^{(t)} \right)^{\lambda_{ij}^{(t)}}, \quad (17)$$

$$\tilde{r}_{ij} = \langle \mu_{ij}, v_{ij} \rangle = \langle \prod_{e=1}^{t} \left(\mu_{ij}^{(e)} \right)^{\lambda_{ij}^{(e)}}, 1 - \prod_{e=1}^{t} \left(1 - v_{ij}^{(e)} \right)^{\lambda_{ij}^{(e)}} \rangle. \quad (18)$$

(3) To provide w_j as weights of evaluation criteria, IF-Entropy is as below [50]:

$$G_j = \frac{1}{m} \sum_{i=1}^{m} \frac{1}{\sqrt{2} - 1} \left(\sin \frac{\pi \left(1 + \mu_{ij} - v_{ij} \right)}{4} + \sin \frac{\pi \left(1 - \mu_{ij} + v_{ij} \right)}{4} - 1 \right), \quad (19)$$

where $0 \leq G_j \leq 1$, $i = 1, 2, \ldots, m$ and $j = 1, 2, \ldots, n$.

Entropy weight of the *j*th criterion is reported as:

$$w_j = \frac{1 - G_j}{n - \sum_{j=1}^{n} G_j},$$

(20)

where $w_j \in [0,1]$, $\sum_{j=1}^{n} w_j = 1$, $j = 1,2,\ldots,n$.

3.2. Ranking of Alternatives by the Model

We can consider different alternatives and compare based on their IF values. Various types of concordance sets as the concordance set, midrange concordance set, and weak concordance set (CS) by ideas of score function and accuracy function are classified. It is also similar to discordance set (DS) [13].

Let $\tilde{X} = (\mu_{\tilde{x}}, v_{\tilde{x}}, \pi_{\tilde{x}})$ be an IF value. The CS C_{kl} of A_k and A_l contains all criteria for which A_k is preferred to A_l. We apply ideas of score function, accuracy function, and hesitancy degree of the IFNs to classify concordance sets. The CS C_{kl} can be provided as follows [13]:

$$C_{kl}^1 = \left\{ j \middle| \mu_{kj} > \mu_{lj} , \ v_{kj} < v_{lj} \ and \ \left(\mu_{kj} + v_{kj} \right) > \left(\mu_{lj} + v_{lj} \right) \right\}$$

(21)

where $j = 1,2,\ldots,n$, and Equation (21) can be more concordant than Equation (22) or Equation (23). The midrange CS C_{kl}^2 is denoted:

$$C_{kl}^2 = \left\{ j \middle| \mu_{kj} > \mu_{lj} , \ v_{kj} < v_{lj} \ and \ \left(\mu_{kj} + v_{kj} \right) \leq \left(\mu_{lj} + v_{lj} \right) \right\}$$

(22)

The main difference between Equations (21) and (22) is the hesitancy degree; it at the *k*th alternative versus the *j*th criterion is regarded higher than the *l*th alternative versus the j*th criterion in the midrange concordance set. Thus, Equation (21) can be more concordant than (22). The weak CS C_{kl}^3 is denoted as:

$$C_{kl}^3 = \left\{ j \middle| \mu_{kj} \geq \mu_{lj} \ and \ v_{kj} \geq v_{lj} \right\}$$

(23)

The degree of non-membership at the *k*th alternative versus the *j*th criterion is regarded higher than the *l*th alternative versus the *j*th criterion in weak concordance set; thus, Equation (22) can be more concordant than (23). The DS includes all criteria for which A_k is not related to A_l by:

$$D_{kl}^1 = \left\{ j \middle| \mu_{kj} < \mu_{lj} , \ v_{kj} \geq v_{lj} \ and \ \left(\mu_{kj} + v_{kj} \right) \leq \left(\mu_{lj} + v_{lj} \right) \right\}$$

(24)

The midrange DS D_{kl}^2 is denoted as follows:

$$D_{kl}^2 = \left\{ j \middle| \mu_{kj} \langle \mu_{lj} , \ v_{kj} \rangle v_{lj} \ and \ \left(\mu_{kj} + v_{kj} \right) > \left(\mu_{lj} + v_{lj} \right) \right\}$$

(25)

Equation (24) can be more discordant than Equation (25). The weak DS D_{kl}^3 is denoted as follows:

$$D_{kl}^3 = \left\{ j \middle| \mu_{kj} < \mu_{lj} \ and \ v_{kj} < v_{lj} \right\}$$

(26)

Equation (25) can be more discordant than Equation (26).

In the proposed new hybrid GRA, IF-ELECTRE and VIKOR model with assessment data, the relative value of CS could be taken through the concordance index. Hence, the concordance index C_{kl} between A_k and A_l in this study is characterized as:

$$\varphi_{kl} = w_{c1} \times \sum_{j \in C_{kl}^1} w_j + w_{c2} \times \sum_{j \in C_{kl}^2} w_j + w_{c3} \times \sum_{j \in C_{kl}^3} w_j, \tag{27}$$

where w_{c1}, w_{c2} and w_{c3} are the weights of the concordance, midrange concordance, and weak concordance sets, respectively, and w_j is the weight of the evaluation criteria.

Concordance matrix Φ could be formed as:

$$\Phi = \begin{bmatrix} - & \varphi_{12} & \varphi_{13} & \cdots & \varphi_{1m} \\ \varphi_{21} & - & \varphi_{23} & \cdots & \varphi_{2m} \\ \vdots & \vdots & - & \ddots & \vdots \\ \varphi_{(m-1)1} & \varphi_{(m-1)2} & \cdots & - & \varphi_{(m-1)m} \\ \varphi_{m1} & \varphi_{m2} & \cdots & \varphi_{m(m-1)} & - \end{bmatrix}, \tag{28}$$

where the maximum and the minimum values of φ_{kl} are denoted by φ^* and, φ^- which are the positive ideal point and negative ideal point, respectively. Also, a higher value of φ_{kl} indicates that A_k would be preferred to A_l and vice versa.

Evaluations of certain A_k are worse than appraisements of a competing A_k. Discordance index is provided as:

$$\varepsilon_{kl} = \frac{\max\limits_{j \in D_{kl}} w_D^* \times d\left(\tilde{x}_{kj}, \tilde{x}_{lj}\right)}{\max\limits_{j \in J} d\left(\tilde{x}_{kj}, \tilde{x}_{lj}\right)}, \tag{29}$$

where $d\left(\tilde{x}_{kj}, \tilde{x}_{lj}\right)$ is determined by Equation (14), and w_D^* is equal to w_{D1}, w_{D2} or w_{D3}. Discordance matrix E is formed as:

$$E = \begin{bmatrix} - & \varepsilon_{12} & \varepsilon_{13} & \cdots & \varepsilon_{1m} \\ \varepsilon_{21} & - & \varepsilon_{23} & \cdots & \varepsilon_{2m} \\ \vdots & \vdots & - & \ddots & \vdots \\ \varepsilon_{(m-1)1} & \varepsilon_{(m-1)2} & \cdots & - & \varepsilon_{(m-1)m} \\ \varepsilon_{m1} & \varepsilon_{m2} & \cdots & \varepsilon_{m(m-1)} & - \end{bmatrix} \tag{30}$$

where the maximum and the minimum values of ε_{kl} are indicated with ε^* and ε^-, which are the negative ideal and positive ideal points, respectively. A higher value of ε_{kl} indicates that A_k would be less favourable than A_l and vice versa.

Steps of the GRA algorithm can be reported as below [51–54]: The grey relational coefficient is calculated. The grey relational coefficient $\gamma\left(x_{0j}, x_{ij}\right)$ is computed by:

$$\gamma\left(x_{0j}, x_{ij}\right) = \frac{\min\limits_i \min\limits_j \left|x_{0j} - x_{ij}\right| + \rho \max\limits_i \max\limits_j \left|x_{0j} - x_{ij}\right|}{\left|x_{0j} - x_{ij}\right| + \rho \max\limits_i \max\limits_j \left|x_{0j} - x_{ij}\right|}, \tag{31}$$

where ρ is the identification coefficient $\rho \in [0,1]$, $i = 1,2,\ldots,m$ and $j = 1,2,\ldots,n$.

The grade $\gamma\left(x_{0j}, x_{ij}\right)$ between x_0 and x_i can be as:

$$\gamma(x_0, x_i) = \sum_{j=1}^{n} w_j \gamma\left(x_{0j}, x_{ij}\right) \text{ and } \sum_{j=1}^{n} w_j = 1, \tag{32}$$

Introduced CD matrix calculation process in this study is according to the compromise solution idea. It means that the alternative must have the shortest grey relational coefficient from PIS and the farthest grey relational coefficient from the NIS; thus, the CD matrix Ψ is formed as:

$$
\Psi = \begin{bmatrix}
- & \psi_{12} & \psi_{13} & \cdots & \psi_{1m} \\
\psi_{1m} & - & p_{23} & \cdots & \psi_{2m} \\
\vdots & \vdots & - & \ddots & \vdots \\
\psi_{(m-1)1} & \psi_{(m-1)2} & \cdots & - & \psi_{(m-1)m} \\
\psi_{m1} & \psi_{m2} & \cdots & \psi_{m(m-1)} & -
\end{bmatrix}
\tag{33}
$$

where

$$
\psi_{kl} = \frac{\xi_{kl}^+}{\xi_{kl}^- + \xi_{kl}^+},
\tag{34}
$$

$$
\xi_{kl}^+ = \frac{\min\limits_{1\le k\le m}\min\limits_{1\le l\le m}\left|\varphi^* - \varphi_{kl}\right| + \rho\max\limits_{1\le k\le m}\max\limits_{1\le l\le m}\left|\varphi^* - \varphi_{kl}\right|}{\left|\varphi^* - \varphi_{kl}\right| + \rho\max\limits_{1\le k\le m}\max\limits_{1\le l\le m}\left|\varphi^* - \varphi_{kl}\right|}, \; k \text{ and } l = 1,2,\ldots,m,
\tag{35}
$$

$$
\xi_{kl}^- = \frac{\min\limits_{1\le k\le m}\min\limits_{1\le l\le m}\left|\varphi^- - \varphi_{kl}\right| + \rho\max\limits_{1\le k\le m}\max\limits_{1\le l\le m}\left|\varphi^- - \varphi_{kl}\right|}{\left|\varphi^- - \varphi_{kl}\right| + \rho\max\limits_{1\le k\le m}\max\limits_{1\le l\le m}\left|\varphi^- - \varphi_{kl}\right|}, \; k \text{ and } l = 1,2,\ldots,m.
\tag{36}
$$

A higher value of ψ_{kl} could indicate that A_k is less favourable than A_l.

In a similar way, the presented DD matrix is formed in this study similar to the proposed CD matrix calculation process; thus, the DD matrix Ω is formed as:

$$
\Omega = \begin{bmatrix}
- & \omega_{12} & \omega_{13} & \cdots & \omega_{1m} \\
\omega_{1m} & - & p_{23} & \cdots & \omega_{2m} \\
\vdots & \vdots & - & \ddots & \vdots \\
\omega_{(m-1)1} & \omega_{(m-1)2} & \cdots & - & \omega_{(m-1)m} \\
\omega_{m1} & \omega_{m2} & \cdots & \omega_{m(m-1)} & -
\end{bmatrix}
\tag{37}
$$

where

$$
\omega_{kl} = \frac{\zeta_{kl}^+}{\zeta_{kl}^- + \zeta_{kl}^+},
\tag{38}
$$

$$
\zeta_{kl}^+ = \frac{\min\limits_{1\le k\le m}\min\limits_{1\le l\le m}\left|\varepsilon^* - \varepsilon_{kl}\right| + \rho\max\limits_{1\le k\le m}\max\limits_{1\le l\le m}\left|\varepsilon^* - \varepsilon_{kl}\right|}{\left|\varepsilon^* - \varepsilon_{kl}\right| + \rho\max\limits_{1\le k\le m}\max\limits_{1\le l\le m}\left|\varepsilon^* - \varepsilon_{kl}\right|}, \; k \text{ and } l = 1,2,\ldots,m,
\tag{39}
$$

$$
\zeta_{kl}^- = \frac{\min\limits_{1\le k\le m}\min\limits_{1\le l\le m}\left|\varepsilon^- - \varepsilon_{kl}\right| + \rho\max\limits_{1\le k\le m}\max\limits_{1\le l\le m}\left|\varepsilon^- - \varepsilon_{kl}\right|}{\left|\varepsilon^- - \varepsilon_{kl}\right| + \rho\max\limits_{1\le k\le m}\max\limits_{1\le l\le m}\left|\varepsilon^- - \varepsilon_{kl}\right|}, \; k \text{ and } l = 1,2,\ldots,m.
\tag{40}
$$

A higher value of ω_{kl} could indicate that A_k is preferred to A_l. According to the VIKOR method idea, the $\mathcal{I}_i$, $\mathcal{R}_i$, $\mathcal{I}_i'$ and $\mathcal{R}_i'$ values are represented by:

$$
\mathcal{I}_i = \sum_{l=1;l\ne k}^{m} \psi_{il},
\tag{41}
$$

$$
\mathcal{R}_i = \max_l(\psi_{il}),
\tag{42}
$$

$$
\mathcal{I}_i' = \sum_{l=1;l\ne k}^{m} \omega_{il},
\tag{43}
$$

$$\mathcal{R}_i' = \max_l(w_{il}).$$ (44)

Then, the values of indices δ_i and ϱ_i are proposed:

$$\delta_i = \left(\frac{\mathcal{I}_i + \mathcal{R}_i}{2}\right)\left(\frac{\mathcal{I}_i - \mathcal{I}^+}{\mathcal{I}^- - \mathcal{I}^+}\right) + \left(\frac{2 - (\mathcal{I}_i + \mathcal{R}_i)}{2}\right)\left(\frac{\mathcal{R}_i - \mathcal{R}^+}{\mathcal{R}^- - \mathcal{R}^+}\right)$$ (45)

and

$$\varrho_i = \left(\frac{\mathcal{I}_i' + \mathcal{R}_i'}{2}\right)\left(\frac{\mathcal{I}_i' - \mathcal{I}'^-}{\mathcal{I}'^+ - \mathcal{I}'^-}\right) + \left(\frac{2 - (\mathcal{I}_i' + \mathcal{R}_i')}{2}\right)\left(\frac{\mathcal{R}_i' - \mathcal{R}'^-}{\mathcal{R}'^+ - \mathcal{R}'^-}\right),$$ (46)

where $\begin{cases} \mathcal{I}^+ = \min_i \mathcal{I}_i \\ \mathcal{I}^- = \max_i \mathcal{I}_i \end{cases}$, $\begin{cases} \mathcal{R}^+ = \min_i \mathcal{R}_i \\ \mathcal{R}^- = \max_i \mathcal{R}_i \end{cases}$, $\begin{cases} \mathcal{I}'^+ = \max_i \mathcal{I}_i' \\ \mathcal{I}'^- = \min_i \mathcal{I}_i' \end{cases}$, $\begin{cases} \mathcal{R}'^+ = \max_i \mathcal{I}_i \\ \mathcal{R}'^- = \min_i \mathcal{I}_i \end{cases}$.

We have the following relation:

$$Q_i = \frac{\delta_i}{\delta_i + \varrho_i}$$ (47)

Q_i is the final value of assessment. All options can be positioned by Q_i. The best option A^* can be characterized as below:

$$A^* = \max\{Q_i\}.$$ (48)

3.3. Algorithm

An algorithm of the proposed decision model can be given as below:

step 1. A group of DMs is established to solve the complicated decision problem by considering conflicting criteria;

step 2. Proper criteria are reported for the selection problem;

step 3. Provide the ratings of each candidate versus each selected criterion for each DM;

step 4. Weight of each DM from the decision matrix is calculated by Equations (15) and (16);

step 5. Construct an aggregated IFS decision matrix by Equations (17) and (18);

step 6. Present the weights of appraisement criteria by Equations (19) and (20);

step 7. Identify the CS and DS. Find C_{kl}^1, C_{kl}^2, C_{kl}^3, D_{kl}^1, D_{kl}^2 and D_{kl}^3 for pair-wise comparisons of candidates by Equations (21)–(26);

step 8. Form the concordance matrix Φ by Equations (27) and (28);

step 9. Calculate the discordance matrix E by Equations (29) and (30);

step 10. Form CD matrix P by Equations (33)–(36);

step 11. Form DD matrix O by Equations (37)–(40);

step 12. Determine the values of $\mathcal{I}_i$, $\mathcal{R}_i$, $\mathcal{I}_i'$ and $\mathcal{R}_i'$ by Equations (41)–(44);

step 13. Compute the values of indices δ_i and ϱ_i are by Equations (45) and (46);

step 14. Calculate values of ranking index (Q_i) using Equation (47). Rank the candidates in decreasing order.

Finally, a flowchart of the proposed model is illustrated in Figure 1.

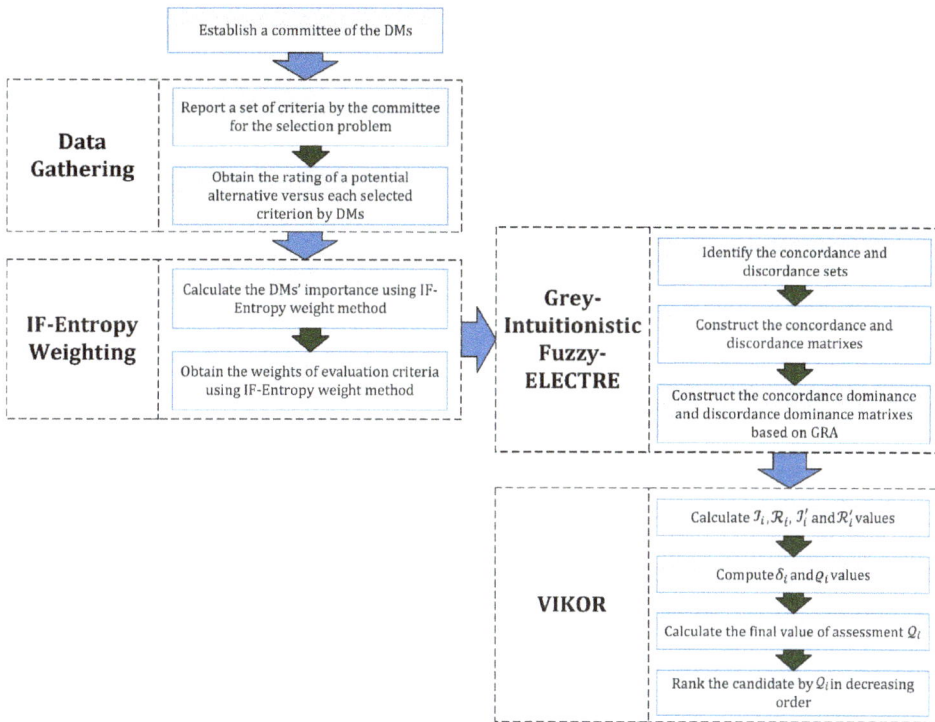

Figure 1. Flowchart of proposed model.

4. Solution of Contractor Assessment Problem

4.1. Implementation and Computational Results

Construction projects are initiated in dynamically changing and complicated environment, which result in circumstances of high uncertainty and risks [55,56]. Choosing the best alternative for a building is of great importance for owners, contractors, and stakeholders [57]. To exhibit the appropriateness of the soft decision model, a case study from the recent literature [58] is presented regarding the construction contractor assessment. This assessment can be via some conflicting criteria. A group of three DMs (DM_1, DM_2, and DM_3) is arranged to appraise the appropriate contractor. In this industrial application, five potential contractors (CO_1, CO_2, ..., CO_5) are chosen, and twenty criteria (CR_1, CR_2, ..., CR_{20}) . are reported for final assessments (steps 1 and 2). By taking DMs' judgments, all ratings of alternatives versus evaluation factors are represented with linguistic variables by Table 1.

Table 1. Linguistic variables for performance ratings.

Linguistic Variables	Intuitionistic Fuzzy Numbers
Verygood (VG)	$\langle 0.90, 0.10 \rangle$
Good (G)	$\langle 0.80, 0.15 \rangle$
Medium good (MG)	$\langle 0.65, 0.25 \rangle$
Fair (F)	$\langle 0.50, 0.40 \rangle$
Medium poor (MP)	$\langle 0.30, 0.60 \rangle$
Poor (P)	$\langle 0.20, 0.75 \rangle$
Verypoor (VP)	$\langle 0.10, 0.90 \rangle$

The performance of alternatives in terms of appraisement criteria is represented by three DMs and then illustrated in Table 2 (step 3).

Table 2. Ratings of contractors.

Criteria	Contractors	Decision Makers		
		DM_1	DM_2	DM_3
CR_1	CO_1	F	MP	MP
	CO_2	G	MG	VG
	CO_3	MP	F	F
	CO_4	MG	G	F
	CO_5	MG	F	F
CR_2	CO_1	F	MP	MP
	CO_2	VG	G	G
	CO_3	MG	G	G
	CO_4	G	MG	VG
	CO_5	MG	F	F
CR_3	CO_1	VG	VG	G
	CO_2	MG	G	G
	CO_3	G	VG	MG
	CO_4	MG	G	G
	CO_5	F	MP	MG
CR_4	CO_1	MP	F	F
	CO_2	G	MG	MG
	CO_3	F	MP	MG
	CO_4	G	VG	MG
	CO_5	MG	G	F
CR_5	CO_1	F	MG	MP
	CO_2	VG	G	G
	CO_3	MG	F	G
	CO_4	MG	F	G
	CO_5	MG	G	G
CR_6	CO_1	F	MG	MP
	CO_2	VG	VG	G
	CO_3	MG	F	F
	CO_4	G	VG	MG
	CO_5	G	VG	MG
CR_7	CO_1	MG	G	G
	CO_2	G	MG	VG
	CO_3	MG	F	F
	CO_4	G	MG	VG
	CO_5	G	MG	MG
CR_8	CO_1	F	MG	MP
	CO_2	MG	G	G
	CO_3	F	MG	MG
	CO_4	MG	F	F
	CO_5	MG	F	F
CR_9	CO_1	MG	G	F
	CO_2	VG	G	G
	CO_3	MG	G	F
	CO_4	G	MG	MG
	CO_5	MG	G	G
CR_{10}	CO_1	F	MG	MG
	CO_2	G	VG	VG
	CO_3	F	MG	MG
	CO_4	G	VG	MG
	CO_5	MG	F	G

Table 2. *Cont.*

Criteria	Contractors	Decision Makers		
		DM_1	DM_2	DM_3
CR_{11}	CO_1	MG	F	F
	CO_2	VG	G	G
	CO_3	MG	F	G
	CO_4	G	VG	MG
	CO_5	MG	F	G
CR_{12}	CO_1	F	MP	MP
	CO_2	MG	F	F
	CO_3	F	MG	MP
	CO_4	F	MP	MG
	CO_5	MG	G	F
CR_{13}	CO_1	MG	G	G
	CO_2	G	VG	VG
	CO_3	MG	F	G
	CO_4	G	VG	VG
	CO_5	G	MG	VG
CR_{14}	CO_1	MP	P	F
	CO_2	F	MG	MP
	CO_3	MP	F	P
	CO_4	F	MP	MP
	CO_5	F	MG	MP
CR_{15}	CO_1	MP	F	P
	CO_2	F	MG	MP
	CO_3	F	MP	MP
	CO_4	F	MP	MP
	CO_5	F	MG	MG
CR_{16}	CO_1	F	MP	MP
	CO_2	MG	F	G
	CO_3	F	MP	MG
	CO_4	MG	F	F
	CO_5	F	MP	MP
CR_{17}	CO_1	F	MG	MG
	CO_2	MG	F	F
	CO_3	MG	G	F
	CO_4	MG	G	G
	CO_5	F	MG	MG
CR_{18}	CO_1	MG	G	G
	CO_2	VG	VG	G
	CO_3	G	MG	VG
	CO_4	G	VG	MG
	CO_5	MG	G	F
CR_{19}	CO_1	F	MP	MP
	CO_2	G	VG	VG
	CO_3	MG	F	G
	CO_4	G	MG	VG
	CO_5	MG	F	F
CR_{20}	CO_1	MG	F	G
	CO_2	G	VG	MG
	CO_3	MG	G	F
	CO_4	G	VG	MG
	CO_5	G	VG	MG

Table 1 Linguistic variables for performance ratings.

The weights of each DM are computed and presented by Equations (15) and (16) and then by Equations (17) and (18), the aggregated IFS decision matrix constructs by the DMs (steps 4 and 5), reported in Table 3.

Table 3. Aggregated IFS decision matrix.

Contractors \ Criteria	CR_1	CR_2	CR_3	CR_4
CO_1	$\langle 0.308, 0.591\rangle$	$\langle 0.308, 0.591\rangle$	$\langle 0.874, 0.113\rangle$	$\langle 0.329, 0.569\rangle$
CO_2	$\langle 0.827, 0.139\rangle$	$\langle 0.841, 0.129\rangle$	$\langle 0.774, 0.167\rangle$	$\langle 0.731, 0.195\rangle$
CO_3	$\langle 0.329, 0.569\rangle$	$\langle 0.774, 0.167\rangle$	$\langle 0.827, 0.139\rangle$	$\langle 0.492, 0.402\rangle$
CO_4	$\langle 0.750, 0.183\rangle$	$\langle 0.827, 0.139\rangle$	$\langle 0.774, 0.167\rangle$	$\langle 0.827, 0.139\rangle$
CO_5	$\langle 0.631, 0.268\rangle$	$\langle 0.631, 0.268\rangle$	$\langle 0.492, 0.402\rangle$	$\langle 0.750, 0.183\rangle$

Contractors \ Criteria	CR_5	CR_6	CR_7	CR_8
CO_2	$\langle 0.841, 0.129\rangle$	$\langle 0.874, 0.113\rangle$	$\langle 0.827, 0.139\rangle$	$\langle 0.774, 0.167\rangle$
CO_3	$\langle 0.750, 0.183\rangle$	$\langle 0.631, 0.268\rangle$	$\langle 0.631, 0.268\rangle$	$\langle 0.645, 0.255\rangle$
CO_4	$\langle 0.750, 0.183\rangle$	$\langle 0.827, 0.139\rangle$	$\langle 0.827, 0.139\rangle$	$\langle 0.631, 0.268\rangle$
CO_5	$\langle 0.774, 0.167\rangle$	$\langle 0.827, 0.139\rangle$	$\langle 0.731, 0.195\rangle$	$\langle 0.631, 0.268\rangle$

Contractors \ Criteria	CR_9	CR_{10}	CR_{11}	CR_{12}
CO_2	$\langle 0.841, 0.129\rangle$	$\langle 0.874, 0.113\rangle$	$\langle 0.841, 0.129\rangle$	$\langle 0.631, 0.268\rangle$
CO_3	$\langle 0.750, 0.183\rangle$	$\langle 0.645, 0.255\rangle$	$\langle 0.750, 0.183\rangle$	$\langle 0.492, 0.402\rangle$
CO_4	$\langle 0.731, 0.195\rangle$	$\langle 0.827, 0.139\rangle$	$\langle 0.827, 0.139\rangle$	$\langle 0.492, 0.402\rangle$
CO_5	$\langle 0.774, 0.167\rangle$	$\langle 0.750, 0.183\rangle$	$\langle 0.750, 0.183\rangle$	$\langle 0.750, 0.183\rangle$

Contractors \ Criteria	CR_{13}	CR_{14}	CR_{15}	CR_{16}
CO_2	$\langle 0.874, 0.113\rangle$	$\langle 0.492, 0.402\rangle$	$\langle 0.492, 0.402\rangle$	$\langle 0.750, 0.183\rangle$
CO_3	$\langle 0.750, 0.183\rangle$	$\langle 0.224, 0.716\rangle$	$\langle 0.308, 0.591\rangle$	$\langle 0.492, 0.402\rangle$
CO_4	$\langle 0.874, 0.113\rangle$	$\langle 0.308, 0.591\rangle$	$\langle 0.308, 0.591\rangle$	$\langle 0.631, 0.268\rangle$
CO_5	$\langle 0.827, 0.139\rangle$	$\langle 0.492, 0.402\rangle$	$\langle 0.645, 0.255\rangle$	$\langle 0.308, 0.591\rangle$

Contractors \ Criteria	CR_{17}	CR_{18}	CR_{19}	CR_{20}
CO_2	$\langle 0.631, 0.268\rangle$	$\langle 0.874, 0.113\rangle$	$\langle 0.874, 0.113\rangle$	$\langle 0.827, 0.139\rangle$
CO_3	$\langle 0.750, 0.183\rangle$	$\langle 0.827, 0.139\rangle$	$\langle 0.750, 0.183\rangle$	$\langle 0.750, 0.183\rangle$
CO_4	$\langle 0.774, 0.167\rangle$	$\langle 0.827, 0.139\rangle$	$\langle 0.827, 0.139\rangle$	$\langle 0.827, 0.139\rangle$
CO_5	$\langle 0.645, 0.255\rangle$	$\langle 0.750, 0.183\rangle$	$\langle 0.631, 0.268\rangle$	$\langle 0.827, 0.139\rangle$

Twenty criteria' weights are established with Equations (19) and (20) and are given in Table 4 (step 6).

Table 4. Aggregated IFS decision matrix by DMs' opinions.

Criteria	CR_1	CR_2	CR_3	CR_4	CR_5
Weights	0.038	0.055	0.063	0.041	0.058
Criteria	CR_6	CR_7	CR_8	CR_9	CR_{10}
Weights	0.059	0.061	0.028	0.064	0.059
Criteria	CR_{11}	CR_{12}	CR_{13}	CR_{14}	CR_{15}
Weights	0.062	0.020	0.082	0.021	0.020
Criteria	CR_{16}	CR_{17}	CR_{18}	CR_{19}	CR_{20}
Weights	0.022	0.040	0.078	0.056	0.073

The CS and DS could be identified (step 7). The relative weights of DMs also are reported as:

$$W' = [w_{C1}, w_{C2}, w_{C3}, w_{D1}, w_{D2}, w_{D3}] = \left[1, \frac{2}{3}, \frac{1}{3}, 1, \frac{2}{3}, \frac{1}{3}\right],$$

The CS can be:

$$C_{kl}^1 = \begin{bmatrix}
- & 3,\ 17 & 3,\ 7,\ 13 & 3,\ 9 & 3,\ 7,\ 18 \\
\begin{matrix}1,\ 2,\ 4,\ 5,\ 6,\ 7,\ 8,\ 9,\ 10,\ 11,\\12,\ 13,\ 16,\ 18,\ 19,\ 20\end{matrix} & - & \begin{matrix}1,\ 2,\ 4,\ 5,\ 6,\ 7,\ 8,\ 9,\ 10,11,\\12,\ 13,\ 16,\ 18,\ 19,\ 20\end{matrix} & \begin{matrix}1,\ 2,\ 5,\ 6,\ 8,\ 9,\ 10,\\11,12,\ 16,\ 18,\ 19\end{matrix} & \begin{matrix}1,\ 2,\ 3,\ 5,\ 6,\ 7,\ 8,\ 9,\ 10,\\11,13,\ 16,\ 18,\ 19\end{matrix} \\
\begin{matrix}2,\ 5,\ 6,\ 8,\ 11,\ 17,\ 18,\ 19\end{matrix} & 3,\ 17 & - & 3,\ 8,\ 9 & \begin{matrix}2,\ 3,\ 8,\ 17,\ 18,\ 19\end{matrix} \\
\begin{matrix}1,\ 2,\ 4,\ 5,\ 6,\ 7,\ 8,\ 10,\ 11,\\13,16,\ 17,\ 18,\ 19,\ 20\end{matrix} & 4,\ 17 & \begin{matrix}1,\ 2,\ 4,\ 6,\ 7,\ 10,\ 11,\\13,16,\ 17,\ 19,\ 20\end{matrix} & - & \begin{matrix}1,\ 2,\ 3,\ 4,\ 7,\ 10,\ 11,\\13,\ 16,\ 17,\ 18,\ 19\end{matrix} \\
\begin{matrix}1,\ 2,\ 4,\ 5,\ 6,\ 8,\ 9,\ 10,\\11,\ 12,\ 13,\ 19,\ 20\end{matrix} & 4,\ 12,\ 15,\ 17 & \begin{matrix}1,\ 4,\ 5,\ 6,\ 7,\ 9,\ 10,\\12,\ 13,\ 15,\ 20\end{matrix} & 5,\ 9,\ 12,\ 15 & -
\end{bmatrix}$$

The midrange CS can be:

$$C_{kl}^2 = \begin{bmatrix}
- & - & - & - & - \\
14,\ 15 & - & 14,\ 15 & 14,\ 15 & \\
1,\ 4,\ 12,\ 15,\ 16 & - & - & - & 16 \\
12,\ 14,\ 15 & - & 14 & - & \\
14,\ 15 & - & 14 & 14 & -
\end{bmatrix}$$

The weak CS can be:

$$C_{kl}^3 = \begin{bmatrix}
- & - & 9,\ 10,\ 14,\ 20 & - & 16,\ 17 \\
- & - & - & 3,\ 7,\ 13,\ 20 & 14,\ 20 \\
9,\ 10,\ 14,\ 20 & - & - & 5,\ 12,\ 15 & 11 \\
- & 3,\ 7,\ 13,\ 20 & 5,\ 12,\ 15,\ 18 & - & 6,\ 8,\ 20 \\
16,\ 17 & 14,\ 20 & 11 & 6,\ 8,\ 20 & -
\end{bmatrix}$$

The DS can be:

$$D_{kl}^1 = \begin{bmatrix}
- & \begin{matrix}1,\ 2,\ 4,\ 5,\ 6,\ 7,\ 8,\ 9,\ 10,\ 11,\\12,\ 13,\ 16,\ 18,\ 19,\ 20\end{matrix} & \begin{matrix}2,\ 5,\ 6,\ 8,\ 11,\ 17,\ 18,\ 19\end{matrix} & \begin{matrix}1,\ 2,\ 4,\ 5,\ 6,\ 7,\ 8,\ 10,\ 11,\ 13,\\16,\ 17,\ 18,\ 19,\ 20\end{matrix} & \begin{matrix}1,\ 2,\ 4,\ 5,\ 6,\ 8,\ 9,\ 10,\ 11,\\12,\ 13,\ 19,\ 20\end{matrix} \\
3,\ 17 & - & 3,\ 17 & 4,\ 17 & 4,\ 12,\ 15,\ 17 \\
3,\ 7,\ 13 & \begin{matrix}1,\ 2,\ 4,\ 5,\ 6,\ 7,\ 8,\ 9,\ 10,\ 11,\\12,\ 13,\ 16,\ 18,\ 19,\ 20\end{matrix} & - & \begin{matrix}1,\ 2,\ 4,\ 6,\ 7,\ 10,\ 11,\ 13,\\16,\ 17,\ 18,\ 19,\ 20\end{matrix} & \begin{matrix}1,\ 4,\ 5,\ 6,\ 7,\ 9,\ 10,\\12,\ 13,\ 15,\ 20\end{matrix} \\
3,\ 9 & \begin{matrix}1,\ 2,\ 5,\ 6,\ 8,\ 9,\ 10,\\11,\ 12,\ 16,\ 18,\ 19\end{matrix} & 3,\ 8,\ 9 & - & 5,\ 9,\ 12,\ 15 \\
3,\ 7,\ 18 & \begin{matrix}1,\ 2,\ 3,\ 5,\ 6,\ 7,\ 8,\ 9,\ 10,\\11,\ 13,\ 16,\ 18,\ 19\end{matrix} & 2,\ 3,\ 8,\ 17,\ 18,\ 19 & \begin{matrix}1,\ 2,\ 3,\ 4,\ 7,\ 10,\ 11,\\13,\ 16,\ 17,\ 18,\ 19\end{matrix} & -
\end{bmatrix}$$

The midrange DS can be:

$$D_{kl}^2 = \begin{bmatrix}
- & 14,\ 15 & 1,\ 4,\ 12,\ 15,\ 16 & 12,\ 14,\ 15 & 14,\ 15 \\
- & - & - & - & - \\
- & 14,\ 15 & - & 14 & 14 \\
- & 14,\ 15 & - & - & 14 \\
- & - & 16 & - & -
\end{bmatrix}.$$

The weak DS can be:

$$D_{kl}^3 = \begin{bmatrix}
- & - & - & - & - \\
- & - & - & - & - \\
- & - & - & - & - \\
- & - & - & - & - \\
- & - & - & - & -
\end{bmatrix}.$$

Then, matrixes of the concordance and discordance can be formed (steps 8 and 9). The respective results can be as:

$$\Phi = \begin{bmatrix} - & 0.241 & 1.069 & 0.241 & 0.694 \\ 3.711 & - & 3.711 & 3.322 & 2.921 \\ 2.922 & 0.241 & - & 1.058 & 1.100 \\ 3.614 & 0.827 & 3.332 & - & 2.636 \\ 3.401 & 1.503 & 2.934 & 1.631 & - \end{bmatrix},$$

$$E = \begin{bmatrix} - & 1.000 & 1.000 & 1.000 & 0.963 \\ 0.165 & - & 0.226 & 0.681 & 0.352 \\ 0.285 & 1.000 & - & 1.000 & 1.000 \\ 0.178 & 0.731 & 0.115 & - & 1.000 \\ 0.780 & 1.000 & 0.909 & 0.960 & - \end{bmatrix}$$

Consequently, matrixes of the CD and DD are constructed (steps 10 and 11). The respective results are as follows:

$$\Psi = \begin{bmatrix} - & 0.250 & 0.369 & 0.250 & 0.315 \\ 0.750 & - & 0.750 & 0.694 & 0.636 \\ 0.636 & 0.250 & - & 0.368 & 0.374 \\ 0.736 & 0.334 & 0.695 & - & 0.595 \\ 0.705 & 0.432 & 0.638 & 0.450 & - \end{bmatrix}$$

and

$$\Omega = \begin{bmatrix} - & 0.750 & 0.750 & 0.750 & 0.729 \\ 0.278 & - & 0.313 & 0.570 & 0.384 \\ 0.346 & 0.750 & - & 0.750 & 0.750 \\ 0.285 & 0.598 & 0.250 & - & 0.750 \\ 0.626 & 0.750 & 0.698 & 0.727 & - \end{bmatrix}$$

$\mathcal{I}_i$, $\mathcal{R}_i$, $\mathcal{I}'_i$ and $\mathcal{R}'_i$ values are determined (step 12).

$$\mathcal{I} = \begin{bmatrix} 0.296 \\ 0.708 \\ 0.407 \\ 0.590 \\ 0.556 \end{bmatrix}, \mathcal{R} = \begin{bmatrix} 0.369 \\ 0.750 \\ 0.636 \\ 0.736 \\ 0.705 \end{bmatrix}, \mathcal{I}' = \begin{bmatrix} 0.745 \\ 0.386 \\ 0.390 \\ 0.471 \\ 0.700 \end{bmatrix}, and \mathcal{R}' = \begin{bmatrix} 0.750 \\ 0.570 \\ 0.750 \\ 0.750 \\ 0.750 \end{bmatrix}.$$

Then, the values of indices δ_i and ϱ_i are computed (step 13).

$$\delta = \begin{bmatrix} 0.000 \\ 1.000 \\ 0.479 \\ 0.812 \\ 0.736 \end{bmatrix}, and \varrho = \begin{bmatrix} 0.000 \\ 1.000 \\ 0.831 \\ 0.562 \\ 0.919 \end{bmatrix}.$$

Finally, the final value of evaluation index (Q_i) s calculated (step 14) as given in Table 5. The optimal ranking order is as $CO_2 > CO_4 > CO_5 > CO_3 > CO_1$ and the best contractors could be the CO_2. In addition, the final value of evaluation index Q_i for appraising alternatives has been taken in comparison with the fuzzy VIKOR method by the previous study [58] in Table 5. The results demonstrate that the same ranking results on the CSP are obtained.

Table 5. Final value of Q_i for ranking order of alternatives.

Contractors	Q_i	Final Ranking (Proposed Model)	Final Ranking (Fuzzy VIKOR by [58])
CO_1	0.000	5	5
CO_2	1.000	1	1
CO_3	0.369	4	4
CO_4	0.599	2	2
CO_5	0.443	3	3

The computational results are given in Table 5; the proposed model (via the GRA, Entropy, IFSs, ELECTRE and new compromise ranking index) versus the modified VIKOR method (via conventional fuzzy sets) by the previous study [58] is compared. Both models can handle complex CSPs with uncertain conditions; however, some main merits of the presented group decision model are provided as below:

- Firstly, this study takes account of key advantages of IFSs and GRA concurrently to handle the uncertain information via the group decision process and to involve more flexibility to illustrate the imprecise and vague data of the several experience DMs.
- Secondly, a new ranking method based on the compromise solution within a new version of classical ELECTRE approach is developed to distinguish potential candidates of the complex CSP as a reasonable way of the optimal ranking, and to introduce stable decisions in the construction industry with uncertain conditions.

4.2. Sensitivity Analysis

A sensitivity analysis is represented on each identification coefficient value ρ. The computational results can be represented in Table 6 and illustrated in Figure 2. According to Table 6, the final ranking orders of contractors with the changes of ρ value ($\rho = 0.1$ to $\rho = 1$) are the same.

Table 6. Sensitivity analysis on each identification coefficient value.

ρ Value		Contractors				
		CO_1	CO_2	CO_3	CO_4	CO_5
$\rho = 0.1$	Q_i	0.000	1.000	0.377	0.618	0.441
	Preference order ranking	5	1	4	2	3
$\rho = 0.2$	Q_i	0.000	1.000	0.374	0.611	0.442
	Preference order ranking	5	1	4	2	3
$\rho = 0.3$	Q_i	0.000	1.000	0.372	0.606	0.442
	Preference order ranking	5	1	4	2	3
$\rho = 0.4$	Q_i	0.000	1.000	0.370	0.602	0.443
	Preference order ranking	5	1	4	2	3
$\rho = 0.5$	Q_i	0.000	1.000	0.369	0.599	0.443
	Preference order ranking	5	1	4	2	3
$\rho = 0.6$	Q_i	0.000	1.000	0.368	0.597	0.444
	Preference order ranking	5	1	4	2	3
$\rho = 0.7$	Q_i	0.000	1.000	0.368	0.595	0.444
	Preference order ranking	5	1	4	2	3
$\rho = 0.8$	Q_i	0.000	1.000	0.367	0.594	0.444
	Preference order ranking	5	1	4	2	3
$\rho = 0.9$	Q_i	0.000	1.000	0.366	0.592	0.444
	Preference order ranking	5	1	4	2	3
$\rho = 1$	Q_i	0.000	1.000	0.366	0.591	0.445
	Preference order ranking	5	1	4	2	3

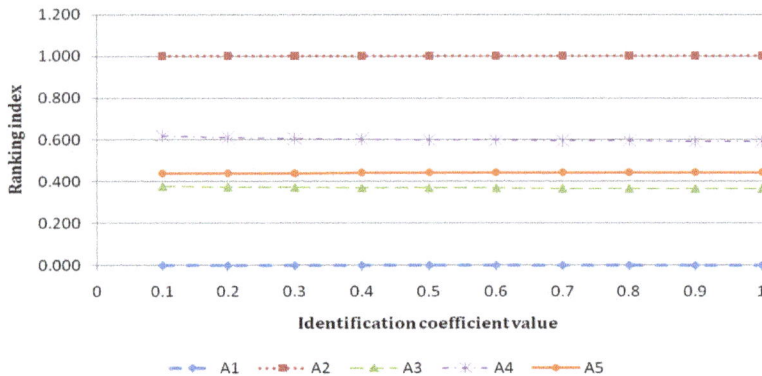

Figure 2. Variation analysis of Q_i for the sensitivity analysis.

New group decision model can take account of the gaps between the Q_i values of various alternatives appear larger when the coefficient cannot change remarkably, and they have enough stability.

5. Concluding Remarks and Future Research

The study introduced a new version of MCGDM model under uncertainty. Major concepts of IFSs theory and GRA were considered in the presented model along with the uncertain ELECTRE and VIKOR methods for selection and assessment problems. For this purpose, linguistic variables denoted by IF-numbers, by regarding the truth-membership and non-truth-membership functions, were utilized to report the importance of each candidate for the complicated problems. Then, a weighting approach was represented for Entropy analysis and IFSs. In addition, a new version of classical ELECTRE method was presented as indicated by the ideas of IFSs and grey theory. Finally, a new ranking index was introduced based on the VIKOR method concept for the appraisement. Furthermore, a case study from the recent literature for construction contractor assessment was indicated to successfully illustrate and validate the proposed model. Comparing with the previous studies, the proposed model assists the DMs or experts with a beneficial way to take fuzzy MCDM complex problems in more generalized methodology because of the way that it applied IFSs rather than conventional fuzzy sets to express the performance ratings of each candidate versus criteria. Although the new decision model provided is demonstrated by a decision problem of the contractor assessment, it is interesting to apply the model in other important management fields, like project selection.

Author Contributions: H.H. and S.M.M. presented the research methodology, performed the development and experiments of this study; E.K.Z. provided extensive advice throughout the study; A.C. and Z.T. assisted with the findings and revised the manuscript. All of the authors have read and approved the final manuscript.

Acknowledgments: The authors are grateful to three anonymous referees for their valuable recommendations that enhanced the quality of the primary version.

Conflicts of Interest: The authors declare no conflict of interest.

References

1. Hatush, Z.; Skitmore, M. Criteria for contractor selection. *Constr. Manag. Econ.* **1997**, *15*, 19–38. [CrossRef]
2. Vahdani, B.; Mousavi, S.M.; Hashemi, H.; Mousakhani, M.; Ebrahimnejad, S. A new hybrid model based on least squares support vector machine for project selection problem in construction industry. *Arab. J. Sci. Eng.* **2014**, *39*, 4301–4314. [CrossRef]
3. Liu, P.; Chen, S.M. Multiattribute group decision making based on intuitionistic 2-tuple linguistic information. *Inf. Sci.* **2018**, *430*, 599–619. [CrossRef]

4. Keshavarz Ghorabaee, M.; Zavadskas, E.K.; Olfat, L.; Turskis, Z. Multi-criteria inventory classification using a new method of evaluation based on distance from average solution (EDAS). *Informatica* **2015**, *26*, 435–451. [CrossRef]

5. Zavadskas, E.K.; Vilutienė, T.; Turskis, Z.; Tamosaitiene, J. Contractor selection for construction works by applying SAW-G and TOPSIS grey techniques. *J. Bus. Econ. Manag.* **2010**, *11*, 34–55. [CrossRef]

6. Zavadskas, E.K.; Turskis, Z.; Antucheviciene, J. Selecting a Contractor by Using a Novel Method for Multiple Attribute Analysis: Weighted Aggregated Sum Product Assessment with Grey Values (WASPAS-G). *Stud. Inform. Control* **2015**, *24*, 141–150. [CrossRef]

7. Zagorskas, J.; Zavadskas, E.K.; Turskis, Z.; Burinskienė, M.; Blumberga, A.; Blumberga, D. Thermal insulation alternatives of historic brick buildings in Baltic Sea Region. *Energy Build.* **2014**, *78*, 35–42. [CrossRef]

8. Zavadskas, E.K.; Kaklauskas, A.; Turskis, Z.; Kalibatas, D. An approach to multi-attribute assessment of indoor environment before and after refurbishment of dwellings. *J. Environ. Eng. Landsc. Manag.* **2009**, *17*, 5–11. [CrossRef]

9. Hashemkhani Zolfani, S.; Zavadskas, E.K.; Turskis, Z. Design of products with both International and Local perspectives based on Yin-Yang balance theory and SWARA method. *Econ. Res.-Ekon. Istraž.* **2013**, *26*, 153–166.

10. Gitinavard, H.; Mousavi, S.M.; Vahdani, B. Soft computing-based new interval-valued hesitant fuzzy multi-criteria group assessment method with last aggregation to industrial decision problems. *Soft Comput.* **2017**, *21*, 3247–3265. [CrossRef]

11. Ebrahimnejad, S.; Hashemi, H.; Mousavi, S.M.; Vahdani, B. A New Interval-valued Intuitionistic Fuzzy Model to Group Decision Making for the Selection of Outsourcing Providers. *J. Econ. Comput. Econ. Cybern. Stud. Res.* **2015**, *49*, 269–290.

12. Khanzadi, M.; Turskis, Z.; Ghodrati Amiri, G.; Chalekaee, A. A model of discrete zero-sum two-person matrix games with grey numbers to solve dispute resolution problems in construction. *J. Civ. Eng. Manag.* **2017**, *23*, 824–835. [CrossRef]

13. Wu, M.C.; Chen, T.Y. The ELECTRE multicriteria analysis approach based on Atanassov's intuitionistic fuzzy sets. *Expert Syst. Appl.* **2011**, *38*, 12318–12327. [CrossRef]

14. Roy, B. Classement et choix en présence de points de vue multiples. Revue française d'automatique, d'informatique et de recherche opérationnelle. *Recherche Opér.* **1968**, *2*, 57–75.

15. Mousavi, M.; Gitinavard, H.; Mousavi, S.M. A soft computing based-modified ELECTRE model for renewable energy policy selection with unknown information. *Renew. Sustain. Energy Rev.* **2017**, *68*, 774–787. [CrossRef]

16. Yu, X.; Zhang, S.; Liao, X.; Qi, X. ELECTRE methods in prioritized MCDM environment. *Inf. Sci.* **2018**, *424*, 301–316. [CrossRef]

17. Hashemi, S.S.; Hajiagha, S.H.R.; Zavadskas, E.K.; Mahdiraji, H.A. Multicriteria group decision making with ELECTRE III method based on interval-valued intuitionistic fuzzy information. *Appl. Math. Model.* **2016**, *40*, 1554–1564. [CrossRef]

18. Azadnia, A.H.; Ghadimi, P.; Saman, M.Z.M.; Wong, K.Y.; Sharif, S. Supplier selection: A hybrid approach using ELECTRE and fuzzy clustering. In *International Conference on Informatics Engineering and Information Science*; Springer: Berlin/Heidelberg, Germany, 2011; pp. 663–676.

19. Sevkli, M. An application of the fuzzy ELECTRE method for supplier selection. *Int. J. Prod. Res.* **2010**, *48*, 3393–3405. [CrossRef]

20. Teixeira de Almeida, A. Multicriteria decision model for outsourcing contracts selection based on utility function and ELECTRE method. *Comput. Oper. Res.* **2007**, *34*, 3569–3574. [CrossRef]

21. Montazer, G.A.; Saremi, H.Q.; Ramezani, M. Design a new mixed expert decision aiding system using fuzzy ELECTRE III method for vendor selection. *Expert Syst. Appl.* **2009**, *36*, 10837–10847. [CrossRef]

22. Marzouk, M. An application of ELECTRE III to contractor selection. In *Construction Research Congress 2010: Innovation for Reshaping Construction Practice*; Ruwanpura, J., Mohamed, Y., Lee, S.H., Eds.; American Society of Civil Engineers: Reston, VA, USA, 2010; pp. 1316–1324.

23. You, X.; Chen, T.; Yang, Q. Approach to multi-criteria group decision-making problems based on the best-worst-method and electre method. *Symmetry* **2016**, *8*, 95. [CrossRef]

24. Zavadskas, E.K.; Vilutienė, T.; Turskis, Z.; Šaparauskas, J. Multi-criteria analysis of Projects' performance in construction. *Arch. Civ. Mech. Eng.* **2014**, *14*, 114–121. [CrossRef]

25. Hashemi, H.; Mousavi, S.M.; Tavakkoli-Moghaddam, R.; Gholipour, Y. Compromise ranking approach with bootstrap confidence intervals for risk assessment in port management projects. *J. Manag. Eng. (ASCE)* **2013**, *29*, 334–344. [CrossRef]

26. Liang, W.; Zhao, G.; Wu, H. Evaluating investment risks of metallic mines using an extended TOPSIS method with linguistic neutrosophic numbers. *Symmetry* **2017**, *9*, 149. [CrossRef]

27. Hashemi, H.; Bazargan, J.; Mousavi, S.M.; Vahdani, B. An extended compromise ratio model with an application to reservoir flood control operation under an interval-valued intuitionistic fuzzy environment. *Appl. Math. Model.* **2014**, *38*, 3495–3511. [CrossRef]

28. Nie, R.X.; Wang, J.Q.; Zhang, H.Y. Solving solar-wind power station location problem using an extended weighted aggregated sum product assessment (WASPAS) technique with interval neutrosophic sets. *Symmetry* **2017**, *9*, 106. [CrossRef]

29. Zavadskas, E.K.; Antucheviciene, J.; Hajiagha, S.H.R.; Hashemi, S.S. Extension of weighted aggregated sum product assessment with interval-valued intuitionistic fuzzy numbers (WASPAS-IVIF). *Appl. Soft Comput.* **2014**, *24*, 1013–1021. [CrossRef]

30. Luo, S.Z.; Cheng, P.F.; Wang, J.Q.; Huang, Y.J. Selecting project delivery systems based on simplified neutrosophic linguistic preference relations. *Symmetry* **2017**, *9*, 151. [CrossRef]

31. Atanassov, K.T. Intuitionistic fuzzy sets. *Fuzzy Sets Syst.* **1986**, *20*, 87–96. [CrossRef]

32. Xu, Z. Intuitionistic preference relations and their application in group decision making. *Inf. Sci.* **2007**, *177*, 2363–2379. [CrossRef]

33. Chen, R.Y. A problem-solving approach to product design using decision tree induction based on intuitionistic fuzzy. *Eur. J. Oper. Res.* **2009**, *196*, 266–272. [CrossRef]

34. Ye, J. Fuzzy decision-making method based on the weighted correlation coefficient under intuitionistic fuzzy environment. *Eur. J. Oper. Res.* **2010**, *205*, 202–204. [CrossRef]

35. Li, D.F.; Chen, G.H.; Huang, Z.G. Linear programming method for multi-attribute group decision making using IF sets. *Inf. Sci.* **2010**, *180*, 1591–1609. [CrossRef]

36. Liu, P. Multiple attribute decision-making methods based on normal intuitionistic fuzzy interaction aggregation operators. *Symmetry* **2017**, *9*, 261. [CrossRef]

37. Fouladgar, M.M.; Yazdani-Chamzini, A.; Zavadskas, E.K.; Yakhchali, S.H.; Ghasempourabadi, M.H. Project portfolio selection using fuzzy AHP and VIKOR techniques. *Transform. Bus. Econ.* **2012**, *11*, 213–231.

38. Hashemi, H.; Bazargan, J.; Mousavi, S.M. A Compromise Ratio Method with an Application to Water Resources Management: An Intuitionistic Fuzzy Set. *Water Res. Manag.* **2013**, *27*, 2029–2051. [CrossRef]

39. Zhao, J.; You, X.Y.; Liu, H.C.; Wu, S.M. An extended VIKOR method using intuitionistic fuzzy sets and combination weights for supplier selection. *Symmetry* **2017**, *9*, 169. [CrossRef]

40. Hosseinzadeh, F.; Sarpoolaki, H.; Hashemi, H. Precursor Selection for Sol-Gel Synthesis of Titanium Carbide Nanopowders by a New Intuitionistic Fuzzy Multi-Attribute Group Decision-Making Model. *Int. J. Appl. Ceram. Technol.* **2014**, *11*, 681–698. [CrossRef]

41. Zavadskas, E.K.; Antucheviciene, J.; Razavi Hajiagha, S.H.; Hashemi, S.S. The interval-valued intuitionistic fuzzy MULTIMOORA method for group decision making in engineering. *Math. Probl. Eng.* **2015**, 560690. [CrossRef]

42. Keshavaraz Ghorabaee, M.K.; Amiri, M.; Sadaghiani, J.S.; Zavadskas, E.K. Multi-Criteria Project Selection Using an Extended VIKOR Method with Interval Type-2 Fuzzy Sets. *Int. J. Inf. Technol. Decis. Mak.* **2015**, *14*, 993–1016. [CrossRef]

43. Shu, M.H.; Cheng, C.H.; Chang, J.R. Using intuitionistic fuzzy sets for fault-tree analysis on printed circuit board assembly. *Microelectron. Reliab.* **2006**, *46*, 2139–2148. [CrossRef]

44. De, S.K.; Biswas, R.; Roy, A.R. Some operations on intuitionistic fuzzy sets. *Fuzzy Sets Syst.* **2000**, *114*, 477–484. [CrossRef]

45. Chen, S.M.; Tan, J.M. Handling multicriteria fuzzy decision-making problems based on vague set theory. *Fuzzy Sets Syst.* **1994**, *67*, 163–172. [CrossRef]

46. Hong, D.H.; Choi, C.H. Multicriteria fuzzy decision-making problems based on vague set theory. *Fuzzy Sets Syst.* **2000**, *114*, 103–113. [CrossRef]

47. Xu, Z.; Yager, R.R. Some geometric aggregation operators based on intuitionistic fuzzy sets. *Int. J. Gen. Syst.* **2006**, *35*, 417–433. [CrossRef]

48. Szmidt, E.; Kacprzyk, J. Distances between intuitionistic fuzzy sets. *Fuzzy Sets Syst.* **2000**, *114*, 505–518. [CrossRef]

49. Shannon, C.E. Communication Theory of Secrecy Systems. *Bell Syst. Tech. J.* **1949**, *28*, 656–715. [CrossRef]

50. Ye, J. Multiple attribute group decision-making methods with completely unknown weights in intuitionistic fuzzy setting and interval-valued intuitionistic fuzzy setting. *Group Decis. Negot.* **2013**, *22*, 173–188. [CrossRef]

51. Celik, E.; Bilisik, O.N.; Erdogan, M.; Gumus, A.T.; Baracli, H. An integrated novel interval type-2 fuzzy MCDM method to improve customer satisfaction in public transportation for Istanbul. *Transp. Res. Part E Logist. Transp. Rev.* **2013**, *58*, 28–51. [CrossRef]

52. Chen, W.H. A grey-based approach for distribution network reconfiguration. *J. Chin. Inst. Eng.* **2005**, *28*, 795–802. [CrossRef]

53. Kuo, M.S.; Liang, G.S. Combining VIKOR with GRA techniques to evaluate service quality of airports under fuzzy environment. *Expert Syst. Appl.* **2011**, *38*, 1304–1312. [CrossRef]

54. Wei, G.W. Gray relational analysis method for intuitionistic fuzzy multiple attribute decision making. *Expert Syst. Appl.* **2011**, *38*, 11671–11677. [CrossRef]

55. Zavadskas, E.K.; Turskis, Z.; Volvačiovas, R.; Kildiene, S. Multi-criteria assessment model of technologies. *Stud. Inform. Control* **2013**, *22*, 249–258. [CrossRef]

56. Turskis, Z.; Juodagalvienė, B. A novel hybrid multi-criteria decision-making model to assess a stairs shape for dwelling houses. *J. Civ. Eng. Manag.* **2016**, *22*, 1078–1087. [CrossRef]

57. Turskis, Z.; Daniūnas, A.; Zavadskas, E.K.; Medzvieckas, J. Multicriteria evaluation of building foundation alternatives. *Comput.-Aided Civ. Infrastruct. Eng.* **2016**, *31*, 717–729. [CrossRef]

58. Vahdani, B.; Mousavi, S.M.; Hashemi, H.; Mousakhani, M.; Tavakkoli-Moghaddam, R. A new compromise solution method for fuzzy group decision-making problems with an application to the contractor selection. *Eng. Appl. Artif. Intell.* **2013**, *26*, 779–788. [CrossRef]

MDPI

Article

Sustainable Construction Supply Chains through Synchronized Production Planning and Control in Engineer-to-Order Enterprises

Patrick Dallasega * and Erwin Rauch

Faculty of Science and Technology, Free University of Bozen-Bolzano, Universitätsplatz 5, 39100 Bolzano, Italy; erwin.rauch@unibz.it
* Correspondence: patrick.dallasega@unibz.it; Tel.: +39-0471-017114

Received: 8 September 2017; Accepted: 18 October 2017; Published: 20 October 2017

Abstract: Sustainability in the supply chain is becoming more and more important for industrial enterprises in different sectors. This research article focuses on construction supply chains (CSCs) in the Engineer-to-Order (ETO) industry, where every product is almost unique based on specific customer needs and requirements. The development of methods and approaches for more sustainable supply chain management in construction is becoming even more important. Engineering, fabrication of parts and their installation on-site are not always well synchronized in ETO supply chains. The results of such supply chains are long lead times, inefficient material transport and high and uncontrolled levels of work-in-progress (WIP). This article describes a conceptual approach to synchronize demand on-site with supply in manufacturing using the CONstant Work In Progress (ConWIP) concept from Lean Management to achieve Just-in-Time (JIT) supply. As a result, sustainable supply chains in ETO enterprises, with optimizations from an economic, ecological and social point of view, can be designed. The approach has been validated in an industrial case study.

Keywords: construction; supply chain management; engineer-to-order; sustainability; resource efficiency; just-in-time; constant work in progress

1. Introduction

A growing number of enterprises are working on the implementation of sustainable manufacturing and supply chain processes. The objective of sustainable supply chain initiatives is the creation of products or objects by means of energy-efficient, resource-saving as well as socially acceptable processes [1]. Customer satisfaction will be achieved in future not only through the creation of the product itself, but also through socially and environmentally responsible as well as economically efficient concepts of manufacturing avoiding negative effects for society [2]. Thus, we need modern organizational models and approaches to design supply chains with a focus on sustainability. In this paper, there is special emphasis on the ETO construction industry focusing on a case study from façade manufacturing industry. Construction supplier companies can be classified according to different market interaction strategies (Figure 1): (1) Make-to-Stock (MTS); (2) Assemble-to-Order (ATO); (3) Production-on-Demand (POD); (4) Build-to-Order (BTO); (5) Configure-to-Order (CTO) and (6) Engineer-to-Order (ETO) [3,4].

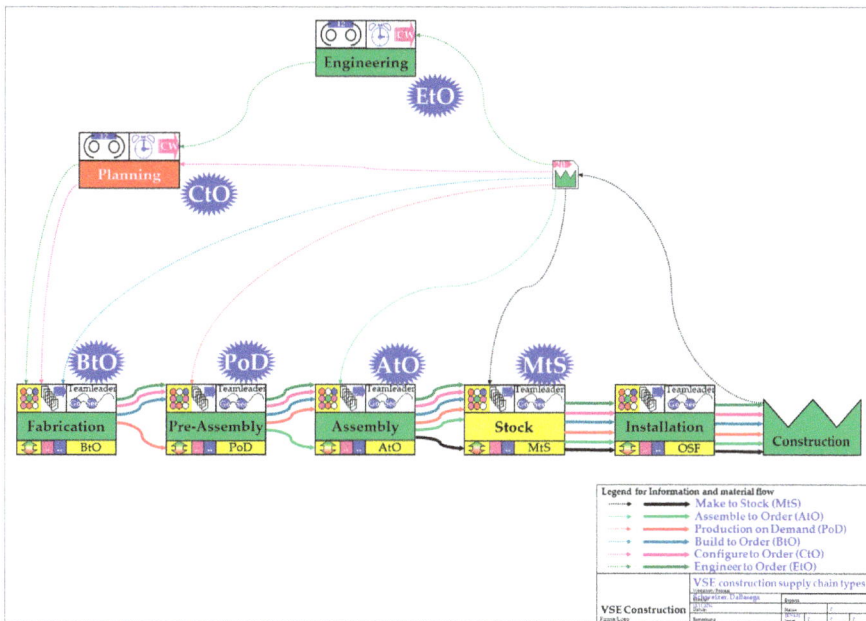

Figure 1. Market interaction strategies for construction suppliers [5].

As in many ETO construction supply chains (CSCs), the problem of traditional façade producers is that manufacturing processes are disconnected from the installation on-site [6]. This is emphasized by considering tier one suppliers, which produce and deliver their products from a fabrication shop to the construction site for installation. As a consequence, economic benefits reached through scale effects in production, are often lost due to an inefficient installation process on-site [7]. ETO companies are characterized by products or singular components that are engineered (developed) and produced according to a specific customer order. This means that every product is unique and therefore different from the other and standardization in production or installation is almost impossible [8]. This customization of products also makes it difficult to gain from learning curves during development, fabrication and installation on-site. Another challenge in the construction industry is the efficient usage of human resources on different projects. Very often ETO enterprises have to manage several projects in different places or countries through accurate multi-project management [9]. Switching personnel frequently between different projects to meet critical deadlines is not recommended, because this means there is an initial learning effort and thus inefficiencies in resource use.

Typical first-tier ETO supply chains in the construction industry consider three different macro-phases [6]:

1. Engineering
2. Fabrication
3. Installation on-site

Traditionally, ETO-projects are not synchronized to the three phases and, therefore, budget overruns occur. The causes for this are manifold. An example could be if fabrication is not informed about the progress of installation on-site. In this case, the fabrication shop could produce too much (overproduction) generating a lot of stock at the production site or at the construction site. In another scenario, faster progress on-site could lead to a bottleneck in material supply, if production is not informed regularly about the construction progress. The same lack of integration and information also occurs between engineering and manufacturing. Installation and/or manufacturing are not able to go on with their work because of missing information or drawings from engineering. ETO companies design and build products to customer specifications. Thus, a significant amount of time and cost goes into the engineering and design stage of the project. On the other side, if changes occur on the shop floor or on the installation site, project planning and scheduling as well as engineering are not always updated automatically. Additional problems occur due to constant changes in the project by manufacturing, installation or customers themselves [10]. In the fabrication shop, often changes in time schedules are due to technical or logistical problems in order processing. Mostly, the customer requires changes before and during the production or installation of components. This leads to capacity bottlenecks and complexity in project management and in supply chain management. All these changes in engineering, manufacturing and installation have to be synchronized and coordinated.

In Lean Management, inventories or buffers stand for waste (muda) and thus they are handled as activities which are not value-adding. For the purposes of a lean process along the entire value chain from engineering to manufacturing to installation, inventory should be reduced to a minimum. Thus, there is a need to close this gap of synchronization between engineering, manufacturing and on-site installation enabling efficient and sustainable supply chain management with low inventory and a Just-in-Time (JIT) delivery of material and information [6].

The paper is structured as follows: after a short introduction in Section 1, Section 2 describes the theoretical background with a review of sustainability challenges in CSC management and as a response recent works in the field of Lean Management, especially models for JIT-oriented production planning and control. Section 3 deals with the analysis of the actual situation of production planning in typical ETO construction suppliers. Section 4 explains the proposed approach for synchronization from manufacturing to on-site installation in ETO supply chains and in Section 5 the practical application and validation in an industrial case study is presented. Finally, Section 6 discusses the benefits and advantages of the proposed approach to increase sustainability in ETO supply chains based on the three key elements: economic, ecological and social sustainability.

2. Theoretical Background

In this section, first sustainability challenges in CSCs are described. As a response, approaches from Lean Management, especially JIT-based models and concepts for production planning and control in construction and especially in the ETO environment are presented.

2.1. Sustainability in CSCs

Usually, CSCs are MTO supply chains, which converge all materials to the site where the building is assembled from incoming materials [11–14]. Moreover, especially in the field of individual construction where ETO products are common, CSCs can usually be considered as temporary initiatives characterized by fragmentation, instability and high inefficiency [11,14]. Therefore, SCM concepts developed and applied in other industries, like the manufacturing industry, cannot be directly applied to the construction industry [15]. As a response, Vrijhoef and Koskela presented four specific roles of SCM in construction focusing on the supply chain, the construction site, or both [14]. Here, by means of the design of new supply chain configurations, the aim is to reach a global reduction of transportation, inventory and production costs. Moreover, by means of industrialization and especially prefabrication, moving activities from an uncontrollable environment (the construction site) to a controllable environment (the fabrication shop), the aim is to achieve improvements in terms of quality, time and cost. Unlike the BTO and MTS manufacturing supply chains (see Figure 1) that have successfully developed lean and sustainable supply chain management [16–20], the ETO CSC is yet to realize sustainability due to many complications from the design, fabrication to installation varying in many supply chain partners. In best practice, sustainability of the supply chain is partially realized. Sustainable construction becomes a concern for many companies in the construction industry [16–20]. There exist guidelines for designing sustainable buildings; however, there are different interests and focuses between Europe, Asia and America, e.g., construction impact, material usage, procurement and transportation [21,22]. Davies and Davies [23] describe drivers and barriers of sustainability in construction. According to their study, typical drivers are client awareness, regulations, financial incentives or tax burdens while typical barriers are affordability, lack of client awareness as well as a lack of proven alternative technologies.

Many sustainability-related concerns are addressed in research but usually in specific activities, e.g., waste management, designing or purchasing [24–26]. Information Lifecycle Management (ILM) and Information Technology (IT) have been widely used to improve information sharing of ETO. However, these technologies are limited to a specific phase of the CSC [27–30]. Total supply chain information sharing and decision support tools are rather ineffective and inefficient. This results in long lead times and overall supply chain inefficiency [31,32]. A study at Kota Kinabalu by Ali et al. found that environmental and sustainability concerns will not be understood if there is lack of knowledge, tools and skills to address the various issues to support a sustainable development plan [33]. The obstacle for sustainable ETO might focus on the social aspects relating to teach sustainability in education and skilled workers to carry out the work. On the other hand, in small construction environments, the decision-making is rather focused on financial and economic perspectives. Supply chain, human resources, environment and risk management are among recent interests. However, the sustainability aspect is yet to be realized [34–36].

In addition to these specific difficulties, there are some general challenges in the ETO industry. In traditional ETO manufacturing, building components are produced mainly according to the push principle. This means, that shop floor drawings are pushed from engineering to fabrication. In the fabrication shop, components are produced according to the available shop floor drawings. Finally, components are pushed from fabrication to the site for installation. Five of the seven types of waste (identified by Taiichi Ohno [37]) are described as problems, which endanger the sustainability of ETO construction suppliers: overproduction, waiting time, motion, inventory and failure correction [38]. Overproduction means that an upstream process, like engineering or fabrication, produces outputs (like drawings or components) in too large a quantity and before they are actually needed by downstream processes. According to practitioners, the justification for large batch sizes is that "this is how we have always done it" or that some of the machines need long setup times. Moreover, materials are ordered from outstanding suppliers more and sooner than needed to make sure they are available when needed. As a result, overproduction is seen as one of the major types of waste because it triggers the other ones mentioned previously. Often, the cause of missing components on-site (creating construction interruptions and so waiting times) is that the fabrication department cannot produce them, due to the missing information about the progress on-site. In addition, the correction of failures shows an economic as well as ecological waste. Errors in the engineering department or manufacturing are usually only discovered on-site. In this case, the defective components have to be engineered and fabricated again and finally reinstalled; eventually, the old parts have to be scrapped as well. Using big batch sizes, in particular, creates cost explosions if correction operations are necessary. The main cause of big levels of inventory and high expenditure for handling and motion operations at the fabrication shop or at the installation site is the aforementioned lack of synchronization between fabrication and installation. The communication and synchronization between the different departments becomes increasingly complex if installation is geographically far away from the fabrication shop, if the company is working in parallel on many different projects (multi-project management) and if the project is extraordinarily extensive and lengthy.

2.2. Review of JIT-Based, Lean Production Planning and Control in ETO CSCs

In recent decades, Lean Principles have been used in manufacturing and later in construction (Lean Construction) to optimize production flows and to reduce waste. However, according to a study performed by Bevilacqua et al. 2017 [39] a wide application of lean principles can be recognized in larger Italian companies. Smaller enterprises often fear that the implementation of lean concepts is costly and time consuming and that it does not contribute to the business growth of a company. Moreover, smaller firms tend to be more agile and flexible to adjusting themselves to changing market conditions and therefore accepting high production costs [39]. This could be especially considered the case for Italian ETO companies working in the CSC industry.

In contrast to traditional push-oriented production planning and control methods, the theory of Lean Construction suggests the adoption of pull-based models [40]. Push systems are those in which production jobs are scheduled, whereas pull systems are those in which the start of one job is triggered by the completion of another. In push systems, an error in demand forecast causes bullwhip effects. However, in JIT ordering systems, amplifications are avoided because the actual demand is used instead of the demand forecast [41]. Two types of JIT control circuits are generally used for production management: the KANBAN system and the CONWIP system. The KANBAN system was developed by the Japanese automobile manufacturer Toyota [42]. Pull production controlled by Kanban requires a steady part flow, which is impractical for small and infrequent orders as it is common in the construction environment [43]. More in detail, by using KANBAN in a small lot size and high product variety environment, unused WIP and unresponsiveness of the system is caused because Kanban cards have to propagate backward through the entire line for releasing new orders [43]. Moreover, because KANBAN requires that cards be assigned to parts, this means that at every station just one standard parts container should be placed allowing that a downstream process is able to pull what needed. As a result, an outstanding space needed for placing part containers and an increased complexity to handle the system for non-repetitive manufacturing would be the case. Moreover, a pull based production with KANBAN negatively affects the product mix variety and especially the reduction of batch sizes [44].

However, production in high volumes contradicts the fundamental principle and JIT performance objective of WIP minimization [45]. In contrast, Spearman et al. developed a pull-based production system called CONWIP (CONstant Work in Progress), which can be used in a wide variety of manufacturing environments [46]. A CONWIP production line sets the WIP levels and measures throughput [46]. As the main difference to a KANBAN system, in CONWIP parts are moved by using standard containers, each containing the same amount of work. In CONWIP cards are assigned to these standard containers. The fundamental difference/advantage is that WIP is directly observable while capacity, which is needed to appropriately release work in a push system, must be estimated. According to Hopp and Spearman [46] and Arbulu [47] in a CONWIP production system, on average less WIP levels are needed to obtain the same throughput as in conventional systems. In a CONWIP system, the whole value chain is just triggered in one point (usually the first process step) which improves drastically the handling of the system. Moreover, CONWIP signals are not product specific, which means that with one signal different product types can be triggered in a flexible way [47]. Arbulu [47] suggests quantifying CONWIP signals in time units as opposed to an amount of product units, because this would be more practicable in industry. In Lean Manufacturing, the consistent number of production orders released at the pacemaker process and simultaneously the taking away of an equal amount of finished goods is called "paced withdrawal". Also, in construction, establishing a constant production pace could create a predictable construction flow that would enable quick corrective action to be taken in case of unforeseen problems [5]. In "lean language", this consistent increment of work is called "Pitch" and is calculated by determining how much work can be done at the bottleneck process at a certain and prior-defined interval (i.e., 1 h, 1 shift, 1 day, etc.) [38]. Therefore, a common unit for the construction flow, independent of the customer order, becomes possible. This "Pitch" becomes the basic unit of the production schedule for the considered product family. For a more detailed explanation of the pitch concept readers could consult the work of Matt [48].

The general practice in multi-project management enables work on an incoming project to start immediately as soon as it is commissioned (provided the first relevant resource is available) [49]. According to Anavi-Isakow and Golany [49] this practice is equivalent to the "push" principle in production management, where there is no control over the number of products in the system. In [49] researchers try to adapt the CONWIP principle to multi-project management. They present two variants of the control circuit, one limiting the number of projects and the other limiting the number of worked hours in the system. The concept focuses on a backlog list where projects, arriving at times when the system is unable to accept them, enter an external queue (the backlog list) where they wait until the load of the system has fallen under the threshold. The first variant Constant Number of Projects in Process (CONPIP) limits the number of projects that are allowed in the system to a fixed number. If there are less than the maximum allowed projects in the system, the backlog list is empty and incoming projects are accepted immediately. When the project is activated, it is broken down into its individual tasks. The completion time of an activated project is dependent on the status of the system, because the task has to be ready (all its predecessor tasks have been completed) and the required resource has to be available (not occupied with another task) [49]. The second variant, Constant Time in Process (CONTIP), controls the total processing time required by all the projects that are active in the system. When one task of an active project is completed by one of the resources in the system, the remaining processing time needed for all active projects is updated. When it falls/goes under a certain limit, a new project is allowed to move out of the backlog list and into the active system. Therefore, the second variant considers and controls the available capacity (in terms of labor resources) when activating a new project. One of the results (mentioned in [49]) states that, by holding projects in the backlog list, when the system has reached its capacity, no flow time performance is lost and no accumulation of overhead costs occurs. Moreover, the mentioned pull approach (CONPIP and CONTIP) allows for easier forecasting of completion times. According to [49], many of the synchronization delays, which consist of a task waiting for its predecessor to be completed and which cannot even start since the relevant resources are faced with queues of tasks from other active projects, disappear. The proposed approaches, CONPIP and CONTIP, address the problem of synchronization between different projects.

Carniel-Perrin et al. [50] tried, in their work, to achieve a pull-oriented ERP system in the ETO environment. The high variety nature of the business and the use of ERP systems impact on the ability to implement pull, a term widely misinterpreted. Using the case of a British ETO company, they analyzed and determined the extent to which an ERP can support an ETO to tend towards a "pure" pull system. In their analyses, they focused on the following three principles: (1) KANBAN; (2) CONWIP and (3) POLCA (Paired-cell Overlapping Loops of Cards) methodology.

Gosling et al. [51] derived principles for the design of ETO supply chains, based on the FORRIDGE principles (that are a combination of principles defined by FORRester and BurbiDGE). Forrester [52] emphasized, in 1961, the role of "*connectance*", feedback and disturbances in manufacturing systems. The more extended the chain, the worse the dynamic response. At around the same time, Burbidge was developing ideas relating to material flow control [53]. The FORRIDGE principles united these different intellectual threads into a succinct set of five principles shown in Table 1 [51]. Gosling et al. analyze how the original principles may be conceived in an ETO environment. A further 'design for X' principle was also added to the original principles shown in Table 1. This is crucial for the ETO supply chain, where companies have to engage in new designs for each customer. This integrates a well-established concept in the design engineering literature with the FORRIDGE principles, thereby expanding and strengthening the principles for use in the ETO sector. Implementing the six principles effectively offers considerable opportunity for competitive advantage for those companies willing to invest. In this way, the paper provides guidance on how to address some of the structural problems outlined in the challenging setting of the ETO sector.

Table 1. Definitions and influences of the FORRIDGE (**FORR**ester and **Burb**I**DGE**) Principles [51].

FORRIDGE Principles	FORRIDGE Definitions	FORRester Inputs	BurbIDGE Inputs
1—Time Compression Principle	Every activity in the chain should be undertaken in the minimum time needed to achieve task goals.	Faster order handling to improve stability and reduction of system time delays.	Minimize the material throughput time.
2—Control System Principle	There is a need to select the most appropriate control system best suited to achieve user targets and take unnecessary guesswork out of the system.	Change inventory policy to adjust the level of inventories and in process orders.	Only make those products that you can quickly dispatch and invoice to customers—only make in one period those components you need for assembly in the next period.
3—Synchronization Principle	All events are synchronized so that orders and deliveries are visible at discrete points in time, and there is continuous ordering synchronized throughout the chain.	Events should be synchronized, so that orders and deliveries are visible at discrete points in time.	Use the shortest planning period—only take deliveries from suppliers in small batches as and when needed for processing or assembly-demand amplification can be reduced by continuous ordering synchronized throughout the chain.
4—Information Transparency Principle	Up-to-the minute data free of "noise" and bias should be accessed by all members in the system.	Ensure correct behavior of information-feedback systems.	Do not rely on long-term forecasts and promote "connectance".
5—Echelon Elimination Principle	There should be the minimum number of echelons appropriate to the goals of the supply chain.	Eliminate distributor level to reduce demand amplification.	Efficiency is inversely proportional to the complexity of its material flow system.

As seen also in the works of Anavi-Isakow and Golany [49], Carniel-Perrin et al. [50] and Gosling et al. [51] pull systems can be adapted and introduced in ETO companies to achieve a CSC oriented to the customer demand on the construction site. However, there is still a lack of methods and tools to synchronize off-site manufacturing of ETO-components with installation on-site. Through the adoption of such pull- and JIT-oriented approaches, not only can efficiency in the supply chain increase, but also sustainability of supply chains increases.

2.3. Research Question

Based on the introduction and the literature review, the authors define the following two hypotheses and research questions in this work:

RQ1: Is it possible to synchronize off-site fabrication of ETO components and on-site installation by the use of CONWIP?

RQ2: What is the impact of such an approach on sustainability of CSCs in ETO companies?

According to several authors, case research is one of the most suitable ways for the development, testing, disproof and/or refining of a theory or hypothesis [48,54,55] as well as for the determination of further research needs [56], especially in a complex and dynamic context [57]. The previously mentioned research questions are analyzed based on a case study from construction industry.

3. Analysis of the Actual Situation in an ETO Construction Company

As mentioned in the introduction, a typical ETO construction project can be subdivided into three main phases: engineering, fabrication and installation. To describe the actual situation in typical ETO CSCs the authors give a short overview of the results of the As-it-is-Analysis at the case study company. The results are categorized into four parts: (1) Project preparation and quotation; (2) Engineering; (3) Fabrication and (4) Installation on-site.

3.1. Project Preparation and Quotation

Before the start of a typical ETO project, the customer or the architect defines the main concept, the customer requirements and the technical specifications. Based on these specifications, an internal project manager elaborates the master plan (usually a macro Gantt chart) where he specifies/estimates the content (work packages), the duration (milestones) for the different phases, and the deadline for the project. Based on this master plan, the ETO-company participates in the bidding process. If the company wins the bidding award, the project enters the multi-project pipeline and the operation work for project realization starts. At this point, the supplier company defines the project team: the project manager, staff from the engineering department, someone responsible for manufacturing and logistics as well as a foreman for the installation team on-site. After the internal project planning, engineering starts with the elaboration of drawings and the technical design. After the realization of the first detailed drawings, manufacturing starts to produce components and parts and the installation crew on-site prepares the construction site for installation.

3.2. Engineering

The technical office or engineering puts together the approval design, which consists of a detailed design of the project for customer approval. Usually, the approval design is a 2D or 3D plan, only specifying the overall geometry of the objects. The approval planning is presented to the customer for agreement. When the customer releases the approval design, engineering goes on with shop floor designs. They specify the different components, their design, material, strength requirements and so on. At the same time, the BOM is elaborated, which contains a first estimate of make or buy of components and parts. Based on the BOM and the technical drawings, the project manager, together with quotations from purchasing, defines whether a part is fabricated internally in the job shop or by an external supplier.

3.3. Fabrication

Once the make or buy decision process is finished and shop floor designs have been created, they are sent to the production planning department. Production planning integrates the project with multi-project planning and tries to achieve efficiency benefits through the creation of bigger lot sizes collecting production orders from different projects. In some cases, this is necessary to achieve rational economic lot size if the fabrication process requires the use of costly machinery with high changeover times. Furthermore, production planning often tends to increase lot sizes in production through the fabrication of the whole lot size of parts for a single project order instead of JIT-oriented production and delivery. Thus, in many cases, production planning accepts higher production lot sizes and therefore a higher stock of semi-finished and finished components for installation, with the aim of increasing efficiency in manufacturing and very often to overcome risks of a supply shortage and thereby necessary extension of the agreed deadline for the project end. The fabrication ensures then, according to the production planning schedule provided, that all necessary information from engineering (e.g., drawings) and from installation (e.g., tolerances to be determined at the construction site) are delivered on time. Furthermore, the logistics department organizes the transport and supply of the necessary components to the construction site.

3.4. Installation On-Site

The installation foreman defines, together with project management, the installation schedule and the installation team (who, where and for how long). Normally, by defining the human resources for projects, learning curve effects should be considered. This means that the uniqueness of every ETO-project requires, at the beginning of every task, a certain time to generate a learning effect for staff. For example, a steel façade installation with a non-repetitive design of façade elements requires some time to optimize the installation tasks for mounting the elements. After some time, the staff become more experienced and achieve higher efficiency in the installation of the remaining façade elements. Given the available resources (labor) and a rough experience-based estimation of the learning curve, the installation foreman schedules the assembly process on-site, communicating this schedule to the project manager and production. In many cases, this sharing of information and schedules occurs only at the beginning of a project, while a continuous or periodic re-alignment is often neglected.

3.5. Findings from Initial Case Study Analysis

Together with the company, the most important challenges for the research were defined after the initial analysis. The analysis showed that synchronization in planning is lacking between engineering and fabrication, between engineering and installation as well as between fabrication and installation. In order to limit the area of investigation, the focus in this research is on the synchronization between manufacturing and on-site installation. Due to planned changes in the IT environment as well as other circumstances, a more accurate and deeper integration of engineering in the synchronization approach was postponed. Furthermore, the main objectives for the synchronization approach were defined as: (a) a reduction of inventory in the supply chain; (b) the reduction of lead times and thus shorter project delivery times; (c) avoiding budget overruns and increasing profit and (d) increasing sustainability in construction projects.

4. Conceptual Approach for Synchronization of Off-Site Manufacturing and On-Site Installation in ETO Supply Chains

The construction industry is generally considered to be some way behind those sectors where effective supply chain management is regarded as key to gaining competitive advantage and dealing with the need to constantly improve operations to satisfy the increasingly sophisticated demands of end users [58]. Since supplier lead times are, for the most part, much greater than the possible accurate foresight regarding work completion on-site, JIT delivery of ETO components from production to the construction site is difficult [59].

In the ETO industry, every product is unique and therefore engineering design, fabrication and installation is made for the specific customer order (batch-of-one production). Even if the final product can involve some standard parts, every customer order requires individual engineering designs and bill of materials, individual production routings and individual installation procedures. Generally, in ETO construction companies, the three phases/departments-engineering, fabrication and installation—are not synchronized with each other. This means that the engineering department produces drawings, which are passed to fabrication. The fabrication department manufactures according to the drawings delivered from the engineering department. Afterwards ETO-components are passed from fabrication to the site for installation. As a result, in the traditional approach, customer orders are pushed from engineering to installation on-site. Therefore, high and uncontrolled levels of WIP occur, which are the main causes of long lead times [60].

The aim of the proposed approach is that ETO-components arrive in the right sequence and quantity needed for installation on-site (Just-in-Sequence (JIS) and JIT). The concept for aligning demand with supply uses the CONWIP regulation circuit (Figure 2).

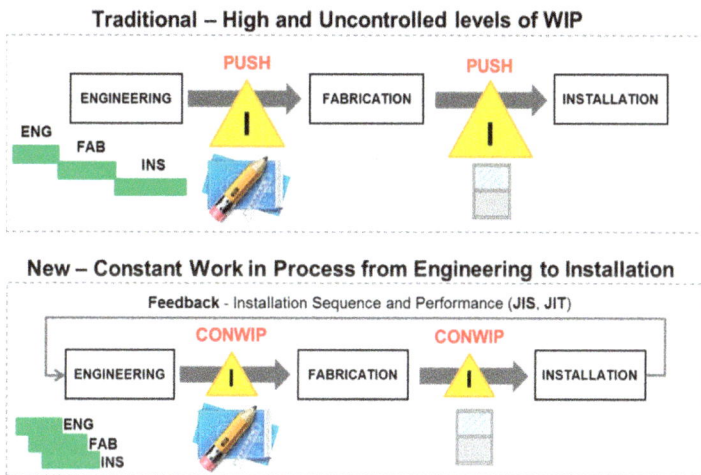

Figure 2. Synchronization approach to align demand with supply (according to [60]).

In this concept, the engineering department defines the optimal installation sequence and performance in collaboration with the installation department. Therefore, the engineering department designs ETO-components in the right sequence needed for installation. Moreover, the right granularity for engineering, fabrication and installation is defined. In detail, the approach consists of two regulation circuits, long- and short-term (Figures 3 and 4) explained in the following Sections 4.1 and 4.2. Here, the customer demand is broken down into small lots of approximately equal size, allowing an optimal capacity saturation and minimal non-productive time. To define those lots with approximately equal size, the so-called pitching concept is used [61]. As shown in Equation (1), for every task, the pitch defines the number of construction areas (CAs), which should be completed by a specific crew in a defined interval. As a practical example, consider façade installation and specifically the task installation of frames, where four CAs, consisting of four rooms, should be completed by a crew composed of two workers in a day.

$$\text{Pitch (task)} = \#\text{CAs}/\text{crew}/\text{time-unit} \tag{1}$$

The pitching concept is used to structure the work in the long-term as well as a short-term control loop (Figures 3 and 4). Pitches that are composed of different services are interpreted in the construction industry like containers in the manufacturing industry incorporating physical products.

4.1. Long-Term Control Loop-Multi-Project Level

The long-term control loop is used to define the optimum number of projects in the production system (Figure 3). In the CONWIP–backlog, the Master Schedules of different projects are combined to calculate the necessary work capacity. Furthermore, the final aim of the CONWIP-backlog is to define the optimal sequence of different projects according to the delivery time. If construction delays occur, the project sequence has to be recalculated in the CONWIP-backlog avoiding a scenario where components are produced too early and must, therefore, be stored throughout the supply chain or on the construction site. The project orders throughout the phases/departments-engineering, fabrication and installation—are structured according to the pitching principle. As a result, one pitch flows from engineering to the site for installation in a continuous way (One-Pitch-Flow). Of course, the granularity to define pitches in the long-term control loop should be at a lower level (e.g., weeks, months). In the lower part of Figure 3, the multi-project capacity planning is shown. The time frame was set at a weekly detail just to demonstrate the concept. In the middle of the figure, the time scheduling is shown. The time schedule contains the pitches from engineering, fabrication and installation, which have to be scheduled to reach the milestones defined in the project contract. Furthermore, pitches visualized in green define project No. 1 and pitches visualized in gray define project No. 2. The optimum level of WIP (in terms of number of pitches) is defined according to the law of little [62]. The best case is when the WIP level consists of 3 pitches and a lead time (LT) of 3 calendar weeks (CW) can be guaranteed to the customer, resulting in a throughput (TH) of 1 pitch/CW (Figure 3). This level of WIP is called the critical WIP (WIP$_C$). If the level of WIP is below the WIP$_C$, the system does not work at full capacity, which means that an output of 2/3 pitches/CW is reached. If the level of WIP is greater than the WIP$_C$, the lead time increases and the promised delivery time cannot be guaranteed to the customer. Therefore, according to the CONWIP principle, when the pitch-Installation-A is finished, the pitch-Engineering-D can enter the system. If the pitch-Installation-A could not be finished in CW 3, then the project sequence in the backlog has to be recalculated and, instead of pitch-Engineering-D, the pitch-Engineering-E should be released. Moreover, in CW 4 not pitch-Fabrication-D, but the pitch-Fabrication E should be released avoiding a scenario where components are produced which cannot be installed immediately, overfilling the buffer for storing material on-site.

As described in [49], every project is broken down into its individual tasks and the necessary labor resources are assigned. Therefore, in comparison to the traditional approach, employees for installation are planned in advance avoiding a switch between different projects and between fabrication and installation. Furthermore, the necessary work capacity (for new projects) can be calculated and compared with the available capacity used in running projects. Important milestones within the three phases (engineering, fabrication and installation) should be set, which take into account the available work capacity. As a result, new projects can be sequenced in an optimal way to reach the defined milestones. The long-term control loop uses the CONWIP approach to determine when a new project can enter the system by taking account of the available work capacity.

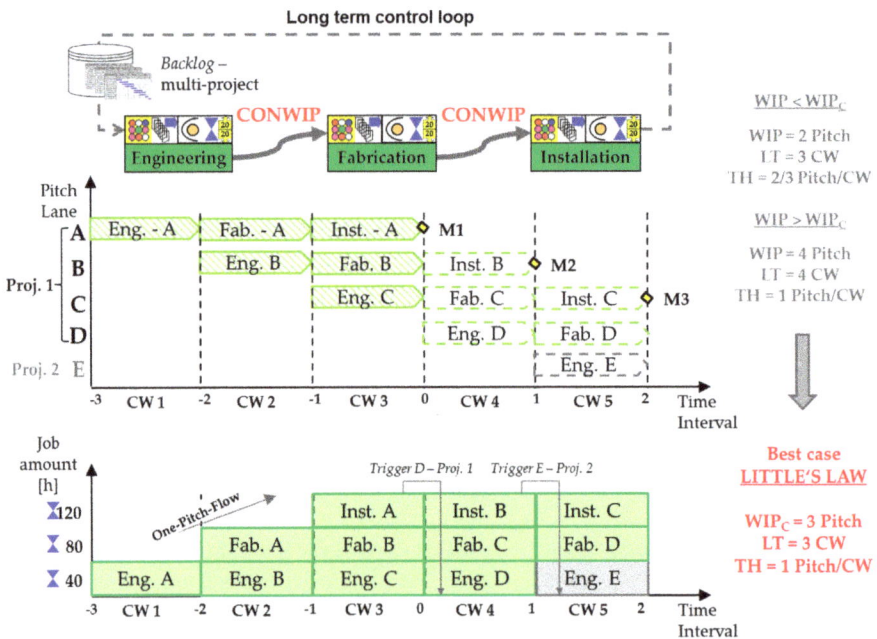

Figure 3. Long-term control loop-multi-project capacity planning.

4.2. Short-Term Control Loop-Project Level

The short-term control loop is used to schedule and control the work to be performed within the supply chain and on the construction site for the upcoming time interval (Figure 4).

In the short-term control loop, an appropriate granularity should be defined in engineering, manufacturing and installation. The same granularity means, in this case, that a Constant Work in Progress (e.g., work for one week or one month) flows from engineering to installation. As visualized in Figure 4, the job amount for engineering, fabrication and installation is different. It depends on the tasks to be performed and, accordingly, on the necessary workforce for each phase. This is different from the conventional assumption where CONWIP is used in manufacturing and parts are moved by using standard containers each containing the same amount of work [46]. In this context, the work to be performed in the three departments is very different (in terms of job type). As a result, the three departments are not balanced in terms of the same job amount avoiding the slowing down of departments due to a different job type. As practical example, if two employees working full-time (FTE) produce shop floor drawings in one week, two employees in production would not be able to produce all the components specified in the shop floor drawings in the following week (usually production takes longer or it needs a larger workforce than engineering). In the proposed approach, the time unit remains constant but the job content for the pitches of the three departments varies (Figure 4). To measure the project performance, the physical part of the building where tasks have to be performed (the CAs), are used. As a practical example, in the case of façade installation (as visualized in Figure 4), CA1 could be the east façade of the fourth floor. In more detail, the engineering department should produce shop floor drawings, which contain a complete CA. Most of the time, due to lack of time, shop floor drawings are produced for the main parts (e.g., east façade of fourth level) and details (e.g., connection east and west façade) are missed and defined at a later stage. Therefore, necessary components for completing a construction area will not be produced and delivered at the right time. This means that there is left over work still to do on-site. In order to complete the remaining work on-site, a higher amount of work in comparison to the normal amount is needed.

Every department works with a self-regulation control loop. This means, if the work for one time interval (e.g., one CW in Figure 4) cannot be completed, the work capacity should be increased. In principal, this could be done in two ways: (a) increasing the size of the workforce by using temporary employees or (b) increasing the working time by introducing overtime. As visualized in Figure 3, pitch A has to be completed before pitch B can be started. As a practical example, if pitch A is not going to be finished in one CW (5 days), overtime will be introduced, which means that the employees responsible for pitch A will also work on Saturday (6 days). As a result, the system will work with a capacity flexibility of ±20%.

To identify in time if a pitch will not be completed according to a specified deadline, a detailed measurement of its tasks is carried out. In frequent time intervals (e.g., daily frequency), the progress of tasks within the three departments is measured (e.g., number of elaborated shop floor drawings, number of produced components and number of completed CAs).

As a result, by limiting the amount of WIP in the system and by using the long-term as well as the short-term control loop, two positive effects can be reached: (1) an over allocation of projects in the system and thus an overloading of employees is prevented and (2) by using the short-term control loop, a delay in predecessor tasks causing a delayed start to successor tasks is avoided.

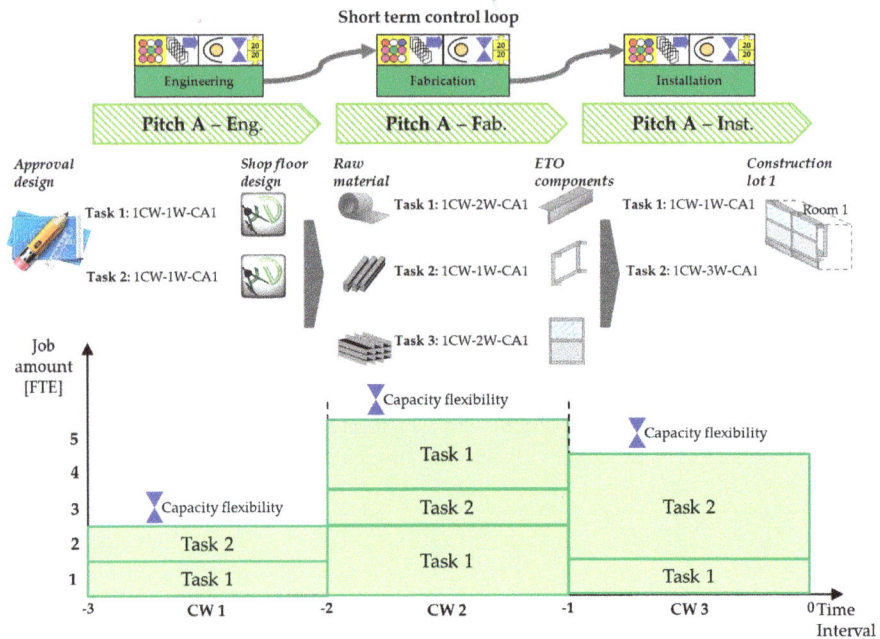

Figure 4. Short-term control loop-synchronization of engineering, fabrication and installation.

5. Validation of the Proposed Synchronization Approach in an Industrial Case Study

In order to answer the research question RQ1 ("Is it possible to synchronize off-site fabrication of ETO components and on-site installation by the use of CONWIP?") the proposed approach was validated in an industrial case study.

The approach was tested in collaboration with the previously mentioned ETO façade supplier company applying it to the extension of a new bedroom block in a hospital construction project. The construction project consists over ground of three wings (A, B and C) with respectively four levels, and a new entrance area. The considered ETO façade supplier company realized as Leader Company in a bidder-consortium the facades of the three wings. The facades were designed according to the conventional building technique, the so-called mullion-and-transom system where every single component is delivered separately to the site and will be assembled on the spot. As a special restriction, a small buffer consisting of around 100 m² for storing material on-site was available to the ETO façade supplier company.

The construction site started on calendar week CW16-2013 and finished on CW50-2014. The researchers were only involved in the project from CW46-2013. This was because the project management recognized an abnormal consumption of budget which was non-compliant with the planned construction progress up to that point in time.

At the beginning, collaborative workshops were performed, in which the project manager, the foreman on-site and the responsible employee from the design office participated. First, the building was structured in CAs where a specific codification was developed (Figure 5). The sector indicates which wing (A, B or C) of the construction project is considered. According to every sector the level (from one to four) and the orientation (east, north, west and south) is specified. The highest level of detail of a CA is defined by its construction units (CUs), specified as the distance between two principal axis of the building. The definition of CUs showed several benefits: (1) every CU has more or less the same size (around 7–8 m); (2) the separation of CUs is visible on-site because every CU is delimited by two pillars and (3) CUs were small enough that a detailed and reliable measurement of the construction progress could be performed every day.

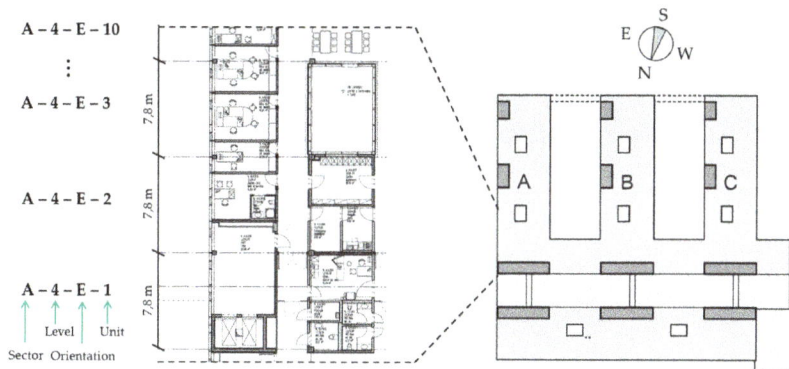

Figure 5. Definition of construction areas (CAs).

Figure 6 shows in detail the applied approach for synchronization of the manufacturing process with the installation on-site. Based on the drawings approved by the customer, the necessary tasks on-site were defined and the foreman estimated the pitch for every task. By taking into account the pitch and the total amount of CAs where the task had to be performed, the duration was calculated and, by taking into account the crew in terms of necessary workforce size, the job amount in man-hours was calculated. As a practical example, consider again the task "*installation of frames*" for which a pitch consisting of 4 CAs of the east façade (four CUs) to be performed by a crew consisting of 2 employees working 8 h/day was estimated. The east façade task "*installation of frames*" should, in this case, be performed in 50 CAs and, as such, a duration of 12.5 days (50 CAs/4 CAs/day) resulting in a job amount of 200 man-hours was calculated (12.5 days × 2 workers × 8 h/day).

This calculation was performed for every task and the job amount was calculated resulting in a budget for value-adding tasks, $Budget_{va}$. To calculate the total budget to complete the installation on-site, the budget needed to include supporting tasks, $Budget_{su}$, which included tasks such as logistics and site management. To calculate the amount of man-hours needed for $Budget_{su}$ the foreman estimated a flat-rate of 15% of $Budget_{va}$. As a result, a total budget of 41,000 man-hours was calculated for the project (see Figure 7).

Furthermore, as shown in Figure 6, the pitch was also used to synchronize the fabrication shop with the construction site. Here, for every task to be performed in a specific CA, the necessary component groups were defined. This was done once the engineering department had produced shop floor drawings. In more detail, in the fabrication shop, the work structure for producing the single components, based on the installation structure on-site, was established. Here, a fundamental distinction between lot-based prefabrication and a final assembly of semi-finished components was made. Consider again the task "*installation of frames*", which requires a certain number of frames per CA, and specific tasks needed to produce single components in the fabrication shop. Here, specifically the cutting of profiles, their welding to frames and the drilling and painting as finishing operations were considered (Figure 6). Based on the assignment of component groups to tasks, a demand-driven release from the construction site could be achieved (upper part of Figure 6). Every week, the foreman on-site scheduled the tasks to be performed for the upcoming CW. This was done based on the progress made in the construction up to that point, in terms of completed CAs. Simultaneously to the weekly work scheduling, the foreman on-site performed a medium-term scheduling of tasks used to release the required component groups from the fabrication shop. This medium-term scheduling period consisted of four weeks, because the foreman stated that he was not able to predict the tasks to be performed on-site more than four weeks ahead. In order to release the component groups that have a lead-time longer than four CWs, a supermarket was installed at the fabrication shop to buffer prefabricated and semi-finished components. These components had been prefabricated according to the master schedule of the company and by taking into account the inventory level in the supermarket.

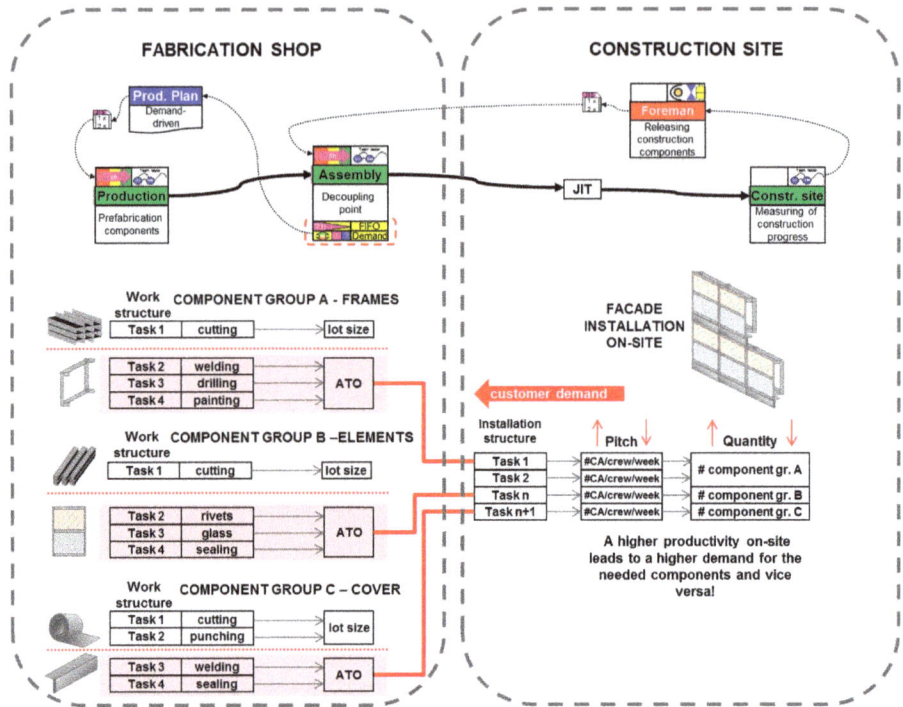

Figure 6. Synchronization of the fabrication shop with the construction site (based on [59]).

As a result, a pull-based production system, taking into account the demand of the construction site could be implemented. To validate the synchronization approach, a specific diagram, displaying the up to date consumed budget in terms of man-hours and the corresponding forecast was developed. It is displayed in Figure 7, where the bars in light gray show the cumulated consumed budget (in man-hours) and the bars in dark gray illustrate the forecast for the necessary job amount in every CW. Equation (2) displays the calculation of the forecast. Every week a comparison of the cumulated consumed budget and the forecast until project completion was computed.

$$Forecast_{(t)} = \left[\left(1 - Closed\ CUs_{(t)} \right) * Budget_{va} \right] + Budget_{su(t)} + Budget_{co(t)} \tag{2}$$

The term $Closed\ CUs_{(t)}$ refers to the amount of completed CAs (until time t) in relation to the total amount of CAs where a specific task has to be completed. As a practical example, consider again task "*installation of frames*" planned to be performed in 50 CAs in total; up until a specific $CW_{(t)}$ a quantity of 30 CAs had been completed resulting in a value of 30/50 $Closed\ CUs_{(t)}$. The term $Budget_{co(t)}$ refers to cumulated reported working hours by the installation crew on-site up to a certain time. As anticipated previously, the $Budget_{va}$ was calculated using the pitch estimated by the foreman during the collaborative workshops (mentioned in the previous paragraphs). If the pitch as performance target is increased, than the *Budget_{va}* and the *Forecast_{(t)}* will be decreased and vice versa. Figure 7 visualizes the juxtaposition of the cumulated consumed budget in man-hours (light gray bars) up to a certain week, the forecast until project completion (dark gray bars) and the total budget limit defined during the collaborative process planning workshops (black line). As visualized in the plot, at the beginning of the validation period, the forecast exceeded the total budget limit. This was because, at the beginning, the research team experienced problems in applying the material request on-site resulting in construction interruptions due to missing materials. From CW03-2014, the synchronization of the fabrication shop to the site began to work and, therefore, the forecast could be steadily reduced. Moreover, the continuous adaptation of pitches to the real conditions on-site allowed reliable material requests and forecasts of the necessary man-hours until project completion to be made if forecast increases happened, as in CW23-2014 or CW28-2014, the plot allowed them to be identified early on and improvement actions to be implemented in time.

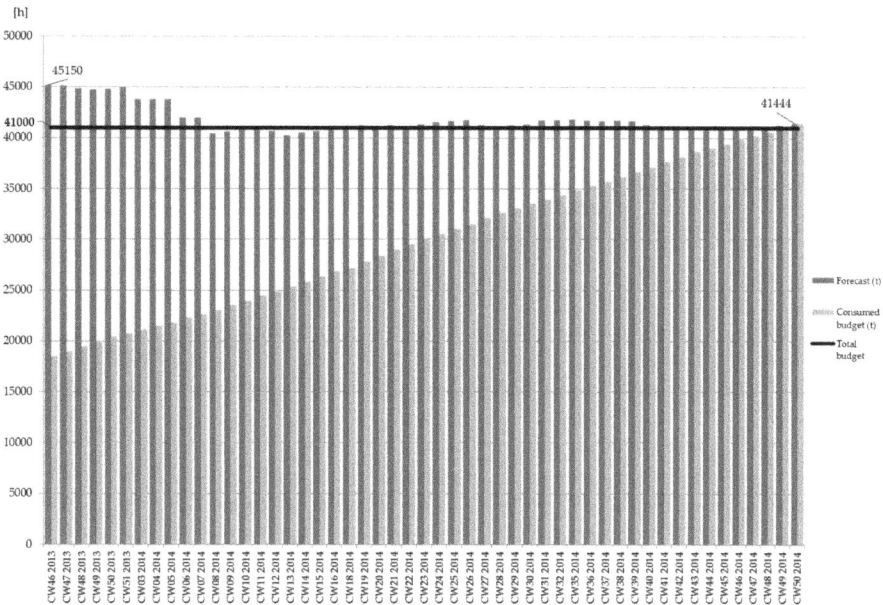

Figure 7. Overview of consumed budget vs. forecast until project completion.

As visualized in Figure 7, the approach allowed the amount of man-hours needed to perform the installation on-site to be reduced from 45,150 h (predicted at CW 46-2013) to 41,444 h actually used up to the end of the project. In other words, by applying the approach for synchronizing manufacturing to the installation on-site, a reduction of the initial forecast of around 8% could be reached. In conclusion, the application of the approach allowed the cost for installation to be kept within the total budget limit. Thus, the answer to RQ1 is that the proposed approach is suitable for synchronizing off-site fabrication of ETO components and on-site installation using CONWIP. Therefore the algorithm had a significant impact to react rapidly on budget overruns and thus to avoid an economic loss in this project. Through a higher accuracy in JIT delivery based on the presented approach the deliverability rate of material for the construction site could be increased by 17%. In the past, the case study company rented very often a production hall with 1000 m² close to the company's fabrication shop and used the space as buffer stock for material waiting for shipping to the construction sites. The management was already thinking about expanding the company's building, using free agricultural land. Through the application of the algorithm, the need to rent these spaces was reduced to sporadic leasing in case of urgent bottlenecks. Therefore, in the case study initially the company planned to lease an intermediate inventory with a capacity of 1850 m². However, the management board decided to apply the presented approach and as such to avoid an intermediate storage and ship directly the needed components to the site. As a result, costs for loading and unloading trucks, costs of insurance of the inventory as well as transportation costs from the inventory to the construction site could be avoided. In total, 5.29% of the budget available for installation could be saved by avoiding an intermediate storage and applying direct deliveries to the site. Such a reduction of space has not only an economic impact, but also an ecological impact with the meaning of lower energy consumption for lighting or heating/cooling and less waste of agricultural space. As another important consideration for economic sustainability, a quantitative evaluation of the effort needed in using the approach was carried out. Here, the effort needed for performing the collaborative planning workshops (to define CAs, tasks and pitches), the effort for scheduling and reporting the work to be performed on-site, as well as the expenditure of time for releasing ETO-components JIT for installation was considered. As a result, an amount of 571 man-hours was calculated and by comparing it with budget a percentage of 1.39% resulted. More in detail, an expenditure of 571 man-hours allowed to decrease the total labor forecast of 3700 man-hours or in other words a saving factor of 6.47 (3700 h/571 h).

Considering the ecological sustainability dimension, the use of express delivery services decreased by more than 70% compared to other projects in the past. As main reason for this positive effect, the project manager stated that the implementation of a weekly release of needed components allowed to assure that they were most of the time available on-site. Express deliveries were just used in the case of engineering and fabrication errors. While in the past rework on the construction site was very common, the amount of rework after the introduction of the synchronization approach could be reduced by 33%. According to the foreman, this reduction was mainly possible because the synchronization of the fabrication shop with the construction site allowed to use smaller lot sizes and as such if errors were identified on-site just the components of the released amount had to be reworked (not the total amount of needed components in the project). As rework of customized parts means in most of the cases that parts have to be substituted (e.g., sealing components or glass elements), the impact on ecologic sustainability is significant.

Considering the social sustainability, an important result was the less fluctuation of the number of assigned operators on-site. Due to a better planning of labor resources at the multi-project level a frequent exchange of operators between different construction sites and between the fabrication shop and the sites was diminished. Throughout the entire lifecycle of the construction site a number of 11 until 13 operators were present on-site. More in detail, as visualized in Figure 7 the bars in light gray, visualizing the consumed budget every week, increased steadily (from 440 h/week to 520 h/week) until project completion. Today, the case study company uses this approach in most of their important and international projects. From a social aspect, the proposed approach increases the know-how of workers and supervisors at the construction site. In order to guarantee a successful roll-out of the approach, the company hired qualified staff, trained these people in the approach and created an own department for synchronizing the different installation sites to the fabrication shop (called as Supply Chain Management department). These trainers now train the project managers and supervisors at the construction site in applying the presented approach. These lead also to a job enrichment of workers and thus to a higher involvement and identification with the employer.

The quantitative results in this case study, having impact on economic, ecological and social sustainability, are summarized in Table 2.

Table 2. Impact of case study results on economic, ecological and social sustainability.

Economic Sustainability	Ecological Sustainability	Social Sustainability
- Reduction of initial labor forecast of 8%	- Reduction of express and special transports of more than 70%	- Less fluctuations of the number of operators on-site (from 11 to 13 workers)
- Increase of material deliverability of 17%	- Reduction of rework of components on-site about 33%	- Creation of 2 new jobs in form of on-site process optimization
- Avoiding of an intermediate storage with an initial calculated cost of 5.29% of installation budget	- Avoiding of an intermediate storage capacity of 1836 m^2 and thus no more need of an extension of the factory on agricultural land	- 6 project manager and 5 foremen trained in the application of the new approach and algorithm
- acceptable effort needed to use the approach in practice (1.39% of the installation budget)		
- Factor effort saving 6.4		

6. Discussion of the Impact on ETO CSC Sustainability

In this section, the findings in Section 4 (proposed synchronization approach) and Section 5 (validation in practice) based on the case study research are generalized and summarized. Before starting with the description of the impact of the proposed approach on sustainability, the term sustainability itself is explained.

Sustainability is part of the optimization of the overall efficiency of enterprises, products and processes. Sustainability traditionally has three dimensions: economy, ecology and social. Costs of energy or materials have an impact on the economic effectiveness. The reduction of resources is a contribution to the economic and ecological effectiveness. The social dimension is represented by work conditions, conditions of education, skills and others [63] (Figure 8). Based on these three dimensions, different aspects and arguments for why the presented approach for synchronization will positively affect the sustainability of construction supply chains in the future are shown.

The application of Lean and JIT in ETO companies leads to more sustainability in their supply chain. Our algorithm described in Section 4 and tested in Section 5 supports ETO manufacturer in the construction industry to overcome difficulties in implementing this concepts in a holistic manner focusing not only lean aspects in the fabrication, but focusing on the synchronization with on-site installation. Previous to this case study, the company tried to implement lean methods in production as well as to conduct Kaizen workshops at the installation site. Both initiatives did not lead to the expected results before synchronizing them and thus introducing a customer pace from installation to manufacturing. In the following Sections 6.1–6.3, the proposed approach will be discussed based on the three dimensions of sustainability. Table 3 summarizes the results and advantages of the approach in terms of a more sustainable CSC in ETO companies in order to answer to research question RQ2 ("What is the impact of such an approach on sustainability of CSCs in ETO companies?").

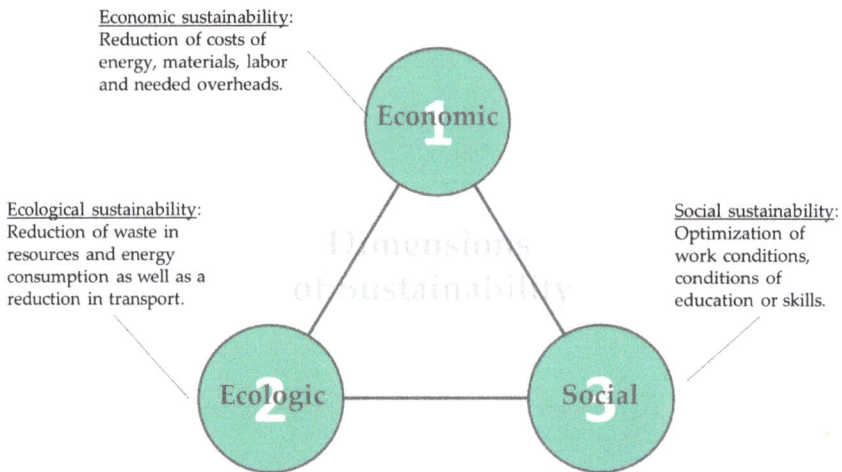

Economic sustainability: Reduction of costs of energy, materials, labor and needed overheads.

Ecological sustainability: Reduction of waste in resources and energy consumption as well as a reduction in transport.

Social sustainability: Optimization of work conditions, conditions of education or skills.

Figure 8. Dimensions of sustainability (adapted from [64]).

6.1. Economic Aspects

The economic aspects in the adoption of the proposed synchronization approach are varied. Firstly, the proposed approach to synchronize manufacturing and on-site installation enforces a cost-efficient coordination, saving time for information sharing and organization and thus costs for supporting activities. As seen before, different types of waste can be reduced saving costs and thus being more attractive to the market and/or increasing profitability for the ETO-company. The reduction of waste in overproduction is reached through a customer-driven triggering of production. In this way, manufacturing produces only the amount needed in a given time period at the construction site. This also leads to a reduction in transportation costs due to an improved coordination of material and logistic flows. At the same time, JIT production and delivery also reduces inventory and material buffers at the production site and at the construction site, saving costs for space and material handling. Especially on the construction site, space is always limited and costly (particularly in urban areas). As shown in Figure 7, the budget overrun identified at the beginning (starting by CW46-2013 until CW03-2014) was mainly caused by non-synchronized material deliveries. The delivery of components *too early* caused an overfilling of the buffer on-site inducing the following types of problems: (a) buffering of components on-site over a longer period caused the risk of damages and therefore *reworks*; (b) *searching* and *moving* of components in the buffer caused an abnormal high amount of supporting work. A delivery of components *too late* caused that the foreman had to *reschedule* the work on-site. In the worst cases, if a rescheduling could not be done, costly construction interruptions happened. A rescheduling of work due to missing material caused that some CAs had to be completed in a second moment, which required a higher amount of work (supporting and value adding) due to a rearranging of tools and initial learning curve effects because of a restarting of the work. As a result, the approach allowed avoiding missing material on-site and as such a reduction of downtimes and waiting times, and therefore increases in construction productivity. This allowed to decrease the initial forecasted budget overruns and a staying on budget until project completion.

6.2. Ecological Aspects

Transport between the fabrication shop and the construction site generate an increase in CO_2 emissions and, therefore, environmental pollution. Bearing in mind the finite resources of fossil fuel and energy, the reduction of unnecessary transport helps protect the environment. The application of the proposed approach in ETO supply chains will reduce traffic on the road, which is a positive aspect for flora, fauna and human society considering noise nuisance and air pollution. A more transparent and better coordinated synchronization of transport between manufacturing and installation on-site not only brings benefits to transport but also in terms of energy consumption for material handling devices. A high inventory level off-site or on-site requires a rearrangement of material due to limited space that could be eliminated or minimized by the introduction of JIT and JIS production and direct supplies to the site. In the case of poorly coordinated manufacturing and installation, parts and components are made to stock. If changes in the geometry of parts are then required due to special situations during the installation, the material stock which has been produced has to be disposed of. This means that waste in material consumption and non-value adding activities not only have an economic, but also an ecological impact. A further aspect is that space is always limited at the fabrication site and at the construction site. A reduction in inventory levels also reduces the need for new space which is very often found in green areas.

6.3. Social Aspects

The proposed approach not only has economic and ecological advantages, but also offers benefits from a social point of view. Better-synchronized coordination leads to increased employee satisfaction and, therefore, lower staff turnover. High transparency in the supply chain also reduces conflict situations between manufacturing and installation on-site. This means there are fewer "emergency" situations, less stress for employees and more time to think about strategic optimization. In addition, the proposed approach also requires highly skilled and qualified personnel for coordination and monitoring of the progress at the construction site. It also enforces the creation of new job profiles, in terms of engineers for process optimization in ETO supply chains and at construction sites. Successful ETO-companies using this approach to optimize their supply chain usually have a competitive position in the market and, therefore, they will create new jobs.

Table 3. Generalized impact of the proposed approach on sustainability in ETO CSCs.

Economic Dimension	Ecological Dimension	Social Dimension	Future Challenges
- Cost-efficient coordination between manufacturing and installation		- Improvement of employee satisfaction through synchronized coordination	■ Education of highly skilled process engineers with a focus on efficient and sustainable supply chains (→ universities and schools)
- Reduction of waste in overproduction	- Reduction of material waste due to bad coordination of manufacturing and installation	- Higher transparency in the supply chain and thus fewer conflicts between manufacturing and the site	
- Reduction of transports caused by a lack in coordination			
- Reduction of inventory and thus inventory costs	- Reduction of energy consumption for material handling devices to handle material stock off-site and on-site due to limited space	- Fewer "emergency" situations in material supply and thus less stress situations for employees	■ Development of an IT tool to implement the proposed approach in the sense of the actual trend in Industry 4.0 and Cyber-Physical-Systems (→ research)
- Reduction of unnecessary handling caused by a high inventory level		- Necessity of highly skilled and qualified personnel for coordination and to monitor the progress on-site	
- Reduction of budget overruns through proactive correction measures	- Reduction of energy consumption due to express transports caused through a lack of coordination	- Job creation in the sense of new job profiles for process optimization in ETO supply chains	■ Extension of the approach to the coordination of different companies on-site and the respective supply chains
- Reduction of waiting times/downtimes caused by missing material supply			
- Increase of efficiency in the fabrication shop due to clear information on the needs on-site	- Reduction of space needed for material buffers off-site and on-site	- Higher employment rate in growing and successful ETO companies using this approach to increase competitiveness in lead times	■ Application of the approach and the IT-tool in practice to evaluate results and to improve it (→ further case studies in industry)
- Higher competitiveness through accurate compliance with deadlines			

7. Conclusions and Outlook

One of the major causes of budget overruns is the lack of ETO-components on-site, which causes wasteful construction downtimes. When construction interruptions occur, tasks on-site have to be rescheduled, which leads consequently to rearrangement of materials or equipment on-site (non-value-adding activities). In addition, a high uncertainty of the progress on-site leads fabrication to anticipate production of components and, therefore, to increase inventory at the fabrication shop or on-site. To avoid such inefficiencies in the supply chain, downtimes at the construction site and high inventory levels, the authors propose an approach to synchronizing manufacturing and installation. Aligning manufacturing with the site could first avoid such non-value-adding activities. Furthermore, through JIT and pull-oriented production and delivery, non-value-adding operations (like searching for components on-site) could be reduced and the chances for early detection of quality problems could be improved. In addition, by pulling manufacturing from a site, a higher degree of capacity and resource utilization can be reached.

In traditional ETO CSCs, the economic benefits of a project, reached through scale effects in production, are lost due to an inefficient installation process on-site. One of the major causes of an inefficient installation process on-site is construction interruptions due to a lack of necessary MTO or ETO components. By limiting the amount of WIP throughout the supply chain and by synchronizing the supply chain to the construction site, all dimensions of sustainability can be addressed positively. The proposed long-term control loop limits the number of projects in the system and thus avoids an accumulation of jobs within the production system and an overloading of employees. As a result, the delivery time can be guaranteed to the customer. The short-term control loop avoids a situation where a delay is propagated throughout the supply chain, by measuring in detail the performance of tasks and introducing, in cases of delay, an appropriate capacity flexibility.

The proposed approach shows that the following dimensions of sustainability can be increased in ETO CSCs. From an economic point of view, overall costs for labor, for material and for resources can be reduced and, therefore, ETO companies can increase their competitiveness in the market. From an ecological point of view, the approach reduces unnecessary transportation and express transportation of materials from the job shop to the construction site. Furthermore, a reduction in material handling reduces the consumption of fuel for motorized material handling devices. A reduced need for storage space also has a positive effect on the consumption of energy for production and storage facilities. According to the third aspect in sustainability research, the social aspect, the proposed approach encourages the use of higher qualified personnel at the construction site and/or training of existing staff in organizational logistics and JIT-oriented monitoring and methods. Furthermore, work conditions improve if the work at the construction site is scheduled accurately and if the right material is available in time, avoiding eventually dangerous corrections with inadequate machinery and equipment on-site.

Critically, it must be added that the proposed approach is not applicable to all construction typologies or supply chains. As such, for example, supply chains dealing with bulk materials need other models for production planning and control. A further critical point is the fact that the proposed approach is based on a single industrial case study. Although the company considered is a good and generalizable example of an ETO manufacturer in CSCs, further validations with other companies should be conducted. A further point of criticism and suggestion for further research is the lack of a quantitative assessment tool for the determination of sustainability performance. Such functionality should be integrated into a future IT-supported tool for sustainable production planning and control.

In future research activities, the proposed approach will be extended to coordinate all companies involved in a construction project and the corresponding supply chains. Furthermore, an IT tool to support and apply the proposed approach in industrial practice in different industrial case studies will be developed. Handling the high variability of construction processes and considering a multi-project environment are future challenges in the realization of such an IT tool.

Acknowledgments: The research presented in this article was started during the project "Modeling and Managing Processes in Construction (MoMaPC)" supported by the Free University of Bozen-Bolzano (Italy) under Grant IN2021 and it was completed within the research project "Collaborative Construction Process Management (Cockpit)" financed by the European Regional Development Fund Investment for Growth and Jobs Programme 2014-2020 under Grant IN2204.

Author Contributions: Patrick Dallasega developed the CONWIP-based approach for synchronization of manufacturing and installation on-site and validated it in the presented case study; Erwin Rauch contributed with expertise in the field of lean methods and production planning and control in manufacturing to synchronize manufacturing and on-site-installation.

Conflicts of Interest: The authors declare no conflict of interest.

References

1. Carter, C.R.; Liane Easton, P. Sustainable supply chain management: Evolution and future directions. *Int. J. Phys. Distrib. Logist. Manag.* **2011**, *41*, 46–62. [CrossRef]
2. Windsperger, A.; Steinlechner, S.; Fischer, M.; Seebacher, U.; Lackner, B.; Hammerl, B.; Kaltenegger, I. *Integriertes Nutzungsmodell Zum Effizienteren Rohstoffeinsatz Im Wirtschaftsbereich (Integrated Usage Model for More Efficient Use of Raw Materials in Economics)*; BMVIT 19/2006; Federal Ministry for Transport, Innovation and Technology: Vienna, Austria, 2006.
3. Browne, J.; Harren, J.; Shivan, J. *Production Management Systems. An Integrated Perspective*; Addison-Wesley: Harlow, UK, 1996.
4. Schweizer, W. *Wertstrom Engineering-Typen-Und Variantenreiche Produktion*; Epubli GmbH: Berlin, Germany, 2013; p. 53.
5. Dallasega, P.; Rauch, E.; Matt, D.T. Sustainability in the supply chain through synchronization of demand and supply in ETO-companies. *Procedia CIRP* **2015**, *29*, 215–220. [CrossRef]
6. Rauch, E.; Dallasega, P.; Matt, D.T. Synchronization of Engineering, Manufacturing and on-site Installation in Lean ETO-Enterprises. *Procedia CIRP* **2015**, *37*, 128–133. [CrossRef]
7. Blismas, N.; Pasquire, C.; Gibb, A. Benefit evaluation for off-site production in construction. *Constr. Manag. Econ.* **2006**, *24*, 121–130. [CrossRef]
8. Matt, D.T.; Rauch, E. Implementing Lean in Engineer-to-Order Manufacturing: Experiences from a ETO Manufacturer. In *Handbook of Research on Design and Management of Lean Production Systems*; Modrák, V., Semančo, P., Eds.; IGI Global: Hershey, PA, USA, 2014; pp. 148–172.
9. Adrodegari, F.; Bacchetti, A.; Pinto, R.; Pirola, F.; Zanardini, M. Engineer-to-order (ETO) production planning and control: An empirical framework for machinery-building companies. *Prod. Plan. Control* **2015**, *26*, 910–932. [CrossRef]
10. Nowotarski, P.; Paslawski, J. Barriers in running construction SME–case study on introduction of agile methodology to electrical subcontractor. *Procedia Eng.* **2015**, *122*, 47–56. [CrossRef]
11. Singleton, T.; Cormican, K. The influence of technology on the development of partnership relationships in the Irish construction industry. *Int. J. Comput. Int. Manuf.* **2013**, *26*, 19–28. [CrossRef]
12. Dong, S.; Feng, C.; Kamat, V.R. Sensitivity analysis of augmented reality-assisted building damage reconnaissance using virtual prototyping. *Autom. Constr.* **2013**, *33*, 24–36. [CrossRef]
13. Kim, B.; Kim, C.; Kim, H. Interactive modeler for construction equipment operation using augmented reality. *J. Comput. Civ. Eng.* **2012**, *26*, 331–341. [CrossRef]
14. Vrijhoef, R.; Koskela, L. The four roles of supply chain management in construction. *Eur. J. Purch. Supply Manag.* **2000**, *6*, 169–178. [CrossRef]
15. Akintoye, A.; McIntosh, G.; Fitzgerald, E. A survey of supply chain collaboration and management in the UK construction industry. *Eur. J. Purch. Supply Manag.* **2000**, *6*, 159–168. [CrossRef]
16. Zailani, S.; Jeyaraman, K.; Vengadasan, G.; Premkumar, R. Sustainable supply chain management (SSCM) in Malaysia: A survey. *Int. J. Prod. Econ.* **2012**, *140*, 330–340. [CrossRef]
17. Vachon, S.; Mao, Z. Linking supply chain strength to sustainable development: A country-level analysis. *J. Clean. Prod.* **2008**, *16*, 1552–1560. [CrossRef]
18. Golicic, S.L.; Smith, C.D. A Meta-Analysis of Environmentally Sustainable Supply Chain Management Practices and Firm Performance. *J. Supply Chain Manag.* **2013**, *49*, 78–95. [CrossRef]

19. Vachon, S.; Klassen, R.D. Extending green practices across the supply chain—The impact of upstream and downstream integration. *Int. J. Oper. Prod. Manag.* **2006**, *26*, 795–821. [CrossRef]

20. Yang, L.R. Key practices, manufacturing capability and attainment of manufacturing goals: The perspective of project/engineer-to-order manufacturing. *Int. J. Proj. Manag.* **2013**, *31*, 109–125. [CrossRef]

21. Bunz, K.R.; Henze, G.P.; Tiller, D.K. Survey of Sustainable Building Design Practices in North America, Europe, and Asia. *J. Archit. Eng.* **2006**, *12*, 33–62. [CrossRef]

22. Shen, L.Y.; Li Hao, J.; Tam, V.W.Y.; Yao, H. A checklist for assessing sustainability performance of construction projects. *J. Civ. Eng. Manag.* **2007**, *13*, 273–281.

23. Davies, R.M.; Davies, I.O.E. Barriers to Implementation of Sustainable Construction Techniques. *MAYFEB J. Environ. Sci.* **2017**, *2*, 1–9.

24. Ofori, G. Greening the construction supply chain in Singapore. *Eur. J. Purch. Supply Manag.* **2000**, *6*, 195–206. [CrossRef]

25. Kofoworola, O.F.; Gheewala, S.H. Estimation of construction waste generation and management in Thailand. *Waste Manag.* **2009**, *29*, 731–738. [CrossRef] [PubMed]

26. Shen, L.Y.; Tam, V.W.Y. Implementation of environmental management in the Hong Kong construction industry. *Int. J. Proj. Manag.* **2002**, *20*, 535–543. [CrossRef]

27. Oh, S.W.; Chang, H.J.; Kim, Y.S.; Lee, Y.B.; Kim, H.S. An application of PDA and barcode technology for the improvement of information management in construction projects. In Proceedings of the 21st International Symposium on Automation and Robotics in Construction, Jeju Island, Korea, 21 September 2004; pp. 518–524.

28. Chen, Z.; Li, H.; Wong, C.T.C. An application of bar-code system for reducing construction wastes. *Autom. Constr.* **2002**, *11*, 521–533. [CrossRef]

29. Cheng, M.Y.; Chen, Y.C. Integrating barcode and GIS for monitoring construction progress. *Autom. Constr.* **2002**, *11*, 23–33. [CrossRef]

30. Tzeng, C.T.; Chiang, Y.C.; Chiang, C.M.; Lai, C.M. Combination of radio frequency identification (RFID) and field verification tests of interior decorating materials. *Autom. Constr.* **2008**, *18*, 16–23. [CrossRef]

31. Hassan, A.S. Towards Sustainable Housing Construction in Southeast Asia. *Agenda* **2002**, *21*, 1–17.

32. Pandit, A.; Zhu, Y. An ontology-based approach to support decision-making for the design of ETO (Engineer-To-Order) products. *Autom. Constr.* **2007**, *16*, 759–770. [CrossRef]

33. Ali, A.N.A.; Jainudin, N.A.; Tawie, R.; Jugah, I. Green Initiatives in Kota Kinabalu Construction Industry. *Proced. Soc. Behav. Sci.* **2016**, *224*, 626–631. [CrossRef]

34. Raftery, J.; Pasadilla, B.; Chiang, Y.H.; Hui, E.C.; Tang, B.S. Globalization and construction industry development: Implications of recent developments in the construction sector in Asia. *Constr. Manag. Econ.* **1998**, *16*, 729–737. [CrossRef]

35. Long, N.D.; Ogunlana, S.; Quang, T.; Lam, K.C. Large construction projects in developing countries: A case study from Vietnam. *Int. J. Proj. Manag.* **2004**, *22*, 553–561. [CrossRef]

36. Chritamara, S.; Ogunlana, S.O. Problems experienced on design and build projects in Thailand. *J. Constr. Procure.* **2001**, *7*, 73–86.

37. Ohno, T. *Toyota Production System: Beyond Large Scale Production*; Productivity Press: Cambridge, MA, USA, 1988.

38. Rother, M.; Shook, J. *Learning to See—Value Stream Mapping to Create Value and Eliminate Muda*; Lean Enterprise Institute: Cambridge, MA, USA, 2009.

39. Bevilacqua, M.; Ciarapica, F.E.; De Sanctis, I. Relationships between Italian companies' operational characteristics and business growth in high and low lean performers. *J. Manuf. Technol. Manag.* **2017**, *28*, 250–274. [CrossRef]

40. Ballard, G.; Howell, G. Toward construction JIT. In *Lean Construction*; Alarco'n, L., Ed.; Taylor & Francis: New York, NY, USA, 1995; pp. 291–300.

41. Takahashi, K.; Hirotani, D. Comparing CONWIP, synchronized CONWIP, and Kanban in complex supply chains. *Int. J. Prod. Econ.* **2005**, *93–94*, 25–40. [CrossRef]

42. Kimura, O.; Terada, H. Design and analysis of pull system: A method of multi-stage production control. *Int. J. Prod. Res.* **1981**, *19*, 241–253. [CrossRef]

43. Hopp, W.J.; Spearman, M.L. *Factory Physics: Foundations of Manufacturing Management*, 2nd ed.; Boston IRWN/McGraw-Hill: Boston, MA, USA, 2001.

44. Bevilacqua, M.; Ciarapica, F.E.; De Sanctis, I. Lean practices implementation and their relationships with operational responsiveness and company performance: An Italian study. *Int. J. Prod. Res.* **2017**, *55*, 769–794. [CrossRef]

45. Papadopoulu, P.C. Application of Lean Scheduling and Production Control in Non-Repetitive Manufacturing Systems Using Intelligent Agent Decision Support. Ph.D. Thesis, The Engineering & Design Brunel University, London, UK, 2013; p. 7.

46. Spearman, M.L.; Woodruff, D.L.; Hopp, W.J. CONWIP: A pull alternative to KANBAN. *Int. J. Prod. Res.* **1990**, *28*, 879–894. [CrossRef]

47. Arbulu, R. Application of Pull and Conwip in Construction Production Systems. In Proceedings of the 14th Annual Conference of the International Group for Lean Construction, Santiago, Chile, 25–27 July 2006; pp. 215–226.

48. Matt, D.T. Adaptation of the value stream mapping approach to the design of lean engineer-to-order production systems. *J. Manuf. Technol. Manag.* **2014**, *25*, 334–350. [CrossRef]

49. Anavi-Isakow, S.; Golany, B. Managing multi-project environments through constant work-in-process. *Int. J. Proj. Manag.* **2003**, *21*, 9–18. [CrossRef]

50. Carniel-Perrin, B.; Dehe, B.; Bamford, D.; Jolley, K. Pull-logic and ERP within Engineering-to-Order (ETO): The case of a British Manufacturer. In Proceedings of the 5th World Production and Operations Management Conference P&OM, Havana, Cuba, 6–10 September 2016.

51. Gosling, J.; Towill, D.R.; Naim, M.M.; Dainty, A.R. Principles for the design and operation of engineer-to-order supply chains in the construction sector. *Prod. Plan. Control* **2015**, *26*, 203–218. [CrossRef]

52. Forrester, J.W. *Industrial Dynamics*; Pegasus Communications: Waltham, MA, USA, 1961.

53. Burbidge, J.L. The new approach to production. *Prod. Eng.* **1961**, *40*, 769–784. [CrossRef]

54. Woodside, A.G.; Wilson, E.J. Case study research methods for theory building. *J. Bus. Ind. Mark.* **2003**, *18*, 493–508. [CrossRef]

55. Gummesson, E. Qualitative research in marketing: Roadmap for a wilderness of complexity and unpredictability. *Eur. J. Mark.* **2005**, *39*, 309–327. [CrossRef]

56. Siggelkow, N. Persuasion with case studies. *Acad. Manag. J.* **2007**, *50*, 20–24. [CrossRef]

57. Perren, L.; Ram, M. Case-study method in small business and entrepreneurial research. *Int. Small Bus. J.* **2004**, *22*, 83–101. [CrossRef]

58. Briscoe, G.; Dainty, A. Construction supply chain integration: An elusive goal? *Supply Chain Manag. Int. J.* **2005**, *10*, 319–326. [CrossRef]

59. Matt, D.T.; Dallasega, P.; Rauch, E. Synchronization of the Manufacturing Process and On-Site Installation in ETO Companies. *Procedia CIRP* **2014**, *17*, 457–462. [CrossRef]

60. Dallasega, P.; Rauch, E.; Matt, D.T.; Fronk, A. Increasing productivity in ETO construction projects through a lean methodology for demand predictability. In Proceedings of the International Conference on Industrial Engineering and Operations Management (IEOM), Dubai, UAE, 3–5 March 2015; pp. 1–11.

61. Dallasega, P.; Frosolini, E.; Matt, D.T. An approach supporting Real-Time Project Management in Plant Building and the Construction Industry. In Proceedings of the XXI Summer School Francesco Turco, Naples, Italy, 13–15 September 2016.

62. Little, J.D.C. A proof for the Formula: L = λW. *Oper. Res.* **1961**, *9*, 383–387. [CrossRef]

63. Westkämper, E. Manufuture and sustainable manufacturing. In *Manufacturing Systems and Technologies for the New Frontier*; Mitsuishi, M., Ueda, K., Kimura, F., Eds.; Springer: London, UK, 2008; pp. 11–14.

64. Rauch, E.; Dallasega, P.; Matt, D.T. Sustainable production in emerging markets through Distributed Manufacturing Systems (DMS). *J. Clean. Prod.* **2016**, *135*, 127–138. [CrossRef]

sustainability

MDPI

Article

Soft Asphalt and Double Otta Seal—Self-Healing Sustainable Techniques for Low-Volume Gravel Road Rehabilitation

Audrius Vaitkus [1], Viktoras Vorobjovas [1], Faustina Tuminienė [1], Judita Gražulytė [1,*] and Donatas Čygas [2]

[1]   Road Research Institute, Vilnius Gediminas Technical University, Linkmenų g. 28, LT-80217 Vilnius, Lithuania; audrius.vaitkus@vgtu.lt (A.V.); viktoras.vorobjovas@vgtu.lt (V.V.); faustina.tuminiene@vgtu.lt (F.T.)

[2]   Department of Roads, Vilnius Gediminas Technical University, Saulėtekio ave. 11, LT-10223 Vilnius, Lithuania; donatas.cygas@vgtu.lt

*   Correspondence: judita.grazulyte@vgtu.lt; Tel.: +37-065-273-1011

Received: 8 January 2018; Accepted: 11 January 2018; Published: 15 January 2018

Abstract: Increased traffic flow on low-volume gravel roads and deficiencies of national road infrastructure, are increasingly apparent in Lithuania. Gravel roads do not comply with requirements, resulting in low driving comfort, longer travelling time, faster vehicle amortization, and dustiness. The control of dustiness is one of the most important road maintenance activities on gravel roads. Another important issue is the assurance of required driving comfort and safety. Soft asphalt and Otta Seal technologies were proposed as a sustainable solution for the improvement of low-volume roads in Lithuania. Five gravel roads were constructed with soft asphalt, and 13 gravel roads were sealed with double Otta Seal, in 2012. The main aim of this research was to check soft asphalt and double Otta Seal's ability to self-heal, on the basis of the results of the qualitative visual assessment of pavement defects and distress. The qualitative visual assessment was carried out twice a year following the opening of the rehabilitated road sections. The results confirmed soft asphalt and double Otta Seal's ability to self-heal. The healing effect was more than 13% and 19% on roads with soft asphalt and double Otta Seal, respectively. In addition, on some roads, all cracks observed in spring self-healed during summer.

Keywords: double Otta Seal; gravel roads; qualitative visual assessment of defects; soft (low viscosity) bitumen; soft asphalt; self-healing; sustainable technique

1. Introduction

According to the Lithuanian Road Administration under the Ministry of Transport and Communication (2015), in Lithuania, more than a third (33.9%) of state roads are gravel roads, and they cause low driving comfort, longer time of travel, faster vehicle amortization, and dustiness.

The control of dustiness is one of the most important maintenance activities on gravel roads. Vehicles travelling at a speed of 75 km/h on gravel roads with annual average daily traffic (AADT) equal to 100 vehicles per day (vpd), throw up about 25 tons of gravel wearing course aggregate annually, per kilometer [1]. This results in approximately 4 mm less thickness of the wearing course for a road 7 m in width [2]. Deficiency in the thickness of the wearing course leads to a corrugated road surface and faster pavement deterioration.

Furthermore, greater visibility is achieved when dustiness is controlled, thus increasing the safety of road users [3]. The less dustiness, the lower the rate of traffic accidents. It also results in lower costs related to vehicle repairs, medical expenses, and loss of life [4]. Dust particles penetrate into the engine and other mechanical components of vehicles, increasing vehicle wear, requiring more frequent maintenance, and hence resulting in higher vehicle operating costs [5,6]. Dust is also a health hazard because particles smaller than 10 μm accumulate in the human respiratory tract and cause allergies and asthma [7,8].

Typically, chemical dust suppressants, such as calcium or magnesium chloride and calcium lignosulphonate, are used to control dustiness. An analysis of the literature reveals that they might reduce dustiness by up to 80% and the total aggregate loss by up to 42–61% [9,10]. However, chemical dust suppressants usually enhance the corrosion of cars and often contain toxic ingredients that negatively influence vegetation and water life [11]. Moreover, their efficiency depends on the amount of precipitation. According to the Lithuanian Hydrometeorological Service under the Ministry of Environment (2013), in Lithuania, the average annual amount of precipitation varies from 600 mm to 900 mm. During the warm period, it varies from 400 mm to 500 mm. Hence, in Lithuania, using chemical dust suppressants to control dustiness is uneconomical, and control of dustiness remains a challenge [12].

Soft asphalt and Otta Seal technologies are a sustainable and economical solution to control dustiness on gravel roads [13–17]. Over the last decade, soft asphalt has been used on the low-volume roads of Nordic countries as it is less sensitive to frost heaves and fatigue, more flexible, and possesses self-healing abilities. Botswana, New Zealand, the United States (Minnesota), Norway, and Sweden, have already observed the benefits of Otta Seal, which can be constructed as a single layer or as double layers, and use it widely [18–22].

Soft asphalt and Otta Seal do not increase bearing capacity, but protect pavements from moisture infiltration and loss of aggregates, and improve driving conditions. Hence, soft asphalt and Otta Seal reduce the maintenance cost of gravel roads and ensure social satisfaction. From the viewpoint of dust minimization alone, these techniques reduce maintenance costs by more than 40% [21].

The suitability of soft asphalt and double Otta Seal for Lithuanian, low-volume gravel roads, was proven by Vaitkus et al. [23]. The main aim of this study is to show soft asphalt and double Otta Seal sustainability and ability to self-heal on the basis of the results of the qualitative visual assessment of pavement defects and distress. The novelty of this study is that these technologies are usually used in the construction of new pavement structures or the rehabilitation of old ones (the resistance of pavement structures to frost is ensured), but in this study, soft asphalt and double Otta Seal are used on the existing pavement structure and only a new unbound base course of crushed stones of 7–15 cm thickness is constructed (the pavement structure is susceptible to frost). Pavement structures that are not resistant to frost are prone to cracking and these cracks are instrumental in causing distress. Consequently, the pavement's ability to self-heal becomes a significant factor in ensuring the desirable performance of these types of pavement structures.

2. Test Road Sections

2.1. Soft Asphalt

In 2012, soft asphalt technology was used on five gravel road sections (Nos. 1716, 2430, 2518, 4028, and 5235; total length 7.39 km). On these road sections, AADT was less than 200 vpd, except on road no. 4028, where AADT was 640 vpd. The number of equivalent single axle loads (ESALs) was less than 0.1 million on all road sections.

The newly constructed pavement structure consisted of 4.5 cm of soft asphalt and a 15 cm unbound base layer of crushed stones. The bearing capacity of the base course had to be at least 120 MPa. If the thickness of the existing gravel pavement was lower than 35 cm, it was increased using unbound materials.

The soft asphalt SA 16-d-V6000 type C was used on the basis of the results of laboratory tests assessed by the Simple Additive Weighting (SAW) method [24,25]. The binder content varied from 4.6% (road No. 5235) to 5% (road No. 2430), air voids content varied from 5.5% (road No. 1716) to

6.9% (road No. 4028), and the indirect tensile strength ratio varied from 60% (road No. 1716) to 99.8% (road No. 2430). More information is given in [26].

2.2. Double Otta Seal

In 2012, double Otta Seal technology was used on the 13 gravel road sections (Nos. 1708, 1717, 4516(1), 4516(2), 3208, 4118, 1235, 2427, 3918, 4726, 5017, 2642, and 5123; total length 11.95 km). On these road sections, AADT was less than 300 vpd, except on road no. 3208, where AADT was 453 vpd. The number of ESALs was less than 0.1 million on all road sections.

Generally, Otta Seal consists of soft (low viscosity) binder sprayed on the surface, followed by the spreading and rolling of graded aggregates. The layer thickness is about 32 mm when double layers are constructed. On test road sections, double Otta Seal was constructed on the 7–10 cm newly constructed unbound base course of crushed stones. The bearing capacity of the base course had to be at least 120 MPa. If the thickness of the existing gravel pavement was lower than 30 cm, it was increased using unbound materials.

There were 0/16 fraction aggregates used, whose gradation limits are shown in Figure 1, and bitumen emulsion C60B1PA-V6000 or C60BF1-PA. The resistance to fragmentation, according to the impact resistance value (SZ), was ≤18, and the Los Angeles coefficient (LA) was ≤20. The recommended sprayed aggregate amount was 14 L/m^2, but it could be corrected according to the aggregate spread test results. Adhesive additives and primes were not used. More information is given in [23].

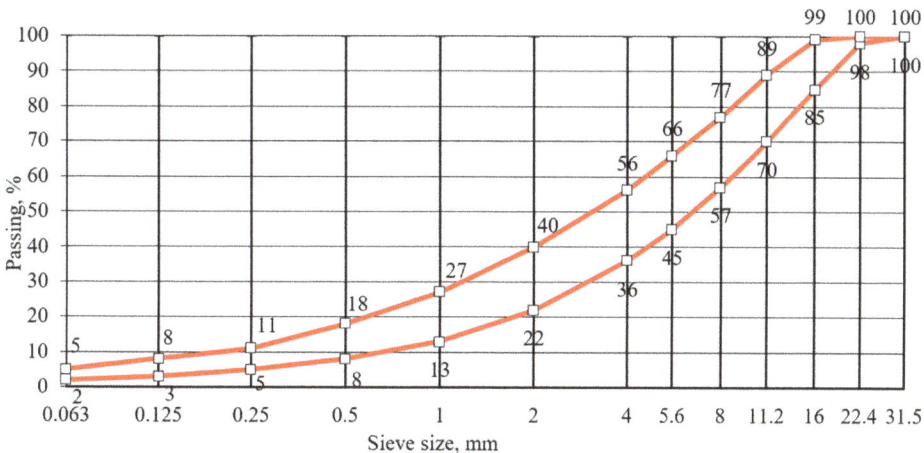

Figure 1. Aggregate gradation limits for double Otta Seal.

3. Research Methodology

In this study, the performance of soft asphalt and double Otta Seal was assessed on the basis of the results of the qualitative visual assessment of defects and distress. This assessment was carried out twice a year. The performance of the soft asphalt was determined in spring and summer. In summer, the qualitative visual assessment of defects and distress was carried out during the hottest period. The performance of double Otta Seal was determined in spring and autumn. In autumn 2016, the qualitative visual assessment of defects and distress on road sections covered with double Otta Seal, was not carried out.

In order to reveal soft asphalt and double Otta Seal's ability to self-heal, the focus was placed mainly on the length of longitudinal and transverse cracks, which are the main result of frost heaves and low temperatures during winter. The severity of longitudinal and transverse cracking was expressed in percentages, as the ratio of the length of the cracks to the length of the road section. The length of longitudinal and transverse cracks was measured with a distance wheel (Figure 2).

The severity of bleeding, potholes, raveling, and seals, was expressed in percentages and represents the distressed road section area.

Figure 2. Measurement of crack length with a distance wheel.

In order to show the sustainability of soft asphalt and double Otta Seal, a life cycle cost analysis was done. Gravel roads with chemical dust suppressants were not analyzed considering life cycle cost since they are less environmentally friendly and sustainable than bituminous materials. The reason is that gravel roads have a negative impact on the corrosion of cars, vegetation, and water life, and their effectiveness depends on the amount of precipitation [11,12]. All of these factors do not occur on roads with bituminous materials. Consequently, a life cycle analysis was conducted for soft asphalt, double Otta Seal, and asphalt concrete.

Dustiness was not directly measured on the test road sections since researchers have already shown that Otta Seal has similar dust emission levels to roads sealed with asphalt concrete (less than 5–10 g/m^2). Thus, bituminous materials on gravel roads reduce dustiness by 2–3 times or even more [21]. The exact soft asphalt and double Otta Seal effect on dustiness will be studied in future research.

4. Results and Discussion

4.1. Soft Asphalt

On road sections with soft asphalt, within five years of operation, longitudinal cracks, transverse cracks, bleeding, potholes, raveling, and seals, were observed. The extent of longitudinal cracks on each road section is shown in Figure 3. The most distressed road section was road no. 2430, where the average severity of longitudinal cracks was 10.00%. The maximum value of 24.10% and the lowest value of 2.16% on road no. 2430, were determined in summer 2016 and summer 2017, respectively. In summer 2013, the longitudinal cracks decreased by 62.98%. This reduction was a consequence of soft asphalt's ability to self-heal under high pavement temperatures. However, in summer 2014, longitudinal cracks showed lower self-healing than in the previous year (only 13.90%), and, in summer 2015, they did not change and even slightly increased (by 6.77%) in summer 2016. This happened because of a less warm summer in 2014, 2015, and 2016, than in 2013. In summer 2017, longitudinal cracks decreased by 87.23%, but the main reason was not soft-asphalt's ability to self-heal, since most of them were sealed during maintenance (Figure 4).

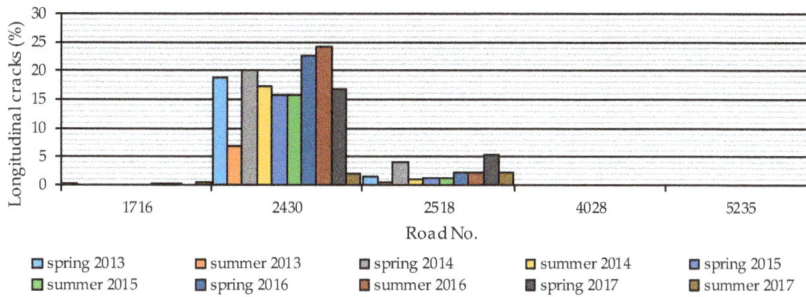

Figure 3. Longitudinal cracks on road sections covered with soft asphalt.

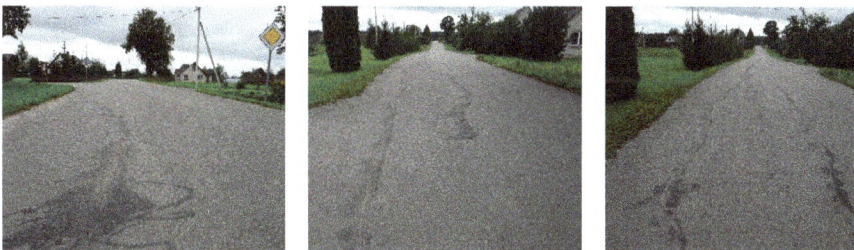

Figure 4. Sealed longitudinal cracks on road No. 2430 in summer 2017.

Similar tendencies for longitudinal crack development were observed on road No. 2518, where the maximum and the minimum values were 5.24% (spring 2017) and 0.39% (summer 2013), respectively. In summer 2013, 2014, and 2017, soft asphalt healed from 57% to 76% of longitudinal cracks. On road No. 1716, longitudinal cracks (0.07%) completely healed in summer 2013 and reappeared only in spring 2016 (0.35%). That summer, the length of longitudinal cracks neither increased nor decreased. However, it increased by 0.6% in summer 2017. On both roads (Nos. 4028 and 5235), longitudinal cracks had not yet formed.

The extent of transverse cracks on each road section is depicted in Figure 5. Transverse cracks appeared one year after soft asphalt construction (after winter 2013/2014). The most distressed road sections were roads No. 2430 and 2518 (Figure 6). The maximum value of 1.88% was determined on road No. 2518 in spring 2017. On this road, in 2014, 2016, and 2017, the length of transverse cracks decreased by 33–98% because of soft asphalt's ability to self-heal under high pavement temperatures. However, in summer 2015, twice as many transverse cracks were observed than in spring. This, as well as the tendency of longitudinal cracks to develop on road No. 2430, is strange because transverse cracks usually occur after winter and do not increase during summer. Sometimes, at high temperatures, they can disappear, but only if the wearing course has self-healing properties. On road No. 5235, transverse cracks (0.35%) completely healed in summer 2014 and reappeared only in spring 2016 (0.45%). That summer, the length of transverse cracks neither increased nor decreased. However, they almost doubled in length in spring 2017 and, in summer, decreased to their previous length (0.45%). On both roads (Nos. 1716 and 4028), transverse cracks had not yet formed.

An analysis of longitudinal and transverse cracks revealed the road sections that performed the best and the worst. Roads No. 1716, 4028, and 5235, were the most resistant to cracking (Figure 7), while roads No. 2430 and 2518 suffered from longitudinal and transverse cracking (Figures 4 and 6) independently of both AADT and ESALs. It was noticed that longitudinal cracking was 10 to 13 times more severe than transverse cracking. Fortunately, soft asphalt's ability to self-heal contributed to the slower development of both longitudinal and transverse cracks and led to lower pavement susceptibility to the frost heaves, fatigue, and low temperatures that directly influence pavement cracking.

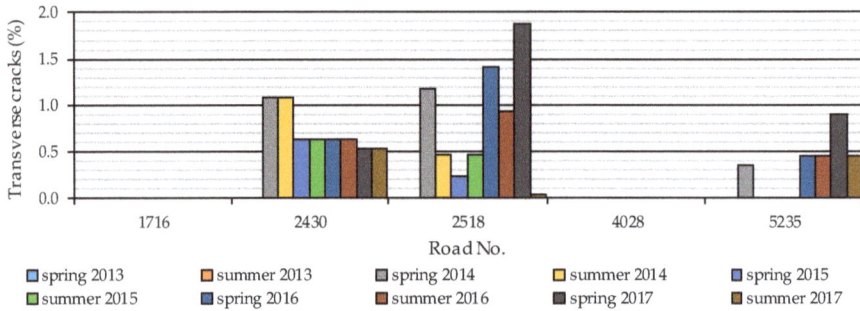

Figure 5. Transverse cracks on road sections covered with soft asphalt.

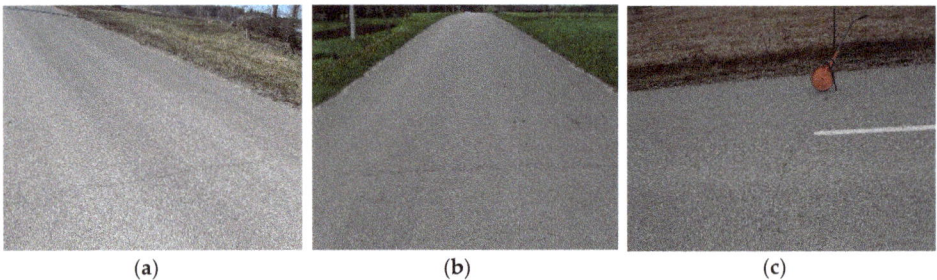

Figure 6. Transverse cracks on road sections covered with soft asphalt: (a) road No. 2430 in spring 2014; (b) road No. 2430 in spring 2016; (c) road No. 2518 in spring 2017.

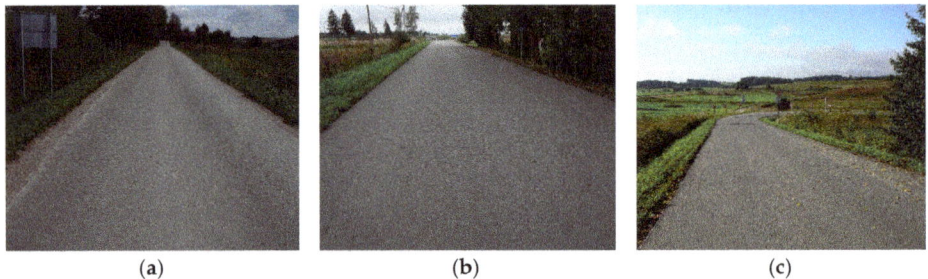

Figure 7. Road sections that showed the best performance after 5 years of operation: (a) road No. 1716; (b) road No. 4028; (c) road No. 5235.

Bleeding, potholes, raveling, and seals are, at least, significant forms of distress. Only on road No. 1716 did bleeding reach up to 4%, while on all other roads it was less than 1%. Potholes, raveling, and seals on each road, at any season, were less than 0.5%.

4.2. Double Otta Seal

On road sections with double Otta Seal, within five years of operation, longitudinal cracks, transverse cracks, bleeding (especially in wheel paths), the loss of both binder and aggregates, and other small defects and distress, were observed. The extent of longitudinal cracks on each road section is shown in Figure 8. This distress was observed from spring 2014. The most distressed road sections were roads no. 1708, 4516(1), and 2642, where the average length of longitudinal cracks on each road, within five years of operation, was 24.62%, 11.83%, and 26.41%, respectively (Figure 9). The maximum value of 51.54% was determined in spring 2017 on road no. 1708. Such a huge development of longitudinal cracks occurred because of frost heaves and low bearing capacity during the spring thaw. On four roads (Nos. 4516(2), 3208, 1235, and 5123) the average length of longitudinal cracks within five years of operation varied from 1% to 7%. Other roads were not prone to the development of longitudinal cracks, because cracks appeared on less than 1% of road length. On some roads (e.g., Nos. 1708, 4516(1), 4516(2), 3208, 2427, 4726, and 5017), in autumn, fewer longitudinal cracks appeared than in spring because a soft bitumen was used for emulsion production, which resulted in pavement self-healing under high pavement temperatures, and thus some cracks healed during summer. However, some anomalies were observed on roads No. 1708, 3208, 2642, and 5123, in 2014 and 2015, since more longitudinal cracks were determined in autumn than in spring, which was unexpected.

The extent of transverse cracks on each road section is shown in Figure 10. This distress was observed from spring 2014. The most distressed road section was road no. 1235, where the average length of transverse cracks within five years of operation was 1.75% (Figure 11). It was emphasized that, on this road section, more than 85% less transverse cracks were determined in autumn than in spring. A similar tendency, of a decrease in transverse cracks, was also observed on other roads (e.g., Nos. 4516(1), 4516(2), 3208, 4726, and 5017). Hence, on the roads covered with double Otta Seal, transverse cracks self-healed under high pavement temperatures. The maximum value of 4.19% of transverse cracks was determined in spring 2014 on road No. 5017. Four roads (Nos. 1708, 1717, 2642, and 5123) were resistant to the development of transverse cracks, since within five years of operation no transverse cracks appeared. On other roads, the average of transverse cracks within five years of operation was less than 1%.

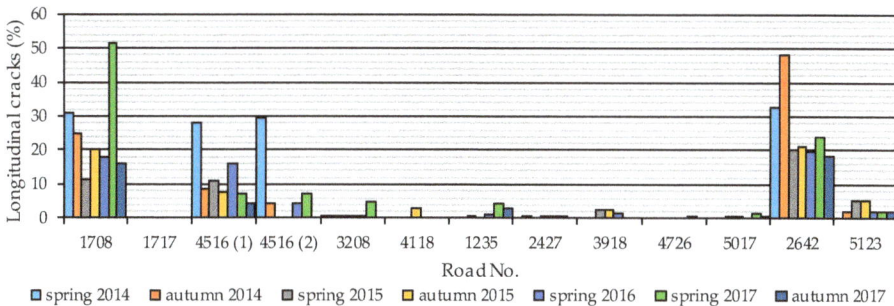

Figure 8. Longitudinal cracks on road sections covered with double Otta Seal.

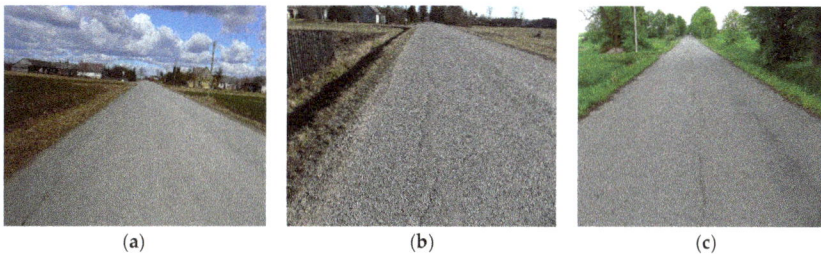

(a) (b) (c)

Figure 9. Longitudinal cracks on road sections covered with double Otta Seal: (a) road No. 1708 in spring 2017; (b) road No. 4516(1) in spring 2014; (c) road No. 2642 in spring 2016.

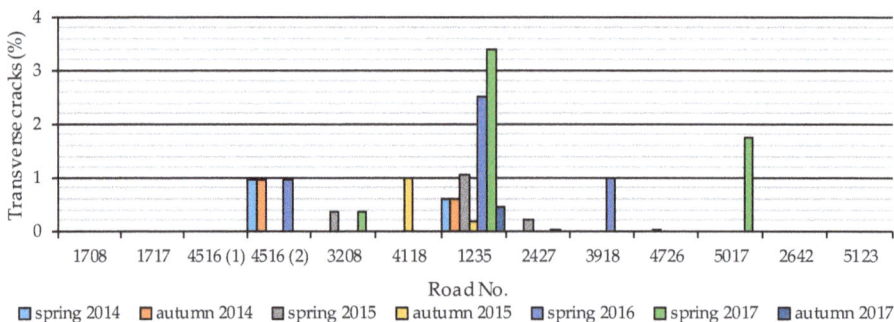

Figure 10. Transverse cracks on road sections covered with double Otta Seal.

(a)   (b)   (c)

Figure 11. Transverse cracks on road No. 1235: (**a**) in spring 2017; (**b**) in spring 2017; (**c**) in autumn 2017.

An analysis of longitudinal and transverse cracks revealed the road sections that performed the best and the worst. Roads No. 1717, 4118, 2427, 3918, 4726, and 5017, were the most resistant to cracking (Figure 12), since the average of both longitudinal and transverse cracks within five years of operation was less than 1%, while roads No. 1708, 4516(1), 1235, and 2642, suffered from longitudinal or transverse cracking (Figures 9 and 11). The severity of cracking did not depend on AADT and ESALs on test roads. It was noticed that longitudinal cracking was 12 to 25 times more severe than transverse cracking. Fortunately, double Otta Seal's ability to self-heal contributed to the slower development of both longitudinal and transverse cracks and led to lower pavement susceptibility to frost heaves, fatigue, and low temperatures.

(a)   (b)   (c)

Figure 12. Road sections that showed the best performance after 5 years of operation: (**a**) road No. 4118; (**b**) road No. 4726; (**c**) road No. 5017.

Qualitative visual assessments of defects and distress revealed that road sections covered with double Otta Seal tended to undergo bleeding (especially in wheel paths). The bleeding, loss of both binder and aggregates, and other small defects and distress, on each road section within three years of operation, are given in [23] and this tendency has not changed significantly.

4.3. Life Cycle Cost

Life cycle analysis for soft asphalt, double Otta Seal, and asphalt concrete, is given in [24]. This revealed that soft asphalt construction was up to 29% cheaper than asphalt concrete, while double Otta Seal saved 44% of asphalt concrete construction cost [24]. It should be noted that maintenance cost was not included in that analysis. However, the expected service life and the actual maintenance needs of the test road sections with soft asphalt and double Otta Seal showed positive results, especially for soft asphalt.

5. Conclusions

Soft asphalt and double Otta Seal can be termed as self-healing, sustainable techniques for low road rehabilitation. These technologies perform as standard asphalt pavements providing smooth, durable, and flexible surfaces, and greatly prolonging maintenance periods. The analysis of the qualitative visual assessments of pavement defects and distress during five years of road operation revealed the following:

- Longitudinal cracking was 10 to 25 times more severe than transverse cracking, depending on the technology (soft asphalt or double Otta Seal) and existing pavement structure composition. An adequate frost resistance of the whole pavement structure and bearing capacity of the base and subgrade (especially during the spring thaw), were vital for satisfactory pavement performance.
- Both technologies were able to self-heal during spring–summer. The healing effect was more than 13% and 19% on roads with soft asphalt and double Otta Seal, respectively. In addition, on some roads, all cracks observed in spring self-healed during summer.
- Three of five roads with soft asphalt (Nos. 1716, 4028, and 5235) and six of 13 roads with double Otta Seal (Nos. 1717, 4118, 2427, 3918, 4726, and 5017) were resistant to both longitudinal and transverse cracking (the average of cracking was less than 1%) independent of both AADT and ESALs. This might be related to the existing pavement structure composition and thickness, since during the design process, neither structure composition nor thickness were considered. Further evaluation and analysis of these factors will be conducted in future research.

The pavement condition of rehabilitated low-volume gravel roads after five years' operation confirmed the suitability of soft asphalt and double Otta Seal technologies for the improvement of gravel roads with annual average daily traffic of less than 300 vpd (except roads No. 3208 and 4028) and 0.1 million ESALs over a 20-year period.

Author Contributions: Audrius Vaitkus initiated the research, set objectives and led the study. Faustina Tuminienė and Judita Gražulytė carried out qualitative visual assessments of defects and distress on these roads. Donatas Čygas, Viktoras Vorobjovas, and Judita Gražulytė analyzed the results and wrote the manuscript. All of the authors contributed to the discussion, refined the final manuscript, and approved it.

Conflicts of Interest: The authors declare no conflict of interest.

References

1. Jones, T.E. *Dust Emission from Unpaved Roads in Kenya*; Laboratory Report 1110; Transport and Road Research Laboratory: Crowthome, UK, 1984.
2. Edvardsson, K. Gravel Roads and Dust Suppression. *Road Mater. Pavement* **2009**, *10*, 439–469. [CrossRef]
3. Behnood, A.; Roshandeh, A.M.; Mannering, F.L. Latent class analysis of the effects of age, gender, and alcohol consumption on driver-injury severities. *Anal. Methods Accid. Res.* **2014**, *3–4*, 56–91. [CrossRef]

4.   Monlux, S.; Mitchell, M. Chloride Stabilization of Unpaved Road Aggregate Surfacing. *Transp. Res. Rec. J. Transp. Res. Board* **2007**, *1989*, 50–58. [CrossRef]

5.   Addo, J.Q.; Sanders, T.G. *Effectiveness and Environmental Impact of Road Dust Suppressants*; Mountain-Plains Consortium; U.S. Department of Commerce: Ft. Collins, CO, USA, 1995.

6.   Thompson, R.J.; Visser, A.T. Selection, performance and economic evaluation of dust palliatives on surface mine haul roads. *J. S. Afr. Inst. Min. Metal.* **2007**, *107*, 435–450.

7.   Gottschalk, K. *Road Dust: A Survey of Particle Size, Elemental Composition, Human Respiratory Deposition and Clearance Mechanisms*; Unpublished Independent Study; Colorado State University: Ft Collins, CO, USA, 1994.

8.   Donaldson, K.; Gilmour, M.I.; MacNee, W. Asthma and PM10. *Respir. Res.* **2000**, *1*, 12–15. [CrossRef] [PubMed]

9.   Hoover, J.M. *Surface Improvement and Dust Palliation of Unpaved Secondary Roads and Streets*; Report ISU-ERI-AMES-72316; Engineering Research Institute, Iowa State University: Ames, IA, USA, 1973.

10.   Sanders, T.G.; Addo, J.; Ariniello, J.; Heiden, W. Relative Effectiveness of Road Dust Suppressants. *J. Transp. Eng.* **1997**, *123*, 393–397. [CrossRef]

11.   Golden, B. Impact of magnesium chloride dust control product on the environment. In Proceedings of the Transportation Association of Canada Annual Conference, Winnipeg, MB, Canada, 15–16 September 1991; pp. 1–6.

12.   Vorobjovas, V. Assurance of the Function of Low-Volume Roads for the Improvement of Driving Conditions. *Balt. J. Road Bridge Eng.* **2011**, *6*, 67–75. [CrossRef]

13.   Roads and Traffic Research Society and Asphalt Roads Working Group. *Supplementary Technical Contract Specifications and Guidelines for the Construction of Carriageway Surfacing from Asphalt ZTV Asphalt—StB 2000*; Road and Transportation Research Association: Cologne, Germany, 2000.

14.   Silfwerbrand, J. Swedish Design of Industrial Concrete Pavements. In Proceedings of the 7th International Conference on Concrete Pavements, Orlando, FL, USA, 9–13 September 2001; p. 16.

15.   Swedish National Road Administration. *General Technical Construction Specifications for Roads*; Chapter 1, Common Prerequisites; Chapter 3, Pavement Design; Chapter 6, Bitumen-Bound Layers; Swedish National Road Administration: Solna, Sweden, 1996.

16.   Joshi, S.G.; Jha, A.K. Otta Seal Experience in Nepal. In Proceedings of the Transportation Research Board 92nd Annual Meeting, Transportation Research Board of the National Academies, Washington, DC, USA, 13–17 January 2013.

17.   Overby, C.; Pinard, M. Development of an Economic and Practical Alternative to Traditional Bituminous Surface Treatments. *Transp. Res. Rec. J. Transp. Res. Board* **2007**, *1989*, 226–233. [CrossRef]

18.   Pinard, M.; Obika, B.; Motswagole, K. Developments in Innovative Low-Volume Road Technology in Botswana. *Transp. Res. Rec. J. Transp. Res. Board* **1999**, *1652*, 68–75. [CrossRef]

19.   Oloo, S.; Lindsay, R.; Mothilal, S. Otta Seals and Gravseals as Low-Cost Surfacing Alternatives for Low-Volume Roads: Experiences in South Africa. *Transp. Res. Rec. J. Transp. Res. Board* **2003**, *1819*, 338–342. [CrossRef]

20.   Johnson, G. Minnesota's Experience with Thin Bituminous Treatments for Low-Volume Roads. *Transp. Res. Rec. J. Transp. Res. Board* **2003**, *1819*, 333–337. [CrossRef]

21.   Waters, J.C. *Long-Term Dust Suppression Using the Otta Seal Technique*; New Zealand Transport Agency: Wellington, New Zealand, 2009.

22.   Overby, C.; Pinard, M.I. The Otta Seal Surfacing: A practical and Economic Alternative to Traditional Bituminous Surface Treatments. *Transp. Res. Rec. J. Transp. Res. Board* **2013**, *2349*, 136–144. [CrossRef]

23.   Vaitkus, A.; Vorobjovas, V.; Tuminienė, F.; Gražulytė, J. Performance of Soft Asphalt and Double Otta Seal within First Three Years. *Adv. Mater. Sci. Eng.* **2016**, 1–12. [CrossRef]

24.   Perveneckas, Z.; Vaitkus, A.; Vorobjovas, V. Soft Asphalt Pavements—Solution for Low Traffic Volume Roads in Lithuania. In Proceedings of the 28th International Baltic Road Conference, Vilnius, Lithuania, 26–28 August 2013.

Sustainability **2018**, *10*, 198

25. Vorobjovas, V.; Andriejauskas, T.; Perveneckas, Z. Selection of Soft Asphalt Pavements for Low-volume roads in Lithuania. In Proceedings of the 9th International Conference on Environmental Engineering, Vilnius, Lithuania, 22–24 May 2014.

26. Gražulytė, J.; Žilionienė, D.; Tuminienė, F. Otta Seal—The New Way to Solve Problems of Maintenance of Gravel Roads in Lithuania. In Proceedings of the 9th International Conference on Environmental Engineering, Vilnius, Lithuania, 22–24 May 2014.

sustainability

MDPI

Article

Computer Vision-Based Bridge Displacement Measurements Using Rotation-Invariant Image Processing Technique

Byung-Wan Jo, Yun-Sung Lee *, Jun Ho Jo and Rana Muhammad Asad Khan

Department of Civil and Environmental Engineering, Hanyang University, 222 Wangsimni-ro, Seongdong-gu, Seoul 04763, Korea; joycon@hanmail.net (B.-W.J.); kpxjuno08@hanyang.ac.kr (J.H.J.); masadkhan87@gmail.com (R.M.A.K.)
* Correspondence: unsaboa@hanyang.ac.kr; Tel.: +82-2220-0327

Received: 20 April 2018; Accepted: 28 May 2018; Published: 29 May 2018

Abstract: Bridges are exposed to various kinds of external loads, including vehicle and hurricanes, during their life cycle. These loads cause structural damage, which may lead to bridge collapse. To ensure bridge safety, it is essential to periodically inspect the physical and functional conditions of bridges. The displacement responses of a bridge have significance in determining the structural behaviors and assessing their safety. In recent years, many researchers have been studying bridge displacement measurements using image processing technologies. Image-processing-based displacement measurements using a computer analysis system can quickly assess bridge conditions and, thus, can be used to enhance the reliability of bridges with high accuracy. This paper presents a method based on multiple-image processing bridge displacement measurements that includes enhanced robustness to image rotation. This study applies template matching and a homography matrix to measure the displacement that works well regardless of the angle between the smartphone camera and the target.

Keywords: bridge; displacement measurement; template matching; homography matrix; smartphone camera

---

1. Introduction

In modern society, bridges serve as the backbone of a nation's economy as they connect separated regions and provide effective means of transportation to accommodate the needs of populations experiencing explosive growth. However, severe weather can badly damage bridge structures, reducing their life and safety. Therefore, systematic inspections and maintenance must be performed to maintain the optimum conditions of bridges. A bridge's displacement responses provide valuable information for structural analysis to assess bridge safety [1,2] and performance. Thus, the monitoring of displacement characteristics is a critical issue for the safety and maintenance of bridges. Displacement response measurements in bridges are normally done using high precision sensors, such as linear-variable differential transformers (LVDTs) [3], accelerometers [4], global positioning systems (GPSs). However, such conventional methods have limitations [5–7]. For instance, an LVDT or accelerometer must be installed with additional equipment, such as data loggers (including cables and power supply equipment), which can introduce inconvenience and extra monitoring-cost for large-scale bridges and structures. Moreover, bridge monitoring based on acceleration measurements is relatively an effortless approach, as it determines the displacement by simply double integrating the acceleration values, however, the biased initial conditions may create unstable results [8]. On the other hand, the GPS-based method has low precision of vertical displacement measurements [9]. The computer vision-based techniques offer significant advantages over conventional sensing systems for measuring structural responses such as (i) computer vision-based techniques can measure displacements of multiple points from a motion image captured by a single camera; (ii) the computer vision-based technique does not require physical access to the structures, as the vision sensors (such as cameras) can be setup at a remote location and can easily alter the measurement points; (iii) computer vision-based techniques can be placed hundreds of meters away from the bridge (when using a zoom high resolution lens) and still achieve satisfactory measurement accuracy. Thus, the computer vision method can effectively reduce time and cost compared to previous sensor-based methods. In past structural inspections, computer vision technology was often used to detect cracks [10]. Crack inspection research focuses primarily on techniques that accurately analyze the length, width, and depth of cracks [11]. Recently, computer vision technique has gradually expanded in the field of dynamic response measurements for various structures, including buildings and bridges [12]. For example, Stephen et al. [13] presented an image-tracking system to measure bridge–deck displacement. This image tracking system was successfully applied at Humber Bridge in the UK. Olaszek [14] utilized photogrammetric principle and introduced a scheme to measure displacement and dynamic characteristics of bridges. Guo [15] proposed computer vision technology including a Lucas–Kanade template tracking algorithm to measure the dynamic displacement of large-scale structures. Shan et al. [16] proposed bridge deflection monitoring technique based on Zhang's [17] camera calibration method. Feng and Feng [18,19] proposed structural modal property identification method utilizing computer vision technique. Similarly, Yoon et al. [20] proposed a computer vision-based method using consumer-grade cameras to identify the dynamic characteristic of structures.

Computer vision-based systems for structural displacement measurements mainly rely on template matching [21]. Template matching approach is an image processing technique to determine the locations of an image matching a predefined template image. To match image, the template image is compared against the source image by moving it one pixel at a time. At each location, a correlation-coefficient is computed indicating the similarity between template image and a specific area of the source image. As all the possible locations of template image are considered and compared with source image, therefore, the area with highest correlation-coefficient becomes the target position. The displacement measurement using the digital image correlation (DIC) approach is a computer vision technology that employs a digital camera and image processing for accurate measurements of changes in images. The use of the DIC method to measure displacements of objects was initially proposed by researchers at the University of South Carolina [22–24]. The DIC technique has shown wide scale powerful and flexible applications in the field of experimental solid mechanics, particularly for measuring the surface deformations [25]. Moreover, it has been studied and applied by several researchers for structural displacement measurement [5,26,27]. In recent years, the displacement measurement method using DIC has been realized as a simple system because it is an image-processing technique correlating the original image and deformation image. Therefore, it has shown high applicability to the various devices. In object tracking, DIC is not an ideal technique as it can be easily affected by image scale and rotation [28]. In order to overcome the limitations of template-based matching using DIC technique, Jeong et al. [29] proposed a pan-tilt-zoom (PTZ) camera-based computer vision system using a sum of squared differences (SSD) and homography-matrix [30], this system is capable of expanded measurable range in monitoring structural displacement. Till present, several correlation-coefficient functions have been introduced including normalized cross-correlation (NCC), sum of absolute differences (SAD), sum of squared differences (SSD), zero-mean normalized cross-correlation (ZNCC), zero-mean sum of absolute differences (ZSAD), zero-mean sum of squared differences (ZSSD), optimized sum of absolute difference (OSAD), and optimized sum of squared difference (OSSD) [31,32]. Among these, NCC, SAD, SSD, and ZNCC have been widely adopted in template matching from the field of objects tracking [32]. The SSD is a popular method for many applications, including template-based matching, due to its efficient computational schemes. However, it is sensitive to illumination changes [33]. On the other hand, NCC has robustness advantages against illumination change. In addition to, NCC is a very simple approach for determining multiple patterns from a source image. Thus, the NCC function is mainly used to measure the displacement of the structure in the civil engineering [34,35].

Present study introduces a computer vision method for bridge displacement measurement, enabled by the robust template matching supported by homography-matrix and NCC function. Additionally, for motion image capture, smartphone camera was used instead of the conventional commercial-scale expensive cameras. The adoption of smartphone would allow the quick safety inspection of bridge because of being cost-effective, mobile, and ease of installation.

2. Literature Review

Jauregui et al. [36] successfully measured the vertical deflections of bridge structures by implementing close-range terrestrial photogrammetry approach. For highly accurate displacement measurements, a sub-pixel edge detection algorithm was introduced by Poudel et al. [37]. Fu and Moosa [38] utilized high resolution CCD camera to develop a cost-effective and efficient technique for the measurement of displacement in civil structures. A vision-based system for the accurate monitoring of absolute displacement was introduced by Wahbeh et al. [39]. The architecture of this system is simply comprised of a LED, a camera, and a computer. Through experiments on the Vincent Thomas Bridge in California, researchers demonstrated the feasibility of this approach for measuring bridge displacement. Kohut et al. [40] conducted a comparative research on radar interferometry and vision-based methods to measure the displacements in civil structures. They demonstrated the precision of vision-based techniques by implementing them on a steel bridge; moreover, they accurately determined the bridge displacement. Lee and Shinozuka [5,41,42] developed an image processing system which is digital enough to monitor and measure the dynamic displacements of flexible bridges in real-time. This system was also shown to have the ability to assess the load carrying capacity of a steel girder bridge located in Korea. Fukuda et al. [6] introduced a vision-based displacement measurement system that utilizes a low-cost digital camcorder and a PC. Similarly, Choi et al. [43] presented a structural dynamic displacement system that could perform multi-measurements using an economical handheld digital camcorder. Ho et al. [44] developed a synchronous vision system implementing an image-processing approach to measure the dynamic displacement of civil infrastructures. This system can simultaneously support several camcorders for real-time multipoint displacement measurements. Tian and Pan [45] proposed a vision-based displacement measurement device using LED targets for real-time measurements of bridge deflection. Yoneyama et al. [27] obtained bridge deflection measurements using a digital image correlation method for bridge load tests. Busca et al. [46] combined pattern matching, digital image correlation, and edge detection and developed a vision-based technique for the monitoring of dynamic response of bridge.

A remote vision-sensor system in combination with advance template matching, to measure the displacement in civil infrastructures was proposed by Feng et al. [47]. Ribeiro et al. [7] introduced a video system based on the RANdom SAmple Consensus (RANSAC) algorithm. This algorithm accurately determines the dynamic displacement of railroad bridges. Feng et al. [48] developed a markerless vision-based system using an advanced orientation code-matching (OCM) algorithm to measure the dynamic displacement of bridges. Jeon et al. [49] presented a paired visual servoing system to measure the six degrees of freedom (6-DOF) displacement of high-rise buildings and long-span bridges. Similarly, Park et al. [50] measured the three-dimensional displacement associated with structures with the utilization of a capture system. This system is capable of high sample rate for the accurate measurement of displacement. Chang and Xiao [51] used a single-camera based video-grammetric approach to measure both three-dimensional translation in concurrence with the rotation of the planer marker installed on the structure. Ji and Chang [52] proposed a novel vision-based method that used an optical flow technique to eliminate the use of a marker. This system used a camcorder to measure the vibrational responses of a small pedestrian bridge. In another study, Ji and Chang [53] presented a markerless stereo vision technique to monitor the dynamic responses of line-like elements, such as cables, in both spatial and temporal domains. Kim and Kim [54] proposed a markerless vision technique that utilized digital

Sustainability **2018**, *10*, 1785

image correlation obtained from a portable and digital camcorder to measure hanger cable tensions of a suspension bridge. Kim [55] proposed a multi-template matching method that rectifies the errors occurring during marker recognition as noise in after-images. This noise is usually caused by the camera shaking and by high-speed motion of the cables. A digital photogrammetry method [56] using multiple camcorders to measure the synchronized ambient vibrations at various points of a stay cable bridge has been proposed, this approach effectively identifies the mode shapes ratios. For the quick inspection of the cable supported bridge, Zhao et al. [57] conducted an experiment on the feasibility and utilization of the computer vision-based cable force estimation method using a smartphone camera. They used a D-viewer and iPhone 6 to measure the vibration characteristics of the cable. Based on a previous study [57], Zhao et al. [58] proposed a computer vision-based approach for identifying the natural frequencies of cable and estimating cable force using a smartphone camera.

3. Computer Vision-Based Bridge Displacement Measurement

3.1. Overview

Figure 1 shows an overview of the proposed computer vision-based displacement measurement methodology. This system consists of a marker, a smartphone camera, and a computer installed with an image processing software. The connection between the smartphone and the computer is conducted by a wireless communication or cable. The proposed computer vision-based displacement measurement method is composed of three main phases: (1) motion image acquisition; (2) image processing for object tracking; and (3) calculation of bridge displacement. The smartphone camera is fixed on a tripod and is located at about 1 m away from the marker. Installation of smartphone camera and its focusing is time-effective, as it takes a few minutes. The smartphone camera gets the motion images of the marker for displacement measurement, and computer analyzes the marker's movement and bridge displacement through image processing software. This image processing relies on the homographic matrix for the planar projections of the captured motion images, and template-based matching algorithms using on the NCC function for object tracking.

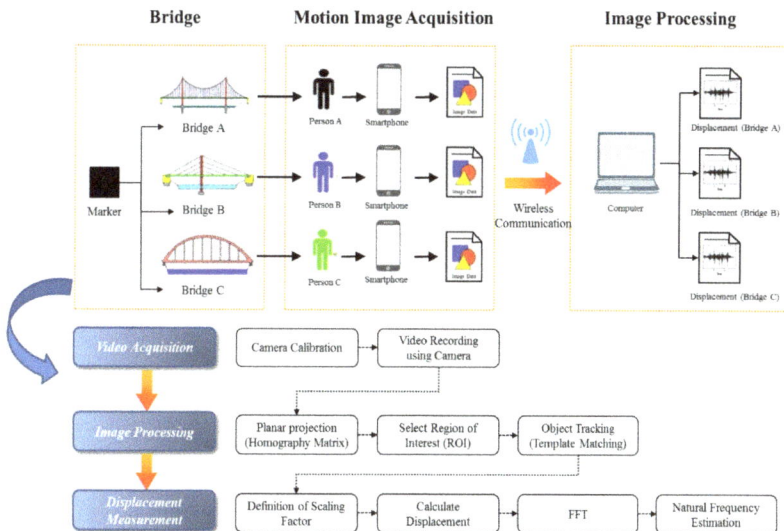

Figure 1. Schematics and Flowchart of the Proposed Methodology.

3.2. Camera Calibration

Accurate calculation of bridge displacement based on measurements from motion images requires the correction of camera distortion. Generally, camera calibration is the most crucial stage for a vision-based system. Because the proposed system relies on motion images from a camera to determine displacement, the accuracy is highly influence by the precision calibration of the camera. In this regard, to eliminate distortion from the camera lens, Zhang [17] proposed a camera calibration algorithm based on the 2D planar pattern. In the present study, the inherent and extraneous characteristics of camera, and distortion coefficient of camera lenses were estimated using Zhang's calibration algorithm (for more details see [17]).

3.3. Template-Based Matching Method

Among the object tracking methods, the template-based matching technique is commonly applied specifically for the measurement of structures. The template-based matching technique is normally implemented by first registering a part of the search image to use as a template image (such as marker). Because the template image is registered, an object can be found in the search image using NCC [28] function. The NCC is a sub-class of DIC is one of the most promising approach to analyze the resemblance of two individual images base on their cross-correlation. The NCC coefficient is defined by Equation (1). The basic concept of the template-matching technique is shown in Figure 2.

$$\gamma(u,v) = \frac{\sum_{x,y}\left\{f(x,y) - \overline{f}\right\}\left\{t(x-u, y-v) - \overline{t}\right\}}{\sqrt{\sum_{x,y}\left\{f(x,y) - \overline{f}\right\}^2 \sum_{x,y}\left\{t(x-u, y-v) - \overline{t}\right\}^2}} \tag{1}$$

$\gamma(u,v)$ = correlation coefficient
$f(x,y)$ = intensity value of search image at point (x,y)
$\overline{f}$ = mean value of $f(x,y)$ within the region of the template
$\overline{t}$ = mean value of the template

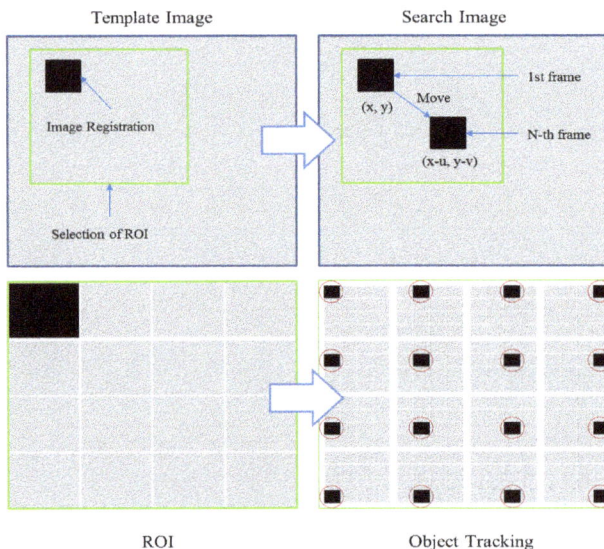

Figure 2. Basic Concept of the Template-based Matching Technique.

3.4. Homography-Matrix

For precise displacement measurements, an image transformation function should be applied to eliminate geometric distortions. Therefore, the present method includes the homography approach [30] to eliminate the geometric distortions of image scale and rotation. Homography simultaneously maps two planes by translating a point P from the *x-y* plane to a point P' in the newly-projected plane *x'-y'*. As shown in Figure 3, the *x-y* plane points P1(*x1, y1*), P2(*x2, y2*), P3(*x3, y3*), and P4(*x4, y4*) are relocated as points P1'(*x1', y1'*), P2'(*x2', y2'*), P3'(*x3', y3'*), and P4'(*x4', y4'*), respectively, on the *x'-y'* plane. These newly oriented points are located at distance R (mm) from the origin. The distances of points P1', P2', P3', and P4' have the same attributes as the origin, and the intersection of each point occurs at 90°; thus, the orthogonality for each new point is maintained. Conclusively, it can be said that, in homography, the standard plane is the distortion-corrected plane in units of millimeters. The homography shows the relationship between points in the form of a 3 × 3 matrix, as illustrated by Equation (2). This relationship remains valid, not only for the direct projection of two planes, but for every plane that is directly or indirectly projected onto a flat plane.

$$
\begin{bmatrix} x_i' \\ y_i' \\ 1 \end{bmatrix} = \begin{bmatrix} h_{11} & h_{12} & h_{13} \\ h_{21} & h_{22} & h_{23} \\ h_{31} & h_{32} & h_{33} \end{bmatrix} \begin{bmatrix} x_i \\ y_i \\ 1 \end{bmatrix}, i = 1, 2, \ldots \tag{2}
$$

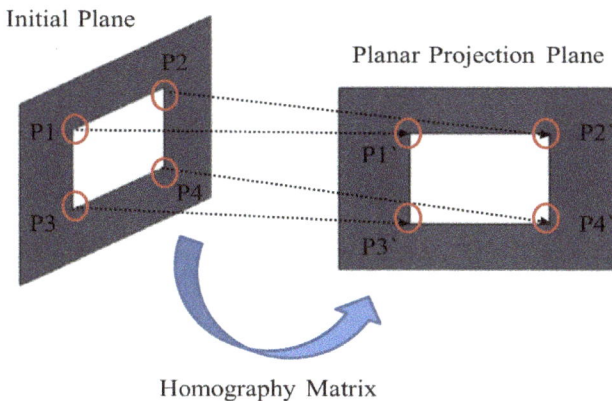

Figure 3. Basic Concept of Planar Projection using a Homography-Matrix.

3.5. Scaling Factor

To obtain the accurate displacement response from the image, it is mandatory to transform the pixel coordinates and physical coordinates. Therefore, a scaling factor is required to determine the transformation coefficient. This scaling factor (mm/pixel) defines as the transformation coefficient between the units on image plane and the physical units. When the image plane is parallel to the object surface, the scaling factor can be calculated using Equation (3).

$$
\text{Scaling factor} = \frac{\text{physical units (such as marker length)}}{\text{pixel units on the image plane}} \tag{3}
$$

3.6. System Implementation

In this study, we developed software to accurately estimate bridge dynamic displacements based on the template-based matching technique and a homography-matrix. This software is operated on a Windows platform, and it is performed in a user-friendly graphical user interface (GUI).

The programming tool for development the image processing software was used Visual Studio 2012 using C++ language. As shown in Figure 4, the developed software can import the motion images acquired from the smartphone camera, register the template image directly using the mouse on the video screen. Also, region of interest (ROI) can be set in the same way. The scaling factor is easily changed by inputting the length of the used marker. The software has been developed to fit the Galaxy S7 Samsung smartphone. However, it can be easily adapted to other smartphone models by changing the focal length through coding. In real time, the users can view the displacement curve through the chart screen. In addition, a fast Fourier transform (FFT) was applied to calculate the natural frequency of the bridge from displacement data. This software is a stand-alone executable program helpful for the measurement of bridge dynamic displacement. Through this software, the bridge dynamic displacements can be obtained easily by recognizing the movement of the pre-designed marker attached on the target using smartphone camera without expensive displacement sensors.

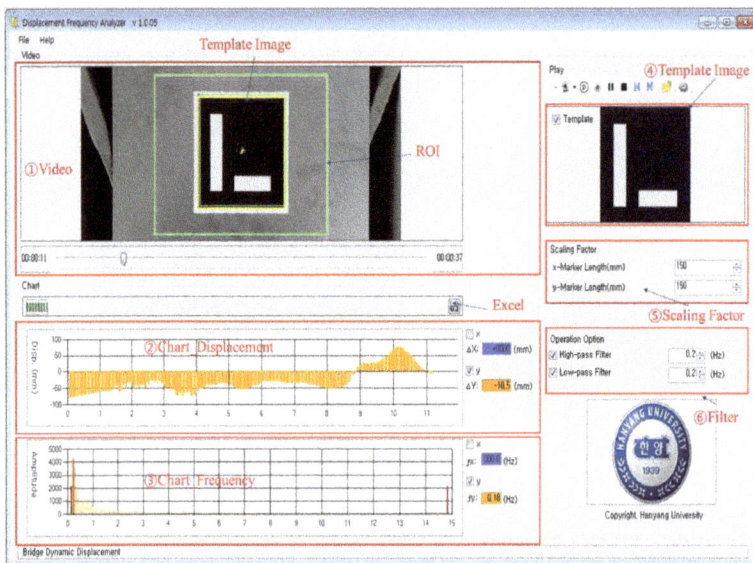

Figure 4. Interface of software for the measurement of bridge displacements.

4. Experimental Validation

4.1. Laboratory Scale Tests

Initially, the performance of proposed computer vision-based bridge displacement measurement technique was evaluated at laboratory scale. In these experimental trials, specially design test equipment was used as shown in Figure 5. A predesigned black and white marker (150 mm × 150 mm) acting as tracking target, was fixed on computer numerical controlled (CNC) movement axis of machine. The linear repeated movement of the axis was fixed at 20 cm from origin point. To evaluate the proposed displacement measurement technique based on homographic matrix, motion images of marker attached movement axis were captured by successively placing the smartphone camera at angles of 0°, 10°, 20°, and 30° with reference to front of marker (shown in Figure 6). The marker's motion images captured by the smartphone camera were recorded at 30 frames per second. The smartphone camera was placed at 1 m from the marker front side so that appropriate installation like field settings without any zoom can be tested. In these experiments, there was no need to install reference sensor to measure the displacement, as the marker is attached CNC movement axis. Thus, the CNC coordinates are more accurate compare to sensor readings and can easily be obtained from the control program. The marker's motion images were recorded and transmitted to computer over wireless communication. The marker displacement readings obtained from the control program were regarded as reference-value and the errors of the proposed approached at various rotation angles were compared. We used Samsung Galaxy S7 smartphone for experimental trials. Table 1 summarizes the hardware specifications of computer vision-based system.

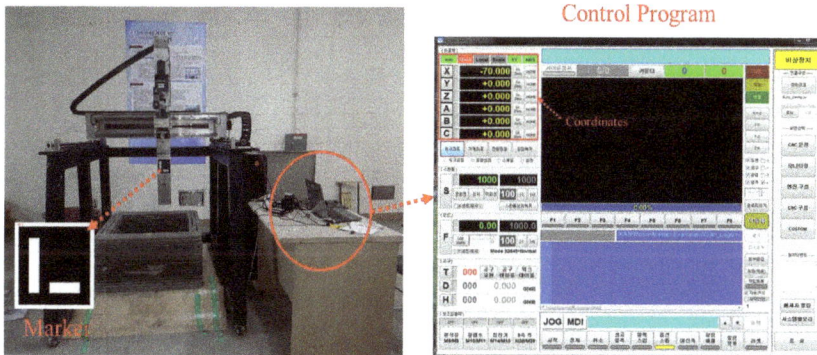

Figure 5. Setup of the Laboratory Tests.

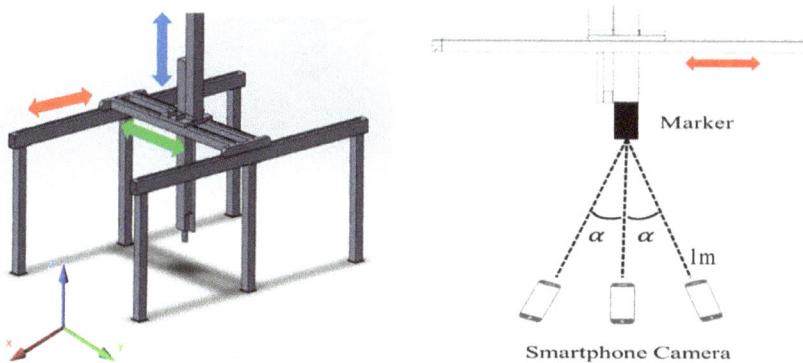

Figure 6. Movement Mechanism of Test Equipment and Angle of Smartphone Camera.

Table 1. The hardware configuration of Computer Vision System.

Types	Models	Specifications
Marker	-	- Two white rectangles with dark background - Length: 150 mm × 150 mm
Smartphone Camera	SM-G930S	Image sensor: Sony IMX260 UHD 4K (3840 × 2160) resolution @30 fps Aperture: f/1.7 Pixel size: 1.4 µm Focal length: 4.20 mm
Computer	Samsung- NT870Z5G	Intel(R) Core(TM) i7-4720HQ CPU 2.60 GHZ 8.00 GB RAM Windows 10 (64-bit)

Figures 7–10 show the measurement results of the displacement at different angles. We conducted error evaluation based on the root mean squared error (RMSE; Equation (4)) to quantify the measurement errors. Table 2 provides the calculated RMSE values for the bridge displacements.

$$\text{RMSE} = \sqrt{\frac{\sum_{i=1}^{n}(x_{1,i} - x_{2,i})^2}{n}} \tag{4}$$

- $x_{1,i}$ = ith set of bridge displacement data at time t_i measured using the proposed method
- $x_{2,i}$ = ith set of bridge displacement data at time t_i measured using LVDT
- n = number of data

The test results show that the proposed method provides a small change according to the angle between the smartphone camera and the marker. When the angle of the smartphone camera and the marker is 0°, RMSE value is 0.037 mm that is similar to the reference value. Although the error is increased by 0.371 mm when the angle is 10°, the magnitude of the error is sufficient to replace the existing displacement sensor. Also, RMSE value is 0.400 mm and 0.398 mm for the angle of 20° and 30°, respectively, that did not show any significant different compare to the angle of 10°. The proposed method maintains a minor error even when the angle increases that can measure accurate displacement of bridge in various range of the angle.

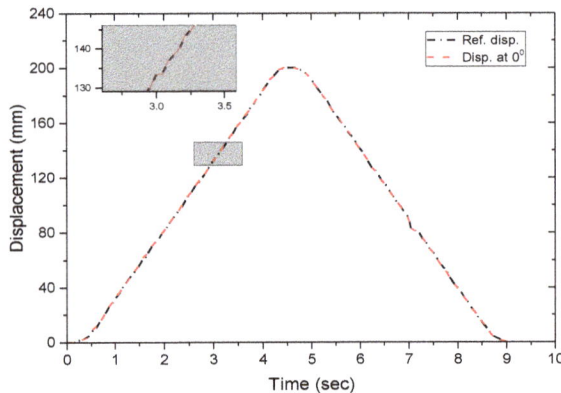

Figure 7. Angle of Smartphone Camera = 0°.

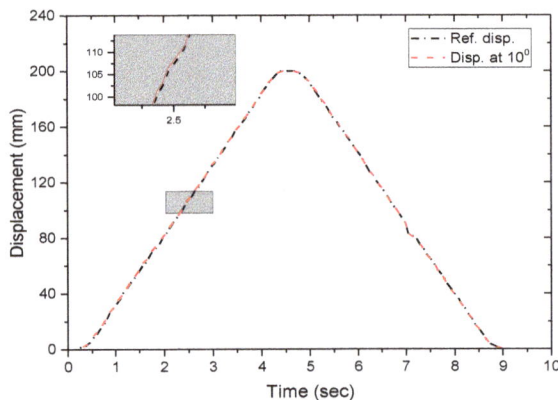

Figure 8. Angle of Smartphone Camera = 10°.

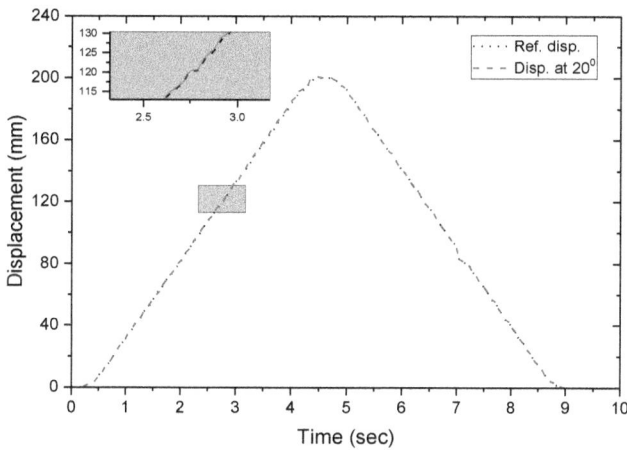

Figure 9. Angle of Smartphone Camera = 20°.

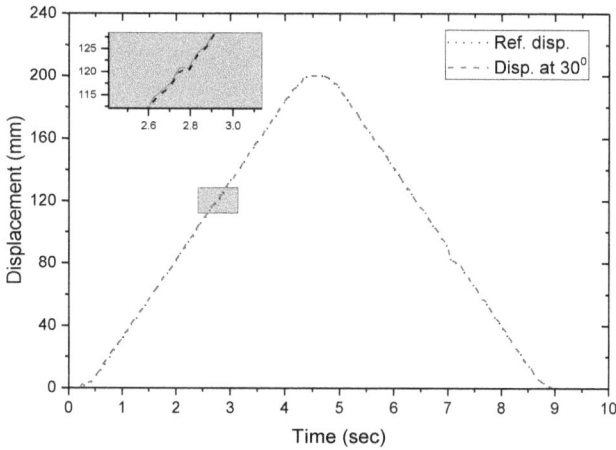

Figure 10. Angle of Smartphone Camera = 30°.

Table 2. Measurement errors (root mean squared error (RMSE)).

	0°	10°	20°	30°
RMSE	0.037	0.371	0.400	0.398

4.2. Field Test

The further experimental verification of proposed system has been performed field test. Like laboratory experiments, the computer vision system for the field test comprised a marker, a smartphone camera (Galaxy S7), and a computer installed with image processing software. In real field experiment, to reduce data transfer time, smartphone and computer were connected using data cable. The marker was an acrylic plate glued to metallic-plate to prevent deformation (Figure 11). The field tests were performed at Seongdong Bridge located in Seoul. In this field experiment, the displacement values obtained from the proposed vision-based approach were compared with conventional displacement sensors (LVDT). The acrylic plate marker was attached to the bridge girder using adhesives. The distance between smartphone camera and marker was set at 1 m and the angle was 0°. The motion images were recorded at 30 fps and transmitted to the computer. In field experimental setup of the proposed vision-based system is shown in Figure 11.

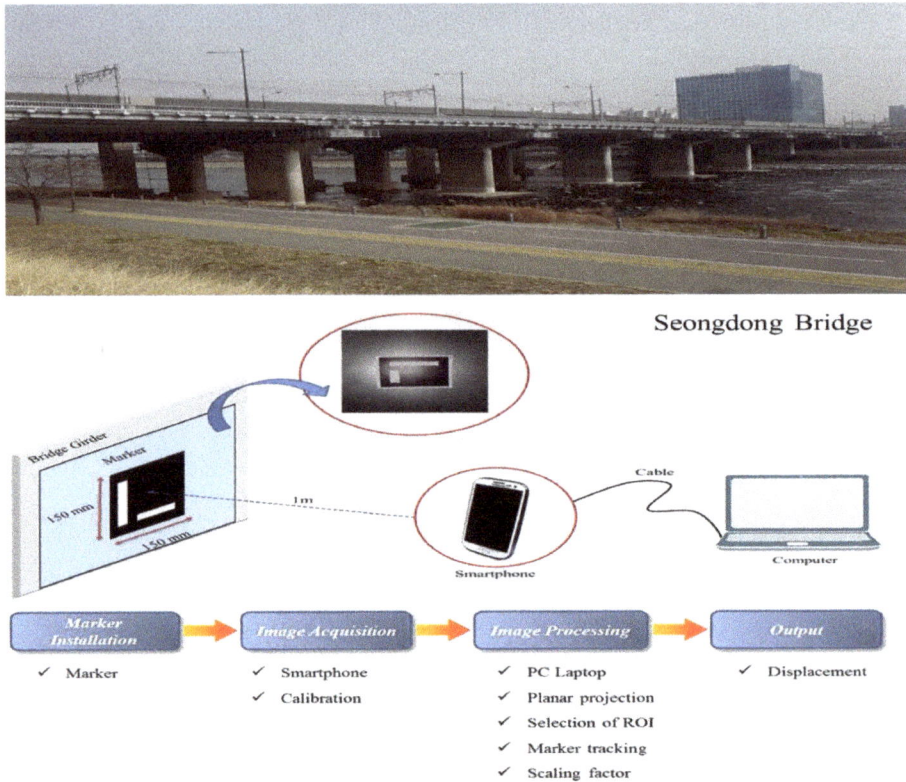

Figure 11. Schematic of the experimental setup of the proposed system.

In the field test, two factors affecting measurement errors should are considered. The first factor to consider is the shaking of the smartphone camera caused by the wind, and the second factor is the influence of vibration around the smartphone camera due to vehicle traffic. Thus, a windless sunny day was selected for the field experimentation to minimize the errors and the trials were performed at the ambient vibration conditions. Using these experimental procedures, we obtained the dynamic displacements-time curve of the Seongdong Bridge by the proposed method and the LVDT. During the two-day test, we measured four sets of displacement data at the same point. The authors compared the displacement data measured by the proposed method and the measured values by the LVDT sensor. As shown in Figure 12, the measurements obtained using the proposed displacement measurement approach agree well with those from the conventional displacement sensor (LVDT).

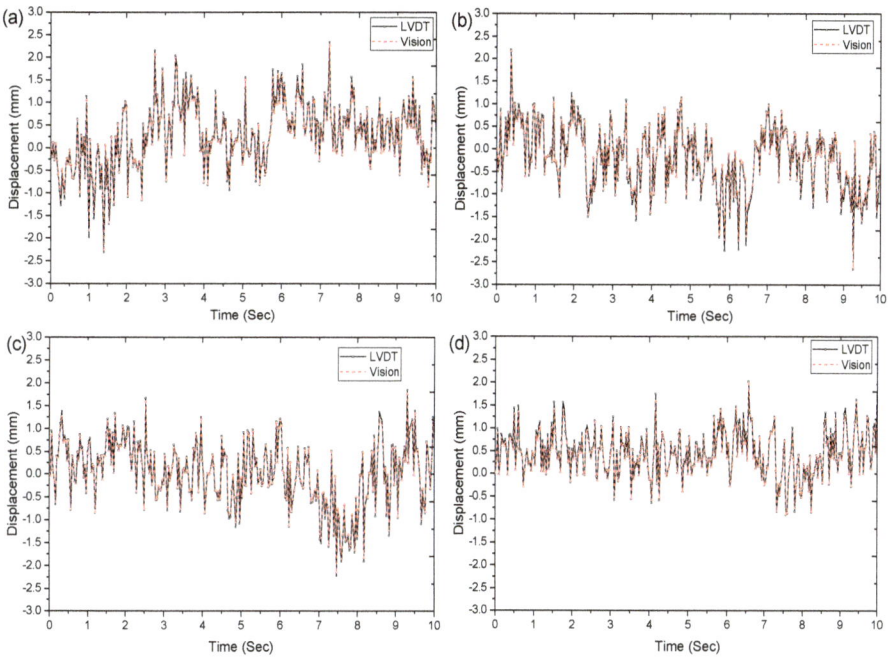

Figure 12. Comparison evaluation of Measured Displacement: (**a**) Data 1; (**b**) Data 2; (**c**) Data 3; (**d**) Data 4.

Similar to the laboratory experiments, error analysis was performed using the RMSE of Equation (4) to quantify the accuracy of proposed method. Table 3 provides the calculated RMSE values that the values were slightly increased compared to the laboratory test. However, the proposed method demonstrated high measurement accuracy with a maximum RMSE value of 0.051 mm in the field. The proposed approach uses motion images obtained by portable devices, such as smartphone camera, to measure the bridge displacement. This technique requires no additional equipment and allows measurements at multiple locations in shorter time than when using conventional methods. Thus, the proposed method can effectively reduce monitoring time and cost compared to the method using conventional displacement sensor.

Table 3. Measurement errors (RMSE).

	(a)	(b)	(c)	(d)	Max.	Min.
RMSE	0.044	0.051	0.040	0.050	0.051	0.040

5. Limitations and Future Research

The accuracy of computer vision-based displacement measurement technique can even compete with conventional displacement sensors. It is also more efficient in terms of sensor installation cost and working time. The accuracy and efficiency of computer vision-based method has been verified by researchers. However, computer vision-based system requires a large-capacity server for data storage and database of images for the long-term monitoring as compared to existing sensors systems. Moreover, the camera performance and image processing algorithm significantly influence the quality of displacement data. In recent years, even though, several displacement measurement algorithms have been developed, yet it is difficult to find a highly suitable algorithm for structures monitoring.

This limitation bounds the utilization for entire civil infrastructure. Another, drawback of computer vision approaches is the possibility of errors due to noise caused by the vibrations around the camera. The utilization of vibration control device such as gimbal may solve this problem.

6. Conclusions

The main objective of this study is to extend the measurable range and to enhance the accuracy of displacement measurement, so that computer vision-based systems can easily be adopted for the civil engineering structures (such as bridges). In this study, we introduced a computer vision-based technique for the measurement of rotation-invariant displacement with the successful utilization of a smartphone camera. The proposed method can transmit motion images acquired by using a smartphone camera to a remote computer using wireless communication. This computer vision-based approach combines two separate image-processing techniques: template-based matching and a homography matrix. As an initial step toward the implementation of this proposed method, we also calibrated the smartphone camera using hang's calibration algorithm to eliminate lens distortion. The next step was marker tracking in the captured motion images. We applied template-based matching using NCC function as the image-processing algorithm to track the marker. To remove the geometric distortion, a planar projection method by homography-matrix was applied. Through this research, we developed GUI-based software for measuring bridge displacements with a user-friendly interface. In addition to laboratory scale testing, field tests were conducted to evaluate the performance of the proposed method. To quantify the measurement error, an error evaluation was performed based on the RMSE. The proposed method showed the accuracy of the displacement measurement even when the angle between the smartphone camera and the marker increased.

Significant advantages of the proposed system include low cost, ease of operation, and flexibility to extract bridge displacements at various angles from a single measurement, making this system highly applicable to civil structures.

Author Contributions: ByungWan Jo conceived the idea and provided the technical support and materials, Yun Sung Lee reviewed literature, carried out the experimentation, writing, and collected data and its management. Jun Ho Jo helped to set up experimental trials and Rana Muhammad Asad Khan helped in writing this paper.

Acknowledgments: We would like to thanks to the anonymous reviews, whose suggestion and reviews comments really helped us to improve the manuscript.

Conflicts of Interest: The authors have no conflicts of interest to declare.

References

1. Chen, Z.; Zhou, X.; Wang, X.; Dong, L.; Qian, Y. Deployment of a smart structural health monitoring system for long-span arch bridges: A review and a case study. *Sensors* **2017**, *17*, 2151. [CrossRef] [PubMed]
2. Vallan, A.; Casalicchio, M.L.; Perrone, G. Displacement and acceleration measurements in vibration tests using a fiber optic sensor. *IEEE Trans. Instrum. Meas.* **2010**, *59*, 1389–1396. [CrossRef]
3. Sung, Y.-C.; Lin, T.-K.; Chiu, Y.-T.; Chang, K.-C.; Chen, K.-L.; Chang, C.-C. A bridge safety monitoring system for prestressed composite box-girder bridges with corrugated steel webs based on in-situ loading experiments and a long-term monitoring database. *Eng. Struct.* **2016**, *126*, 571–585. [CrossRef]
4. Park, J.-W.; Sim, S.-H.; Jung, H.-J. Wireless displacement sensing system for bridges using multi-sensor fusion. *Smart Mater. Struct.* **2014**, *23*, 045022. [CrossRef]
5. Lee, J.-J.; Shinozuka, M. Real-time displacement measurement of a flexible bridge using digital image processing techniques. *Exp. Mech.* **2006**, *46*, 105–114. [CrossRef]
6. Fukuda, Y.; Feng, M.Q.; Shinozuka, M. Cost-effective vision-based system for monitoring dynamic response of civil engineering structures. *Struct. Control Health Monit.* **2010**, *17*, 918–936. [CrossRef]
7. Ribeiro, D.; Calcada, R.; Ferreira, J.; Martins, T. Non-contact measurement of the dynamic displacement of railway bridges using an advanced video-based system. *Eng. Struct.* **2014**, *75*, 164–180. [CrossRef]
8. Park, K.-T.; Kim, S.-H.; Park, H.-S.; Lee, K.-W. The determination of bridge displacement using measured acceleration. *Eng. Struct.* **2005**, *27*, 371–378. [CrossRef]

9. Psimoulis, P.; Pytharouli, S.; Karambalis, D.; Stiros, S. Potential of global positioning system (GPS) to measure frequencies of oscillations of engineering structures. *J. Sound Vib.* **2008**, *318*, 606–623. [CrossRef]

10. Abdel-Qader, L.; Abudayyeh, O.; Kelly, M.E. Analysis of edge-detection techniques for crack identification in bridges. *J. Comput. Civ. Eng.* **2003**, *17*, 255–263. [CrossRef]

11. Koch, C.; Georgieva, K.; Kasireddy, V.; Akinci, B.; Fieguth, P. A review on computer vision based defect detection and condition assessment of concrete and asphalt civil infrastructure. *Adv. Eng. Inform.* **2015**, *29*, 196–210. [CrossRef]

12. Ye, X.; Dong, C.; Liu, T. A review of machine vision-based structural health monitoring: Methodologies and applications. *J. Sens.* **2016**, *2016*. [CrossRef]

13. Stephen, G.; Brownjohn, J.; Taylor, C. Measurements of static and dynamic displacement from visual monitoring of the humber bridge. *Eng. Struct.* **1993**, *15*, 197–208. [CrossRef]

14. Olaszek, P. Investigation of the dynamic characteristic of bridge structures using a computer vision method. *Measurement* **1999**, *25*, 227–236. [CrossRef]

15. Guo, J. Dynamic displacement measurement of large-scale structures based on the lucas–kanade template tracking algorithm. *Mech. Syst. Signal Process.* **2016**, *66*, 425–436. [CrossRef]

16. Shan, B.; Wang, L.; Huo, X.; Yuan, W.; Xue, Z. A bridge deflection monitoring system based on CCD. *Adv. Mater. Sci. Eng.* **2016**, *2016*. [CrossRef]

17. Zhang, Z. A flexible new technique for camera calibration. *IEEE Trans. Pattern Anal. Mach. Intell.* **2000**, *22*, 1330–1334. [CrossRef]

18. Feng, D.; Feng, M.Q. Identification of structural stiffness and excitation forces in time domain using noncontact vision-based displacement measurement. *J. Sound Vib.* **2017**, *406*, 15–28. [CrossRef]

19. Feng, D.; Feng, M.Q. Experimental validation of cost-effective vision-based structural health monitoring. *Mech. Syst. Signal Process.* **2017**, *88*, 199–211. [CrossRef]

20. Yoon, H.; Elanwar, H.; Choi, H.; Golparvar-Fard, M.; Spencer, B.F. Target-free approach for vision-based structural system identification using consumer-grade cameras. *Struct. Control Health Monit.* **2016**, *23*, 1405–1416. [CrossRef]

21. Brunelli, R. *Template Matching Techniques in Computer Vision: Theory and Practice*; John Wiley & Sons: New York, NY, USA, 2009.

22. Peters, W.; Ranson, W. Digital imaging techniques in experimental stress analysis. *Opt. Eng.* **1982**, *21*, 213427. [CrossRef]

23. Peters, W.; Ranson, W.; Sutton, M.; Chu, T.; Anderson, J. Application of digital correlation methods to rigid body mechanics. *Opt. Eng.* **1983**, *22*, 226738. [CrossRef]

24. Sutton, M.; Mingqi, C.; Peters, W.; Chao, Y.; McNeill, S. Application of an optimized digital correlation method to planar deformation analysis. *Image Vis. Comput.* **1986**, *4*, 143–150. [CrossRef]

25. Pan, B.; Li, K.; Tong, W. Fast, robust and accurate digital image correlation calculation without redundant computations. *Exp. Mech.* **2013**, *53*, 1277–1289. [CrossRef]

26. Chen, J.; Jin, G.; Meng, L. Applications of digital correlation method to structure inspection. *Tsinghua Sci. Technol.* **2007**, *12*, 237–243. [CrossRef]

27. Yoneyama, S.; Kitagawa, A.; Iwata, S.; Tani, K.; Kikuta, H. Bridge deflection measurement using digital image correlation. *Exp. Tech.* **2007**, *31*, 34–40. [CrossRef]

28. Lewis, J. *Fast normalized cross-correlation, In Vision Interface*; Canadian Image Processing and Pattern Recognition Society: Quebec City, QC, Canada, 1995; Volume 10, pp. 120–123.

29. Jeong, Y.; Park, D.; Park, K.H. Ptz camera-based displacement sensor system with perspective distortion correction unit for early detection of building destruction. *Sensors* **2017**, *17*, 430. [CrossRef] [PubMed]

30. Hartley, R.; Zisserman, A. *Multiple View Geometry in Computer Vision*; Cambridge University Press: Cabridge, UK, 2003.

31. Luo, J.; Konofagou, E.E. A fast normalized cross-correlation calculation method for motion estimation. *IEEE Trans. Ultrason. Ferroelectr. Freq. Control* **2010**, *57*, 1347–1357. [PubMed]

32. Abdulfattah, G.; Ahmad, M. Face localization-based template matching approach using new similarity measurements. *J. Theor. Appl. Inf. Technol.* **2013**, *57*, 424–431.

33. Dawoud, N.N.; Samir, B.B.; Janier, J. Fast template matching method based optimized sum of absolute difference algorithm for face localization. *Int. J. Comput. Appl.* **2011**, *18*, 30–34.

34. Kim, S.-W.; Lee, S.-S.; Kim, N.-S.; Kim, D.-J. Numerical model validation for a prestressed concrete girder bridge by using image signals. *KSCE J. Civ. Eng.* **2013**, *17*, 509. [CrossRef]

35. Kim, S.-W.; Jeon, B.-G.; Kim, N.-S.; Park, J.-C. Vision-based monitoring system for evaluating cable tensile forces on a cable-stayed bridge. *Struct. Health Monit.* **2013**, *12*, 440–456. [CrossRef]

36. Jáuregui, D.V.; White, K.R.; Woodward, C.B.; Leitch, K.R. Noncontact photogrammetric measurement of vertical bridge deflection. *J. Bridge Eng.* **2003**, *8*, 212–222. [CrossRef]

37. Poudel, U.; Fu, G.; Ye, J. Structural damage detection using digital video imaging technique and wavelet transformation. *J. Sound Vib.* **2005**, *286*, 869–895. [CrossRef]

38. Fu, G.; Moosa, A.G. An optical approach to structural displacement measurement and its application. *J. Eng. Mech.* **2002**, *128*, 511–520. [CrossRef]

39. Wahbeh, A.M.; Caffrey, J.P.; Masri, S.F. A vision-based approach for the direct measurement of displacements in vibrating systems. *Smart Mater. Struct.* **2003**, *12*, 785. [CrossRef]

40. Kohut, P.; Holak, K.; Uhl, T.; Ortyl, Ł.; Owerko, T.; Kuras, P.; Kocierz, R. Monitoring of a civil structure's state based on noncontact measurements. *Struct. Health Monit.* **2013**, *12*, 411–429. [CrossRef]

41. Lee, J.J.; Shinozuka, M. A vision-based system for remote sensing of bridge displacement. *NDT E Int.* **2006**, *39*, 425–431. [CrossRef]

42. Lee, J.J.; Shinozuka, M.; Lee, C.G. Evaluation of bridge load carrying capacity based on dynamic displacement measurement using real-time image processing techniques. *Int. J. Steel Struct.* **2006**, *6*, 377–385.

43. Choi, H.-S.; Cheung, J.-H.; Kim, S.-H.; Ahn, J.-H. Structural dynamic displacement vision system using digital image processing. *NDT E Int.* **2011**, *44*, 597–608. [CrossRef]

44. Ho, H.-N.; Lee, J.-H.; Park, Y.-S.; Lee, J.-J. A synchronized multipoint vision-based system for displacement measurement of civil infrastructures. *Sci. World J.* **2012**, *2012*. [CrossRef] [PubMed]

45. Tian, L.; Pan, B. Remote bridge deflection measurement using an advanced video deflectometer and actively illuminated led targets. *Sensors* **2016**, *16*, 1344. [CrossRef] [PubMed]

46. Busca, G.; Cigada, A.; Mazzoleni, P.; Zappa, E. Vibration monitoring of multiple bridge points by means of a unique vision-based measuring system. *Exp. Mech.* **2014**, *54*, 255–271. [CrossRef]

47. Feng, D.M.; Feng, M.Q.; Ozer, E.; Fukuda, Y. A vision-based sensor for noncontact structural displacement measurement. *Sensors* **2015**, *15*, 16557–16575. [CrossRef] [PubMed]

48. Feng, M.Q.; Fukuda, Y.; Feng, D.; Mizuta, M. Nontarget vision sensor for remote measurement of bridge dynamic response. *J. Bridge Eng.* **2015**, *20*, 04015023. [CrossRef]

49. Jeon, H.; Bang, Y.; Myung, H. A paired visual servoing system for 6-dof displacement measurement of structures. *Smart Mater. Struct.* **2011**, *20*, 045019. [CrossRef]

50. Park, S.; Park, H.; Kim, J.; Adeli, H. 3D displacement measurement model for health monitoring of structures using a motion capture system. *Measurement* **2015**, *59*, 352–362. [CrossRef]

51. Chang, C.; Xiao, X. Three-dimensional structural translation and rotation measurement using monocular videogrammetry. *J. Eng. Mech.* **2009**, *136*, 840–848. [CrossRef]

52. Ji, Y.; Chang, C. Nontarget image-based technique for small cable vibration measurement. *J. Bridge Eng.* **2008**, *13*, 34–42. [CrossRef]

53. Ji, Y.; Chang, C. Nontarget stereo vision technique for spatiotemporal response measurement of line-like structures. *J. Eng. Mech.* **2008**, *134*, 466–474. [CrossRef]

54. Kim, S.-W.; Kim, N.-S. Dynamic characteristics of suspension bridge hanger cables using digital image processing. *NDT E Int.* **2013**, *59*, 25–33. [CrossRef]

55. Kim, B.H. Extracting modal parameters of a cable on shaky motion pictures. *Mech. Syst. Signal Process.* **2014**, *49*, 3–12. [CrossRef]

56. Chen, C.-C.; Wu, W.-H.; Tseng, H.-Z.; Chen, C.-H.; Lai, G. Application of digital photogrammetry techniques in identifying the mode shape ratios of stay cables with multiple camcorders. *Measurement* **2015**, *75*, 134–146. [CrossRef]

57. Zhao, X.; Ri, K.; Han, R.; Yu, Y.; Li, M.; Ou, J. Experimental research on quick structural health monitoring technique for bridges using smartphone. *Adv. Mater. Sci. Eng.* **2016**, *2016*. [CrossRef]
58. Zhao, X.; Ri, K.; Wang, N. Experimental verification for cable force estimation using handheld shooting of smartphones. *J. Sens.* **2017**, *2017*. [CrossRef]

sustainability

MDPI

Article

Decision-Aiding Evaluation of Public Infrastructure for Electric Vehicles in Cities and Resorts of Lithuania

Vytautas Palevičius [1], Askoldas Podviezko [2], Henrikas Sivilevičius [3] and Olegas Prentkovskis [3,*]

[1]   Department of Roads, Faculty of Environmental Engineering, Vilnius Gediminas Technical University, Saulėtekio al. 11, LT-10223 Vilnius, Lithuania; vytautas.palevicius@vgtu.lt
[2]   Department of Economics Engineering, Faculty of Business Management, Vilnius Gediminas Technical University, Saulėtekio al. 11, LT-10223 Vilnius, Lithuania; askoldas@gmail.com
[3]   Department of Mobile Machinery and Railway Transport, Faculty of Transport Engineering, Vilnius Gediminas Technical University, Plytinės g. 27, LT-10105 Vilnius, Lithuania; henrikas.sivilevicius@vgtu.lt
*   Correspondence: olegas.prentkovskis@vgtu.lt; Tel.: +370-5-2744784

Received: 2 February 2018; Accepted: 17 March 2018; Published: 21 March 2018

Abstract: In the National Communication Development of 2014–2022 Program and Guidelines of the Development of the Public Electric Vehicle Charging Infrastructure confirmed by the Government of the Republic of Lithuania, it is planned that, until the year of 2025, among newly registered vehicles, electric ones should make at least 10%. Analysis of the trend of electric vehicles makes evident that the target does not have a real chance to be achieved without targeted efforts. In order to improve the infrastructure of electric vehicles in major cities and resorts of Lithuania, we have carried out a comparative analysis of public infrastructure for electric vehicles in 18 Lithuanian cities and resorts. For the quantitative analysis, we proposed eight criteria describing such an infrastructure. As perception of the infrastructure by owners of electric cars depends on complex factors, we used multiple criteria evaluation methods (MCDM) for evaluation of the current state of its development by four such methods: EDAS, SAW, TOPSIS, and PROMETHEE II. Based on the evaluation results, prominent and lagging factors were understood, and proposals for effective development of public infrastructure of electric vehicles were proposed for improvement of the infrastructure.

Keywords: development; electric vehicle; public infrastructure; MCDM; EDAS; SAW; TOPSIS; PROMETHEE II; COIN; decision-making

1. Introduction

Presently, properly developed public infrastructure of electric vehicles can rarely be found even in major cities of the world. Potential buyers of such vehicles usually primarily focus their concern on technical and operating parameters [1], infrastructure of charging facilities and ease of access to such facilities in the city of residence [2–4], costs and benefits for drivers of electric vehicles [5,6], and on other economic and social factors [7].

Countries of the European Union, which are committed to implementing European transport space development plans in time, should achieve the level of about 10% sales of new cars to be electric ones in the year 2025 [8]. As the global market of electric vehicles is not fully developed, car manufacturers could make their own promotion of electric vehicles. Nevertheless, promotion by governments of various countries of ecological vehicles in major cities and resorts, emphasizing that the electric vehicle is such a technological alternative that carries a high potential for reducing pollution and energy dependence in the city [9,10], creates a more realistic potential for increasing the number of such vehicles in the roads of each country. Analysis of the latest literature suggests that development of electric vehicles in a city heavily depends on both infrastructure of electric vehicles together with considerable environmental and energy benefits of such vehicles comparing to traditional vehicles [11–13].

Compared to neighbors, Lithuania is lagging in the level of development of public infrastructure for electric vehicles. For example, presently there are four times as many electric vehicles in Estonia than can be found in Lithuania. The strongest inducement was caused by the decision of the Estonian Government to use funds received for emission allowances sold for the purchase of electric vehicles. A considerable amount of electric vehicles was provided to public servants along with compensation to the first 500 buyers of private electric vehicles. In addition, a substantial network of charging stations was installed using the funds, which is currently considered to be the second largest network, related to the density of population, after Norway. There are currently 384 charging stations installed in Estonia; 193 of which are fast-loading; 72 charging stations in Latvia; 12 of which are fast-charging; and 329 charging stations in Poland; 39 of which are fast-charging [14–16].

Lithuanian cities have excellent conditions for testing perspectives of development of the electric vehicle penetration due to rather high temperature fluctuations (from -30 °C to $+35$ °C) and reasonable distances between cities and resort areas (the very maximum distance in Lithuania is 356 km). From this observation naturally stems the purpose of the present article to evaluate possibilities of development of public infrastructure for drivers of electric vehicles and to offer effective solutions for infrastructure improvement based on the results obtained. We have to mention, though, related risks, which are beyond the scope of our research: capacity of electric supply, and efficiency of its present network. A situation with planned construction of the Ignalina Nuclear Power Plant is not yet clear, while all its reactors were closed on 31 December 2009. Nevertheless, the recently stretched two links "NordBalt" with Sweden over the bottom of the Baltic Sea, and connection with Poland "LitPol Link" should most probably ensure uninterrupted transmission of electricity after some adjusting steps will have been successfully accomplished.

As perception of the infrastructure by owners of electric cars is a complex one, we used multiple criteria evaluation methods (MCDM) of the current state of development of the infrastructure by several such MCDM methods as EDAS, SAW, TOPSIS, and PROMETHEE II.

2. Promotion of Electric Vehicles Policy Tools in Lithuania

Increase of use of electric vehicles is explained by the desire to save environment, non-renewable energy sources, and reduce fuel costs [17–19]. As of 1 February 2018 according to the local registrar of vehicles, the State Enterprise "Regitra" (Vilnius, Lithuania) 661 electric vehicles were registered in Lithuania (Figure 1). The most popular among them were Nissan Leaf (329 cars), Tesla's model S (78 cars), and BMW i3 (57 cars), while 8590 cars of the M1 class hybrid vehicles (petrol/electric and diesel/electric) were registered. The most popular among them in the class diesel/electric were Toyota (5186 units), Lexus (2446 units), and Honda (403 units), and in the class petrol/electric were Chevrolet Volt (102 units), and Toyota Prius (24 units).

At present, in Lithuania, no electric vans and trucks have been registered. Only passenger electric vehicles have been registered. It is planned that the first Tesla electric truck will appear in Lithuania this year as it was ordered from the Tesla company by one of the largest European transport companies—Girteka Logistics (Vilnius, Lithuania).

We distinguish two types of promotional activities of electric vehicles: soft and hard. In Lithuania, only soft measures were undertaken so far, such as allowing the use of the dedicated for the public transport A-lane for electric vehicles starting from 19 January 2013; free parking of electric vehicles in city centers; some special road signs allocated indicating that road signs are not valid for electric vehicles, e.g., parking. However, based on current tendencies of using electric vehicles in this country, it can be forecasted that additional measures are required to sufficiently motivate the population to increase using electric vehicles. Consequently, the Ministry of Transport and Communications together with the Lithuanian Road Administration has prepared an incentive project for development hard promotional activities, estimated by 3 million euros investment. In addition, the EU provides financing possibilities for development of normal and high-power charging posts on major roads and within cities of the country, with over 25,000 in population or have the resort status.

Figure 1. The number of electric vehicles and hybrids in Lithuania (2014–2018).

The current development is apparently insufficient: in 2018, 12 electric high-power charging posts were installed on the major highway of the country. It is planned that 14 additional electric vehicle-charging posts will be installed on Lithuanian highways in the beginning of 2019, and it is planned that on Lithuanian highways at least 28 high-power access points will be created. In addition, according to the project, by the year 2022, at least 150 normal- and high-power stations for electric vehicles will be installed in cities and resorts, and around 30 high power stations on highways.

3. Choosing Criteria of Evaluation

A list of eight factors influencing the development of public infrastructure of electric vehicles in the city was created by authors of this paper using the expert examination method and by consulting a group of experts consisting of transport, road and civil engineers working at relevant ministries, researchers in the field of energy and environment. They were chosen by the duration of their work experience of no less than 10 years, occupied positions of no less than the head of a department, and by their academic degree of no less than a master's.

In 2011, the White Paper on European Transport Policy for the development of a Single European Transport System was published by the European Commission. It aims to create an agenda for European Commission initiatives in the field of transport policy by the year 2050. The White Paper presents 40 specific initiatives to be taken by each European Union member state over the next decade to achieve the goals outlined in the document. The following eight criteria were created in accordance with the named Policy that are influencing the development of public infrastructure of electric vehicle (Table 1).

Table 1. Factors influencing development of public electric vehicle infrastructure in the city.

Factor	Dimension	Reference
1. Streets in the city with the dedicated A lane for public transport allowed to be used also by electric vehicles	Part of all streets in the city	National legislation
2. Parking places and areas exempt from the parking fee for electric vehicles	Number of parking places per 1000 inhabitants	National legislation
3. Development of high-power charging posts	The number of access points for 1000 inhabitants	European Union legislation adopted in the national legislation
4. Development of electric charging posts	The number of access points for 1000 inhabitants	European Union legislation adopted in the national legislation
5. Investment of state institutions to the infrastructure for electric vehicles	Euro per 1000 inhabitants	Created by authors of this paper
6. Investment of private institutions to the infrastructure for electric vehicles	Euro per 1000 inhabitants	Created by authors of this paper
7. Integrated electric vehicle infrastructure development projects	Construction of access roads and its infrastructure, in euro per 1000 inhabitants	Created by authors of this paper
8. Installation of high-power charge posts on roads of national importance within 50 km distance from major city center	The number of access points for 1000 inhabitants	European Union legislation adopted in the national legislation

We have selected 18 major cities and resorts of Lithuania for the evaluation. The data reflecting each city and resort by every criterion is formed into a decision-matrix $R = \|r_{ij}\|$, of the size (m, n), where m is the number of chosen criteria (8), and n is a number of cities participating in the evaluation (18).

4. Eliciting of Weights of Criteria and Gauging the Level of Concordance of Opinions of Experts

MCDM methods require using weights ω_i, which express importance of each criterion (where i is an index for denoting criteria).

For the chosen task, it was decided to elicit opinions of importance of criteria from experts, and, in addition, to employ the entropy method, which finds weights based on data only. Two absolutely different methods were gathered in the way that possible flaws of one method were outweighed with another method. Unlike previous attempts, where geometric mean for comprising results of weights elicited from experts and the ones, which were elicited from data, the arithmetic mean was used because of the following logic. In case there is small entropy observed in the data, the geometric mean would make the final weight negligible in spite of opinions of experts. Consequently, as our purpose was to smoothen possible flaws of used methods, we chose the arithmetic mean, as it will reflect both the structure of data and opinions of experts.

Experts {E1, E8} were asked to fill in forms, stating weights of importance of criteria in percent, so that aggregate weights in a group of either criteria or categories make up 100%. The summary of opinions of eight experts used for the study on significance of weights is presented in Table 2. Final weights, which are averages of weights of experts, are presented in the right-hand column of the table.

Table 2. Summary of expert opinions on weights.

Experts / Criteria	E_1	E_2	E_3	E_4	E_5	E_6	E_7	E_8	Final Weights w_i
1	0.15	0.2	0.08	0.06	0.11	0.2	0.3	0.13	0.154
2	0.14	0.12	0.1	0.1	0.15	0.2	0.2	0.08	0.136
3	0.11	0.1	0.2	0.25	0.16	0.1	0.02	0.14	0.135
4	0.21	0.21	0.17	0.18	0.16	0.1	0.18	0.19	0.175
5	0.09	0.09	0.06	0.18	0.09	0	0.19	0.11	0.101
6	0.12	0.15	0.12	0.08	0.11	0.2	0.01	0.12	0.114
7	0.11	0.08	0.13	0.05	0.13	0.1	0.1	0.13	0.104
8	0.07	0.05	0.14	0.1	0.09	0.1	0	0.1	0.081

In order to gauge the level of concordance of opinions of experts, magnitudes of provided weights were ranked in order to apply the theory of concordance by Kendall [20–22]. Such ranks we denoted as e_{ik}, where $i = 1, 2, \ldots, m$ is the index for criteria (m is 8 in our case), while $k = 1, 2, \ldots, r$ is the index to denote responded experts (r—is also 8 in our case). The Kendall variable W, which is used in the chi-squared test statistics for gauging the level of concordance, depends on the squared deviations of sums of all ranks e_{ik} by all experts (1):

$$e_i = \sum_{k=1}^{r} e_{ik} \tag{1}$$

from the mean of such sums (2):

$$\bar{e} = \frac{\sum_{i=1}^{m} e_i}{m}. \tag{2}$$

In the case when there are no equal ranks of criteria, Kendall variable W equals the ratio between the sum S mentioned above, which is calculated by Formula (3):

$$S = \sum_{i=1}^{m} (e_i - \bar{e})^2 \tag{3}$$

and its largest deviation, denoted by S_{max}, observed in the case of absolute concordance of opinions of experts, in terms of ranks of importance of criteria (4):

$$S_{max} = \frac{r^2 m(m^2 - 1)}{12}. \tag{4}$$

As we found eight sets of equal ranks within expert estimates, we use the adjusted formula for calculating Kendall's variable [23] (5):

$$W = \frac{12S}{r^2 m(m^2 - 1) - r\sum_\phi \left(t_\phi^3 - t_\phi\right)}, \tag{5}$$

where ϕ denotes sets of equal ranks, and t_ϕ denotes the number of equal ranks within a set within ϕ. Chi-squared test statistics for this variable is the following (6):

$$\chi^2 = Wr(m - 1) \tag{6}$$

for the number of degrees of freedom $v = m - 1 = 7$. For the test statistics, we chose the level of significance $\alpha = 0.05$. The critical level of χ^2 distribution for the chosen threshold and the number of degrees of freedom $v = 7$ is $\chi^2_{crit} = 14.07$. Calculations of the adjusted Kendall's variable produced the result $W = 0.326$, while test statistics for this result appeared to be beyond the critical threshold $\chi^2 = 18.26 > \chi^2_{crit} = 14.07$, and we may reject the hypothesis that opinions of experts are non-concordant.

5. Weights Obtained Using Entropy

The method uses structure of data instead of eliciting perception of importance of each chosen criterion from experts. Nevertheless, we note that entropy may yield too high or too low differences between weights and a method of softening such influences is plausible to use, which will be described in the next section. For example, it was proposed in [24,25] to outweigh these effects by applying the CILOS weight estimation method instead of entropy based on criterion impact losses proposing the IDOCRIW method [26,27], a combination of two methods [28,29].

The degree of entropy, E_i, for $i = 1, 2, \ldots, m$ for each criterion is calculated as follows (7):

$$E_i = -\frac{1}{\ln m} \sum_{j=1}^{n} \tilde{r}_{ij} \cdot \ln \tilde{r}_{ij}, \ (i = 1, 2, \ldots, m;\ 0 \le E_i \le 1), \tag{7}$$

where $\tilde{r}_{ij}$ are normalized values of the i-th criterion for the j-th alternative (8):

$$\tilde{r}_{ij} = \frac{r_{ij}}{\sum_{j=1}^{n} r_{ij}}. \tag{8}$$

The degrees of variation, d_i, i.e., non-normalized values of the weights determined by the entropy method, are calculated for each criterion (9):

$$d_i = 1 - E_i. \tag{9}$$

Entropy weights W_i are normalized values of d_i calculated as follows (10):

$$W_i = \frac{d_i}{\sum_{i=1}^{m} d_i}. \tag{10}$$

Obtained entropy weights are presented in Table 3:

Table 3. Obtained entropy weights.

Criteria	1	2	3	4	5	6	7	8
Weights	0.001	0.079	0.019	0.132	0.162	0.341	0.212	0.053

Whenever there is a small dispersion between normalised values obtained using Formula (10), entropy weights appear to be small (e.g., the weight for the criterion 1), and vice versa; whenever there is a large dispersion between normalised values, entropy weights are large (e.g., the weight for the criterion 6).

6. A Combined COIN (COmpensating INfluences) Method of Obtaining Weights

The idea of the COIN method is suggested by the second author and is presented in this paper for the first time. It suggests comprising both ways of eliciting weights: from data by using the entropy method, and from experts that differ considerably. We will opt to choose average values between entropy weights w_i^{en} and weights elicited from experts w_i^{ex} instead of the geometric mean as proposed in [24,30] (11):

$$w_i^C = \frac{w_i^{en} + w_i^{ex}}{2}, \tag{11}$$

where w_i^C are the ultimate weights obtained by the COIN method, which will be used in this paper for making evaluations.

The choice in favour of the average values is made because the geometric mean does not reflect the idea of compensating differences between considerably different methods and overestimates the influence of very small values. Consequently, the geometric mean would under-value such criteria as 1 and 3 where values obtained by the entropy method appear to be small, and where weights elicited from experts appeared to be considerable. Weights obtained using different methods are presented in Table 4. Such weights equally reflect the structure of data and opinions of experts.

Table 4. Weights obtained using different methods.

Criteria	1	2	3	4	5	6	7	8
Weights (entropy), ω_i^{en}	0.001	0.079	0.019	0.132	0.162	0.341	0.212	0.053
Weights (elicited from experts), ω_i^{ex}	0.154	0.136	0.135	0.175	0.101	0.114	0.104	0.081
Weights (COIN), ω_i^C	0.078	0.108	0.077	0.154	0.132	0.227	0.158	0.067

7. Evaluation Using MCDM Methods

In order to evaluate conditions of availability and growth of public infrastructure for electric vehicles in eighteen cities of Lithuania and to make conclusions on their progress and proposals on future opportunities, a multiple criteria evaluation was performed. The choice in favour of multiple criteria decision aid methods was determined both because of the structure of data [31,32] and clear necessity to comprise opinions of experts on how the city should develop [33].

Statistical methods are useful only in such cases when data is appropriate and there is a sufficient number of entries, in our case sufficient number of alternatives. Moreover, data should be normally distributed. However, this is not the case: our data represents only 18 cities and is not suitable for statistical analysis due to a small scope of the data. Moreover, in our case, MCDM methods are perfectly suitable as the methods provide the result of quantitative evaluation of all alternatives, which are real (evaluation of the state in 18 cities in our case). Consequently, the methods reveal attractiveness in broad terms providing ranking of alternatives. Moreover, the methods also provide a powerful tool for analysis of causes of prominence and lagging of the alternatives.

The major idea of any MCDM method is to create a cumulative criterion for each alternative, reflecting the attractiveness of the alternative in quantitative terms, expressed in a single value related to each alternative. Such a cumulative criterion comprises both weights of importance of criteria chosen for evaluation (Sections 4–6) and values criteria in a way that the more attractive alternative outranks a weaker alternative in case the cumulative criterion of this alternative appears to be larger. Several methods must be used, as there is no single best MCDM method that guarantees precision of evaluation. In the paper, we will use four different MCDM methods, as it appears to be a popular option used by researchers that use MCDM methods. Such an approach mitigates distortions, which inevitably are introduced by applying different types of normalisation [34]. Each chosen MCDM method of the four has not only different, specific only to this method, principles and logic but also uses different types of normalisation, based on different principles. It is known that values of criteria could be normalized in many different ways; it should also be borne in mind that some MCDM methods use transformation proprietary for that method. Such an approach of integrating results obtained by different methods reduces discrepancies within the results.

We use the following MCDM methods: the EDAS (Evaluation Based on Distance from Average Solution) method was proposed in 2015; a very popular raw MCDM method SAW (Simple Additive Weighing); the TOPSIS (Technique for Order Preference by Similarity to an Ideal Solution) method as a popular contemporary method, and the PROMETHEE II (Preference Ranking Organisation Method for Enrichment Evaluation) method as the most thorough method, which performs pairwise comparison of all alternatives.

As mentioned above, choice of the methods was determined by the fact that they employ both different logic and normalisation. Even if the EDAS method uses a similar idea, as it was proposed in the TOPSIS method: to measure a distance of an alternative to be evaluated to a benchmark solution (or benchmark solutions as in the TOPSIS method), there is a considerable difference between the benchmarks used. In the TOPSIS method, artificial worst and best alternatives with normalised weighted coordinates are used, while the EDAS method uses the average artificially created benchmark solution "*AV*" and measures weighted sums of "positive" and "negative" distances. We note that proprietary normalisations for the two methods are different. The TOPSIS method uses the Euclidean distance, while the EDAS method uses summation of coordinates, and the sums are finally normalised. The SAW method also differs from the previous two methods by its logic even if some similarity with the EDAS method, in terms of the SAW using weighted sum of normalised values, could be observed. However, first, the EDAS normalises weighted sums at the final stage, while the SAW normalises each value. Second, normalisation for the SAW method is different: each value of the data was divided by the sum of values by each alternative. The PROMETHEE II method is found in the different class of MCDM methods as it uses pairwise comparison in all pairs of alternatives. In addition, instead of normalization, a preference function is used. Unlike the previous three methods, distances between coordinates are normalised instead of values of criteria or weighted sums. The authors believe that such a mixture of different normalisation formulae, as well as different MCDM approaches, will reduce discrepancies. Considering an MCDM method as a random variable, the average of solutions produced by random MCDM methods with different logic and normalisation should considerably reduce variance of the average estimation in accordance with the logic of statistical inference from a random sample similarly as it happens to the estimation of the mean of the population by taking a random sample.

In all MCDM methods, the same decision matrix is used. The matrices contain statistical data $R = \|r_{ij}\|$, which describe the objects being evaluated. Weights of criteria are denoted as ω_i ($\sum_{i=1}^{n} \omega_i = 1$), $i = 1, 2, \ldots, m; j = 1, 2, \ldots, n$, where m is the number of criteria, and n is the number of the evaluated objects or alternatives. All criteria in our case are maximizing. The larger is the value of a maximizing criterion, the better it is in terms of attractiveness; the smaller is the value of a minimizing criterion for an alternative, the more attractive it becomes.

8. Evaluation by the EDAS Method

The idea and prominence of the EDAS method are reflected in the name of the method. In contrast to the TOPSIS method, the EDAS method uses the solution with average values of criteria as benchmark solution "*AV*" [35,36] (12):

$$AV_i = \frac{\sum_{j=1}^{m} r_{ij}}{m}. \tag{12}$$

At the next step, positive and negative distances from *AV* are calculated for each alternative and each criterion as follows, separately for maximizing Labels (13) and (14) and minimizing criteria Labels (15) and (16):

$$PDA_{ij} = \frac{\max\left[0, (r_{ij} - AV_i)\right]}{AV_i}, \tag{13}$$

$$NDA_{ij} = \frac{\max\left[0, (AV_i - r_{ij})\right]}{AV_i}, \tag{14}$$

$$PDA_{ij} = \frac{\max\left[0, (AV_i - r_{ij})\right]}{AV_i}, \tag{15}$$

$$NDA_{ij} = \frac{\max\left[0, (r_{ij} - AV_i)\right]}{AV_i}. \tag{16}$$

At the next step, weights are incorporated to find NSP_j and NSN_j (17) and (18):

$$NSP_j = \frac{\sum_{i=1}^{n} \omega_i PDA_{ij}}{\max_j \sum_{i=1}^{n} \omega_i PDA_{ij}}, \tag{17}$$

$$NSN_j = 1 - \frac{\sum_{i=1}^{n} \omega_i NDA_{ij}}{\max_j \sum_{i=1}^{n} \omega_i NDA_{ij}}.$$ (18)

Finally, the cumulative criterion of the method is found by the Formula (19):

$$AS_j = \frac{1}{2}(NSP_j + NSN_j).$$ (19)

The results of the evaluation by the EDAS method using COIN weights from Table 4 are presented in Table 5.

Table 5. Results of the evaluation by the EDAS method using COIN weights.

Alternatives	1	2	3	4	5	6	7	8	9
EDAS	0.462	0.208	0.051	0.063	0.220	0.243	0.300	0.023	0.004
Rank	2	6	15	13	5	4	3	17	18
Alternatives	10	11	12	13	14	15	16	17	18
EDAS	0.906	0.195	0.184	0.196	0.061	0.080	0.034	0.206	0.075
Rank	1	9	10	8	14	11	16	7	12

Notes: Alternatives are cities of Lithuania: 1—Birštonas, 2—Druskininkai, 3—Alytus, 4—Jonava, 5—Kaunas, 6—Kėdainiai, 7—Klaipėda, 8—Marijampolė, 9—Mažeikiai, 10—Neringa, 11—Palanga, 12—Panevėžys, 13—Šiauliai, 14—Tauragė, 15—Telšiai, 16—Utena, 17—Vilnius, 18—Visaginas, authors' calculations.

9. Evaluation by the SAW Method

The method reflects the core idea of MCDM methods: it comprises both normalized values of criteria and weights by using a simple additive sum of both, by all criteria chosen [37,38] (20). This operation is performed for each considered alternative:

$$S_j = \sum_{i=1}^{m} \omega_i \tilde{r}_{ij},$$ (20)

where $\tilde{r}_{ij}$ are normalized values of the i-th criterion for the j-th alternative, and ω_i are weights of the i-th criterion.

The cumulative criterion S_j, similarly to all other MCDM methods considered in the paper, reflects attractiveness of each alternative by its magnitude: the larger is the criterion, the more attractive appears to be the alternative. Final values of the cumulative criterion S_j are presented in Table 6.

Table 6. The results of evaluation by the SAW method using COIN weights.

Alternatives	1	2	3	4	5	6	7	8	9
SAW	0.175	0.037	0.015	0.019	0.045	0.036	0.080	0.014	0.012
Rank	2	7	15	13	5	9	3	16	18
Alternatives	10	11	12	13	14	15	16	17	18
SAW	0.329	0.042	0.030	0.036	0.020	0.022	0.013	0.057	0.017
Rank	1	6	10	8	12	11	17	4	14

Notes: Alternatives are cities of Lithuania: 1—Birštonas, 2—Druskininkai, 3—Alytus, 4—Jonava, 5—Kaunas, 6—Kėdainiai, 7—Klaipėda, 8—Marijampolė, 9—Mažeikiai, 10—Neringa, 11—Palanga, 12—Panevėžys, 13—Šiauliai, 14—Tauragė, 15—Telšiai, 16—Utena, 17—Vilnius, 18—Visaginas, authors' calculations.

10. Evaluation by the TOPSIS Method

The TOPSIS method [39,40] is one of the most popular and interesting contemporary MCDM methods among researchers [41]. The method uses the Euclidean proximity to the best and the worst hypothetical solutions [42,43]. The smaller is the distance to the best hypothetical solution, and the greater it is to the worst hypothetical solution, the greater appears the cumulative criterion of the method. The method requires a proprietary normalization of values of criteria [44], in accordance with Formula (21):

$$\tilde{r}_{ij} = \frac{r_{ij}}{\sqrt{\sum\limits_{j=1}^{n} r_{ij}^2}} \qquad (i = 1, \ldots, m; j = 1, \ldots n). \tag{21}$$

Denote the best hypothetical alternative as V^*. It is found in accordance with the following Formula (22):

$$V^* = \{V_1^*, V_2^*, \ldots, V_m^*\} = \{(\max_j \omega_i \tilde{r}_{ij}/i \in I_1, (\min_j \omega_i \tilde{r}_{ij}/i \in I_2)\}, \tag{22}$$

where I_1 is the set of indices of the maximizing criteria, I_2 is the set of indices of the minimizing criteria.

Denote the worst hypothetical alternative as V^-. It is found in accordance with the following Formula (23):

$$V^- = \{V_1^-, V_2^-, \ldots, V_m^-\} = \{(\min_j \omega_i \tilde{r}_{ij}/i \in I_1), (\max_j \omega_i \tilde{r}_{ij}/i \in I_2)\}. \tag{23}$$

The Euclidean distance to the best and the worst alternatives is calculated in accordance to the following Formulas (24) and (25):

$$D_j^* = \sqrt{\sum_{i=1}^{m} (\omega_i \tilde{r}_{ij} - V_i^*)^2}, \tag{24}$$

$$D_j^- = \sqrt{\sum_{i=1}^{m} (\omega_i \tilde{r}_{ij} - V_i^-)^2}. \tag{25}$$

Such distances are used for finding the cumulative criterion of the method C_j^* (26). It becomes as close to 1 as the alternative is closer to the best hypothetical alternative while its distance to the worst hypothetical solution is as great as possible. Furthermore, in the case of the opposite, the cumulative solution approaches zero. The ranking of attractiveness of alternatives is based on values of the cumulative criterion of the method C_j^* (26) and is carried in decreasing order:

$$C_j^* = \frac{D_j^-}{D_j^* + D_j^-}(j = 1, 2, \ldots, n), \quad (0 \le C_j^* \le 1). \tag{26}$$

The results of the evaluation by the TOPSIS method using COIN weights (Table 4) are presented in Table 7:

Table 7. The results of the evaluation by the TOPSIS method using COIN weights.

Alternatives	1	2	3	4	5	6	7	8	9
TOPSIS	0.431	0.124	0.041	0.082	0.153	0.107	0.223	0.061	0.063
Rank	2	8	18	13	6	9	3	15	14
Alternatives	**10**	**11**	**12**	**13**	**14**	**15**	**16**	**17**	**18**
TOPSIS	0.749	0.163	0.100	0.124	0.095	0.096	0.048	0.186	0.052
Rank	1	5	10	7	12	11	17	4	16

Notes: Alternatives are cities of Lithuania: 1—Birštonas, 2—Druskininkai, 3—Alytus, 4—Jonava, 5—Kaunas, 6—Kėdainiai, 7—Klaipėda, 8—Marijampolė, 9—Mažeikiai, 10—Neringa, 11—Palanga, 12—Panevėžys, 13—Šiauliai, 14—Tauragė, 15—Telšiai, 16—Utena, 17—Vilnius, 18—Visaginas, authors' calculations.

11. Evaluation by the PROMETHEE II Method

The PROMETHEE II method [45] uses some general ideas of MCDM framework; nevertheless, it discerns to a high extent from other MCDM methods. The preference functions of chosen shapes $p_t(d_i(A_j, A_k))$ (the shape is represented by the index t) normalize values of criteria, or, more precisely, of differences d_i between values of criteria in pairs (A_j, A_k) compared. The method compares all such pairs and creates a cumulative criterion based on such comparisons. Choice of parameters q and s enhances the evaluation by providing more options and has influence on the result [34]. The cumulative criterion is calculated in two steps. First, for every alternative A_j and all remaining alternatives A_k, two inward and backward aggregated preference indices $\pi(A_j, A_k)$ and $\pi(A_k, A_j)$ are calculated in accordance with the following Formula (27):

$$\pi(A_j, A_k) = \sum_{i=1}^{m} \omega_i p_t(d_i(A_j, A_k)) \tag{27}$$

by multiplication of values of preference function with weights, quite similar as in the SAW method. $\pi(A_j, A_k)$ shows the level of preference of the alternative A_j over A_k; conversely, $\pi(A_k, A_j)$ shows the level of preference of the alternative A_k over A_j. At the next step, positive and negative outranking flows are calculated by summing inward and backward aggregated preference indices over all alternatives (28) and (29):

$$F_j^+ = \sum_{k=1}^{n} \pi(A_j, A_k)(j = 1, 2, \ldots, n), \tag{28}$$

$$F_j^- = \sum_{k=1}^{n} \pi(A_k, A_j)(j = 1, 2, \ldots, n). \tag{29}$$

The positive flow comprises only magnitudes of preference of a chosen alternative relative to all other alternatives otherwise including nil values, while the negative flow comprises only magnitudes of outranking of other alternatives over the chosen alternative A_j. The larger is F_j^+ and the smaller is F_j^- the better alternative is evaluated. The cumulative criterion incorporates both flows (30):

$$F_j = F_j^+ - F_j^-. \tag{30}$$

We chose the following preference function $p_0(x)$ (Figure 2), which linearly maps differences between values of criteria:

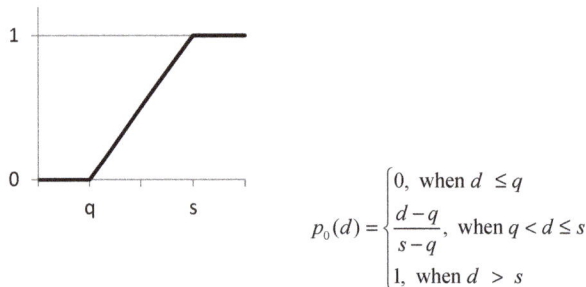

$$p_0(d) = \begin{cases} 0, & \text{when } d \leq q \\ \dfrac{d-q}{s-q}, & \text{when } q < d \leq s \\ 1, & \text{when } d > s \end{cases}$$

Figure 2. The analytical expression and shape of the preference function $p_0(x)$.

The result of the evaluation by the PROMETHEE II method using COIN weights (Table 4) is presented in Table 8:

Table 8. The results of the evaluation by the PROMETHEE II method using COIN weights.

Alternatives	1	2	3	4	5	6	7	8	9
F_j^+	6.641	1.287	0.240	0.612	1.717	0.887	2.643	0.356	0.359
F_j^-	1.634	1.766	2.444	2.106	1.831	1.806	1.581	2.221	2.263
F_j	5.007	−0.479	−2.204	−1.494	−0.114	−0.919	1.062	−1.865	−1.904
Rank	2	7	18	13	6	9	3	14	16
Alternatives	**10**	**11**	**12**	**13**	**14**	**15**	**16**	**17**	**18**
F_j^+	10.989	2.024	0.820	0.981	0.714	0.771	0.260	2.294	0.357
F_j^-	0.710	1.440	1.885	1.856	2.069	2.048	2.239	1.798	2.256
F_j	10.279	0.584	−1.065	−0.875	−1.355	−1.277	−1.979	0.496	−1.899
Rank	1	4	10	8	12	11	17	5	15

Notes: Alternatives are cities of Lithuania: 1—Birštonas, 2—Druskininkai, 3—Alytus, 4—Jonava, 5—Kaunas, 6—Kėdainiai, 7—Klaipėda, 8—Marijampolė, 9—Mažeikiai, 10—Neringa, 11—Palanga, 12—Panevėžys, 13—Šiauliai, 14—Tauragė, 15—Telšiai, 16—Utena, 17—Vilnius, 18—Visaginas, authors' calculations.

12. Results

For the purpose of increasing reliability of the evaluation, we combine results obtained by four MCDM methods similarly as in [22,42]. First, averages of obtained rankings by all the four MCDM methods presented in Tables 5–8 are calculated. Second, final rankings are found based on the averages calculated at the previous step. Such rankings are presented in Table 9. For the purpose of increasing descriptiveness, some other augmented reporting tools to decision-makers are recommended to be used.

Table 9. Final rankings of alternatives.

Alternatives	1	2	3	4	5	6	7	8	9
Average rank	2.0	7.0	16.5	13.0	5.5	7.8	3.0	15.5	16.5
The final rank of the evaluation	2	7	16.5	13	5	8.5	3	15	16.5
Alternatives	**10**	**11**	**12**	**13**	**14**	**15**	**16**	**17**	**18**
Average rank	1.0	6.0	10.0	7.8	12.5	11.0	16.8	5.0	14.3
The final rank of the evaluation	1	6	10	8.5	12	11	18	4	14

Notes: Alternatives are cities of Lithuania: 1—Birštonas, 2—Druskininkai, 3—Alytus, 4—Jonava, 5—Kaunas, 6—Kėdainiai, 7—Klaipėda, 8—Marijampolė, 9—Mažeikiai, 10—Neringa, 11—Palanga, 12—Panevėžys, 13—Šiauliai, 14—Tauragė, 15—Telšiai, 16—Utena, 17—Vilnius, 18—Visaginas, authors' calculations.

Correlation between aggregate criteria of the multiple criteria methods used in the paper appeared to be high. Test statistics for correlation coefficients between all groups of methods appeared to be well above the critical value t_{cr} = 2.120 for the t-distribution for 16 degrees of freedom at the chosen 5% level of significance. Namely, the correlation coefficient between TOPSIS and SAW: 0.990 (test statistics 28.57 > 2.120); between TOPSIS and EDAS: 0.972 (test statistics 16.60 > 2.120); between SAW and EDAS: 0.973 (test statistics 16.98 > 2.120); between TOPSIS and PROMETHEE: 0.998 (test statistics 67.06 > 2.120); between EDAS and PROMETHEE: 0.973 (test statistics 16.72 > 2.120); and between SAW and PROMETHEE: 0.992 (test statistics 30.96 > 2.120) appeared to reveal a high degree of correlation.

Normalised values of aggregate criteria of all four methods for 18 alternatives are presented in Figure 3.

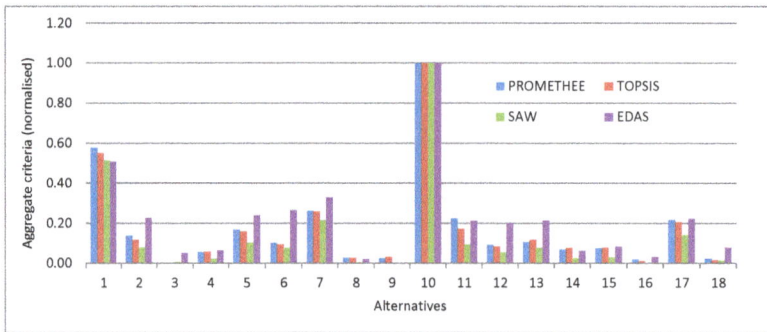

Figure 3. Normalised values of aggregate criteria of all four methods for 18 alternatives. Notes: Alternatives are cities of Lithuania: 1—Birštonas, 2—Druskininkai, 3—Alytus, 4—Jonava, 5—Kaunas, 6—Kėdainiai, 7—Klaipėda, 8—Marijampolė, 9—Mažeikiai, 10—Neringa, 11—Palanga, 12—Panevėžys, 13—Šiauliai, 14—Tauragė, 15—Telšiai, 16—Utena, 17—Vilnius, 18—Visaginas, authors' calculations.

The formula of normalisation is as follows (31):

$$\tilde{K}_f = \frac{K_f}{\max_f K_f},$$ (31)

where f is the index representing a method of evaluation (ranging from 1 to 4); K_f are values of the aggregate criteria of a method f; $\tilde{K}_f$ are normalized values of the aggregate criteria of a method f.

It can be observed that results among the methods are related.

In Figure 4, relative positions of attractiveness of cities and resorts of Lithuania in terms of the public infrastructure for electric vehicles can be observed, where final ranks obtained by the evaluation are depicted. The ranks resulted after comprising both data reflecting the state at every place as well as opinions of experts on importance of criteria of evaluation.

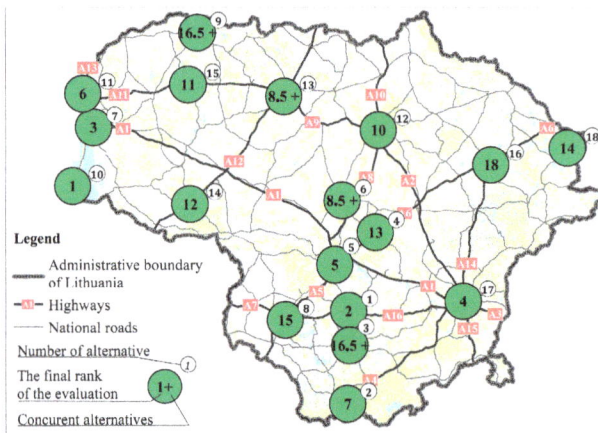

Figure 4. The state of the public infrastructure for electric vehicles in cities and resorts of Lithuania. Notes: Alternatives are cities of Lithuania: 1—Birštonas, 2—Druskininkai, 3—Alytus, 4—Jonava, 5—Kaunas, 6—Kėdainiai, 7—Klaipėda, 8—Marijampolė, 9—Mažeikiai, 10—Neringa, 11—Palanga, 12—Panevėžys, 13—Šiauliai, 14—Tauragė, 15—Telšiai, 16—Utena, 17—Vilnius, 18—Visaginas, authors' calculations.

13. Conclusions

Evaluation of public infrastructure for electric vehicles in cities and resorts of Lithuania revealed that the highest positions are currently attained by resorts. Namely, Neringa attained the 1st place; Birštonas—the 2nd place; Palanga the 6th place, and Druskininkai the 7th place. It could be concluded that, in general, that cities are lagging behind and require additional investment. We also point out the importance of private investment for the listed resorts as the corresponding factor "Investment of private institutions to the infrastructure for electric vehicles" has the largest weight of 0.227. Only Neringa may be distinguished by good scores in terms of said factor.

More uniform development of infrastructure would ensure attractiveness of electric vehicles among citizens of Lithuania. Extension of facilities for electric vehicles in lagging cities, among which only Klaipėda and Vilnius have good positions (the 3rd and the 5th positions, respectively), is required. Other cities are lacking streets with the dedicated A lane; development of electric charging posts; investment of both state and private institutions to the infrastructure for electric vehicles; integrated electric vehicle infrastructure development projects; installation of high-power charge posts on roads of national importance within 50 km distance from major city center. Such development would make a significant contribution to reduction of ecological effect in towns, namely, lowering air pollution, noise levels and greenhouse gas emissions in cities.

A combination of techniques used for estimation of weights is used in the paper. Two sources for such estimation were used: experts and data itself. Opinions of experts were estimated in terms of their concordance based on ranks of importance of criteria, using Kendall theory. A balancing combination of experts' estimations of weights and of the entropy weights, the method COIN, was proposed.

Four considerably different MCDM methods were used in the paper. The SAW method comprises normalized values of criteria multiplied by weights into the aggregate criterion of the method, the TOPSIS uses distances in the m-dimensional space (where m is the number of criteria) to the hypothetical worst and best alternatives, the EDAS method uses distances of each criterion from its average value, and the PROMETHEE II method makes pairwise comparison in all pairs of alternatives also by using a special type of normalization by applying a preference function. Such a combination of intrinsically different methods enhanced reliability of results. Especially it was observed in our case, when test statistics for correlation coefficients between all groups of methods appeared to be well above the critical value $t_{cr} = 2.120$ for the t-distribution for 16 degrees of freedom at the chosen 5% level of significance. Namely, the correlation coefficient between TOPSIS and SAW: 0.990 (test statistics 28.57 > 2.120); between TOPSIS and EDAS: 0.972 (test statistics 16.60 > 2.120); between SAW and EDAS: 0.973 (test statistics 16.98 > 2.120); between TOPSIS and PROMETHEE: 0.998 (test statistics 67.06 > 2.120); between EDAS and PROMETHEE: 0.973 (test statistics 16.72 > 2.120); and between SAW and PROMETHEE: 0.992 (test statistics 30.96 > 2.120) appeared to reveal a high degree of correlation. Thus, we can be more confident in the obtained results.

The research carried out shows that the current promotion measures for the development of Lithuanian electric vehicles access infrastructure need to be reviewed since most of the funds for the development of electric vehicle's access infrastructure are directed to the big cities of Lithuania. For making infrastructure for electric vehicles uniform within the country, incentives should be directed to resort cities. For the purpose of implementing the development plan of the infrastructure for electric vehicles in Lithuania and achieving more coordinated distribution of incentives between municipalities, the model of MCDM evaluation is presented in this paper, which could be used by interested institutions. The theoretical model can be easily used in reality, which would provide suitable conditions for the development of electric vehicle access infrastructure in cities of different municipalities all around the country.

Acknowledgments: All sources of funding of the study should be disclosed. Please clearly indicate grants that you have received in support of your research work. Clearly state if you received funds for covering the costs to publish in open access.

Author Contributions: All authors contributed equally to this work. Vytautas Palevičius developed the project idea, analysed data, conducted the expert interviews, created graphs and figures, outlined conclusions and recommendations, and contributed to drafting the paper. Askoldas Podviezko developed the model of quantitative and qualitative data collection, created the method COIN, performed calculations of weights and multiple criteria analysis, created tables, and contributed to drafting the paper. Henrikas Sivilevičius developed the project idea, led the development of the methodology and contributed to drafting the paper. Olegas Prentkovskis analysed data, contributed to the revision of the draft version and improvement of the final paper. All authors discussed the results and commented on the paper.

Conflicts of Interest: The authors declare no conflict of interest.

References

1. Han, X.; Ouyang, M.; Lu, L.; Li, J.; Zheng, Y.; Li, Z. A comparative study of commercial lithium ion battery cycle life in electrical vehicle: Aging mechanism identification. *J. Power Sources* **2014**, *251*, 38–54. [CrossRef]
2. Neubauer, J.; Wood, E. The impact of range anxiety and home, workplace, and public charging infrastructure on simulated battery electric vehicle lifetime utility. *J. Power Sources* **2014**, *257*, 12–20. [CrossRef]
3. Dong, J.; Liu, C.; Lin, Z. Charging infrastructure planning for promoting battery electric vehicles: An activity-based approach using multiday travel data. *Transp. Res. Part C Emerg. Technol.* **2014**, *38*, 44–55. [CrossRef]
4. Wang, J.; Shan, L.; Dai, Y.; Ming, L.; Yu, D. A systematic planning method for the electric vehicles charging service network. *J. Clean Energy Technol.* **2015**, *3*, 155–158. [CrossRef]
5. Chung, S.H.; Kwon, C. Multi-period planning for electric car charging station locations: A case of Korean Expressways. *Eur. J. Oper. Res.* **2015**, *242*, 677–687. [CrossRef]
6. Salah, F.; Ilg, J.P.; Flath, C.M.; Basse, H.; Van Dinther, C. Impact of electric vehicles on distribution substations: A Swiss case study. *Appl. Energy* **2015**, *137*, 88–96. [CrossRef]
7. Pasaoglu, G.; Zubaryeva, A.; Fiorello, D.; Thiel, C. Analysis of European mobility surveys and their potential to support studies on the impact of electric vehicles on energy and infrastructure needs in Europe. *Technol. Forecast. Soc. Chang.* **2014**, *87*, 41–50. [CrossRef]
8. Raslavičius, L.; Azzopardi, B.; Keršys, A.; Starevičius, M.; Bazaras, Ž.; Makaras, R. Electric vehicles challenges and opportunities: Lithuanian review. *Renew. Sustain. Energy Rev.* **2015**, *42*, 786–800. [CrossRef]
9. Naor, M.; Bernardes, E.S.; Druehl, C.T.; Shiftan, Y. Overcoming barriers to adoption of environmentally-friendly innovations through design and strategy: Learning from the failure of an electric vehicle infrastructure firm. *Int. J. Oper. Prod. Manag.* **2015**, *35*, 26–59. [CrossRef]
10. Delucchi, M.A.; Yang, C.; Burke, A.F.; Ogden, J.M.; Kurani, K.; Kessler, J.; Sperling, D. An assessment of electric vehicles: Technology, infrastructure requirements, greenhouse-gas emissions, petroleum use, material use, lifetime cost, consumer acceptance and policy initiatives. *Philos. Trans. R. Soc. Lond. Math. Phys. Eng. Sci.* **2014**, *372*, 1–27. [CrossRef] [PubMed]
11. Aghaei, J.; Nezhad, A.E.; Rabiee, A.; Rahimi, E. Contribution of plug-in hybrid electric vehicles in power system uncertainty management. *Renew. Sustain. Energy Rev.* **2016**, *59*, 450–458. [CrossRef]
12. Nie, Y.M.; Ghamami, M.; Zockaie, A.; Xiao, F. Optimization of incentive polices for plug-in electric vehicles. *Transp. Res. Part B Methodol.* **2016**, *84*, 103–123. [CrossRef]
13. Shafie-Khah, M.; Neyestani, N.; Damavandi, M.Y.; Gil, F.A.S.; Catalão, J.P.S. Economic and technical aspects of plug-in electric vehicles in electricity markets. *Renew. Sustain. Energy Rev.* **2016**, *53*, 1168–1177. [CrossRef]
14. Sierzchula, W.; Bakker, S.; Maat, K.; Van Wee, B. The influence of financial incentives and other socio-economic factors on electric vehicle adoption. *Energy Policy* **2014**, *68*, 183–194. [CrossRef]
15. Andriukaitis, D.; Bagdanavicius, N.; Jokuzis, V.; Kilius, S. Investigation of prospects for electric vehicle development in Lithuania. *Prz. Elektrotech.* **2014**, *90*, 101–104.
16. Raslavičius, L.; Starevičius, M.; Keršys, A.; Pilkauskas, K.; Vilkauskas, A. Performance of an all-electric vehicle under UN ECE R101 test conditions: A feasibility study for the city of Kaunas, Lithuania. *Energy* **2013**, *55*, 436–448. [CrossRef]
17. Huo, H.; Cai, H.; Zhang, Q.; Liu, F.; He, K. Life-cycle assessment of greenhouse gas and air emissions of electric vehicles: A comparison between China and the US. *Atmos. Environ.* **2015**, *108*, 107–116. [CrossRef]
18. Sternberg, A.; Bardow, A. Power-to-What?—Environmental assessment of energy storage systems. *Energy Environ. Sci.* **2015**, *8*, 389–400. [CrossRef]

19. Von Rosenstiel, D.P.; Heuermann, D.F.; Hüsig, S. Why has the introduction of natural gas vehicles failed in Germany?—Lessons on the role of market failure in markets for alternative fuel vehicles. *Energy Policy* **2015**, *78*, 91–101. [CrossRef]

20. Kendall, M.G.; Gibbons, J.D. *Rank Correlation Methods*, 5th ed.; Oxford University Press: New York, NY, USA, 1990.

21. Onat, N.C.; Gumus, S.; Kucukvar, M.; Tatari, O. Application of the TOPSIS and intuitionistic fuzzy set approaches for ranking the life cycle sustainability performance of alternative vehicle technologies. *Sustain. Prod. Consum.* **2016**, *6*, 12–25. [CrossRef]

22. Çolak, M.; Kaya, İ. Prioritization of renewable energy alternatives by using an integrated fuzzy MCDM model: A real case application for Turkey. *Renew. Sustain. Energy Rev.* **2017**, *80*, 840–853. [CrossRef]

23. Burinskiene, M.; Bielinskas, V.; Podviezko, A.; Gurskiene, V.; Maliene, V. Evaluating the Significance of Criteria Contributing to Decision-Making on Brownfield Land Redevelopment Strategies in Urban Areas. *Sustainability* **2017**, *9*, 759. [CrossRef]

24. Zavadskas, E.K.; Govindan, K.; Antuchevičienė, J.; Turskis, Z. Hybrid multiple criteria decision-making methods: A review of applications for sustainability issues. *Econ. Res. Ekon. Istraž.* **2016**, *29*, 857–887. [CrossRef]

25. Mardani, A.; Jusoh, A.; Nor, K.M.D.; Khalifah, Z.; Zakwan, N.; Valipour, A. Multiple criteria decision-making techniques and their applications—A review of the literature from 2000 to 2014. *Econ. Res. Ekon. Istraž.* **2015**, *28*, 516–571. [CrossRef]

26. Zavadskas, E.; Cavallaro, F.; Podvezko, V.; Ubarte, I.; Kaklauskas, A. MCDM Assessment of a Healthy and Safe Built Environment According to Sustainable Development Principles: A practical neighborhood approach in Vilnius. *Sustainability* **2017**, *9*, 702. [CrossRef]

27. Huang, C.-Y.; Hung, Y.-H.; Tzeng, G.-H. Using hybrid MCDM methods to assess fuel cell technology for the next generation of hybrid power automobiles. *J. Adv. Comput. Intell. Intell. Inform.* **2011**, *15*, 406–417. [CrossRef]

28. Podvezko, V.; Kildiene, S.; Zavadskas, E. Assessing the performance of the construction sectors in the Baltic states and Poland. *Panoeconomicus* **2017**, *64*, 493–512. [CrossRef]

29. Čereska, A.; Zavadskas, E.; Cavallaro, F.; Podvezko, V.; Tetsman, I.; Grinbergienė, I. Sustainable assessment of aerosol pollution decrease applying multiple attribute decision-making methods. *Sustainability* **2016**, *8*, 586. [CrossRef]

30. Hwang, C.-L.; Yoon, K. *Multiple Attribute Decision Making Methods and Applications*; Springer: Berlin, Germany, 1981.

31. Mardani, A.; Zavadskas, E.K.; Khalifah, Z.; Jusoh, A.; Nor, K.M.D. Multiple criteria decision-making techniques in transportation systems: A systematic review of the state of the art literature. *Transport* **2016**, *31*, 359–385. [CrossRef]

32. Wu, Y.; Yang, M.; Zhang, H.; Chen, K.; Wang, Y. Optimal site selection of electric vehicle charging stations based on a cloud model and the PROMETHEE method. *Energies* **2016**, *9*, 157. [CrossRef]

33. Yavuz, M.; Oztaysi, B.; Onar, S.C.; Kahraman, C. Multi-criteria evaluation of alternative-fuel vehicles via a hierarchical hesitant fuzzy linguistic model. *Expert Syst. Appl.* **2015**, *42*, 2835–2848. [CrossRef]

34. Wu, S.-M.; Liu, H.-C.; Wang, L.-E. Hesitant fuzzy integrated MCDM approach for quality function deployment: A case study in electric vehicle. *Int. J. Prod. Res.* **2017**, *55*, 4436–4449. [CrossRef]

35. Keshavarz Ghorabaee, M.; Zavadskas, E.K.; Olfat, L.; Turskis, Z. Multi-criteria inventory classification using a new method of evaluation based on distance from average solution (EDAS). *Informatica* **2015**, *26*, 435–451. [CrossRef]

36. Keshavarz Ghorabaee, M.; Amiri, M.; Zavadskas, E.K.; Turskis, Z.; Antucheviciene, J. A new hybrid simulation-based assignment approach for evaluating airlines with multiple service quality criteria. *J. Air Transp. Manag.* **2017**, *63*, 45–60. [CrossRef]

37. Zhao, H.; Li, N. Optimal siting of charging stations for electric vehicles based on fuzzy Delphi and hybrid multi-criteria decision making approaches from an extended sustainability perspective. *Energies* **2016**, *9*, 270. [CrossRef]

38. Dičiūnaitė-Rauktienė, R.; Gurskienė, V.; Burinskienė, M.; Malienė, V. The usage and perception of pedestrian zones in Lithuanian cities: Multiple criteria and comparative analysis. *Sustainability* **2018**, *10*, 818. [CrossRef]

39. Opricovic, S.; Tzeng, G.-H. Compromise solution by MCDM methods: A comparative analysis of VIKOR and TOPSIS. *Eur. J. Oper. Res.* **2004**, *156*, 445–455. [CrossRef]
40. Guo, S.; Zhao, H. Optimal site selection of electric vehicle charging station by using fuzzy TOPSIS based on sustainability perspective. *Appl. Energy* **2015**, *158*, 390–402. [CrossRef]
41. Palevicius, V.; Sivilevicius, H.; Podviezko, A.; Griskeviciute-Geciene, A.; Karpavicius, T. Evaluation of park and ride facilities at communication corridors in a middle-sized city. *Econ. Comput. Econ. Cybern. Stud. Res.* **2017**, *51*, 231–248.
42. Wątróbski, J.; Małecki, K.; Kijewska, K.; Iwan, S.; Karczmarczyk, A.; Thompson, R.G. Multi-criteria analysis of electric vans for city logistics. *Sustainability* **2017**, *9*, 1453. [CrossRef]
43. Stević, Ž.; Pamučar, D.; Zavadskas, E.K.; Ćirović, G.; Prentkovskis, O. The selection of wagons for the internal transport of a logistics company: A novel approach based on rough BWM and rough SAW methods. *Symmetry* **2017**, *9*, 264. [CrossRef]
44. Behzadian, M.; Otaghsara, S.K.; Yazdani, M.; Ignatius, J. A state-of the-art survey of TOPSIS applications. *Expert Syst. Appl.* **2012**, *39*, 13051–13069. [CrossRef]
45. Brans, J.-P.; Mareschal, B. PROMETHEE methods. In *Multiple Criteria Decision Analysis: State of the Art Surveys. International Series in Operations Research & Management Science*; Springer: New York, NY, USA, 2005; Volume 78, pp. 163–186.

sustainability

MDPI

Article

Research on Construction Engineering Project Risk Assessment with Some 2-Tuple Linguistic Neutrosophic Hamy Mean Operators

Shengjun Wu [1], Jie Wang [1], Guiwu Wei [1],* and Yu Wei [2],*

[1] School of Business, Sichuan Normal University, Chengdu 610101, China; wsjn382@sina.com (S.W.); JW970326@163.com (J.W.)
[2] School of Finance, Yunnan University of Finance and Economics, Kunming 650221, China
* Correspondence: weiguiwu1973@sicnu.edu.cn (G.W.); weiyusy@126.com (Y.W.); Tel.: +86-28-8448-0719 (G.W.)

Received: 14 April 2018; Accepted: 4 May 2018; Published: 11 May 2018

Abstract: In this paper, we expand the Hamy mean (HM) operator, weighted Hamy mean (WHM), dual Hamy mean (DHM) operator, and weighted dual Hamy mean (WDHM) operator with 2-tuple linguistic neutrosophic numbers (2TLNNs) to propose a 2-tuple linguistic neutrosophic Hamy mean (2TLNHM) operator, 2-tuple linguistic neutrosophic weighted Hamy mean (2TLNWHM) operator, 2-tuple linguistic neutrosophic dual Hamy mean (2TLNDHM) operator, and 2-tuple linguistic neutrosophic weighted dual Hamy mean (2TLNWDHM) operator. Then, the multiple attribute decision-making (MADM) methods are proposed with these operators. Finally, we utilize an applicable example in risk assessment for construction engineering projects to prove the proposed methods.

Keywords: multiple attribute decision-making (MADM); neutrosophic numbers; 2-tuple linguistic neutrosophic numbers set (2TLNNSs); 2TLNHM operator; 2TLNWHM operator; 2TLNDHM operator; 2TLNWDHM operator; construction engineering projects; risk assessment

1. Introduction

Neutrosophic sets (NSs), which were proposed originally by Smarandache [1,2], have attracted the attention of many scholars, and NSs have acted as a workspace in depicting indeterminate and inconsistent information. A NS has more potential power than other modeling mathematical tools, such as fuzzy set [3], intuitionistic fuzzy set (IFS) [4] and interval-valued intuitionistic fuzzy set (IVIFS) [5]. But, it is difficult to apply NSs to solve real life problems. Therefore, Wang et al. [6,7] defined single valued neutrosophic sets (SVNSs) and interval neutrosophic sets (INS), which are characterized by a truth membership, an indeterminacy membership and a falsity membership. Hence, SVNSs and INSs can express much more information than fuzzy sets, IFSs and IVIFSs. Ye [8] proposed a multiple attribute decision-making (MADM) method with correlation coefficients of SVNSs. Broumi and Smarandache [9] defined the correlation coefficients of INSs. Biswas et al. [10] proposed the Technique for Order Preference by Similarity to an Ideal Solution(TOPSIS) method with SVNNs. Liu et al. [11] defined the generalized neutrosophic number Hamacher aggregation for SVNSs. Sahin and Liu [12] defined the maximizing deviation model under a neutrosophic environment. Ye [13] developed some similarity measures of INS. Zhang et al. [14] defined some aggregating operators with INNs. Ye [15] defined a simplified neutrosophic set (SNS). Peng et al. [16] developed aggregation operators under SNS. Peng et al. [17] investigated the outranking approach with SNS, and then Zhang et al. [18] extended Peng's approach. Liu and Liu [19] proposed a power averaging operator with SVNNs. Deli and Subas [20] discussed a novel method to rank SVNNs. Peng et al. [21] proposed multi-valued neutrosophic sets. Zhang et al. [22] gave the improved weighted correlation coefficient for interval neutrosophic sets. Chen and Ye [23] proposed Dombi operations for SVNSs. Liu and Wang [24] proposed the MADM method based on a SVN-normalized weighted Bonferroni mean. Wu et al. [25] proposed a cross-entropy and prioritized an aggregation operator with SNSs in MADM problems. Li et al. [26] developed SVNN Heronian mean operators in MADM problems. Zavadskas et al. [27] proposed a model for residential house elements and material selection using the neutrosophic MULTIMOORA method. Zavadskas et al. [28] studied the sustainable market valuation of buildings using the SVN MAMVA method. Bausys and Juodagalviene [29] investigated the garage location selection for residential houses using the WASPAS-SVNS method. Wu et al. [30] proposed some Hamacher aggregation operators under an SVN 2-tuple linguistic environment for MAGDM.

Although SVNS theory has been successfully applied in some areas, the SVNS is also characterized by truth membership degree, indeterminacy membership degree, and falsity membership degree information. However, all the above approaches are unsuitable for describing the truth membership degree, indeterminacy membership degree, and falsity membership degree information of an element of a set by linguistic variables on the basis of the given linguistic term sets, which can reflect a decision maker's confidence level when they are making an evaluation. In order to overcome this limit, we propose the concept of a 2-tuple linguistic neutrosophic numbers set (2TLNNSs) to solve this problem based on SVNS [6,7] and a 2-tuple linguistic information processing model [31]. Thus, how to aggregate these 2-tuple linguistic neutrosophic numbers is an interesting topic. To solve this issue, in this paper, we develop aggregation operators with 2TLNNs based on the traditional operator [32]. In order to do so, the remainder of this paper is set out as follows. In the next section, we propose the concept of 2TLNNSs. In Section 3, we propose Hamy mean (HM) operators with 2TLNNs. In Section 4, we give a numerical example for risk assessment of a construction engineering projects. Section 5 concludes the paper with some remarks.

2. Preliminaries

In this section, we propose the concept of using 2-tuple linguistic neutrosophic sets (2TLNSs) based on SVNSs [6,7] and 2-tuple linguistic sets (2TLSs) [31].

2.1. 2TLSs

Definition 1. *Let* $S = \{s_i | i = 0, 1, \ldots, t\}$ *be a linguistic term set with an odd cardinality. Any label,* s_i, *represents a possible value for a linguistic variable, and S can be defined as:*

$$S = \left\{ \begin{array}{l} s_0 = extremely\ poor,\ s_1 = very\ poor,\ s_2 = poor,\ s_3 = medium, \\ s_4 = good,\ s_5 = very\ good,\ s_6 = extremely\ good. \end{array} \right\} \tag{1}$$

Herrera and Martinez [27,28] developed the 2-tuple fuzzy linguistic representation model based on the concept of symbolic translation. It is used for representing the linguistic assessment information by means of a 2-tuple (s_i, ρ_i), where s_i is a linguistic label for predefined linguistic term set S and ρ_i is the value of symbolic translation, and $\rho_i \in [-0.5, 0.5)$.

2.2. SVNSs

Let X be a space of points (objects) with a generic element in a fixed set, X, denoted by x. An SVNS, A, in X is characterized as the following [6,7]:

$$A = \{(x, T_A(x), I_A(x), F_A(x)) | x \in X\} \tag{2}$$

where the truth membership function, $T_A(x)$, indeterminacy-membership, $I_A(x)$, and falsity membership function, $F_A(x)$, are single subintervals/subsets in the real standard $[0,1]$, that is,

$T_A(x) : X \to [0,1]$, $I_A(x) : X \to [0,1]$ and $F_A(x) : X \to [0,1]$. In addition, the sum of $T_A(x)$, $I_A(x)$ and $F_A(x)$ satisfies the condition $0 \leq T_A(x) + I_A(x) + F_A(x) \leq 3$. Then, a simplification of A is denoted by $A = \{(x, T_A(x), I_A(x), F_A(x)) | x \in X\}$, which is a SVNS.

For a SVNS $\{(x, T_A(x), I_A(x), F_A(x)) | x \in X\}$, the ordered triple components, $(T_A(x), I_A(x), F_A(x))$, are described as a single-valued neutrosophic number (SVNN), and each SVNN can be expressed as $A = (T_A, I_A, F_A)$, where $T_A \in [0,1]$, $I_A \in [0,1]$, $F_A \in [0,1]$ and $0 \leq T_A + I_A + F_A \leq 3$.

2.3. 2TLNSs

Definition 2. *Assume that* $\varphi = \{\varphi_0, \varphi_1, \ldots, \varphi_t\}$ *is a 2TLSs with an odd cardinality,* $t + 1$. *If* $\varphi = \langle (s_T, \alpha), (s_I, \beta), (s_F, \gamma) \rangle$ *is defined for* $(s_T, \alpha), (s_I, \beta), (s_F, \gamma) \in \varphi$ *and* $\alpha, \beta, \gamma \in [0, t]$, *where* $(s_T, \alpha), (s_I, \beta)$ *and* (s_F, γ) *express independently the truth degree, indeterminacy degree, and falsity degree by 2TLSs, then 2TLNSs is defined as follows:*

$$\varphi_j = \left\langle \left(s_{T_j}, \alpha_j\right), \left(s_{I_j}, \beta_j\right), \left(s_{F_j}, \gamma_j\right) \right\rangle \tag{3}$$

where $0 \leq \Delta^{-1}\left(s_{T_j}, \alpha_j\right) \leq t, 0 \leq \Delta^{-1}\left(s_{I_j}, \beta_j\right) \leq t, 0 \leq \Delta^{-1}\left(s_{F_j}, \gamma_j\right) \leq t$, *and* $0 \leq \Delta^{-1}\left(s_{T_j}, \alpha_j\right) + \Delta^{-1}\left(s_{I_j}, \beta_j\right) + \Delta^{-1}\left(s_{F_j}, \gamma_j\right) \leq 3t$.

Definition 3. *Let* $\varphi_1 = \left\langle \left(s_{T_1}, \alpha_1\right), \left(s_{I_1}, \beta_1\right), \left(s_{F_1}, \gamma_1\right) \right\rangle$ *be a 2TLNN in* φ. *Then, the score and accuracy functions of* φ_1 *are defined as follows:*

$$S(\varphi_1) = \Delta \left\{ \frac{\left(2t + \Delta^{-1}\left(s_{T_1}, \alpha_1\right) - \Delta^{-1}\left(s_{I_1}, \beta_1\right) - \Delta^{-1}\left(s_{F_1}, \gamma_1\right)\right)}{3} \right\}, S(\varphi_1) \in [0, t] \tag{4}$$

$$H(\varphi_1) = \Delta \left\{ \Delta^{-1}\left(s_{T_1}, \alpha_1\right) - \Delta^{-1}\left(s_{F_1}, \gamma_1\right) \right\}, H(\varphi_1) \in [-t, t]. \tag{5}$$

Definition 4. *Let* $\varphi_1 = \left\langle \left(s_{T_1}, \alpha_1\right), \left(s_{I_1}, \beta_1\right), \left(s_{F_1}, \gamma_1\right) \right\rangle$ *and* $\varphi_2 = \left\langle \left(s_{T_2}, \alpha_2\right), \left(s_{I_2}, \beta_2\right), \left(s_{F_2}, \gamma_2\right) \right\rangle$ *be two 2TLNNs, then*

(1) if $S(\varphi_1) < S(\varphi_2)$, then $\varphi_1 < \varphi_2$;
(2) if $S(\varphi_1) > S(\varphi_2)$, then $\varphi_1 > \varphi_2$;
(3) if $S(\varphi_1) = S(\varphi_2), H(\varphi_1) < H(\varphi_2)$, then $\varphi_1 < \varphi_2$;
(4) if $S(\varphi_1) = S(\varphi_2), H(\varphi_1) > H(\varphi_2)$, then $\varphi_1 > \varphi_2$;
(5) if $S(\varphi_1) = S(\varphi_2), H(\varphi_1) = H(\varphi_2)$, then $\varphi_1 = \varphi_2$.

Definition 5. *Let* $\varphi_1 = \left\langle \left(s_{T_1}, \alpha_1\right), \left(s_{I_1}, \beta_1\right), \left(s_{F_1}, \gamma_1\right) \right\rangle$ *and* $\varphi_2 = \left\langle \left(s_{T_2}, \alpha_2\right), \left(s_{I_2}, \beta_2\right), \left(s_{F_2}, \gamma_2\right) \right\rangle$ *be two 2TLNNs,* $\zeta > 0$, *then*

$$(1) \quad \varphi_1 \oplus \varphi_2 = \left\{ \begin{array}{l} \Delta\left(t\left(\frac{\Delta^{-1}\left(s_{T_1}, \alpha_1\right)}{t} + \frac{\Delta^{-1}\left(s_{T_2}, \alpha_2\right)}{t} - \frac{\Delta^{-1}\left(s_{T_1}, \alpha_1\right)}{t} \cdot \frac{\Delta^{-1}\left(s_{T_2}, \alpha_2\right)}{t}\right)\right), \\ \Delta\left(t\left(\frac{\Delta^{-1}\left(s_{I_1}, \beta_1\right)}{t} \cdot \frac{\Delta^{-1}\left(s_{I_2}, \beta_2\right)}{t}\right)\right), \Delta\left(t\left(\frac{\Delta^{-1}\left(s_{F_1}, \gamma_1\right)}{t} \cdot \frac{\Delta^{-1}\left(s_{F_2}, \gamma_2\right)}{t}\right)\right) \end{array} \right\};$$

$$(2) \quad \varphi_1 \otimes \varphi_2 = \left\{ \begin{array}{l} \Delta\left(t\left(\frac{\Delta^{-1}\left(s_{T_1}, \alpha_1\right)}{t} \cdot \frac{\Delta^{-1}\left(s_{T_2}, \alpha_2\right)}{t}\right)\right), \\ \Delta\left(t\left(\frac{\Delta^{-1}\left(s_{T_1}, \beta_1\right)}{t} + \frac{\Delta^{-1}\left(s_{T_2}, \beta_2\right)}{t} - \frac{\Delta^{-1}\left(s_{T_1}, \beta_1\right)}{t} \cdot \frac{\Delta^{-1}\left(s_{T_2}, \beta_2\right)}{t}\right)\right), \\ \Delta\left(t\left(\frac{\Delta^{-1}\left(s_{F_1}, \gamma_1\right)}{t} + \frac{\Delta^{-1}\left(s_{F_2}, \gamma_2\right)}{t} - \frac{\Delta^{-1}\left(s_{F_1}, \gamma_1\right)}{t} \cdot \frac{\Delta^{-1}\left(s_{F_2}, \gamma_2\right)}{t}\right)\right) \end{array} \right\};$$

$$
(3) \quad \zeta\varphi_1 = \left\{ \begin{array}{l} \Delta\left(t\left(1-\left(1-\frac{\Delta^{-1}\left(s_{T_1},\alpha_1\right)}{t}\right)^{\zeta}\right)\right), \Delta\left(t\left(\frac{\Delta^{-1}\left(s_{I_1},\beta_1\right)}{t}\right)^{\zeta}\right), \\ \Delta\left(t\left(\frac{\Delta^{-1}\left(s_{F_1},\gamma_1\right)}{t}\right)^{\zeta}\right) \end{array} \right\}, \zeta > 0;
$$

$$
(4) \quad (\varphi_1)^{\zeta} = \left\{ \begin{array}{l} \Delta\left(t\left(\frac{\Delta^{-1}\left(s_{T_1},\alpha_1\right)}{t}\right)^{\zeta}\right), \Delta\left(t\left(1-\left(1-\frac{\Delta^{-1}\left(s_{I_1},\beta_1\right)}{t}\right)^{\zeta}\right)\right), \\ \Delta\left(t\left(1-\left(1-\frac{\Delta^{-1}\left(s_{F_1},\gamma_1\right)}{t}\right)^{\zeta}\right)\right) \end{array} \right\}, \zeta > 0.
$$

2.4. HM Operator

Definition 6 [32]. *The Hamy mean (HM) operator is defined as follows:*

$$
HM^{(x)}(\varphi_1, \varphi_2, \cdots, \varphi_n) = \frac{\sum\limits_{1 \leq i_1 < \ldots < i_x \leq n} \left(\prod\limits_{j=1}^{x} \varphi_{i_j}\right)^{\frac{1}{x}}}{C_n^x}, \tag{6}
$$

where x is a parameter and $x = 1, 2, \ldots, n$, $i_1, i_2, \ldots, i_x$ are x integer values taken from the set $\{1, 2, \ldots, n\}$ of k integer values, C_n^x denotes the binomial coefficient and $C_n^x = \frac{n!}{x!(n-x)!}$.

3. Some 2TLNHM Operators

3.1. 2TLNHM Operator

In this section, we will combine HM and 2TLNNs and propose the 2-tuple linguistic neutrosophic Hamy mean (2TLNHM) operator.

Definition 7. *Let $\varphi_j = \left\langle \left(s_{T_j}, \alpha_j\right), \left(s_{I_j}, \beta_j\right), \left(s_{F_j}, \gamma_j\right) \right\rangle (j = 1, 2, \ldots, n)$ be a set of 2TLNNs. The 2TLNHM operator is*

$$
2TLNHM^{(x)}(\varphi_1, \varphi_2, \cdots, \varphi_n) = \frac{\bigoplus\limits_{1 \leq i_1 < \ldots < i_x \leq n} \left(\bigotimes\limits_{j=1}^{x} \varphi_{i_j}\right)^{\frac{1}{x}}}{C_n^x}. \tag{7}
$$

Theorem 1. *Let $\varphi_j = \left\langle \left(s_{T_j}, \alpha_j\right), \left(s_{I_j}, \beta_j\right), \left(s_{F_j}, \gamma_j\right) \right\rangle (j = 1, 2, \ldots, n)$ be a set of 2TLNNs. The aggregated value from the 2TLNHM operators is also a 2TLNN where*

$$2\text{TLNHM}^{(x)}(\varphi_1, \varphi_2, \cdots, \varphi_n) = \frac{\underset{1 \leq i_1 < \ldots < i_x \leq n}{\overset{\oplus}{}} \left(\underset{j=1}{\overset{x}{\otimes}} \varphi_{i_j} \right)^{\frac{1}{x}}}{C_n^x}$$

$$= \left\{ \begin{array}{l} \Delta\left(t\left(1 - \left(\displaystyle\prod_{1 \leq i_1 < \ldots < i_x \leq n} \left(1 - \left(\displaystyle\prod_{j=1}^{x} \left(\frac{\Delta^{-1}\left(s_{T_j}, \alpha_j\right)}{t} \right)\right)^{\frac{1}{x}} \right)\right)^{\frac{1}{C_n^x}} \right)\right), \\[20pt] \Delta\left(t\left(\displaystyle\prod_{1 \leq i_1 < \ldots < i_x \leq n} \left(1 - \left(\displaystyle\prod_{j=1}^{x} \left(1 - \frac{\Delta^{-1}\left(s_{I_j}, \beta_j\right)}{t} \right)\right)^{\frac{1}{x}} \right)\right)^{\frac{1}{C_n^x}} \right)\right), \\[20pt] \Delta\left(t\left(\displaystyle\prod_{1 \leq i_1 < \ldots < i_x \leq n} \left(1 - \left(\displaystyle\prod_{j=1}^{x} \left(1 - \frac{\Delta^{-1}\left(s_{F_j}, \gamma_j\right)}{t} \right)\right)^{\frac{1}{x}} \right)\right)^{\frac{1}{C_n^x}} \right)\right) \end{array} \right\}. \tag{8}$$

Proof:

$$\underset{j=1}{\overset{x}{\otimes}} \varphi_{i_j} = \left\{ \begin{array}{l} \Delta\left(t\displaystyle\prod_{j=1}^{x} \left(\frac{\Delta^{-1}\left(s_{T_j}, \alpha_j\right)}{t} \right)\right), \Delta\left(t\left(1 - \displaystyle\prod_{j=1}^{x} \left(1 - \frac{\Delta^{-1}\left(s_{I_j}, \beta_j\right)}{t} \right)\right)\right), \\[20pt] \Delta\left(t\left(1 - \displaystyle\prod_{j=1}^{x} \left(1 - \frac{\Delta^{-1}\left(s_{F_j}, \gamma_j\right)}{t} \right)\right)\right) \end{array} \right\}. \tag{9}$$

Thus,

$$\left(\underset{j=1}{\overset{x}{\otimes}} \varphi_{i_j} \right)^{\frac{1}{x}} = \left\{ \begin{array}{l} \Delta\left(t\left(\displaystyle\prod_{j=1}^{x} \left(\frac{\Delta^{-1}\left(s_{T_j}, \alpha_j\right)}{t} \right)\right)^{\frac{1}{x}}\right), \Delta\left(t\left(1 - \left(\displaystyle\prod_{j=1}^{x} \left(1 - \frac{\Delta^{-1}\left(s_{I_j}, \beta_j\right)}{t} \right)\right)^{\frac{1}{x}}\right)\right), \\[20pt] \Delta\left(t\left(1 - \left(\displaystyle\prod_{j=1}^{x} \left(1 - \frac{\Delta^{-1}\left(s_{F_j}, \gamma_j\right)}{t} \right)\right)^{\frac{1}{x}}\right)\right) \end{array} \right\} \tag{10}$$

Thereafter,

$$\underset{1 \leq i_1 < \ldots < i_x \leq n}{\overset{\oplus}{}} \left(\underset{j=1}{\overset{x}{\otimes}} \varphi_{i_j} \right)^{\frac{1}{x}} = \left\{ \begin{array}{l} \Delta\left(t\left(1 - \displaystyle\prod_{1 \leq i_1 < \ldots < i_x \leq n} \left(1 - \left(\displaystyle\prod_{j=1}^{x} \left(\frac{\Delta^{-1}\left(s_{T_j}, \alpha_j\right)}{t} \right)\right)^{\frac{1}{x}} \right)\right)\right), \\[20pt] \Delta\left(t\left(\displaystyle\prod_{1 \leq i_1 < \ldots < i_x \leq n} \left(1 - \left(\displaystyle\prod_{j=1}^{x} \left(1 - \frac{\Delta^{-1}\left(s_{I_j}, \beta_j\right)}{t} \right)\right)^{\frac{1}{x}} \right)\right)\right), \\[20pt] \Delta\left(t\left(\displaystyle\prod_{1 \leq i_1 < \ldots < i_x \leq n} \left(1 - \left(\displaystyle\prod_{j=1}^{x} \left(1 - \frac{\Delta^{-1}\left(s_{F_j}, \gamma_j\right)}{t} \right)\right)^{\frac{1}{x}} \right)\right)\right) \end{array} \right\}. \tag{11}$$

Therefore,

$$2\text{TLNHM}^{(x)}(\varphi_1, \varphi_2, \cdots, \varphi_n) = \frac{\overset{\oplus}{\underset{1 \le i_1 < \ldots < i_x \le n}{}} \left(\overset{x}{\underset{j=1}{\otimes}} \varphi_{ij} \right)^{\frac{1}{x}}}{C_n^x}$$

$$= \left\{ \begin{array}{l} \Delta \left(t \left(1 - \left(\displaystyle\prod_{1 \le i_1 < \ldots < i_x \le n} \left(1 - \left(\displaystyle\prod_{j=1}^{x} \left(\frac{\Delta^{-1}(s_{T_j}, \alpha_j)}{t} \right) \right)^{\frac{1}{x}} \right) \right)^{\frac{1}{C_n^x}} \right) \right), \\[30pt]
\Delta \left(t \left(\displaystyle\prod_{1 \le i_1 < \ldots < i_x \le n} \left(1 - \left(\displaystyle\prod_{j=1}^{x} \left(1 - \frac{\Delta^{-1}(s_{I_j}, \beta_j)}{t} \right) \right)^{\frac{1}{x}} \right) \right)^{\frac{1}{C_n^x}} \right) \right), \\[30pt]
\Delta \left(t \left(\displaystyle\prod_{1 \le i_1 < \ldots < i_x \le n} \left(1 - \left(\displaystyle\prod_{j=1}^{x} \left(1 - \frac{\Delta^{-1}(s_{F_j}, \gamma_j)}{t} \right) \right)^{\frac{1}{x}} \right) \right)^{\frac{1}{C_n^x}} \right) \right) \end{array} \right\} \tag{12}$$

Hence, (7) is kept.

Then we need to prove that (7) is a 2TLNN. We need to prove two conditions, as follows:

① $0 \le \Delta^{-1}(s_T, \alpha) \le t, 0 \le \Delta^{-1}(s_I, \beta) \le t, 0 \le \Delta^{-1}(s_F, \gamma) \le t.$

② $0 \le \Delta^{-1}(s_T, \alpha) + \Delta^{-1}(s_I, \beta) + \Delta^{-1}(s_F, \gamma) \le 3t.$

Let

$$\frac{\Delta^{-1}(s_T, \alpha)}{t} = 1 - \left(\prod_{1 \le i_1 < \ldots < i_x \le n} \left(1 - \left(\prod_{j=1}^{x} \left(\frac{\Delta^{-1}(s_{T_j}, \alpha_j)}{t} \right) \right)^{\frac{1}{x}} \right) \right)^{\frac{1}{C_n^x}}$$

$$\frac{\Delta^{-1}(s_I, \beta)}{t} = \left(\prod_{1 \le i_1 < \ldots < i_x \le n} \left(1 - \left(\prod_{j=1}^{x} \left(1 - \frac{\Delta^{-1}(s_{I_j}, \beta_j)}{t} \right) \right)^{\frac{1}{x}} \right) \right)^{\frac{1}{C_n^x}}$$

$$\frac{\Delta^{-1}(s_F, \gamma)}{t} = \left(\prod_{1 \le i_1 < \ldots < i_x \le n} \left(1 - \left(\prod_{j=1}^{x} \left(1 - \frac{\Delta^{-1}(s_{F_j}, \gamma_j)}{t} \right) \right)^{\frac{1}{x}} \right) \right)^{\frac{1}{C_n^x}}$$

□

Proof: ① Since $0 \le \frac{\Delta^{-1}(s_{T_j}, \alpha_j)}{t} \le 1$, we get

$$0 \le \prod_{j=1}^{x} \left(\frac{\Delta^{-1}(s_{T_j}, \alpha_j)}{t} \right) \le 1 \text{ and } 0 \le 1 - \left(\prod_{j=1}^{x} \left(\frac{\Delta^{-1}(s_{T_j}, \alpha_j)}{t} \right) \right)^{\frac{1}{x}} \le 1 \tag{13}$$

Then,

$$0 \le \prod_{1 \le i_1 < \ldots < i_x \le n} \left(1 - \left(\prod_{j=1}^{x} \left(\frac{\Delta^{-1}(s_{T_j}, \alpha_j)}{t} \right) \right)^{\frac{1}{x}} \right) \le 1, \tag{14}$$

$$0 \le 1 - \left(\prod_{1 \le i_1 < \ldots < i_x \le n} \left(1 - \left(\prod_{j=1}^{x} \left(\frac{\Delta^{-1}(s_{T_j}, \alpha_j)}{t} \right) \right)^{\frac{1}{x}} \right) \right)^{\frac{1}{C_n^x}} \le 1. \tag{15}$$

That means $0 \le \Delta^{-1}(s_T, \alpha) \le t$. Similarly, we can get $0 \le \Delta^{-1}(s_I, \beta) \le t, 0 \le \Delta^{-1}(s_F, \gamma) \le t$ so ① is maintained. ② Since $0 \le \Delta^{-1}(s_T, \alpha) \le t, 0 \le \Delta^{-1}(s_I, \beta) \le t, 0 \le \Delta^{-1}(s_F, \gamma) \le t.,$ $0 \le \Delta^{-1}(s_T, \alpha) + \Delta^{-1}(s_I, \beta) + \Delta^{-1}(s_F, \gamma) \le 3t. \square$

Example 1. *Let* $\langle(s_5,0),(s_2,0),(s_1,0)\rangle, \langle(s_4,0),(s_3,0),(s_4,0)\rangle, \langle(s_2,0),(s_5,0),(s_1,0)\rangle$ *and* $\langle(s_5,0),(s_1,0),(s_3,0)\rangle$ *be four 2TLNNs, and suppose* $x = 2$*, then according to (4), we have*

$$2TLNHM^{(2)}(\varphi_1,\varphi_2,\cdots,\varphi_n) = \frac{\underset{1\leq i_1<...<i_x\leq n}{\oplus}\left(\overset{x}{\underset{j=1}{\otimes}}\varphi_{ij}\right)^{\frac{1}{x}}}{C_n^x}$$

$$= \left\{ \begin{array}{l} \Delta\left(6\times\left(1-\left(\begin{array}{l}\left(1-\left(\frac{5}{6}\times\frac{4}{6}\right)^{\frac{1}{2}}\right)\times\left(1-\left(\frac{5}{6}\times\frac{2}{6}\right)^{\frac{1}{2}}\right)\times\left(1-\left(\frac{5}{6}\times\frac{5}{6}\right)^{\frac{1}{2}}\right)\\ \times\left(1-\left(\frac{4}{6}\times\frac{2}{6}\right)^{\frac{1}{2}}\right)\times\left(1-\left(\frac{4}{6}\times\frac{5}{6}\right)^{\frac{1}{2}}\right)\times\left(1-\left(\frac{2}{6}\times\frac{5}{6}\right)^{\frac{1}{2}}\right)\end{array}\right)^{\frac{1}{C_4^2}}\right)\right), \\ \Delta\left(6\times\left(\begin{array}{l}\left(1-((1-\frac{2}{6})\times(1-\frac{3}{6}))^{\frac{1}{2}}\right)\times\left(1-((1-\frac{2}{6})\times(1-\frac{5}{6}))^{\frac{1}{2}}\right)\times\left(1-\left((1-\frac{2}{6})\times\left(1-\frac{1}{6}\right)\right)^{\frac{1}{2}}\right)\\ \times\left(1-((1-\frac{3}{6})\times(1-\frac{5}{6}))^{\frac{1}{2}}\right)\times\left(1-\left((1-\frac{3}{6})\times\left(1-\frac{1}{6}\right)\right)^{\frac{1}{2}}\right)\times\left(1-\left((1-\frac{5}{6})\times\left(1-\frac{1}{6}\right)\right)^{\frac{1}{2}}\right)\end{array}\right)^{\frac{1}{C_4^2}}\right), \\ \Delta\left(6\times\left(\begin{array}{l}\left(1-\left(\left(1-\frac{1}{6}\right)\times\left(1-\frac{4}{6}\right)\right)^{\frac{1}{2}}\right)\times\left(1-\left(\left(1-\frac{1}{6}\right)\times\left(1-\frac{1}{6}\right)\right)^{\frac{1}{2}}\right)\times\left(1-\left(\left(1-\frac{1}{6}\right)\times(1-\frac{3}{6})\right)^{\frac{1}{2}}\right)\\ \times\left(1-\left(\left(1-\frac{4}{6}\right)\times\left(1-\frac{1}{6}\right)\right)^{\frac{1}{2}}\right)\times\left(1-\left((1-\frac{4}{6})\times(1-\frac{3}{6})\right)^{\frac{1}{2}}\right)\times\left(1-\left(\left(1-\frac{1}{6}\right)\times(1-\frac{3}{6})\right)^{\frac{1}{2}}\right)\end{array}\right)^{\frac{1}{C_4^2}}\right) \end{array} \right\}$$

$$= \langle(s_4,0.0235),(s_3,-0.1556),(s_2,0.2489)\rangle$$

Now, we will give some properties of a 2TLNHM operator.

Property 1. *(Idempotency) If* $\varphi_j = \left\langle\left(s_{T_j},\alpha_j\right),\left(s_{I_j},\beta_j\right),\left(s_{F_j},\gamma_j\right)\right\rangle(j=1,2,\ldots,n)$ *are equal, then*

$$2TLNHM^{(x)}(\varphi_1,\varphi_2,\cdots,\varphi_n) = \varphi. \tag{16}$$

Proof: Since $\varphi_j = \varphi = \langle(s_T,\alpha),(s_I,\beta),(s_F,\gamma)\rangle$, then

$$2TLNHM^{(x)}(\varphi_1,\varphi_2,\cdots,\varphi_n) = \frac{\underset{1\leq i_1<...<i_x\leq n}{\oplus}\left(\overset{x}{\underset{j=1}{\otimes}}\varphi_{ij}\right)^{\frac{1}{x}}}{C_n^x}$$

$$= \left\{ \begin{array}{l} \Delta\left(t\left(1-\left(\underset{1\leq i_1<...<i_x\leq n}{\prod}\left(1-\left(\overset{x}{\underset{j=1}{\prod}}\left(\frac{\Delta^{-1}(s_T,\alpha)}{t}\right)\right)^{\frac{1}{x}}\right)\right)^{\frac{1}{C_n^x}}\right)\right), \Delta\left(t\left(\underset{1\leq i_1<...<i_x\leq n}{\prod}\left(1-\left(\overset{x}{\underset{j=1}{\prod}}\left(1-\frac{\Delta^{-1}(s_I,\beta)}{t}\right)\right)^{\frac{1}{x}}\right)\right)^{\frac{1}{C_n^x}}\right), \\ \Delta\left(t\left(\underset{1\leq i_1<...<i_x\leq n}{\prod}\left(1-\left(\overset{x}{\underset{j=1}{\prod}}\left(1-\frac{\Delta^{-1}(s_F,\gamma)}{t}\right)\right)^{\frac{1}{x}}\right)\right)^{\frac{1}{C_n^x}}\right) \end{array} \right\}$$

$$= \left\{ \begin{array}{l} \Delta\left(t\left(1-\left(\left(1-\left(\left(\frac{\Delta^{-1}(s_T,\alpha)}{t}\right)^x\right)^{\frac{1}{x}}\right)^{\frac{1}{C_n^x}}\right)^{C_n^x}\right)\right), \Delta\left(t\left(\left(1-\left(1-\frac{\Delta^{-1}(s_I,\beta)}{t}\right)^x\right)^{\frac{1}{x}}\right)^{\frac{1}{C_n^x}}\right)^{C_n^x}\right), \\ \Delta\left(t\left(\left(1-\left(1-\frac{\Delta^{-1}(s_F,\gamma)}{t}\right)^x\right)^{\frac{1}{x}}\right)^{\frac{1}{C_n^x}}\right)^{C_n^x}\right) \end{array} \right\}$$

$$= \langle(s_T,\alpha),(s_I,\beta),(s_F,\gamma)\rangle = \varphi$$

□

Property 2. *(Monotonicity) Let* $\varphi_{a_j} = \left\langle\left(s_{T_{a_j}},\alpha_{a_j}\right),\left(s_{I_{a_j}},\beta_{a_j}\right),\left(s_{F_{a_j}},\gamma_{a_j}\right)\right\rangle(j=1,2,\ldots,n)$ *and* $\varphi_{b_j} = \left\langle\left(s_{T_{b_j}},\alpha_{b_j}\right),\left(s_{I_{b_j}},\beta_{b_j}\right),\left(s_{F_{b_j}},\gamma_{b_j}\right)\right\rangle(j=1,2,\ldots,n)$ *be two sets of 2TLNNs. If* $\Delta^{-1}\left(s_{T_{a_j}},\alpha_{a_j}\right) \leq \Delta^{-1}\left(s_{T_{b_j}},\alpha_{b_j}\right), \Delta^{-1}\left(s_{I_{a_j}},\beta_{a_j}\right) \geq \Delta^{-1}\left(s_{I_{b_j}},\beta_{b_j}\right)$ *and* $\Delta^{-1}\left(s_{F_{a_j}},\gamma_{a_j}\right) \geq \Delta^{-1}\left(s_{F_{b_j}},\gamma_{b_j}\right)$ *hold for all j, then*

$$2TLNHM^{(x)}(\varphi_{a_1},\varphi_{a_2},\cdots,\varphi_{a_n}) \leq 2TLNHM^{(x)}(\varphi_{b_1},\varphi_{b_2},\cdots,\varphi_{b_n}) \tag{17}$$

Proof: Let $\varphi_{a_j} = \left\langle \left(s_{T_{a_j}}, \alpha_{a_j} \right), \left(s_{I_{a_j}}, \beta_{a_j} \right), \left(s_{F_{a_j}}, \gamma_{a_j} \right) \right\rangle$ and $\varphi_{b_j} = \left\langle \left(s_{T_{b_j}}, \alpha_{b_j} \right), \left(s_{I_{b_j}}, \beta_{b_j} \right), \left(s_{F_{b_j}}, \gamma_{b_j} \right) \right\rangle$,
given that $\Delta^{-1}\left(s_{T_{a_j}}, \alpha_{a_j} \right) \leq \Delta^{-1}\left(s_{T_{b_j}}, \alpha_{b_j} \right)$, we can obtain

$$\prod_{j=1}^{x} \left(\frac{\Delta^{-1}\left(s_{T_{a_j}}, \alpha_{a_j} \right)}{t} \right) \leq \prod_{j=1}^{x} \left(\frac{\Delta^{-1}\left(s_{T_{b_j}}, \alpha_{b_j} \right)}{t} \right), \tag{18}$$

$$1 - \left(\prod_{j=1}^{x} \left(\frac{\Delta^{-1}\left(s_{T_{a_j}}, \alpha_{a_j} \right)}{t} \right) \right)^{\frac{1}{x}} \geq 1 - \left(\prod_{j=1}^{x} \left(\frac{\Delta^{-1}\left(s_{T_{b_j}}, \alpha_{b_j} \right)}{t} \right) \right)^{\frac{1}{x}}. \tag{19}$$

Thereafter,

$$\left(\prod_{1 \leq i_1 < \ldots < i_x \leq n} \left(1 - \left(\prod_{j=1}^{x} \left(\frac{\Delta^{-1}\left(s_{T_{a_j}}, \alpha_{a_j} \right)}{t} \right) \right)^{\frac{1}{x}} \right) \right)^{\frac{1}{C_n^x}} \geq \left(\prod_{1 \leq i_1 < \ldots < i_x \leq n} \left(1 - \left(\prod_{j=1}^{x} \left(\frac{\Delta^{-1}\left(s_{T_{b_j}}, \alpha_{b_j} \right)}{t} \right) \right)^{\frac{1}{x}} \right) \right)^{\frac{1}{C_n^x}}. \tag{20}$$

Furthermore,

$$1 - \left(\prod_{1 \leq i_1 < \ldots < i_x \leq n} \left(1 - \left(\prod_{j=1}^{x} \left(\frac{\Delta^{-1}\left(s_{T_{a_j}}, \alpha_{a_j} \right)}{t} \right) \right)^{\frac{1}{x}} \right) \right)^{\frac{1}{C_n^x}} \leq 1 - \left(\prod_{1 \leq i_1 < \ldots < i_x \leq n} \left(1 - \left(\prod_{j=1}^{x} \left(\frac{\Delta^{-1}\left(s_{T_{b_j}}, \alpha_{b_j} \right)}{t} \right) \right)^{\frac{1}{x}} \right) \right)^{\frac{1}{C_n^x}}. \tag{21}$$

That means $\Delta^{-1}\left(s_{T_a}, \alpha_a \right) \leq \Delta^{-1}\left(s_{T_b}, \alpha_b \right)$. Similarly, we can obtain $\Delta^{-1}\left(s_{I_a}, \beta_a \right) \geq \Delta^{-1}\left(s_{I_b}, \beta_b \right)$ and $\Delta^{-1}\left(s_{F_a}, \gamma_a \right) \geq \Delta^{-1}\left(s_{F_b}, \gamma_b \right)$.

If $\Delta^{-1}\left(s_{T_a}, \alpha_a \right) < \Delta^{-1}\left(s_{T_b}, \alpha_b \right), \Delta^{-1}\left(s_{I_a}, \beta_a \right) \geq \Delta^{-1}\left(s_{I_b}, \beta_b \right)$ and $\Delta^{-1}\left(s_{F_a}, \gamma_a \right) \geq \Delta^{-1}\left(s_{F_b}, \gamma_b \right)$,
$2TLNHM^{(x)}(\varphi_a, \varphi_a, \cdots, \varphi_a) < 2TLNHM^{(x)}(\varphi_b, \varphi_b, \cdots, \varphi_b)$
If $\Delta^{-1}\left(s_{T_a}, \alpha_a \right) = \Delta^{-1}\left(s_{T_b}, \alpha_b \right), \Delta^{-1}\left(s_{I_a}, \beta_a \right) > \Delta^{-1}\left(s_{I_b}, \beta_b \right)$ and $\Delta^{-1}\left(s_{F_a}, \gamma_a \right) > \Delta^{-1}\left(s_{F_b}, \gamma_b \right)$,
$2TLNHM^{(x)}(\varphi_a, \varphi_a, \cdots, \varphi_a) < 2TLNHM^{(x)}(\varphi_b, \varphi_b, \cdots, \varphi_b)$
If $\Delta^{-1}\left(s_{T_a}, \alpha_a \right) = \Delta^{-1}\left(s_{T_b}, \alpha_b \right), \Delta^{-1}\left(s_{I_a}, \beta_a \right) = \Delta^{-1}\left(s_{I_b}, \beta_b \right)$ and $\Delta^{-1}\left(s_{F_a}, \gamma_a \right) = \Delta^{-1}\left(s_{F_b}, \gamma_b \right)$,
$2TLNHM^{(x)}(\varphi_a, \varphi_a, \cdots, \varphi_a) = 2TLNHM^{(x)}(\varphi_b, \varphi_b, \cdots, \varphi_b)$
So, Property 2 is right. $\square$

Property 3. *(Boundedness) Let* $\varphi_j = \left\langle \left(s_{T_j}, \alpha_j \right), \left(s_{I_j}, \beta_j \right), \left(s_{F_j}, \gamma_j \right) \right\rangle (j = 1, 2, \ldots, n)$ *be a set of 2TLNNs. If* $\varphi_i^+ = \left(max_i \left(s_{T_j}, \alpha_j \right), min_i \left(s_{I_j}, \beta_j \right), min_i \left(s_{F_j}, \gamma_j \right) \right)$ *and* $\varphi_i^+ = \left(max_i \left(s_{T_j}, \alpha_j \right), min_i \left(s_{I_j}, \beta_j \right), min_i \left(s_{F_j}, \gamma_j \right) \right)$, *then*

$$\varphi^- \leq 2TLNHM^{(x)}(\varphi_1, \varphi_2, \cdots, \varphi_n) \leq \varphi^+. \tag{22}$$

From Property 1,

$$2TLNHM^{(x)}\left(\varphi_1^-, \varphi_2^-, \cdots, \varphi_n^- \right) = \varphi^-$$
$$2TLNHM^{(x)}\left(\varphi_1^+, \varphi_2^+, \cdots, \varphi_n^+ \right) = \varphi^+ .$$

From Property 2,

$$\varphi^- \leq 2TLNHM^{(x)}(\varphi_1, \varphi_2, \cdots, \varphi_n) \leq \varphi^+.$$

3.2. The 2TLNWHM Operator

In an actual MADM, it is important to consider attribute weights. This section proposes a 2-tuple linguistic neutrosophic weighted Hamy mean (2TLNWHM) operator as follows.

Definition 8. *Let* $\varphi_j = \left\langle \left(s_{T_j}, \alpha_j\right), \left(s_{I_j}, \beta_j\right), \left(s_{F_j}, \gamma_j\right) \right\rangle (j = 1, 2, \ldots, n)$ *be a set of 2TLNNs with a weight vector,* $w_i = (w_1, w_2, \ldots, w_n)^T$, *thereby satisfying* $w_i \in [0, 1]$ *and* $\sum_{i=1}^n w_i = 1$. *Then, we can define the 2TLNWHM operator as follows:*

$$2\text{TLNWHM}_w^{(x)}(\varphi_1, \varphi_2, \ldots, \varphi_n) = \frac{\overset{\oplus}{\underset{1 \leq i_1 < \ldots < i_x \leq n}{}} \left(\overset{x}{\underset{j=1}{\otimes}} \left(\varphi_{i_j}\right)^{w_{i_j}} \right)^{\frac{1}{x}}}{C_n^x}. \tag{23}$$

Theorem 2. *Let* $\varphi_j = \left\langle \left(s_{T_j}, \alpha_j\right), \left(s_{I_j}, \beta_j\right), \left(s_{F_j}, \gamma_j\right) \right\rangle (j = 1, 2, \ldots, n)$ *be a set of 2TLNNs. The aggregated value determined using a 2TLNWHM operator is also a 2TLNN, where*

$$2\text{TLNWHM}_w^{(x)}(\varphi_1, \varphi_2, \ldots, \varphi_n) = \frac{\overset{\oplus}{\underset{1 \leq i_1 < \ldots < i_x \leq n}{}} \left(\overset{x}{\underset{j=1}{\otimes}} \left(\varphi_{i_j}\right)^{w_{i_j}} \right)^{\frac{1}{x}}}{C_n^x}$$

$$= \left\{ \begin{array}{l} \Delta\left(t\left(1 - \left(\displaystyle\prod_{1 \leq i_1 < \ldots < i_x \leq n}\left(1 - \left(\displaystyle\prod_{j=1}^x \left(\frac{\Delta^{-1}\left(s_{T_j}, \alpha_j\right)}{t}\right)^{w_{i_j}}\right)^{\frac{1}{x}}\right)\right)^{\frac{1}{C_n^x}}\right)\right), \\[4ex] \Delta\left(t\left(\displaystyle\prod_{1 \leq i_1 < \ldots < i_x \leq n}\left(1 - \left(\displaystyle\prod_{j=1}^x \left(1 - \frac{\Delta^{-1}\left(s_{I_j}, \beta_j\right)}{t}\right)^{w_{i_j}}\right)^{\frac{1}{x}}\right)\right)^{\frac{1}{C_n^x}}\right), \\[4ex] \Delta\left(t\left(\displaystyle\prod_{1 \leq i_1 < \ldots < i_x \leq n}\left(1 - \left(\displaystyle\prod_{j=1}^x \left(1 - \frac{\Delta^{-1}\left(s_{F_j}, \gamma_j\right)}{t}\right)^{w_{i_j}}\right)^{\frac{1}{x}}\right)\right)^{\frac{1}{C_n^x}}\right) \end{array} \right\}. \tag{24}$$

Proof: From Definition 5, we can obtain,

$$\left(\varphi_{i_j}\right)^{w_{i_j}} = \left\{ \begin{array}{l} \Delta\left(t\left(\frac{\Delta^{-1}\left(s_{T_j}, \alpha_j\right)}{t}\right)^{w_{i_j}}\right), \Delta\left(t\left(1 - \left(1 - \frac{\Delta^{-1}\left(s_{I_j}, \beta_j\right)}{t}\right)^{w_{i_j}}\right)\right), \\[3ex] \Delta\left(t\left(1 - \left(1 - \frac{\Delta^{-1}\left(s_{F_j}, \gamma_j\right)}{t}\right)^{w_{i_j}}\right)\right) \end{array} \right\} \tag{25}$$

Thus,

$$\overset{x}{\underset{j=1}{\otimes}} \left(\varphi_{i_j}\right)^{w_{i_j}} = \left\{ \begin{array}{l} \Delta\left(t\left(\displaystyle\prod_{j=1}^x \left(\frac{\Delta^{-1}\left(s_{T_j}, \alpha_j\right)}{t}\right)^{w_{i_j}}\right)\right), \Delta\left(t\left(1 - \displaystyle\prod_{j=1}^x \left(1 - \frac{\Delta^{-1}\left(s_{I_j}, \beta_j\right)}{t}\right)^{w_{i_j}}\right)\right), \\[3ex] \Delta\left(t\left(1 - \displaystyle\prod_{j=1}^x \left(1 - \frac{\Delta^{-1}\left(s_{F_j}, \gamma_j\right)}{t}\right)^{w_{i_j}}\right)\right) \end{array} \right\} \tag{26}$$

Therefore,

$$\left(\overset{x}{\underset{j=1}{\otimes}} \left(\varphi_{i_j}\right)^{w_{i_j}}\right)^{\frac{1}{x}} = \left\{ \begin{array}{l} \Delta\left(t\left(\displaystyle\prod_{j=1}^x \left(\frac{\Delta^{-1}\left(s_{T_j}, \alpha_j\right)}{t}\right)^{w_{i_j}}\right)^{\frac{1}{x}}\right), \Delta\left(t\left(1 - \left(\displaystyle\prod_{j=1}^x \left(1 - \frac{\Delta^{-1}\left(s_{I_j}, \beta_j\right)}{t}\right)^{w_{i_j}}\right)^{\frac{1}{x}}\right)\right), \\[3ex] \Delta\left(t\left(1 - \left(\displaystyle\prod_{j=1}^x \left(1 - \frac{\Delta^{-1}\left(s_{F_j}, \gamma_j\right)}{t}\right)^{w_{i_j}}\right)^{\frac{1}{x}}\right)\right) \end{array} \right\}. \tag{27}$$

Thereafter,

$$
\bigoplus_{1\le i_1 <...<i_x \le n} \left(\bigotimes_{j=1}^{x} \left(\varphi_{i_j}\right)^{w_{i_j}} \right)^{\frac{1}{x}} =
\left\{
\begin{array}{l}
\Delta\left(t\left(1 - \prod_{1\le i_1<...<i_x\le n} \left(1 - \left(\prod_{j=1}^{x} \left(\frac{\Delta^{-1}\left(s_{T_j},\alpha_j\right)}{t} \right)^{w_{i_j}} \right)^{\frac{1}{x}} \right) \right) \right), \\[4mm]
\Delta\left(t\left(\prod_{1\le i_1<...<i_x\le n} \left(1 - \left(\prod_{j=1}^{x} \left(1 - \frac{\Delta^{-1}\left(s_{I_j},\beta_j\right)}{t} \right)^{w_{i_j}} \right)^{\frac{1}{x}} \right) \right) \right), \\[4mm]
\Delta\left(t\left(\prod_{1\le i_1<...<i_x\le n} \left(1 - \left(\prod_{j=1}^{x} \left(1 - \frac{\Delta^{-1}\left(s_{F_j},\gamma_j\right)}{t} \right)^{w_{i_j}} \right)^{\frac{1}{x}} \right) \right) \right)
\end{array}
\right\}
\tag{28}
$$

Furthermore,

$$
2\text{TLNWHM}_w^{(x)}(\varphi_1,\varphi_2,\ldots,\varphi_n) = \frac{\bigoplus_{1\le i_1 <...<i_x \le n} \left(\bigotimes_{j=1}^{x} \left(\varphi_{i_j}\right)^{w_{i_j}} \right)^{\frac{1}{x}}}{C_n^x}.
$$

$$
= \left\{
\begin{array}{l}
\Delta\left(t\left(1 - \left(\prod_{1\le i_1<...<i_x\le n} \left(1 - \left(\prod_{j=1}^{x} \left(\frac{\Delta^{-1}\left(s_{T_j},\alpha_j\right)}{t} \right)^{w_{i_j}} \right)^{\frac{1}{x}} \right) \right)^{\frac{1}{C_n^x}} \right) \right), \\[4mm]
\Delta\left(t\left(\prod_{1\le i_1<...<i_x\le n} \left(1 - \left(\prod_{j=1}^{x} \left(1 - \frac{\Delta^{-1}\left(s_{I_j},\beta_j\right)}{t} \right)^{w_{i_j}} \right)^{\frac{1}{x}} \right) \right)^{\frac{1}{C_n^x}} \right), \\[4mm]
\Delta\left(t\left(\prod_{1\le i_1<...<i_x\le n} \left(1 - \left(\prod_{j=1}^{x} \left(1 - \frac{\Delta^{-1}\left(s_{F_j},\gamma_j\right)}{t} \right)^{w_{i_j}} \right)^{\frac{1}{x}} \right) \right)^{\frac{1}{C_n^x}} \right)
\end{array}
\right\}.
\tag{29}
$$

Hence, (23) is kept.

Then we need to prove that (23) is a 2TLNN. We need to prove two conditions as follows:

① $\quad 0 \le \Delta^{-1}(s_T,\alpha) \le t, 0 \le \Delta^{-1}(s_I,\beta) \le t, 0 \le \Delta^{-1}(s_F,\gamma) \le t.$

② $\quad 0 \le \Delta^{-1}(s_T,\alpha) + \Delta^{-1}(s_I,\beta) + \Delta^{-1}(s_F,\gamma) \le 3t.$

Let

$$
\frac{\Delta^{-1}(s_T,\alpha)}{t} = 1 - \left(\prod_{1\le i_1<...<i_x\le n} \left(1 - \left(\prod_{j=1}^{x} \left(\frac{\Delta^{-1}\left(s_{T_j},\alpha_j\right)}{t} \right)^{w_{i_j}} \right)^{\frac{1}{x}} \right) \right)^{\frac{1}{C_n^x}}
$$

$$
\frac{\Delta^{-1}(s_I,\beta)}{t} = \left(\prod_{1\le i_1<...<i_x\le n} \left(1 - \left(\prod_{j=1}^{x} \left(1 - \frac{\Delta^{-1}\left(s_{I_j},\beta_j\right)}{t} \right)^{w_{i_j}} \right)^{\frac{1}{x}} \right) \right)^{\frac{1}{C_n^x}}
$$

$$
\frac{\Delta^{-1}(s_F,\gamma)}{t} = \left(\prod_{1\le i_1<...<i_x\le n} \left(1 - \left(\prod_{j=1}^{x} \left(1 - \frac{\Delta^{-1}\left(s_{F_j},\gamma_j\right)}{t} \right)^{w_{i_j}} \right)^{\frac{1}{x}} \right) \right)^{\frac{1}{C_n^x}}.
$$

□

Proof. ① Since $0 \le \dfrac{\Delta^{-1}\left(s_{T_j}, \alpha_j\right)}{t} \le 1$, we get

$$0 \le \prod_{j=1}^{x}\left(\frac{\Delta^{-1}\left(s_{T_j}, \alpha_j\right)}{t}\right)^{w_{ij}} \le 1 \text{ and } 0 \le 1 - \left(\prod_{j=1}^{x}\left(\frac{\Delta^{-1}\left(s_{T_j}, \alpha_j\right)}{t}\right)^{w_{ij}}\right)^{\frac{1}{x}} \le 1 \tag{30}$$

Then,

$$0 \le \prod_{1 \le i_1 < \ldots < i_x \le n}\left(1 - \left(\prod_{j=1}^{x}\left(\frac{\Delta^{-1}\left(s_{T_j}, \alpha_j\right)}{t}\right)^{w_{ij}}\right)^{\frac{1}{x}}\right) \le 1 \tag{31}$$

$$0 \le 1 - \left(\prod_{1 \le i_1 < \ldots < i_x \le n}\left(1 - \left(\prod_{j=1}^{x}\left(\frac{\Delta^{-1}\left(s_{T_j}, \alpha_j\right)}{t}\right)^{w_{ij}}\right)^{\frac{1}{x}}\right)\right)^{\frac{1}{C_n^x}} \le 1 \tag{32}$$

That means $0 \le \Delta^{-1}(s_T, \alpha) \le t$. Similarly, we can get $0 \le \Delta^{-1}(s_I, \beta) \le t, 0 \le \Delta^{-1}(s_F, \gamma) \le t$. So, ① is maintained; ② Since $0 \le \Delta^{-1}(s_T, \alpha) \le t, 0 \le \Delta^{-1}(s_I, \beta) \le t, 0 \le \Delta^{-1}(s_F, \gamma) \le t$. $0 \le \Delta^{-1}(s_T, \alpha) + \Delta^{-1}(s_I, \beta) + \Delta^{-1}(s_F, \gamma) \le 3t$. □

Example 2. *Let* $\langle (s_5, 0), (s_2, 0), (s_1, 0) \rangle, \langle (s_4, 0), (s_3, 0), (s_4, 0) \rangle, \langle (s_2, 0), (s_5, 0), (s_1, 0) \rangle$ *and* $\langle (s_5, 0), (s_1, 0), (s_3, 0) \rangle$ *be four 2TLNNs,* $w = (0.2, 0.3, 0.4, 0.1)$ *and suppose* $x = 2$, *then according to (23), we have*

$$2\text{TLNHM}^{(2)}(\varphi_1, \varphi_2, \cdots, \varphi_n) = \frac{\overset{\oplus}{\underset{1 \le i_1 < \ldots < i_x \le n}{\bigoplus}}\left(\overset{x}{\underset{j=1}{\otimes}}\left(\varphi_{ij}\right)^{w_{ij}}\right)^{\frac{1}{x}}}{C_n^x}$$

$$= \left\{ \begin{array}{l}
\Delta\left\{6 \times \left\{1 - \left\{\begin{array}{l} \left(1 - \left(\left(\frac{5}{6}\right)^{0.2} \times \left(\frac{4}{6}\right)^{0.3}\right)^{\frac{1}{2}}\right) \times \left(1 - \left(\left(\frac{5}{6}\right)^{0.2} \times \left(\frac{2}{6}\right)^{0.4}\right)^{\frac{1}{2}}\right) \times \left(1 - \left(\left(\frac{5}{6}\right)^{0.2} \times \left(\frac{5}{6}\right)^{0.1}\right)^{\frac{1}{2}}\right) \\ \times \left(1 - \left(\left(\frac{4}{6}\right)^{0.3} \times \left(\frac{2}{6}\right)^{0.4}\right)^{\frac{1}{2}}\right) \times \left(1 - \left(\left(\frac{4}{6}\right)^{0.3} \times \left(\frac{5}{6}\right)^{0.1}\right)^{\frac{1}{2}}\right) \times \left(1 - \left(\left(\frac{2}{6}\right)^{0.4} \times \left(\frac{5}{6}\right)^{0.1}\right)^{\frac{1}{2}}\right) \end{array}\right\}^{\frac{1}{C_4^2}}\right\}\right\}, \\
\\
\Delta\left\{6 \times \left\{\begin{array}{l} \left(1 - \left(\left(1 - \left(\frac{2}{6}\right)^{0.2}\right) \times \left(1 - \left(\frac{3}{6}\right)^{0.3}\right)\right)^{\frac{1}{2}}\right) \times \left(1 - \left(\left(1 - \left(\frac{2}{6}\right)^{0.2}\right) \times \left(1 - \left(\frac{5}{6}\right)^{0.4}\right)\right)^{\frac{1}{2}}\right) \\ \times \left(1 - \left(\left(1 - \left(\frac{2}{6}\right)^{0.2}\right) \times \left(1 - \left(\frac{1}{6}\right)^{0.1}\right)\right)^{\frac{1}{2}}\right) \times \left(1 - \left(\left(1 - \left(\frac{3}{6}\right)^{0.3}\right) \times \left(1 - \left(\frac{5}{6}\right)^{0.4}\right)\right)^{\frac{1}{2}}\right) \\ \times \left(1 - \left(\left(1 - \left(\frac{3}{6}\right)^{0.3}\right) \times \left(1 - \left(\frac{1}{6}\right)^{0.1}\right)\right)^{\frac{1}{2}}\right) \times \left(1 - \left(\left(1 - \left(\frac{5}{6}\right)^{0.4}\right) \times \left(1 - \left(\frac{1}{6}\right)^{0.1}\right)\right)^{\frac{1}{2}}\right) \end{array}\right\}^{\frac{1}{C_4^2}}\right\}, \\
\\
\Delta\left\{6 \times \left\{\begin{array}{l} \left(1 - \left(\left(1 - \left(\frac{1}{6}\right)^{0.2}\right) \times \left(1 - \left(\frac{4}{6}\right)^{0.3}\right)\right)^{\frac{1}{2}}\right) \times \left(1 - \left(\left(1 - \left(\frac{1}{6}\right)^{0.2}\right) \times \left(1 - \left(\frac{1}{6}\right)^{0.4}\right)\right)^{\frac{1}{2}}\right) \\ \times \left(1 - \left(\left(1 - \left(\frac{1}{6}\right)^{0.2}\right) \times \left(1 - \left(\frac{3}{6}\right)^{0.1}\right)\right)^{\frac{1}{2}}\right) \times \left(1 - \left(\left(1 - \left(\frac{4}{6}\right)^{0.3}\right) \times \left(1 - \left(\frac{1}{6}\right)^{0.4}\right)\right)^{\frac{1}{2}}\right) \\ \times \left(1 - \left(\left(1 - \left(\frac{4}{6}\right)^{0.3}\right) \times \left(1 - \left(\frac{3}{6}\right)^{0.1}\right)\right)^{\frac{1}{2}}\right) \times \left(1 - \left(\left(1 - \left(\frac{1}{6}\right)^{0.4}\right) \times \left(1 - \left(\frac{3}{6}\right)^{0.1}\right)\right)^{\frac{1}{2}}\right) \end{array}\right\}^{\frac{1}{C_4^2}}\right\}
\end{array} \right\}.$$

$$= \langle (s_5, 0.3604), (s_1, -0.0344), (s_1, -0.3963) \rangle$$

Now, we will discuss some properties of the 2TLNWHM operator.

Property 4. *(Monotonicity) Let* $\varphi_{a_j} = \left\langle \left(s_{T_{a_j}}, \alpha_{a_j}\right), \left(s_{I_{a_j}}, \beta_{a_j}\right), \left(s_{F_{a_j}}, \gamma_{a_j}\right) \right\rangle (j = 1, 2, \ldots, n)$ *and* $\varphi_{b_j} = \left\langle \left(s_{T_{b_j}}, \alpha_{b_j}\right), \left(s_{I_{b_j}}, \beta_{b_j}\right), \left(s_{F_{b_j}}, \gamma_{b_j}\right) \right\rangle (j = 1, 2, \ldots, n)$ *be two sets of 2TLNNs. If* $\Delta^{-1}\left(s_{T_{a_j}}, \alpha_{a_j}\right) \leq \Delta^{-1}\left(s_{T_{b_j}}, \alpha_{b_j}\right), \Delta^{-1}\left(s_{I_{a_j}}, \beta_{a_j}\right) \geq \Delta^{-1}\left(s_{I_{b_j}}, \beta_{b_j}\right)$ *and* $\Delta^{-1}\left(s_{F_{a_j}}, \gamma_{a_j}\right) \geq \Delta^{-1}\left(s_{F_{b_j}}, \gamma_{b_j}\right)$ *hold for all j, then*

$$2\text{TLNWHM}^{(x)}(\varphi_{a_1}, \varphi_{a_2}, \cdots, \varphi_{a_n}) \leq 2\text{TLNWHM}^{(x)}(\varphi_{b_1}, \varphi_{b_2}, \cdots, \varphi_{b_n}) \tag{33}$$

The proof is similar to 2TLNWHM; it is omitted here.

Property 5. *(Boundedness) Let* $\varphi_j = \left\langle \left(s_{T_j}, \alpha_j\right), \left(s_{I_j}, \beta_j\right), \left(s_{F_j}, \gamma_j\right) \right\rangle (j = 1, 2, \ldots, n)$ *be a set of 2TLNNs. If* $\varphi_i^+ = \left(\max_i\left(s_{T_j}, \alpha_j\right), \min_i\left(s_{I_j}, \beta_j\right), \min_i\left(s_{F_j}, \gamma_j\right)\right)$ *and* $\varphi_i^+ = \left(\max_i\left(s_{T_j}, \alpha_j\right), \min_i\left(s_{I_j}, \beta_j\right), \min_i\left(s_{F_j}, \gamma_j\right)\right)$, *then*

$$\varphi^- \leq 2\text{TLNWHM}^{(x)}(\varphi_1, \varphi_2, \cdots, \varphi_n) \leq \varphi^+. \tag{34}$$

From theorem 2, we get

$$2\text{TLNWHM}_w^{(x)}\left((\varphi_1^-, \varphi_2^-, \cdots, \varphi_n^-)\right) = \frac{\overset{\oplus}{\underset{1 \leq i_1 < \ldots < i_x \leq n}{}}\left(\overset{x}{\underset{j=1}{\otimes}}\left(\min \varphi_{i_j}\right)^{w_{i_j}}\right)^{\frac{1}{x}}}{C_n^x}$$

$$= \left\{ \begin{array}{l} \Delta\left(t\left(1 - \left(\prod_{1 \leq i_1 < \ldots < i_x \leq n}\left(1 - \left(\prod_{j=1}^{x}\left(\frac{\min \Delta^{-1}\left(s_{T_j}, \alpha_j\right)}{t}\right)^{w_{i_j}}\right)^{\frac{1}{x}}\right)\right)^{\frac{1}{C_n^x}}\right)\right), \\[4mm] \Delta\left(t\left(\prod_{1 \leq i_1 < \ldots < i_x \leq n}\left(1 - \left(\prod_{j=1}^{x}\left(1 - \frac{\max \Delta^{-1}\left(s_{I_j}, \beta_j\right)}{t}\right)^{w_{i_j}}\right)^{\frac{1}{x}}\right)\right)^{\frac{1}{C_n^x}}\right), \\[4mm] \Delta\left(t\left(\prod_{1 \leq i_1 < \ldots < i_x \leq n}\left(1 - \left(\prod_{j=1}^{x}\left(1 - \frac{\max \Delta^{-1}\left(s_{F_j}, \gamma_j\right)}{t}\right)^{w_{i_j}}\right)^{\frac{1}{x}}\right)\right)^{\frac{1}{C_n^x}}\right) \end{array} \right\}, \tag{35}$$

$$2\text{TLNWHM}_w^{(x)}\left((\varphi_1^+, \varphi_2^+, \cdots, \varphi_n^+)\right) = \frac{\overset{\oplus}{\underset{1 \leq i_1 < \ldots < i_x \leq n}{}}\left(\overset{x}{\underset{j=1}{\otimes}}\left(\max \varphi_{i_j}\right)^{w_{i_j}}\right)^{\frac{1}{x}}}{C_n^x}$$

$$= \left\{ \begin{array}{l} \Delta\left(t\left(1 - \left(\prod_{1 \leq i_1 < \ldots < i_x \leq n}\left(1 - \left(\prod_{j=1}^{x}\left(\frac{\max \Delta^{-1}\left(s_{T_j}, \alpha_j\right)}{t}\right)^{w_{i_j}}\right)^{\frac{1}{x}}\right)\right)^{\frac{1}{C_n^x}}\right)\right), \\[4mm] \Delta\left(t\left(\prod_{1 \leq i_1 < \ldots < i_x \leq n}\left(1 - \left(\prod_{j=1}^{x}\left(1 - \frac{\min \Delta^{-1}\left(s_{I_j}, \beta_j\right)}{t}\right)^{w_{i_j}}\right)^{\frac{1}{x}}\right)\right)^{\frac{1}{C_n^x}}\right), \\[4mm] \Delta\left(t\left(\prod_{1 \leq i_1 < \ldots < i_x \leq n}\left(1 - \left(\prod_{j=1}^{x}\left(1 - \frac{\min \Delta^{-1}\left(s_{F_j}, \gamma_j\right)}{t}\right)^{w_{i_j}}\right)^{\frac{1}{x}}\right)\right)^{\frac{1}{C_n^x}}\right) \end{array} \right\}. \tag{36}$$

From property 4, we get

$$\varphi^- \le 2\text{TLNWHM}^{(x)}(\varphi_1, \varphi_2, \cdots, \varphi_n) \le \varphi^+ \tag{37}$$

It is obvious that the 2TLNWHM operator lacks the property of idempotency.

3.3. The 2TLNDHM Operator

Based on the Hamy mean (HM) operator [32], we propose the dual Hamy mean (DHM) operator.

Definition 9. *The DHM operator is defined as follows:*

$$\text{DHM}^{(x)}(\varphi_1, \varphi_2, \cdots, \varphi_n) = \left(\prod_{1 \le i_1 < \ldots < i_x \le n} \left(\frac{\sum\limits_{j=1}^{x} \varphi_{i_j}}{x} \right) \right)^{\frac{1}{C_n^x}}. \tag{38}$$

where x is a parameter and $x = 1, 2, \ldots, n$, $i_1, i_2, \ldots, i_x$ are x integer values taken from the set $\{1, 2, \ldots, n\}$ of k integer values, C_n^x denotes the binomial coefficient and $C_n^x = \frac{n!}{x!(n-x)!}$.

In this section, we propose the 2-tuple linguistic neutrosophic DHM (2TLNDHM) operator.

Definition 10. *Let $\varphi_j = \left\langle \left(s_{T_j}, \alpha_j \right), \left(s_{I_j}, \beta_j \right), \left(s_{F_j}, \gamma_j \right) \right\rangle (j = 1, 2, \ldots, n)$ be a set of 2TLNNs. The 2TLNDHM operator is:*

$$2\text{TLNDHM}^{(x)}(\varphi_1, \varphi_2, \cdots, \varphi_n) = \left(\bigotimes_{1 \le i_1 < \ldots < i_x \le n} \left(\frac{\bigoplus\limits_{j=1}^{x} \varphi_{i_j}}{x} \right) \right)^{\frac{1}{C_n^x}}. \tag{39}$$

Theorem 3. *Let $\varphi_j = \left\langle \left(s_{T_j}, \alpha_j \right), \left(s_{I_j}, \beta_j \right), \left(s_{F_j}, \gamma_j \right) \right\rangle (j = 1, 2, \ldots, n)$ be a set of 2TLNNs. The aggregated value determined using 2TLNDHM operators is also a 2TLNN where*

$$2\text{TLNDHM}^{(x)}(\varphi_1, \varphi_2, \cdots, \varphi_n) = \left(\bigotimes_{1 \le i_1 < \ldots < i_x \le n} \left(\frac{\bigoplus\limits_{j=1}^{x} \varphi_{i_j}}{x} \right) \right)^{\frac{1}{C_n^x}}$$

$$= \left\{ \begin{array}{l} \Delta \left(t \left(\prod\limits_{1 \le i_1 < \ldots < i_x \le n} \left(1 - \left(\prod\limits_{j=1}^{x} \left(1 - \frac{\Delta^{-1}\left(s_{T_j}, \alpha_j \right)}{t} \right) \right)^{\frac{1}{x}} \right) \right)^{\frac{1}{C_n^x}} \right), \\[3mm] \Delta \left(t \left(1 - \left(\prod\limits_{1 \le i_1 < \ldots < i_x \le n} \left(1 - \left(\prod\limits_{j=1}^{x} \left(\frac{\Delta^{-1}\left(s_{I_j}, \beta_j \right)}{t} \right) \right)^{\frac{1}{x}} \right) \right)^{\frac{1}{C_n^x}} \right) \right), \\[3mm] \Delta \left(t \left(1 - \left(\prod\limits_{1 \le i_1 < \ldots < i_x \le n} \left(1 - \left(\prod\limits_{j=1}^{x} \left(\frac{\Delta^{-1}\left(s_{F_j}, \gamma_j \right)}{t} \right) \right)^{\frac{1}{x}} \right) \right)^{\frac{1}{C_n^x}} \right) \right) \end{array} \right\} \tag{40}$$

Proof:

$$
\overset{x}{\underset{j=1}{\oplus}} \varphi_{ij} =
\left\{
\begin{array}{l}
\Delta\left(t\left(1 - \prod\limits_{j=1}^{x}\left(1 - \frac{\Delta^{-1}\left(s_{T_j},\alpha_j\right)}{t}\right)\right)\right), \Delta\left(t\left(\prod\limits_{j=1}^{x}\left(\frac{\Delta^{-1}\left(s_{I_j},\beta_j\right)}{t}\right)\right)\right), \\[4mm]
\Delta\left(t\left(\prod\limits_{j=1}^{x}\left(\frac{\Delta^{-1}\left(s_{F_j},\gamma_j\right)}{t}\right)\right)\right)
\end{array}
\right\}
\tag{41}
$$

Thus,

$$
\frac{\overset{x}{\underset{j=1}{\oplus}} \varphi_{ij}}{x} =
\left\{
\begin{array}{l}
\Delta\left(t\left(1 - \left(\prod\limits_{j=1}^{x}\left(1 - \frac{\Delta^{-1}\left(s_{T_j},\alpha_j\right)}{t}\right)\right)^{\frac{1}{x}}\right)\right), \Delta\left(t\left(\prod\limits_{j=1}^{x}\left(\frac{\Delta^{-1}\left(s_{I_j},\beta_j\right)}{t}\right)\right)^{\frac{1}{x}}\right), \\[4mm]
\Delta\left(t\left(\prod\limits_{j=1}^{x}\left(\frac{\Delta^{-1}\left(s_{F_j},\gamma_j\right)}{t}\right)\right)^{\frac{1}{x}}\right)
\end{array}
\right\}
\tag{42}
$$

Thereafter,

$$
\underset{1\leq i_1<\ldots<i_x\leq n}{\otimes}\left(\frac{\overset{x}{\underset{j=1}{\oplus}} \varphi_{ij}}{x}\right) =
\left\{
\begin{array}{l}
\Delta\left(t\left(\prod\limits_{1\leq i_1<\ldots<i_x\leq n}\left(1 - \left(\prod\limits_{j=1}^{x}\left(1 - \frac{\Delta^{-1}\left(s_{T_j},\alpha_j\right)}{t}\right)\right)^{\frac{1}{x}}\right)\right)\right), \\[4mm]
\Delta\left(t\left(1 - \prod\limits_{1\leq i_1<\ldots<i_x\leq n}\left(1 - \left(\prod\limits_{j=1}^{x}\left(\frac{\Delta^{-1}\left(s_{I_j},\beta_j\right)}{t}\right)\right)^{\frac{1}{x}}\right)\right)\right), \\[4mm]
\Delta\left(t\left(1 - \prod\limits_{1\leq i_1<\ldots<i_x\leq n}\left(1 - \left(\prod\limits_{j=1}^{x}\left(\frac{\Delta^{-1}\left(s_{F_j},\gamma_j\right)}{t}\right)\right)^{\frac{1}{x}}\right)\right)\right)
\end{array}
\right\}
\tag{43}
$$

Therefore,

$$
\text{2TLNDHM}^{(x)}(\varphi_1, \varphi_2, \cdots, \varphi_n) = \left(\underset{1\leq i_1<\ldots<i_x\leq n}{\otimes}\left(\frac{\overset{x}{\underset{j=1}{\oplus}} \varphi_{ij}}{x}\right)\right)^{\frac{1}{C_n^x}}
$$

$$
= \left\{
\begin{array}{l}
\Delta\left(t\left(\prod\limits_{1\leq i_1<\ldots<i_x\leq n}\left(1 - \left(\prod\limits_{j=1}^{x}\left(1 - \frac{\Delta^{-1}\left(s_{T_j},\alpha_j\right)}{t}\right)\right)^{\frac{1}{x}}\right)\right)^{\frac{1}{C_n^x}}\right), \\[4mm]
\Delta\left(t\left(1 - \left(\prod\limits_{1\leq i_1<\ldots<i_x\leq n}\left(1 - \left(\prod\limits_{j=1}^{x}\left(\frac{\Delta^{-1}\left(s_{I_j},\beta_j\right)}{t}\right)\right)^{\frac{1}{x}}\right)\right)^{\frac{1}{C_n^x}}\right)\right), \\[4mm]
\Delta\left(t\left(1 - \left(\prod\limits_{1\leq i_1<\ldots<i_x\leq n}\left(1 - \left(\prod\limits_{j=1}^{x}\left(\frac{\Delta^{-1}\left(s_{F_j},\gamma_j\right)}{t}\right)\right)^{\frac{1}{x}}\right)\right)^{\frac{1}{C_n^x}}\right)\right)
\end{array}
\right\}.
\tag{44}
$$

Hence, (39) is kept.

Then, we need to prove that (39) is a 2TLNN. We need to prove two conditions as follows:

① $0 \leq \Delta^{-1}(s_T, \alpha) \leq t, 0 \leq \Delta^{-1}(s_I, \beta) \leq t, 0 \leq \Delta^{-1}(s_F, \gamma) \leq t.$

② $0 \leq \Delta^{-1}(s_T, \alpha) + \Delta^{-1}(s_I, \beta) + \Delta^{-1}(s_F, \gamma) \leq 3t.$

Let

$$\frac{\Delta^{-1}(s_T,\alpha)}{t} = \left(\prod_{1\le i_1<...<i_x\le n} \left(1 - \left(\prod_{j=1}^{x} \left(1 - \frac{\Delta^{-1}\left(s_{T_j},\alpha_j\right)}{t} \right) \right)^{\frac{1}{x}} \right) \right)^{\frac{1}{C_n^x}}$$

$$\frac{\Delta^{-1}(s_I,\beta)}{t} = 1 - \left(\prod_{1\le i_1<...<i_x\le n} \left(1 - \left(\prod_{j=1}^{x} \left(\frac{\Delta^{-1}\left(s_{I_j},\beta_j\right)}{t} \right) \right)^{\frac{1}{x}} \right) \right)^{\frac{1}{C_n^x}}$$

$$\frac{\Delta^{-1}(s_F,\gamma)}{t} = 1 - \left(\prod_{1\le i_1<...<i_x\le n} \left(1 - \left(\prod_{j=1}^{x} \left(\frac{\Delta^{-1}\left(s_{F_j},\gamma_j\right)}{t} \right) \right)^{\frac{1}{x}} \right) \right)^{\frac{1}{C_n^x}}$$

□

Proof. ① Since $0 \le \dfrac{\Delta^{-1}\left(s_{T_j},\alpha_j\right)}{t} \le 1$, we get

$$0 \le \prod_{j=1}^{x} \left(1 - \frac{\Delta^{-1}\left(s_{T_j},\alpha_j\right)}{t} \right) \le 1 \text{ and } 0 \le 1 - \left(\prod_{j=1}^{x} \left(1 - \frac{\Delta^{-1}\left(s_{T_j},\alpha_j\right)}{t} \right) \right)^{\frac{1}{x}} \le 1. \qquad (45)$$

Then,

$$0 \le \prod_{1\le i_1<...<i_x\le n} \left(1 - \left(\prod_{j=1}^{x} \left(1 - \frac{\Delta^{-1}\left(s_{T_j},\alpha_j\right)}{t} \right) \right)^{\frac{1}{x}} \right) \le 1 \qquad (46)$$

$$0 \le \left(\prod_{1\le i_1<...<i_x\le n} \left(1 - \left(\prod_{j=1}^{x} \left(1 - \frac{\Delta^{-1}\left(s_{T_j},\alpha_j\right)}{t} \right) \right)^{\frac{1}{x}} \right) \right)^{\frac{1}{C_n^x}} \le 1 \qquad (47)$$

That means $0 \le \Delta^{-1}(s_T,\alpha) \le t$. Similarly, we can get $0 \le \Delta^{-1}(s_I,\beta) \le t, 0 \le \Delta^{-1}(s_F,\gamma) \le t$. So, ① is maintained. ② Since $0 \le \Delta^{-1}(s_T,\alpha) \le t, 0 \le \Delta^{-1}(s_I,\beta) \le t, 0 \le \Delta^{-1}(s_F,\gamma) \le t$. $0 \le \Delta^{-1}(s_T,\alpha) + \Delta^{-1}(s_I,\beta) + \Delta^{-1}(s_F,\gamma) \le 3t$. □

Example 3. Let $\langle(s_5,0),(s_2,0),(s_1,0)\rangle, \langle(s_4,0),(s_3,0),(s_4,0)\rangle, \langle(s_2,0),(s_5,0),(s_1,0)\rangle$ and $\langle(s_5,0),(s_1,0),(s_3,0)\rangle$ be four 2TLNNs, and suppose $x = 2$, then according to (39), we have

$$2TLNDHM^{(2)}(\varphi_1,\varphi_2,\cdots,\varphi_n) = \left(\bigotimes_{1\le i_1<...<i_x\le n} \left(\frac{\bigoplus_{j=1}^{x}\varphi_{i_j}}{x} \right) \right)^{\frac{1}{C_n^x}}$$

$$= \left\{ \begin{array}{l} \Delta\left(6 \times \left(\begin{array}{l} \left(1 - \left((1-\frac{5}{6})\times(1-\frac{4}{6})\right)^{\frac{1}{2}}\right) \times \left(1 - ((1-\frac{5}{6})\times(1-\frac{2}{6}))^{\frac{1}{2}}\right) \times \left(1 - ((1-\frac{5}{6})\times(1-\frac{5}{6}))^{\frac{1}{2}}\right) \\ \times \left(1 - \left((1-\frac{4}{6})\times(1-\frac{2}{6})\right)^{\frac{1}{2}}\right) \times \left(1 - \left((1-\frac{4}{6})\times(1-\frac{5}{6})\right)^{\frac{1}{2}}\right) \times \left(1 - ((1-\frac{2}{6})\times(1-\frac{5}{6}))^{\frac{1}{2}}\right) \end{array} \right)^{\frac{1}{C_4^2}} \right), \\ \Delta\left(6 \times \left(1 - \left(\begin{array}{l} \left(1 - (\frac{2}{6}\times\frac{3}{6})^{\frac{1}{2}}\right) \times \left(1 - (\frac{2}{6}\times\frac{5}{6})^{\frac{1}{2}}\right) \times \left(1 - (\frac{2}{6}\times\frac{1}{6})^{\frac{1}{2}}\right) \\ \times \left(1 - (\frac{3}{6}\times\frac{5}{6})^{\frac{1}{2}}\right) \times \left(1 - (\frac{3}{6}\times\frac{1}{6})^{\frac{1}{2}}\right) \times \left(1 - (\frac{5}{6}\times\frac{1}{6})^{\frac{1}{2}}\right) \end{array} \right)^{\frac{1}{C_4^2}} \right) \right), \\ \Delta\left(6 \times \left(1 - \left(\begin{array}{l} \left(1 - (\frac{1}{6}\times\frac{4}{6})^{\frac{1}{2}}\right) \times \left(1 - (\frac{1}{6}\times\frac{1}{6})^{\frac{1}{2}}\right) \times \left(1 - (\frac{1}{6}\times\frac{3}{6})^{\frac{1}{2}}\right) \\ \times \left(1 - (\frac{4}{6}\times\frac{1}{6})^{\frac{1}{2}}\right) \times \left(1 - (\frac{4}{6}\times\frac{3}{6})^{\frac{1}{2}}\right) \times \left(1 - (\frac{1}{6}\times\frac{3}{6})^{\frac{1}{2}}\right) \end{array} \right)^{\frac{1}{C_4^2}} \right) \right) \end{array} \right\}$$

$$= \langle(s_4,0.1802),(s_3,-0.4123),(s_2,0.0680)\rangle$$

Similar to the 2TLNHM operator, we can get the properties, as follows.

Property 6. *(Idempotency) If* $\varphi_j = \left\langle \left(s_{T_j}, \alpha_j \right), \left(s_{I_j}, \beta_j \right), \left(s_{F_j}, \gamma_j \right) \right\rangle (j = 1, 2, \ldots, n)$ *are equal, then*

$$2\text{TLNDHM}^{(x)}(\varphi_1, \varphi_2, \cdots, \varphi_n) = \varphi. \tag{48}$$

Property 7. *(Monotonicity) Let* $\varphi_{a_j} = \left\langle \left(s_{T_{a_j}}, \alpha_{a_j} \right), \left(s_{I_{a_j}}, \beta_{a_j} \right), \left(s_{F_{a_j}}, \gamma_{a_j} \right) \right\rangle (j = 1, 2, \ldots, n)$ *and* $\varphi_{b_j} = \left\langle \left(s_{T_{b_j}}, \alpha_{b_j} \right), \left(s_{I_{b_j}}, \beta_{b_j} \right), \left(s_{F_{b_j}}, \gamma_{b_j} \right) \right\rangle (j = 1, 2, \ldots, n)$ *be two sets of 2TLNNs. If* $\Delta^{-1}\left(s_{T_{a_j}}, \alpha_{a_j} \right) \leq \Delta^{-1}\left(s_{T_{b_j}}, \alpha_{b_j} \right)$, $\Delta^{-1}\left(s_{I_{a_j}}, \beta_{a_j} \right) \geq \Delta^{-1}\left(s_{I_{b_j}}, \beta_{b_j} \right)$ *and* $\Delta^{-1}\left(s_{F_{a_j}}, \gamma_{a_j} \right) \geq \Delta^{-1}\left(s_{F_{b_j}}, \gamma_{b_j} \right)$ *hold for all j, then*

$$2\text{TLNDHM}^{(x)}(\varphi_{a_1}, \varphi_{a_2}, \cdots, \varphi_{a_n}) \leq 2\text{TLNDHM}^{(x)}(\varphi_{b_1}, \varphi_{b_2}, \cdots, \varphi_{b_n}) \tag{49}$$

Property 8. *(Boundedness) Let* $\varphi_j = \left\langle \left(s_{T_j}, \alpha_j \right), \left(s_{I_j}, \beta_j \right), \left(s_{F_j}, \gamma_j \right) \right\rangle (j = 1, 2, \ldots, n)$ *be a set of 2TLNNs. If* $\varphi_i^+ = \left(\max_i \left(s_{T_j}, \alpha_j \right), \min_i \left(s_{I_j}, \beta_j \right), \min_i \left(s_{F_j}, \gamma_j \right) \right)$ *and* $\varphi_i^+ = \left(\max_i \left(s_{T_j}, \alpha_j \right), \min_i \left(s_{I_j}, \beta_j \right), \min_i \left(s_{F_j}, \gamma_j \right) \right)$ *then*

$$\varphi^- \leq 2\text{TLNDHM}^{(x)}(\varphi_1, \varphi_2, \cdots, \varphi_n) \leq \varphi^+. \tag{50}$$

3.4. The 2TLNWDHM Operator

In an actual MADM, it is important to consider attribute weights; in this section we shall propose the 2-tuple linguistic neutrosophic weighted DHM (2TLNWDHM) operator.

Definition 11. *Let* $\varphi_j = \left\langle \left(s_{T_j}, \alpha_j \right), \left(s_{I_j}, \beta_j \right), \left(s_{F_j}, \gamma_j \right) \right\rangle (j = 1, 2, \ldots, n)$ *be a set of 2TLNNs with the weight vector,* $w_i = (w_1, w_2, \ldots, w_n)^T$, *thereby satisfying* $w_i \in [0, 1]$ *and* $\sum_{i=1}^{n} w_i = 1$. *If*

$$2\text{TLNWDHM}^{(x)}(\varphi_1, \varphi_2, \cdots, \varphi_n) = \left(\bigotimes_{1 \leq i_1 < \ldots < i_x \leq n} \left(\frac{\overset{x}{\underset{j=1}{\oplus}} w_{i_j} \varphi_{i_j}}{x} \right) \right)^{\frac{1}{C_n^x}}. \tag{51}$$

Theorem 4. *Let* $\varphi_j = \left\langle \left(s_{T_j}, \alpha_j \right), \left(s_{I_j}, \beta_j \right), \left(s_{F_j}, \gamma_j \right) \right\rangle (j = 1, 2, \ldots, n)$ *be a set of 2TLNNs. The aggregated value by using 2TLNWDHM operators is also a 2TLNN where*

$$
2TLNWDHM^{(x)}(\varphi_1, \varphi_2, \cdots, \varphi_n) = \left(\underset{1 \le i_1 < \ldots < i_x \le n}{\otimes} \left(\frac{\overset{x}{\underset{j=1}{\oplus}} w_{i_j} \varphi_{i_j}}{x} \right) \right)^{\frac{1}{C_n^x}}
$$

$$
= \left\{
\begin{array}{l}
\Delta \left(t \left(\prod_{1 \le i_1 < \ldots < i_x \le n} \left(1 - \left(\prod_{j=1}^{x} \left(1 - \frac{\Delta^{-1}\left(s_{T_j}, \alpha_j\right)}{t} \right)^{w_{i_j}} \right)^{\frac{1}{x}} \right) \right)^{\frac{1}{C_n^x}} \right), \\[3em]
\Delta \left(t \left(1 - \left(\prod_{1 \le i_1 < \ldots < i_x \le n} \left(1 - \left(\prod_{j=1}^{x} \left(\frac{\Delta^{-1}\left(s_{I_j}, \beta_j\right)}{t} \right)^{w_{i_j}} \right)^{\frac{1}{x}} \right) \right)^{\frac{1}{C_n^x}} \right) \right), \\[3em]
\Delta \left(t \left(1 - \left(\prod_{1 \le i_1 < \ldots < i_x \le n} \left(1 - \left(\prod_{j=1}^{x} \left(\frac{\Delta^{-1}\left(s_{F_j}, \gamma_j\right)}{t} \right)^{w_{i_j}} \right)^{\frac{1}{x}} \right) \right)^{\frac{1}{C_n^x}} \right) \right)
\end{array}
\right\}
\tag{52}
$$

Proof. From Definition 5, we can obtain that

$$
w_{i_j} \varphi_{i_j} = \left\{
\begin{array}{l}
\Delta \left(t \left(1 - \left(1 - \frac{\Delta^{-1}\left(s_{T_j}, \alpha_j\right)}{t} \right)^{w_{i_j}} \right) \right), \\[2em]
\Delta \left(t \left(\frac{\Delta^{-1}\left(s_{I_j}, \beta_j\right)}{t} \right)^{w_{i_j}} \right), \Delta \left(t \left(\frac{\Delta^{-1}\left(s_{F_j}, \gamma_j\right)}{t} \right)^{w_{i_j}} \right)
\end{array}
\right\}
\tag{53}
$$

Then,

$$
\overset{x}{\underset{j=1}{\oplus}} w_{i_j} \varphi_{i_j} = \left\{
\begin{array}{l}
\Delta \left(t \left(1 - \prod_{j=1}^{x} \left(1 - \frac{\Delta^{-1}\left(s_{T_j}, \alpha_j\right)}{t} \right)^{w_{i_j}} \right) \right), \\[2em]
\Delta \left(t \left(\prod_{j=1}^{x} \left(\frac{\Delta^{-1}\left(s_{I_j}, \beta_j\right)}{t} \right)^{w_{i_j}} \right) \right), \Delta \left(t \left(\prod_{j=1}^{x} \left(\frac{\Delta^{-1}\left(s_{F_j}, \gamma_j\right)}{t} \right)^{w_{i_j}} \right) \right)
\end{array}
\right\}
\tag{54}
$$

Thus,

$$
\frac{\overset{x}{\underset{j=1}{\oplus}} w_{i_j} \varphi_{i_j}}{x} = \left\{
\begin{array}{l}
\Delta \left(t \left(1 - \left(\prod_{j=1}^{x} \left(1 - \frac{\Delta^{-1}\left(s_{T_j}, \alpha_j\right)}{t} \right)^{w_{i_j}} \right)^{\frac{1}{x}} \right) \right), \\[2em]
\Delta \left(t \left(\prod_{j=1}^{x} \left(\frac{\Delta^{-1}\left(s_{I_j}, \beta_j\right)}{t} \right)^{w_{i_j}} \right)^{\frac{1}{x}} \right), \Delta \left(t \left(\prod_{j=1}^{x} \left(\frac{\Delta^{-1}\left(s_{F_j}, \gamma_j\right)}{t} \right)^{w_{i_j}} \right)^{\frac{1}{x}} \right)
\end{array}
\right\}
\tag{55}
$$

Therefore,

$$
\underset{1\leq i_1<...<i_x\leq n}{\otimes}\left(\frac{\overset{x}{\underset{j=1}{\oplus}}w_{i_j}\varphi_{i_j}}{x}\right)=\left\{
\begin{array}{c}
\Delta\left(t\left(\underset{1\leq i_1<...<i_x\leq n}{\prod}\left(1-\left(\overset{x}{\underset{j=1}{\prod}}\left(1-\frac{\Delta^{-1}\left(s_{T_j},\alpha_j\right)}{t}\right)\right)^{w_{i_j}}\right)^{\frac{1}{x}}\right)\right)\right),\\[20pt]
\Delta\left(t\left(1-\underset{1\leq i_1<...<i_x\leq n}{\prod}\left(1-\left(\overset{x}{\underset{j=1}{\prod}}\left(\frac{\Delta^{-1}\left(s_{I_j},\beta_j\right)}{t}\right)^{w_{i_j}}\right)^{\frac{1}{x}}\right)\right)\right),\\[20pt]
\Delta\left(t\left(1-\underset{1\leq i_1<...<i_x\leq n}{\prod}\left(1-\left(\overset{x}{\underset{j=1}{\prod}}\left(\frac{\Delta^{-1}\left(s_{F_j},\gamma_j\right)}{t}\right)^{w_{i_j}}\right)^{\frac{1}{x}}\right)\right)\right)
\end{array}
\right\}
\tag{56}
$$

Therefore,

$$
2\text{TLNWDHM}^{(x)}(\varphi_1,\varphi_2,\cdots,\varphi_n)=\left(\underset{1\leq i_1<...<i_x\leq n}{\otimes}\left(\frac{\overset{x}{\underset{j=1}{\oplus}}w_{i_j}\varphi_{i_j}}{x}\right)\right)^{\frac{1}{C_n^x}}
$$

$$
=\left\{
\begin{array}{c}
\Delta\left(t\left(\underset{1\leq i_1<...<i_x\leq n}{\prod}\left(1-\left(\overset{x}{\underset{j=1}{\prod}}\left(1-\frac{\Delta^{-1}\left(s_{T_j},\alpha_j\right)}{t}\right)\right)^{w_{i_j}}\right)^{\frac{1}{x}}\right)^{\frac{1}{C_n^x}}\right),\\[20pt]
\Delta\left(t\left(1-\left(\underset{1\leq i_1<...<i_x\leq n}{\prod}\left(1-\left(\overset{x}{\underset{j=1}{\prod}}\left(\frac{\Delta^{-1}\left(s_{I_j},\beta_j\right)}{t}\right)^{w_{i_j}}\right)^{\frac{1}{x}}\right)\right)^{\frac{1}{C_n^x}}\right)\right),\\[20pt]
\Delta\left(t\left(1-\left(\underset{1\leq i_1<...<i_x\leq n}{\prod}\left(1-\left(\overset{x}{\underset{j=1}{\prod}}\left(\frac{\Delta^{-1}\left(s_{F_j},\gamma_j\right)}{t}\right)^{w_{i_j}}\right)^{\frac{1}{x}}\right)\right)^{\frac{1}{C_n^x}}\right)\right)
\end{array}
\right\}
\tag{57}
$$

Hence, (51) is kept.

Then, we need to prove that (51) is a 2TLNN. We need to prove two conditions as follows:

① $0\leq\Delta^{-1}(s_T,\alpha)\leq t,0\leq\Delta^{-1}(s_I,\beta)\leq t,0\leq\Delta^{-1}(s_F,\gamma)\leq t.$

② $0\leq\Delta^{-1}(s_T,\alpha)+\Delta^{-1}(s_I,\beta)+\Delta^{-1}(s_F,\gamma)\leq 3t.$

Let

$$
\frac{\Delta^{-1}(s_T,\alpha)}{t}=\left(\underset{1\leq i_1<...<i_x\leq n}{\prod}\left(1-\left(\overset{x}{\underset{j=1}{\prod}}\left(1-\frac{\Delta^{-1}\left(s_{T_j},\alpha_j\right)}{t}\right)\right)^{w_{i_j}}\right)^{\frac{1}{x}}\right)^{\frac{1}{C_n^x}}
$$

$$
\frac{\Delta^{-1}(s_I,\beta)}{t}=1-\left(\underset{1\leq i_1<...<i_x\leq n}{\prod}\left(1-\left(\overset{x}{\underset{j=1}{\prod}}\left(\frac{\Delta^{-1}\left(s_{I_j},\beta_j\right)}{t}\right)^{w_{i_j}}\right)^{\frac{1}{x}}\right)\right)^{\frac{1}{C_n^x}}
$$

$$
\frac{\Delta^{-1}(s_F,\gamma)}{t}=1-\left(\underset{1\leq i_1<...<i_x\leq n}{\prod}\left(1-\left(\overset{x}{\underset{j=1}{\prod}}\left(\frac{\Delta^{-1}\left(s_{F_j},\gamma_j\right)}{t}\right)^{w_{i_j}}\right)^{\frac{1}{x}}\right)\right)^{\frac{1}{C_n^x}}
$$

□

Proof. ① Since $0 \leq \frac{\Delta^{-1}\left(s_{T_j}, \alpha_j\right)}{t} \leq 1$, we get

$$0 \leq \prod_{j=1}^{x}\left(1 - \frac{\Delta^{-1}\left(s_{T_j}, \alpha_j\right)}{t}\right) \leq 1 \text{ and } 0 \leq 1 - \left(\prod_{j=1}^{x}\left(1 - \frac{\Delta^{-1}\left(s_{T_j}, \alpha_j\right)}{t}\right)^{w_{ij}}\right)^{\frac{1}{x}} \leq 1 \quad (58)$$

Then,

$$0 \leq \prod_{1 \leq i_1 < \ldots < i_x \leq n}\left(1 - \left(\prod_{j=1}^{x}\left(1 - \frac{\Delta^{-1}\left(s_{T_j}, \alpha_j\right)}{t}\right)^{w_{ij}}\right)^{\frac{1}{x}}\right) \leq 1 \quad (59)$$

$$0 \leq \left(\prod_{1 \leq i_1 < \ldots < i_x \leq n}\left(1 - \left(\prod_{j=1}^{x}\left(1 - \frac{\Delta^{-1}\left(s_{T_j}, \alpha_j\right)}{t}\right)^{w_{ij}}\right)^{\frac{1}{x}}\right)\right)^{\frac{1}{C_n^x}} \leq 1. \quad (60)$$

That means $0 \leq \Delta^{-1}(s_T, \alpha) \leq t$. Similarly, we can get $0 \leq \Delta^{-1}(s_I, \beta) \leq t, 0 \leq \Delta^{-1}(s_F, \gamma) \leq t$ so ① is maintained. ② Since $0 \leq \Delta^{-1}(s_T, \alpha) \leq t, 0 \leq \Delta^{-1}(s_I, \beta) \leq t, 0 \leq \Delta^{-1}(s_F, \gamma) \leq t$. $0 \leq \Delta^{-1}(s_T, \alpha) + \Delta^{-1}(s_I, \beta) + \Delta^{-1}(s_F, \gamma) \leq 3t$. $\square$

Example 4. *Let* $\langle(s_5, 0), (s_2, 0), (s_1, 0)\rangle, \langle(s_4, 0), (s_3, 0), (s_4, 0)\rangle, \langle(s_2, 0), (s_5, 0), (s_1, 0)\rangle$ *and* $\langle(s_5, 0), (s_1, 0), (s_3, 0)\rangle$ *be four 2TLNNs,* $w = (0.2, 0.3, 0.4, 0.1)$ *and suppose* $x = 2$, *then according to (51), we have*

$$2\text{TLNWDHM}^{(2)}(\varphi_1, \varphi_2, \cdots, \varphi_n) = \left(\underset{1 \leq i_1 < \ldots < i_x \leq n}{\otimes}\left(\frac{\overset{x}{\underset{j=1}{\oplus}} w_{ij}\varphi_{ij}}{x}\right)\right)^{\frac{1}{C_n^x}}$$

$$= \left\{ \begin{array}{l} \Delta\left\{6 \times \left[\begin{array}{l} \left(1 - \left(\left(1 - \frac{5}{6}\right)^{0.2} \times \left(1 - \frac{4}{6}\right)^{0.3}\right)^{\frac{1}{2}}\right) \times \left(1 - \left(\left(1 - \frac{5}{6}\right)^{0.2} \times \left(1 - \frac{2}{6}\right)^{0.4}\right)^{\frac{1}{2}}\right) \\ \times \left(1 - \left(\left(1 - \frac{5}{6}\right)^{0.2} \times \left(1 - \frac{5}{6}\right)^{0.1}\right)^{\frac{1}{2}}\right) \times \left(1 - \left(\left(1 - \frac{4}{6}\right)^{0.3} \times \left(1 - \frac{2}{6}\right)^{0.4}\right)^{\frac{1}{2}}\right) \\ \times \left(1 - \left(\left(1 - \frac{4}{6}\right)^{0.3} \times \left(1 - \frac{5}{6}\right)^{0.1}\right)^{\frac{1}{2}}\right) \times \left(1 - \left(\left(1 - \frac{2}{6}\right)^{0.4} \times \left(1 - \frac{5}{6}\right)^{0.1}\right)^{\frac{1}{2}}\right) \end{array} \right]^{\frac{1}{C_4^2}} \right\}, \\ \Delta\left\{6 \times \left\{1 - \left\{ \begin{array}{l} \left(1 - \left(\left(\frac{2}{6}\right)^{0.2} \times \left(\frac{3}{6}\right)^{0.3}\right)^{\frac{1}{2}}\right) \times \left(1 - \left(\left(\frac{2}{6}\right)^{0.2} \times \left(\frac{5}{6}\right)^{0.4}\right)^{\frac{1}{2}}\right) \times \left(1 - \left(\left(\frac{2}{6}\right)^{0.2} \times \left(\frac{1}{6}\right)^{0.1}\right)^{\frac{1}{2}}\right) \\ \left(1 - \left(\left(\frac{3}{6}\right)^{0.3} \times \left(\frac{5}{6}\right)^{0.4}\right)^{\frac{1}{2}}\right) \times \left(1 - \left(\left(\frac{3}{6}\right)^{0.3} \times \left(\frac{1}{6}\right)^{0.1}\right)^{\frac{1}{2}}\right) \times \left(1 - \left(\left(\frac{5}{6}\right)^{0.4} \times \left(\frac{1}{6}\right)^{0.1}\right)^{\frac{1}{2}}\right) \end{array} \right\}^{\frac{1}{C_4^2}} \right\}\right\}, \\ \Delta\left\{6 \times \left\{1 - \left\{ \begin{array}{l} \left(1 - \left(\left(\frac{1}{6}\right)^{0.2} \times \left(\frac{4}{6}\right)^{0.3}\right)^{\frac{1}{2}}\right) \times \left(1 - \left(\left(\frac{1}{6}\right)^{0.2} \times \left(\frac{1}{6}\right)^{0.4}\right)^{\frac{1}{2}}\right) \times \left(1 - \left(\left(\frac{1}{6}\right)^{0.2} \times \left(\frac{3}{6}\right)^{0.1}\right)^{\frac{1}{2}}\right) \\ \times \left(1 - \left(\left(\frac{4}{6}\right)^{0.3} \times \left(\frac{1}{6}\right)^{0.4}\right)^{\frac{1}{2}}\right) \times \left(1 - \left(\left(\frac{4}{6}\right)^{0.3} \times \left(\frac{3}{6}\right)^{0.1}\right)^{\frac{1}{2}}\right) \times \left(1 - \left(\left(\frac{1}{6}\right)^{0.4} \times \left(\frac{3}{6}\right)^{0.1}\right)^{\frac{1}{2}}\right) \end{array} \right\}^{\frac{1}{C_4^2}} \right\}\right\} \end{array} \right\}$$

$= \langle(s_1, 0.3339), (s_5, 0.0807), (s_5, -0.4164)\rangle$

Now, we will discuss some properties of the 2TLNWDHM operator.

Property 9. *(Monotonicity) Let* $\varphi_{a_j} = \left\langle \left(s_{T_{a_j}}, \alpha_{a_j}\right), \left(s_{I_{a_j}}, \beta_{a_j}\right), \left(s_{F_{a_j}}, \gamma_{a_j}\right) \right\rangle (j = 1, 2, \ldots, n)$ *and* $\varphi_{b_j} = \left\langle \left(s_{T_{b_j}}, \alpha_{b_j}\right), \left(s_{I_{b_j}}, \beta_{b_j}\right), \left(s_{F_{b_j}}, \gamma_{b_j}\right) \right\rangle (j = 1, 2, \ldots, n)$ *be two sets of 2TLNNs. If* $\Delta^{-1}\left(s_{T_{a_j}}, \alpha_{a_j}\right) \leq \Delta^{-1}\left(s_{T_{b_j}}, \alpha_{b_j}\right), \Delta^{-1}\left(s_{I_{a_j}}, \beta_{a_j}\right) \geq \Delta^{-1}\left(s_{I_{b_j}}, \beta_{b_j}\right)$ *and* $\Delta^{-1}\left(s_{F_{a_j}}, \gamma_{a_j}\right) \geq \Delta^{-1}\left(s_{F_{b_j}}, \gamma_{b_j}\right)$ *hold for all j, then*

$$2\text{TLNWDHM}^{(x)}\left(\varphi_{a_1}, \varphi_{a_2}, \cdots, \varphi_{a_n}\right) \leq 2\text{TLNWDHM}^{(x)}\left(\varphi_{b_1}, \varphi_{b_2}, \cdots, \varphi_{b_n}\right) \tag{61}$$

Property 10. *(Boundedness) Let* $\varphi_j = \left\langle \left(s_{T_j}, \alpha_j\right), \left(s_{I_j}, \beta_j\right), \left(s_{F_j}, \gamma_j\right) \right\rangle (j = 1, 2, \ldots, n)$ *be a set of 2TLNNs. If* $\varphi_i^+ = \left(\max_i\left(s_{T_j}, \alpha_j\right), \min_i\left(s_{I_j}, \beta_j\right), \min_i\left(s_{F_j}, \gamma_j\right)\right)$ *and* $\varphi_i^+ = \left(\max_i\left(s_{T_j}, \alpha_j\right), \min_i\left(s_{I_j}, \beta_j\right), \min_i\left(s_{F_j}, \gamma_j\right)\right)$, *then*

$$\varphi^- \leq 2\text{TLNWDHM}^{(x)}\left(\varphi_1, \varphi_2, \cdots, \varphi_n\right) \leq \varphi^+ \tag{62}$$

From Theorem 4,

$$2\text{TLNWDHM}^{(x)}\left(\varphi_1^+, \varphi_2^+, \cdots, \varphi_n^+\right)$$

$$= \left\{ \begin{array}{l} \Delta\left(t\left(\displaystyle\prod_{1\leq i_1 <\ldots<i_x\leq n}\left(1-\left(\prod_{j=1}^{x}\left(1-\frac{\max\Delta^{-1}\left(s_{T_j},\alpha_j\right)}{t}\right)^{w_{ij}}\right)^{\frac{1}{x}}\right)\right)^{\frac{1}{C_n^x}}\right), \\[3em] \Delta\left(t\left(1-\left(\displaystyle\prod_{1\leq i_1 <\ldots<i_x\leq n}\left(1-\left(\prod_{j=1}^{x}\left(\frac{\min\Delta^{-1}\left(s_{I_j},\beta_j\right)}{t}\right)^{w_{ij}}\right)^{\frac{1}{x}}\right)\right)^{\frac{1}{C_n^x}}\right)\right), \\[3em] \Delta\left(t\left(1-\left(\displaystyle\prod_{1\leq i_1 <\ldots<i_x\leq n}\left(1-\left(\prod_{j=1}^{x}\left(\frac{\min\Delta^{-1}\left(s_{F_j},\gamma_j\right)}{t}\right)^{w_{ij}}\right)^{\frac{1}{x}}\right)\right)^{\frac{1}{C_n^x}}\right)\right) \end{array} \right\} \tag{63}$$

$$2\text{TLNWDHM}^{(x)}\left(\varphi_1^-, \varphi_2^-, \cdots, \varphi_n^-\right)$$

$$= \left\{ \begin{array}{l} \Delta\left(t\left(\displaystyle\prod_{1\leq i_1 <\ldots<i_x\leq n}\left(1-\left(\prod_{j=1}^{x}\left(1-\frac{\min\Delta^{-1}\left(s_{T_j},\alpha_j\right)}{t}\right)^{w_{ij}}\right)^{\frac{1}{x}}\right)\right)^{\frac{1}{C_n^x}}\right), \\[3em] \Delta\left(t\left(1-\left(\displaystyle\prod_{1\leq i_1 <\ldots<i_x\leq n}\left(1-\left(\prod_{j=1}^{x}\left(\frac{\max\Delta^{-1}\left(s_{I_j},\beta_j\right)}{t}\right)^{w_{ij}}\right)^{\frac{1}{x}}\right)\right)^{\frac{1}{C_n^x}}\right)\right), \\[3em] \Delta\left(t\left(1-\left(\displaystyle\prod_{1\leq i_1 <\ldots<i_x\leq n}\left(1-\left(\prod_{j=1}^{x}\left(\frac{\max\Delta^{-1}\left(s_{F_j},\gamma_j\right)}{t}\right)^{w_{ij}}\right)^{\frac{1}{x}}\right)\right)^{\frac{1}{C_n^x}}\right)\right) \end{array} \right\} \tag{64}$$

From Property 9,

$$\varphi^- \leq 2\text{TLNWDHM}^{(x)}\left(\varphi_1, \varphi_2, \cdots, \varphi_n\right) \leq \varphi^+ \tag{65}$$

It is obvious that 2TLNWDHM operator lacks the property of idempotency.

4. Numerical Example and Comparative Analysis

4.1. Numerical Example

The constructional engineering projects have the following characteristics: large investment, many participants, complex project environment, and a wide range of risk factors on the basis of the Engineering Procurement Construction (EPC) mode. Therefore, it is necessary to analyze and assess the constructional engineering project's risks during the life cycle; a risk assessment is good for implementing projects and completing project goals. During the process of implementation, constructional engineering projects face aspects of risk—political, economic, social and natural and other aspects of risks. These risks have a great influence on our construction company, and produce many factors with high probability that are difficult to estimate and quantify. Thus, we shall give a numerical example for construction engineering project risk assessment with 2TLNNs in order to illustrate the method proposed in this paper. There are five possible construction engineering projects $A_i(i = 1, 2, 3, 4, 5)$ to evaluate. The expert selects four attributes to evaluate the five possible construction engineering projects: ① G_1 is the construction work environment; ② G_2 is the construction site safety protection measures; ③ G_3 is the safety management ability of the engineering project management; and ④ G_4 is the safety production responsibility system. The five possible construction engineering projects, $A_i(i = 1, 2, 3, 4, 5)$, will be evaluated using the 2TLNNs by the decision maker using the above four attributes (whose weighting vector is $\omega = (0.2, 0.1, 0.5, 0.2)$ and expert weighting vector is $\omega = (0.2, 0.4, 0.4)$, which are listed in Tables 1–3.

Table 1. 2-tuple linguistic neutrosophic numbers (2TLNN) decision matrix (R_1).

	G_1	G_2	G_3	G_4
A_1	<(s₄, 0), (s₃, 0) (s₂, 0)>	<(s₅, 0), (s₃, 0) (s₁, 0)>	<(s₄, 0), (s₁, 0) (s₂, 0)>	<(s₂, 0), (s₃, 0) (s₂, 0)>
A_2	<(s₃, 0), (s₂, 0) (s₄, 0)>	<(s₄, 0), (s₂, 0) (s₂, 0)>	<(s₃, 0), (s₂, 0) (s₂, 0)>	<(s₄, 0), (s₃, 0) (s₃, 0)>
A_3	<(s₅, 0), (s₄, 0) (s₃, 0)>	<(s₄, 0), (s₄, 0) (s₃, 0)>	<(s₂, 0), (s₁, 0) (s₂, 0)>	<(s₄, 0), (s₃, 0) (s₂, 0)>
A_4	<(s₂, 0), (s₁, 0) (s₂, 0)>	<(s₅, 0), (s₁, 0) (s₂, 0)>	<(s₄, 0), (s₃, 0) (s₅, 0)>	<(s₃, 0), (s₁, 0) (s₁, 0)>
A_5	<(s₄, 0), (s₃, 0) (s₁, 0)>	<(s₅, 0), (s₂, 0) (s₂, 0)>	<(s₃, 0), (s₂, 0) (s₁, 0)>	<(s₃, 0), (s₂, 0) (s₂, 0)>

Table 2. 2TLNN decision matrix (R_2).

	G_1	G_2	G_3	G_4
A_1	<(s₃, 0), (s₂, 0) (s₃, 0)>	<(s₃, 0), (s₃, 0) (s₂, 0)>	<(s₃, 0), (s₁, 0) (s₂, 0)>	<(s₄, 0), (s₁, 0) (s₃, 0)>
A_2	<(s₂, 0), (s₃, 0) (s₃, 0)>	<(s₃, 0), (s₃, 0) (s₃, 0)>	<(s₃, 0), (s₂, 0) (s₂, 0)>	<(s₃, 0), (s₄, 0) (s₃, 0)>
A_3	<(s₂, 0), (s₃, 0) (s₃, 0)>	<(s₃, 0), (s₂, 0) (s₂, 0)>	<(s₂, 0), (s₃, 0) (s₁, 0)>	<(s₃, 0), (s₂, 0) (s₄, 0)>
A_4	<(s₃, 0), (s₂, 0) (s₂, 0)>	<(s₂, 0), (s₂, 0) (s₃, 0)>	<(s₃, 0), (s₄, 0) (s₂, 0)>	<(s₃, 0), (s₁, 0) (s₂, 0)>
A_5	<(s₃, 0), (s₂, 0) (s₁, 0)>	<(s₃, 0), (s₄, 0) (s₃, 0)>	<(s₄, 0), (s₁, 0) (s₁, 0)>	<(s₂, 0), (s₃, 0) (s₂, 0)>

Table 3. 2TLNN decision matrix (R_3).

	G_1	G_2	G_3	G_4
A_1	<(s₃, 0), (s₃, 0) (s₁, 0)>	<(s₄, 0), (s₂, 0) (s₁, 0)>	<(s₄, 0), (s₄, 0) (s₃, 0)>	<(s₄, 0), (s₁, 0) (s₃, 0)>
A_2	<(s₂, 0), (s₂, 0) (s₂, 0)>	<(s₄, 0), (s₄, 0) (s₄, 0)>	<(s₃, 0), (s₂, 0) (s₃, 0)>	<(s₂, 0), (s₁, 0) (s₃, 0)>
A_3	<(s₂, 0), (s₁, 0) (s₂, 0)>	<(s₃, 0), (s₂, 0) (s₂, 0)>	<(s₄, 0), (s₅, 0) (s₂, 0)>	<(s₂, 0), (s₄, 0) (s₄, 0)>
A_4	<(s₃, 0), (s₁, 0) (s₂, 0)>	<(s₂, 0), (s₁, 0) (s₂, 0)>	<(s₃, 0), (s₄, 0) (s₅, 0)>	<(s₅, 0), (s₃, 0) (s₁, 0)>
A_5	<(s₃, 0), (s₃, 0) (s₂, 0)>	<(s₃, 0), (s₂, 0) (s₂, 0)>	<(s₃, 0), (s₂, 0) (s₃, 0)>	<(s₅, 0), (s₃, 0) (s₄, 0)>

Then, we utilize the approach developed to select the best construction engineering projects.

Definition 12. *Let* $\varphi_j = \left\langle \left(s_{T_j}, \alpha_j \right), \left(s_{I_j}, \beta_j \right), \left(s_{F_j}, \gamma_j \right) \right\rangle (j = 1, 2, \ldots, n)$ *be a set of 2TLNNs with the weight vector,* $w_i = (w_1, w_2, \ldots, w_n)^T$, *thereby satisfying* $w_i \in [0, 1]$ *and* $\sum_{i=1}^{n} w_i = 1$, *then we can obtain*

$$2TLNNWAA(\varphi_1, \varphi_2, \ldots, \varphi_n) = \sum_{j=1}^{n} w_j \varphi_j$$

$$= \begin{cases} \Delta \left(t \left(1 - \prod_{j=1}^{n} \left(1 - \frac{\Delta^{-1}\left(s_{T_j}, \alpha_j \right)}{t} \right)^{w_j} \right) \right), \\ \Delta \left(t \left(\prod_{j=1}^{n} \left(\frac{\Delta^{-1}\left(s_{I_j}, \beta_j \right)}{t} \right)^{w_j} \right) \right), \Delta \left(t \left(\prod_{j=1}^{n} \left(\frac{\Delta^{-1}\left(s_{F_j}, \gamma_j \right)}{t} \right)^{w_j} \right) \right) \end{cases} \tag{66}$$

$$2TLNNWGA(\varphi_1, \varphi_2, \ldots, \varphi_n) = \sum_{j=1}^{n} (\varphi_j)^{w_j}$$

$$= \begin{cases} \Delta \left(t \left(\prod_{j=1}^{n} \left(\frac{\Delta^{-1}\left(s_{T_j}, \alpha_j \right)}{t} \right)^{w_j} \right) \right), \Delta \left(t \left(1 - \prod_{j=1}^{n} \left(1 - \frac{\Delta^{-1}\left(s_{I_j}, \beta_j \right)}{t} \right)^{w_j} \right) \right), \\ \Delta \left(t \left(1 - \prod_{j=1}^{n} \left(1 - \frac{\Delta^{-1}\left(s_{F_j}, \gamma_j \right)}{t} \right)^{w_j} \right) \right) \end{cases} \tag{67}$$

Step 1. According to the 2TLNNs, $r_{ij}(i = 1, 2, 3, 4, 5, j = 1, 2, 3, 4)$, we can calculate all 2TLNNs r_{ij} by using the 2-tuple linguistic neutrosophic numbers weighted average (2TLNNWA) operator and 2-tuple linguistic neutrosophic numbers weighted geometric (2TLNNWG) operator to get the overall 2TLNNs $A_i(i = 1, 2, 3, 4, 5)$ of the construction engineering projects. Then, the results are shown in Table 4.

Table 4. The aggregating results by the 2TLNNWAA operator.

	G_1	G_2
A_1	<(s_3, 0.2337), (s_3, −0.4492), (s_2, −0.2174)>	<(s_4, −0.0477), (s_3, −0.4492), (s_1, 0.3195)>
A_2	<(s_2, 0.2236), (s_2, 0.3522), (s_3, −0.2981)>	<(s_4, −0.3522), (s_3, 0.1037),(s_3, 0.1037)>
A_3	<(s_3, −0.0314), (s_2, 0.0477), (s_3, −0.4492)>	<(s_3, 0.2337), (s_2, 0.2974), (s_2, 0.1689)>
A_4	<(s_3, −0.1777), (s_1, 0.3195), (s_2, 0.0000)>	<(s_3, −0.0314), (s_1, 0.3195), (s_2, 0.3522)>
A_5	<(s_3, 0.2337), (s_3, −0.4492), (s_1, 0.3195)>	<(s_4, −0.4082), (s_3, −0.3610), (s_2, 0.3522)>

	G_3	G_4
A_1	<(s_4, −0.3522), (s_2, −0.2589), (s_2, 0.3522)>	<(s_4, −0.2974), (s_1, 0.2457), (s_3, −0.2337)>
A_2	<(s_3, 0.000), (s_2, 0.0000), (s_2, 0.3522)>	<(s_3, −0.1037), (s_2,0.1689), (s_3, 0.0000)>
A_3	<(s_3, −0.0314), (s_3, −0.0458), (s_2, −0.4843)>	<(s_3, -−0.1037), (s_3, −0.1381), (s_3,0.4822)>
A_4	<(s_3, 0.2337), (s_4, −0.2236), (s_3, 0.4657)>	<(s_4, 0.0668), (s_2, −0.4482), (s_1,0.3195)>
A_5	<(s_3, 0.4492), (s_2, −0.4843), (s_2, −0.4482)>	<(s_4, −0.1689), (s_3, −0.2337),(s_3, −0.3610)>

Step 2. According to Table 4, we can calculate the r_{ij} of all 2TLNNs by using the 2TLNWHM (2TLNWDHM) operator to get the overall 2TLNNs $A_i(i = 1, 2, 3, 4, 5)$ of the construction engineering projects, A_i. Suppose that $x = 2$, then the aggregating results are as shown in Table 5.

Table 5. The aggregating results of the construction engineering projects by the 2TLNWHM (2TLNWDHM) operator.

	2TLNWHM	2TLNWDHM
A_1	<(s_5, 0.3257), (s_1, −0.4776), (s_1, −0.4012)>	<(s_1, 0.1564), (s_5, −0.4119), (s_5, −0.2877)>
A_2	<(s_5, 0.0468), (s_1, −0.3632), (s_1, −0.2177)>	<(s_1, −0.1456), (s_5, −0.2124), (s_5, −0.0171)>
A_3	<(s_5, 0.1002), (s_1, −0.2491), (s_1, −0.3474)>	<(s_1, −0.1092), (s_5, −0.0603), (s_5, −0.1946)>
A_4	<(s_5, 0.2143), (s_1, −0.3927), (s_1, −0.3263)>	<(s_1, 0.0275), (s_5, −0.2639), (s_5, −0.1713)>
A_5	<(s_5, 0.2941), (s_1, −0.3870), (s_1, −0.4981)>	<(s_1, −0.1127), (s_5, −0.2367), (s_5, −0.4415)>

Step 3. In accordance with the aggregating results shown in Table 5, the score functions of the construction engineering projects are shown in Table 6.

Table 6. The score functions of the construction engineering projects.

	2TLNWHM	2TLNWDHM
A_1	$(s_5, 0.4015)$	$(s_1, 0.2853)$
A_2	$(s_5, 0.2092)$	$(s_1, 0.0280)$
A_3	$(s_5, 0.2323)$	$(s_1, 0.0486)$
A_4	$(s_5, 0.3111)$	$(s_1, 0.1542)$
A_5	$(s_5, 0.3930)$	$(s_1, 0.2636)$

Step 4. In accordance with the scores shown in Table 6 and the comparison formulas of the score functions, the ordering of the construction engineering projects are shown in Table 7. The best construction engineering project is A_1.

Table 7. Ordering of the construction engineering projects.

	Ordering
2TLNWHM	$A_1 > A_5 > A_4 > A_3 > A_2$
2TLNWDHM	$A_1 > A_5 > A_4 > A_3 > A_2$

4.2. Influence of the Parameter on the Final Result

In order to show the effects on the ranking results by changing parameters of x in the 2TLNWHM (2TLNWDHM) operators, all the results are shown in Tables 8 and 9.

Table 8. Ranking results for different operational parameters of the 2TLNWHM operator.

	$s(A_1)$	$s(A_2)$	$s(A_3)$	$s(A_4)$	$s(A_5)$	Ordering
$x = 1$	0.9134	0.8799	0.8865	0.9070	0.9065	$A_1 > A_4 > A_5 > A_3 > A_2$
$x = 2$	0.9003	0.8682	0.8720	0.8852	0.8988	$A_1 > A_5 > A_4 > A_3 > A_2$
$x = 3$	0.8953	0.8642	0.8661	0.8696	0.8958	$A_5 > A_1 > A_4 > A_3 > A_2$
$x = 4$	0.8927	0.8621	0.8627	0.8571	0.8942	$A_5 > A_1 > A_2 > A_3 > A_4$

Table 9. Ranking results for different operational parameters of the 2TLNWDHM operator.

	$s(A_1)$	$s(A_2)$	$s(A_3)$	$s(A_4)$	$s(A_5)$	Ordering
$x = 1$	0.1922	0.1489	0.1557	0.1796	0.1817	$A_1 > A_5 > A_4 > A_3 > A_2$
$x = 2$	0.2142	0.1713	0.1748	0.1924	0.2106	$A_1 > A_5 > A_4 > A_3 > A_2$
$x = 3$	0.2239	0.1815	0.1852	0.1965	0.2252	$A_5 > A_1 > A_4 > A_3 > A_2$
$x = 4$	0.2293	0.1878	0.1922	0.1985	0.2342	$A_5 > A_1 > A_4 > A_3 > A_2$

4.3. Comparative Analysis

Then, we compare our proposed method with the LNNWAA operator and LNNWGA operator [33] and cosine measures of linguistic neutrosophic numbers [34]. The comparative results are shown in Table 10.

Table 10. Ordering of the construction engineering projects.

	Ordering
LNNWAA [33]	$A_5 > A_1 > A_4 > A_3 > A_2$
LNNWGA [33]	$A_5 > A_1 > A_3 > A_2 > A_4$
$C^{w_1}_{LNNs}$ [34]	$A_5 > A_1 > A_4 > A_2 > A_3$
$C^{w_2}_{LNNs}$ [34]	$A_5 > A_1 > A_4 > A_2 > A_3$

From above, we determine that the optimal construction engineering projects to show the practicality and effectiveness of the proposed approaches. However, the LNNWAA operator and LNNWGA operator do not consider the information about the relationship between the arguments being aggregated and thus, cannot eliminate the influence of unfair arguments on the decision result. Our proposed 2TLNWHM and 2TLNWDHM operators consider the information about the relationship among arguments being aggregated.

5. Conclusions

In this paper, we investigated the MADM problems with 2TLNNs. Then, we utilized the Hamy mean (HM) operator, weighted Hamy mean (WHM) operator, dual Hamy mean (DHM) operator and weighted dual Hamy mean (WDHM) operator to develop some Hamy mean aggregation operators with 2TLNNs: 2-tuple linguistic neutrosophic Hamy mean (2TLNHM) operator, 2-tuple linguistic neutrosophic weighted Hamy mean (2TLNWHM) operator, 2-tuple linguistic neutrosophic dual Hamy mean (2TLNDHM) operator, and 2-tuple linguistic neutrosophic weighted dual Hamy mean (2TLNWDHM) operator. The prominent characteristics of these proposed operators were studied. Then, we utilized these operators to develop some approaches to solve MADM problems with 2TLNNs. Finally, a practical example for construction engineering project risk assessment was given to show the developed approach. In the future, the application of the 2TLNNs needs to be investigated under uncertain [35–46] and fuzzy environments [47–54].

Author Contributions: S.W., J.W., G.W. and Y.W. conceived and worked together to achieve this work, J.W. compiled the computing program by Matlab and analyzed the data, J.W. and G.W. wrote the paper. Finally, all the authors have read and approved the final manuscript.

Acknowledgments: The work was supported by the National Natural Science Foundation of China under Grant No. 71571128 and the Humanities and Social Sciences Foundation of Ministry of Education of the People's Republic of China (16XJA630005) and the Construction Plan of Scientific Research Innovation Team for Colleges and Universities in Sichuan Province (15TD0004).

Conflicts of Interest: The authors declare no conflict of interest.

References

1. Smarandache, F. *Neutrosophy: Neutrosophic Probability, Set, and Logic: Analytic Synthesis & Synthetic Analysis*; American Research Press: Rehoboth, DE, USA, 1998.
2. Smarandache, F. *A Unifying Field in Logics: Neutrosophic Logic. Neutrosophy, Neutrosophic Set, Neutrosophic Probability and Statistics*, 3rd ed.; Phoenix: Xiquan, China, 2003.
3. Zadeh, L.A. Fuzzy sets. *Inf. Control* **1965**, *8*, 338–353. [CrossRef]
4. Atanassov, K. Intuitionistic fuzzy sets. *Fuzzy Sets Syst.* **1986**, *20*, 87–96. [CrossRef]
5. Atanassov, K.; Gargov, G. Interval-valued intuitionistic fuzzy sets. *Fuzzy Sets Syst.* **1989**, *31*, 343–349. [CrossRef]
6. Wang, H.; Smarandache, F.; Zhang, Y.Q.; Sunderraman, R. Single valued neutrosophic sets. *Multispace Multistruct.* **2010**, *4*, 410–413.
7. Wang, H.; Smarandache, F.; Zhang, Y.Q.; Sunderraman, R. *Interval Neutrosophic Sets and Logic: Theory and Applications in Computing*; Hexis: Phoenix, AZ, USA, 2005.
8. Ye, J. Multicriteria decision-making method using the correlation coefficient under single-valued neutrosophic environment. *Int. J. General Syst.* **2013**, *42*, 386–394. [CrossRef]
9. Broumi, S.; Smarandache, F. Correlation coefficient of interval neutrosophic set. *Appl. Mech. Mater.* **2013**, *436*, 511–517. [CrossRef]
10. Biswas, P.; Pramanik, S.; Giri, B.C. TOPSIS method for multi-attribute group decision-making under single-valued neutrosophic environment. *Neural Comput. Appl.* **2016**, *27*, 727–737. [CrossRef]
11. Liu, P.D.; Chu, Y.C.; Li, Y.W.; Chen, Y.B. Some generalized neutrosophic number Hamacher aggregation operators and their application to Group Decision Making. *Int. J. Fuzzy Syst.* **2014**, *16*, 242–255.
12. Sahin, R.; Liu, P.D. Maximizing deviation method for neutrosophic multiple attribute decision making with incomplete weight information. *Neural Comput. Appl.* **2016**, *27*, 2017–2029. [CrossRef]

13. Ye, J. Similarity measures between interval neutrosophic sets and their applications in multicriteria decision-making. *J. Intell. Fuzzy Syst.* **2014**, *26*, 165–172.
14. Zhang, H.Y.; Wang, J.Q.; Chen, X.H. Interval neutrosophic sets and their application in multicriteria decision making problems. *Sci. Word J.* **2014**, *2014*, 1–15. [CrossRef] [PubMed]
15. Ye, J. A multicriteria decision-making method using aggregation operators for simplified neutrosophic sets. *J. Intell. Fuzzy Syst.* **2014**, *26*, 2459–2466.
16. Peng, J.J.; Wang, J.Q.; Wang, J.; Zhang, H.Y.; Chen, X.H. Simplified neutrosophic sets and their applications in multicriteria group decision-making problems. *Int. J. Syst. Sci.* **2016**, *47*, 2342–2358. [CrossRef]
17. Peng, J.J.; Wang, J.Q.; Zhang, H.Y.; Chen, X.H. An outranking approach for multi-criteria decision-making problems with simplified neutrosophic sets. *Appl. Soft Comput.* **2014**, *25*, 336–346. [CrossRef]
18. Zhang, H.; Wang, J.Q.; Chen, X.H. An outranking approach for multi-criteria decision-making problems with interval-valued neutrosophic sets. *Neural Comput. Appl.* **2016**, *27*, 615–627. [CrossRef]
19. Liu, P.D.; Xi, L. The neutrosophic number generalized weighted power averaging operator and its application in multiple attribute group decision making. *Int. J. Mach. Learn. Cybernet.* **2016**. [CrossRef]
20. Deli, I.; Şubaş, Y. A ranking method of single valued neutrosophic numbers and its applications to multiattribute decision making problem. *Int. J. Mach. Learn. Cybernet.* **2017**, *8*, 1309–1322. [CrossRef]
21. Peng, J.J.; Wang, J.Q.; Wu, X.H.; Wang, J.; Chen, X.H. Multi-valued neutrosophic sets and power aggregation operators with their applications in multi-criteria group decision-making problems. *Int. J. Comput. Intell. Syst.* **2015**, *8*, 345–363. [CrossRef]
22. Zhang, H.Y.; Ji, P.; Wang, J.Q.; Chen, X.H. An improved weighted correlation coefficient based on integrated weight for interval neutrosophic sets and its application in multi-criteria decision-making problems. *Int. J. Comput. Intell. Syst.* **2015**, *8*, 1027–1043. [CrossRef]
23. Chen, J.Q.; Ye, J. Some Single-Valued Neutrosophic Dombi Weighted Aggregation Operators for Multiple Attribute Decision-Making. *Symmetry* **2017**, *9*, 82. [CrossRef]
24. Liu, P.D.; Wang, Y.M. Multiple attribute decision making method based on single-valued neutrosophic normalized weighted Bonferroni mean. *Neural Comput. Appl.* **2014**, *25*, 2001–2010. [CrossRef]
25. Wu, X.H.; Wang, J.Q.; Peng, J.J.; Chen, X.H. Cross-entropy and prioritized aggregation operator with simplified neutrosophic sets and their application in multi-criteria decision-making problems. *J. Intell. Fuzzy Syst.* **2016**, *18*, 1104–1116. [CrossRef]
26. Li, Y.; Liu, P.; Chen, Y. Some Single Valued Neutrosophic Number Heronian Mean Operators and Their Application in Multiple Attribute Group Decision Making. *Informatica* **2016**, *27*, 85–110. [CrossRef]
27. Zavadskas, E.K.; Bausys, R.; Juodagalviene, B.; Garnyte-Sapranaviciene, I. Model for residential house element and material selection by neutrosophic MULTIMOORA method. *Eng. Appl. Artif. Intell.* **2017**, *64*, 315–324. [CrossRef]
28. Zavadskas, E.K.; Bausys, R.; Kaklauskas, A.; Ubarte, I.; Kuzminske, A.; Gudiene, N. Sustainable market valuation of buildings by the single-valued neutrosophic MAMVA method. *Appl. Soft Comput.* **2017**, *57*, 74–87. [CrossRef]
29. Bausys, R.; Juodagalviene, B. Garage location selection for residential house by WASPAS-SVNS method. *J. Civ. Eng. Manag.* **2017**, *23*, 421–429. [CrossRef]
30. Wu, Q.; Wu, P.; Zhou, L.G.; Chen, H.Y.; Guan, X.J. Some new Hamacher aggregation operators under single-valued neutrosophic 2-tuple linguistic environment and their applications to multi-attribute group decision making. *Comput. Ind. Eng.* **2017**, *116*. [CrossRef]
31. Herrera, F.; Martinez, L. A 2-tuple fuzzy linguistic representation model for computing with words. *IEEE Trans. Fuzzy Syst.* **2000**, *8*, 746–752.
32. Hara, T.; Uchiyama, M.; Takahasi, S.E. A refinement of various mean inequalities. *J. Inequal. Appl.* **1998**, *2*, 387–395. [CrossRef]
33. Fang, Z.; Ye, J. Multiple Attribute Group Decision-Making Method Based on Linguistic Neutrosophic Numbers. *Symmetry* **2017**, *9*, 111. [CrossRef]
34. Shi, L.; Ye, J. Cosine Measures of Linguistic Neutrosophic Numbers and Their Application in Multiple Attribute Group Decision-Making. *Information* **2017**, *8*, 117.
35. Chen, T. The inclusion-based TOPSIS method with interval-valued intuitionistic fuzzy sets for multiple criteria group decision making. *Appl. Soft Comput.* **2015**, *26*, 57–73. [CrossRef]

36. Pérez-Fernández, R. Monotonicity-based consensus states for the monometric rationalisation of ranking rules with application in decision making. *Int. J. Approx. Reason.* **2018**, *16*, 109–110. [CrossRef]
37. Gao, H.; Wei, G.; Huang, Y. Dual Hesitant Bipolar Fuzzy Hamacher Prioritized Aggregation Operators in Multiple Attribute Decision Making. *IEEE Access* **2018**, *6*, 11508–11522. [CrossRef]
38. Krishnamurthy, M.; Marcinek, P.; Malik, K.M.; Afzal, M. Representing Social Network Patient Data as Evidence-Based Knowledge to Support Decision Making in Disease Progression for Comorbidities. *IEEE Access* **2018**, *6*, 12951–12965. [CrossRef]
39. Rahman, M.A.; Mezhuyev, V.; Bhuiyan, M.Z.A.; Sadat, S.N.; Zakaria, S.A.B.; Refat, N. Reliable Decision Making of Accepting Friend Request on Online Social Networks. *IEEE Access* **2018**, *6*, 9484–9491. [CrossRef]
40. Ma, X.; Zhan, J.; Ali, M.I.; Mehmood, N. A survey of decision making methods based on two classes of hybrid soft set models. *Artif. Intell. Rev.* **2018**, *49*, 511–529. [CrossRef]
41. Garg, H.; Arora, R. Generalized and group-based generalized intuitionistic fuzzy soft sets with applications in decision-making. *Appl. Intell.* **2018**, *48*, 343–356. [CrossRef]
42. Tang, X.Y.; Wei, G.W. Models for green supplier selection in green supply chain management with Pythagorean 2-tuple linguistic information. *IEEE Access* **2018**, *6*, 18042–18060. [CrossRef]
43. Wei, G.W.; Lu, M. Pythagorean Fuzzy Maclaurin Symmetric Mean Operators in multiple attribute decision making. *Int. J. Intell. Syst.* **2018**, *33*, 1043–1070. [CrossRef]
44. Jiang, F.; Ma, Q. Multi-attribute group decision making under probabilistic hesitant fuzzy environment with application to evaluate the transformation efficiency. *Appl. Intell.* **2018**, *48*, 953–965. [CrossRef]
45. Wu, P.; Liu, S.; Zhou, L.; Chen, H. A fuzzy group decision making model with trapezoidal fuzzy preference relations based on compatibility measure and COWGA operator. *Appl. Intell.* **2018**, *48*, 46–67. [CrossRef]
46. Kamacı, H.; Atagün, A.O.; Sönmezoğlu, A. Row-products of soft matrices with applications in multiple-disjoint decision making. *Appl. Soft Comput.* **2018**, *62*, 892–914. [CrossRef]
47. Wei, G.W.; Gao, H.; Wei, Y. Some q-Rung Orthopair Fuzzy Heronian Mean Operators in Multiple Attribute Decision Making. *Int. J. Intell. Syst.* **2018**. [CrossRef]
48. Wei, G.W.; Lu, M.; Tang, X.Y.; Wei, Y. Pythagorean Hesitant Fuzzy Hamacher Aggregation Operators and Their Application to Multiple Attribute Decision Making. *Int. J. Intell. Syst.* **2018**, 1–37. [CrossRef]
49. Liao, H.C.; Yang, L.Y.; Xu, Z.S. Two new approaches based on ELECTRE II to solve the multiple criteria decision making problems with hesitant fuzzy linguistic term sets. *Appl. Soft Comput.* **2018**, *63*, 223–234. [CrossRef]
50. Liu, F.; Liu, Z.-L.; Wu, Y.-H. A group decision making model based on triangular fuzzy additive reciprocal matrices with additive approximation-consistency. *Appl. Soft Comput.* **2018**, *65*, 349–359. [CrossRef]
51. Liang, H.; Xiong, W.; Dong, Y. A prospect theory-based method for fusing the individual preference-approval structures in group decision making. *Comput. Ind. Eng.* **2018**, *117*, 237–248. [CrossRef]
52. Wu, T.; Liu, X.; Qin, J. A linguistic solution for double large-scale group decision-making in E-commerce. *Comput. Ind. Eng.* **2018**, *116*, 97–112. [CrossRef]
53. Wu, H.; Xu, Z.; Ren, P.; Liao, H. Hesitant fuzzy linguistic projection model to multi-criteria decision making for hospital decision support systems. *Comput. Ind. Eng.* **2018**, *115*, 449–458. [CrossRef]
54. Xu, Y.; Wen, X.; Zhang, W. A two-stage consensus method for large-scale multi-attribute group decision making with an application to earthquake shelter selection. *Comput. Ind. Eng.* **2018**, *116*, 113–129. [CrossRef]

sustainability

MDPI

Article

Application of Fuzzy DEMATEL Method for Analyzing Occupational Risks on Construction Sites

Sukran Seker [1,*] **and Edmundas Kazimieras Zavadskas** [2]

[1] Department of Industrial Engineering, Yildiz Technical University, 34349 Besiktas, Turkey
[2] Faculty of Civil Engineering, Vilnius Gediminas Technical University, LT-10223 Vilnius, Lithuania;
    edmundas.zavadskas@vgtu.lt
* Correspondence: sseker@yildiz.edu.tr; Tel.: +90-212-383-2875

Received: 7 September 2017; Accepted: 9 November 2017; Published: 13 November 2017

Abstract: The construction industry is known as a hazardous industry because of its complexity and strategic nature. Therefore, it is important to know the main causes of occupational accidents to prevent fatal occupational accidents in construction industry. At building construction sites, workers performing tasks are continuously exposed to risks, not only emerging from their own mistakes but also from the mistakes of their co-workers. A great deal of studies investigating risks and preventing occupational hazards for the construction industry has been carried out in the literature. The quantitative conventional methods mostly use either probabilistic techniques or statistics, or both, but they have limitations dealing with the ambiguity and fuzziness in information. In this study, to overcome these limitations, an applicable and improved approach, which helps construction managers to propose preventive measures for accidents on construction sites, is proposed to simplify the risk assessment. It is shown that the Fuzzy Decision Making Trial and Evaluation Laboratory (DEMATEL) method can evaluate causal factors of occupational hazards by a cause–effect diagram and improve certain safety measures on construction sites. In addition, sensitivity analysis is conducted to verify the robustness of the results.

Keywords: risk evaluation; construction sites; occupational accidents; fuzzy sets; DEMATEL

1. Introduction

The construction industry is infamous for the highest accident rates compared to any other industry in many parts of the world [1]. The amount of work accidents and injuries has been increasing drastically for years in construction industry. This occurs because the construction industry comprises higher percentage of self-employed workers, and large number of seasonal and migrant workers [2]. Hence, the importance of accident and injury prevention, which requires a knowledge of accidents' causal factors and how the factors increase the probability of risks that can cause accidents, has arisen [3,4].

The literature on occupational risk assessment reveals that accidents are caused from a wide range of factors such as unsafe tools, conditions related with the work site, the industry specific problems, unsafe methods related with the work, human factors and management issues [5]. Physical hazards on construction sites occur because of continuing exposure to mechanical process or work activity. As a result, physical hazards can cause various types of injuries, from minor and requiring first aid only, to disabling and/or fatal.

Physical hazards involve conditions such as working at height, falling objects, exposure to electricity, etc. Falls from height have been regarded as the most frequent cause of injury or death among the accidents on construction sites [6]. Slips and trips that cause fall are also assumed as the most prevalent occupational hazards and lead to one third of all serious injuries. Being struck and crushed by equipment, fires and explosions related with the ignition of flammable materials are other common occupational risks on construction sites. For example, Pipitsupaphol and Watanabe [7] represented that falling, workers being struck by falling objects, stepping on or striking against objects are the three most frequently occurring type of accidents in Thailand. In addition, OSHA (Occupational Safety and Health Administration) and Huang and Hinze [8] informed that falls and struck by falling objects also have been the cause of the highest number of injuries and fatalities in the U.S. construction industry.

As for occupation type, particularly, falling off machines and machines overturning when travelling up or down slopes are commonly encountered major hazards for construction site workers [9]. In addition, electrical devices such as cables, circuit breaker panels and cords present a high risk for workers in areas exposed to electricity. Common sources of these physical injuries on construction sites occur due to technical or human errors [10]. Dumrak and Mostafa [11] and Jackson and Loomis [12] claimed that truck drivers, plant operators, electricians are highly susceptible to fatal accidents.

These may arise from lack of safety knowledge, training, supervision, uncontrolled working environment, inability to carry out a task safely, and error of judgments, carelessness, apathy or reckless operations. Unsafe behaviors, which are the results of a poor safety culture, are other significant factors in the cause of site accidents [13]. In addition, safety is considered as part of Total Quality Management (TQM); poor safety practices are also accepted as the cause of accidents and subsequent injuries [2].

According to Toole's [14] study in the USA, the causes of accidents include: unsafe methods or sequencing; deficient enforcement of safety; lack of proper training; safety equipment not provided; unsafe site conditions; poor attitude toward safety; not using provided safety equipment; and remote and deviation from regular behavior. Similarly, in their study, Tam et al. [15] addressed that the main factors affecting safety performance are reckless operations, lack of training, and poor safety awareness of supervisors and top management.

Tam et al. [15] represented that trained or skilled workers ensure improvement of the site safety. However, high mobility of workers on construction sites and frequent move from one construction site to another makes it difficult to train workers. Dester and Blockley [16] and Zhou et al. [17] pointed that the poor safety in the construction industry occurs because of poor safety culture rather than the inherent (nature) hazards of the industry. In addition, Agvu and Olele [18] represented that poor safety culture cause increased rate of unsafe acts/fatalities in the Nigerian construction industry. Barofsky and Legro [19] and Folkard and Tucker [20] concluded that fatigue is a versatile and complex occurrence, including physical, mental and emotional stress and other behavioral points, all of which require additional examination. The importance of management commitment to safety and safety regulations, which influence organizational safety performance in a good way, are indicated by Ismail et al. [21]. In addition, they represented that there is a relationship between safe behavior, safety rules, and legalization, and management commitment.

Causes of accidents in construction industry are obtained in many ways using variety of ORA (Occupational Risk Assessment) methods.

In this study, a better and more practical approach is recommended to simplify the risk assessment process for construction industry. The DEMATEL method is commonly used to obtain a cause–effect diagram of interdependent factors. This method is superior to conventional techniques due to exposing the relationships between criteria, ranking the criteria relating to the type of relationships and revealing intensity of their effects on each criterion. Since a single method is not sufficient to identify occupational risks under uncertainty and vagueness, there exists a need to apply an integrated approach to solve the problem considered. Therefore, fuzzy linguistic modeling is utilized to represent and handle flexible information [22].

Accordingly, the DEMATEL method is used to reveal the effect and cause criteria, and to increase the model applicability in terms of linguistic variables combined with triangular fuzzy numbers. As a multi criteria decision making method, DEMATEL contributes to risk assessment literature a different point of view by providing an evaluation that enables modeling cause and effect relationships among the risk factors and exposing the degree of relation or the strength of influence analytically [23].

The proposed approach presents the following advantages compared to traditional methods for ORA:

(1) The proposed method illustrates the interrelationships among critical occupational hazards by constructing causal relationship among construction activities.
(2) Identifying each hazard using triangular fuzzy numbers gives better and more reliable results, as the uncertainty and vagueness of the data can be managed with a fuzzy approach.
(3) The proposed method offers highly accurate and effective material to support the risk assessment procedure because occupational hazards can be better ranked and well evaluated to prevent critical hazards in construction industry.

The remainder of the paper is organized as follows: Section 2 discusses the limitations of traditional methods for ORA in construction industry and common Multi Criteria Decision Making (MCDM) methods used for risk assessment under fuzzy environment. Section 3 formally describes proposed framework steps and applied techniques, and presents the analysis of the proposed approach for risk assessment on construction sites. Section 4 reports the results of aspects, criteria and data analysis, which are based on the proposed steps. In addition, a sensitivity analysis has been carried out to indicate the verification of the results. Finally, Section 5 concludes the paper by summarizing obtained results and discussing strategic decisions.

2. Literature Review: Risk Analysis on Construction Sites

There is comprehensive literature about occupational risk assessment in the construction industry. In addition, there are many approaches for risk analysis in construction industry and other operating facilities [24–27]. In common, the application procedure is presented as follows:

(1) Project managers prepare a corporate safety program, which includes performance standards in a number of classes, and then they compare these classes due to their importance and propose a new safety profile [28].
(2) Decision is made about the strengths and weaknesses of a recent safety program using a safety audit, which is part of the company's safety program [29].
(3) Injury rate of recurrence, which is the number of lost-time injuries per million hours worked and assumed as a method of evaluating safety performance, is computed [30,31].

However, many authors [32–37] presented the shortcomings of traditional methods for ORA because these methods are formed from incomplete information, which is based on uncertainty, vagueness and imprecision. On the other hand, sources of imprecision usually require data obtained from expert judgment, which cannot be evaluated easily by traditional (probabilistic) methods [38]. In addition, Pinto et al. [39] investigated traditional methods for occupational risk assessment in the construction industry pointing out limitations and benefits of using fuzzy sets approaches to cope with imprecise situations. Thus, several methods have been presented using fuzzy principles for analyzing risks [40–44].

However, in recent years, numerous studies have been carried out using MCDM methods to analyze risks in construction industry [45]. Dejus and Antucheviciene [46] proposed MCDM technique for assessment and selection of appropriate solutions for occupational safety. Furthermore, in construction industry, Efe et al. [47] suggested an integrated intuitionistic fuzzy multi-criteria decision making method and a linear programming for risk evaluation in three firms. The paper aimed to overcome the limitations of traditional Failure Mode and Effects Analysis (FMEA) for risk evaluation. Tamošaitienė and Zavadskas [41] proposed risk assessment method using the Technique for Order of Preference by Similarity to Ideal Solution with fuzzy information (TOPSIS-F) method for project of commercial center. Basahel and Taylan [48] suggested a model that can be used to evaluate the most important factors of SMS (safety management system). The significance of these factors and their sub-factors was obtained using the fuzzy Analytic Hierarchy Process (AHP) technique, and the effectiveness of the four construction companies' SMSs was obtained using fuzzy TOPSIS. Liu and Tsai [49] proposed a fuzzy risk assessment method to decrease or prevent occupational hazards. The method includes two-stage quality function deployment (QFD) tables to represent the relationships among construction items, hazard types and causes and a fuzzy Analytic Network Process (ANP) method to classify important hazard types and hazard causes and Failure Modes and Effect Analysis (FMEA) to evaluate the risk value of hazard causes based on the fuzzy inference approach. Janackovic et al. [50] proposed the expert evaluation method and the fuzzy AHP to represent the factors, performance, and indicators of occupational safety with ranking at a Serbian road construction company in Serbia. Kim et al. [51] presented a decision support model, which is based on fuzzy AHP, to quantify the failure risk and to show experts' and practitioners' subjectivities. Using results as an input for fuzzy comprehensive operations, the quantitative failure risks were found. Wu and Shen [52] suggested an assessment model based on the fuzzy grey relational analysis theory for the factors influencing highway construction safety. In addition, the construction safety evaluation index system was built and the weight of each index was found using AHP. Li et al. [53] proposed an improved AHP Method (IAHP) for risk identification. It was conducted to open-cut subway construction to show performance of the proposed approach. The application results show that IAHP is predominant to AHP in terms of comparison matrix (CM) consistency. Yuan et al. [54] improved and employed an effective method to evaluate safety risks on construction projects using the Fuzzy Analytic Network Process (FANP). Zamri et al. [55] proposed a Fuzzy TOPSIS (FTOPSIS) with Z-numbers to handle uncertainty in the construction problems. The proposed approach was used to determine the causes of workers' accident at the construction sites.

In addition, Leonavičiūtė and Dėjus [56] introduced a new MCDM method of the Weighted Aggregated Sum Product Assessment with Grey numbers (WASPAS-G) to determine the best alternative in safety measures to prevent accidents on construction sites. Mohsen and Fereshteh [57] used Z numbers to overcome uncertainty in the experts' judgments for risk assessment in a geothermal power plant. In addition, they used the fuzzy VIseKriterijumska Optimizacija I Kompromisno Resenje (VIKOR) technique to rank and prioritize the failure modes based on the minimum individual regret and the maxi group utility.

The goal of this paper is to depict how Fuzzy DEMATEL can be used as a valuable managerial tool for managers of construction sites to develop effective precautions strategies to avoid construction accidents, and to evaluate causal factors of accidents for construction site of buildings.

3. Fuzzy DEMATEL Approach

Although conventional quantification methods present exact solutions, they are not useful to resolve people-centered problems due to the complexities arising from human factors [58]. Therefore, the concepts of fuzzy set theory introduced by Zadeh [59] who is commonly used in these types of real-world problems where there is uncertainty and fuzziness related with the environment. In real world applications, decision making problems need to be carried out under uncertainty because goals, constraints and possible actions are ambiguous [58].

DEMATEL method is a well-known and comprehensive method to obtain a structural model that provides casual relationships between complex real-world factors. The DEMATEL method is superior to other techniques such as Analytic Hierarchy Process (AHP) since it accounts for the interdependence among the factors of a system via causal diagram, which is overlooked in traditional techniques [60]. The basis of the DEMATEL method comprises the following steps [58,61]:

(1) In this step, the elements related to the problem and degree of influence between elements is formed. The influential factors of the complex system are defined based on data from literature reviews, brainstorming, or expert opinions.

(2) A direct relation matrix is constructed. Then, a questionnaire survey method is carried out after the importance of the measurement scales is defined.

$$
\begin{bmatrix}
0 & x_{12} & \cdots & x_{1n} \\
x_{21} & 0 & \cdots & x_{2n} \\
\vdots & \vdots & \vdots & \vdots \\
x_{n1} & x_{n2} & \cdots & 0
\end{bmatrix}
\tag{1}
$$

(3) A normalized direct relation matrix is built on the basis of direct relation matrix using Equation (2).

$$
\lambda = \frac{1}{\max\limits_{1 \le i \le n} \left(\sum_{j=1}^{n} x_{ij} \right)}, i, j = 1, 2, \ldots, n
\tag{2}
$$

$$
N = \lambda X
$$

(4) The total relation matrix (T) is obtained through Formula (3).

$$
T = \lim_{k \to \infty} \left(N + N^2 + \ldots + N^k \right) = N(1 - N)^{-1}
\tag{3}
$$

(5) In this step, the sum of the values in each column and each row is calculated in the total relation matrix. Thus, D_i shows the sum of the ith row and R_j shows the sum of the jth column. The direct and indirect influences between factors are shown with D_i and R_j, respectively.

$$
D_i = \sum_{j=1}^{n} t_{ij}(i = 1, 2, \ldots, n)
\tag{4}
$$

$$
R_i = \sum_{i=1}^{n} t_{ij}(j = 1, 2, \ldots, n)
\tag{5}
$$

(6) In this stage, cause and effect diagram is built. Horizontal axis $(D + R)$ is made summing R and D, the vertical axis $(D - R)$ is made subtracting R from D. While $(D + R)$ is defined as "prominence", which shows the importance degree of the criterion, $(D - R)$ is defined as "relation", which shows the extent of the influence. If the $(D - R)$ is negative, the criterion is grouped into the effect group. It means that it is influenced by other criteria. If $(D - R)$ is positive, it means that it has a significant impact. It should be improved first.

In this study, Fuzzy DEMATEL method is used to assess causal relations of accidents for construction processes. This combination is used for the imprecise and subjective nature of human judgments. Interval sets are used rather than real numbers in fuzzy set theory. Linguistics terms are converted to fuzzy numbers. The proposed method is favorable to reveal the relationships among factors and ranking the criteria related to the type of relationships and the impact of severe degree on each criterion.

The analysis procedures of Fuzzy DEMATEL method are explained as follows [61,62].

Step 1. Define the evaluation criteria.

Step 2. Select a group of experts who have knowledge and experience about problem to evaluate the effect between factors using pairwise comparison.

Step 3. Define the fuzzy linguistic scale for dealing with the vagueness of human assessments, the linguistic variable "influence" is used with a five-level scale containing the following scale items in the group decision-making proposed by Li [63]: No influence, Very low influence, Low influence, High influence, and Very high influence. The fuzzy numbers for these linguistic terms are given in Table 1.

Step 4. Obtain an initial direct relation matrix with pair wise comparison. Develop the initial fuzzy direct-relation matrix Z^k by having evaluators introduce the fuzzy pair-wise influence relationships between the components in a $n \times n$ matrix where k is the number of experts. Accordingly, the direct-relation matrix is established as $Z^k = [z^k{}_{ij}]$ where Z is a $n \times n$ non-negative matrix; z_{ij} represents the direct impact of factor i on factor j; and, when $i = j$, the diagonal elements $z_{ij} = 0$.

For simplicity, denote Z^k as

$$z^k_{ij} = (l_{ij}, m_{ij}, u_{ij})$$

$$Z^k = \begin{array}{c} C_1 \\ C_2 \\ \vdots \\ C_n \end{array} \left[\begin{array}{cccc} [0,0] & \otimes z^k_{12} & \cdots & \otimes z^k_{1n} \\ \otimes z^k_{21} & [0,0] & \cdots & \otimes z^k_{2n} \\ \vdots & \vdots & \vdots & \vdots \\ \otimes z^k_{n1} & \otimes z^k_{n2} & \cdots & [0,0] \end{array} \right] \tag{6}$$

Step 5. Obtain the normalized fuzzy direct-relation matrix "D" using Expressions (7) related to the overall fuzzy direct-relation matrix Z.

$$D = \frac{Z^k}{\max\limits_{1 \leq i \leq n} \sum_{j=1}^{n} z_{ij}}, i, j = 1, 2, \ldots, n \tag{7}$$

Step 6. Compute the total-relation matrix T using Expression (8), where $n \times n$ identity matrix is represented with I. Upper and lower values are calculated separately.

$$T = D(I - D)^{-1} \tag{8}$$

where $T = D + D^2 + \ldots + = \sum\limits_{i=1}^{\infty} D^i$.

Step 7. Determine row (r_i) and column (c_j) sums for each row i and column j from the T matrix, respectively, with following equations.

$$T = [t_{ij}]_{nxn} \qquad\qquad i, j = 1, 2, \ldots n \tag{9}$$

$$r_i = \sum_{1 \leq j \leq n}^{n} t_{ij} \qquad\qquad \forall i \tag{10}$$

$$c_j = \sum_{1 \leq i \leq n}^{n} t_{ij} \qquad\qquad \forall j \tag{11}$$

Step 8. The causal diagram is built with the horizontal axis $(r_i + c_j)$ and the vertical axis $(r_i - c_j)$. The horizontal axis "Prominence" refers the importance degree of the factor, whereas the vertical axis "Relation" shows the extent of the influence. If the $(r_i - c_j)$ axis is positive, the factor is in the cause group. Otherwise, if the $(r_i - c_j)$ axis is negative, the factor is in the effect group. Causal diagrams can convert complex relationships of factors into an easy to understand structural model, providing awareness for problem solving.

Table 1. The fuzzy linguistic scale for the respondents' evaluations [63].

Linguistic Terms	Corresponding Triangular Fuzzy Numbers (TFNs)
No influence (NO)	(0, 0, 0.25)
Very low influence (VL)	(0, 0.25, 0.5)
Low influence (L)	(0.25, 0.5, 0.75)
High influence (H)	(0.5, 0.75, 1)
Very high influence (VH)	(0.75, 1, 1)

The Application Procedures of Fuzzy DEMATEL for Construction Sites—Analysis of the Proposed Approach

Based on risk assessment standard process, this study utilizes the Fuzzy DEMATEL approach to identify risk assessment elements and evaluate comprehensive influence relations for construction sites. Implementation of Fuzzy DEMATEL approach for critical occupational hazards on construction sites is introduced as follows (see Figure 1).

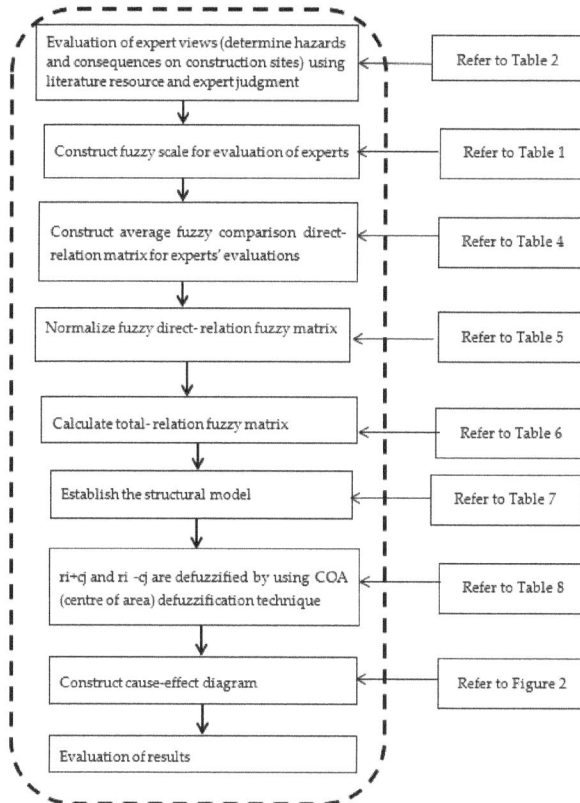

Figure 1. Implementation of a Fuzzy DEMATEL approach.

Step 1. Determine risk factors: In risk identification, 14 safety causal factors related to occupational hazards were constructed using an extensive literature review and witnesses of evaluators.

Step 2. Before the implementation of this approach, five evaluators, who actively work on construction sites, evaluated the causal factors of accidents. Evaluators with varying ages (30–50 years) and career lengths (5–30 years), as an indicator of their level of experience, were selected. The characteristics of the five decision-making evaluators are given in Table 2.

The evaluators expressed their thoughts relating to their knowledge, experience and expertise. In this step, significant causal factors of accidents in construction sites and potential occupational hazards are identified in Table 3. The experience and the knowledge level of the evaluators were different. However, their understanding of the causes of accidents in construction industry was adequate to be selected for this study.

Table 2. The characteristics of the five decision-making evaluators.

	Age	Education Level	Experience (Years)	Job Title	Job Responsibility
Evaluator 1	35	Bachelor in Civil Engineering and a getting certificate in occupational health and safety	>10	Construction manager	directing construction projects, planning site investigations, conducting feasibility studies, providing construction safety
Evaluator 2	50	Bachelor in Civil Engineering and a getting certificate in occupational health and safety	>25	Construction Manager and Construction Safety Expert	directing construction projects, planning site investigations, conducting feasibility studies, providing construction safety
Evaluator 3	40	Bachelor in Civil Engineering and PhD in occupational health and safety	>15	Construction Safety Expert	providing a safe working environment for workers, preparing safety procedures, training of workers on safety topics.
Evaluator 4	30	Bachelor in Civil Engineering and a getting certificate in occupational health and safety	>5	Construction Safety Expert	providing a safe working environment for workers, preparing safety procedures, training of workers on safety topics.
Evaluator 5	30	Bachelor in Civil Engineering and a getting certificate in occupational health and safety	>5	Construction Safety Expert	providing a safe working environment for workers, preparing safety procedures, training of workers on safety topics.

Table 3. Significant causal factors of accidents and Potential critical occupational hazards on construction sites.

Code	Causal Factors of Accidents	Potential Critical Occupational Hazards
T1	Worker actions/behavior	Improper positioning and posture while working, not wearing personal protective equipment (PPE) provided, falling from height, getting foreign objects into the eye such as during welding, cutting, grinding, etc.
T2	Worker capabilities (knowledge, skills, experience)	Exceeding the operator's lifting capacity, lack of experience of using vehicles bulldozers, diggers, excavator, etc., lack of experience on operations welding, cutting, etc., cave-ins (while or after excavation), transport accidents, exposure to fumes, gases, fire and explosions, etc.
T3	The lack of proper communication among workers or between workers and employees.	Confusion because of the physical distance between workers or high levels of background noise or poor line of communication among safety officer and employees, heavy equipment accidents, exposure to electricity, etc.
T4	Worker health/fatigue	Unsafe behavior due to fatigue, falls from vehicles (bulldozers, diggers, excavator), transport accidents on site, loss of balance (while working on the scaffold), etc.

Table 3. *Cont.*

Code	Causal Factors of Accidents	Potential Critical Occupational Hazards
T5	Site conditions	Working at crowded space, poor warning signage, working at high level (expose to high level), tripping over cables or falling into holes, crushed, jammed or pinched in or between objects, caught between machinery part, etc.
T6	Work scheduling	Irregular work schedules of workers (extended work hours), bad planning between tasks and workers, and using the wrong equipment, power tool and machinery accidents, cave-ins (while or after excavation), etc.
T7	Unsuitability of materials	Not meeting specification requirements of materials, not disposing all surplus and unsuitable materials; such as brush, grass, weeds, tripped on rubble, exposure to gases, fumes, smoke, fire, etc.
T8	Unsuitability of equipment	Shortcomings of equipment, including PPE, etc.; inappropriate use of equipment for the tasks such as carrying, lifting heavy equipment accidents, crushed, jammed or pinched in or between objects, etc.
T9	Safety culture	Not making safety an integral part of the job, not having pre-qualified of contractors for safety, not training workers on use of safety equipment, safety expectations, and any safety risks and precautions relevant to their job duties, falls, tripped on brick on scaffold, struck scaffold, etc.
T10	Construction process	Manual handling, exposure to hazardous materials, scaffolding, ground working, struck by formwork on crane, building/structure collapse, etc.
T11	Shortage of safety management	Lack of personal protection equipment, regular safety meetings, and safety training, fall accidents, struck by lorry platform whilst attaching it, struck by scaffold, etc.
T12	Poor education of laborers, inadequate safety training	Lack of proper education, not receiving proper safety training and, lack of understanding of the job, not training on the accident prevention, risk and work hazard identification, reporting near-miss falls from height, falling objects, unsafe lifting, unsafe carrying, unsafe placing, etc.
T13	Poor safety conscientiousness of laborers	Lack of safety awareness and conscious of construction workers on the job-related safety and health issues, lack of conscious on the wearing of personal protective equipment, lack of enforcement of safety regulations, exposure to electricity, caught between machinery parts, etc.
T14	Poor site management	Not providing sufficient PPE and safety equipment of management, lack of education (safety training and orientation), fall accidents (fall from height and falling objects), crushed by plasterboard whilst removing from trolley, exposure to high level vibration, etc.

Step 3. Evaluators used five level scales containing the following scale item factor influence relationships: Very low, Low, Medium, High, and Very high (see Table 1).

Step 4. The pair wise comparison was made by using linguistics variables. The average linguistic scores of the evaluator opinions were shown in Table 4. Using the fuzzy scale shown in Table 1, the initial direct relation matrix was obtained.

Step 5. Using presence of the initial direct relation matrix, the normalized fuzzy direct-relation matrix "*N*" was built. The normalized fuzzy direct-relation matrix can be calculated using Expressions (7) (see Table 5).

Step 6. Total-relation fuzzy matrix was derived, after having obtained normalized direct-relation fuzzy matrix. This can be obtained by using Expression (8), where it is represented an $n \times n$ identity matrix. Total relation fuzzy matrix was shown in Table 6.

Step 7. The structural model was established. After having built matrix T, $r_i + c_j$ and $r_i - c_j$ were determined. In Expressions (9)–(11), r_i and c_j are obtained with the sum of the rows and columns of matrix T. While $r_i + c_j$ donates the importance of factor I, $r_i - c_j$ indicates the net effect of factor i. Results were presented in Table 7.

Step 8. Using Centre of Area (COA) defuzzification technique, $r_i + c_j$ and $r_i - c_j$ were defuzzified and obtained Best Non-fuzzy Performance (BNP) values [64]. In the COA process, the defuzzified factor of risk is represented by the geometric center of the area limited by the curve that represents its membership function. Crisp values of r_i, c_j, $r_i + c_j$ and $r_i - c_j$ are shown in Table 8.

Table 4. Linguistic assessment of the evaluators' opinion (average).

Code	T1	T2	T3	T4	T5	T6	T7	T8	T9	T10	T11	T12	T13	T14
T1	NO	L	H	VH	NO	NO	NO	NO	H	H	H	NO	NO	H
T2	VH	NO	VH	H	L	VH	NO	NO	H	VH	L	L	L	L
T3	H	NO	NO	NO	L	NO	NO	NO	NO	H	H	NO	H	H
T4	H	VH	H	NO	NO	L	NO	NO	NO	H	H	NO	H	H
T5	H	VH	H	H	NO	L	L	L	L	H	H	NO	L	L
T6	H	L	NO	H	NO	NO	NO	NO	NO	VH	NO	L	L	H
T7	VH	L	NO	VH	NO	H	NO	NO	L	VH	H	NO	H	H
T8	VH	L	NO	VH	L	H	NO	NO	L	VH	H	NO	H	H
T9	VH	L	H	VH	L	NO	H	H	NO	H	VH	H	VH	H
T10	H	NO	NO	H	L	H	H	H	NO	NO	L	L	L	H
T11	VH	L	L	VH	H	H	H	H	VH	H	NO	H	VH	VH
T12	H	L	H	VH	L	H	H	H	H	VH	H	NO	VH	H
T13	H	L	VH	H	NO	H	NO	NO	H	VH	VH	H	NO	VH
T14	VH	L	H	VH	VH	VH	VH	VH	VH	H	H	H	VH	NO

Table 5. Normalized initial direct-relation fuzzy matrix.

Code	T1	T2	T3	T4	T5	...	T13	T14
T1	(0.00, 0.00, 0.02)	(0.04, 0.06, 0.08)	(0.04, 0.06, 0.08)	(0.06, 0.08, 0.08)	(0.00, 0.00, 0.02)	...	(0.00, 0.00, 0.02)	(0.04, 0.06, 0.08)
T2	(0.00, 0.00, 0.02)	(0.00, 0.00, 0.02)	(0.06, 0.08, 0.08)	(0.04, 0.06, 0.08)	(0.02, 0.04, 0.06)	...	(0.02, 0.04, 0.06)	(0.02, 0.04, 0.06)
T3	(0.00, 0.00, 0.02)	(0.00, 0.00, 0.02)	(0.00, 0.00, 0.02)	(0.00, 0.00, 0.00)	(0.04, 0.06, 0.06)	...	(0.04, 0.06, 0.08)	(0.04, 0.06, 0.08)
T4	(0.00, 0.00, 0.02)	(0.08, 0.08, 0.08)	(0.04, 0.06, 0.08)	(0.00, 0.00, 0.00)	(0.00, 0.00, 0.02)	...	(0.04, 0.06, 0.08)	(0.04, 0.06, 0.08)
T5	(0.00, 0.00, 0.02)	(0.08, 0.08, 0.08)	(0.04, 0.06, 0.08)	(0.04, 0.06, 0.08)	(0.00, 0.00, 0.02)	...	(0.02, 0.04, 0.06)	(0.02, 0.04, 0.06)
⋮								
T13	(0.04, 0.06, 0.08)	(0.04, 0.06, 0.06)	(0.06, 0.08, 0.08)	(0.04, 0.06, 0.08)	(0.00, 0.00, 0.02)	...	(0.00, 0.00, 0.02)	(0.06, 0.08, 0.08)
T14	(0.06, 0.08, 0.08)	(0.04, 0.06, 0.06)	(0.04, 0.06, 0.08)	(0.06, 0.08, 0.08)	(0.08, 0.08, 0.08)	...	(0.06, 0.08, 0.08)	(0.00, 0.00, 0.02)

Table 6. Total-relation fuzzy matrix.

Code	T1	T2	T3	T4	T5	...	T13	T14
T1	(0.02, 0.18, 0.24)	(0.03, 0.15, 0.24)	(0.05, 0.19, 0.24)	(0.07, 0.26, 0.28)	(0.01, 0.09, 0.16)	...	(0.02, 0.17, 0.21)	(0.05, 0.22, 0.28)
T2	(0.08, 0.48, 0.33)	(0.01, 0.25, 0.33)	(0.07, 0.38, 0.28)	(0.06, 0.49, 0.32)	(0.03, 0.24, 0.22)	...	(0.04, 0.42, 0.28)	(0.04, 0.42, 0.31)
T3	(0.05, 0.21, 0.27)	(0.01, 0.09, 0.27)	(0.01, 0.12, 0.17)	(0.02, 0.16, 0.21)	(0.03, 0.12, 0.18)	...	(0.05, 0.20, 0.24)	(0.05, 0.20, 0.27)
T4	(0.06, 0.26, 0.30)	(0.06, 0.19, 0.30)	(0.05, 0.21, 0.25)	(0.02, 0.21, 0.24)	(0.01, 0.10, 0.16)	...	(0.05, 0.24, 0.27)	(0.05, 0.25, 0.30)
T5	(0.06, 0.30, 0.35)	(0.07, 0.22, 0.35)	(0.05, 0.23, 0.28)	(0.06, 0.31, 0.33)	(0.01, 0.12, 0.19)	...	(0.04, 0.26, 0.29)	(0.04, 0.26, 0.32)
⋮								
T13	(0.07, 0.58, 0.34)	(0.03, 0.36, 0.34)	(0.07, 0.47, 0.28)	(0.07, 0.61, 0.33)	(0.01, 0.27, 0.19)	...	(0.02, 0.50, 0.25)	(0.08, 0.56, 0.34)
T14	(0.10, 0.67, 0.40)	(0.04, 0.40, 0.40)	(0.06, 0.50, 0.32)	(0.10, 0.70, 0.38)	(0.07, 0.37, 0.27)	...	(0.09, 0.63, 0.35)	(0.04, 0.55, 0.33)

Table 7. Fuzzy values of r_i, c_i, $r_i + c_i$ and $r_i - c_i$.

Code	r_i	c_i	$r_i + c_i$	$r_i - c_i$
T1	(0.436, 3.097, 3.076)	(0.957, 5.753, 4.635)	(1.393, 8.85, 7.712)	(−0.521, −2.656, −1.559)
T2	(0.613, 7.058, 3.81)	(0.458, 3.487, 4.635)	(1.07, 10.545, 8.445)	(0.155, 3.571, −0.826)
T3	(0.351, 2.653, 2.913)	(0.586, 3.487, 4.635)	(0.937, 6.14, 7.549)	(−0.235, −0.834, −1.722)
T4	(0.489, 3.52, 3.313)	(0.933, 5.998, 4.39)	(1.422, 9.518, 7.703)	(−0.444, −2.478, −1.077)
T5	(0.57, 4.157, 3.91)	(0.341, 2.806, 2.848)	(0.911, 6.963, 6.758)	(0.228, 1.351, 1.062)
T6	(0.367, 5.535, 2.917)	(0.605, 4.316, 3.69)	(0.971, 9.851, 6.607)	(−0.238, 1.219, −0.773)
T7	(0.577, 4.078, 3.54)	(0.374, 3.164, 2.776)	(0.951, 7.242, 6.316)	(0.202, 0.914, 0.763)
T8	(0.606, 4.272, 3.725)	(0.374, 6.137, 2.776)	(0.981, 10.409, 6.501)	(0.232, −1.865, 0.949)
T9	(0.8, 9.342, 4.362)	(0.522, 3.9, 3.344)	(1.322, 13.242, 7.707)	(0.279, 5.442, 1.018)
T10	(0.489, 6.246, 3.652)	(0.95, 6.137, 4.635)	(1.439, 12.383, 8.288)	(−0.462, 0.109, −0.983)
T11	(0.888, 9.908, 4.59)	(0.728, 4.926, 4.206)	(1.616, 14.834, 8.796)	(0.161, 4.982, 0.384)
T12	(0.816, 5.372, 4.572)	(0.35, 26.012, 2.927)	(1.166, 31.384, 7.499)	(0.466, −20.64, 1.646)
T13	(0.709, 9.015, 3.901)	(0.719, 5.275, 4.003)	(1.428, 14.289, 7.903)	(−0.01, 3.74, −0.102)
T14	(0.988, 10.167, 4.66)	(0.801, 5.289, 4.489)	(1.789, 15.456, 9.149)	(0.187, 4.879, 0.171)

Table 8. Crisp values of r_i, c_j, $r_i + c_j$ and $r_i - c_j$.

Code	r_i	c_j	$r_i + c_j$	$r_i - c_j$
T1	2.20	3.78	5.99	−1.58
T2	3.83	2.86	6.69	0.97
T3	1.97	2.90	4.88	−0.93
T4	2.44	3.77	6.21	−1.33
T5	2.88	2.00	4.88	0.88
T6	2.94	2.87	5.81	0.07
T7	2.73	2.10	4.84	0.63
T8	2.87	3.10	5.96	−0.23
T9	4.83	2.59	7.42	2.25
T10	3.46	3.91	7.37	−0.45
T11	5.13	3.29	8.42	1.84
T12	3.59	9.76	13.35	−6.18
T13	4.54	3.33	7.87	1.21
T14	5.27	3.53	8.80	1.75

The cause–effect diagram was drawn after obtaining horizontal axis $(r_i + c_j)$ and vertical axis $(r_i - c_j)$. While $(r_i + c_j)$ refers to the strength of influence among criteria, $(r_i - c_j)$ refers to the influence relation among criteria. The cause–effect diagram is shown in Figure 2.

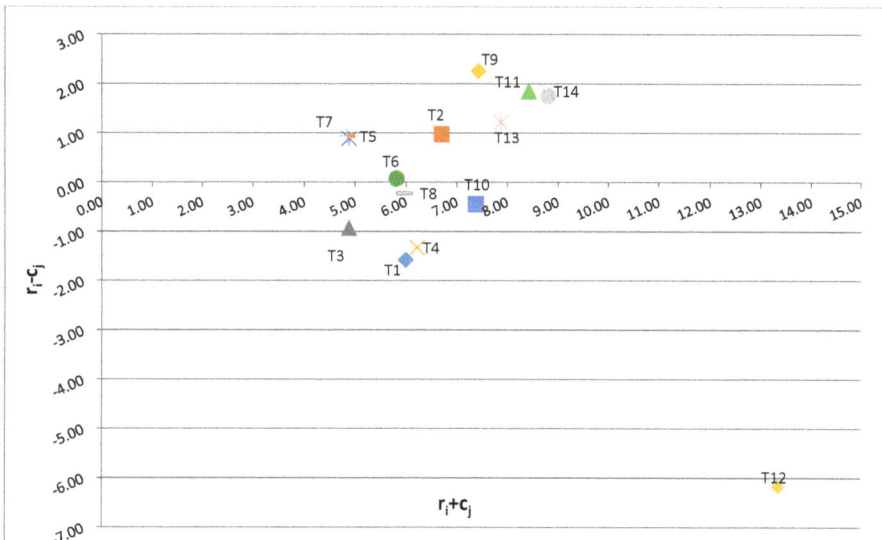

Figure 2. Cause and effect diagram.

4. Results

This study combines Fuzzy System Theory and DEMATEL method to develop a systematic risk assessment methodology for potential occupational hazards on construction sites. The results are summarized based on the causal diagram as follows. The assessment criteria Worker capabilities (knowledge, skills) (T2), Site conditions (excluding equipment, weather, materials) (T5), Work scheduling (T6), Unsuitability of materials (T7) Safety culture (T9), Shortage of safety management (T11), Poor safety conscientiousness of laborers (T13), and Poor site management (T14) are classified into the cause criteria group, while effect criteria group includes Worker actions/behavior (T1),

Communication (T3), Worker health/fatigue (T4), Unsuitability of equipment (T8), Construction process (T10), and Poor of education of laborers (T12) , which need to be improved. Since cause factors influence the effect group criteria, they should be the focus. The cause group criteria refer to the implication of the influencing criteria, while the effect group criteria refer to the implication of the influenced criteria. Considering the interdependence among factors, much attention should be paid to the cause group criteria related to their influence on the effect group criteria [65]. Therefore, by improving cause factors, effect factors are developed simultaneously. Therefore, T9, T11, T14, T12 and T10 are critical occupational factors of accidents to be considered in construction industry based on evaluators' wide experiences and knowledge.

The most significant causal factor of occupational hazards that cause accident is "Safety culture (T9)" has the highest $(r_i - c_j)$ value with 2.25, which means (T9) should be given more consideration on the overall system of critical occupational hazards on construction sites. Besides, Table 7 shows that influential impact degree of (T9) is 4.83, which is ranked the third highest degree among all causal factors. In common, (T9) is a main factor that requires more consideration in the construction industry process. "Shortage of safety management (T11)" has significant impact on other cause group factors with the second highest $(r_i - c_j)$ degree. Furthermore, (T11) has the second highest ri value (5.13) among the causal factors in terms of prominent impact degree. Likewise, "Poor site management (T14)" is another significant factor because the $r_i - c_j$ value is in the third place (1.75). Besides, T14 has the highest r_i value (5.27).

If the value of $(r_i - c_j)$ is negative, such perspective is classified in the effect group (hazards), and is largely influenced by others. In this study, "Poor education of laborers (T12)" has the highest $(r_i + c_j)$ value among the whole process. However, their $(r_i - c_j)$ values are very high when compared to other factors in effect group (−6.18). This means that it has a significant impact on the other factors. However, its $(r_i + c_j)$ score is the second highest in the other effect group criteria. "Construction process (T10)" has impact to improve the system as its $r_i - c_j$ value is very low (−0.45). It is easily affected by the other factors. "Worker health/fatigue (T4)" has the third highest $(r_i + c_j)$ in the whole process. The remaining factors have moderate $(r_i + c_j)$ values. Their $(r_i - c_j)$ values are comparatively low, which point as a strong influenced degree.

Considering the results, managers are able to define regular precautions that must be taken for critical causal factors of occupational hazards. The precautions against the most critical causal factors in construction industry are given in Table 9.

Table 9. The precautions against the most critical causal factors on construction sites.

Code	Critical Causal Factors	Safety Precautions
T9	Safety culture	• Examine individual and group values, attitudes, competencies and patterns of behavior. • Provide safety knowledge, worker/co-workers • Interpersonal skills and appropriate attitudes and beliefs among workers. • Establish a robust safe system of work.
T11	Shortage of safety management	• Implement effective emergency management. • Identify hazard and develop emergency management procedures. • Worker participation in managing safety is important to generate ideas and to build ownership and responsibility.
T14	Poor site management	• Safety needs to be owned by all workers (project team, designers, engineers and operators). • Raise awareness of hazards and safety training, hazard recognition and prepare behavioral-based safety (BBS) programs, toolbox meetings, etc. • Overlook safety in the context of heavy workloads and other priorities. • Make sure all workers take regular breaks to reduce the chances of accidents due to exhaustion.
T12	Poor of education of laborers	• Educate all construction workers and train on safety at each construction processes. • Supervise work crews.
T10	Construction process	• Give greater attention to the design and selection of tools, equipment and materials. • Provide all workers with high-visibility apparel including reflective vests. • Prepare organization's health and safety programs to suit the particular hazards in construction processes.

Sensitivity Analysis of Results

Sensitivity analysis is performed to test the reliability of decisions made by evaluators. To understand the effects of using various combinations of decision criteria weight (keeping equal weightings and adjust more weight to any evaluator), sensitivity analysis is conducted [66]. To verify the results, sensitivity analysis is presented in this study. Initially, equal weights are assigned to each evaluator (Scenario 1). Then, weights of each evaluator are changed according to their years of experience and job responsibilities to analyze how much the cause–effect relations vary. Accordingly, considering that Evaluator 2, Evaluator 3 and Evaluator 1, respectively, faced more occupational accidents during their job experiences, their weights are assigned higher than other evaluators. Scenarios based on evaluator weights are given in Table 10.

Table 10. Different important weights of evaluators in sensitivity analysis.

	Scenario 1	Scenario 2	Scenario 3	Scenario 4	Scenario 5
Evaluator 1	0.2	0.15	0.15	0.2	0.25
Evaluator 2	0.2	0.4	0.4	0.4	0.35
Evaluator 3	0.2	0.3	0.35	0.3	0.3
Evaluator 4	0.2	0.05	0.05	0.05	0.05
Evaluator 5	0.2	0.1	0.05	0.05	0.05

The results of Scenarios 1–5 are presented in Table 11 and Figure 3.

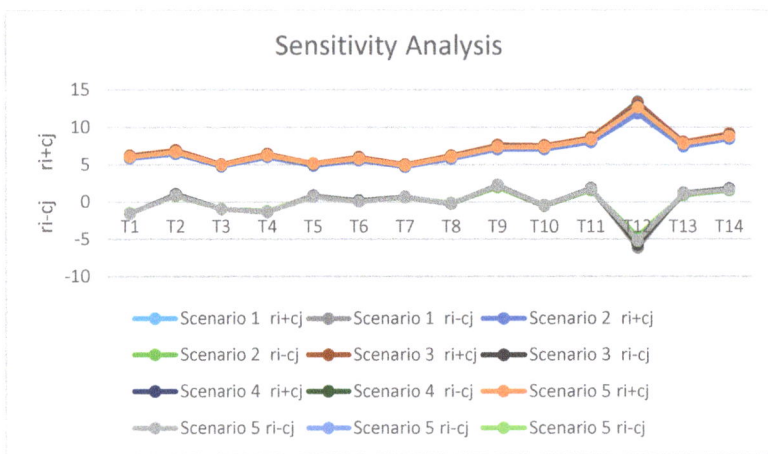

Figure 3. Causal diagram of sensitivity analysis.

Table 11. Cause and effect parameters obtained in sensitivity analysis.

Scenario 1		Scenario 2		Scenario 3		Scenario 4		Scenario 5	
$r_i + c_j$	$r_i - c_j$	$r_i + c_j$	$r_i - c_j$	$r_i + c_j$	$r_i - c_j$	$r_i + c_j$	$r_i - c_j$	$r_i + c_j$	$r_i - c_j$
5.99	−1.58	5.83	−1.51	6.19	−1.60	6.03	−1.58	6.02	−1.60
6.69	0.97	6.42	0.83	6.90	1.03	6.69	0.92	6.67	0.89
4.88	−0.93	4.74	−0.95	4.98	−0.98	4.90	−0.97	4.90	−0.97
6.21	−1.33	6.00	−1.27	6.38	−1.37	6.24	−1.33	6.24	−1.34
4.88	0.88	4.82	0.73	5.11	0.77	5.04	0.73	5.08	0.71
5.81	0.07	5.53	0.06	5.97	0.171	5.79	0.069	5.77	0.016
4.84	0.63	4.67	0.63	4.96	0.66	4.86	0.65	4.86	0.65
5.96	−0.23	5.76	−0.16	6.16	−0.23	6.01	−0.20	6.01	−0.19
7.42	2.25	7.01	1.94	7.57	2.15	7.36	2.12	7.34	2.12
7.37	−0.45	7.01	−0.57	7.51	−0.55	7.34	−0.54	7.33	−0.53
8.42	1.84	7.95	1.55	8.55	1.76	8.35	1.71	8.34	1.70
13.35	−6.18	11.79	−4.73	13.23	−5.75	12.72	−5.37	12.63	−5.25
7.87	1.21	7.36	0.89	7.94	1.05	7.76	1.03	7.76	1.04
8.8	1.75	8.39	1.53	9.05	1.77	8.79	1.66	8.75	1.63

The results show that the ranking of cause and effect factors remain unchanged in all scenarios. The sensitivity analysis has shown robust and valid results that are close to real preferences of the consulted evaluators. As a result, the understanding of evaluators about the causes of accidents in construction industry is adequate for this study.

5. Conclusions

This study aims to improve Fuzzy DEMATEL approach to causal factors of critical occupational hazards, which are implemented through 14 criteria, for construction industry. Thus, this study presents a novel occupational risk assessment approach for evaluating critical casual factors of accidents for construction industry, which can help managers of construction industry to make proper precautionary strategies for accidents. The proposed method is superior to conventional techniques because of exposing the relationships between factors and ranking the criteria relating to the type of relationships and intensity of their effects on each criterion. In addition, by using fuzzy linguistic scale, imprecise and inaccurate information has been handled. Due to these advantages, DEMATEL is used to reveal a better knowledge of the influences of the analysis of cause and effect criteria, and to increase the model applicability. Thus, the proposed method has capability to represent the causal relationship of criteria and is favorable to handle group decision making in fuzzy environment.

According to the findings, several precautions can be suggested for potential occupational hazards. Firstly, it can be proposed to concentrate on the cause group criteria because of their influences on the effect group criteria. Arrangement of cause group criteria are much more difficult than the effect group criteria.

Moreover, managers should focus on critical causal factors in construction industry, which are Worker capabilities (knowledge, skills) (T2), Site conditions (excluding equipment, weather, materials) (T5), Work scheduling (T6), Unsuitability of materials (T7) Safety culture (T9), Shortage of safety management (T11), Poor safety conscientiousness of laborers (T13), Poor site management (T14). Worker actions/behavior (T1), Communication (T3), Worker health/fatigue (T4), Unsuitability of equipment (T8), Construction process (T10), and Poor of education of laborers (T12).

The sensitivity analysis is also introduced to reveal robust and valid results that are close to real preferences of evaluators. The Fuzzy DEMATEL method is a useful tool and widely used in all industry sectors to handle problems that need group decision-making in a fuzzy environment. Therefore, the proposed framework can be enhanced in further studies to test the research findings presented in this study by applying them to a real construction site. In addition, in future research, more evaluator opinions can be evaluated.

Author Contributions: Authors have worked together on this manuscript.

Conflicts of Interest: The authors declare no conflicts of interest.

References

1. Solomon, O.; Lijian, N.; Eucharia, E. Hazards in Building Construction Sites and Safety Precautions in Enugu Metropolis, Enugu State, Nigeria. *Imp. J. Interdiscip. Res.* **2016**, *2*, 282–289.
2. Vitharana, V.H.P.; De Silva, S.; De Silva, S. Health Hazards, Risk and Safety Practices in Construction Sites—A Review Study. *Engineer* **2015**, *48*, 35–44. [CrossRef]
3. Manu, P.A.; Ankrah, A.; Proverbs, D.G.; Suresh, S. Investigating the multi-causal and complex nature of the accident causal influence of construction project features. *Accid. Anal. Prev.* **2012**, *48*, 126–133. [CrossRef] [PubMed]
4. Li, R.Y.M.; Poon, S.W. A Literature Review on the Causes of Construction Accidents. *Constr. Saf. Risk Eng.* **2013**. [CrossRef]
5. Hamid, A.; Majid, M.; Singh, B. Causes of accidents at construction Sites. *Malays. J. Civ. Eng.* **2008**, *20*, 242–259.

6. Occupational Health and Safety APRIL 30. Environmental, Health, and Safety (EHS) Guidelines. General EHS Guidelines: Occupational Health and Safety. 2007. Available online: https://www.ifc.org/wps/wcm/connect/554e8d80488658e4b76af76a6515bb18/Final+General+EHS+Guidelines.pdf?MOD=AJPERES (accessed on 20 July 2017).

7. Pipitsupaphol, T.; Watanabe, T. Identification of Root Causes of Labor Accidents in the Thai Construction Industry. In Proceedings of the 4th Asia Pacific Structural Engineering and Construction Conference (APSEC 2000), Kuala Lumpur, Malaysia, 13–25 September 2000; pp. 193–202.

8. Huang, X.; Hinze, J. Analysis of Construction Worker Fall Accidents. *J. Constr. Eng. Manag.* **2003**, 262–271. [CrossRef]

9. Health and Safety Executive (HSE). Health and Safety Executive: Improving Health and Safety in the Construction Industry: House of Commons Committee of Public Accounts, Fifty-Second Report of Session. 2004. Available online: https://publications.parliament.uk/pa/cm200304/cmselect/cmpubacc/627/627.pdf (accessed on 9 November 2017).

10. Phoya, S. Health and Safety Risk Management in Building Construction Sites in Tanzania: The Practice of Risk Assessment, Communication and Control. Master's Thesis, Department of Architecture, Chalmers University of Technology Gothenburg, Göteborg, Sweden, 2012.

11. Dumrak, J.; Mostafa, S.; Kamardeen, I.; Rameezdeen, R. Factors associated with the severity of construction accidents: The case of South Australia. *Australas. J. Constr. Econ. Build.* **2013**. [CrossRef]

12. Jackson, S.A.; Loomis, D. Fatal occupational injuries in the North Carolina construction industry, 1978–1994. *Appl. Occup. Environ. Hyg.* **2002**, *17*, 27–33. [CrossRef] [PubMed]

13. Saiman, B.T. Managing Physical Hazards on Construction Site Sıtı Norlizah. Bachelor's Thesis, Faculty of Civil Engineering & Earth Resources, Universiti Malaysia Pahang, Pekan, Malaysia, 2010.

14. Toole, T.M. Construction Site Safety Roles. *J. Constr. Eng. Manag.* **2002**, *12*, 203–210. [CrossRef]

15. Tam, C.M.; Zeng, S.X.; Deng, Z.M. Identifying elements of poor construction safety management in China. *Saf. Sci.* **2004**, *42*, 569–586. [CrossRef]

16. Dester, W.S.; Blockley, D.I. Safety behavior and culture in construction. Engineering. *Constr. Archit. Manag.* **1995**, *2*, 17–26. [CrossRef]

17. Zhou, Z.; Goh, Y.M.; Li, Q. Overview and analysis of safety management studies in the construction industry. *Saf. Sci.* **2015**, *72*, 337–350. [CrossRef]

18. Agvu, O.; Olele, H. Fatalities in the Nigerian Construction Industry: A Case of Poor Safety Culture. *Br. J. Econ. Manag. Trade* **2014**, *4*, 431–452.

19. Barofsky, I.; Legro, M. Definition and measurement of fatigue. *Rev. Infect. Dis.* **1991**, *13*, 94–97. [CrossRef]

20. Folkard, S.; Tucker, P. Shift work, safety productivity. *Occup. Med.* **2003**, *53*, 95–101. [CrossRef]

21. Ismail, Z.; Doostdar, S.; Harun, Z. Factors influencing the implementation of a safety management system for construction sites. *Saf. Sci.* **2012**, *50*, 418–423. [CrossRef]

22. Wu, W.W. Choosing knowledge management strategies by using a combined ANP and DEMATEL approach. *Expert Syst. Appl.* **2008**, *35*, 828–835. [CrossRef]

23. Mentes, A.; Akyildiz, H.; Yetkin, M.; Türkoğlu, N. A FSA based fuzzy DEMATEL approach for risk assessment of cargo ships at coasts and open seas of Turkey. *Saf. Sci.* **2015**, *79*, 1–10. [CrossRef]

24. Aneziris, O.N.; Topali, E.; Papazoglou, I.A. Occupational risk of building construction. *Reliab. Eng. Syst. Saf.* **2012**, *105*, 36–46. [CrossRef]

25. Parida, R.; Ray, P. Study and analysis of occupational risk factors for ergonomic design of construction worksystems. *Work* **2012**, *41*, 3788–3794. [PubMed]

26. Dabrowski, A. An investigation and analysis of safety issues in Polish small construction plants. *Int. J. Occup. Saf. Ergon.* **2015**, *21*, 498–511. [CrossRef] [PubMed]

27. Zavadskas, E.K.; Vaidogas, E.R. Bayesian reasoning in managerial decisions on the choice of equipment for the prevention of industrial accidents. *Inz. Ekon. Eng. Econ.* **2008**, *60*, 32–40.

28. Fletcher, J. *The Industrial Environment*; National Profile Ltd.: Willowdale, ON, Canada, 1972.

29. Kavianian, H.R.; Wentz, C.A. *Occupational and Environmental Safety Engineering and Management*; Van Nostrand Reinhold: New York, NY, USA, 1990; ISBN 978-0-471-28912-8.

30. Jannadi, M.O.; Al-Sudairi, A. Safety management in the construction industry in Saudi Arabia. *Build. Res. Inf.* **1995**, *29*, 15–24. [CrossRef]

31. Priyadarshani, K.; Karunasena, G.; Jayasuriya, S. Construction Safety Assessment Framework for Developing Countries: A Case Study of Sri Lanka. *J. Constr. Dev. Ctries.* **2013**, *18*, 33–51.

32. Pender, S. Managing incomplete knowledge: Why risk management is not sufficient. *Int. J. Proj. Manag.* **2001**, *19*, 79–87. [CrossRef]

33. Sii, H.S.; Wang, J.; Ruxton, T. Novel risk assessment techniques for maritime safety management system. *Int. J. Qual. Reliab. Manag.* **2001**, *18*, 982–999.

34. Tixier, J.; Dusserre, G.; Salvi, O.; Gaston, D. Review of 62 risk analysis methodologies of industrial plants. *J. Loss Prev. Process Ind.* **2002**, *15*, 291–303. [CrossRef]

35. Faber, M.H.; Stewart, M.G. Risk assessment for civil engineering facilities: Critical overview and discussion. *Reliab. Eng. Syst. Saf.* **2003**, *80*, 173–184. [CrossRef]

36. Nilsen, T.; Aven, T. Models and model uncertainty in the context of risk analysis. *Reliab. Eng. Syst. Saf.* **2003**, *79*, 309–317. [CrossRef]

37. Kentel, E.; Aral, M.M. Probabilistic-fuzzy health risk modeling. *Stoch. Environ. Res. Risk Assess.* **2004**, *18*, 324–338. [CrossRef]

38. Pinto, A.; Ribeiro, R.; Isabel, L. Qualitative Occupational Risk Assessment model—An introduction. In Proceedings of the World Congress on Risk 2012: Risk and Development in a Changing World, Sydney, Australia, 17–20 July 2012.

39. Pinto, A.; Nunes, I.L.; Ribeiro, R.A. Occupational risk assessment in construction industry-overview and reflection. *Saf. Sci.* **2011**, *49*, 616–624. [CrossRef]

40. Zavadskas, E.; Turkis, Z.; Tamošaitiene, J. Risk assessment of construction projects. *J. Civ. Eng. Manag.* **2010**, *16*, 33–46. [CrossRef]

41. Tamošaitienė, J.; Zavadskas, E.; Turskis, Z. Multi-criteria Risk Assessment of a Construction Project. *Procedia Comput. Sci.* **2013**, *17*, 129–133. [CrossRef]

42. Azadeh, A.; Fam, I.M.; Khoshnoud, M.; Nikafrouz, M. Design and implementation of fuzzy expert system for performance assessment of an integrated health, safety, environment (HSE) and ergonomics system: The case of a gas refinery. *Inf. Sci.* **2008**, *178*, 4280–4300. [CrossRef]

43. Cuny, X.; Lejeune, M. Occupational risks and the value and modelling of a measurement of severity. *Saf. Sci.* **1999**, *31*, 213–229. [CrossRef]

44. Rafat, H. *Machinery Safety: The Risk Based Approach*; Technical Communications (Publishing) Ltd.: Hertfordshire, UK, 1995.

45. Mardani, A.; Jusoh, A.; Nor, K.; Khalifah, Z.; Zakwan, N.; Valipour, A. Multiple criteria decision-making techniques and their applications—A review of the literature from 2000 to 2014. *Econ. Res. Ekon. Istraž.* **2015**. [CrossRef]

46. Dejus, T.; Antucheviciene, J. Assessment of health and safety solutions at a construction site. *J. Civ. Eng. Manag.* **2013**, *19*, 728–737. [CrossRef]

47. Efe, B.; Kurt, M.; Efe, Ö. An integrated intuitionistic fuzzy set and mathematical programming approach for an occupational health and safety polic. *Gazi Univ. J. Sci.* **2017**, *30*, 73–95.

48. Basahel, A.; Taylan, O. Using Fuzzy Ahp And Fuzzy Topsis Approaches For Assessing Safety Conditions At Worksites In Construction Industry. *Int. J. Saf. Secur. Eng.* **2016**, *6*, 728–745.

49. Liu, H.; Tsai, Y. A fuzzy risk assessment approach for occupational hazards in the construction industry. *Saf. Sci.* **2012**, *50*, 41067–41078. [CrossRef]

50. Janackovic, G.L.; Savic, S.M; Stankovic, M.S. Selection and Ranking of Occupational Safety Indicators Based on Fuzzy-AHP: A Case Study in Road Construction Companies. *S. Afr. J. Ind. Eng.* **2013**, *24*, 175–189. [CrossRef]

51. Kim, D.I.; Yoo, W.S.; Cho, H.; Kang, K.I. A Fuzzy AHP-Based Decision Support Model for Quantifying Failure Risk of Excavation Work. *KSCE J. Civ. Eng.* **2014**, *18*, 1966–1976. [CrossRef]

52. Wu, Z.; Shen, R. Safety Evaluation Model of Highway Construction based on Fuzzy Grey Theory. *Procedia Eng.* **2012**, *45*, 64–69. [CrossRef]

53. Li, F.; Phoon, K.; Du, X.; Zhang, M. Improved AHP Method and Its Application in Risk Identification. *J. Constr. Eng. Manag. ASCE* **2013**, *139*, 312–320. [CrossRef]

54. Yuan, P.; Patrick, Z.; Jimmie, H.; Rinker, M. Assessing Safety Risks On Construction Projects Using Fuzzy Analytic Network Process (Anp): A Proposed Model. In *International Conference Evolution of and Direction in*

Construction Safety and Health—Gainesville, United States; In House Publishing: Rotterdam, The Netherlands, 2008; pp. 599–610.

55. Zamri, N.; Ahmad, F.; Rose, A.; Makhtar, M. Fuzzy TOPSIS with Z-Numbers Approach for Evaluation on Accident at the Construction Site. *Int. Conf. Soft Comput. Data Min.* **2016**, *549*, 41–50.

56. Leonavičiūtė, G.; Dėjus, T.; Antuchevičienė, J. Analysis and prevention of construction site accidents. *Građevinar* **2016**. [CrossRef]

57. Mohsen, O.; Fereshteh, N. An extended VIKOR method based on entropy measure for the failure modes risk assessment – A case study of the geothermal power plant (GPP). *Saf. Sci.* **2017**, *92*, 160–172. [CrossRef]

58. Tsai, S.; Chien, M.; Xue, Y.; Li, L.; Jiang, X.; Chen, Q.; Zhou, J.; Wang, L. Using the Fuzzy DEMATEL to Determine Environmental Performance: A Case of Printed Circuit Board Industry in Taiwan. *PLoS ONE* **2015**, *10*, e0129153. [CrossRef] [PubMed]

59. Zadeh, L. Fuzzy Sets. *Inf. Control* **1965**, *8*, 338–353. [CrossRef]

60. Mentes, A.; Akyildiz, H.; Helvacioğlu, I. A Grey Based Dematel Technique for Risk Assessment of Cargo Ships. In Proceedings of the 7th International Conference on Model Transformation ICMT 2014, Glasgow, UK, 7–9 July 2014.

61. Akyuz, E.; Celik, E. A fuzzy DEMATEL method to evaluate critical operational hazards during gas freeing process in crude oil tankers. *J. Loss Prev. Process Ind.* **2015**, *38*, 243–253. [CrossRef]

62. Lin, R. Using fuzzy DEMATEL to evaluate the green supply chain management practices. *J. Clean. Prod.* **2013**, *40*, 32–39. [CrossRef]

63. Li, R.J. Fuzzy Method in Group Decision Making. *Comput. Math. Appl.* **1999**, *38*, 91–101. [CrossRef]

64. Ross, T.J. *Fuzzy Logic with Engineering Applications*; McGraw-Hill, Inc.: New York, NY, USA, 1995.

65. Fontela, E.; Gabus, A. *The Dematel Observer, Dematel 1976 Report*; Battelle Geneva Research Center: Geneva, Switzerland, 1976.

66. Emovon, I.; Norman, R.; Murphy, A. An integration of multi-criteria decision making techniques with a delay time model for determination of inspection intervals for marine machinery systems. *Appl. Ocean Res.* **2016**, *59*, 65–82. [CrossRef]

![sustainability logo] *sustainability*

MDPI

Article

Project Portfolio Risk Identification and Analysis, Considering Project Risk Interactions and Using Bayesian Networks

Foroogh Ghasemi [1,*]**, Mohammad Hossein Mahmoudi Sari** [1]**, Vahidreza Yousefi** [2]**, Reza Falsafi** [1] **and Jolanta Tamošaitienė** [3]

[1] Project and Construction Management, Faculty of Architecture and Urban Planning, University of Art, Tehran 1136813518, Iran; mahmoudi@art.ac.ir (M.H.M.S.); r.falsafi@art.ac.ir (R.F.)
[2] Construction and Project Management, University of Tehran, Tehran 1417614418, Iran; vr.yousefi@ut.ac.ir
[3] Civil Engineering Faculty, Vilnius Gediminas Technical University, Saulėtekio al. 11, LT 2040 Vilnius, Lithuania; jolanta.tamosaitiene@vgtu.lt
* Correspondence: ghasemiforoogh@gmail.com; Tel.: +98-935-849-7254

Received: 18 April 2018; Accepted: 9 May 2018; Published: 17 May 2018

Abstract: An organization's strategic objectives are accomplished through portfolios. However, the materialization of portfolio risks may affect a portfolio's sustainable success and the achievement of those objectives. Moreover, project interdependencies and cause–effect relationships between risks create complexity for portfolio risk analysis. This paper presents a model using Bayesian network (BN) methodology for modeling and analyzing portfolio risks. To develop this model, first, portfolio-level risks and risks caused by project interdependencies are identified. Then, based on their cause–effect relationships all portfolio risks are organized in a BN. Conditional probability distributions for this network are specified and the Bayesian networks method is used to estimate the probability of portfolio risk. This model was applied to a portfolio of a construction company located in Iran and proved effective in analyzing portfolio risk probability. Furthermore, the model provided valuable information for selecting a portfolio's projects and making strategic decisions.

Keywords: project portfolio risk; risk interactions; risk analysis; risk identification; Bayesian networks

---

1. Introduction

A project portfolio is a collection of projects that are managed coordinately to achieve an organization's strategic objectives [1–3]. However, portfolio success and achieving those objectives is affected by risk [1]. Therefore, portfolio risk management can be remarkably effective in a portfolio's alignment with strategic objectives. Furthermore, portfolio risk management improves organizational learning and prevents a risk of one project from occurring in other projects. Thus, even a project's negative effects can have positive implications for a portfolio in the distant future [4]. Moreover, the concept of sustainability, which nowadays is of growing importance, has risk management as one of its recognized 'impact areas' [5]. Nevertheless, in the field of risk management, there are gaps to be filled before it can provide such benefits [6].

Risk analysis is a crucial part of portfolio risk management. It provides information about the effects of choosing a project based on portfolio risk, and so facilitates making decisions about project selection. Furthermore, applying the concept of sustainability through risk analysis makes this concept a standard part of project-related decisions [7]. However, most of today's frameworks for portfolio risk analysis are not designed for achieving strategic objectives. Moreover, they are not written based on portfolio characteristics and are adapted from generic frameworks [8].

Prior to risk analysis is risk identification, the first and the most critical activity in risk management [1,9], upon which the integration of sustainability depends [5]. However, gaps in the literature call for the identification of portfolio risks, beyond single project risks [9,10]. Nowadays, understanding interdependencies between projects and their following implications have a considerable effect on making the right decisions [11–15]. Therefore, in risk identification, the effect of interdependencies on portfolio risk also needs to be considered [16].

This paper introduces a framework for portfolio risk analysis that addresses the aforementioned issues and contributes to a clearer and more sustainable decision-making process. The objectives of this paper are:

1. To identify portfolio risks, beyond single project risks: portfolio risks are put in three categories, (1) project risks; (2) portfolio-level risks (risks that are created specifically in the portfolio); and (3) project interdependency risks (risks that arise from interdependencies between projects). Risks of the second and the third categories are identified and then validated by experts.

2. To develop a model for analyzing portfolio risk, the cause–effect relationships and interdependencies between its components should be considered: for this purpose, using Bayesian networks, a graph of risks based on their cause–effect relationships is constructed and the probability of portfolio risk is calculated.

The data collected from a company active in the construction industry were used to evaluate the model. The model proved effective and could provide valuable information for project portfolio selection, which is a subject undergoing intense study in portfolio management [17–19]. It could show how the decision to add or eliminate one or more specific projects affected the portfolio risk.

This paper is one of the first attempts to identify the aforementioned risks and to develop a framework based on Bayesian networks for analyzing portfolio risk. The proposed framework focuses on risks interrelations, portfolio management objectives, and portfolio success factors.

2. Literature Review

2.1. Risk Identification

Risk identification refers to a series of actions with the ultimate goal of sustainably detecting uncertain events or conditions that, if they occur, have positive or negative effects on one or more objectives [20,21]. Since project risks (first category) are well documented in the literature, this paper does not focus on identifying them and readers are referred to studies that have focused on construction risks [22–25].

Studies on the second and third risk categories are quite limited. Sanchez et al. [8] presented, as far as we know, the only framework for identifying portfolio risks. Using this framework, the user had to find risks among numerous factors that could lead to making key decisions. Although Sanchez et al. considered some types of project interdependencies, outcome interdependency—to use the end result of one project in another one—[26,27], and accomplishment interdependency—the increase in a project's probability of success as a result of undertaking another project—[26,28] were not mentioned in their framework.

2.2. Project Interdependencies

Gear and Cowie [26] were the first to identify different types of project interdependencies. Later Killen and Kjaer [27] presented a comprehensive categorization, in which they separated interdependencies from dependencies. In Table 1, different types of project interdependencies are presented. It should be considered that in some cases different authors have used different names for them.

Table 1. Project interdependencies.

Interdependency	Description	References
Value interdependency	Total value of two projects being greater or less than the sum of their individual values (synergism or antagonism)	Gear, Cowie [26]; Killen, Kjaer [27]; Schmidt [28]; Bhattacharyya et al. [29]
Resource interdependency	Sharing resources among projects	Killen, Kjaer [27]; Schmidt [28]; Bhattacharyya et al. [29]; Verma, Sinha [30]
Technology interdependency	Using a specific technology in several projects	Gear, Cowie [26]; Bhattacharyya et al. [29]; Verma, Sinha [30]
Accomplishment dependency	Increase in project A's probability of success as a result of undertaking project B	Gear, Cowie [26]; Schmidt [28]
Outcome dependency	Using the end result, knowledge or capabilities gained from project B, in project A	Gear, Cowie [26]; Killen, Kjaer [27]

2.3. Bayesian Networks

Bayesian networks are directed acyclic graphs, where nodes correspond to random variables and arcs specify direct causal relations between the linked nodes [31–34]. A BN has conditional probability distributions for all possible combinations of variable values [35], and each node comes with a conditional probability table (CPT). Therefore, despite most of simulation models, such as artificial neural networks (ANNs) that can be time-consuming due to high computational complexity [36], a compiled BN model can quickly calculate a specific probability distribution. Hence, the user can combine expert knowledge with available data and update the CPTs and the associated results, as new evidence comes in [31,36–38] or new causal relations between the nodes become recognized.

2.4. Bayesian Networks in Risk Management

Bayesian networks contribute to risk analysis in different ways. BNs facilitate the "what–if" analysis and help to estimate how the elimination of one or some risks affects the others. Moreover, the visual representation of BN contributes to the identification of risk resources. Thus, it is an effective method for supporting strategic decisions [27,36,37]. On the other hand, unlike ANNs that are praised for their objectivity [39], constructing BN graphs and CPTs based on expert knowledge are criticized for subjectivity [33]. However, it should be noted that "subjective" is not a synonym for "uninformative" [40]. Furthermore, projects and portfolios are unique; therefore, gathering risk-related data from all of them may take a considerable amount of time and money [38]. In developing countries like Iran, expert knowledge can remedy this deficiency [40].

Bayesian networks have been applied to project risk management. Odimabo et al. [41] proposed a methodology based on BNs to assess the risk of construction projects in developing countries. Hu et al. [37] used BNs with causality constraints to develop a model for software project risk analysis. Vitabile et al. [42] presented a BN-based model to measure risk and assess coastal sustainability, which can be used as a decision support system in coastal zone management. Woodberry et al. [43] and Pollino et al. [44] proposed a methodology for parameterization and evaluation of a BN to assess sustainability-related risks. Wang et al. [45] proposed a model based on Bayesian networks to assess the risk of hazmat transportation. This model included seven risk factors and its qualitative part was organized using the adjacency matrix. Yet et al. [38] presented a BN framework to compute costs, benefits, and return on investment (ROI) of a project, considering the effect of risk factors as one of the causal factors. A minor part of the model was assigned to risk factors and the model did not take into account the possible interdependency among these factors.

Fan and Yu [33] introduced a BN-based procedure for software projects. This procedure applied a feedback loop to identify potential erroneous assumptions. Lee et al. [46] proposed a BN-based scheme for managing the risk of engineering projects. In their model, the level of each risk was determined using Equation (1) and a risk matrix. Risk matrices provide qualitative categorization of risks and eliminate the quantitative information. So, in a situation where one has to choose between two risks, which are in the same qualitative category, there is no way to choose a risk based on the risk matrix, and the error probability would be 50% [47]. In order to overcome this deficiency, Yousefi et al. [48] used risk matrices along with an AHP (analytical hierarchy process) method to achieve a more precise ranking of project risks and to identify the most critical ones.

$$\text{Risk} = (\text{the degree of loss}) \times (\text{the probability of occurrence}) \qquad (1)$$

Despite the vast use of BNs in project risk management, their application in portfolio risk management has been quite limited. To the best of our knowledge, Du-juan and Pen's [49] paper was the only study that used BNs in portfolio risk management. They used BN structure learning to construct a network of interdependent portfolio projects' risks. In their study, the effect of value and accomplishment outcome of project risks was explained. However, in order to simplify the calculation, value interaction was not considered in developing the network. Considering the clear advantages of BNs, in the remainder of this research BNs are used to develop a model for portfolio risk analysis.

3. Risk Identification

In this section, portfolio-level risks and project interdependency risks are identified and then validated by experts. Since this paper is one of the first studies to identify the aforementioned risks, a general approach was taken and the results can apply to different industries. Identifying these risks contributes to a more precise and sustainable risk analysis and makes managers aware of the consequences of their decisions.

3.1. Identification of Portfolio-Level Risks

The literature on portfolio-level risks is poor and incomplete [8,9]; however, to conduct this part of the study, we needed to find a valid and well-founded basis. According to the Standard for Portfolio Management, portfolio risk is an uncertain event or condition which, should it occur, may positively or negatively affect one or more project objectives and one or more portfolio success criteria [1]. Based on this definition, it can be said that portfolio risk can cause consequences, such as an increase or decrease in the probability of meeting portfolio success criteria and achieving project objectives. Since portfolio success criteria and project objectives are well documented in previous studies, we chose them as the basis. Then, we tried to identify the causes (portfolio-level risks) by recognizing their consequences (failure to meet portfolio success criteria and to achieve project objectives).

Portfolio success criteria were recognized by studying: portfolio management objectives—since once they are achieved the portfolio is considered successful—and portfolio success factors, because their presence leads to meeting portfolio success criteria. The main portfolio management objectives, as described in detail in Table 2, are: (1) portfolio value maximization; (2) strategic alignment; (3) portfolio balance; and (4) right number of projects [50,51]. It can also be deduced from the literature that the main portfolio success factors are relevant to stakeholders, organizational changes, information, and some of portfolio management knowledge areas. These factors are fully reported in Table 3.

Then, based on Tables 2 and 3, portfolio-level risks were proposed considering that: (1) if portfolio-level risk is considered as a threat and as the only factor that affects portfolio success criteria, then failure to meet these criteria indicates that this risk has happened; (2) hence, the probability of portfolio-level risk equals the probability of failure to meet the criteria.

For example, *achieving the objective of "strategic alignment"* is a portfolio success criterion. Based on this, events or conditions like error in a project portfolio selection, external changes that lead to changing the organizational strategy, some projects' lack of alignment with the new strategy, and a governance review board's reluctance to kill projects off when they are no longer aligned with organizational strategy were proposed as portfolio-level risks because they keep us from achieving the aforementioned objective. These risks and other proposed portfolio-level risks are presented in Table 4.

Table 2. Portfolio management objectives.

Portfolio Management Objective	Description	References
Value maximization	To achieve the average success of single projects in terms of economic objective.	Heising [51]; Cooper et al. [52]; Project Management Institute [1]
	To achieve the average success of single projects in time, cost, quality and customer satisfaction objectives.	Meskendahl [53]
	To use the cross-project synergies in the portfolio.	Meskendahl [53]
Strategic alignment	To align portfolio projects with organizational strategy.	Cooper et al. [52]; Heising [51]; Project Management Institute [1]
Balance	To balance the constant use of resources during project implementation and the constant generation of cash flow.	Heising [51]
	To balance high risk projects and low risk ones.	Archer, Ghasemzadeh [54]; Cooper et al. [52]; Killen et al. [55]; Jonas [56]
	To balance large projects and small ones.	Archer, Ghasemzadeh [54]
	To balance long term projects and short term ones.	Archer, Ghasemzadeh [54]; Cooper et al. [52]; Killen et al. [55]
	To balance projects of new areas of application and projects of old areas of application.	Jonas [56]
	To balance the use of new and current technologies in projects.	Cooper et al. [52]; Killen et al. [55]
	To balance projects across various markets.	Cooper et al. [52]; Killen et al. [55]
	To balance project types (including new product production, improvement, maintenance and repair, research and development (R&D), cost-cutting, etc.).	Cooper et al. [52]
Right number of projects	To balance available resources and resources needed for ongoing projects.	Cooper et al. [52]

Table 3. Portfolio success factors.

Portfolio Success Factor	Description	References
Cooperation quality	Project managers' collaboration for resolving conflicts, and their preparedness to help fellow project managers.	Jonas [56]; Unger et al. [57]
	Mutual assistance of project teams.	Jonas [56]; Unger et al. [57]
Inter-project abilities	Cross-trained staff who can easily switch from project to project.	Fricke, Shenhar [58]
Competency	Governance review board's competency.	Authors
	Portfolio manager's competency.	Authors
Top manager's appropriate engagement intensity	Top manager's non-interference in governance review board's decisions.	Authors
Appropriate pace and frequency of organizational changes	Appropriate pace and frequency of changes in positions, responsibilities, and organizational structure to maintain continuity in work.	Elonen, Artto [3]
Stakeholder management	Clarity in stakeholders' roles and the intensity of their engagement.	Beringer et al. [59]
Management of portfolio risk	Managing portfolio risks (uncertain events or conditions that, if they occur, positively or negatively affect portfolio objectives) and their interdependencies.	Teller[4]
Information sharing	Sharing information and transparency in information.	Martinsuo, Lehtonen [60]; Killen, Kjaer [27]; Unger et al. [57]
Information quality	Information quality and accuracy.	Jonas [56]; Unger et al. [57]

Table 4. Portfolio-Level Proposed Risks.

Risk	Description
Portfolio's imbalance	Imbalance between high-risk projects and low-risk ones.
	Imbalance between large projects and small ones.
	Imbalance between long-term projects and short-term ones.
	Imbalance between projects of new areas of application and projects of old areas of application.
	Imbalance between the use of new and current technologies in projects.
	Imbalance in projects of various markets.
	Imbalance in terms of project types (including new product production, improvement, maintenance and repair, R&D, cost-cutting, etc.).
Strategic lack of alignment	Choosing projects that are not aligned with strategic objectives of the organization.
	Political, social or legislative changes which lead to changing the organizational strategy, and project's objectives lack of alignment with the new strategy.
	Governance review board's reluctance to kill poor projects during their implementation and when they are no longer aligned with organizational strategy.
Wrong number of projects	Choosing too many projects for the available resources.
	Governance review board reluctance to kill off or suspend projects when their required resources are no longer available.
Incompetency	Governance review board's incompetency.
	Portfolio manager's incompetency.
Top manager's interference	Top manager's interference in governance review board's decisions (which leads to choosing projects whose required resources are not available or that are not aligned with strategic objectives of the organization).
Lack of quality in cooperation	Conflict among project managers.
	Lack of quality in cooperation among project teams.
Lack of inter-project abilities	Not having cross-trained staff who can easily switch from project to project
Recurrent organizational changes	Recurrent and rapid changes in positions, responsibilities and organizational structure, which hampers continuity in work.
Insufficient stakeholder management	Lack of clarity in stakeholders' roles and the intensity of their engagement.
Insufficient portfolio risk management	Insufficient portfolio risk management (which leads to more risk materialization and consequently more unexpected and undesirable events).
Lack of sharing and transparency in information	Lack of sharing or transparency in information (which leads to the making of wrong decisions).
Lack of quality in information	Inaccuracy and lack of quality in information (which leads to the making of wrong decisions).

3.2. Identification of Project Interdependency Risks

Project interdependencies are praised for creating synergies and having a positive effect on portfolio success. However, the complexities that they create and their negative implications are less frequently discussed. In this section, risks that arise from these complexities are presented in Table 5. For example, with regards to resource interdependency, projects are interrelated because of shared resources, and risks may arise from inappropriate management of these resources. Therefore, *lack of or delay in supply of shared resources, error in resource allocation,* and *conflict between the managers of these projects* are proposed as the project interdependency risks arising from resource interdependency.

Table 5. Project Interdependencies Proposed Risks.

Arising From	Risk	Description
Value interdependency	Losing the potential value.	When the total value of two projects is greater than the sum of their individual values, if the success of one of them is at risk, then the potential value of value interdependency may be at risk too.
Resource interdependency	Lack of or delay in supply of shared resources.	If two projects share resources, then lack of or delay in supply of shared resources may pose a threat to the success of both projects.
	Error in resource allocation.	If two projects share resources, then error in resource allocation may be a threat to the success of both projects.
	Conflict between project managers.	If two projects share resources, then conflict between project managers may present a threat to success of both projects.
Accomplishment dependency	Cancellation of project B.	When the probability of project A's success increases as a result of undertaking project B, if project B is cancelled, then the success of project A may be at risk.
Outcome dependency	Failure of project B.	When project A is dependent on the end result, knowledge or capability gained through project B, if project B's success is at risk, then the project A's success may be at risk too.

3.3. Validity Estimation for Proposed Risks

In the next step, using 6-point Likert scale questions, experts were asked about the validity of the proposed risks. These experts were in the field of project, program or portfolio management. The questionnaire's validity was estimated and supported as well using face validity, and its reliability was corroborated by Cronbach's alpha measure, which was 0.7939. Then, using a content validity ratio (CVR) with the minimum of 0.33 for 35 experts [61], 22 valid risks were distinguished, which are shown in Tables 6 and 7. Table 6 shows that from the portfolio-level risks category, risks related to strategic unalignment, lack of sharing of information and deficiencies in portfolio management are the most approved risks. It can also be deduced from Table 7 that risks caused by resource interdependency are the most confirmed risks from the third category.

Table 6. Portfolio proposed valid risks.

Abbr.	Portfolio-Level Risk	CVR
PLR1	Choosing projects that are not aligned with strategic objectives of the organization.	
PLR2	Lack of sharing or transparency in information.	0.94
PLR3	Insufficient portfolio risk management.	
PLR4	Portfolio manager's incompetency.	
PLR5	Portfolio's imbalance in terms of high-risk projects versus low risk ones.	0.88
PLR6	Political, social or legislative changes which leads to changing the organizational strategy, and project's objectives lack of alignment with the new strategy.	0.82
PLR7	Top manager's interference in governance review board's decisions.	
PLR8	Choosing too many projects for the available resources.	0.77
PLR9	Inaccuracy and lack of quality in information.	
PLR10	Portfolio's imbalance between long-term projects and short-term ones.	0.71
PLR11	Governance review board's incompetency.	0.60
PLR12	Frequent changes in roles, responsibilities and organizational structure.	0.54
PLR13	Lack of clarity in stakeholders' roles and the intensity of their engagement.	
PLR14	Governance review board reluctance to kill poor projects during their implementation, when they are no longer aligned with business strategy.	0.48
PLR15	Governance review board's reluctance to kill or suspend projects when their required resources are no longer available.	
PLR16	Portfolio's imbalance across various markets.	0.42
PLR17	Portfolio's imbalance in terms of project types.	

Table 7. Project interdependencies proposed valid risks.

Arising From	Project Interdependency Risk	CVR
Resource interdependency	Lack of or delay in supply of shared resources.	1
Resource interdependency	Error in resource allocation.	0.88
Outcome dependency	Failure of project B (see Table 4).	0.71
Value interdependency	Losing the potential value.	0.65
Resource interdependency	Conflict between project managers.	0.60

4. Portfolio Risk Analysis Model

4.1. Development of the Model

After identifying risks, the portfolio risk analysis model is developed. The model includes three types of risks within the portfolio. For analyzing these risks, their relationships must be visually represented. This is because their effect on portfolio risk is not always simple and direct, and sometimes they increase this effect by raising the probability of or creating other risks.

Despite these complexities, it is clear that these risks have cause–effect relationships with the portfolio total risk, because they all can be a reason for an increase in portfolio risk.

Figure 1 plainly shows these relationships, based on which the BN graph is constructed: "a project is a temporary endeavor undertaken to create a unique product, service, or result" [1] and based on whether or not it has interdependency with other projects, its risk falls under the category of either dependent project risk or independent project risk. Dependent project risk is affected by each type of interdependency risk in a unique way, which is explained in the remainder of this section. In Figure 1, one single box embodies all portfolio-level risks, which is also further expanded. At the end of this section all these risks are connected to construct a Bayesian network, in this network each node represents a risk and every arc specifies a cause–effect relationship.

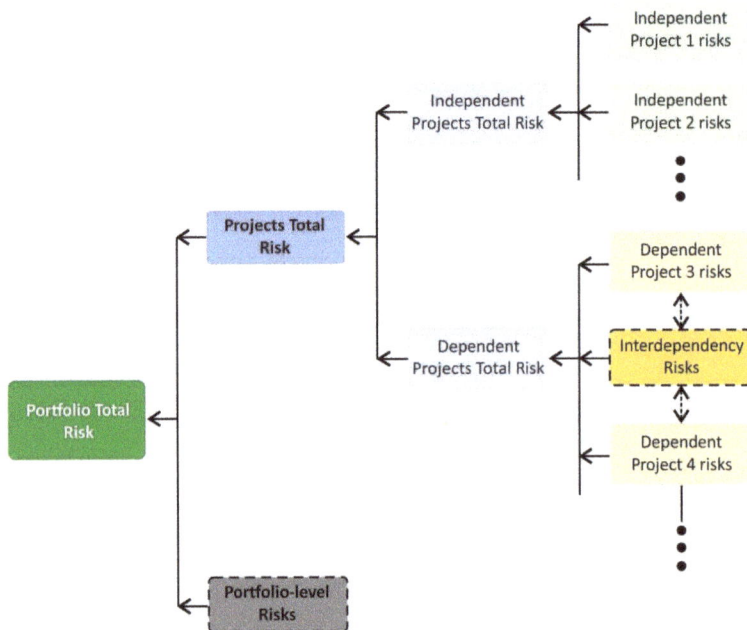

Figure 1. Project portfolio total risk content.

4.1.1. Risks of an Independent Project in a Bayesian Network

Since sustainably analyzing the risk of the portfolio is the main goal of this paper, project risk must be seen from a viewpoint that reveals its effects on the portfolio risk. Since for studying portfolio risks we focused on portfolio management objectives and portfolio success factors, in this section we stay in line with this approach too. In one of the portfolio objectives, that is value maximization, project success is mentioned and divided into two dimensions: (1) success of project in economic objective [1,51,52]; and (2) success of project in time, budget, quality and customer satisfaction objectives [53]. Therefore, we define project risk as an event or condition, which keeps us from achieving these objectives and as shown in Figure 2, in the BN graph, we connect the node for project risk factors (P_n Risk Factors) to P_n E Risk (risk of not achieving the average success in economic objective) and P_n T.B.Q.S Risk (risk of not achieving the average success in time, cost, quality and customer satisfaction objectives).

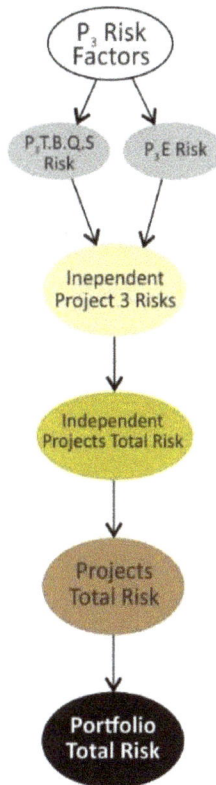

Figure 2. An independent project risk in portfolio risk Bayesian network.

4.1.2. Risks of Projects with Resource Interdependency in a Bayesian Network

In the BN, the risk of a dependent project is shown to be similar to an independent one, except with interdependency risks added to it. If two projects have resource interdependency, then risks arising from this interdependency pose threats to the success of both projects. Therefore, in the BN, these risks are connected to both of them, as shown in Figure 3.

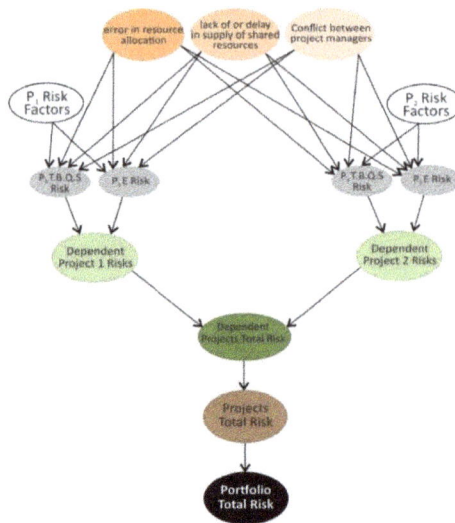

Figure 3. Projects with resource interdependency in portfolio risk Bayesian network.

4.1.3. Risks of Projects with Outcome Dependency in a Bayesian Network

When project 1 is dependent on the outcome of project 2, if project 2's success is at risk, then project 1's success may be at risk too. Therefore, as shown in Figure 4, the risk of project 2 is considered as a new risk for project 1.

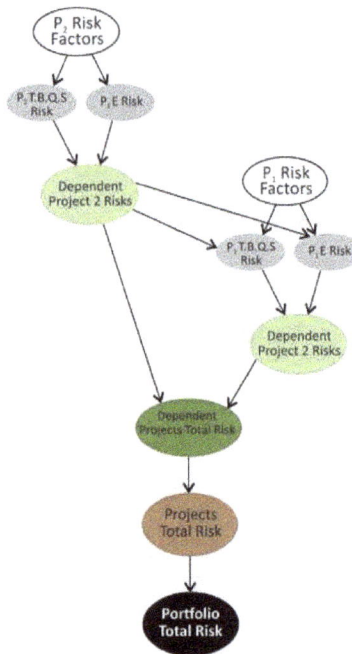

Figure 4. Projects with outcome dependency in portfolio risk Bayesian network.

4.1.4. Risks of Projects with Value Interdependency in a Bayesian Network

When two projects have value interdependency, if one of them is at risk, then the potential value of this interdependency and consequently the success of both projects will be at risk. Thus, as shown in Figure 5, a new risk called "losing potential value of value interdependency" is added to the BN. The risks of dependent projects 1 and 2 are assigned as its parent nodes and it has an effect on the total risk of dependent projects.

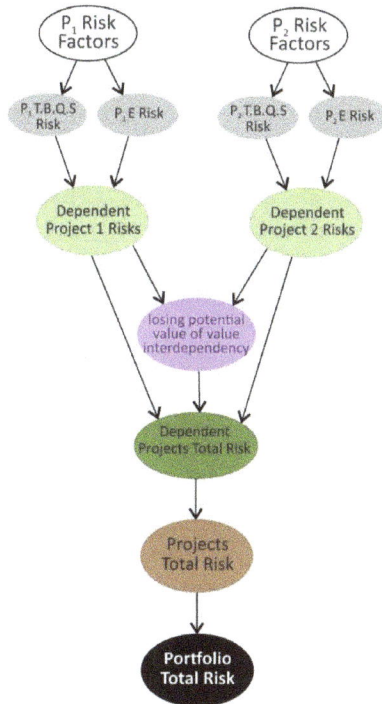

Figure 5. Projects with value interdependency in portfolio risk Bayesian network.

4.1.5. Portfolio-Level Risks in a Bayesian Network

Apart from portfolio-level risks' effect on portfolio total risk, they have cause–effect relationships with each other as well. Based on field research and expert consultation, these relationships are constructed in a BN that is shown in Figure 6. In addition to the risks shown in Table 6, two nodes are added to the BN for having a more precise and clear analysis: "portfolio imbalance" (PLR18) that specifies the total imbalance of the portfolio as a consequence of different imbalance risks, and "other risks" (PLR19) that represent all portfolio-level risks that are not mentioned in this model.

It can be deduced from Figure 6 that:

- "Top manager's interference in governance review board's decisions" (PLR7) and "governance review board's incompetency" (PLR11), by having six child nodes, are the most important causes (parent nodes) among portfolio risks.
- "Inaccuracy and lack of quality in information" (PLR9) and "lack of sharing or transparency in information" (PLR2), by having five child nodes, are on the next level of important causes (parent nodes) among portfolio risks.

- "Choosing projects that are unaligned with strategic objectives of the organization" (PLR1), having six parent nodes, is the most important consequence among portfolio risks.
- "Choosing too many projects for the available resources" (PLR8) and "portfolio's imbalance in terms of high-risk project versus low-risk ones" (PLR5), by having five parent nodes, are on the next level of important consequences among portfolio risks.

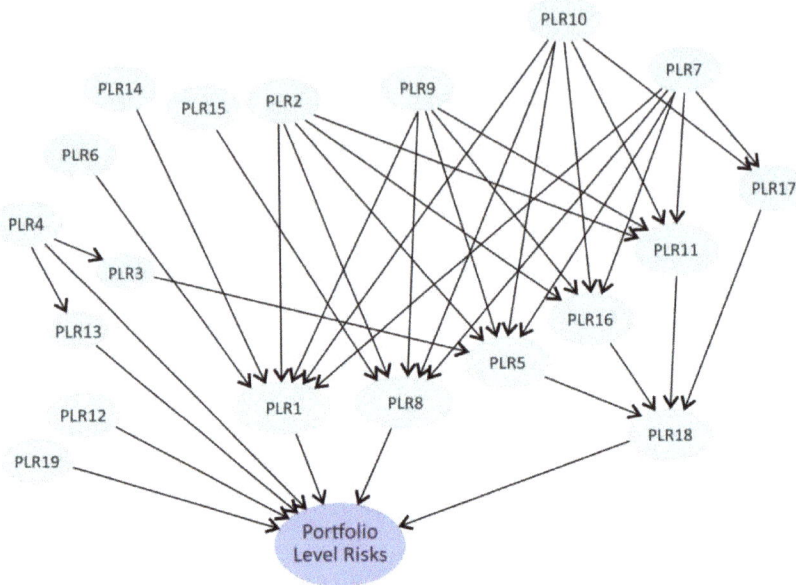

Figure 6. Portfolio-level risks in a Bayesian network.

These points can contribute to portfolio risk management success by facilitating the recognition of: (1) "upstream" risks with numerous child nodes, since by managing them the occurrence of many "downstream" risks can be prevented; (2) "downstream" risks with numerous parent nodes, because the probability of their occurrence may be higher than other risks [62].

4.1.6. Compiled Bayesian Network for Portfolio Risk Analysis

By putting the previously explained parts of the network together, the qualitative part of the BN for portfolio risk analysis is constructed. Figure 7 shows this part for a portfolio that has two independent projects and two projects with outcome dependency and value interdependency. Afterward, the quantitative part is assigned to the BN. This part includes the conditional probabilities of nodes, which are binary variables and can take either the low or the high state. Constructing the quantitative part begins with assigning the probability of root nodes (nodes with no parents). Then, based on the cause–effect relationships, other nodes are associated with conditional probability tables (CPTs). As an example, Figure 8 shows the CPT of PLR8, which has PLR7 as its only parent.

In this model, numbers for CPTs are elicited from expert's subjective assessments. However, it cannot be neglected that a node like PLR16 with 6 parent nodes requires $2^7 = 128$ conditional probabilities to complete its CPT. Commenting on these numbers, which may have slight differences, while keeping the structure of reasoning in mind, could be a complicated and time-consuming task for the expert. These cases can be handled by Pearl's noisy OR-gate [32] that reduces the number of required conditional probabilities to the number of a node's parents.

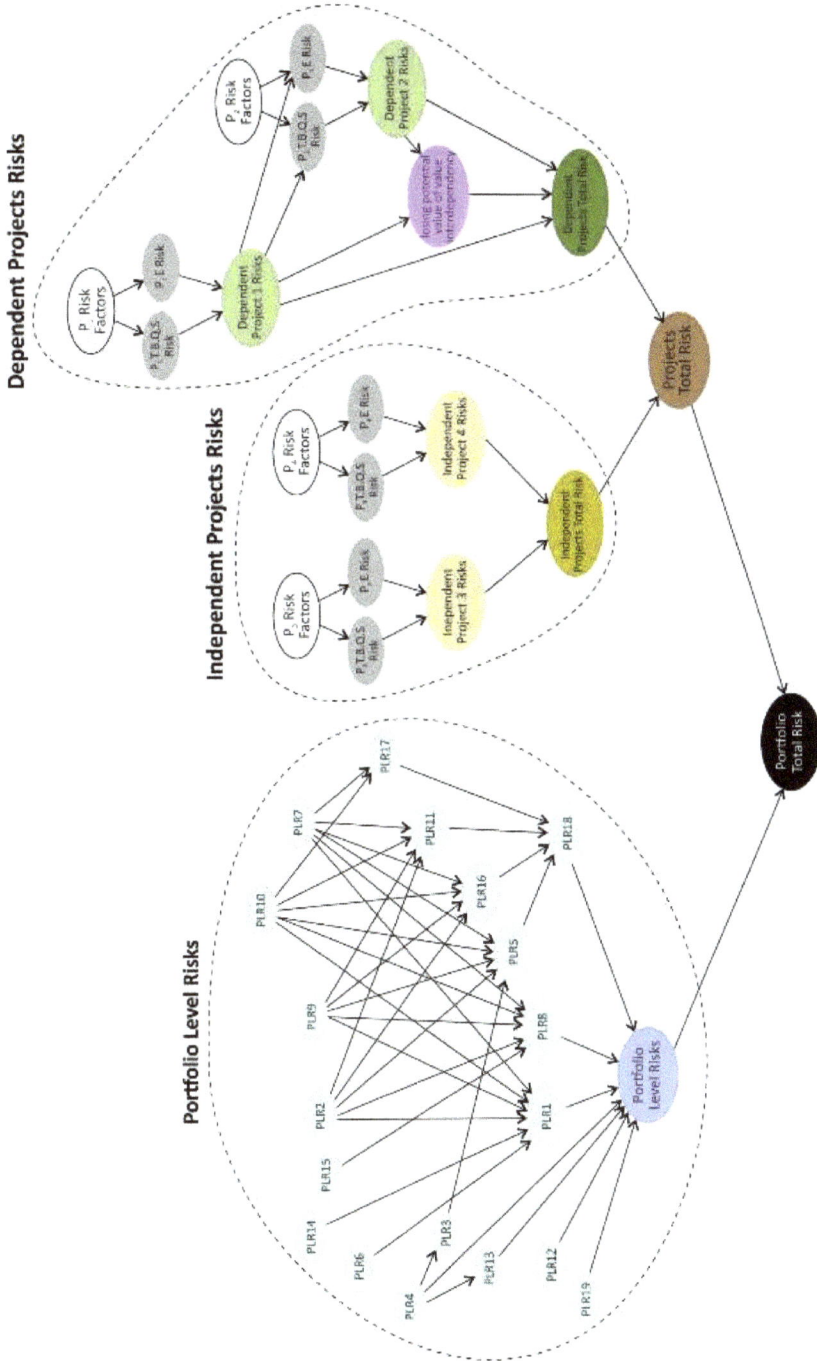

Figure 7. Example of a Bayesian network for portfolio risk analysis.

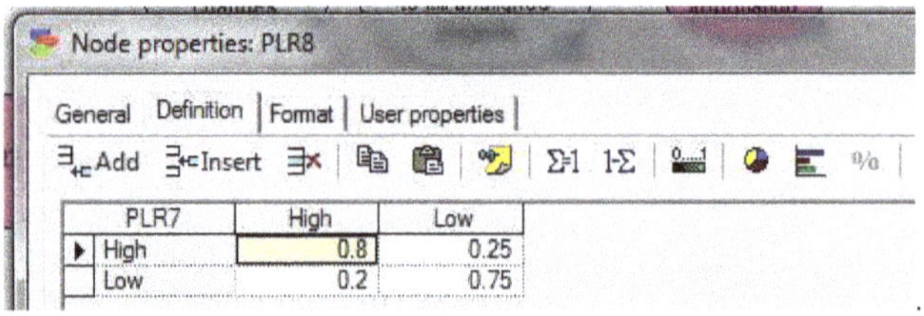

Figure 8. Conditional probability table (CPT) of PLR8.

The noisy OR-gate is based on two assumptions: (1) accountability and (2) exception independence [32]. With regard to the subject of the paper, the first assumption requires that a risk R be presumed LOW (i.e., $P(R = HIGH) = 0$) if all of its causes are LOW. The second assumption asserts that if a risk R is a consequence of either one of two risks r1 and r2, then the mechanism that may inhibit R being high because of r1's high status (I1 inhibitory mechanism) is independent of I2. Considering these two assumptions to be true, using Equation (2), the numbers for a node's CPT can be calculated.

$$P\left(x|u\right) = \begin{cases} \prod\limits_{i \in H_i} q_i & \text{if} \quad x = 0 \quad X = \text{LOW} \\ 1 - \prod\limits_{i \in H_i} q_i & \text{if} \quad x = 1 \quad X = \text{HIGH} \end{cases} \tag{2}$$

H_i represents the subset of risk parents of a node that are HIGH and q_i denote the probability that the i-th inhibitor is active. If n is the only parent of x that is HIGH, x will be HIGH if q_n is not active. Thus, we have:

$$P(x = \text{HIGH} \mid n = \text{HIGH}) = 1 - q_n \tag{3}$$

The left part of Equation (3) is the number that the expert specifies for the noisy OR-gate model. Table 8 shows the typical CPT of node x, while Tables 9 and 10 show that CPT, using the noisy OR-gate model.

Table 8. CPT of node x.

i	High		Low	
n	High	Low	High	Low
High	$P(x = \text{HIGH} \mid i = \text{HIGH},$ $n = \text{HIGH})$	$P(x = \text{HIGH} \mid i = \text{HIGH},$ $n = \text{LOW})$	$P(x = \text{HIGH} \mid i = \text{LOW},$ $n = \text{HIGH})$	$P(x = \text{HIGH} \mid i = \text{LOW},$ $n = \text{LOW})$
Low	$P(x = \text{LOW} \mid i = \text{HIGH},$ $n = \text{HIGH})$	$P(x = \text{LOW} \mid i = \text{HIGH},$ $n = \text{LOW})$	$P(x = \text{LOW} \mid i = \text{LOW}, n$ $= \text{HIGH})$	$P(x = \text{LOW} \mid i = \text{LOW}, n$ $= \text{LOW})$

Table 9. CPT of node x, using the noisy OR-gate model (1).

i	High		Low	
n	High	Low	High	Low
High	$1 - q_i\, q_n$	$1 - q_i$	$1 - q_n$	0
Low	$q_i\, q_n$	q_i	q_n	1

After constructing the BN graph of risks, the network and the conditional probabilities are inserted as inputs to GeNIe [63], which is a software used for Bayesian network analysis. Then, for a more accurate analysis, if available, the evidence suggesting whether a risk is in the high state or the low state is set in the software. Finally, based on the risk graph and the CPTs associated with it, the probability of total portfolio risk is calculated. Figure 9 shows the procedure of portfolio risk analysis in this model.

Table 10. CPT of node *x*, using the noisy OR-gate model (2).

i	High		Low	
n	High	Low	High	Low
High	$1 - [(1 - P(x = \text{HIGH} \mid i = \text{HIGH}))*(1 - P(x = \text{HIGH} \mid i = \text{HIGH}))]$	$P(x = \text{HIGH} \mid i = \text{HIGH})$	$P(x = \text{HIGH} \mid n = \text{HIGH})$	0
Low	$(1 - P(x = \text{HIGH} \mid i = \text{HIGH}))*(1 - P(x = \text{HIGH} \mid n = \text{HIGH}))$	$1 - P(x = \text{HIGH} \mid i = \text{HIGH})$	$1 - P(x = \text{HIGH} \mid n = \text{HIGH})$	1

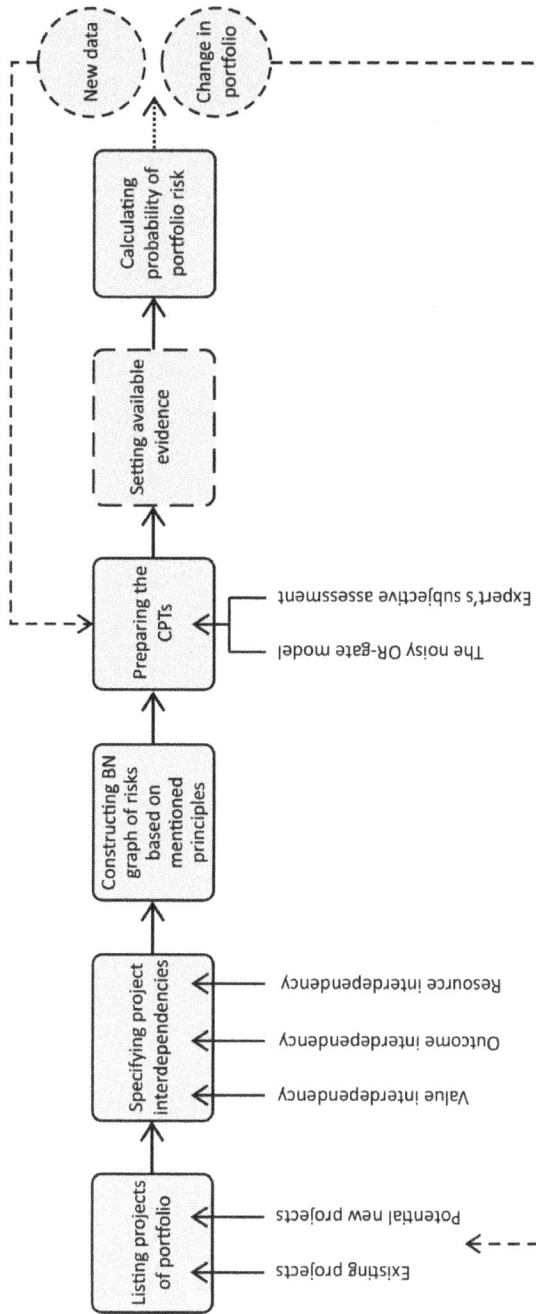

Figure 9. Portfolio Risk analysis procedure.

4.2. Evaluation of the Model

Evaluation of the model's performance and practicality was assessed with real data and in the real portfolio environment. A portfolio from a subsidiary of a Swedish company, located in Iran and active in the construction industry, was chosen to be analyzed. The company had around 50 employees and an operating profit of 8.7%. It has provided consultancy and engineering services and has seven years of experience. Its portfolio included six projects: three independent projects (projects 1 to 3) and three projects with value interdependency (projects 4 to 6). Project 4 was the greatest project, area-wise, that they had ever had and it was going to be delivered in partnership with another company. Furthermore, adding another independent project to the portfolio was under review. A part of project 4 was implemented by another firm but it was suspended after the contract's termination.

First, the probability of portfolio risk was calculated considering the six existing projects. To do so, the risk graph was constructed based on the principles mentioned (Figure 10) and the CPTs were completed with the CEO's assessments. Finally, the model showed the result as presented in Figure 11, from which it can be deduced that:

- Project 4, which had value interdependency and was being implemented through partnership with another company, had the highest risk probability of 70% among all projects. Since the company had no experience of a project at this scale and based on the interviews with the CEO, aspects of their partnership were not quite clarified, this probability of risk is not surprising.
- PLR14 and PLR15, with probabilities of 75% and 60%, respectively, were the highest portfolio-level risks. This is because the board of directors believed that project termination involved a very complex and time-consuming legal process, and was damaging for the company's reputation. Therefore, they avoided killing off a project, even when it did not reflect the organizational strategy, or its required resources were hardly available.
- The probability of high portfolio total risk was 39%. Furthermore, a look at its parent and grandparent nodes showed that the dependent projects' total risk with the probability of 59% was the highest risk among them. This piece of information could contribute to more effective risk-response planning since the model revealed that reducing the risk of dependent projects would have a considerable effect on portfolio risk reduction.

Afterwards, the fourth independent project was added to the BN and the following steps were taken for re-analyzing the network:

- Probabilities and conditional probabilities for the four new nodes, which this project added to the BN, were inserted as inputs to the tables.
- The CPT of independent projects total risk which by adding this project had a new parent, was reassessed.
- Those portfolio-level risks that could change because of adding a new project were re-assessed as well; especially risks concerned with portfolio's imbalance (PLR5, PLR10, PLR16, and PLR17), lack of resources (PLR8), strategic unalignment (PLR1) and lack of clarity in stakeholders' roles (PLR13).

The network was then updated and showed the result presented in Figure 12. The model revealed that by adding this project to the portfolio, the total risk of independent projects would increase from 38% to 48% but the portfolio total risk would increase only by 2%. This is while, based on the CEO's assessment, the new project would not affect the portfolio-level risks. Therefore, the slight change in portfolio total risk could mean that the effect of independent projects on portfolio risk was minor. Thus, from the portfolio risk management point of view, although selecting the fourth independent project increased the portfolio total risk, the project only marginally affected this risk.

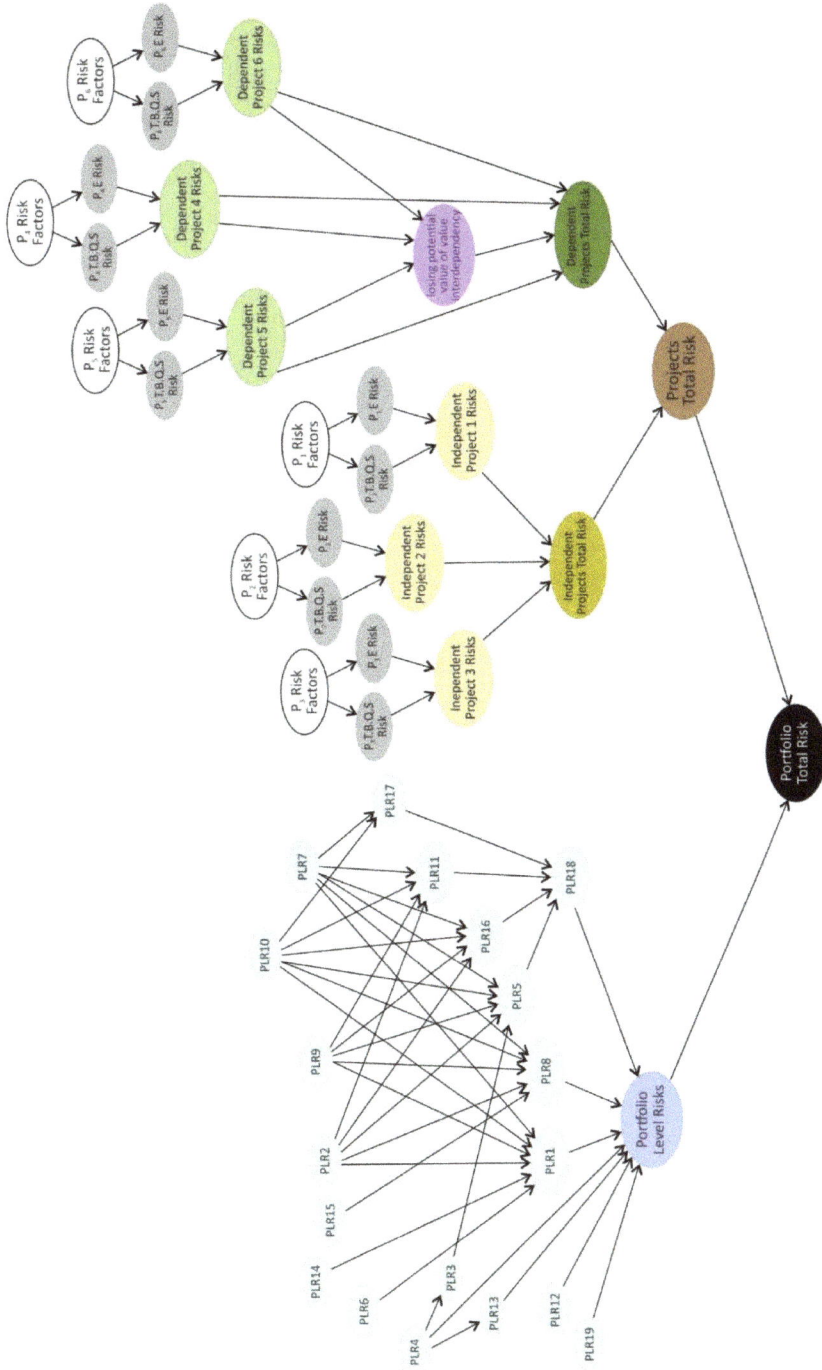

Figure 10. Bayesian network for the analyzed portfolio.

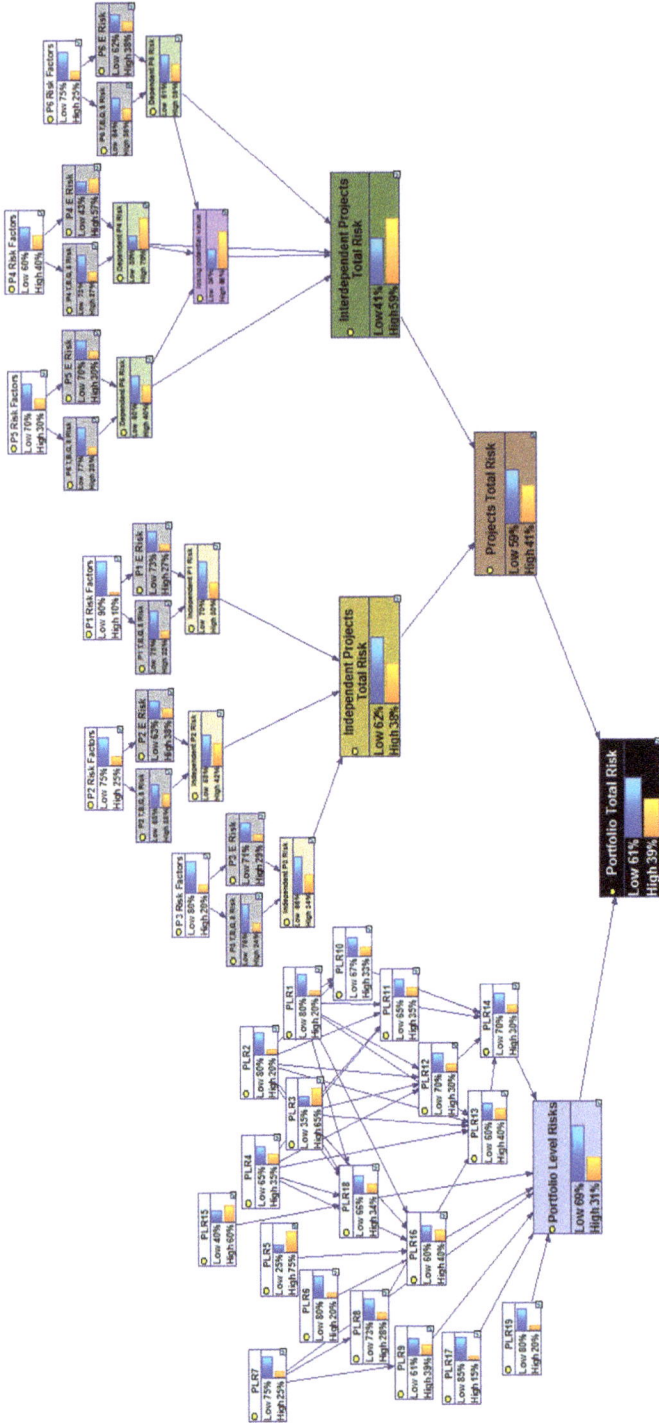

Figure 11. The portfolio risk analysis result.

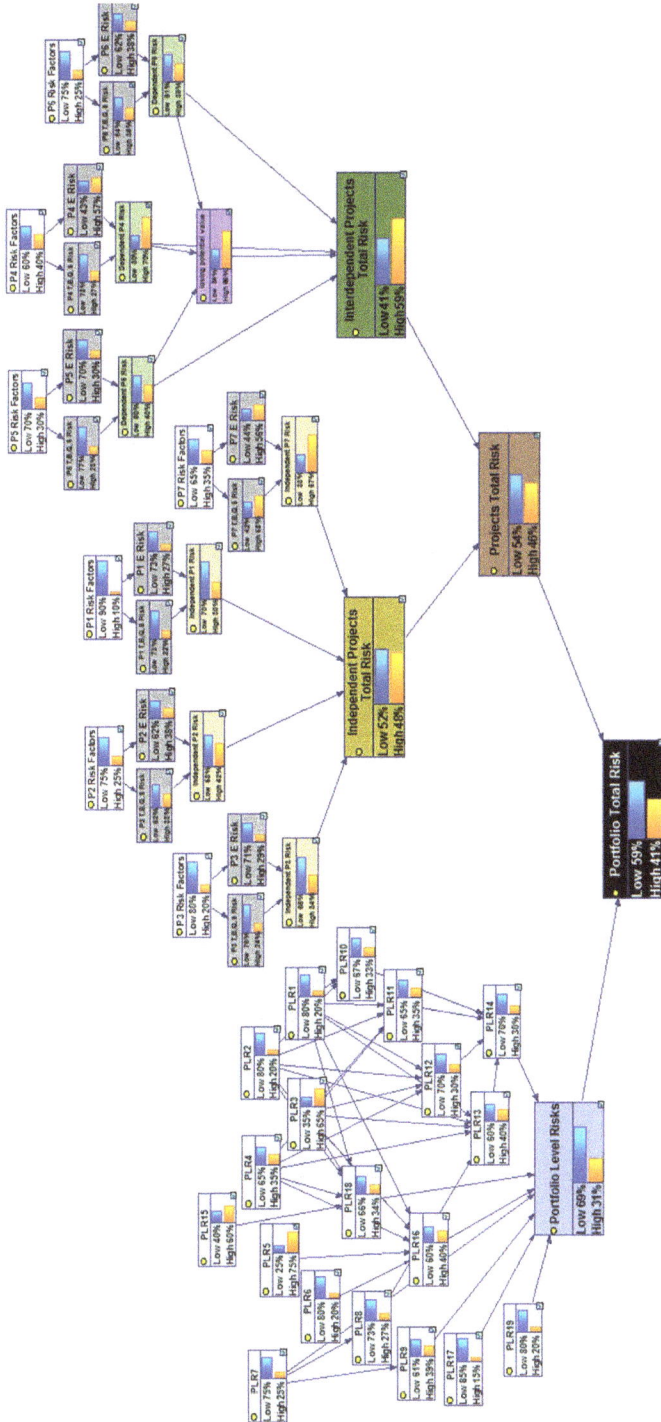

Figure 12. The portfolio risk analysis result after adding the fourth independent project.

5. Conclusions

This paper presented a BN modeling framework for analyzing the risk of a portfolio in order to achieve sustainability. This framework includes a Bayesian network of portfolio risks, which clarifies their cause–effect relationships. Besides single-project risks, the network involves risks that arise from project interdependencies and portfolio-level risks, which were identified in this paper as well.

The results supported the idea that a way of identifying portfolio-level risks and integrating sustainability is to study portfolio success criteria, and marked risks such as "choosing projects that are unaligned with strategic objectives of the organization", "lack of sharing or transparency in information", "deficiency in portfolio risk management" and "portfolio manager's incompetency" as the most approved portfolio-level risks. Furthermore, the results confirmed that although project interdependencies create synergies, they pose threats to the portfolio and projects' success.

After identifying the risks, constructing the Bayesian network visualized how the risks of interdependent projects affect the other risks and the total risk of the portfolio. Moreover, this network highlighted the most important upstream ("top manager's interference in governance review board's decisions" and "governance review board's incompetency") and downstream ("choosing projects that are unaligned with strategic objectives of the organization", "portfolio imbalance in terms of high-risk projects versus low-risk ones", and "choosing too many projects for the available resources") portfolio-level risks.

Having these characteristics, the presented BN-based framework was evaluated with the real data of a portfolio containing construction projects. In addition to calculating the probability of portfolio total risk, the model proved effective in quickly estimating the change in portfolio risk probability, in the case of adding or removing projects. Therefore, the framework showed that it could contribute to more accurate and sustainable portfolio project selection and decision-making process. Furthermore, it could discriminate between portfolio-level and project-level risks, proving beneficial in the development of diversified portfolios regarding the projects' associated risks. In other words, the presented framework confirmed the approach that a diversified portfolio with a desired level of risk is not a set of projects with that risk level, but a group of projects with different levels of risk that, through interrelations, lead to a portfolio with the intended risk level. Moreover, as presented in Figures 11 and 12, since the result of the model showed the probability of different categories of risks, the most effective risks in increasing the portfolio total risk could be recognized. This type of information can help managers and the decision makers in conducting a more successful risk-response planning and managing a more sustainable portfolio.

Author Contributions: F.G. and V.F., designed the research, set the objectives, studied the literature, analyzed data, designed and developed the portfolio risk analysis model based on Bayesian networks and using GeNIe software, and wrote the paper. R.F. and M.H.M.S. gathered data through conducting interviews with the experts, contributed to developing the research idea, and provided extensive advice on the literature review, risk identification, model development and its evaluation. J.T. collected and analyzed the data and the obtained results, provided extensive advice throughout the study, and revised the manuscript, methodology and findings. All authors discussed the model evaluation results and commented on the paper.

Funding: This research received no external funding.

Conflicts of Interest: The authors declare no conflict of interest.

References

1. Project Management Institute. *The Standard for Portfolio Management*, 3rd ed.; Project Management Institute: Newtown Square, PA, USA, 2013.
2. Sanchez, H.; Robert, B.; Pellerin, R. A project portfolio risk-opportunity identification framework. *Proj. Manag. J.* **2008**, *39*, 97–109. [CrossRef]
3. Elonen, S.; Artto, K.A. Problems in managing internal development projects in multi-project environments. *Int. J. Proj. Manag.* **2003**, *21*, 395–402. [CrossRef]

4. Teller, J. Portfolio Risk Management and Its Contribution to Project Portfolio Success: An Investigation of Organization, Process, and Culture. *Proj. Manag. J.* **2013**, *44*, 36–51. [CrossRef]

5. Silvius, G. Integrating sustainability into project risk management. In *Global Business Expansion: Concepts, Methodologies, Tools, and Applications*; IGI Global: Hershey, PA, USA, 2018; pp. 330–352.

6. Iqbal, S.; Choudhry, R.M.; Holschemacher, K.; Ali, A.; Tamošaitienė, J. Risk management in construction projects. *Technol. Econ. Dev. Econ.* **2015**, *21*, 65–78. [CrossRef]

7. MacAskill, K.; Guthrie, P. Risk-based approaches to sustainability in civil engineering. *Eng. Sustain.* **2013**, *166*, 181–190. [CrossRef]

8. Sanchez, H.; Robert, B.; Bourgault, M.; Pellerin, R. Risk management applied to projects, programs, and portfolios. *Int. J. Manag. Proj. Bus.* **2009**, *2*, 14–35. [CrossRef]

9. Teller, J.; Kock, A. An empirical investigation on how portfolio risk management influences project portfolio success. *Int. J. Proj. Manag.* **2013**, *31*, 817–829. [CrossRef]

10. Eik-Andresen, P.; Johansen, A.; Landmark, A.D.; Sørensen, A.Ø. Controlling a Multibillion Project Portfolio—Milestones as Key Performance Indicator for Project Portfolio Management. *Proced. Soc. Behav. Sci.* **2016**, *226*, 294–301. [CrossRef]

11. Olsson, R. Risk management in a multi-project environment. An approach to manage portfolio risks. *Int. J. Qual. Reliab. Manag.* **2008**, *25*, 60–71. [CrossRef]

12. Zhang, Y. Selecting risk response strategies considering project risk interdependence. *Int. J. Proj. Manag.* **2016**, *34*, 819–830. [CrossRef]

13. Canbaz, B.; Marle, F. Construction of project portfolio considering efficiency, strategic effectiveness, balance and project interdependencies. *Int. J. Proj. Organ. Manag.* **2016**, *8*, 103–126. [CrossRef]

14. Bathallath, S.; Smedberg, Å.; Kjellin, H. Managing project interdependencies in IT/IS project portfolios: A review of managerial issues. *Int. J. Inf. Syst. Proj. Manag.* **2016**, *4*, 67–82.

15. Dehdasht, G.; Mohamad Zin, R.; Ferwati, M.S.; Mohammed Abdullahi, M.; Keyvanfar, A.; McCaffer, R. DEMATEL-ANP Risk Assessment in Oil and Gas Construction. *Sustainability* **2017**, *9*, 1420. [CrossRef]

16. Gónzalez, M.P.; De La Rosa, C.G.B.; Moran, F.J.C. Fuzzy cognitive maps and computing with words for modeling project portfolio risks interdependencies. *Int. J. Innov. Appl. Stud.* **2016**, *15*, 737.

17. Tofighian, A.A.; Moezzi, H.; Barfuei, M.K.; Shafiee, M. Multi-period project portfolio selection under risk considerations and stochastic income. *J. Ind. Eng. Int.* **2018**, 1–14. [CrossRef]

18. Panadero, J.; Doering, J.; Kizys, R.; Juan, A.A.; Fito, A. A variable neighborhood search simheuristic for project portfolio selection under uncertainty. *J. Heuristics* **2018**, 1–23. [CrossRef]

19. Sefair, J.A.; Méndez, C.Y.; Babat, O.; Medaglia, A.L.; Zuluaga, L.F. Linear solution schemes for Mean-SemiVariance Project portfolio selection problems: An application in the oil and gas industry. *Omega* **2017**, *68*, 39–48. [CrossRef]

20. Marle, F.; Vidal, L.A.; Bocquet, J.C. Interactions-based risk clustering methodologies and algorithms for complex project management. *Int. J. Prod. Econ.* **2013**, *142*, 225–234. [CrossRef]

21. Tamošaitienė, J.; Zavadskas, E.K.; Turskis, Z. Multi-criteria risk assessment of a construction project. *Proced. Comput. Sci.* **2013**, *17*, 129–133. [CrossRef]

22. Boateng, P.; Chen, Z.; Ogunlana, S.O. An Analytical Network Process model for risks prioritisation in megaprojects. *Int. J. Proj. Manag.* **2015**, *33*, 1795–1811. [CrossRef]

23. Diab, M.F.; Nassar, K. Using Risk Assessment to Improve Highway Construction Project Performance. In Proceedings of the ASC Annual 48th Annual International Conference, Birmingham, UK, 11–14 April 2012.

24. Levine, H.A. *Project Portfolio Management: A Practical Guide to Selecting Projects, Managing Portfolios, and Maximizing Benefits*; John Wiley & Sons: San Francisco, CA, USA, 2007.

25. Quresh, A.A.; Jeswani, H. Qualitative study on construction project risk. *Int. J. Eng. Technol. Sci. Res.* **2018**, *5*, 40–48.

26. Gear, T.E.; Cowie, G.C. A note on modeling project interdependence in research and development. *Decis. Sci.* **1980**, *11*, 738–748. [CrossRef]

27. Killen, C.P.; Kjaer, C. Understanding project interdependencies: The role of visual representation, culture and process. *Int. J. Proj. Manag.* **2012**, *30*, 554–566. [CrossRef]

28. Schmidt, R.L. A model for R&D project selection with combined benefit, outcome and resource interactions. *IEEE Trans. Eng. Manag.* **1993**, *40*, 403–410.

29. Bhattacharyya, R.; Kumar, P.; Kar, S. Fuzzy R&D portfolio selection of interdependent projects. *Comput. Math. Appl.* **2011**, *62*, 3857–3870.

30. Verma, D.; Sinha, K.K. Toward a theory of project interdependencies in high tech R&D environments. *J. Oper. Manag.* **2002**, *20*, 451–468.

31. Charniak, E. Bayesian Networks without Tears. *AI Mag.* **1991**, *12*, 50–63.

32. Pearl, J. *Probabilistic Reasoning in Intelligent Systems: Networks of Plausible Inference*; Morgan Kaufmann: San Mateo, CA, USA, 1988.

33. Fan, C.F.; Yu, Y.C. BBN-based software project risk management. *J. Syst. Softw.* **2004**, *73*, 193–203. [CrossRef]

34. Heckerman, D. Bayesian Networks for Data Mining. *Data Min. Knowl. Discov.* **1997**, *1*, 79–119. [CrossRef]

35. Jain, A.K.; Mao, J.; Mohiuddin, K.M. Artificial Neural Networks: A Tutorial. *Computer* **1996**, *29*, 31–44. [CrossRef]

36. Uusitalo, L. Advantages and challenges of Bayesian networks in environmental modelling. *Ecol. Model.* **2007**, *203*, 312–318. [CrossRef]

37. Hu, Y.; Zhang, X.; Ngai, E.W.T.; Cai, R.; Liu, M. Software project risk analysis using Bayesian networks with causality constraints. *Decis. Support Syst.* **2013**, *56*, 439–449. [CrossRef]

38. Yet, B.; Constantinou, A.; Fenton, N.; Neil, M.; Luedeling, E.; Shepherd, K. A Bayesian network framework for project cost, benefit and risk analysis with an agricultural development case study. *Exp. Syst. Appl.* **2016**, *60*, 141–155. [CrossRef]

39. Hu, Y.; Huang, J.; Chen, J.; Liu, M.; Xie, K. Software Project Risk Management Modeling with Neural Network and Support Vector Machine Approaches. In Proceedings of the Third International Conference on Natural Computation, Haikou, China, 24–27 August 2007; pp. 358–362.

40. Shepherd, K.; Hubbard, D.; Fenton, N.; Claxton, K.; Luedeling, E.; Leeuw, J. Policy: Development goals should enable decision-making. *Nature* **2015**, *523*, 152–154. [CrossRef] [PubMed]

41. Odimabo, O.O.; Oduoza, C.; Suresh, S. Methodology for Project Risk Assessment of Building Construction Projects Using Bayesian Belief Networks. *Int. J. Constr. Eng. Manag.* **2017**, *6*, 221–234.

42. Vitabile, S.; Farruggia, A.; Pernice, G.; Gaglio, S. Assessing coastal sustainability: A Bayesian approach for modeling and estimating a global index for measuring risk. *J. Telecommun. Inf. Technol.* **2013**, *4*, 5–15.

43. Woodberry, O.; Nicholson, A.E.; Korb, K.B.; Pollino, C.A. Parameterising Bayesian networks. In Proceedings of the 17th Australian Joint Conference on Artificial Intelligence, Cairns, Australia, 4–6 December 2004; pp. 1101–1107.

44. Pollino, C.A.; Woodberry, O.; Nicholson, A.; Korb, K.; Hart, B.T. Parameterisation and evaluation of a Bayesian network for use in an ecological risk assessment. *Environ. Model. Softw.* **2007**, *22*, 1140–1152. [CrossRef]

45. Wang, X.; Zhu, J.; Ma, F.; Li, C.; Cai, Y.; Yang, Z. Bayesian network-based risk assessment for hazmat transportation on the Middle Route of the South-to-North Water Transfer Project in China. *Stoch. Environ. Res. Risk Assess.* **2016**, *30*, 841–857. [CrossRef]

46. Lee, E.; Park, Y.; Shin, J.G. Large engineering project risk management using a Bayesian belief network. *Exp. Syst. Appl.* **2009**, *36*, 5880–5887. [CrossRef]

47. Cox, L.A. What's wrong with risk matrices? *Risk Anal.* **2008**, *28*, 497–512. [PubMed]

48. Yousefi, V.; Yakhchali, S.H.; Khanzadi, M.; Mehrabanfar, E.; Šaparauskas, J. Proposing a neural network model to predict time and cost claims in construction projects. *J. Civ. Eng. Manag.* **2016**, *22*, 967–978. [CrossRef]

49. Du-Juan, G.; Pen, G. Constructing interdependent risks network of project portfolio based on bayesian network. In Proceedings of the International Conference on Management Science & Engineering (21th), Helsinki, Finland, 17–19 August 2014.

50. Cooper, R.; Edgett, S.; Kleinschmidt, E. New product portfolio management: Practices and performance. *J. Prod. Innov. Manag.* **1999**, *16*, 333–351. [CrossRef]

51. Heising, W. The integration of ideation and project portfolio management—A key factor for sustainable success. *Int. J. Proj. Manag.* **2012**, *30*, 582–595. [CrossRef]

52. Cooper, R.; Edgett, S.; Kleinschmidt, E. *Portfolio Management: Fundamental for New Product Success*; John Wiley & Sons: New York, NY, USA, 2002.

53. Meskendahl, S. The influence of business strategy on project portfolio management and its success—A conceptual framework. *Int. J. Proj. Manag.* **2010**, *28*, 807–817. [CrossRef]

54. Archer, N.P.; Ghasemzadeh, F. An integrated framework for project portfolio selection. *Int. J. Proj. Manag.* **1999**, *17*, 207–216. [CrossRef]

55. Killen, C.P.; Hunt, R.A.; Kleinschmidt, E.J. Project portfolio management for product innovation. *Int. J. Qual. Reliab. Manag.* **2008**, *25*, 24–38. [CrossRef]

56. Jonas, D. Empowering project portfolio managers: How management involvement impacts project portfolio management performance. *Int. J. Proj. Manag.* **2010**, *28*, 818–831. [CrossRef]

57. Unger, B.N.; Gemünden, H.G.; Aubry, M. The three roles of a project portfolio management office: Their impact on portfolio management execution and success. *Int. J. Proj. Manag.* **2012**, *30*, 608–620. [CrossRef]

58. Fricke, S.E.; Shenhar, A.J. Managing multiple engineering projects in a manufacturing support environment. *IEEE Trans. Eng. Manag.* **2000**, *47*, 258–268. [CrossRef]

59. Beringer, C.; Jonas, D.; Kock, A. Behavior of internal stakeholders in project portfolio management and its impact on success. *Int. J. Proj. Manag.* **2013**, *31*, 830–846. [CrossRef]

60. Martinsuo, M.; Lehtonen, P. Role of single-project management in achieving portfolio management efficiency. *Int. J. Proj. Manag.* **2007**, *25*, 56–65. [CrossRef]

61. Lawshe, C.H. A quantitative approach to content validity. *Pers. Psychol.* **1975**, *28*, 563–575. [CrossRef]

62. Fang, C.; Marle, F. A simulation-based risk network model for decision support in project risk management. *Decis. Support Syst.* **2012**, *52*, 635–644. [CrossRef]

63. *GeNIe. Version 2.0*; A Decision Systems Laboratory, School of Information Sciences, University of Pittsburgh: Pittsburgh, PA, USA, 2015.

![sustainability logo] *sustainability*

MDPI

Article

Research on Factors Affecting Public Risk Perception of Thai High-Speed Railway Projects Based on "Belt and Road Initiative"

Sangsomboon Ploywarin [1,2], Yan Song [1,*] and Dian Sun [1]

[1]   School of Economics and Management, Harbin Engineering University, Harbin 150001, China;
       sangsombun@hotmail.com (S.P.); sundian900621@163.com (D.S.)
[2]   International Strategy Division, Rajamangala University of Technology, Thanyaburi 12110,
       Pathum Thani, Thailand
*   Correspondence: songyan@hrbeu.edu.cn

Received: 25 April 2018; Accepted: 8 June 2018; Published: 13 June 2018

Abstract: Studies on the factors affecting public risk perception of high-speed railway projects in Thailand are very limited. The aim of this study was to assess the influencing factors of public railway project risk perception, which described the public trust degrees of government, enterprise, media and experts with a combination of variables. Therefore, the study used the widely accepted influential factors and proposed a comprehensive framework to clarify the mechanism among various factors in the public risk perception. Dataset of 675 samples was collected from Don Muang area Bangkok, Pak Thong Chai, Pak Chong, Kaengkhoi area, and Nakhon Ratchasima Province, Thailand through questionnaire. Rationality of the questionnaire was ensured through its high reliability and efficiency. The dimension hypothesis of the second-order factor was validated by confirmatory factor analysis, and the relationship among information acquisition, trust, emotion and risk perception was analyzed through the structural equation model. The results show that, within the factors that affect risk perception, the public has a more direct effect on the factors of social emotion of railway projects compared with information acquirement and the factors of trust level of each subject. This study exerts practical implications to reduce public risk perception of railway projects and promote the development of railway in Thailand.

Keywords: Thai-Sino high-speed railway project; the Belt and Road; public risk perception; affecting factors

---

1. Introduction

The International Summit Forum at Beijing in May 2017 represented the remarkable success of the contributions of the past three years of "Belt and Road Initiative". Meanwhile, it will play a greater role in the world economy. During September and October 2013, President Xi Jinping visited Central and Southeast Asian countries, where he proposed a major initiative to build the "Silk Road Economic Belt" and "21st-Century Maritime Silk Road" (referred to as "the Belt and Road Initiative"), which received a positive response from Thai government. In November 2014, when Thai Prime Minister Yasushi Akira attended APEC leaders' meeting in Beijing, he said that [1], "Thailand is exploring a path of development which meets its own national realities and hopes to make exchanges and deepen cooperation with China. With the help of 'the Belt and Road Initiative' in particular, Thai will promote agriculture, railway cooperation and regional interconnectivity, expand the export of its own agricultural products to China, facilitate non-governmental exchanges as well as strengthen personnel training". In December the same year, Prime Minister Paryuth Chan-ocha visited China again and reiterated that Thailand is willing to actively participate in "the Belt and Road Initiative" proposed by President Xi Jinping. Thailand would deepen cooperation in the fields of railways, communications and tourism; promote regional interconnectivity; and march to the goal of a free trade zone in Asia-Pacific region [2]. "The belt and Road initiative" is both a strategic opportunity and a realistic challenge for Thai-Sino relations.

Because of the decrease in Thailand's GDP since 2014, Thai government was eager to pour driving force into Thailand's weak economy through high-speed rail project and other infrastructure. In April 2016, the Thai National Railway Administration approved to construct two new railroad tracks, with a total investment of 143.5 billion baht (about 4.1 billion US dollars), which showed that the railway development has entered a new stage after the stagnation for up to half a century. The negotiations for Thai-Sino High-speed Railway project lasted for three years and plans to begin building the longest Bangkok-Nakhon Ratchasima section in 2017 were made. The reasons for such a long negotiation were the differences in funds, the design of blueprints, and the delay in the railway development, which raised suspicion of the railway project and conflicting emotions among Thai people. Many other experiences in large-scale projects such as the Thailand Coal Power Plant project showed that the public's attitude towards the project depended on their subjective awareness of the project, that is, risk perception, rather than scientific and rational judgment. Thus, it is necessary to propose a public risk perception of high-speed railways in Thailand.

Scholars have studied the influencing factors of risk perception from different aspects. It is common to analyze the impact factors of risk perception in large-scale projects. Qualitative analysis is widely used compared to quantitative empirical research methods. In the process of railway project safety management, the judgments and reactions of people can be investigated when people assess the potential risks of railway projects [3]. Therefore, this paper constructs the high-order variable trust to describe the overall trust level of the government, experts, media and enterprises from the public. It also establishes a theoretical framework for the impact mechanism of risk communication, information acquisition, social emotion and trust on the public perception of risk. Structural equation and hierarchical regression analysis method were used to analyze the influence of variables to reduce public opposition to Thailand high-speed rail projects and provide useful suggestions for the administrative department to formulate relevant policies and regulations of railway projects.

2. Theoretical Background and Research Hypotheses

Study of the influencing factors of public risk perception includes not only the characteristics of risk itself and the sources of information, but also the analysis of trust and emotion as well as value orientation [4–7]. Elsewhere, scholars have explored the influencing factors for risk perception from different perspectives. Lupton [8] suggested that the level of public risk knowledge, risk perceptions, and the acceptability of risks depended on the social environment in which the public was located. Different social backgrounds and cultural heritage formulate different public understanding for risk and the formation of risk perception is more a process of social development. Connor et al. [9] found that in the United States of America the knowledge of risk, risk assessment and behavioral intention are all different, which directly affects the public acceptance and judgment for the risk. The results showed that the factors influencing risk judgment not only include the knowledge reasons discussed by predecessors, but also other reasons such as politics, economy, environment, culture and family relations. The public perception of risk will take all characteristics affecting public risk perception which is important in the index system of major project stability evaluation. Similarly, various psychological factors should be considered in the public risk perception of railway project.

According to the particularity of Thailand high-speed railway project and the harmful consequence that technical backwardness may bring, the technology acceptability model [10] put forward by scholars was used. The model draws on the important role of trust factors and combines the widely-accepted emotion factors in the field of risk perception. This paper integrates the influential factors which have been widely accepted by scholars and attempts to adopt a comprehensive framework to clarify the mechanism among various factors in the public risk perception.

2.1. Important Role of Trust

Trust is the main factor that affects public risk perception [11]. Previous research, whether in terms of the model building or the empirical conclusion, reflects the inverse relationship between trust factor and risk perception [12]. Media is a major link in the process of social amplification of risk and a main way for people to obtain information [13]. When people do not have sufficient knowledge of the risks, the public often takes information directly from the media. Thus, the media plays an important role in the public's risk perception in Thailand railway project. From the point of the related responsibility of the railway project in Thailand, nuclear power plant research experience [14] is applied to study the effect of public trust factor in terms of the media, the government, the experts and the enterprises.

Earle and Cvetkovich [15] pointed out that public decision-making and judgments are often guided by social trust because of the lack of knowledge. They prefer to choose experts based on trust, rather than making rational judgments on unfamiliar risks. Studies by Siegrist and Cvetkovich show that, when public knowledge of risk programs is scarce, public judgments about risk and income will depend on social trust [16]. In the field of power plant, a lot of research shows that trust has a significant effect on the perception of specific technologies [17–20]. Compared with the public, experts have a comprehensive knowledge and information about technology which can objectively evaluate the feasibility of technology [21]. However, in Thailand's social realities, experts cannot make concerted agreement on the railway project due to the reluctance to accommodate varied opinions. Hence, the credibility of experts is compromised, resulting in increased public awareness of the railway risk and the assessment of risk is far higher than the expert evaluation.

Based on the above analysis, this paper puts forward the following hypotheses:

Hypothesis H1: *Trust from the public is a high-order variable of the government, experts, media and enterprises.*

Hypothesis H2: *Trust has an inverse correlation with the public perception of risk in railway projects.*

Hypothesis H3: *Trust and social emotion are inversely related, that is, the higher the degree of trust, the more positive the mood, and vice versa.*

2.2. Social Sentiment and Public Perception of Risk

Previous studies show that public choice behavior in crisis situations is influenced by perceptions of risk [22–24]. A too high-risk perception will lead to an overreaction of the public, resulting in unnecessary irrational behavior. The difference of public risk perception often influences the effect of response measures of governmental crisis. Loewenstein et al. [25] established the emotional hypothesis which pointed out that risk perception depends not only on rational cognitive judgment, but also on the mood of decision makers. The idea that emotion has a significant impact on risk perception has been widely accepted by scholars. Studies have shown that people with positive emotions are at a lower risk level than those with social emotions. Due to the particularity of railway projects, its construction will take up large amount of land which will naturally arouse public anxiety and fear. Therefore, it is necessary to explore the impact of social emotion on project risk perception.

Based on the above analysis, this paper puts forward the following hypothesis:

Hypothesis H4: *Social emotion is positively correlated with the public risk perception of the railway project.*

2.3. Information Acquisition

Numerous studies have shown that the dissemination of information is very important to the public's risk perception in risk events. The methods of sourcing for information and the presentation of information content will greatly influence and even determine the public's risk perception of the event [26].

Midden [27] suggested that trust depends on the quality of information, the source of the information, and the organization and performance of the institution in the public risk perception. The public trust in government mainly comes from the historical performance and current behavior of the government's ability which is the long-term construction process. Especially, the effectiveness of the government and experts' risk communication in each crisis events will directly affect the trust level of the public. Therefore, the acquisition of information is very important for building trust in railway projects.

Based on the above analysis, this paper puts forward the following hypotheses:

Hypothesis H5: *There is a positive correlation between information acquisition and trust.*

Hypothesis H6: *Information acquisition is positively correlated with social emotion.*

2.4. Technology Risk and Public Risk Perception

Based on Baker's analysis of the ubiquity of risk, the uncertainty of modern science technology, the failure of national society management, and the technological risk of the expansion of globalization have aroused general risk awareness due to the serious consequences. However, there is no significant consistency between the actual risk and the perception of risk in risk event. The risk communication expert Sandman has introduced the concept of "anger" to explain the sources of differences between technology risk and risk perceptions. As a result, public outrage can result in a cognitive amplification of actual risks. In other words, for those projects or events with relatively low technical risk, they will be considered as high-risk projects or events if they encounter outrageous crowd. In the Thai-Sino high-speed railway project, the Chinese railway technology will be adopted. Although China's high-speed railway has developed rapidly in recent years, Thai people have little knowledge about high-speed railway, because of the stagnation in the development of Thai railway in recent decades, which will inevitably raise suspicion of China high-speed railway technology.

Based on the above analysis, this paper puts forward the following hypothesis:

Hypothesis H7: *The technical risk is positively correlated with the public risk perception of railway project.*

2.5. Conceptual Model

A conceptual model of the impact factors in high-speed railway project risk perception can be developed based on the above analysis, as shown in Figure 1.

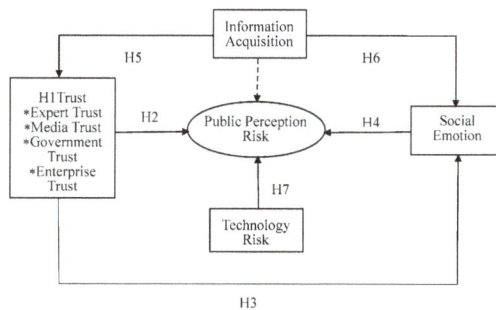

Figure 1. Conceptual model of influencing factors of risk perception in high-speed railway projects.

3. Research and Design

The variables data for this study were collected and counted through a wide range of questionnaires. Rationality of the questionnaire was ensured through its high reliability and efficiency. The dimension hypothesis of the second-order factor was validated by confirmatory factor analysis, and the relationship between information acquisition, trust, emotion and risk perception was analyzed through the structural equation model.

3.1. Questionnaire Design and Variable Processing

The operational definition and measure index of the variable were sorted. Combined with the construction of railway projects, the measuring scale of four latent variables of risk perception, trust, emotion and information were designed, and the variables were measured using a five-point Likert scale, from 1 to 5 representing: 1, "very disagree"; 2, "disagree"; 3, "general"; 4, "consent"; and 5, "very agreeable". For the negative effect involved in the questionnaire, the negative score was set for the needs of the research.

The selection of indicators and questionnaires was discussed in interviews with 12 engineering experts at a Thai university, and the scale was revised and improved. Before the formal investigation, 60 college students were selected to carry out a small-scale preliminary survey. According to the recommendations, the problems in the questionnaire were clarified, which enhanced its operability. The revised scale is shown in Table 1.

Table 1. Sample population description analysis.

Statistical Content	Category	Sample Number	Proportion
Gender	Male	301	44.59%
	Female	374	55.41%
Age	Under 20	88	13.03%
	20~30	196	29.04%
	30~40	167	24.74%
	40~50	113	16.74%
	50~60	87	12.89%
	60 and above	24	3.56%
Education	Junior High School	132	19.56%
	High School	154	51.56%
	Undergraduate	348	8.94%
	Postgraduate	41	6.07%
Occupation	Party and Government Officer	49	7.26%
	Institution Officer	94	13.93%
	Enterprise officer	146	21.63%
	Professional and Technical Personnel (Teachers, Doctors, Engineers)	67	9.93%
	Student	219	32.45%
	Retired	29	4.30%
	Freelance	45	6.65%
	Others	26	3.85%

3.2. Overall Sample Description and Data Collection

According to the analysis of 675 effective questionnaires, the overall description of the sample was summarized and analyzed which includes gender, age, education and occupation. The results of the overall description of the samples are shown in Table 1.

Because the current research does not involve regional differences, the study analyzed the influencing factors of railroad project risk perception from the individual microcosmic angle.

This research adopted the quantitative investigation method to obtain the data. According to the first-phase in Thai-Sino high-speed railway project, Bangkok and Nakhon Ratchasima mansion would be connected with a length of 252.5 km and the design speed of 250 km/h.

Data were collected through questionnaire from 16 August 2017 to 25 December 2017 in Bangkok and Nakhon Ratchasima mansion. During the period of this investigation, 800 questionnaires were issued and 728 questionnaires were collected, among which 675 questionnaires were effective, and the effective recovery rate was 92.71%. From the gender perspective, the male and female proportions of the sample were almost the same: 374 were female (55.41%), and 301 were male (44.59%). From the age perspective, 88 were <20 years old (13.03%), 196 were 20–30 years old (29.04%), 167 were 30–40 years old (24.74%), 113 were 40–50 years old (16.74%), 87 were 50–60 years old (12.89%), and 24 were >60 years old (3.56%). For the level of education, 132 had junior high school education or below, 154 had college/undergraduate education, 348 had graduate education, and 41 had postgraduate education or above. From the professional point of view, there were students, business staff and institutions staff: 219 from students, 146 from enterprise staff and 94 from units which accounted for 68.01%. Professional and technical personnel and the number of party workers were relatively large, accounting for 9.93% and 7.26%, respectively. Freelancers, farmers and others accounted for 6.65%, 4.3%, and 3.85%, respectively. From two urban population characteristics, the sample had a good representation, and the data could meet the needs of further research.

3.3. Reliability of Scale and Analysis of Validity

Reliability and validity analysis are the precondition of model evaluating. This research carried out the screening of the questionnaire, and then selected data were input to SPSS23.0 and AMOS21.0 statistic software (see Table 2). Cronbach's α was used to evaluate the reliability of data. The Cronbach's α value of the whole scale is 0.994 (>0.70). The α value of the corresponding component table of each variable is also greater than 0.7 which shows that the scale has high reliability. Validity analysis includes two parts: content validity and structure validity. For content validity, the problem of measuring variables is obtained based on analysis and collation of domestic literature and completed through the discussion and confirmation with experts as well as small sample to be modified and tested. Hence, the variables have good content validity. As for structural validity, by applying the factor analysis method in statistical analysis which can test the factor load of the variables corresponding to each index, each topic has a good factor load on its corresponding factors which reflects the relative importance of the factor load in public factors. $p \leq 0.01$ indicated that it has strong statistical significance and sound convergence validity. Through the above results, each topic can clearly explain the corresponding variables, so the scale has good structural validity. For sample and data test results, the KMO (Kaiser–Meyer–Olkin) value is 0.945 and Bartlett's Spherical test significance is 0.000 which means the questionnaire data are suitable for application factor analysis. Through the correlation coefficient range of the variable index in Table 2, we can see that the variable index correlation coefficient does not contain the value 1, and the corresponding p value is significant ($p < 0.01$ or 0.05) which indicates that the validity of the scale is verified by statistics.

Table 2. Variable reliability and validity analysis.

Variable	Item	Factor Loading	Cronbach's α	Variable Index Correlation Coefficient (Range)
Information Acquisition (IA)	Railway Accident Information (IA1)	0.680	0.933	0.523–0.597
	Railway Safety Information (IA2)	0.448		
	Reducing Railway Hazard Information (IA3)	0.480		
	Railway Project Operation Supervision Support Information (IA4)	0.557		
	Railway Transportation Efficiency Guarantee Information (IA5)	0.309		
Risk Perception (RP)	Visibility (RP1)	0.960	0.975	0.521–0.644
	Frightening (RP2)	0.662		
	Possibility (RP3)	0.604		
	Seriousness (RP3)	0.608		
	Controllability (RP4)	0.962		
Social Emotion (SE)	Unconvinced (SE1)	0.849	0.973	0.507–0.654
	Worry (SE 2)	0.875		
	Tense or Anxious (SE3)	0.875		
	Discontent or Dislike (SE4)	0.477		
	Scared or Frightened (SE5)	0.855		
	Anger (SE6)	0.888		
Technology Risk (TR)	Technical Capability (TR1)	0.854	0.982	0.566–0.671
	Risk Control Ability(TR2)	0.854		
	System Technical Performance (TR3)	0.791		
	Technology Risk Management (TR4)	0.655		
Expert Trust (ExT)	Evaluation Ability (ExT1)	0.749	0.984	0.575–0.691
	Speech Authenticity (ExT2)	0.709		
	Accountability (ExT3)	0.633		
	Reliability (ExT4)	0.672		
Media Trust (MT)	Reporting Authenticity (MT1)	0.778	0.947	0.540–0.691
	Report Timely (MT2)	0.745		
	Responsibility (MT3)	0.772		
Government Trust (GT)	Political Reliability (MT4)	0.602	0.985	0.478–0.551
	Possibility of Supervision and Management System (GT1)	0.564		
	Feasibility of Decision Making (GT2)	0.941		
	Emergency Management Practicability (GT3)	0.672		
Enterprise Trust (ET)	Managerial Competence and Empirical Reliability (ET1)	0.728	0.976	0.484–0.569
	Rationality of Decision Process (ET2)	0.762		
	Fund Reliability (ET3)	0.778		

4. Verification and Modification of Model

4.1. Trust Second-Order Factor Model

To test the measurement rationality of the observational variables, the study used AMOS6 to analyze confirmatory factor of each variable. The test trust factor is the higher-order factor assumption by the public composed by four dimensions including experts, media, government and enterprise. An overall model suitability test for second-order confirmatory factor analysis is between 2 and 5. The smaller is the numerical value, the better is the fitting degree of the model. Incremental Fit Index (IFI), Normed Fit Index (NFI), Comparative Fit Index (CFI), and Goodness of Fit Index (GFI) are more than 0.90. Root Mean Square of approximation (RMSEA) is less than 0.05 which should be less than 0.08 (Table 3). To verify the dimensions, polymerization validity, difference validity of trust measurement, and the reliability of second-order factor construction, this study constructed four correlative first-order factors formed by trust items and high-order second-order factor models formed by the corresponding first-order factors (see Figure 2). As shown in Table 3, the associated first- and second-order factor models are less than 5, and the two RMSEA are less than 0.08. IFI, NFI, CFI, and GFI values are greater than 0.9. The model fitted the corresponding criteria. The factor load of trust is above 0.5, and the path of the item to the factor is very significant, which shows that the public has excellent polymerization validity for the structure model. To further verify that the trust factor dimension is a higher-order factor, we should prove the rationality of the second-order factor not only through the comparison value, but also the value of the goal coefficient T. The value of the target coefficient T is the ratio of chi square value about the low-order factor model and the higher-order factor model [28]. The closer is the T value to 1, the more reasonable is the higher-order factor. When the T value is higher than 0.90, the higher-order factor model can effectively explain the correlation between the lower-order factor [29]. Here, the target coefficient T = 38/40 = 0.95 indicates that the trust factor interprets 95% relationship of four dimensions feature and the second-order factor loads are very significant. Therefore, assuming that H1 is validated, the trust is a higher-order variable composed of the public to the media, government, experts, and enterprises. It is more reasonable to make further investigation on the public perception of trust with different subjects. In the four trust dimensions, the government trust has the most obvious contribution to the trust factor according to the factor load coefficient, followed by the trust of the railway enterprise, media, and finally the expert trust. This shows that the importance of different dimensions is different in the consideration of the relationship between trust factors and other variables. For instance, the degree of trust in the government is significantly higher than others, that is, when raising public confidence in government (government credibility), the role of trust factors can be significantly improved.

Table 3. Fitting indexes of each factor model (*n* = 675).

Model	Chi-Square	df	Chi-Square/df	GFI	RMSEA	CFI	AGFI	NFI
Associated First-order Factor	139.337	71	1.962	0.971	0.038	0.949	0.958	0.903
second-order Factor	159.238	73	2.181	0.939	0.058	0.955	0.925	0.934

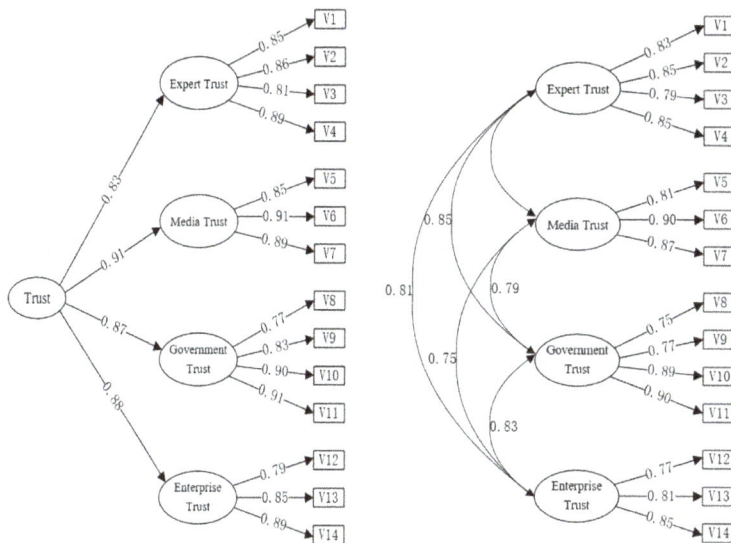

Figure 2. Confidence analyses of second-order factor confirmation.

4.2. Inspection and Correction of the Whole Model

Based on the above analysis, the trust variable is a higher-order variable composed of four dimensions of media trust, government trust, expert trust and enterprise trust. The high dimension brings great inconveniences to the calculation, and there may be complex nonlinear correlation between each index. In the past, most studies used the average value of the measurement index of each variable to substitute each variable, which could lead to the deviation of the estimation result. Therefore, the weighted arithmetic mean was used to calculate the five-dimension variables, which are the five measurements of trust. The weights are determined by the measurement of the variables and the load factors of potential variables.

After confirming the reliability and validity of the measurement model, the AMOS6 software was used to analyze the path based on the maximum likelihood estimation method. The model was modified preliminarily according to the modification indices of the model output to enhance the fitting of the model. By modifying the model, the fitting result of the whole model is shown in Figure 3. The χ^2/df value is 1.779 < 5, RMSEA is 0.34 < 0.08. CFI is 0.944 and Adjusted Goodness of Fit Index (AGFI) is 0.943 which is greater than 0.9 (see Table 4). Parsimonious Normed Fit Index (PNFI) is 0.785, which fits well. Thus, the model reached an acceptable level. Based on the criterion of fitting, the model was tested and the results are shown in Table 5.

Table 4. Comparison of fitting indexes of each model.

Model	Chi-Square	df	Chi-Square/df	GFI	RMSEA	CFI	AGFI	NFI
Original Model	1477.009	202	7.312	0.863	0.097	0.536	0.829	0.504
Modified Model	352.311	198	1.779	0.955	0.034	0.944	0.943	0.882

Table 5. Parameter estimation results of the modified model.

			Estimate	S.E.	C.R.	*p*	Label
Trust	<—	Information Acquisition	0.271	0.125	1.923	0.016	par_1
Social Emotion	<—	Information Acquisition	0.315	0.108	2.925	0.002	par_3
Social Emotion	<—	Trust	0.315	0.108	2.925	0.002	par_6
Perception Risk	<—	Information Acquisition	0.314	0.119	2.628	0.031	par_2
Perception Risk	<—	Trust	0.183	0.105	1.742	0.082	par_4
Perception Risk	<—	Social Emotion	0.372	0.174	2.965	***	par_5
Perception Risk	<—	Technology Risk	0.355	0.150	3.132	***	par_23
S1	<—	Information Acquisition	1.000				
S2	<—	Information Acquisition	0.478	0.201	2.376	0.017	par_7
S3	<—	Information Acquisition	0.465	0.220	2.115	0.034	par_8
S4	<—	Information Acquisition	0.573	0.219	2.616	0.009	par_9
S5	<—	Information Acquisition	0.146	0.108	1.353	0.176	par_10
S6	<—	Trust	1.000				
S7	<—	Trust	0.020	0.084	0.237	0.812	par_11
S8	<—	Trust	0.920	0.223	4.118	***	par_12
S9	<—	Trust	0.274	0.069	3.988	***	par_13
S10	<—	Perception Risk	1.000				
S11	<—	Perception Risk	1.100	0.160	6.862	***	par_14
S12	<—	Perception Risk	1.344	0.177	7.584	***	par_15
S13	<—	Perception Risk	0.850	0.129	6.594	***	par_16
S14	<—	Perception Risk	0.790	0.134	5.901	***	par_17
S15	<—	Social Emotion	1.000				
S16	<—	Social Emotion	1.811	0.389	4.838	***	par_18
S17	<—	Social Emotion	1.547	0.341	4.934	***	par_19
S18	<—	Social Emotion	0.843	0.252	3.350	0.003	par_20
S19	<—	Social Emotion	0.636	0.244	2.612	0.009	par_21
S20	<—	Social Emotion	0.021	0.050	0.414	0.679	par_24
S21	<—	Technology Risk	1.000				
S22	<—	Technology Risk	1.605	0.561	4.890	***	par_22
S23	<—	Technology Risk	0.791	0.231	4.631	***	par_25
S24	<—	Technology Risk	0.655	0.145	3.547	***	par_26

Notes: Standard errors in indicates. *** *p* value is less than 0.001.

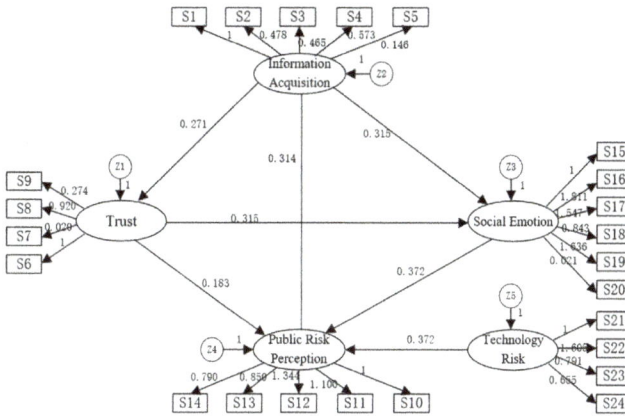

Figure 3. Overall model.

Based on Figure 3, it is obvious that Information acquisition has a significant and positive effect on Trust (b = 0.271, *p* = 0.016), Social emotion (b = 0.315, *p* = 0.002) and Public Risk Perception (b = 0.314, *p* = 0.031). Thus, H1, H5, and H6 are accepted, while H2 is not (b = 0.183, *p* = 0.082). If the direct path between trust and risk perception, information acquisition and social emotion is set to be freely evaluated, the chi square value will be greatly reduced. Then, the fitting degree of the model and the corresponding parameter estimation value will be improved. To improve the data fitting of the existing model and to find the potential optimal model, we construct the competition model by comparing their fitting.

4.3. Relationship among Social Emotion, Trust and Risk Perception

Public trust and social sentiment are relevant. From the above path analysis, it can be concluded that social emotion has a significant positive effect on the public perception of risk (b = 0.372), but trust has no significant effect (b = 0.183, *p* = 0.082), which contradicts with the hypothesis of this article. Then, it is necessary to analyze their relationship. Theoretically, trust can serve as mediation variable as well as mediate variable. The degree of public trust in related subjects will change the impact of social sentiment on risk perception. When the public extremely trusts government decision-making, the project risk perception of railways will not be high even if there is a certain fear of railway projects. On the contrary, the lower is the degree of trust, the greater is the impact of social emotions. On the other hand, the public's social sentiment towards the railway project will affect the public's trust, and the degree of trust will have a certain impact on public perception of risk. Therefore, it is necessary to do further analysis to explore the role of trust.

According to the test steps to adjust variable and intermediary variable proposed by Wen Zhonglin [30], the score of latent variable is centralized. The effect by multicollinearity is eliminated, and the relationship between variables and the moderating effect of trust are established by the multi-level analytic hierarchy process. The first step is to establish a regression model with risk perception as the dependent variable, and trust and social emotion as the independent variable in Model 1. The second step is to add the product of trust and social emotion as the independent variable in Model 1, and then form Model 2. The concrete regression results (see Table 6 for details).

In Models 1 and 2, the correlation coefficient between social emotion and public perception of risk is positive, so the social emotion will have positive effect on public risk perception considering the main effect variable and adjusting the variable in a single or simultaneous way. With the introduction of the intersection of trust and social emotion, the explanatory ability of Model 2 has been improved compared to Model 1 (the change of R^2 is about 7%), which suggests that the moderating effect of trust is conducive to a more comprehensive explanation to the process of social emotion. The regression coefficient of trust and social emotion is positive which shows that the moderating effect of trust is significant. With the increase of public trust, the influence of social mood on public risk perception level will be reduced (see Figure 4).

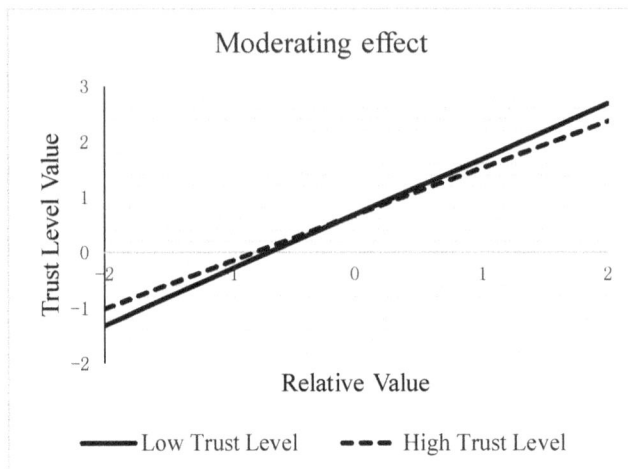

Figure 4. The moderating effect of trust level.

Table 6. Results of hierarchical regression analysis.

	Dependent Variable	Public Perception Risk	
		Model-1	Model-2
Intercept Term		11.646	−0.853
Independent Variable	Trust	−0.003	0.266 *
	Social Emotion	0.253 **	0.923 **
	Trust × Social Emotion		−0.014 **
	R²	0.057	0.063
	ΔR²		0.007 *

Notes: Standard errors in indicates. *, **: 0.1 and 0.05, respectively.

5. Conclusions

This study aimed to assess the influencing factors of public railway project risk perception and described the trust degree of public to government, enterprise, media and experts with a combination of variables. Structure equation model and hierarchical regression analysis method were used to explore the mechanism of trust, information acquirement and social emotion of the public perception of risk. The results show that, among the factors that affect risk perception, the public has a more direct influence on the factors of social emotion of railway projects compared with information acquirement and the factors of trust level of each subject. This research has practical significance to reduce the public's perception of the risk in railway projects. The policy-making departments should popularize and educate about relevant railway related knowledge, reduce the public's anxiety, and raise attention to the public emotion factor. For instance, different strategies are formulated according to the differences of public mood characteristics at different levels. Then, the risk perception level of the public can be directly reduced. On the other hand, the serious consequences of railway accidents in the past have brought a shadow to the people around the world. If the government departments want to improve the public's social mood, they must pay attention to the promotion of their own credibility, and promote the ability of media, enterprises and experts to reduce negative impact brought by other Thai railway construction accidents.

The significant effect of information acquisition factors on the level of trust provided the possibility to raising public confidence in the media, government, experts and enterprises. The Thai government should improve not only the release mechanism of railway project information and the transparency of project construction information, but also the information mechanism of railway project risk communication so that the public can better understand railway technology and consequently the public's trust level to the related subjects can be raised.

Supplementary Materials: The following are available online at http://www.mdpi.com/2071-1050/10/6/1978/s1.

Author Contributions: S.P. and Y.S. conceived and designed the research questions, the conceptual model, the questionnaire and the methodology. S.P. collected and analyzed the data. Y.S. revised the paper. All authors read and approved the final manuscript.

Funding: The study was supported by the Chinese National Nature Science Foundation (Grant No. 71771061) and the Major Strategic Research Plan of Harbin Engineering University (Grant No. HEUCFW170903).

Conflicts of Interest: The authors declare no conflict of interest.

References

1. Xinhua News Agency. Xi Jinping meets with Prime Minister of Thailand, Ba Yu. *Xinhua*, 9 November 2014.
2. Xinhua News Agency. Xi Jinping meets with Prime Minister of Thailand, Ba Yu. *Xinhua*, 23 December 2014.
3. Xie, X.F. Several psychological problems in risk studies. *Psychol. Sci.* **1994**, *2*, 104–108.
4. Cox, D. Risk Handling in Consumer Behavior: An Intensive Study of Two Cases. In *Risk-Taking and Information Handling in Consumer Behavior*; Harvard University Press: Boston, MA, USA, 1967; pp. 34–81.
5. Starr, C. Social Benefit versus Technological Risk. *Science* **1969**, *165*, 1232–1238. [CrossRef] [PubMed]
6. Flynn, J.; Slovic, P.; Mertz, C. Gender, Race, and Perception of Environmental Health Risks. *Risk Anal.* **1994**, *14*, 1101–1108. [CrossRef] [PubMed]
7. Lai, J.; Tao, J. Perception of Environmental Hazards in Hong Kong Chinese. *Risk Anal.* **2003**, *23*, 669–684. [PubMed]
8. Lupton, D. *Risk*; Routledge: New York, NY, USA, 1999; pp. 28–33.
9. O'Connor, R.E.; Bord, R.J.; Fisher, A. The curious impact of knowledge about climate change on risk perceptions and willingness to sacrifice. *Risk Decis. Policy* **1998**, *3*, 145–155. [CrossRef]
10. Siegrist, M. The influence of trust and perceptions of risks and benefits on the acceptance of gene technology. *Risk Anal.* **2000**, *20*, 195–204. [CrossRef] [PubMed]
11. Rosa, E.A.; Tuler, S.P.; Fischhoff, B.; Webler, T.; Friedman, S.M.; Sclove, R.E.; Shrader-Frechette, K.; English, M.R.; Kasperson, R.E.; Goble, R.L.; et al. Nuclear waste: Knowledge waste? *Science (Washington)* **2010**, *329*, 762–763. [CrossRef] [PubMed]
12. Slovic, P.; Flynn, J.H.; Layman, M. Perceived risk, trust, and the politics of nuclear waste. *Science* **1991**, *254*, 1603–1607. [CrossRef] [PubMed]
13. Bu, Y.M. Social amplification of risk: Framework and empirical research and implications. *Study Pract.* **2009**, *12*, 120–125.
14. Quan, S.-W.; Zeng, Y.-C.; Huang, B. Public Perception and Acceptance of Nuclear Power. *Beijing Soc. Sci.* **2012**, *5*, 55–60.
15. Earle, T.C.; Cvetkovich, G. *Social Trust: Toward a Cosmopolitan Society*; Greenwood Publishing Group: Westport, CT, USA, 1995.
16. Siegrist, M.; Cvetkovich, G.; Roth, C. Salient value similarity, social trust, and risk/benefit perception. *Risk Anal.* **2000**, *20*, 353–362. [CrossRef] [PubMed]
17. Bord, R.J.; O'Connor, R.E. Determinants of risk perceptions of a hazardous waste site. *Risk Anal.* **1992**, *12*, 411–416. [CrossRef]
18. Flynn, J.; Burns, W.; Mertz, C.K.; Slovic, F. Trust as a determinant of opposition to a high-level radioactive waste repository: Analysis of a structural model. *Risk Anal.* **1992**, *12*, 417–429. [CrossRef]

19. Groothuis, P.A.; Miller, G. The role of social distrust in risk-benefit analysis: A study of the siting of a hazardous waste disposal facility. *J. Risk Uncertain.* **1997**, *15*, 241–257. [CrossRef]

20. Jungermann, H.; Pfister, H.R.; Fischer, K. Credibility, information preferences, and information interests. *Risk Anal.* **1996**, *16*, 251–261. [CrossRef]

21. SiegrisL, M.; CveLkovich, G. Perception of Hazards: The Role of Social Trust and Knowledge. *Risk Anal.* **2000**, *20*, 713–720. [CrossRef]

22. Sandman, P.M. Risk Communication: Facing Public Outrage. *EPA J.* **1987**, *13*, 21–22. [CrossRef]

23. Finucane, M.L.; Alhakami, A.; Slovic, P.; Johnson, S.M. The affect heuristic in judgments of risks and benefits. *J. Behav. Decis. Mak.* **2000**, *13*, 1–17. [CrossRef]

24. Epstein, S. Integration of the cognitive and psychodynamic unconscious. *Am. Psychol.* **1994**, *49*, 709–724. [CrossRef] [PubMed]

25. Loewenstein, G.F.; Weber, E.U.; Hsee, C.K.; Welch, N. Risk as feelings. *Psychol. Bull.* **2001**, *127*, 267–286. [CrossRef] [PubMed]

26. Wang, J.; Qi, L.; Zhou, Q. Review of Research on Influencing Factor of the Public' Risk Perception. In Proceedings of the Conference on Psychology and Social Harmony (CPSH2012), Shanghai, China, 20–22 May 2012; pp. 485–488.

27. Midden, C. Credibility and risk communication. In Proceedings of the International Workshop on Risk Communication, Juelich, Germany, 17–20 October 1988; Volume 17, p. 21.

28. Lichtenthaler, U.; Lichtenthaler, E. A Capability-Based Framework for Open Innovation: Complementing Absorptive Capacity. *J. Manag. Stud.* **2009**, *46*, 1315–1338. [CrossRef]

29. Marsh, H.W.; Hocevar, D. Application of confirmatory factor analysis to the study of self-concept: First-and higher order factor models and their invariance across groups. *Psychol. Bull.* **1985**, *97*, 562. [CrossRef]

30. Wen, Z.; Hau, K.-T.; Chang, L. A Compar Ison Ofmoderator Andmed Iator and Their Applications. *Acta Psychol. Sin.* **2005**, *37*, 268–274.

sustainability

MDPI

Article

Sustainable Construction Risk Perceptions in the Kuwaiti Construction Industry

Dalya Ismael * and Tripp Shealy

Department of Civil and Environmental Engineering, Virginia Tech, Blacksburg, VA 24061, USA; tshealy@vt.edu
* Correspondence: dalya88@vt.edu

Received: 30 April 2018; Accepted: 1 June 2018; Published: 3 June 2018

Abstract: Sustainable construction is fundamentally different than traditional construction because it requires whole systems thinking, early collaboration across stakeholders, and core principles like reducing resource consumption, eliminating toxins, and applying life cycle costing. Construction professionals unfamiliar with this mindset and approach may perceive sustainable construction as risky. One of the global regions in need of more sustainable construction is the Middle Eastern and North African (MENA) region. The MENA region is one of the fastest developing in the world. However, it is the slowest one in implementing sustainable construction practices. Kuwait, in particular, contributes 53% more carbon emissions per capita than the United States. To understand how the Kuwaiti construction industry perceives risks associated with more sustainable construction, a survey was developed with 52 risk elements in which 131 industry professionals responded. The results indicate that industry professionals perceive a lack of public awareness as the risk element with the highest probability of occurrence. The risk element with the highest possible negative impact on future projects is designers' and contractors' inexperience with sustainable construction. Other risks were found to include a high initial cost for materials and overall project costs. Educational interventions, changes in risk allocation, and behavioral science to reframe upfront costs as long-term savings are offered as possible solutions.

Keywords: sustainable construction; risk perceptions; risk management; Kuwait

---

1. Introduction

Sustainable development goals usually focus on broad problems like climate change, energy reduction, and clean air and water. While these broad objectives are necessary, their abstractness can make it challenging for construction professionals to know how to achieve them [1]. This is because the techniques and approaches that are optimal for specific projects vary depending on the geographic location, regional energy sources, community characteristics, stakeholder priorities, and many other variables. In addition, sustainable construction, which refers to achieving social, financial, and environmental sustainability throughout a building's whole life-cycle [2] requires a higher level of collaboration between stakeholders when compared to traditional construction projects due to increased uncertainty [3,4]. This uncertainty is due to the complexity of project decisions, which are often made using prior held judgments and heuristics [5,6]. Without prior experience, the outcomes of sustainable choices can appear risky [7]. When adopting new techniques for sustainable construction, previously held judgments and heuristics must also change.

Currently held judgments and heuristics about sustainable construction do not always represent reality [8]. For example, the long-held perception that sustainable construction costs more financially upfront does not hold true today [9]. Cognitive biases such as risk aversion and status quo bias can lead to these overly generalized assumptions [10,11]. Better understanding of these perceptions, judgments, and heuristics can add value by offering opportunities to educate, increase awareness, and, most importantly, help stakeholders make more informed project level decisions.

Making more informed project level decisions means knowing what factors cause risk and which risks are worth taking [12]. Sustainable construction can appear overly risky when outcomes of value appear not to align with stated stakeholder objectives [13]. Prior research finds differences in stakeholder objectives can increase risk perceptions [14]. For instance, transaction costs are a real risk for general contractors when adopting sustainable construction practices, but this risk is often not a concern from the owners' perspective [15].

How risk and uncertainties vary between stakeholders, industry sectors (e.g., private or public), and global regions are still not well understood [16]. Risks about sustainable construction are particularly challenging to manage because much of the risk occurs upfront while the value comes later over time. The success of sustainable construction (e.g., eliminating impacts on the environment and natural resources, enhancing the health, well-being and productivity of occupants, creating new economic development, and applying a lifecycle approach during planning) heavily relies on contractors' willingness to adopt this new mindset and associated means and methods [17].

Regions of the world that were early to adopt sustainable design and construction practices continue to lead in global sustainable development. The number of new buildings in the United States certified by the Leadership in Energy and Environmental Design (LEED) rating system has grown exponentially in the last decade [18]. As more countries begin to adopt similar rating systems and tools for industry professionals, these industries will face similar barriers experienced by the United States construction market nearly a decade ago [9,19].

One of the global regions in need of sustainable construction is the Middle Eastern and North African (MENA) region. The MENA region is one of the fastest developing in the world. However, this region is slow in developing and implementing sustainable design and construction practices [20]. An early adopter of sustainable construction in this region who experiences a loss of profits or increased burden may have a lasting negative impact in the region's adoption of sustainable design and construction practices in the future [21]. Therefore, understanding the risks associated with sustainable construction prior to adopting these techniques can have a positive impact on the entire region.

Sustainable Construction in Kuwait and the MENA Region

Increasing adoption of sustainable construction practices in the MENA region is necessary for global sustainable development goals. For instance, greenhouse gas emissions are not isolated to merely one region or country but have a negative effect globally. In the past 20 years, the MENA region has increased carbon dioxide emissions by around 114% and per capita by 44% [22]. While this region currently only contributes approximately 6% of the world's carbon dioxide emissions, the trend suggests the region will become a larger contributor to global greenhouse gas emissions in the next century [23].

In comparison to its neighboring countries in the MENA region, Kuwait holds the lowest commitment to sustainable construction [20]. Per capita, residents contribute 53% more carbon dioxide emissions than residents of the United States [22]. This is partly because the country relies on energy-intensive desalination to produce potable water [24] and nearly 85% of electricity costs are subsidized [20]. Furthermore, Kuwait's construction industry has doubled their annual landfill waste in the past five years [25].

One explanation for the lack of commitment among the construction industry is the lack of awareness. Few professionals have experience in sustainable construction techniques and associated technology [20]. Lack of experience means there are fewer projects that demonstrate the benefits and, therefore, less motivation among industry professionals to try something new [26,27]. When awareness about the benefits of sustainable construction increase, the demand for more sustainable buildings and infrastructure also grow [28]. This demand helps drive further adoption and innovation helping these early adopters of sustainable design and construction reach even higher achievement in the future [29].

The lack of political support and incentives to adopt the design and construction techniques that promote sustainability from the Kuwaiti government is likely another reason for slow adoption of sustainable construction. Kuwait engineering design and construction professionals agree that government intervention through new standards and policy is necessary to accelerate adoption [20]. External involvement and additional incentives may help offset perceived risks and encourage market demand.

To help Kuwait and the MENA region more quickly adopt sustainable construction techniques, the purpose of the research reported in this paper is to assess current perceived risks within the industry. Knowing how sustainable construction is perceived can help design interventions, tools, and processes to help this industry overcome these perceived barriers or more efficiently manage them in the future.

2. Background

Sustainable construction is fundamentally different than traditional construction [30]. The purpose of sustainable construction is to create and operate a building based on core principles across the building's life cycle. Sustainable construction should reduce resource consumption, reuse resources, integrate recyclable resources, protect nature, eliminate toxins, apply life cycle costing, and focus on quality [31]. Achieving such high standards requires those involved in the design and construction process to "begin with the end in mind" [17]. This means setting specific goals and building features early during the feasibility stage that align with core sustainability principles.

Those involved in the design and construction process must also take a whole systems approach [32]. A whole systems approach encourages the consideration of interrelated components and people to optimize the performance of the entire building rather than an individual part [33]. For example, Rocky Mountain Institute, which is a consulting firm in the United States, uses a whole systems approach to optimize thermal mass of building envelopes and windows and other components to produce the most cost-effective passive building [34]. Whole systems requires establishing common goals that align incentives, encourage mutual learning, and sharing of information across stakeholder groups [35].

Adopting a whole systems approach and applying sustainable practices can appear risky compared to traditional construction especially for those that are new to sustainable construction. A broadly defined risk is the combination of the probability of an event and its outcomes [36]. Risk management is then the process by which risks are identified, quantified, and used to inform decision making and planning future events [37]. Risk management includes the ability to recognize risks with low probability and low impact compared to risks with high probability and high impact. For example, recognizing the likelihood of failure, delay of schedule, or increased cost by changing construction techniques to reduce the disruption of soils or the probability of success in installing more sensors in a building for enhanced monitoring and control.

Without prior experience, for instance, in changing commonly used materials for those with less embodied energy, errors can occur in judgment and lead to overly weighing probabilities of risks [38]. For example, sustainable construction can appear more expensive when construction professionals are unaware of these possible risks and, as a result, assign higher contingencies [15]. The opposite can also be true. The pseudo-certainty effect occurs when a decision maker perceives an outcome as certain while, in fact, it is uncertain [39]. The success of sustainable construction depends on the judgment of perceived risks and the development of an appropriate risk management plan [40,41]. Understanding the unique variation between conventional and sustainable projects is also essential for beginning to develop risk management techniques and interventions for sustainable construction that has neither overweight nor underweight risks.

Perceived risks to sustainable construction continue to emerge globally as developing countries begin to explore and adopt new techniques and technologies [26,27]. Countries in the MENA region like Kuwait that lack prior experience with sustainable construction techniques may fall into a trap of assigning higher contingencies, which increases the overall cost of the project and creates negative barriers to more sustainable construction projects in the future [42]. Overlooking possible risks leading to negative outcomes for the project team, also known as the pseudo-certainty effect, is also a possibility, which results in less incentive to adopt new techniques as industry norms in the future. Prior research in the MENA region discuss construction risks (e.g., in Kuwait see Reference [43], the U.A.E. see Reference [44], Qatar see Reference [45], and Bahrain see Reference [46]) but fall short in covering, discussing, or outlining possible risks related to new means and methods that incorporate principles of sustainable construction.

Improved ability for assessing risks can help shift industry professionals' focus to appropriately reduce the risk of failure for sustainable construction [47]. Prior research about sustainable construction practices in developed countries like Australia, the United States, and Europe do outline increased risks and recommendations [26] and novel contract structures to share the risk burden [48]. The lists of possible risks from these prior works were used to develop a survey instrument detailed in the methods section of this paper. The weighting of these risks may be different as a result of cultural values, regional or national economic incentives, and political interests [49]. Therefore, identifying the potential risk factors and weights of perceived probability plays a crucial role in enhancing the performance and accomplishing the successful delivery of the project.

Synthesis of Research about Risks Associated with Sustainable Construction

Risks associated with sustainable construction were gathered from prior literature and synthesized into categories broadly defined within the design and construction process of new buildings. The purpose of this literature review was to identify potential sustainable risks during a project's life-cycle. The review consisted of an extensive literature search of recently identified sustainable construction risks in other countries including, more specifically, the work from Reference [43] and Reference [46]. Inclusion criteria required research to be within 10 years (articles published from 2005 to 2015) since modern risk management has greatly evolved over the last decade. This review process follows a similar process of prior synthesized literature reviews [50].

More than 20 papers were included and used to develop the synthesized list presented in Table 1. Nine categories and 52 risk elements were organized from the literature. The nine categories include design, management, construction, material, technology, labor and equipment, external factors, finance, and certification. Risk elements associated with design include both inexperience when dealing with sustainable construction and changes as a result of sustainable construction. Management risks are related to design. Management includes risk elements upstream (clients) and downstream (subcontractors) from the general construction process. Lack of communication, lack of dispute resolution, and general planning are particular risk elements that were identified from prior studies [17,51,52]. Construction techniques, defects, and inexperience were also included as risk elements in the construction category. The categories material, technology, and labor and equipment include risk elements about prior knowledge, lack of experience, and non-compliance. Cost is always a concern and represented in both the external and finance categories through a lack of market demand and associated with payback period and cost overruns. Project certification was also identified as an incentive for pursuing sustainable construction and a possible barrier [53]. Energy models that do not align with actual energy performance are increasingly problematic in new sustainable buildings [54].

Table 1 was used to develop both the research questions and survey instruments to measure both perceived probability of occurrence and possible impact of these risk elements during construction projects in Kuwait that include sustainable design and construction principles.

Table 1. Synthesized list of construction risks associated with adopting techniques and technologies that promote sustainability.

Design	External
Design changes during construction [55]	Lack of market demand [56]
Slow response to meet design changes [26]	Lack of political support and incentives [49]
Design-team inexperience [57]	Lack of public awareness and knowledge [58]
Design defects which could result in failure to achieve certification [53]	Uncertain governmental policies [19]
Management	**Finance**
Lack of quantitative evaluation tools [59]	Cost estimation inaccuracy [46]
Not achieving client expectations [55]	Payback period is too long [60]
Difficulty in the selection of subcontractors who provide sustainable construction services [51]	Performance problems since sustainable building projects face a greater potential in failure (causing liabilities) [53]
Poor interrelationships between supply chain partners [26]	Increased soft costs due to delays in sustainable building completion [53]
Lack of upfront planning by all parties [17]	High cost of sustainable materials and equipment [51]
Sustainability measures not considered early by stakeholders [52]	Cost overrun due to lack of sustainable building knowledge [61]
Delays in resolving disputes [26]	High initial sustainable construction costs [26]
Slow approval processes due to sustainable specifications [26]	Investor cannot fund the high sustainability measure costs [61]
Outdated contractual agreements [62]	Costs of investment in skills development [57]
	High sustainable construction premiums [17]
Material	**Labor and Equipment**
Unavailability of sustainable building materials [63]	Handling recycled materials puts construction workers at safety risks [43]
Poor material quality [26]	Unavailability of specific equipment [15]
Uncertainty in the performance of sustainable materials [59]	Additional responsibilities for construction maintenance [59]
Non-complying products and materials [26]	Lack of practical experience [15]
Change in material types and specifications during construction [43]	Uncertainty with specialized sustainable equipment [51]
Technology	**Certification**
Challenges for operating renewable energy systems [63]	An event that causes the loss of certification [53]
Unacceptable performance of modern technologies [61]	Lower certification than what was expected due to design defects [53]
Technological failures [61]	Changing certification procedures [61]
Misunderstanding of sustainable technological operations [59]	Loss of financing or losing loans for not achieving certification [53]
Construction	
Unforeseen circumstances in execution of the sustainable project [51]	More complex construction techniques [60]
Safety issues [61]	Project delay [55]
Contractors' inexperience with sustainable buildings [57]	Incremental time caused by sustainable construction [59]
Construction defects [53]	

3. Research Objective and Questions

The objective of the research presented in this paper is to understand what perceived risks are associated with sustainable construction techniques of new buildings in Kuwait and what possible methods industry professionals are currently using to overcome these perceived risks to avoid cost increases, time overruns, and long-term quality issues. The results of this research can be used by local and international industry professionals to develop a better understanding of the critical risk factors that are perceived to influence cost, time, and quality of construction projects. This understanding will lead to the development of risk management processes not only during the construction stages but also during the evolution of the design phase. Ultimately, by better understanding the risks associated with sustainable construction, the adoption rate of sustainable construction techniques and subsequent performance of the construction industry in the MENA will increase.

The three research questions are:

1. What sustainable construction risks do professionals in Kuwait believe have the highest probability of occurrence?

2. What sustainable construction risks do professionals in Kuwait believe have the highest negative impact on project outcomes?

3. How do perceptions of sustainable construction risks differ between those working in private and public sectors, across project types (residential, commercial, and industrial), across professions (design engineer, contractor, sub-contractor), and across years of experience?

The results offer answers to these questions and the discussion offers relevant risk management strategies from past research to support future implementation of sustainable design and construction.

4. Research Methodology

The survey to professionals in the Kuwaiti construction industry consisted of two sections. The first section asked general information about the respondents such as their professional experience, type of organization, and their familiarity with construction techniques and technologies that contribute to sustainability. The purpose of this section was to gather descriptive statistics about the sample population. The second section included the 52 risks identified from the literature review.

Respondents were asked to "Please evaluate the probability of the following risks based on the outcomes of sustainable construction projects". Respondents were given a Likert scale (1 = very low probability and 5 = very high probability). Respondents were asked again to "Please evaluate the impact of the following risks based on how they negatively affect the outcomes of sustainable construction projects." Respondents were given a Likert scale (1 = very low impact and 5 = very high impact).

Prior to distribution of the survey, the survey was given to a focus group of five construction professionals in which each had 10 or more years of experience for content validity and to review and provide feedback about the questions. Changes made to the survey helped clarify the meaning of specific risks and certain wording was adjusted to more clearly communicate the meaning. For example, some of the risks were combined such as "design changes during construction" and "changes in work" since they were closely related to each other.

The survey was distributed to a national sample of professionals currently working in the construction industry in Kuwait. Professionals were selected randomly from a list of all construction companies in the country. A total of 195 surveys were sent to construction professionals and 131 surveys were returned (67% response rate).

Risk Assessment

The probability and impact of each risk was evaluated using a weighted score approach. This method was adopted from previous literature [43]. The weighted score approach in Equation (1) shows that, for every identified risk, the weighted score was calculated by adding the product of the number of respondents, x, with their corresponding selected Likert ranking, r.

$$S_{Wj} = \Sigma \, (x \times r) \tag{1}$$

where S_{Wj} is the weighted score, x is the number of respondents for each Likert rank, r is the corresponding Likert scale ranking, and j is a subscript index that represents p for probability or i for impact. A sample collected from the data in Table 2 shows that 17 out of 128 respondents ranked the identified risk element for "design changes during construction" with a Likert scale ranking of 5 (most probable) and the value ($17 \times 5 = 85$) is the product. Similarly, nine individuals responded with a ranking of 1 (least probable) and the value ($9 \times 1 = 9$) is the product. The total weighted score for this particular risk is 416, which is the summation of all the product values. Table 2 represents a sample on the probability of risk. The same equation was applied for the responses about the impacts of risk.

Table 2. Sample calculation of weighted scores for risk probability of design changes during construction.

Variable	Design Changes during Construction				
Likert Scale (r)	1	2	3	4	5
Number of respondents for each Likert Scale ranking (x)	9	21	40	40	17
Number of respondents multiplied by the Likert Scale ranking ($x \times r$)	9	42	120	160	85
Weighted Score, $S_{Wp} = \Sigma (x \times r)$			416		

The probability P and impact I are scaled weighted scores between 0 and 1 obtained by taking the percentage of the score with respect to maximum possible points for each category. In the sample calculation shown in Table 2, the maximum possible points for that risk is 640, which is obtained by multiplying the total number of respondents ($x = 128$) with the maximum Likert scale ($r = 5$). Equations (2) and (3) express this formulation below.

$$P = S_{Wp}/(x \times 5) \qquad (2)$$

$$I = S_{Wi}/(x \times 5) \qquad (3)$$

where S_{Wp} is the weighted score for risk probability and S_{Wi} is the weighted score of risk impact.

Equation (4) was used to calculate the product of both probability and impact of occurrence. The purpose was to quantify the degree of risk. This equation was adopted from Reference [64].

$$R = P \times I \qquad (4)$$

where R is the degree of perceived risk measured between 0 and 1, P is the probability of the risk occurring measured between 0 and 1, and I is the degree of impact of the risk measured between 0 and 1. This method scales risks from high (close to 1) and low (close to 0) by considering both weighted probability and impact.

5. Results

The group of responses ($n = 131$) were contractors (27%), owners/clients (25%), construction consultants (18%), and subcontractors and suppliers (30%). Out of 131 surveyed, 45% indicated more than 10 years of experience in the construction industry, 29% had between five to 10 years of experience, and 27% had less than five years of experience. Most of the respondents came from organizations that have more than 100 employees (53%), 17% from organizations that have 50–100 employees, and 30% from organizations with less than 50 employees. Nearly 40% of respondents indicated their level of sustainability awareness was "poor" and only 19% perceived "good" awareness of sustainability concepts, procedures, and technologies. Nearly half of respondents (47%) reported that the current percentage of construction projects that include sustainable construction practices or technologies in their organization was between 0% and 19%. Half of the respondents (50%) indicated they work with the public sector, 32% with the private sector, and 18% are quasi-public sectors. Details of the main research findings are listed in the sub-sections below.

5.1. Risks That Have the Highest Probability and Impact of Occurrence

To identify the risks that have the highest probability and highest impact of occurrence, the respondents were asked to evaluate the risks based on their probability and impact of occurrence in construction projects in Kuwait. Tables 3–5 present the highest probability of perceived occurrence, expected impact, and their combined rankings (highest degree of perceived risk). Table 3 presents the top 10 risks with the highest perceived probability in ascending order based on their total weighted scores. Lack of public awareness about the benefits of sustainable design and construction and high costs of sustainable material and equipment are perceived as the risks having the highest probability of occurrence in construction projects in Kuwait.

Table 3. The top 10 risks based on the highest expected probability.

Risks with Highest Probability of Occurrence	Weighted Score, S_{Wp}
1 Lack of public awareness about the benefits for sustainability	492
2 High costs of sustainable materials & equipment	488
3 High initial sustainable construction costs	472
4 Lack of market demand	471
5 Lack of practical experience	470
6 Lack of political support & incentives	455
7 Contractors inexperience with sustainable buildings	448
8 Unavailability of specific equipment	448
9 Unavailability of sustainable building materials in the market	446
10 Uncertain governmental policies	444

Table 4 includes the top 10 risks with perceived high expected impact. Notably, contractors' inexperience with sustainable construction practices and technologies had the highest potential risk according to respondents. The second risk associated with adopting sustainable construction practices was design team inexperience, which was followed by the unavailability of sustainable building materials and lack of practical experience. These top four risks are 4% greater (about 20 points on average) in total weighted score compared to the bottom six. In other words, while these top 10 are close in value, the top four appear most critical for the respondents. Contractor's inexperience with sustainable construction is nearly 8% greater from the fifth-ranked highest cost of sustainable materials and equipment.

Table 4. The top 10 risks based on the highest expected impact.

Risks with Highest Impact of Occurrence	Weighted Score, S_{Wi}
1 Contractors inexperience with sustainable construction	519
2 Design team inexperience	499
3 Unavailability of sustainable building materials in the market	496
4 Lack of practical experience	491
5 High costs of sustainable materials & equipment	481
6 Lack of political support & incentives	481
7 Design changes during construction	479
8 Non-complying products & materials	479
9 Poor material quality	476
10 Lack of public awareness	471

Table 5 presents the risks perceived as having a high degree of risk, *R* (combined high probability and high impact). The degree of risk is considered high if the value is closer to 1. The top five risks with similar total weighted scores (within 1% to 2% of each other), are high costs of sustainable materials and equipment, contractor's inexperience with sustainable construction, lack of practical experience, lack of public awareness, and high initial sustainable construction costs. The perceptions about higher upfront costs for sustainable materials and equipment are likely true in a country like Kuwait that is still early in the adoption of these materials and practices. However, increased awareness of the benefits may help offset or balance these higher up-front expenditures.

Table 5. The top 10 combined high probability and impact risks.

Risks with Highest Degree of Perceived Risk		Degree of Perceived Risk, R
1	High costs of sustainable materials & equipment	0.61
2	Contractor's inexperience with sustainable construction	0.59
3	Lack of practical experience	0.58
4	Lack of public awareness	0.579
5	High initial sustainable construction costs	0.575
6	Unavailability of sustainable building materials in the market	0.55
7	Lack of market demand	0.548
8	Lack of political support & incentives	0.547
9	Cost estimation inaccuracy	0.536
10	Difficulty in the selection of subcontractors who provide sustainable construction practices	0.532

5.2. Risk Categories that Have the Highest Probability and Impact of Occurrence

The risk categories from Table 1 with the average highest perceived probability were external risks (e.g., public awareness and knowledge, government incentives, market demand) (mean Likert score = 3.7; 1 = very low probability/impact and 5 = very high probability/impact) and finance-related risks (e.g., cost of sustainable materials, schedule delays, payback period) (mean Likert score = 3.5). Scores were determined using the mean value of the Likert scale responses to each of the individual risks. An average of the Likert scale was calculated to compare between categories where there are several risks in each category, which can reveal the statistical differences. The perceived probability of risk occurrence in all risk categories was above average (more than 50%, mean scores above 3.0) except for the Certification category (mean score less than 3.0). The difference in mean scores between the categories (probability and impact) was statistically significant ($p < 0.001$). The one-way ANOVA test was used since it can determine the statistical difference between the means of two or more independent groups. The mean scores of the probability of each risk category are shown in Figure 1.

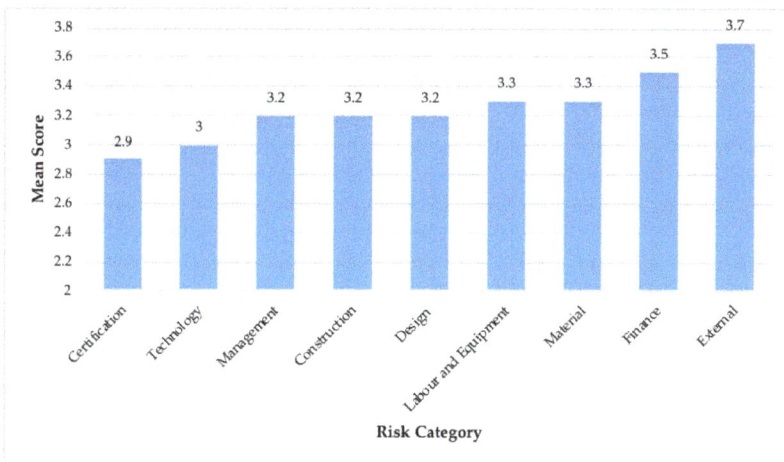

Figure 1. Mean scores for probability of risk categories (scores range from 1 to 5).

Respondents perceive risk impacts related to the materials such as unavailability of sustainable building materials, uncertainty of quality, and change orders for material as the highest possible impact for project outcomes. Design risks were also perceived as having a high potential impact including design changes during construction, design defects, and inexperience with sustainable design. The perceived risk impact in all risk categories was above average (more than 50%, mean scores above 3.0). The mean scores of the impact of each risk category are shown in Figure 2.

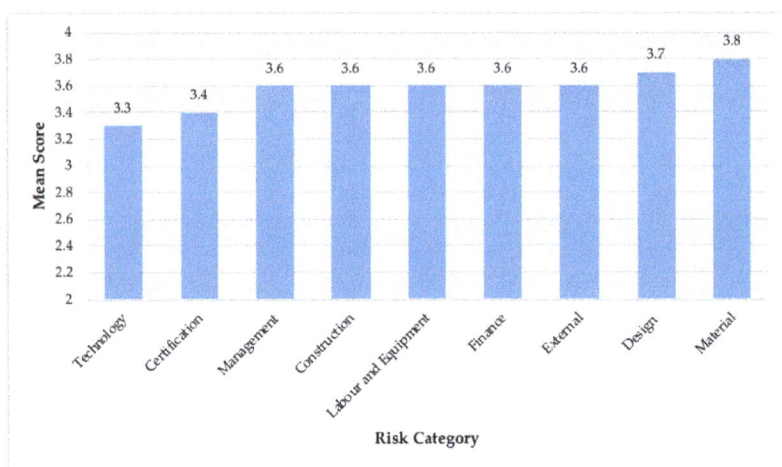

Figure 2. Mean scores for impact of risk categories (scores range from 1 to 5).

The results of each of the probability and impact of risk categories are significantly different. However, the top five categories are similar. This means the nine associated risk categories include risks associated with material, design, external factors, finance, labor, and equipment. The risks have the highest probability of occurrence and would cause the greatest impact on project outcomes. In terms of both probability and impact, the certification and technology categories are of the least concern for the respondents.

5.3. Perceptions of Construction Professionals in the Private Sector Compared to the Public Sector

The findings indicate no significant differences in the mean scores for all risk categories between the different titles of the professionals (project manager, site engineer, or architect), their practical experience (years), or their typical project types (residential, commercial, and industrial).

The only factor with a significant difference in response was whether their client base was private or public. The mean scores for the perceived probability of risk between private and public sectors were significantly different in the following categories: design ($p = 0.004$), construction ($p < 0.001$), management ($p = 0.017$), finance ($p = 0.013$), and technology ($p = 0.013$). Professionals working with the public sector perceive higher probabilities of risk in construction, management, finance, and technology but lower probabilities of risk compared to the private sector related to design. Risks associated with management were perceived as significantly higher by the public sector ($p = 0.023$) than the private sector. The other four categories had similar perceptions between respondents with public and private clients. The one-way ANOVA test was used to generate the p-values. Table 6 summarizes these results.

Table 6. Comparison of risk categories by sector types.

Risk Categories	Public Sector		Private Sector		p-Value
	Mean	(SD)	Mean	(SD)	
Design related risks					
Probability of risk occurrence	3.2	(0.7)	3.5	(0.6)	**0.004**
Expected impact of risk	3.8	(0.8)	3.7	(0.6)	0.905
Construction related risks					
Probability of risk occurrence	3.4	(0.7)	3.1	(0.7)	**<0.001**
Expected impact of risk	3.7	(0.8)	3.6	(0.7)	0.477
Management related risks					
Probability of risk occurrence	3.4	(0.7)	3.0	(0.8)	**0.017**
Expected impact of risk	3.7	(0.6)	3.3	(0.7)	**0.023**
Finance related risks					
Probability of risk occurrence	3.7	(0.8)	3.3	(0.7)	**0.013**
Expected impact of risk	3.7	(0.8)	3.7	(0.7)	0.560
Technology related risks					
Probability of risk occurrence	3.3	(1.0)	2.7	(0.9)	**0.013**
Expected impact of risk	3.5	(1.0)	3.2	(0.9)	0.142

The difference in perceived probability and impact between construction professionals with public sector clients and private sector clients appear to relate to their level of awareness to sustainability in Kuwait. Between the two sectors, the private sector has a higher level of perceived sustainability awareness, which is illustrated in Figure 3. None of the respondents (whether public or private) believe that their knowledge of sustainable construction is "very good".

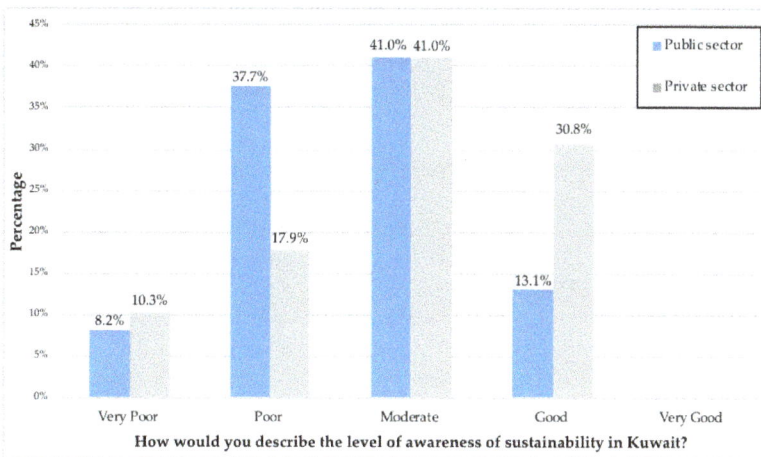

Figure 3. Level of awareness about sustainability among construction professionals in Kuwait that work with clients from the private and public sectors.

6. Discussion

Construction professionals in Kuwait that serve clients in the private sector appear to perceive less probability of risks related to construction, management, and finance when adopting sustainable techniques and technologies. These results are somewhat surprising compared to other regions and countries like the United States and Europe where government institutions were the early adopters of sustainable construction, which mandates the use of LEED and BREEAM, respectively, nearly two decades ago. The private sector may contribute to more innovative techniques and technologies with contract structures that help distribute risks among multiple stakeholder groups. The culture and ability to innovate because of distributed risk may explain why these differences occur between client types [64].

Another reason for the differences in perceptions between construction professionals that work with public or private sector clients is that the private sector in Kuwait tends to hire more non-Kuwaitis than the public sector. Industry professionals hired from outside of the country might bring experience and understanding of sustainability. The public sector predominately hires Kuwaiti citizens who, due to the currently low adoption rate of sustainable design and construction, are less likely familiar with sustainability principles and applications for construction [65]. This gap in sustainable awareness and experience may contribute to why construction professionals that work with private sector clients are less concerned about probabilities of risk related to sustainability compared to those that represent clients from the public sector.

Even with respondents in the private sector indicating "moderate" to "good" understanding of sustainable construction, risks associated with designer and contractors' inexperience with sustainable construction rank highest and could negatively impact project outcomes. A tested solution to overcome these barriers is clients or project owners that recognize the benefits of incorporating sustainable design and construction techniques into their buildings and infrastructure and are motivated to request contractors and design teams with experience in sustainable projects [7]. Another strategy is through contracting sustainability experts. Experts may help general contractors recognize specialty issues and can help facilitate new markets or products that meet sustainability criteria. For example, risks identified in the results with high probability and impact such as unavailability of specific materials and equipment in the market can occur due to unforeseen procurement issue and lag times. Experts can help estimate additional times, delays, and scheduling issues that may arise because they have experience with these products [26].

Respondents also recognize the high costs of sustainable materials and equipment as high in probability and negative impact on future projects that adopt sustainability. Since this industry and region are new toward adopting sustainable construction, the higher cost of materials and equipment is similar to the greater cost of sustainable buildings in the United States and Europe nearly two decades ago when they were early in the adoption phase [9]. To bridge this gap, sharing knowledge across suppliers about what new standards and requirements mean and must be able to document is a possible strategy to reduce perceived risks over time [26,55]. Toyota, for example, groups suppliers together that use similar production processes to ensure that the information can be shared between them and is relevant for all [66]. Other interventions can be applied such as education and awareness programs for engineering professionals and public awareness programs. If these approaches are implemented, the major obstacle of lack of information on sustainable construction can be overcome.

Similar to higher costs of materials and equipment, broad categories of risk related to external factors such as the lack of market demand and lack of regional incentives as well as financial factors such as payback period and performance uncertainties are of greatest concern for these construction professionals. Prior research finds that material scarcity and availability to meet sustainability standards is a predominant factor and can directly impact costs [4]. Contingency premiums can work to reduce the cost for contractors and shift risk to clients and project owners [67].

Inherently, projects that include design and construction techniques for sustainability will incur more risks in the Kuwait construction industry because of the novelty and lack of experience among construction professionals. Traditional risk management strategies such as coordination with subcontractors, increasing workforce and equipment, producing programs using subjective decisions, and producing schedules that offer realistic resource procurement timelines [43] are helpful in traditional projects but may fall short when adopting sustainable construction practices. Achieving project goals for sustainability will require new technologies and strategies unfamiliar to the current workforce. Risk management programs should begin by addressing the barriers identified in the results of this paper. Regular training can help address the lack of awareness [26]. Innovative contract structures and employing those with experience in sustainable design and construction, especially in the public sector, can also help. The upfront cost is a real barrier especially for early adopters [60]. More focus on the benefits of sustainable design and construction may help balance the perceived high cost by educating suppliers and subcontractors and creating a network of professionals that supply the resources and materials that meet the standards for more sustainable materials.

Incorporating more nuanced techniques through behavioral science may also have an effect on construction professionals across cultures and regions, which are worth exploring [8]. For example, representing risks as embedded characteristics of engineering options can change the propensity of decision makers to take risks [68]. Framing risks as loss or gain can raise uncertainty awareness of decision-makers and can nudge construction professionals away from riskier, more uncertain options and towards less risky and certain options [69]. Including a feasibility example or a role model project for construction professionals to use as a guide can encourage higher levels of sustainability achievement [70]. However, these behavioral interventions have not been tested across cultures or with those less aware of sustainable design and construction.

The strength of the effect of framing interventions related to risk is limited to the values of the decision maker [71]. In other words, the effects of framing are larger when the concern or knowledge is low [72]. Participants with relevant experiences and more information on the subject may answer differently than those with limited awareness [12]. Therefore, professionals in Kuwait may be more influenced by framing or other behavioral interventions about risk than professionals in the United States or parts of Europe that have decades of experience and formed heuristics about sustainable construction. Future research can now begin to explore these possible interventions to shift perceptions and nudge contractors to incorporate sustainable design and construction techniques and technologies into building and infrastructure projects.

There are a few limitations to this study. First, the list of synthesized risks is not a comprehensive one since it includes risks from prior research published within the last 10 years only. The rationale for this choice is because risk management has evolved substantially over the last decade, but there could be some sustainability risks identified earlier or not included in this list that the construction industry faces today. Second, literature about sustainable construction risks in Kuwait and the MENA region is limited. There were only a few direct sources of reference from those countries. However, risks identified globally have been used in the literature review. The professionals did perceive many of these risk elements as having a high probability and impact of occurrence in Kuwait. Regardless of these limitations, this research identifies perceived risks in the Kuwaiti construction industry, which is a step forward in understanding and adopting more sustainable construction techniques.

7. Conclusions

The limited adoption rate for sustainable design and construction practices in the MENA region especially Kuwait is troublesome given that per capita residents in this region of the world produce 53% more greenhouse gas emissions than in the United States [22]. The lack of experience in sustainable construction appears to increase perceived risks among construction professionals. Industry professionals perceive that most risks of sustainable construction have high probabilities and impacts of risk occurrence. This perceived risk and higher cost for sustainable materials and equipment likely act as a barrier to adoption of new techniques and technologies. Other perceived risks include the lack of public awareness and practical experience, which are both related to knowledge and expertise. Differences in perceptions between sector types are significant. Construction professionals with clients in the private sector are less concerned with the probabilities and impacts of risk compared to the public sector specifically in risks related to finance, management, construction, and technology. Interestingly, project managers, site engineers, and architects showed no variation in their risk perceptions.

However, as important as the high probability and impact risks, are those with low probability and low impact. Time focused on these risks are potentially limiting the attention to risk with much higher negative effects. The construction professionals represented in this research overwhelming agreed risks related to technology and sustainability certification were low. Several sustainable risk management strategies can encourage more sustainable adoption and reduction of perceived risk. Contingency plans and shared risks with innovative contract structures can work to reduce the cost for contractors [57]. Sustainability experts can also help facilitate new markets or products that meet sustainability criteria and sharing knowledge across suppliers and subcontracts can spur industry support that, over time, reduces procurement costs and time [26,66]. Behavioral science approaches such as framing risks as gains in value instead of a loss or providing a role model project for teams to follow may also help nudge the industry forward in the adoption of sustainable construction techniques. Future research can now begin to measure the effect of new risk management strategies and behavioral interventions to change the perceptions identified in this paper and measure the adoption rate of sustainability in Kuwait and the entire MENA region [73].

Author Contributions: D.I. took part in conceptualization, formal analysis, methodology, and original draft writing. T.S. conducted results validation, writing the review, and editing the manuscript.

Conflicts of Interest: The authors declare no conflicts of interest.

References

1. Holden, E.; Linnerud, K.; Banister, D. The Imperatives of Sustainable Development. *Sustain. Dev.* **2017**, *25*, 213–226. [CrossRef]
2. Hill, R.C.; Bowen, P.A. Sustainable construction: Principles and a framework for attainment. *Constr. Manag. Econ.* **1997**, *15*, 223–239. [CrossRef]
3. Reed, W.G.; Gordon, E.B. Integrated design and building process: What research and methodologies are needed? *Build. Res. Inf.* **2000**, *28*, 325–337. [CrossRef]
4. Klotz, L.; Horman, M. Counterfactual Analysis of Sustainable Project Delivery Processes. *J. Constr. Eng. Manag.* **2010**, *136*, 595–605. [CrossRef]
5. Beamish, T.D.; Biggart, N. *Social Heuristics: Decision Making and Innovation in a Networked Production Market*; Social Science Research Network: Rochester, NY, USA, 2010.
6. Klotz, L.; Mack, D.; Klapthor, B.; Tunstall, C.; Harrison, J. Unintended anchors: Building rating systems and energy performance goals for U.S. buildings. *Energy Policy* **2010**, *38*, 3557–3566. [CrossRef]
7. Demaid, A.; Quintas, P. Knowledge across cultures in the construction industry: Sustainability, innovation and design. *Technovation* **2006**, *26*, 603–610. [CrossRef]
8. Shealy, T.; Klotz, L. Choice Architecture as a Strategy to Encourage Elegant Infrastructure Outcomes. *J. Infrastruct. Syst.* **2017**, *23*, 4016023. [CrossRef]
9. Ahn, Y.H.; Pearce, A.R. Green Construction: Contractor Experiences, Expectations, and Perceptions. *J. Green Build.* **2007**, *2*, 106–122. [CrossRef]
10. Samuelson, W.; Zeckhauser, R. Status quo bias in decision making. *J. Risk Uncertain.* **1988**, *1*, 7–59. [CrossRef]
11. Kahneman, D.; Knetsch, J.L.; Thaler, R.H. Anomalies: The Endowment Effect, Loss Aversion, and Status Quo Bias. *J. Econ. Perspect.* **1991**, *5*, 193–206. [CrossRef]
12. Fischhoff, B.; Kadvany, J.D. *Risk: A Very Short Introduction*; Oxford University Press: Oxford, UK; New York, NY, USA, 2011; ISBN-13 9780199576203; ISBN-10 0199576203.
13. Yang, R.; Zou, P.; Wang, J. Modelling stakeholder-associated risk networks in green building projects. *Int. J. Proj. Manag.* **2015**, *34*. [CrossRef]
14. Andi. The importance and allocation of risks in Indonesian construction projects. *Constr. Manag. Econ.* **2006**, *24*, 69–80. [CrossRef]
15. Qian, Q.K.; Chan, E.H.W.; Khalid, A.G. Challenges in Delivering Green Building Projects: Unearthing the Transaction Costs (TCs). *Sustainability* **2015**, *7*, 3615–3636. [CrossRef]
16. Bryde, D.J.; Volm, J.M. Perceptions of owners in German construction projects: Congruence with project risk theory. *Constr. Manag. Econ.* **2009**, *27*, 1059–1071. [CrossRef]
17. Robichaud, L.B.; Anantatmula, V.S. Greening Project Management Practices for Sustainable Construction. *J. Manag. Eng.* **2011**, *27*, 48–57. [CrossRef]
18. Cidell, J. Building Green: The Emerging Geography of LEED-Certified Buildings and Professionals. *Prof. Geogr.* **2009**, *61*, 200–215. [CrossRef]

19. Zou, P.X.W.; Zhang, G.; Wang, J. Understanding the key risks in construction projects in China. *Int. J. Proj. Manag.* **2007**, *25*, 601–614. [CrossRef]

20. AlSanad, S. Awareness, drivers, actions, and barriers of sustainable construction in Kuwait. *Procedia Eng.* **2015**, *118*, 969–983. [CrossRef]

21. Omran, A.; Mohd Shafie, M.W.; Osman Kulaib, H.M. Identifying Environmental Risk in Construction Projects in Malaysia: Stakeholder Perspective. *Ann. Fac. Eng. Hunedoara* **2015**, *13*, 89–92.

22. The World Bank Kuwait Data. Available online: http://data.worldbank.org/country/kuwait (accessed on 7 February 2017).

23. World Energy Council World Energy Perspective: Energy Efficiency Policies—What Works and What Does Not. Available online: https://www.worldenergy.org/wp-content/uploads/2013/09/WEC-Energy-Efficiency-Policies-executive-summary.pdf (accessed on 11 December 2017).

24. Darwish, M.A.; Al-Awadhi, F.M.; Darwish, A.M. Energy and water in Kuwait Part I. A sustainability view point. *Desalination* **2008**, *225*, 341–355. [CrossRef]

25. Kuwait Central Statistical Bureau. Available online: http://www.csb.gov.kw/Socan_Statistic_EN.aspx?ID=18 (accessed on 13 October 2016).

26. Zou, P.X.W.; Couani, P. Managing risks in green building supply chain. *Arch. Eng. Des. Manag.* **2012**, *8*, 143–158. [CrossRef]

27. AlSanad, S.; Gale, A.; Edwards, R. Challenges of sustainable construction in Kuwait: Investigating level of awareness of Kuwait stakeholders. *World Acad. Sci. Eng. Technol. Int. J. Environ. Chem. Ecol. Geol. Geophys. Eng.* **2011**, *5*, 753–760.

28. Eichholtz, P.; Kok, N.; Quigley, J.M. Doing Well by Doing Good? Green Office Buildings. *Am. Econ. Rev.* **2010**, *100*, 2492–2509. [CrossRef]

29. Mang, P.; Haggard, B. Regenesis Regenerative Development and Design: A Framework for Evolving Sustainability. Available online: http://www.wiley.com/WileyCDA/WileyTitle/productCd-1118972864.html (accessed on 1 September 2017).

30. Rafindadi, A.D.; Mikić, M.; Kovačić, I.; Cekić, Z. Global Perception of Sustainable Construction Project Risks. *Procedia Soc. Behav. Sci.* **2014**, *119*, 456–465. [CrossRef]

31. Kibert, C.J. *Sustainable Construction: Green Building Design and Delivery*; John Wiley & Sons: New York, NY, USA, 2016; ISBN 978-1-119-05517-4.

32. Harris, N.; Shealy, T.; Klotz, L. Choice Architecture as a Way to Encourage a Whole Systems Design Perspective for More Sustainable Infrastructure. *Sustainability* **2016**, *9*, 54. [CrossRef]

33. Lovins, A.; Bendewald, M.; Kinsley, M.; Bony, L.; Hutchinson, H.; Pradhan, A.; Sheikh, I.; Acher, Z. Factor Ten Engineering Design Principles. Available online: http://www.10xe.orwww.10xe.org/Knowledge-Center/Library/2010-10_10xEPrinciples (accessed on 18 May 2018).

34. Yardi, R.; Archambault, T.; Wang, K.; Eubank, H. Home Energy Briefs: #9 Whole System Design. Available online: http://www.10xe.orwww.10xe.org/Knowledge-Center/Library/2004-21_HEB9WholeSystemDesign (accessed on 18 May 2018).

35. Blizzard, J.L.; Klotz, L. A framework for sustainable whole systems design. *Des. Stud.* **2012**, *33*, 456–479. [CrossRef]

36. Smith, N.J.; Merna, T.; Jobling, P. *Managing Risk: In Construction Projects*; John Wiley & Sons: New York, NY, USA, 2009; ISBN 978-1-4051-7274-5.

37. Westland, J. *The Project Management Life Cycle: A Complete Step-By-Step Methodology for Initiating, Planning, Executing & Closing a Project Successfully*; Kogan Page Publishers: London, UK, 2007; ISBN 978-0-7494-4808-0.

38. Keller, C.; Siegrist, M.; Gutscher, H. The Role of the Affect and Availability Heuristics in Risk Communication. *Risk Anal.* **2006**, *26*, 631–639. [CrossRef] [PubMed]

39. Tversky, A.; Kahneman, D. Choices, values, frames. *Am. Psychol.* **1984**, *39*, 341–350. [CrossRef]

40. Hwang, B.; Ng, W.J. Are Project Managers Ready for Green Construction?—Challenges, Knowledge Areas, and Skills. In Proceedings of the CIB World Building Congress 2013, St Lucia, QLD, Australia, 5–9 May 2013.

41. Kerur, S.; Marshall, W. Identifying and managing risk in international construction projects. *Int. Rev. Law* **2012**, *1*. [CrossRef]

42. Pearce, A.R. Sustainable capital projects: Leapfrogging the first cost barrier. *Civ. Eng. Environ. Syst.* **2008**, *25*, 291–300. [CrossRef]

43. Kartam, N.A.; Kartam, S.A. Risk and its management in the Kuwaiti construction industry: A contractors' perspective. *Int. J. Proj. Manag.* **2001**, *19*, 325–335. [CrossRef]

44. El-Sayegh, S.M. Risk assessment and allocation in the UAE construction industry. *Int. J. Proj. Manag.* **2008**, *26*, 431–438. [CrossRef]

45. Jarkas, A.M.; Haupt, T.C. Major construction risk factors considered by general contractors in Qatar. *J. Eng. Des. Technol.* **2015**, *13*, 165–194. [CrossRef]

46. Altoryman, A.S. Identification and Assessment of Risk Factors Affecting Construction Projects in the Gulf Region: Kuwait and Bahrain. Ph.D Thesis, The University of Manchester, Manchester, UK, 30 April 2014.

47. Van Buiten, M.; Hartmann, A. Public-private partnerships: Cognitive biases in the field. In Proceedings of the Engineering Project Organization (EPOC 2013), Winter Park, CO, USA, 9–11 July 2013.

48. Shen, L.Y. Project risk management in Hong Kong. *Int. J. Proj. Manag.* **1997**, *15*, 101–105. [CrossRef]

49. Douglas, P.M. *Risk and Blame*; Routledge: London, UK, 2013; ISBN 978-1-136-49004-0.

50. Alvaro, S.-C.; Alfalla-Luque, R.; Irimia Diéguez, A.I. Risk Identification in Megaprojects as a Crucial Phase of Risk Management: A Literature Review. *Proj. Manag. J.* **2017**, *47*, 75.

51. Hwang, B.-G.; Ng, W.J. Project management knowledge and skills for green construction: Overcoming challenges. *Int. J. Proj. Manag.* **2013**, *31*, 272–284. [CrossRef]

52. Bal, M.; Bryde, D.; Fearon, D.; Ochieng, E. Stakeholder Engagement: Achieving Sustainability in the Construction Sector. *Sustainability* **2013**, *5*, 695–710. [CrossRef]

53. Tollin, H.M. Green Building Risks: It's Not Easy Being Green. *Environ. Claims J.* **2011**, *23*, 199–213. [CrossRef]

54. Swan, L.G.; Ugursal, V.I. Modeling of end-use energy consumption in the residential sector: A review of modeling techniques. *Renew. Sustain. Energy Rev.* **2009**, *13*, 1819–1835. [CrossRef]

55. O'Connor, H. Architect's Professional Liability Risks in the Realm of Green Buildings. Available online: https://uk.perkinswill.com/research/architects-professional-liability-risks-in-the-realm-of-green-buildings.html (accessed on 28 April 2018).

56. Hwang, B.-G.; Tan, J.S. Green building project management: Obstacles and solutions for sustainable development. *Sustain. Dev.* **2012**, *20*, 335–349. [CrossRef]

57. Azizi, M.; Fassman, E.; Wilkinson, S. Risks Associated in Implementation of Green Buildings. Available online: http://www.thesustainabilitysociety.org.nz/conference/2010/papers/Mokhtar-Azizi-Fassman-Wilkinson.pdf (accessed on 4 May 2017).

58. Aminu Umar, U.; Khamidi, D.M.F. Determined the Level of Green Building Public Awareness: Application and Strategies. In Proceedings of the International Conference on Civil, Offshore and Environmental Engineering, Kuala Lumpur, Malaysia, 12–14 June 2012.

59. Lam, P.T.; Chan, E.H.; Poon, C.S.; Chau, C.K.; Chun, K.P. Factors affecting the implementation of green specifications in construction. *J. Environ. Manag.* **2010**, *91*, 654–661. [CrossRef] [PubMed]

60. Turner Construction 2014 Green Building Market Barometer. Available online: http://www.turnerconstruction.com/download-document/turner2014greenbuildingmarketbarometer.pdf (accessed on 22 April 2017).

61. Arashpour, M.; Arashpour, M. A collaborative perspective in green construction risk management. In Proceedings of the 37th Annual Conference of Australasian Universities Building Educators Association (AUBEA), Sydney, Australia, 4–6 July 2012; UTS Publishing/University of New South Wales: Sydney, Australia, 2012; pp. 1–11.

62. Anderson, M.K.; Bidgood, J.K.; Heady, E.J. Hidden Legal Risks of Green Building. *Fla. Bar J.* **2010**, *84*, 35–41.

63. Dagdougui, H. Decision Support Systems for Sustainable Renewable Energy Systems and Hydrogen Logistics: Modelling, Control and Risk Analysis. Ph.D. Thesis, École Nationale Supérieure des Mines de Paris, Università degli studi di Genova, Genova, Italy, 2011.

64. Zhi, H. Risk management for overseas construction projects. *Int. J. Proj. Manag.* **1995**, *13*, 231–237. [CrossRef]

65. Gulseven, O. Challenges to Employing Kuwaitis in the Private Sector. Available online: http://www.oxgaps.org/files/analysis_gulseven.pdf (accessed on 19 May 2017).

66. Dyer, J.H.; Nobeoka, K. Creating and Managing a High-Performance Knowledge-Sharing Network: The Toyota Case. *Strateg. Manag. J.* **2000**, *21*, 345–367. [CrossRef]

67. Lapinski, A.R.; Horman, M.J.; Riley, D.R. Lean Processes for Sustainable Project Delivery. *J. Constr. Eng. Manag.* **2006**, *132*, 1083–1091. [CrossRef]

68. Van Buiten, M.; Hartmann, A.; van der Meer, J.P. Nudging for smart construction: Tackling uncertainty by changing design engineers' choice architecture. In Proceedings of the Engineering Project Organization Conference 2016, Cle Elum, DC, USA, 28–30 June 2016.

69. Shealy, T.; Ismael, D.; Hartmann, A.; van Buiten, M. Removing Certainty from the Equation: Using Choice Architecture to Increase Awareness of Risk in Engineering Design Decision Making. In Proceedings of the Engineering Project Organization Conference, Stanford, CA, USA, 5–7 June 2017.

70. Harris, N.; Shealy, T.; Klotz, L. How Exposure to "Role Model" Projects Can Lead to Decisions for More Sustainable Infrastructure. *Sustainability* **2016**, *8*, 130. [CrossRef]

71. McClure, J.; White, J.; Sibley, C.G. Framing effects on preparation intentions: Distinguishing actions and outcomes. *Disaster Prev. Manag. Int. J.* **2009**, *18*, 187–199. [CrossRef]

72. Newman, C.L.; Howlett, E.; Burton, S.; Kozup, J.C.; Heintz Tangari, A. The influence of consumer concern about global climate change on framing effects for environmental sustainability messages. *Int. J. Advert.* **2012**, *31*, 511–527. [CrossRef]

73. Tang, O.; Nurmaya Musa, S. Identifying risk issues and research advancements in supply chain risk management. *Int. J. Prod. Econ.* **2011**, *133*, 25–34. [CrossRef]

MDPI

St. Alban-Anlage 66

4052 Basel

Switzerland

Tel. +41 61 683 77 34

Fax +41 61 302 89 18

www.mdpi.com

Sustainability Editorial Office

E-mail: sustainability@mdpi.com

www.mdpi.com/journal/sustainability

MDPI
St. Alban-Anlage 66
4052 Basel
Switzerland

Tel: +41 61 683 77 34
Fax: +41 61 302 89 18

www.mdpi.com

MDPI

ISBN 978-3-03897-167-2